Meals in science and practice

Related titles

Understanding consumers of food products
(ISBN 978-1-84569-009-0)
It is very important for food businesses, scientists and policy makers to understand
consumers of food products: in the case of businesses to develop successful products
and in the case of policy makers to gain and retain consumer confidence. Consumers'
requirements and desires are affected by issues such as culture, age and gender. Other
issues that have recently become important to consumers, such as diet and health, or
GM foods, will not always be so significant. Therefore food businesses and policy
makers need to understand consumers' attitudes and the influences upon them in order
to respond effectively. Edited by two distinguished experts, this book is an essential
guide for food businesses, food scientists and policy makers.

Case studies in food product development
(ISBN 978-1-84569-260-5)
Product development is vital to the food industry but a successful outcome is often
elusive. Many publications cover product development principles and techniques, but
much can still be learnt from those engaging in product development in industry. The
study of actual projects can generate new ideas on the philosophies, systems,
organisations and techniques of food product development. Edited by leading
authorities on the subject, and with an international team of contributors, *Case studies
in food product development* describes participants' involvement in developing new
products and improving existing ones, and discusses what others can gain from their
experiences.

Food for the ageing population
(ISBN 978-1-84569-193-6)
The world's ageing population is increasing. Food professionals will have to address
the needs of older generations more closely in the future. This unique volume reviews
the characteristics of the ageing population as food consumers, the role of nutrition in
healthy ageing and the design of food products and services for the elderly. In Part I,
aspects of the elderly's relationship with food, such as appetite and ageing and the
social significance of meals, are discussed. In Part II, the role of nutrition in
conditions such as Alzheimer's and eye-related disorders is reviewed. Concluding
chapters address issues such as food safety and the elderly and nutrition education
programmes.

Details of these books and a complete list of Woodhead's titles can be obtained by:

- visiting our web site at www.woodheadpublishing.com
- contacting Customer Services (e-mail: sales@woodheadpublishing.com;
  fax: +44 (0) 1223 893694; tel.: +44 (0) 1223 891358 ext.130; address:
  Woodhead Publishing Limited, Abington Hall, Granta Park, Great Abington,
  Cambridge CB21 6AH, UK)

Meals in science and practice

Interdisciplinary research and business applications

**Edited by
Herbert L. Meiselman**

CRC

CRC Press
Boca Raton Boston New York Washington, DC

WOODHEAD PUBLISHING LIMITED
Oxford   Cambridge   New Delhi

Published by Woodhead Publishing Limited, Abington Hall, Granta Park,
Great Abington, Cambridge CB21 6AH, UK
www.woodheadpublishing.com

Woodhead Publishing India Private Limited, G-2, Vardaan House, 7/28 Ansari Road,
Daryaganj, New Delhi – 110002, India

Published in North America by CRC Press LLC, 6000 Broken Sound Parkway, NW,
Suite 300, Boca Raton, FL 33487, USA

First published 2009, Woodhead Publishing Limited and CRC Press LLC
© 2009, Woodhead Publishing Limited
The authors have asserted their moral rights.

British Library Cataloguing in Publication Data
A catalogue record for this book is available from the British Library.

Library of Congress Cataloging in Publication Data
A catalog record for this book is available from the Library of Congress.

Woodhead Publishing ISBN 978-1-84569-403-6 (book)
Woodhead Publishing ISBN 978-1-84569-571-2 (e-book)
CRC Press ISBN 978-1-4398-0106-2
CRC Press order number: N10007

The publishers' policy is to use permanent paper from mills that operate a
sustainable forestry policy, and which has been manufactured from pulp
which is processed using acid-free and elemental chlorine-free practices.
Furthermore, the publishers ensure that the text paper and cover board used
have met acceptable environmental accreditation standards.

Typeset by Replika Press Pvt Ltd, India
Printed by TJ International, Padstow, Cornwall, UK

To Deborah, with whom I share my meals.

Contents

Part VII Meals in practice/meals as art

Part VIII Further perspectives on meals

Contributor contact details

(* = main contact)

Editor and Chapters 1 and 2
Herbert L. Meiselman
Herb Meiselman Training and
    Consulting Services
Rockport
MA 01966
USA

E-mail:
herbertlmeiselman@verizon.net
herb@herbmeiselman.com

Chapter 3
Dr Johanna Mäkelä
National Consumer Research Centre
PO Box 3
00531 Helsinki
Finland

E-mail: johanna.makela@ncrc.fi

Chapter 4
Associate Professor Peter G.
    Williams
Smart Foods Centre
School of Health Sciences
University of Wollongong
Wollongong NSW
Australia 2522

E-mail: peter_williams@uow.edu.au

Chapter 5
Dr Unni Kjærnes*
The National Institute for
    Consumer Research (SIFO)
PO Box 4682 Nydalen
N-0405 Oslo
Norway

E-mail: Unni.kjarnes@sifo.no

Professor Lotte Holm
Department of Human Nutrition
Rolighedsvej 30
DK-1958 Frederiksberg C
Denmark

E-mail: loho@life.ku.dk

Professor Jukka Gronow
Department of Sociology
Uppsala University
P.O. Box 624
SE - 75126 Uppsala
Sweden

E-mail:Jukka.Gronow@soc.uu.se

Johanna Mäkelä
National Consumer Research
  Centre
PO Box 3
00531 Helsinki
Finland

E-mail: johanna.makela@ncrc.fi

Associate Professor Marianne
  Pipping Ekström
Department of Culinary Arts and
  Meal Science
Örebro University
Sörälgsvägen 2
SE-712 60 Grythyttan
Sweden

Chapter 6
Dr Øydis Ueland
Nofima Food
Osloveien 1
1430 Ås
Norway

E-mail: oydis.ueland@nofima.no

Chapter 7
Professor John S. A. Edwards* and
  Dr Heather J. Hartwell
Foodservice and Applied Nutrition
  Research Group
School of Services Management
Bournemouth University
Poole
Dorset
BH12 5BB
UK

E-mail: edwardsj@bournemouth.ac.uk
       hhartwell@bournemouth.ac.uk

Chapter 8
Dr Isabelle Boutrolle*
Danone Research, Consumer Science
Route Départementale 128
91 767 Palaiseau
France

E-mail: isabelle.boutrolle@danone.com

Dr Julien Delarue
AgroParisTech, Laboratoire de
  Perception Sensorielle et
  Sensométrie
1 avenue des Olympiades
91 744 Massy
France

E-mail: julien.delarue@agroparistech.fr

Chapter 9
Professor Patricia Pliner*
Department of Psychology
University of Toronto Mississauga
3359 Mississauga Rd
Mississauga
ON L5L 1C6
Canada

E-mail: patricia.pliner@utoronto.ca

Rick Bell (deceased)
Natick Army Soldier Center
Kansas Street
Natick
MA
01760-5020
USA

Chapter 10
Professor W. Alex McIntosh*
Department of Sociology
4351 TAMU
Texas A&M University
College Station
TX 77843-4351
USA

E-mail: w-mcintosh@neo.tamu.edu

Dr Wesley Dean
Department of Sociology
4351 TAMU
Texas A&M University
College Station
TX 77843-4351
USA

E-mail: wdean@tamu.edu

Professor Cruz C. Torres
Department of Recreation, Park and
    Tourism Sciences
2261 TAMU
Texas A&M University
College Station
TX 77843-2261, USA

Professor Jenna Anding
Department of Nutrition and Food
    Sciences
2471 TAMU
Texas A&M University
College Station TX 77843-2471
USA

Professor Karen S. Kubena
College of ArgiLife
2402 TAMU
Texas A&M University
College Station
TX 77843-2402
USA

Professor Rodolfo Nayga
Department of Agricultural
    Economics
2124 TAMU
Texas A&M University
College Station
TX 77843-2124

E-mail: rnayga@tamu.edu

Chapter 11
Professor Christina Fjellström
Department of Food, Nutrition and
    Dietetics
Box 560
Uppsala University
Husargatan 3
SE-753 09 Uppsala
Sweden

E-mail: christina.fjellstrom@ikv.uu.se

Chapter 12
Dr Inger M. Jonsson* and
    Associate Professor Marianne
    Pipping Ekström
Department of Culinary Arts and
    Meal Science
Örebro University
Sörälgsvägen 2
SE-712 60 Grythyttan
Sweden

E-mail: inger.m.jonsson@oru.se

Chapter 13
Dr Janet Chrzan
Department of Anthropology
University of Pennsylvania
Philadelphia
323 Museum
33rd and Spruce Streets
Pennsylvania, PA 19104
USA

E-mail: jchrzan@sas.upenn.edu

Chapter 14
Professor Inga-Britt Gustafsson,*
   Associate Professor Åsa Öström,
   Guest Professor Judith Annett
School of Hospitality Culinary Arts
   and Meal Science
Örebro University
Sörälgsvägen 2
SE-712 60 Grythyttan
Sweden

E-mail: inga-britt.gustafsson@oru.se

Chapter 15
Gerald Darsch* and Stephen
   Moody
Natick Army Soldier Center
Development & Engineering Center
Kansas Street
Natick
MA 01760-5020
USA

E-mail: gerald.darsch@us.army.mil
         Stephen.Moody@us.army.mil

Chapter 16
Claude Grignon* and Christiane
   Grignon
Maison des Sciences de l'Homme
   Paris
6 rue Vulpian
F-75013 Paris
France

E-mail: cgrignon@msh-paris.fr

Chapter 17
Associate Professor Erminio
   Monteleone* and Dr Caterina
   Dinnella
Dipartimento di Biotecnologie
   Agrarie
University of Florence
Via Donizetti, 6
52144 Firenze
Italy

E-mail: erminio.monteleone@unifi.it

Chapter 18
Dr Rosires Deliza*
Embrapa Food Technology
Sensory and Instrumental
   Evaluation Lab
Av. das Américas, 29501
23.020-470 Rio de Janeiro–RJ
Brazil

E-mail: rodeliza@ctaa.embrapa.br

Dr Letícia Casotti
Coppead Business School
Federal University of Rio de
   Janeiro (UFRJ)
Caixa Postal 68514
21.945-970 Rio de Janeiro–RJ
Brazil

Chapter 19
Dr Colleen Taylor Sen
2557 West Farwell Avenue
Chicago
Il 60645
USA

E-mail: ctsen2001@yahoo.com

Chapter 20
Associate Professor Sam-ang
    Seubsman* and P. Suttinan
Room 101, Tresorn Building
Sukhothai Thammathirat Open
    University (STOU)
Bangpood
Pakkret
Nonthaburi
Bangkok
Thailand 11120

E-mail: sam-ang.seubsman@anu.edu.au

Dr Jane Dixon and Dr Cathy
    Banwell
Australian National University
Australia

Chapter 21
Dr J. A. Klein
Department of Anthropology and
    Sociology
School of Oriental and African
    Studies (SOAS)
University of London
Thornhaugh Street
Russell Square
London WC1H 0XG
UK

E-mail: jk2@soas.ac.uk

Chapter 22
Dr David N. Cox
CSIRO Human Nutrition
PO Box 10041
Adelaide BC
SA 5000
Australia

E-mail: david.cox@csiro.au

Chapter 23
Kevan Vetter
Executive Chef / Manager
Culinary Product Development
McCormick and Company
204 Wight Avenue
Hunt Valley, MD 21031
USA

E-mail: kevan_vetter@mccormick.com

Chapter 24
Dr Howard R. Moskowitz
President
Moskowitz Jacobs Inc.
1025 Westchester Ave. – Suite 400
White Plains
NY 10604
USA

E-mail: mjihrm@sprynet.com

Michele Reisner
Vice President
Moskowitz Jacobs, Inc.
1025 Westchester Ave. – Suite 400
White Plains
NY 10604
USA

Gwen Ishmael
Decision Analyst, Inc.
Arlington
TX 76011
USA

Chapter 25
Dr K. E. D'Anci and Professor R.
    B. Kanarek*
Department of Psychology
Tufts University
490 Boston Ave.
Medford
MA 02155
USA

E-mail: robin.kanarek@tufts.edu

Chapter 26
Associate Professor Jordan L.
    LeBel*
Marketing Department
John Molson School of Business
Concordia University
1455 de Maisonneuve Blvd West,
    GM 300-7
Montreal
Canada
H3G 1M8

E-mail: jlebel@jmsb.concordia.ca

Associate Professor Rhona
    Richman Kenneally
Department of Design and
    Computation Arts
Concordia University
1455 de Maisonneuve Blvd West,
    EV 6-753
Montreal
Quebec
Canada, H3G1M8

E-mail: mk@alcor.concordia.ca

Chapter 27
Eric Clay
Shared Journeys
832 North Aurora Street
Ithaca
NY 14850
USA

E-mail: wericclay@aol.com

Gil Marks
Cookbook Author and Rabbi
208 W. 80 Street #1A
New York
NY 10024
USA

E-mail: gilmarks@hotmail.com

Muhammad Munir Chaudry
President of the Islamic Food and
    Nutrition Council of America
IFANCA Halal Research Center
777 Busse Highway
Park Ridge
IL 60068
USA

E-mail: m.chaudry@ifanca.org
            drchaudry@gmail.com

Mian Riaz
Head, Food Protein Center and
    Professor of Food Science
Food Protein R&D Center
2476 TAMU
Texas A&M University
College Station
TX 77843-2476
USA

Huma Siddiqui
Cookbook Author
White Jasmine LLC
PO Box 2561
Madison
WI 53701
USA

E-mail: huma@whitejasmine.com

Joe M. Regenstein*
Head, Cornell Kosher and Halal
    Food Initiative and Professor of
    Food Science
Stocking Hall
Cornell University
Ithaca
NY 14853-7201
USA

E-mail: jmr9@cornell.edu

Chapter 28
Professor David Marshall*
University of Edinburgh Business
    School
50 George Square
Edinburgh
EH8 9JY
UK

E-mail: d.w.marshall@ed.ac.uk

Dr Clare Pettinger
School of Health Professions
Peninsula Allied Health Centre
Derriford Road
University of Plymouth
PL6 9BH
UK

E-mail: clare.pettinger@plymouth.ac.uk

Preface

The subject of meals is addictive to anyone interested in food and eating and people. It is still a constant surprise to me that many people who are interested in food express their interest outside of the context of meals. This is true of researchers and academics, but also of environmentalists, health proponents, and many others who recognize the importance of eating as a human activity. Most food is consumed as part of a meal, and that makes the meal the proper setting or context for all of these concerns about food. Without considering the meal, one cannot have a major impact on foods and eating.

That is why I have embraced the study of meals in two books, *Dimensions of the meal* (Meiselman, 2000) and the present book *Meals in science and practice*. In the process of editing these two books, I have learned a great deal about meals. But the main lesson for me has been how much I do not know about meals. Each of us lives in one subculture, characterized by its own meal customs. Many of us also belong to a 'foodie' group, with a high interest in foods and meals. But around us there are many other types of meals, leading to the enormous variety worldwide of how we eat.

No matter what your interest in food, you need to be aware of this enormous variety of meals. As business, including the food business, becomes more global, we need to see this meal variety and understand its implications. 'Breakfast' means different things around the world, and means even more things when the history of breakfast is considered. When we think of developing new products for the world market or feeding the hungry or dealing with obesity, all of these issues and more, we need to understand these problems and address them within the context of meals. The meal is how people eat. People do not eat individual foods, nor do they eat random combinations of foods. The combinations of foods which we call meals are the result of history and custom and availability and many other factors. Many of these factors are addressed in this book.

I hope that this book on meals will help to spread the word on the richness of meals, as objects of study and as subjects for discussion. Each of the authors in this book has taken a topic and has presented compelling information on meals, on their enormous variety, on their individual, family and social implications, and on their health and business effects.

I hope that this book will be a serious addition to the literature on meals. It provides readers with a range of topics, a range of skilled authors, and an almost overwhelming amount of information. But the subject of meals is both addictive and endless – for those interested in food, eating and meals there is an almost endless amount and variety of information.

Herbert L. Meiselman

Part I

Introduction

1

Dimensions of the meal: a summary

H.L. Meiselman, formerly Natick Soldier Research Development and Engineering Center, and Herb Meiselman Training and Consulting, USA

Abstract: In this chapter, an overview is presented of *Dimensions of the meal*, the first book which brought together information on meals from a broad range of disciplines and specialties, both technical and non-technical; related materials in other, more recent, meal books are identified. *Dimensions of the meal* was interdisciplinary, and included the basic scientific disciplines of social science (psychology and sociology), biological science (biology and physiology) and nutrition. Various meal cultures were described: China, Japan, USA, Scandinavia, Netherlands and UK. Foodservice aspects of meals were addressed from the restaurant and institutional perspectives, and from operator and customer perspectives.

Key words: meals, psychology, sociology, biology, physiology, nutrition, culture, foodservice.

1.1 Introduction

Dimensions of the meal (Meiselman, 2000) was the first book which brought together information on meals from a broad range of disciplines and specialties, both technical and non-technical. The purpose of this chapter in this current book is to present an overview of *Dimensions of the meal*, and identify related materials in other, more recent, meal books. Two recent books are highly relevant, the book coming from the Oxford Symposium *The Meal* (Walker, 2002) and the summary of the Nordic study of meals (Kjærnes, 2001).

Dimensions of the meal was interdisciplinary, and included the basic scientific disciplines of social science (psychology and sociology), biological science (biology and physiology) and nutrition, as well as foodservice and culinary considerations. Also included were specialized areas relating to sensory studies and cultural studies. Meals in a number of specific cultures were described: China, Japan, United States (American Holidays), Scandinavia, Netherlands and Britain. Foodservice aspects of meals were addressed from the restaurant and institutional perspectives, and from operator and customer perspectives. In the final chapter 'An Integrative Summary' the editor Herb Meiselman stated '…two of our goals were to appreciate the complexity of the meal and to appreciate the need for more interdisciplinary approaches.' (p. 311). *Dimensions of the meal* was aimed at filling the void in interdisciplinary texts about meals. This paucity of books on meals stems from several trends:

1. **Discipline specialization**. Research on meals, when it occurs, tends to fall within a specialized discipline such as nutrition, physiology, anthropology, sociology, psychology, food science and technology, foodservice and business, and other fields as well. Researchers in one area do not tend to be familiar with research in other areas. This is very noticeable in definitions of what the word 'meal' means; these definitions make sense within one field but ignore the perspectives of other fields. Some extremes result: sociologically the meal is a social event, and meals eaten alone are not considered to be meals. Physiologists prefer to avoid the culturally laden term 'meal' and use terms like 'eating episode' and 'ingestion'.

2. **Product development**. Much of the interest in food products is focused on products in isolation. Although most food is consumed as parts of meals, food products are often developed in isolation. The product development industry is a major sponsor and customer for all types of food research; their focus on products and relatively low interest in meals has failed to stimulate more research on meals.

3. **Food service as a business.** The food service or catering field is interested in meals, but foodservice is mainly approached from a business perspective and much less as a technical field. Thus, meals research which might have come from foodservice has not had adequate stimulus. As foodservice gradually becomes more technical, interest in meals from technical perspectives is growing within that industry.

4. **Nutrition and meals**. The nutrition field itself has been part of this problem. Although people eat meals, nutritionists continue to be mainly interested in food products and in macro- and micro-nutrient contents. Nutritionists consider calories, protein and fat, or good foods and bad foods, but rarely meals, even though people eat meals.

5. **Normal and abnormal eating**. Within the US at least, a large amount of research on eating has focused on abnormal eating – this research has received major funding for the past 30–40 years, while normal eating has received relatively less support. Combined with the limited interest from the food product industry and the limited research by the foodservice industry, this has produced a lack of information on how most people eat on a daily basis. The major sources of data on normal eating come from Europe, and will be noted later in this chapter.

1.2   The challenge of interdisciplinary research on meals

Not only has study of the meal suffered for all of the following reasons, but the interdisciplinary nature of meals research is itself a challenge. Information about meals exists in so many disparate sources, in so many disciplines, and in both historical and contemporary sources. One's discipline to some degree determines how you define the term 'meal' (Table 1.1). Product developers tend to focus on single products as noted above, but also might consider how combining foods might impact their products. Within foodservice, meals represent critical issues of food sequences and food compatibilities – what foods follow each other and what foods go well with other foods. The biologist might emphasize food intake at meals with special emphasis on meal timing and the pattern of how much is consumed over the day. Psychologists and sociologists look at the interaction of people and meals, focusing on their individual and group significance. The social meaning and importance of meals is a vast area of importance to many people in society. Marketing examines satisfaction provided by meals and this is affected by pricing and branding. Health sciences look at the major health issues of overeating and under-eating in the meal context. In fact, some of the earliest published research on regular (non-laboratory) meals was within the health community (e.g., Stunkard, 1968).

The question of meal definition was addressed by a number of authors in *Dimensions of the meal*. Mäkelä stresses cultural aspects of meal definition: 'Culture defines how possible nutrition is coded into acceptable food.' and '…every culture has its own definitions of a meal' (Mäkelä, 2000, p. 7). There is a fundamental truth that meals have a cultural basis, but many people forget this in pronouncements about meals. Mäkelä goes on to discuss three dimensions of meals: meal format (main course, etc.), daily eating pattern (timing of meals and snacks), and social organization (where and with whom). The social aspects of eating are covered in detail by Sobal.

Table 1.1   Definitions of meals based on scientific or technical discipline/orientation

History – pattern of meals
Product development – food combinations
Foodservice – food sequences, food compatibilities, sensory themes
Designer – meal locations, environments, physical settings (fast food-luxury food)
Sensory – combination of sensory experiences
Biology – food intake timing and pattern (grazing vs. meals)
Physiology – internal satiety signals for meals
Nutrition/dietetics – food intake, macro/micronutrients
Sociology – commensality, social rules
Anthropology – cultural differences
Psychology – the personal meaning of meals
Marketing – price/value, brand, satisfaction
Abnormal psychology/health – undereating, overeating

1.3   The psychological perspective

The psychological perspective on meal definition was presented by Pliner and Rozin (2000). They also make several fundamental statements about meals: 'The meal is a very real psychological entity. It is a virtually universal physical and behavioral feature of human life. (Pliner and Rozin, 2000, p. 19). People 'do most of their eating in relatively short periods of time, separated by periods of minimal if any consumption'. This is a basic definition of a meal, and they go on to say that 'the meal is a privileged unit' of eating. Related to this status of the meal, Pliner and Rozin reported an earlier study by one of the authors studying meal terminology with native speakers of 18 languages. They reported that 17 out of 18 languages had a word for meal, 16 had unique words for breakfast and lunch, and all have a word for a main meal later in the day, and 16 have a word for snack. Only half of the languages have words for specific meal components of the major meal such as appetizer or dessert. Pliner and Rozin also address social interaction at meals and they present a detailed review of psychological factors which affect meal initiation, meal termination and meal size. These factors include meal visibility/ availability, effort, palatability, mood, variety, learning/experience, social factors, cultural factors, and location. This is an exhaustive treatment of the psychological variables affecting meals. Pliner and Rozin also present treatment of breakfast as an anomalous meal, different from the other meals.

1.4   The nutritional perspective

De Graaf (2000) treats the definition of meals from a nutritional perspective as follows: '…the definition and description of the meal from a nutritional perspective refers to the frequency, distribution, and variability of energy and nutrient intake across the day' (de Graaf, 2000, p. 47). While acknowledging the environmental, psychological, and physiological factors, de Graaf addresses the nutritional issues, citing data from the USA and Europe. A number of authors in the book speak of three meals per day as a common or universal pattern of eating, but de Graaf presents US and Dutch data showing more widespread eating. Only half of US respondents report three meals per day, with close to 20% showing both 2 and 4. Similarly, a study of elderly people across Europe showed 3–7 'eating occasions (including snacks)/day'. Evening dinner accounted for the largest percentage of total daily energy intake in the Netherlands across many age groups but the mid-day meal was larger for the elderly in many European countries including France, Italy, Switzerland, Poland and Northern Ireland. Snacks also contribute substantially to daily intake in many countries, although it is not clear whether differences among countries reflect actual snack practice or methods used in the different studies. De Graaf emphasizes the anomalous nutritional nature of breakfast, the meal with the greatest carbohydrate content and lowest protein and fat content. Breakfast is a meal which is sometimes skipped by people. Finally, de Graaf emphasizes that nutritional variables such as energy and nutrient intake are variable between subjects and within subjects, across meals and days.

Criteria for definitions of the meal have also been discussed by Oltersdorf (1999) and are presented in Table 1.2. These varying definitions of the term 'meal' include the disciplines of nutrition (energy content) and sociology (social interaction).

1.5 The social perspective

Finally, definitions of the meal were discussed in the report on the Nordic study of meals, *Eating Patterns* (Kjærnes, 2001). In the introductory chapter of the book, Kjærnes, Ekström, Gronow, Holm and Mäkelä address the question 'What is a meal?', and begin their discussion with the work of the earlier British authority Mary Douglas (1975). Douglas identified a food event when food is eaten and a structured social event; she defined a meal as a structured social event where food is eaten. This definition leaves unclear whether meals eaten alone are meals, or whether small amounts of food eaten with other people are meals. Kjærnes *et al.* address this issue by distinguishing four types of food events: individual and social meals, and private and public environments. This 2 × 2 matrix yields four possible food events. Kjærnes *et al.* emphasize that the variety of eating within the Nordic countries makes it difficult to use meal terminology such as breakfast, lunch and dinner. This is an important historical point that has been lost among many people discussing meals. Dinner should refer to the main meal of the day (Lehmann, 2002) but is sometimes used to refer to the evening hot meal. This leads us to some history of meals.

Table 1.2 Criteria for the definition of a meal*

Criteria*	Meal defined as	Author
Time of consumption	Eating events in the morning, at midday and in the evening	Fabry *et al.*, 1964 and others
Energy content	Consumption of >375 kcal	Bernstein *et al.*, 1981
Social interaction	Presence of fellow eaters	Rotenberg, 1981
Food quality	More than one single food	Skinner *et al.*, 1985
Energy content + interval time since last eating event	210 or 420 or 840 kJ + 15 or 45 min interval time	De Castro, 1993
No predefined concept	Self reported by subject	Gatenby, 1997

*Oltersdorf *et al.* (1999)

1.6   The historical perspective

Dimensions of the meal does not contain separate historical chapters, although much history is present in many of the chapters. Interestingly, most of the history of meals in *Dimensions of the meal* is within the foodservice/catering chapters. With foodservice there appears to be more awareness of, and more sensitivity to, history of meals. One of the most surprising things about meals is how little people know about meal history and contemporary food culture. Meals change frequently over history, and meals differ from culture to culture at any given time. Simply put, the way we eat today is but a dot on the historical development of meals, and the way we eat today, is different from the way people eat in other parts of the world. There are very few rules of meals that hold over centuries and over cultures. There are many historical chapters in the Oxford Symposium *The Meal* (Walker, 2002), which help to put contemporary meals in many different cultures within the historical context of change.

One of the most striking pieces of history is the comparatively recent development of restaurants and more formal meals as we know them. Formal service was *service a la francaise* from the 17th century until the 19th century; foods were loaded onto the table, and people ate what was within reach. Meals were not served by waiters; meals might have three courses but, in each course, many items for the course were placed on the table. This buffet-like service was finally replaced by *service a la russe*, in which a series of courses was served by waiters. Kaufman (2002) describes these two service types in great detail, and discusses reasons for the shift such as a greater focus on food taste rather than on food appearance.

Edwards (2000) traces the history of meals from the Middle Ages when there were two daily meals, mid-day dinner and later supper. Breakfast was established by the 18th century, although questions remained about its healthfulness. As dinner moved later in the day, a midday meal emerged called 'nuncheon' which eventually became luncheon. Lehmann (2002) in *The Meal* has a good discussion of meals, with the following premise: '...the pattern of three meals a day as an immutable part of daily life in a civilized society, whereas in fact this is simply not true' (p. 140). Lehmann goes on to point out that the only constant has been dinner as the main meal. Supper has arguably been a meal, while breakfast was only a meal for a brief period in history. A more complete historical review of breakfast is presented by Albala (2002) in *The Meal*.

Restaurants did not really emerge until the early 19th century, although the first restaurant is attributed to 18th-century Spain. Schafheitle (2000) presents historical multi-course meals of earlier times, with as many as 16 or more courses, and also more 'modern shortened menu(s)' of five courses. Today, the typical French meal served in a restaurant has three courses.

1.7   The biological perspective

Dimensions of the meal treats a number of meal issues from a more biological perspective. I have already noted de Graaf's treatment of meal nutrition above. Kissileff (2000) addresses physiological controls of single meals, which he prefers to call 'eating episodes' in keeping with the discomfort experienced by many biologists/physiologists with concepts which have large social/psychological/anthropological dimensions (recall that the Nordic study used the term food events rather than meals). Kissileff states that 'An eating episode is any amount of eating in any time frame separated from other such episodes by unrelated behavior' (pp. 63–64) and '...it is easier to study eating episodes when eating is the only event taking place and when only a single food is being consumed' (p. 64). Hence, we see very clearly that we mean different things by meals. Kissileff goes on to present an outline of methods for studying how much is eaten during eating episodes. He specifically considers models for studying homeostasis of meal size. And then goes on to analyze meal initiation and termination, which are probably the areas in which physiologists believe that physiology has a major role in eating and in meals.

Lawless (2000) presents another important biological dimension of meals, the sensory dimension. While the sensory dimension is important to consumers, to chefs, and to just about everyone involved with food, one needs to come back to sensory physiology and psychology to find out how the senses work in the complex stimulus environment of the meal. Lawless notes that 'Probably no other common experience assaults all of the senses as does the consumption of foods' (p. 93). Lawless uses the sensory phenomena of mixture interactions, taste and smell interactions, sensory adaptation and release from suppression to show sensory complexities, and he emphasizes the importance of the multi-sensory experience of meals involving taste and smell, touch, texture, temperature and pain.

If we are being bombarded by all of these sensations during a meal, how does the body react? One answer is that people seek variety, and meals provide a variety of sensations. In fact all animals and even human infants respond to variety. Beyond that, however, foods tend to decrease in appeal as we eat them; this is the phenomenon of sensory specific satiety, which has been extensively researched by Rolls and colleagues and is presented in *Dimensions of the meal* (Rolls, 2000). Sensory specific satiety is demonstrated by showing that the foods you have just finished eating are rated lower than they were at the beginning of their consumption, but new foods show no decrement. It is called sensory specific because it applies to new tastes and flavors, not all tastes and flavors. Thus, sensory specific satiety ensures that we switch from food to food as we progress through a meal, and encourages variety in the meal.

1.8   The cultural perspective

One of the most important aspects of meals is their cultural/social component. Meals are one of the main points of interaction within a culture and within a family, and meal time serves as a major point of socialization for children in the culture. But much more is linked to the social setting of the meal. Sobal (2000) discusses this in *Dimensions of the meal*, and begins by suggesting that sociability really defines a meal: 'A "proper" or "ideal" meal is typically eaten with others ... with eating alone not considered a "real" meal for many people' (p. 119). Sobal discusses eating alone, which is more fully addressed in this volume by Pliner and Bell (Chapter 9). Another social phenomenon in eating is social facilitation of eating in which more food is eaten in the presence of other people. This was researched in a series of studies by de Castro and colleagues, who showed how social facilitation of eating holds over a broad range of different situations (de Castro *et al.* 1990). Sobal also discusses the important social phenomenon of commensality – eating together – and what social rules define with whom we eat. As noted above, meals provide the opportunity for social interaction and, hence, sociability and socialization. One of the largest literatures concerning meals is the social science literature on meals including sociology and anthropology. Many interesting cultural perspectives on meals can be found in the Oxford Symposium entitled *The Meal*. Many of these are from non-Western countries and are an important lesson for those used to Western meals. As many of the authors in *The Meal* point out, meals in other cultures can be totally different from western meals, including their location, who is present, the timing, and what is eaten (for example, see Iddison, 2002). The Nordic study on meals (Kjærnes, 2002) presents a good cultural view of these four North European countries which have many cultural similarities and differences.

If culture is so important in meals, then how is culture communicated through cooking meals. Elizabeth Rozin (2000) showed us how this works with an analysis of basic rules which apply to all cuisines. Cuisines are differentiated in three ways; (1) basic foods, (2) culinary techniques, and (3) flavor. Basic foods such as butter and olive oil, rice and potato, lamb and pork differentiate cuisines. Culinary techniques reinforce these differences. One of the most important culinary techniques is heating or cooking itself; cuisines differentially use dry heat baking or roasting, liquid boiling or steaming, or cooking with fat as in sautéing and frying. Fermentation techniques also differ across cuisines producing dairy products, alcohol products, condiments (soy sauce), and others. Flavor 'seems to be the ... part of culinary practice that is most capable of evoking a particular ethnicity...' (p. 135). We can identify cuisines from flavors because cultures tend to use the same combinations of flavors repeatedly, for example the lemon and oregano of Greece or the Chinese combination of soy and sesame, or the Neapolitan combination of garlic, basil and oregano. Elizabeth Rozin refers to these recurring flavor combinations as flavor principles, which have been the basis for an entire *Flavor Principle Cookbook* (Rozin, 1973).

Cultural meals probably are at their most profound in holiday meals, which are discussed by Long (2000) using the American Thanksgiving as an example. Long points out that holiday meals represent the crossing of three domains: food, family, and holidays. Long takes the perspective of folklore which addresses holidays, food, and family. Long views meals as 'the tangible expression or enactment of the beliefs, customs, history and aesthetics surrounding a culture's eating habits' (p. 144). Folklore refers to the activities surrounding food as 'foodways'. Long uses the American Thanksgiving holiday meal to demonstrate foodways. For those readers not familiar with the American Thanksgiving meal, it is a unique holiday and a unique meal which has changed very little. There is no gift giving at Thanksgiving. The meal is roast turkey with accompanying vegetable dishes and baked pies for dessert. Long discusses the rituals surrounding Thanksgiving and other holiday meals, and the (family) politics of these events.

Elizabeth Rozin's presentation on cuisines and flavor principles is a good lead in to presentation of several cuisines. *Dimensions of the meal* presents four examples of cuisines: Chinese, Japanese, Nordic, and British.

Jacqueline Newman (2000) describes Chinese meals including sample Chinese menus for breakfast, the noon or lesser meal, and dinner or main meal. Newman also presents a typical Chinese banquet meal. Newman identifies the common practice of making larger pieces of food into smaller pieces through chopping and other techniques in order to reduce fuel use and reduce cooking times. Interestingly, the Chinese eat little or no raw food, and meals always include grains. While the Chinese eat three times a day, the bulk of their mid-day and evening meals is carbohydrate. The most common carbohydrate is not rice, and includes a long list of grains. Rice is more favored in the south, and wheat and other grains in the north. The main carbohydrate is called 'fan', translating as being 'rice' and 'meal'. Dishes accompanying the fan are called 'cai', which literally translates as vegetables. Chinese breakfasts differ between north and south; in the south breakfast might be a porridge or gruel called 'congee', and in the north it might be deep fried wheat crullers (doughnut-like pastries) with warm soy milk and sugar. Newman emphasizes the very long history of Chinese food and cooking, and the many traditions and rituals which surround every aspect of Chinese meals. Klein covers Chinese meals in Chapter 21 of this volume.

Sigeru Otsuka (2000) presents Japanese meals. Otsuka also includes typical Japanese meals, providing a sample menu of meals for an entire week. Otsuka begins by discussing the imperial ban on eating meat from the 7th–19th century, making fish an important part of Japanese meals. The main flavorings in traditional Japanese food preparation were 'shoyu' (soy sauce) and 'miso' (fermented bean paste). Soy sauce is used in most dishes and is the basis for the flavor of Japanese foods. Otsuka points out the impact on modern Japanese meals of the school lunch program which introduced Japanese children, now adults, to western foods including wheat and bread. The school lunch menu which Otsuka presents contains many items which would be quite foreign in a western school lunch, such as fried squid and beef udon (noodle soup), but it also contains the very western hamburger.

Ritva Prattala (2000) presents meals from North Europe, including Denmark, Finland, Norway and Sweden. These Nordic countries have a history of studying meals, with an excellent summary available in English (Kjærnes, 2002). The historical meal pattern was one of multiple hot meals per day with hot breakfast, midday and evening meals plus other snacks. While the number of hot meals has decreased, there are still usually three meals per day. Interestingly the traditional breakfast of boiled potatoes, herring and porridge did not differ from the midday or evening meals, which also featured potatoes, salted fish, and porridge or gruel. Today, the breakfast is the same throughout the Nordic countries and is based on bread and cheese or spreads. In Norway and Denmark, the midday meal is also a sandwich meal, and the evening meal is usually the main meal. In Sweden and Finland, the midday meal is often the main hot meal of the day. The difference in meal patterns among countries which are very close in many ways probably follows from different social policies and school lunch programs in the four countries. However, the main meal, whether mid-day or evening, is similar in all Nordic countries with hot meat and boiled potatoes. There is often a vegetable, salad, and bread but rarely a dessert.

British meals have been the subject of a great deal of meals research, going back to the pioneering writing of Mary Douglas. David Marshall (2000) presents British meals in *Dimensions of the meal*. Marshall presents historical and contemporary British meal patterns, again showing the pattern of more meals for the working class and fewer meals for the upper classes. The meal pattern changed in the mid-20th century to the hot meal after work. Marshall uses the approach of Mary Douglas in viewing meals as structured events containing food. Marshall (see Table 13.1) categorizes British meals into the following categories: celebratory/festive meals, main meals (weekend/Sunday), main meals (weekday), light meals, and snacks. Marshall goes on to specify the meal structure of these different meal types.

1.9   The foodservice industry perspective

At the end of the book, *Dimensions of the meal* considers the foodservice perspectives of meals. After all, people go to foodservice outlets for meals, and foodservice is perhaps the best vantage point from which to examine meals. Edwards (2000) includes both restaurant and institutional foodservice; this is an important point because both restaurants and institutions are major providers of meals worldwide, and meals in the two types of outlets differ in their goals and practice. Edwards presents the term 'menu' as used interchangeably with the word meal in foodservice. Menu includes not only a list of dishes but some sense of sequence of foods, which is important in catered meals. Edwards points out that many domestic meals are composed of one course, often the multi-item plated meal. Many catered meals have a starter, a main course, and a dessert. Edwards categorizes catered meals into three occasions: for pleasure, for work or business, and through necessity. Edwards presents a comprehensive model (Fig. 14.1) of the numerous factors which affect all meals. One of the main differences between home meals and catered meals is food preparation; Edwards reviews foodservice food preparation techniques in use today and finds that this is a rapidly changing situation as technology changes. One of the most interesting aspects of eating meals outside the home is the growth in this part of the meals business. Meals away from home now account for more than half of the food dollars in the USA, and this trend to eat out more is happening in many other countries. Studies of meals should consider both home meals and also meals consumed away from home.

Howard Moskowitz (2000) presents an approach for developing meal concepts in industry. His approach uses the following stages:

1. define the requirements,
2. develop the overarching concept,
3. identify the components of the meal,
4. identify off the shelf products within the meal,
5. optimize the meat component, and
6. optimize the package.

Moskowitz explains that developing the concept for the meal can be done by experts or by consumers. Of course, Moskowitz is well known for his consumer approach using conjoint analysis. First, the consumer develops elements, that is, parts of concepts. Then consumers judge these elements combined into concepts. Moskowitz presents an example of this approach for a lunch meal.

Finally, we hear from meal practitioners, the chefs. We were fortunate in *Dimensions of the meal* to get interviews with four British chefs conducted by chef and university lecturer Joachim Schafheitle (2000). Schafheitle emphasizes menu development, as noted by Edwards above. Schafheilte includes sample menus from these four chefs, and asks each chef what factors they consider when developing menus. The result is a fascinating documentary of how chefs design and prepare meals. Often these world-class meals demonstrate phenomena known in the various sciences and technologies that underlie meals. Indeed, there are links between meals research and meals in practice.

1.10 References

Albala, K. (2002) 'Hunting for breakfast in medieval and early modern Europe.' In: Walker, H. (Ed.) *The Meal*, Totnes, Devon, England: Prospect Books. Pp. 20–30.
de Castro, J. M., Brewer, E. M., Elmore, D. K. and Orozco, S. (1990) 'Social facilitation

of the spontaneous meal size of humans is independent of time, place, alcohol, or snacks.' *Appetite*, **15**, 89–101.

De Graff, C. (2000) 'Nutritional definitions of the meal'. In: Meiselman, H. L. (Ed.) *Dimensions of the Meal: The Science, Culture, Business and Art of Eating*. Gaithersburg, Maryland: Aspen Publishers, Inc., pp. 47–59.

Douglas, M. (1975) *Implicit Meanings. Essay in Anthropology*. London: Routledge and Kegan Paul.

Edwards, J. S. A. (2000) 'Food service/catering restaurant and institutional perspectives of the meal'. In: Meiselman, H. L. (Ed.) *Dimensions of the meal: The Science, Culture, Business and Art of Eating*. Gaithersburg, Maryland: Aspen Publishers, Inc., pp. 223–244.

Iddison, P. (2002) 'Perpetual picnics – the meal in the UAE.' In: Walker, H. (Ed.) *The Meal*, Totnes, Devon, England: Prospect Books, pp. 113–122.

Kissileff, H. R. (2000) 'Physiological controls of single meals (eating episodes). In: Meiselman, H. L. (Ed.) *Dimensions of the meal: The Science, Culture, Business and Art of Eating*. Gaithersburg, Maryland: Aspen Publishers, Inc., pp. 63–91.

Kjærnes U. (Ed.) (20012), *Eating Patterns: A Day in the Lives of Nordic Peoples. Report No. 7-2001*. Lysaker Norway: National Institute for Consumer Research.

Lawless, H. T. (2000) 'Sensory combinations in the meal'. In: Meiselman, H. L. (Ed.) *Dimensions of the meal: The Science, Culture, Business and Art of Eating*. Gaithersburg, Maryland: Aspen Publishers, Inc., pp. 92–106.

Lehmann, G. (2002) 'Meals and mealtimes, 1600–1800.' In: Walker, H. (Ed.) *The Meal*, Totnes, Devon, England: Prospect Books, pp. 139–154.

Long, L. M. (2000) 'Holiday meals: rituals of family tradition'. In: Meiselman, H. L. (Ed.) *Dimensions of the meal: The Science, Culture, Business and Art of Eating*. Gaithersburg, Maryland: Aspen Publishers, Inc., pp. 143–159.

Marshall, D. W. (2000) 'British meals and food choice'. In: Meiselman, H. L. (Ed.) *Dimensions of the meal: The Science, Culture, Business and Art of Eating*. Gaithersburg, Maryland: Aspen Publishers, Inc., pp. 202–220.

Mäkelä, J. (2000), Cultural definitions of the meal. In Meiselman, H. L. (Ed.) *Dimensions of the meal*, Gaithersburg, MD: Aspen, 7–18.

Meiselman H. L. (Ed.) (2000), *Dimensions of the meal*. Gaithersburg, Maryland: Aspen Publishers.

Meiselman, H. L. (2000) 'The meal: an integrative summary'. In: Meiselman, H. L. (Ed.) *Dimensions of the meal: The Science, Culture, Business and Art of Eating* Gaithersburg, Maryland: Aspen Publishers, Inc. pp. 311–333.

Moskowitz, H. R. (2000) 'Integrating consumers, developers, designers, and researchers into the development and optimization of meals'. In: Meiselman, H. L. (Ed.) *Dimensions of the meal: The Science, Culture, Business and Art of Eating*. Gaithersburg, Maryland: Aspen Publishers, Inc., pp. 245–269.

Newman, J. M. (2000) 'Chinese Meals'. In: Meiselman, H. L. (Ed.) *Dimensions of the meal: The Science, Culture, Business and Art of Eating*. Gaithersburg, Maryland: Aspen Publishers, Inc., pp. 163–177.

Oltersdorf, U. Schlettwein-Gsell, D. and Winkler, G. (1999) 'Assessing eating patterns – an emerging research topic in nutritional sciences: introduction to the symposium'. *Appetite*, **32**, 1–7.

Otsuka, S. (2000) 'Japanese meals'. In: Meiselman, H. L. (Ed.) *Dimensions of the meal: The Science, Culture, Business and Art of Eating*. Gaithersburg, Maryland: Aspen Publishers, Inc., pp. 178–190.

Pliner, P. and Rozin, P. (2000) 'The psychology of the meal'. In: Meiselman, H. L. (Ed.) *Dimensions of the meal: The Science, Culture, Business and Art of Eating*. Gaithersburg, Maryland: Aspen Publishers, Inc., pp. 9–46.

Prättälä, R. (2000) 'North European meals: observations from Denmark, Finland, Norway, and Sweden'. In: Meiselman, H. L. (Ed.) *Dimensions of the meal: The Science, Culture,*

Business and Art of Eating. Gaithersburg, Maryland: Aspen Publishers, Inc., pp. 191–201.

Rolls, B. J. (2000) 'Sensory-specific satiety and variety in the meal'. In: Meiselman, H. L. (Ed.) *Dimensions of the meal: The Science, Culture, Business and Art of Eating*. Gaithersburg, Maryland: Aspen Publishers, Inc., pp. 107–116.

Rozin, E. (1976) *Ethnic Cuisine: The Flavor Principle Cookbook*. Lexington, MA: The Stephen Green Press.

Rozin, E. (2000) 'The role of flavor in the meal and the culture'. In: Meiselman, H. L. (Ed.) *Dimensions of the meal: The Science, Culture, Business and Art of Eating*. Gaithersburg, Maryland: Aspen Publishers, Inc., pp. 134–142.

Sobal, J. (2000) 'Sociability and meals: facilitation, commensality, and interaction'. In: Meiselman, H. L. (Ed.) *Dimensions of the meal: The Science, Culture, Business and Art of Eating*. Gaithersburg, Maryland: Aspen Publishers, Inc., pp. 119–133.

Schafheitle, J. M. (2000) 'Meal design: a dialogue with four acclaimed chefs'. In: Meiselman, H. L. (Ed.) *Dimensions of the meal: The Science, Culture, Business and Art of Eating*. Gaithersburg, Maryland: Aspen Publishers, Inc., pp. 270–310.

Stunkard, A. J. (1968) 'Environment and obesity: recent advances in our understanding of regulation of food intake in man'. *Federation Proceedings*, **27**, 1367–1373.

Walker, H. (2002) (Ed.) *The Meal*, Totnes, Devon, England: Prospect Books.

2

Meals in science and practice: an overview and summary

H. L. Meiselman, formerly Natick Soldier Research Development and Engineering Center and Herb Meiselman Training and Consulting, USA

Abstract: An overview of the coverage of *Meals in science and practice* is given. The focus is on the current status of meals and their future. The authors have moved beyond the definition of meals to give a broad interdisciplinary examination of meals, from many perspectives and many cultures. The goals of this second meals book are outlined, i.e. to showcase the richness of meals. The intention is to help the reader appreciate where meals research is at the present and where it should be heading in the future.

Key words: meals, meals research, meals science, meals cultures.

2.1 Introduction

In *Dimensions of the meal* (Meiselman, 2000) one of the aims of the book was defining the word 'meal'. This was an attempt to present the complexity of the meal topic. Presentations from the perspectives of social science, nutrition, foodservice, the art of chefs, sensory science, culinary history and other perspectives tried to make it clear to the reader that the subject of meals is very broad, and that meals can be seen from many different perspectives. The word 'meal' means different things to different people.

In *Meals in science and practice*, we take meals as the starting point and investigate their current status and their future. Here, the authors have, on their own and without direction from me, moved beyond the theme of definition to the extremely rich subject matter of meals. This book is an incredibly broad examination of meals from many disciplines, many perspectives and many cultures. Thus, the goals of the first meals book and this second meals book are the same: to showcase the richness of meals. The reasons for this are several:

(1) As one interested in eating research, I have dedicated my career to developing research methodology which is a valid representation of how we eat. I have argued for years (Meiselman 1992a,b) that we should study eating real foods in real places, in other words meals in natural locations.

(2) Those conducting meals research should be more aware of the very different approaches for studying meals. As this book demonstrates, the task of fully describing meals is almost impossible because of the diversity of meals research. Interdisciplinary books like *Dimensions of the meal* and *Meals in science and practice* provide opportunities to tell the meals story across discipline lines. This is important because many conferences and many books work within one or very few disciplines – the meals story needs to be viewed in an interdisciplinary way.

(3) The interdisciplinary approach is especially helpful when we look at the future of meals. This is one of the main points of this book – are meals changing and, if so, where are they heading? To answer these questions, we need to examine evidence from many fields and look for agreement/disagreement.

My goal in this chapter is to present an overview of some key topics in this volume. The task is overwhelming. I am humbled by the enormous detail, color, and subtlety which the authors have brought to bear on the subject of meals. It would be easy to shrink back from this and argue that an overview of such a vast amount of material is not possible.

The purpose of my overview should not be seen as an executive summary of the book – an executive summary would miss the point of the book which is meal complexity and meal variety. This overview should provide a stimulus to read more, and get into the details. This overview should also help the reader to begin to form questions as to where meals research is heading at the present and where it should be heading in the future. It is important to note that my overview naturally presents some of my own biases, both professional and personal. I hope each reader will try to recognize their own cultural and professional biases in their own reading, and that is one of the reasons why reading the whole book is important.

The following summary is divided into key topics which appear and re-appear throughout this book. One strength of the book is gaining an interdisciplinary view of these topics.

2.2   Styles of meals: home, restaurant, institution

2.2.1   Home meals

For many people, the home represents the typical meal situation. But meals are consumed in a variety of environments. Both Edwards and Hartwell (Chapter 7) and Williams (Chapter 4) deal with the vast field of foodservice, which is called catering in some countries. We will deal with foodservice below. First, let us look at how different chapters deal with the home. Kjærnes *et al.* (Chapter 5) confirm that in the four Nordic countries, the majority of meals, or eating events as they prefer to call them, take place at home. Other than eating at work, no other location was a major location contributing to meal consumption in their Nordic study. This is probably different from some people's expectation of where the relatively affluent Nordic people eat meals. Specifically, they report that 'Frequent eating at restaurants, cafés, bars and fast-food outlets, often seen as a characteristic of modern life, does not show in our data.'

Deliza and Casotti (Chapter 18) focus on home meals in their treatment of Brazilian meals. However, they note that their '…decision to focus on the home cooking of large Brazilian cities brought the additional difficulty of separating home-cooked meals from the growing habit of eating out, a trend that occurs for a multitude of reasons and is associated with many recent changes in family, work and leisure environments'. Many specific examples of eating at home are present in the sections on national and regional meals below.

If the home is the basic location of eating, then should foods being developed for home meals be tested in the home? When new food products are tested, either alone or in a meal context, they are usually tested in one of two places: (1) Central location tests: tests under controlled conditions, or (2) Home tests: tests under natural conditions. Boutrolle and Delarue (Chapter 8) explore how differences in test location may affect differences in consumer judgments about whether the food is liked or disliked. These decisions by companies affect what foods are offered to the consumer.

Boutrolle and Delarue present the many contextual variables that may influence food liking. This is important in considering how meals are appreciated in both testing and in actual consumption situations. Boutrolle and Delarue focus their attention on the contrast between so called central location tests, used by industry for food testing under controlled conditions, and home use testing, where meals are both tested and actually consumed under natural conditions. Boutrolle and Delarue consider how these two types of meal tests might differ; these variables include the following:

(1) the quantity of food eaten during tasting (normal portion at home),
(2) tested food preparation (food temperature; controlled individually or centrally),
(3) food environment (other meal items and companion foods),
(4) consumption time (appropriate-inappropriate),
(5) social environment (social facilitation; commensality; meal duration; natural groups), and
(6) the scope for choosing the food tasted (choice).

The authors emphasize the role of expectations (through assimilation and contrast) on the impact of context on product ratings. The authors end with practical suggestions for conducting improved food product testing, and with a call for increased research on meal testing.

2.2.2 Restaurant meals

We now turn to meals in restaurants. Restaurant meals form the basis of one of the two applications of meals to education in this book. Gustafsson *et al.* (Chapter 14) use the teaching of restaurant meals within their unique Doctoral program at Orebro University in Culinary Arts and Meal Science (CAMS). The Orebro approach emphasizes multidisciplinarity in science, in training of practical skills in handicraft, and in creativity to prepare aesthetically pleasing meals. Their chapter presents both the theoretical underpinnings of the CAMS approach along with their curriculum.

The Culinary Arts and Meal Science (CAMS) program is based on a Five Aspects Meal Model (FAMM), developed from the assessment of restaurants performed by Guide Michelin. FAMM begins with a visit to a restaurant starting with the '*room, meeting* a headwaiter, being given a table and thereafter receiving some food and beverages (here called the *product*). In addition, there is a surrounding *atmosphere,* by which we mean the guest's perception of the total situation, and a *control management system*, which encompasses the overall business planning…'. The authors explain that '*The room, meeting, product and control management system* together create *the atmosphere*'. This well thought out model of restaurant dining forms the basis for their education in restaurant dining.

Performing at the level of a chef involves a combination of education, intuition, and experience as Chef Kevan Vetter points out. 'For many, this craft is truly a process of developing new flavors and menu items through feeling and intuition. Culinary skills are honed through not only years of formal education at universities and culinary schools but mostly through hands on mentoring by experienced chefs, constant collaboration with fellow chefs and most importantly first hand experiences with ingredients, cuisines and cultures.' Vetter argues that it is the intuitive part of cooking which is unique to the chef.

Restaurant meals are one of the meal venues considered by Chef Kevan Vetter in his treatment of meals. He reviews the process used by large chain restaurants, in which menu decisions can take nine to twelve months of 'in-depth consumer research, product or category immersions, ideation and proto-cept development, concept testing and validation, product development, consumer validation of the finished dish and operations'. Smaller restaurants use much simpler approaches, and many menu decisions in restaurants are made on the spot by chefs for both practical and creative reasons. A practical reason might be to use an over-supply of ingredients, and a creative reason might be to highlight a wonderful ingredient found in the market.

2.2.3   Institutional meals

Meals in science and practice includes some areas of foodservice not addressed
in detail in *Dimensions of the meal*, and one of these areas is Institutional
Meals which is covered by Williams (Chapter 4) and by Edwards and Hartwell
(Chapter 7). Williams divides institutional meals into the following meal
categories: morale-centred meals, manners-centered meals, and medicine-
centered meals. He describes a number of institutional meal service types:
cafeteria, delivered meal trays, military ration packs, and supplementary
feeding.

'In the morale-centred meal services, there is a particular emphasis on
planning the meal service to prevent boredom, provide familiar and perhaps
comforting foods to people in otherwise deprived circumstances, or to
demonstrate that the employer cares for the wellbeing of the clients.' Williams
points out that morale-centered meals apply to some military meals, workplace
canteens in isolated locations, and prisons. Military meals are also covered
by Edwards and Hartwell (Chapter 7), and by Darsch and Moody (Chapter
15)

'In a manners-centred meal service one of the articulated roles of the meal
time is to ensure appropriate behaviour is taught and good behaviour reinforced,
while inappropriate behaviour is corrected.' This applies to childcare and
school settings. Medicine-centered meals apply to healthcare (including
hospital) settings.

Edwards and Hartwell point out that institutional meals are one of the
major sources of meals world-wide encompassing meals in 'universities,
schools, armed forces, hospitals, prisons and employee feeding'. Institutional
meals also go by the names 'Not for Profit Sector, Cost Sector, Subsidised or
Welfare'. They suggest that the very large variety of institutional meals
should be described and grouped by their characteristics rather than by their
location. They go on to describe typical meals in a number of different
settings: hospital and welfare, school, university, prison, military, and business.
Edwards and Hartwell describe the categories of institutional meals and the
characteristics of institutional meals, and present a fairly comprehensive
look at meal variables.

Darsch and Moody present institutional meals from the military perspective.
Within the USA, military meals have been the subject of extensive research
and development since the 1940s. Darsch and Moody present military rations,
designed for feeding soldiers in the field (not in cafeterias), from both an
historical and from a contemporary perspective. They note the technology
advances that underlie ration improvements over the years, and they detail
both technological limitations put on military rations, such as very long shelf
life, compared with civilian meals, and the same consumer requirements for
good flavor and variety.

One of the important characteristics of institutional meals is the perception
that they are of lower quality. This was shown some time ago by Cardello
et al. (1996) and has been the basis for explaining the differences in expectations

of how good meals will be in different service locations (Meiselman *et al.*, 2000; Edwards *et al.*, 2003). Edwards and Hartwell state: 'In many cases, individuals still have negative attitudes, either real or imaginary, towards institutional meals and perceive them being inferior in quality to those found in the commercial sector. These perceptions are, however, changing as the quality and standard of meals improve ...'

2.3   Meal patterns: breakfast, lunch, dinner

Most chapters in this book work on the model of the three meal pattern. McIntosh *et al.* (Chapter 10) note that 'The three meals a day pattern appears to have arisen in Western Europe as a result of industry pressure on workers to spend less time eating and more at work. Historically, four meals a day were eaten'. Fjellström (Chapter 11) confirms this daily three meal pattern in the West. Yet many cultures have eating patterns of main meals and lighter meals or snacks. Monteleone and Dinnella point out that breakfast is not a historical meal in Italy and, today, two morning snacks better describe morning eating in Italy. And even within the four Nordic countries reported by Kjærnes *et al.* (Chapters 5), one can discern similarities and differences in meal pattern. Kjærnes *et al.* look at the pattern of eating in the Nordic countries and observe 'distinct rhythms which are relatively uniform within each country and, for the first part of the day, across the countries. Different national patterns are identifiable in terms of the degrees to which people eat at the same time and the same type of food'.

Grignon and Grignon (Chapter 16) report very high regularity of the three meal pattern in France with more than 90% reporting eating each of the three daily meals. Kjærnes *et al.* observed much greater variability in the Nordic countries. Breakfast had a clear peak of eating. They found no hours when more than 35 per cent of the respondents had eaten in Norway, and only peaks of 40–50% at lunch time in Finland and Sweden. There were also peaks of eating mid-morning, mid-afternoon and in the evening.

Deliza and Casotti (Chapter 18) provide an interesting view of the traditional/historical meal pattern in Brazil up to the beginning of the 20th century: '...the first meal, a "lunch" at around seven o' clock in the morning; the second, the "dinner" at around midday; later followed by the "snack", a short meal around three in the afternoon and the 'supper' at around six o'clock'. This is four meals as suggested by McIntosh *et al.* for the traditional pattern. The contemporary Brazilian meal pattern has shifted to the more common western pattern of breakfast, lunch, and dinner. Breakfast, especially in cities is not a family meal and is usually coffee, milk and bread. Lunch is larger and usually includes rice, beans, meat and vegetables. Dinner can be either lighter and almost snack-like, or similar to lunch. Another example is Thailand, which has a three-meal pattern, but breakfast can be either a hot meal or a lighter cold meal. So there is not only between-country variation in meal pattern, but also within-country variation.

Cox notes that there are scant national data on meals in Australia. He describes Australian meals by weekday and weekend, based on a dietary survey of educated Australian women. The Australian meal pattern is breakfast, lunch, dinner. Breakfasts are composed of mainly cold cereal and fall within a pre-work time slot. Breakfast consumption on weekends is more varied in time but similar in composition, still with an emphasis on cold cereal. Sandwiches and bread accompanying savory foods dominate weekday and Saturday lunches. For many people, the Sunday cooked lunch still exists in Australia, followed in the evening by a cold dinner. For lunches and dinners, Asian foods appear to be growing in popularity in Australia. The variety of foods consumed is very large. The 65 women surveyed collectively ate between 35 and 53 different food types at weekday and weekend lunches and dinners.

Clay *et al.* (Chapter 27) describes meal practices for Muslims (halal food) and Jews (kosher food) and the prayers which are said both before and after meals. For Jews, the prayers before meals depend on whether the meal contains bread, which defines a meal, and determines the content of the prayer. Jewish law further distinguishes simple food meals (arucha) and feasts (seudah) and then further breaks those down into *aruchat boker* (breakfast), *arucha mispachtit* (family meal), *arucha iskit* (business meal), and *arucha meshutaf* (joint meal/meal for everyone).

Many other chapters in this book deal with what we eat – Kanarek and D'anci (Chapter 25) deal with the effects of what we eat on the individual, not on social and family issues. They report on the effects of meals on human cognition, performance, and mood. Not surprisingly, eating meals is important for mood and performance. They begin with breakfast and note: 'Individuals who eat breakfast on a regular basis, on average, consume a more nutritious diet, have micronutrient intakes which better match the recommended daily allowances and are less likely to be overweight than those who forgo the morning meal'. More specifically, 'Children and adults who consume breakfast perform better on tasks measuring problem-solving abilities, logical reasoning, short- and long-term memory, and attention than those who skip breakfast'. Breakfast is also associated with increased vigor and reductions in fatigue. For the elderly with poor memories, eating a complex carbohydrate for breakfast improves cognitive performance and, for children, eating a high-calorie breakfast improves performance in creativity, physical endurance, and mathematical reasoning.

Most people are familiar with the post-lunch slump and, not surprisingly, eating lunch usually leads to decreased energy and fatigue, especially for high carbohydrate meals. Research shows reduced mood and energy following either a high carbohydrate or a high protein meal. Greater increases in fatigue were seen two hours after the high-carbohydrate, supporting the idea that high-carbohydrate meals promote sleepiness. Eating a lunch with balanced proportions of carbohydrate and fat was associated with better cognitive performance, less fatigue or drowsiness, and better mood in comparison to meals high in either fat or carbohydrate. People fed a high-fat lunch performed more slowly, but also more accurately, on a selective attention task, and reported feeling calmer. Thus, although lunch-time meals are followed by drowsiness or sleepiness, attention to the nutrient composition of the meal can benefit mental performance. This postlunch slump effect may be due to normal circadian rhythms since reports of decreased mood and energy are often seen in the afternoon.

It is interesting to note that there is not the usual distinction between breakfast lunch and dinner for the military rations that are eaten at meal-times. Of course, for soldiers operating in the field, there are no fixed meal times. Eating occurs when it is possible, and is often rushed rather than leisurely. The US military ration does not have a real breakfast ration, and there is no distinction made between lunch and dinner rations. Thus, certain circumstances dictate that meal designations vary from the norm.

2.4 Snacking and grazing

One of the main issues in this book, and in general discussion of meals, is whether meals are declining and snacks or grazing increasing. There is a tendency to assume the decline of meals and the increase of snacking without supporting data. In the discussion of meals above, we have already encountered examples of data on snacking, for example in the Nordic eating pattern reported by Kajærnes *et al.*

Deliza and Casotti (Chapter 18) report on snacking in Brazil: interestingly they have a number of different ways of referring to this, making it more difficult to ask about this behavior in questionnaires. Expressions include "'snacking", "have a little something to eat", "eating junk" or simply resort to referring to what they are eating using the diminutive form: "um cafezinho" ("a small-coffee"), "um pãozinho" ("a little-bread-roll"), um "lanchinho" ("a little-snack")'. The international language for snacking and grazing is truly amazing as the reader will see in other chapters.

Kjærnes *et al.* (Chapter 5) discuss the prevalence of grazing in the Nordic countries which they refer to as 'eating in passing'. They measured eating at a fast-food outlet, in the street and in the home in places other than the kitchen or at a dinner table. Perhaps surprising to many people who feel that grazing is overtaking traditional eating, they found that grazing outside the home 'constituted a small proportion of daily eating (average 4–6%). At home, very little of the recorded eating took place at other sites (at a work desk, in bed etc.) or without sitting down at all'. Mäkelä (2001) has pointed out previously that grazing does not seem to be a major trend in eating, and this has been confirmed by Mestdag (2005) with data from the Low Countries. In *Meals in science and practice*, the reader will see many other data on the composition of snacks and the pattern of snacking.

In addition to reporting behavioural effects of meals, Kanarek and D'anci (Chapter 25) report on the effects of snacking. They note that: 'Although confection-based (sugary) snacks may be perceived as unhealthy, they do have positive effects on mental performance'. For example, college-aged students perform better following a confectionary snack or fruit-flavored yogurt. In other examples, both college-aged and school-aged children showed improved cognitive performance with confectionary snacks.

Kanarek and D'anci distinguish short-term effects of foods and snacks, i.e. acute effects, from longer term or chronic effects. They note that chronic intake of snack foods can have a large effect on mental function in both adults and in children. It appears that we need a more reasoned view of snacking.

2.5   Gender

The subject of gender fits into many chapters of this book. Within the meals realm, there is strong reason to believe that gender differences will exist because gender roles with respect to food have traditionally been different, and some chapters point out both how that continues and how it is changing. As more people begin to examine eating alone, as Pliner does in this book, I expect more studies looking at women.

Gender continues to be an important topic with respect to meals. Gender is important for meal planning and preparation, and is also a factor with respect to meal preferences. Gender is also an important factor with respect to health, for example with respect to dietary restraint. For generations, females typically planned and prepared meals in the home and, for generations, males were traditionally chefs in restaurants. Both traditions are changing, but females continue to prepare most meals in the home. In restaurants, females are now chefs in many countries.

While Pliner's entire chapter (Chapter 9) addresses eating alone, Jonsson and Ekstrom (Chapter 12) also discuss eating alone, specifically addressing females eating alone in restaurants. They discuss the social rules of eating out, the difficulties of eating out alone, and the groups (Communal Tables in the USA, Talking Tables in the UK, and Tables for Sitting Together in Sweden) that have sprung up to address the issue. Jonsson and Ekstrom present reasons why people prefer to eat with other people in restaurants, and present data on eating dinner alone in restaurants. As dining out continues to increase in many countries, the issues of eating out alone will also increase, and the role of the lone woman diner might change.

Ueland (Chapter 6) also discusses the role of gender in meals. Given the emphasis on the social aspect of meals, it is not surprising that the variable of gender presents some interesting differences focusing on food preferences, food-related attitudes and behaviors. Ueland emphasizes the traditional role of women in meal preparation, which continues to the present day behavior in women's deference to men's preferences, and women's role as the major drivers for healthy meals. Men who live with women have healthier diets! Today women, more than men, choose different foods depending on whether they are eating alone, with their children, and/or with men. Women defer to men's preferences, and even more to children's. Fjellström (Chapter 11), in her discussion of European family meals, questions who really controls what food is served: children, father, or mother. Even in families where the father's role is dominant, do the parents cater to children, or does the mother choose based on her own preferences, either consciously or unconsciously? As with other topics concerning meals, we seem to be entering a period when some traditional assumptions are being questioned, and other alternatives are being considered.

The traditional picture of the meal continues with respect to male and female meal preferences, with women preferring vegetables, white meat and fish, and men preferring meat and potatoes. Ueland notes that the different meal preferences of males and females align with their hierarchy of foods based on importance and status. One can still list 'male foods' and 'female foods' although this distinction is affected by the level of education. There are fewer obvious gender differences at breakfast, other than the fact that more females skip breakfast. Dinner preferences reflect the overall gender differences.

Kjærnes *et al.* report from their study of Nordic meals that gender received considerable attention. They observed very few gender differences in meal patterns, acknowledging the large gender differences for health issues. They warn that it might be lone diners who differ – in this case lone women diners. They also confirm that across all four Nordic countries more women cooked the hot meals. One of the more humorous gender observations is that 80% of the female respondents claimed that they had cooked the meal, but over 50% of males said they had cooked the meal. Kjærnes *et al.* are kind enough to suggest that the problem might lie in different ideas of what constitutes cooking. Klein (Chapter 21) also reports on gender and meals in China, noting the change in women's role in the labor market but no change in women's role with respect to domestic work such as cooking meals. As noted with Muslim halal meals, men and women sometimes eat separately, and meals do not represent the family meals discussed in this book with respect to western countries, although this is changing in some areas. The separation of genders extends to restaurant eating, which can be male dominated. Another demographic issue discussed with respect to Chinese meals is age; traditionally the elderly received special attention in many spheres of life including meals. This is now changing, and food, especially in the cities of China is becoming more attuned to youth as it is in the West.

2.6   Eating together/eating alone

2.6.1   Family meals

Just as the terminology of meals and snacks can be confusing, so the terminology of family meals can be confusing as McIntosh *et al.* and Fjellström discuss in Chapters 10 and 11, respectively. McIntosh *et al.* distinguish the family meal, which is the items consumed (the food), from the habitual patterns in the preparation and consumption of meals (the family meal ritual).

There might be some confusion between home meals and family meals; as McIntosh *et al.* point out, a home meal 'does not guarantee' a family meal, since one cannot assume that the whole family is present. In fact, McIntosh *et al.* make it clear that there are no 'precise data' on family meal participation in the USA. Part of the reason for this is that many nutritional studies of eating focus on the individual as the unit of study rather than the family. McIntosh *et al.* and Fjellström agree on the difficulties of studying families in terms of whom you study and with what methods. McIntosh *et al.* note that the members of the family have changed and the structure and content of meals have changed. Those making pronouncements on family meals would do well to read these sections, as well as Chapter 3 by Mäkelä, and understand the many decisions about methods that have to be made in order to study family meals. This seemingly simple concept, the family meal, is not simple to investigate.

Fjellström agrees on the difficulty of studying families because it is not clear how to define family membership and families as a whole. Fjellström asks: 'We may ask, however, who we define as a family member. Is it the partner we live with, the person we have children with? Is it the nuclear family we associate with the family dinner or are other relatives included? Can an au pair living with a family for longer periods be included …'? What is a family today? Fjellström goes on to question the perception that the family meal 'is the most ideal, optimal and universal eating occasion.' This reminds me of the finding some time ago of the many family disputes which occur at holiday times, which are thought to be periods of great joy.

Fjellström questions the historical validity of family meals: 'Do we actually know if the family meal has been a natural part of human everyday life throughout history…' She cites Murcott (1997), who also questioned the assumption that historically family members had a structured family meal.

In their large study of Nordic meals, Kjærnes *et al.* report that most breakfasts were eaten alone, lunch was with family or colleagues at work, and dinner was with family if one lived with family. They also report that the ideal of the daily family meal is not met; about half of families report everyone present at a meal. The norm appears to be everyone present approximately every-other-day for both households with children and in households without children.

Researchers have not only looked at the prevalence of family meals and their patterns, but they have looked at their consequences. McIntosh *et al.* note that a lot of expectation is placed on the (American) family meal. The family meal is seen as a symptom of societal health, and also as a medicine for any lack of societal health. It is not clear whether the concern is about family health or greater individuation among family members. Kjærnes *et al.* reflect this same issue (family–individual) seeing meals as a matter of compromise between individual and social concerns. This issue is addressed in great depth by Mäkelä, with a broad discussion of the current balancing of family/social meals and individual meals or individual expression at meals. Many chapters in this book emphasize that meals are training grounds for young children, but Fjellström questions this assumption; children maybe have a greater influence on other family members, especially in their food preferences if the children's food preferences dominate at the table.

Thus, the topic of family meals brings in topics which are well beyond the scope of this book (What is a family?) but includes topics which are within the scope of this book (What is a meal?). Fjellström, Mäkelä, Pettinger and Marshall (Chapter 28), McIntosh *et al.* and others contribute to this discussion. Many chapters address the related topics of working parents, working mothers, television viewing, and other contemporary topics within the home.

The family meal can also serve as a basis for nutrition education. Chrzan (Chapter 13) also points out the perceived importance of the family meal 'as a symbol of family functioning' and a 'pre-eminent space for the cultural training of children in manners, social skills and nutrition'. She points out, however, that 'relatively few nutrition education programs explicitly use the meal as a tool of education…'.

Chrzan describes the use of meal concepts to teach nutrition education to at-risk teenagers in the USA. She compares her approach, based on nutritional anthropology, with traditional dietary counseling and its emphasis upon 'component and nutrient-based curricula'. In this approach, the emphasis is on eating more of certain nutrients and foods containing them. She claims that this approach loses the meal context of eating, and she argues that the meal context is a better way to teach nutrition. By focusing on family dietary intakes, and relating that to cuisine and cultural meal patterns, Chrzan specifically aims at increasing fruit and vegetable intake.

2.6.2   Eating meals alone

We now turn from family meals to eating alone which is covered in Chapter 9 by Pliner and Bell. They begin by acknowledging the great emphasis on eating meals with others, which is called commensality. This emphasis goes back to the important writings of Douglas and continues to this day. Douglas's work is reviewed in a number of chapters in this book, especially by Mäkelä and by Pettinger and Marshall. Pliner and Bell make the important point that solitary dining is not a unitary concept; people dine alone for many reasons and under many conditions. Two main points of this book are that family dining is not a simple concept and dining alone is also not a simple concept.

Pliner and Bell try to gauge the scope of living alone and dining alone, and report limited data and limited agreement on trends among the data, recalling the conclusion by McIntosh *et al.* on the difficulty of finding normative data for family meals. Pliner and Bell also report a lack of information on lay people's views of eating alone, despite the widespread views of academics on the subject. Limited questionnaire data from students and soldiers both support the academic view of meals as social. A questionnaire produced feelings of 'boredom, loneliness, and self-consciousness' as reasons for not eating alone. Jonsson and Ekstrom in Chapter 12 also add the reason of poor service in restaurants.

Pliner and Bell make the interesting suggestion to revise the term 'social facilitation' to 'solitary inhibition', and to look for reasons why people eat less when alone rather than why they eat more when in groups. It might be the aversiveness of eating alone, which is covered by Jonsson and Ekstrom.

2.7   National meals

Meals within different cultures represent two different and interesting phenomena, the pattern of uniformity of the meal for a culture combined with the pattern of regional differences within that culture. Cultures present us with both uniformity and diversity. Below, I review some of the regional diversity presented in this volume when describing meals of different cultures. First, I will examine some of the findings in this book on the unifying themes within these cultures. While reviewing these typical national meals, we need to keep in mind Mäkelä's argument that these typical national meals are changing; for example, Western meals which once centered on a meat dish, might now center on a fish, chicken, pasta, or something else. Some of the concern expressed over the disappearance or decline in meals, especially family meals, might be related to the changing composition of meals. A number of chapters in this book present sample meals in different cultures. These examples give a clear picture of meal composition, both traditional and contemporary.

Grignon and Grignon (Chapter 16) emphasize that French meals display a common daily schedule and common meal composition. The data on the very high frequency of French meal consumption is quite startling – over 90% of French people report eating the three meals of the day. This shows a high degree of uniformity in the French meal culture, and this alone might distinguish it from many other cultures.

Main Italian meals, as discussed by Monteleone and Dinella (Chapter 17), continue to have a clear structure of sequential courses: '... a first course of complex carbohydrates (*primo piatto*, e.g. pasta or rice), a second course that serves as source of protein (*secondo piatto*) and includes a side dish (*contorno*, e.g. salad or cooked vegetable), and as a final course a fruit or dessert ...' And the authors point out that pasta, which is eaten today throughout Italy and perhaps throughout the world, only became an important part of the Italian meal in Italy in the 19th century.

Similarly, the Brazilian meal almost always includes rice and beans. The presentation of the Brazilian meal differs from that in many western and Asian countries where dishes are served individually and sequentially. As

Deliza and Casotti (Chapter 18) point out '... several types of food are not only served together at the table, but are also placed on the same plate.' They further point out that mixtures typify the dishes of the Brazilian meal. '*Cozido, feijoada* and *moqueca*, for example, all seem to typify the Brazilian preference for one-pot meals that are neither liquid nor solid and which combine "something of everything".' *Feijoada* typifies the country and its cooking by representing three different races and food cultural backgrounds.

Whereas Brazil's colonization is one of the influences on its meal structure, colonization is also an influence on the meal pattern of Thailand, along with religion, the monarchy and national economic directions. The beans and rice foundation of Brazilian meals is replaced with 'four fundamental ingredients...rice, fish, dipping sauce (*Namprik*) and herbs'. But Seubsman *et al.* in Chapter 20 point out 'The importance of rice to the Thai people is economic, symbolic and nutritional.' Rice is much more than a foodstuff, and rice is the basis of Thai meals across all age and income groups. The complementary dishes served with rice are called *kap khao*, and the most common is fish, eaten along with a variety of vegetables, spices and herbs. A Thai meal usually consist of three to five dishes varying in color and flavor. Along with rice, fish sauce is a daily ritual in Thai cuisine, and is made by fermenting small fish in a heavy amount of salt for one year, after which it is distilled, bottled, and added to virtually all Thai dishes.

The meal pattern within the two most populous countries in the world demonstrates a largely vegetarian or agrarian diet (discussed by Klein in Chapter 21). The Chinese traditional meal pattern is based on a relatively larger portion of cooked grain (called fan) and a relatively smaller portion of side dishes, which are vegetables, pulses or meat (called cai). As Klein points out, 'The fan/cai distinction serves as the basis for an entire "grammar" of food.' The proper balance between fan and cai determine whether a food occasion is a meal or a snack, and determines whether new western fast food is considered a meal (it is not a meal but a snack). The Chinese usually eat two fan/cai meals per day at mid-day and evening. The Indian meal pattern is similar in that it is centered in a cereal (wheat, rice, millet, sorghum or corn), served with lentils or pulses. These are augmented with a small amount of meat, fish, and vegetable (like cai in Chinese cuisine) and the meal is usually accompanied by flavorings in the form of dairy, chutney or pickles. After the meal, they sometimes serve buttermilk or yoghurt. These Chinese and Indian meal patterns cover the whole of each country and represent truly national cuisine meal compositions for these two vast countries.

A number of chapters discuss religious influences on national meals. Chapter 19 Indian meals (Sen), Chapter 20 on Thai meals (Seubsman *et al.*), and, to a lesser degree, Chapter 21 on Chinese meals (Klein), present religious influences within their national meals. It is interesting to note that there is generally very little influence of Christian religions on meal structure and composition, so that most meals in the West are relatively little influenced by religion as compared with meals in the East and Middle-East. In this book,

Clay *et al.* present kosher and halal meals, which are obviously much broader than national meals. They describe the spread and resulting variety of Jewish food around the world in countries 'each possessing its own unique history, customs, and cuisine'. They also point out the different Sephardic and Ashkenazic foods, and their influence on the Sephardic migrants into North Africa and Western Europe, and the Askenazic peoples in Germany and further east. The most familiar Jewish food in the USA is of the Askenazic form coming from Eastern Europe, and includes bagels, borchst (beet soup), gefilte fish (chopped carp and other fish that made it edible despite the floating 'Y' bones in the fillets), grebenis (rendered poultry skin), schmaltz (rendered poultry fat), kugel (a noodle, potato, or matzos baked sweet main meal starch), tzimmes (vegetables and some sweet fruits with cheap meat), and cholent (as mentioned earlier a bean and cheap meat (e.g., ribs) dish).

Clay *et al.* describe kosher and halal food laws concerning which animals are permitted as food (for example, no pork) and the prohibition of animal blood in food. In addition, the kosher laws prohibit mixing of milk and meat and have special rules for Passover meals. The halal laws also prohibit alcohol. These laws make it very difficult for Jews and Muslims to share meals prepared by non-observant people, and even to eat in restaurants and other venues. For Jews, the separation of meat and milk requires two separate food preparation areas and equipment in the home, and in kosher hospitals and campus dining halls. Kosher restaurants often function as either dairy or meat establishments.

2.8   Convenience meals

When discussing family meals or national meals, one cannot ignore the important topic of convenience meals. Convenience products and convenience food sources are contributing to changes in meals world-wide. Many chapters deal both with changes which have occurred from use of convenience foods and possible trends in the future. Convenience foods include the following sub-topics: prepared foods/meals, take-away foods, cooking meals vs. buying meals. In some countries, there are now large industries providing both foods and entire meals to homes, businesses, and institutions.

As with many other trends, there might be a danger in over-estimating how much change has occurred in the direction of convenience meal. For example McIntosh *et al.* present conflicting data that between $1/3$ and $1/2$ of US meals are made 'from scratch'. Some use convenience products, but some use take-out food from a restaurant or market, or simply eat in a restaurant. Eating in restaurants account for about half of the food dollars in the USA. This figure is growing in the USA and in many other countries.

In their discussion of American family meals, McIntosh *et al.* also note that whether food is cooked from scratch or served from one of the above convenience categories has implications, since children participate less in family meals when mothers use takeout foods more frequently. Mäkelä also comments on the growing use of convenience products as part of the balancing of social issues and individual issues in meal preparation. The chapters on national meals present the current situation and trend for convenience foods in India, China, Brazil, Thailand and other countries.

2.9   Regional meals within countries

The chapters on meals from different countries and cultures demonstrate the importance of regional cuisines in those countries. It is important to emphasize that all of these countries are large countries with large populations. Several of the countries are among the largest countries in the world in both size and population. Nevertheless, the importance of regional meals is important to understand.

In Italy, Monteleone and Dinella (Chapter 17) point out that the meal sequence of three courses and the general meal composition is retained throughout the country. Many of the same dishes appear throughout the country, but their composition reflects local ingredients – thus there is variability in dishes with the same name. For example, three different regions of Italy have three different recipes for lasagna, with lasagne in Emilia Romagna (in the north) including the local prosciutto and parmigiano. In Campagna, the lasagne might feature mozzarella and ricotta, and in Liguria the same dish is lightened with mushrooms. Different regions use different seasonings for the same named dish. For the all important pasta, there are regional differences as well. Compared with what we will see below in some other cuisines, the Italian meals seem more similar containing more of the same foods in name with some variation in ingredients. Perhaps this is what is meant by a national cuisine, if indeed such a thing exists (see below).

In Klein's presentation of Chinese meals (Chapter 21), he points out that China has very different agricultural environments which led to different products in different parts of China. China can be divided into at least four regional cuisines, although greater division is possible. The rice regions of China include the Eastern region associated with Jiangsu meals, the Western region, and the Southern region associated with Cantonese and Sichuan meals. The non-rice region in the North is associated with Shandong meals. Klein raises the intriguing question of whether the gradual blurring of regional cuisines will lead to one national cuisine. This, of course, raises the question of whether other countries with long traditions of cuisine have national cuisines – do the French have one cuisine?

Sen presents the regional Indian meals (Chapter 19), again stressing the enormous variety in this large and heavily populated country with an extensive variety of environments. Sen presents four examples of regional meals: Uttar Pradesh, West Bengal, Gujarat, and Andhra Pradesh. These four regions all use rice and or bread as the meal staple, but which is used, and for which meal, and with what accompaniments varies a great deal. Sen points out that the Southern Indian breakfast meal is now served worldwide in many Indian restaurants as lunch or dinner!

Considering another large and developing country, Deliza and Casotti (Chapter 18) describe the regional meals of Brazil, each with its unique foods. The authors point out that in addition to different products coming from the vastly different geographical areas which influence meal variety, that subsequent immigration within Brazil moved these traditions around, resulting in even more complexity. Of course, the unique Brazilian staple dish is beans and rice which appears everywhere. In the North, fruit is especially used in meals. The Amazon region presents a veritable 'confrontation' of indigenous peoples, Europeans and Africans resulting in a unique meal situation. In the Northeast (Bahia) there is also a strong African influence in foods, and a strong use of street foods. In Middle Brazil (Pantanal), there is an emphasis on the use of meats and sausages, while in the Southeast (Minas Gerais) there is the unique 'table of abundance' presented not according to the usual norms of food combinations, and also the prevalence of dairy products. In the South (Rio Grande do Sul) they have a prevalence of meats in great variety, and a unique bitter tea drink.

While discussing American family meals, McIntosh et al. (Chapter 10) present regional differences in American foods and meals. They point out that in a country based on immigration that ethnic, cultural and regional influences are all apparent in meal selection. Simplistic discussions of what Americans eat miss the point that Americans eat differently regionally and culturally. These cultural influences continue with Italian, Mexican and Chinese restaurants being very popular, and their foods are often eaten in the home. Sen also talks about the growing influence of foreign restaurant food in urban India. Regionally, the USA can be divided into the two coasts (East Coast, West Coast), the South, the Midwest or Plains. In addition, the Mid-Atlantic States are often separated out, and the Deep South is separated from the broader South. For example, an important staple food in the South, seen especially at breakfast is grits, which one will rarely see in the North.

Thai meals also show regional differences. Seubsman et al. (Chapter 20) describe the varieties of meals in four Thai regions. Central Thailand shows both an Indian influence and a Royal Household influence in their meals, Northeast Thailand uses a lot of herbs and spices but less liquid, while the North serves meals that are less spicy based on less use of chili and the South favors seafood dishes. A good example of regional variation is the common fish sauce which varies from region to region; in the Northeast it might be made with fermented fish, while in the North it might contain crab essence producing rather different flavors.

2.10   Questions for further discussion and study

I am not ending this summary and review with a statement of clear conclusions. This book is far too rich and varied to permit simple conclusions. Rather, I will end this review with a number of questions which, I hope, will stimulate further discussion and study:

1. Do we eat three meals a day, or two meals a day? Should we consider breakfast a meal, and is breakfast becoming more or less meal-like? Albala (2002) presents an interesting discussion of the history of breakfast, and perhaps we need to revisit this history and this discussion. Breakfast is often solitary rather than social, breakfast often includes very few food items, and breakfast is more often skipped than other meals. Is breakfast a meal? And, if it is still it a meal, is it becoming less meal-like?

2. Are meals disappearing or just changing in composition? Whether one defines meals by their social nature or by their composition, it is clear that many meals still occur. However, it also appears that meal composition has changed during the 20th century, and continues to change in the 21st century. Do we need new concepts of meals that include these changes in composition? Perhaps 'meat and two veg' might be too traditional for a meal definition in the 21st century.

3. Are family meals disappearing or just changing in frequency and composition of who is included? As pointed out in this book, we are not certain that family meals with all members present were the norm throughout history. As we enter the 21st century with changes in families and in households, we need to again redefine family meals. Family meals appear to occur frequently and perhaps continue to be the norm, but family meals well into the 21st century might require a further redefinition of the family with respect to the family meal.

4. Do national cuisines exist, and if so, are they continuing to develop or are they gradually disappearing? Perhaps no question is more provoking with audiences than the question of national cuisines. People feel strongly about their national foods and national cuisine. Are national cuisines continuing to develop, or are they being weakened and eroded? As with meals, and family meals, do we need a redefinition of national meals and national cuisines?

The broad question is what is in the future for meals? Meals are a combination of the traditional, the contemporary, and the future, and they will remain an exciting area for study and discussion.

2.11   References

Albala, K. (2002) 'Hunting for breakfast in medieval and early modern Europe'. In Walker, H. (Ed.) *The Meal*, Totnes, Devon: Prospect Books, pp. 20–30.

Cardello A. V., Bell R. and Kramer F. M. (1996), 'Attitudes of consumers toward military and other institutional foods'. *Food Quality and Preference*, **7**, 7–20.

Douglas, M. (1976), 'Culture and food'. Russell Sage Foundation Annual Report 1976–77 (pp. 51–58). Reprinted in M. Freilich (Ed) *The Pleasures of Anthropology*, New York: Mentor Books, 1983, 74–101.

Douglas, M. and Nicod, M. (1974), 'Taking the biscuit: the structure of British meals'. *New Society*, **19**, 744–747.

Edwards, J. S. A., Meiselman, H. L., Edwards, A. and Lesher. L. (2003) 'The influence of eating location on the acceptability of identically prepared foods'. *Food Quality and Preference*, **14**, 647–652.

Mäkelä J. (2001), 'The meal format', In Kjærnes, U. (Ed.), *Eating Patterns: A Day in the Lives of Nordic Peoples*. Report No. 7–2001. Lysaker Norway: National Institute for Consumer Research, 125–158.

Meiselman H. L. (1992a), 'Methodology and theory in human eating research'. *Appetite*, **19**, 49–55.

Meiselman H. L. (1992b), 'Obstacles to studying real people eating real meals in real situations: Reply to Commentaries'. *Appetite*, **19**, 84–86.

Meiselman H. L. (Ed.) (2000), *Dimensions of the Meal*. Gaithersburg: Aspen.

Meiselman, H. L., Johnson, J. L., Reeve, W., and Crouch, J. E. 'Demonstration of the influence of the eating environment on food acceptance'. *Appetite*, 2000, 35, 231–237.

Mestdag, I. (2005) 'Disappearance of the traditional meal: temporal, social and spatial destructuration'. *Appetite*, **45**, 62–74.

Murcott A (1997), 'Family meals: a thing of the past?', in Caplan P. (Ed.), *Food, Health and Identity*, London and New York, Routledge, 32–49.

Part II

Defining meals: definitions of the meal reconsidered

3

Meals: the social perspective

J. Mäkelä, National Consumer Research Centre, Finland

Abstract: Sociability of meals is studied from a social science point of view and the influence of classic British meal definitions on research on meals is discussed. Commensality is essential to people as human beings and an inherent core of meals. Therefore, the chapter explores the ongoing debate on individualisation and de-structuration of meal eating. Are these trends really happening or are new types of meal patterns emerging? The central question in relation to the future of meal eating is whether eating meals together continues to be meaningful in the light of historic and contemporary visions about future food.

Key words: meals, sociability, commensality, de-structuration, individualisation.

3.1 Introduction

Studies of food and eating from a social science perspective have often focused on meals. The reason is that meals and practices that have evolved around their preparation and eating are a telling story about the importance of eating habits to culture and modern societies. The social aspects of eating have especially been addressed in sociological and anthropological studies of meals (Wood, 1995). Many of the explorations have sought inspiration from the influential study by Mary Douglas (1975) in which she explored the structure of meals in detail.

The sociability of eating is essential to people as human beings. Therefore, commensality – the act of eating at the same table – is often seen as an inherent core of meals (Sobal, 2000; Mäkelä, 2000; Visser, 1996). A meal is social action regulated by certain rules and behaviour (Mäkelä, 1991). Meals eaten alone can be seen as something unwanted (Sobal, 2000) or even as 'an oxymoron', (Pliner and Bell, Chapter 9) although solitary meals regularly occur (Sobal and Nelson, 2003). Meals mark social relations and they have a vital role in everyday life and in festive occasions. The notion of eating as reproduction of social relationships is crystallised in meals that allow people to eat the same food at the same time and therefore to share the ideas of commensality attached to meals.

This chapter is about sociability of meals from a social science point of view. It starts with a description of the classic British meal definitions and how they have influenced research on meals. The following section gives examples of the commensality and reciprocity related to meals. The next part addresses the ongoing debate on individualisation and de-structuration of meal eating. Are these trends really happening or are they just hype? After that, there is a deliberation on new types of sociability in contemporary meal patterns. In the final section, the future of meal eating is examined. The central question is whether eating together still remains meaningful in the light of historic and contemporary visions about future food.

3.2   Classic meal definitions

One of the main concerns of social scientific research on food and eating has been the dissection of meals and their structure. Since the 1970s, British scholars in particular have been both active and influential in the analysis of the meal structure (Marshall, 2000; for a review of cultural definitions of a meal, see Mäkelä, 2000). However, the idea of connecting meals to a wider social system in which a meal is one ordered system related to other ordered systems (Douglas, 1975) has been the dominant paradigm of the sociology of food for the past 30 years.

The approach has been eminently clear in the work of Mary Douglas, who together with Nicod analysed the structure of British working-class meals (Douglas and Nicod, 1974). Their criteria for the classification of meals were complexity, copiousness, ceremoniousness, and the structural and sensory qualities embedded in the binary oppositions savoury/sweet, hot/cold, liquid/dry. To Douglas, meals and drinks are two important opposite categories, as in the relation between solids and liquids. A meal has both solid and liquid components and it has to have a dimension of bland–sweet–sour. The meaning of a meal is based on a system of repeated analogies (Douglas, 1975).

A meal is not, however, the only type of eating. Douglas and Nicod presented four terms to describe different types of eating. 1. A 'food event' is an occasion when food is eaten. 2. A 'structured event' is a social occasion organised by rules concerning time, place and sequence of action. 3. Food eaten as part of a structured event is a 'meal'. A meal is connected to the rules of combination and sequence. 4. A 'snack' is an unstructured food event without any rules of combination and sequence.

The meal system proposed by Douglas and Nicod consists of three types of meals: (A) a major meal/the main meal, (B) a minor meal/the second meal, and (C) an even less significant meal/the third meal (biscuits and a hot drink) (Douglas, 1983; Douglas and Gross, 1981; Douglas and Nicod, 1974). Thus, the model also contains complementary classifications. The first course of a meal (A) has, both on Sundays and weekdays, the same basic structure based on a staple (potato), a centre (meat, fish or egg), trimmings (vegetables), and dressing (gravy). Everything is savoury and hot. The second course has the same structure except that everything is sweet. The staple is cereal, the centre is fruit and the dressing is liquid custard or cream. Meal (B) follows the structure of meal (A), but the staple is cereal (bread) and not potato.

A few years later, Murcott (1982, 1983 and 1986) studied people's concept of 'a proper meal' in South Wales. In this qualitative study, a proper meal always proved to be a cooked dinner of a certain type. The creation of a proper meal requires transforming ingredients by cooking and combining them into a meal. A proper meal consists of one course only, which is always a variation on meat and two vegetables. 'Meat' must be (fresh) meat; sausages or offal do not qualify. Poultry will do as a last resort, but fish is definitely ruled out. Potatoes and other vegetables are necessary and at least one of the vegetables must be green. Finally, gravy gives the finishing touch to this plateful and combines all the ingredients into a proper meal.

In the late 1980s, Charles and Kerr (1988) still identified a similar British proper meal as in the studies of Murcott or Douglas and Nicod: a cooked meal with meat (fish), potatoes and vegetables as opposed to a snack, which is bread-based and merely prepared instead of cooked. Murcott and Charles and Kerr extended Douglas and Nicod's ideas by involving the aspect of gender, division of labour, and food preparation as important elements in proper meals [see also Ekström, 1990; and Ueland (Chapter 6) on gender; and Jonsson and Ekström (Chapter 12) on gender in restaurants].

Interestingly, a study on changes in food choice and eating habits in Scotland during the transition from single to married or cohabiting suggests that the notions of a proper British meal have changed in the 1990s (Kemmer et al., 1998). Even though the idea of a proper meal as 'meat and two veg' was self-evident to young couples, this traditional format was seen as somewhat boring. Influences from non-British food cultures have changed the food habits of the younger generation. To them, a salad, pasta, chilli, or curry meets the criteria of a proper meal (see also Marshall, 2000).

It seems that, especially in the Nordic countries, the research on meals has primarily been inspired by the British approach, even though it is also commonly referred to in American studies (e.g. Beck, 2007). In a quantitative Swedish study, Ekström (1990) identified four components in a cooked meal. The main dish gives the name to the course and is usually animal protein. The first trimming is the starchy base of the meal, the second trimming consists of vegetables. Extra trimmings are vegetables or different condiments. Bugge and Døving (2000) used qualitative data to define a Norwegian proper meal as a cooked one with vegetable trimmings. In Denmark, similar concepts of proper meals have been found in qualitative studies, and surveys of dietary practices show that the most typical hot meal consists of meat, potatoes, vegetables, and sauce (Haraldsdottir et al., 1987). A qualitative study on eating in Finland concluded that a proper meal could be defined as 'a hot dish' (i.e. cooked food) accompanied by 'a salad' (i.e. uncooked vegetables) (Mäkelä, 1996). In a Nordic study (Kjærnes, 2001), the eating system included the eating pattern and the social organisation of eating the meal format, which refers to both the composition of the main course and the sequence of the whole meal.

However, the idea of a proper meal as a plateful is not the convention in all European countries. In Italy, a traditional proper meal has four or five courses (see e.g. Fischler, 1990, p.160 and Chapter 17 on Italian food in this volume). In France, the proper or real meal ('un vrai repas') with a three-course structure is well established in the everyday routines of the Parisian bourgeoisie (see e.g. Sjögren-de Beauchaine, 1988). In fact, Jean-Pierre Poulain (2002) describes a French 'proper or full meal' as including four courses: a starter, a main course with vegetables, cheese, and a dessert (see also Grignon, 1996, and Grignon and Grignon Chapter 16 in this volume).

3.3   Commensality and reciprocity of meals

The scholars mentioned above do not only analyse the structure of a meal, they are also especially interested in food as a marker of social relations. Mary Douglas (1975) seeks the connection between food and social relationships. According to her, meals are reserved for family members and close friends while beverages could be offered to other people as well. For Anne Murcott (1982) and Robert Rotenberg (1981), a meal is essentially a social affair. Annick Sjögren-de Beauchaine (1988) describes a meal as ordered social action that includes ritualised sharing of food.

German sociologist Georg Simmel (1910) pointed out the role of meals in the socialisation process already in the beginning of the 20th century. The sociability of eating is related to the refinement of social forms of interaction. People gather together in order to eat meals (ibid.). Eating in company shows how important it is to belong to an 'eating community' (Falk, 1994) or 'commensal units and circles' (Sobal, 2000).

From a social scientist's point of view, a family meal is the meal par excellence (Holm, 2001; see also Fjellström on European family meals and McIntosh *et al.* on American family meals in this volume). The commensality of meals is crystallised in family meals on Sundays and Christmas. These meals are key symbols of the family (Charles and Kerr, 1988). Each year we celebrate the chain of family ties and honour the past generations while preparing the same traditional dishes that our grandmothers used to cook for seasonal feasts. A proper family meal is not only nutritionally important, but vital for the reproduction of social relationships within the family. Eating together means staying together (Charles and Kerr, 1988; DeVault, 1991; Brombach, 2000). Sharing the same table is important for newly married and cohabiting couples (Marshall and Anderson, 2002; Bove *et al.*, 2003). A proper meal is eaten with company (Mäkelä, 1996.) As meals knit people together, they mark and structure both everyday life and festive occasions (Holm, 2003).

The family meals act as an arena for socialisation. Children learn about food and how to behave at family meals (McIntosh, 1996) and discussion during dinners is an important part of communication within families (Holm, 2003). At the table, information is shared as family members talk together (Coveney, 2000). Furthermore, meals civilise the children in a certain culture. Parents usually have to make a project out of promoting and requiring table manners. Good table manners are considered an integral part of desirable social behaviour, especially among middle-class families. Children learn the limits of edible and inedible food, but, in addition, they learn to manage their bodies and to appreciate the proper way to both cook and eat food in general (Bell and Valentine, 1997; Fischler, 1986; Grieshaber, 1997; see also Fjellström, Chapter 11, and McIntosh et al., Chapter 10, in this volume on European and American family meals).

The provision of food is an important part of the social organisation of households. Cooking has been and still is a female chore (Murcott, 1983; Charles and Kerr, 1988; Ekström and Fürst, 2001; Bahr Bugge and Almås, 2006). A (British) meal symbolises home and the relationship between spouses (Murcott, 1982; 1986). Even though cooking can be either a pleasant or an unpleasant chore, the point of it is that people usually cook for others. As Anne Murcott (1983) aptly writes in her article, 'It's a pleasure to cook for him'. A self-cooked meal at home is a gift while a meal consumed in a restaurant is more like a commodity (Fürst, 1997). An essential part of this gift is the very idea of sharing it with other people, family or friends (Sidenvall et al., 2000).

Furthermore, the social nature of eating is also present in other types of food exchange and reciprocity. The forms of exchange include hospitality, sharing other eating activities than meals, exchanging foods, giving speciality foods as gifts or payments for services, and cooperative cooking (Thephano and Curtis, 1991). Dinner or lunch parties are usually based on inherent reciprocity. If the invitation is not returned in appropriate time, there is a threat of being left outside the chain of hospitality.

3.4   De-structuration and individualisation of meals

In the era of individualisation, meal-eating patterns are changing. This has raised discussion among scholars about the demise of the family meal as an essence of sociability and concern for the general public since the beginning of the 1990s (see e.g. Mennell et al., 1992, p. 116; Whit, 1995, p. 146, Warde, 1997; Sobal and Nelson 2003; Bahr Bugge and Almås, 2006). The view is based on the alleged diminishing number of shared family meals and the increase of solitary meals in addition to a growing tendency to eat snacks and fast food.

Market research coined the expression 'grazing' in order to grasp this phenomenon. It has been used to describe the irregular eating patterns that are seen as individualised behaviour, seeking instant satisfaction like cows on the meadows (see Caplan, 1997). Other wordings used in this context include e.g. 'nibbling' (Poulain, 2002) and 'snacking' (Murcott, 1997), which in French is referred to with the term *'grignotage'* (Poulain, 2005; Diasio, 2008; see also Grignon and Grignon Chapter 16 in this volume). In Sweden, the de-structuration of meal eating has been called 'breakfastication' (*frukostisering*) (Ekström, 1990, p. 73) and in Denmark 'eating on the go' (*spisning i forbifarten*) (Holm, 2003, p. 25). So this dear child of the ongoing debate on de-structuration and individualisation of eating has many names and is addressed as a challenge especially to healthy eating habits (see also Kjærnes *et al.* Chapter 5).

If the origin of meal studies lies in the idea of a certain structure with rules as a basis for proper meals, the discourse on individualisation is epitomised – for those who do not see individualisation as a blessing to humans – by the disappearance of these very rules. Fischler (1980; 1990) has repeatedly pointed out that, in an industrialised, urbanised and individualised world, the rules and norms concerning food and eating begin to dissolve as the social practices related to them change. Individuals lose their sense of what, how and when they should eat. The result is a state that Fischler calls 'gastroanomy'. Poulain (2002) has continued this discussion with his concept 'vagabond feeding' as an opposite (or alternative) to commensalism based on structured meals. Falk (1994) predicts that the eating community will cease to be a principle structuring social life. Nowadays, meals are about communication rather than communion. Although meals are marginalized, non-ritual oral consumption, for example, chewing gum, eating snacks and sweets, or smoking, is increasing.

In addition to the present individualisation and de-structuration of eating, the disappearance of seasonal variations is considered to be ruining the traditional annual rhythm of eating and sense of seasons with distinctive raw ingredients (see Grignon, 1992). The range of convenience foods available in the Western world is steadily widening. These developments seem to evoke different reactions. Some see them as a blessing for busy parents, whereas others regard convenience foods as a threat to traditional home cooking and the microwave oven as a medium for individualising meals. The new possibilities either to eat outside the home or to buy convenience foods, home meal replacements (HMR) and take-away foods influence our meal patterns.

It is evident that the societal changes, abundance of food available and industrialisation of food production have changed the structure of our eating. Often the concern regarding this change is epitomised in contentions on the perils of a loss of family meals. However, it seems that this argument is based on idealistic and nostalgic ideas of a past that never actually existed at large. Therefore, the importance of the family meal and concern for its decline are part of a powerful myth. As Murcott (1997) points out, the interesting feature of this idea of a decline is the very contention itself. If we look at meals from the point of view of gender, age, and class, the idealistic picture of family meals starts to fall apart. Even in the past, the children of upper-class families did not share their meals with their parents but ate in their own rooms, and working-class wives merely played the part of a waiter instead of sharing the meal with their husbands.

The importance of meals is often seen as related to the cultural and social aspects of eating, but also the nutritional and health consequences of de-structuration of meals should be acknowledged (Fjellström, 2004). However, the commensality of eating may either promote health or be a risk to healthy eating (Sobal and Nelson, 2003). There is evidence that social facilitation increases the food intake in many cases (Hetherington *et al.*, 2006), but not always (Pliner and Bell, Chapter 9). Therefore, studies on obesity, on the one hand, and disordered eating, on the other, nowadays often focus on meal patterns (Bertéus Forslund *et al.*, 2002; Neumark-Sztainer, 2004). An American study on differences between weight-stable and weight-gaining persons focuses on meal patterns, with the social facilitation of food intake as one of the key aspects (Pearcey and Castro, 2002). A Finnish study concludes that among schoolchildren healthy food choices are associated with the tendency to eat together (Haapalahti *et al.*, 2003). Furthermore, a Danish study suggests that the social context of eating influences the experience of satiety (Kristensen *et al.*, 2002). As the social dimension of eating is taken into account in studies on healthy food habits, the multifaceted nature of our eating becomes recognised.

Whereas the tendencies of de-structuration and individualisation are widely discussed, there is, however, surprisingly little empirical evidence to back up the demise of family meals in particular. Actually, several studies and many of the articles in this volume demonstrate that (family) meals are not disappearing (see e.g. Mestdag, 2005; Holm, 2003; Bugge and Døving, 2000; Kjærnes, 2001; and Fjellström, Grignon and Grignon, and McIntosh *et al.*, Chapters 11, 16 and 10, respectively, in this volume). The meal patterns prove to be more resistant than perhaps has been anticipated by the ongoing debate. Even though people are eating alone as well, they still eat commensal meals surprisingly often. Is this all much ado about nothing or is it time to rethink the sociability of eating?

3.5 New interpretations of the sociability of meal preparation and eating

Even though the discussion on the commensality of meals is often related to family meals, a 'family meal' is only one type of shared meal. Eating together creates a feeling of community and solidarity among people who do not share family ties. For example, people commonly share their meals at work or school canteens. Presumably, a kind of togetherness resembling the family meal may emerge in such fairly stable groups. A lovers' dinner at a restaurant, a teenagers' hamburger meal at a fast food place, and a lunch with colleagues are all meals, but they are set outside the home and the family. Furthermore, the claim of a demise of family meals tends to forget that the traditional family is no longer the representative household pattern in real life. The amount of single-person households is increasing.

As it is evident that our meal patterns are gradually changing, it is maybe time to reconsider how we regard the sociability of meals. Research on meals has been criticised for using a very traditional or even stereotypical notion of a nuclear family with two parents and children. It is apparent that it is hard to squeeze all households into the mould of traditional families. To overcome this dilemma, the term 'postmodern family' has been suggested to describe the variety of households today (Bell and Valentine, 1997).

Preparing meals is an essential part of the social organisation and sociability of meals. What happens when people are freed from the necessity of preparing meals? Interestingly, what seems to occur is that the sociability of meals takes new forms. In a world of increasing variety of ready-made meals, HMRs and convenience foods, the preparation of meals from scratch gets new meanings. The younger generations who have become accustomed to easy and convenient food preparation seem to enjoy cooking as a precious and stimulating hobby. Nowadays, preparing food can be a joint event where hosts and guests cook together adventurous dishes. The shared event of meal preparation extends the sociability of eating together to cooking together. Cooking food together might not be the new bee, such a groundbreaking form of social gathering, but the practice effectively blurs the boundaries of division of labour by gathering both genders into the same kitchen at the same time. It seems that meal preparation and eating take new forms in new types of eating communities or commensal circles extending the nuclear family.

Even though the category of a meal seems to be both theoretically and empirically identifiable, the definition changes over time. During the 1990s, informants in Finnish studies regarded the hot temperature of food as a definite marker of a meal (Mäkelä, 1996) but today a salad – on the wave of increasing awareness of health concerns – would qualify as a meal. Fischler (1990) claims that the traditional order is replaced by disorder and unpredictability, while the findings by Kemmer et al. (1998) indicate the emergence of new types of order. The traditional 'meat and two veg' is not the only structure for a proper meal in a world where Britons consume curries, pastas, and salads with a gusto. It should, however, be noted that most earlier studies have mainly been concerned with characterising typical or even ideal proper meals rather than with describing the heterogeneity of everyday eating in modern times.

The focus has been on meals in general. However, it is interesting to ponder whether the changing ideas and practices of meals are crystallised in how we regard lunches and dinner in relation to commensality. Many studies show that people do eat alone in practice but still consider a shared meal as the ideal (see Fjellstöm, Pliner and Bell, and McIntosh *et al.*, Chapters 11, 9 and 10, respectively, in this volume). Nevertheless, people recognise and accept that all their eating events cannot, within the frame of contemporary life, be commensal meals. Therefore, it seems that dinner has become the quintessential meal. Dinner is *the* family meal (Sobal and Nelson, 2003; Fjellström, Chapter 11 in this volume) and couples also stress the importance of shared evening meals (Marshall and Anderson, 2002).

Meals remain a vital social institution that structures our everyday life. It seems that in real life the change has not been as drastic as anticipated. Our eating patterns have proved to be quite enduring. Yet, we live within continuous change: our food habits are all the time gradually changing, as they always have been. The dialogue between continuity and change moulds our meal patterns, which must adjust to contemporary life with its possibilities and restraints. This might mean a simplification of the meal structure, more snacking and fewer meals, but still the quest for commensal eating is a vivid part of our daily lives. Even though shared meals are seen as important, they do not have to take place every day. Actually, the coexistence of commensalism and individualism is very present in our modern eating habits. Sometimes we need to nurture our bodies, sometimes we want to reproduce our social relationships. For some people a meal is a gift, for others it is a constraint (see Fischler and Masson, 2008). We may have entered an era of 'culinary plurality' (Bahr Bugge and Almås, 2006) that allows us to creatively organise our eating according to our needs and wishes.

3.6   Are we still going to eat meals together in the future?

Commensality still is, despite the continuous and apparent changes, an indispensable denominator of a proper meal. The interesting questions are: Are we going to eat meals together with other people and what will we mean by a meal in the future? The answer to the first question is probably 'yes', but the answer to the second question needs reflection as to how our meal patterns will change and transform. However, trying to forecast the future is a tough task. Belasco's (2006) funny and intelligent book on the history of future food gives us interesting insights to how past visions of future have not been fulfilled. Meal pills have been offered as a solution to food security, the ethical and environmental problems of meat eating, and women's liberation. However, those who foresaw meals-in-a-pill as saviours did not count on how strong the traditional food patterns and values can be.

The reoccurring idea of meal pills, however, is today more appealing than ever. The present discussion and research on nutriceuticals and nutrigenomics again blurs the boundaries between food and medicine (Niva, 2008). Therefore, one serious challenge to the commensality of meals is the idea of tailored diets crafted according to individuals' needs. These promises relate to the present dimensions of the current discourses: health on the one side and individuality on the other side.

The changes in Western societies have an effect on our meal eating. We do eat differently in different situations. The changes in our eating habits are related to changes in our life. Even though the focus here has been on the social aspect of meals, it is evident that meal patterns and their social organisation are influenced by the changes in the local and global food systems surrounding us. The trends and possibilities in eating are many and often even contradictory. On the one hand, convenience of meal preparation and eating is preferred. On the other hand, the origin of foods interests consumers. The heated discussion on climate change pushes forward increasing awareness of sustainability of both the production and consumption of food. The principles of healthy eating are known to almost everybody. Yet, the enactment of healthy eating habits is not easy in the context of busy everyday life.

Nevertheless, it seems that it is relatively safe to foresee that the commensality of meals is not going to be lost very easily. The contents of our plates are probably going to change more than the idea of sharing food with other human beings. Pollan (2006) in his quest for a perfect meal among different types of food chains always relishes these meals in the company of other people. Perhaps the various types of meals in the end still partake of this vital feature of how different their ways to the table have been. Even though a meal might mean many different things to people, commensality is the essence of our culture.

3.7   References

Bahr Bugge A and Almås R (2006) 'Domestic dinner. Representations and practices of a proper meal among young suburban mothers', *Journal of Consumer Culture* **6**(2), 203–228.

Beck ME (2007), 'Dinner preparation in the modern United States', *British Food Journal* **109**(7), 531–547.

Belasco W (2006), *Meals to come. A history of the future of food,* Berkeley, Los Angeles, London, University of California Press.

Bell D and Valentine G (1997), *Consuming geographies: we are where we eat,* London and New York, Routledge.

Bertéus Forslund H, Lindroos AK, Sjöström L and Lissner L (2002), 'Meal pattern and obesity in Swedish women – a simple instrument describing usual meal types, frequency and temporal distribution', *European Journal of Clinical Nutrition* **56**(8), 740–747.

Bove CF, Sobal J and Rauchenbach BS (2003), 'Food choices among newly married couples: convergence, conflict, individualism, and project', *Appetite* **40**(1), 25–41.

Brombach C (2000), 'Mahlzeit – (H)Ort der Familie?! eine empirische Deskription von Familien-mahlzeiten', *Internationaler Arbeitskreis für Kulturforschung des Essens* 7, 2–13.

Bugge A and Døving R (2000), *Det norske måltidsmønsteret – ideal og praksis*, Forskningsrapport nr. 2. Lysaker, SIFO.

Caplan A (1997) 'Approaches to the study of food, health and identity', in Caplan P (ed.), *Food health and identity*, London and New York, Routledge, 1–31.

Charles N and Kerr M (1988), *Women, food and families*, Manchester and New York, Manchester University Press.

Coveney J (2000), *Food, morals and meaning. The pleasure and anxiety of eating*, London, Routledge.

DeVault ML (1991), *Feeding the family: the social organization of caring as gendered work*, Chicago and London, The University of Chicago Press.

Diasio N (2008), *Grignotages et jeux avec les normes. Une ethnographie des comportements alimentaires enfantins à Paris et à Rome*, Mise en ligne 18 mars 2008. Available at http://www.lemangeur-ocha.com/fileadmin/images/enfants/Diasio-Grignotages-jeux-normes.pdf.

Douglas M (1975), *Implicit meanings: essays in anthropology*, London, Melbourne, and Henley, Routledge & Kegan Paul.

Douglas M (1983), 'Culture and food', in Freilich M (ed.), *The pleasures of anthropology*. Merton, New American Library, 74–101.

Douglas M and Gross J (1981), 'Food and culture: Measuring the intricacy of rule systems', *Social Science Information* 20(1), 1–35.

Douglas M and Nicod M (1974), 'Taking the biscuit: the structure of British meals', *New Society* 30(637), 744–747.

Ekström M (1990), *Kost, klass och kön*, Umeå Studies in Sociology No. 98. Umeå, Umeå Universitet.

Ekström MP and Fürst EL (2001), 'The Gendered Division of Cooking', in Kjærnes U (ed.) *Eating Patterns. A Day in the Lives of Nordic Peoples*, Rapport no. 7, Lysaker, SIFO, 213–233.

Falk P (1994), *The consuming body*, London, Thousand Oaks, New Delhi, Sage Publications.

Fischler C (1980), 'Food habits, social change and the nature/culture dilemma', *Social Science Information* 19(6), 937–953.

Fischler C (1986), 'Learned versus "spontaneous" dietetics: French mothers' views of what children should eat', *Social Science Information* 25(4), 945–965.

Fischler C (1990), *L'Homnivore. La goût, la cuisine et le corps*, Paris, Editions Odile Jacob.

Fischler C and Masson E (2008), *Manger. Français, Européens et Américains face à l'alimentation*, Paris, Editions Odile Jacob.

Fjellström, C (2004), 'Mealtime and meal patterns from a cultural perspective', *Appetite* 48(4), 161–164.

Fürst EL (1997), 'Cooking and Feminity', *Women's Studies International Forum* 20(3), 441–449.

Grieshaber S (1997), 'Mealtime rituals: power and resistance in the construction of mealtime rules', *The British Journal of Sociology* 48(4), 649–666.

Grignon C (1992), 'Manger en temps et en heure: la popularisation d'une discipline dominante', *Social Science Information* 31(4), 643–668.

Grignon C (1996), 'Rule, fashion, work: the social genesis of the contemporary French pattern of meals', *Food and Foodways* 6(3–4), 205–241.

Haapalahti M, Mykkanen H, Tikkanen S and Kokkonen J (2003), 'Meal patterns and food use in 10- to 11-year-old Finnish children', *Public Health Nutrition* 6(4), 365–370.

Haraldsdottir J, Holm L, Jensen JH and Møller A (1987), *Danskernes kostvaner 1985. 2. Hvem spiser hvad?*, København, Levnedsmiddelstyrelsen.

Hetherington MM, Anderson AS, Norton GNM and Newson L (2006), 'Situational effects on meal intake: a comparison of eating alone and eating with others', *Physiology and Behavior* 88(4–5), 498–505.

Holm L (2001), 'The Family Meal', in Kjærnes U (ed.) *Eating Patterns. A Day in the Lives of Nordic Peoples*, Rapport no. 7, Lysaker, SIFO, 199–212.

Holm L (2003), 'Måltidet som socialt fælleskap', in Holm L, *Mad, mennesker og måltide – samfundsvidenskaplige perspektiver,* Copenhagen, Munksgaard Danmark, 21–34.

Kemmer D, Anderson AS and Marshall DW (1998), 'Living together and eating together: Changes in food choice and eating habits during the transition from single to married/ cohabiting', *The Sociological Review* **46**(1), 48–72.

Kjærnes U (ed.) (2001), *Eating Patterns. A Day in the Lives of Nordic Peoples,* Rapport no. 7, Lysaker, SIFO.

Kristensen ST, Holm L, Raben A and Astrup A (2002) 'Achieving "proper" satiety in different social contexts – qualitative interpretations from a cross-disciplinary project, sociomæt', *Appetite* **39**(3), 207–215.

Marshall DW (2000), 'British Meals and Food Choice', in Meiselman H (ed), *Dimensions of the meal: the science, culture, business, and art of eating,* Gaithersburg, Aspen Publishers, Inc., 202–220.

Marshall DW and Anderson AS (2002), 'Proper meals in transition: young married couples on the nature of eating together', *Appetite* **39**(3), 193–206.

Mennell S, Murcott A and van Otterloo AH (1992), *The sociology of food: Eating, diet and culture,* London, Newbury Park, New Delhi, Sage Publications.

Mestdag I (2005), 'Disappearance of the traditional meal: temporal, social and spatial destructuration', *Appetite* **45**(1), 62–74.

McIntosh WA (1996), *Sociologies of food and nutrition,* New York and London, Plenum Press.

Murcott A (1982), 'On the social significance of "cooked dinner" in South Wales', *Social Science Information* **21**(4/5), 677–696.

Murcott A (1983), 'It's a pleasure to cook for him: food, mealtimes and gender in some South Wales households', in Gamarnikow E (ed.), *The public and the private,* Portsmouth, NH, Heinemann, 78–90.

Murcott A (1986), 'Opening the "black box": food, eating and household relationships', *Sosiaalilääketieteellinen Aikakauslehti* **23**(2), 85–92.

Murcott A (1997), 'Family meals: A thing of the past?', in Caplan P (ed.), *Food, health and identity,* London and New York, Routledge, 32–49.

Mäkelä J (1991), 'Defining a meal', in Fürst EL, Prättälä R, Ekström M, Holm L and Kjaernes U (eds.), *Palatable worlds: Sociocultural food studies.* Oslo, Solum, 87–95.

Mäkelä J (1996), 'Kunnon ateria: Pääkaupunkiseudun perheellisten naisten käsityksiä' (in Finnish. A proper meal: exploring the views of women with families), *Sosiologia* **33**(1), 12–22.

Mäkelä J (2000), 'Cultural Definitions of the Meal', in Meiselman H (ed.), *Dimensions of the meal. The science, culture, business and art of eating,* Gaithersburg, Aspen Publishers Inc., 7–18.

Neumark-Sztainer D, Wall M, Story M and Fulkerson JA (2004), 'Are family meal patterns associated with disordered eating behaviors among adolescents?', *Journal of Adolescent Health* **35**(5), 350–359.

Niva M (2008), *Consumers and the conceptual and practical appropriation of functional foods,* Helsinki, National Consumer Research Centre.

Pearcey SM and de Castro J (2002), 'Food intake and meal patterns of weight-stable and weight-gaining persons', *American Journal of Clinical Nutrition* **76**(1), 107–112.

Pollan M (2006), *The omnivore's dilemma. The natural history of four meals,* New York, The Penguin Press.

Poulain JP (2002), 'The contemporary diet in France: "de-structuration" or from commensalism to "vagabond feeding"', *Appetite* **39**(1), 43–55.

Poulain JP (2005), *Méthodologies d'étude des pratiques alimentaires. Les descripteurs,* Mise en ligne 6 Juillet 2005. Available at http://www.lemangeur-ocha.com/uploads/ tx_smilecontenusocha/05_descripteurs.pdf (accessed 18 July 2008).

Rotenberg R (1981), 'The impact of industrialization on meal patterns in Vienna, Austria', *Ecology of Food and Nutrition* **11**(1), 25–35.

Sidenvall B, Nydahl M and Fjellström C (2000), 'The meal as a gift. The meaning of cooking among retired women', *The Journal of Applied Gerontology* **19**(4), 405–423.

Simmel G (1910), 'Soziologie der Mahlzeit', *Berliner Tageblatt* 10 Oktober 1910.

Sjögren-de Beauchaine A (1988), *The bourgeoisie in the dining-room: Meal ritual and cultural process in Parisian families of today,* Stockholm, Institutet för folklivsforskning vid Nordiska museet och Stockholms universitet.

Sobal J (2000), 'Sociability and the meal: facilitation, commensality, and interaction', in Meiselman H (ed), *Dimensions of the meal: the science, culture, business, and art of eating,* Gaithersburg, Aspen Publishers Inc., 119–133.

Sobal J and Nelson MK (2003), 'Commensal eating patterns: a community study', *Appetite* **41**(2), 181–190.

Thephano J and Curtis K (1991), 'Sisters, mother, and daughters: Food exchange and reciprocity in an Italian–American community', in Sharman A, Theophano J, Curtis K and Messer E (eds.), *Diet and domestic life in society,* Philadelphia, Temple University Press, 147–171.

Visser, M (1996), *The rituals of dinner: the origins, evolution, eccentricities and the meaning of table manners,* London, Penguin Books Ltd.

Warde A (1997), *Consumption, food and taste: culinary antinomies and commodity culture,* London, Sage Publications.

Whit WC (1995), *Food and society: a sociological approach,* Dix Hills, General Hall, Inc.

Wood RC (1995), *The sociology of the meal,* Edinburgh, Edinburgh University Press.

4

Foodservice perspective in institutions

P. G. Williams, University of Wollongong, Australia

Abstract: The different types of meals and foodservice systems used in institutional settings, such as hospitals, nursing homes, prisons, schools, military settings and workplace canteens. are described. The menus used, nutritional standards, food waste, meals times, methods of counting meals and possible future trends are discussed.

Key words: hospital meals, prison food, military rations, school meals, food waste, food service.

4.1   Introduction

Throughout the western world today, more and more meals are being consumed away from the home. Edwards (2000) has pointed out that this can be for pleasure (e.g., in restaurants) or through necessity, in settings where individuals, given a choice, would perhaps choose not to be. Batstone (1983) has made a similar distinction between 'domestic meal provision', where meals are provided to meet principally social goals and personal needs, tastes and comforts, and 'functional meal provision', where meals are provided in a context or rules governing work and especially time constraints.

This latter category encompasses a wide range of foodservice, which can together be considered as institutional settings, including:

- Healthcare settings (hospitals, nursing homes);
- Prisons;
- Schools and child care organisations;
- Military settings (canteens and combat rations);
- Meals on Wheels; and
- Workplace canteens.

There are no comprehensive international data on the size of the institutional foodservice market, but it was estimated to be worth £3.3 billion in the UK in 2003 (IGD, 2004) and $64.1 billion in the USA in 2000 (Price, 2002). In western countries, the institutional sector provides between 10 and 15% of all foodservice meals. Over the decade from 1987 to 1997 in the USA, non-commercial foodservice sales grew 46% (Price, 1998). Some of the greatest growth areas were childcare facilities (186%) and educational institutions (72%), driven by large number of baby boomers' children making their way through the education system. Sales declined in only one sector: hospital foodservice dropped 7%, which could be the result of a trend to more day surgery and shorter lengths of stay. However, despite this growth, as a proportion of all foodservice, institutional meals in the USA have been progressively declining over the past 50 years, from 30.8% in 1955 to 14.6% in 2005 (USDA, 2007). This pattern is likely to be worldwide because of the much greater growth in non-institutional meals from fast food outlets and the general trend to more out-of-home recreational dining.

The meal experience is significantly shaped by the individual living arrangements in institutions (Sydner and Fjellström, 2005) and it has even been suggested that the word 'meal' may be inappropriate to some experiences (like Meals on Wheels), where food is provided, but the social and emotional contexts of eating are missing (de Raeve, 1994). Nonetheless, in all of these settings one can distinguish two goals that they have in common with all other meal service settings – (1) meeting customer expectations and needs (e.g., safety, taste, price, service), and (2) providing physical sustenance (e.g., satiation and nourishment). However, in the institutional settings, there are three other important roles that may inform the goals and objectives of the meal service, which may be considered under the headings of 3 'M's: morale, manners, or medicine.

4.1.1 Morale-centred meals

In the morale-centred meal services, there is a particular emphasis on planning the meal service to prevent boredom, provide familiar and perhaps comforting foods to people in otherwise deprived circumstances, or to demonstrate that the employer cares for the wellbeing of the clients. Food provided to military staff serving in combat zones is an example of this type of service, as are some workplace canteens, especially in isolated locations (e.g., offshore oil platforms, or remote mining camps) where there are few or no alternative sources of meals other than those provided in the workplace.

Prisons also demonstrate some of the aspects of the morale-based service. Meals become very important social occasions in prison as an escape from the boredom of daily routine, and the ability to prepare some home-made and culturally specific food is highly prized (Godderis, 2006a). Complaints about food can be a significant focus of unrest in prisons. Most prison riots begin at meal times in canteens because they are occasions when inmates can congregate and interact (Valentine and Longstaff, 1998), and meeting minimum expectations of service quality is important to help maintain a harmonious environment.

4.1.2   Manners-centred meals

In a manners-centred meal service, one of the articulated roles of the meal time is to ensure appropriate behaviour is taught and good behaviour reinforced, while inappropriate behaviour is corrected. Childcare and school settings provide examples of this, where the importance of providing children with opportunities to try a wide variety of foods, to learn and display appropriate social interactions with other children, and even learn some food preparation and service skills, can be part of the explicit aims of the meal occasion.

In prisons, inmates are often employed in the preparation and service of meals and, as in schools, there can be some socialisation and rehabilitation activities based on meal-time interactions. Conversely, the lack of control over meals by inmates can be seen as part of the process of reinforcing their lack of power and identity within the institution (Godderis, 2006b). One of the complaints that women's prisons in particular have made at times is that they can lose important domestic management skills and confidence if they are not involved in the service of meals (Smith, 2002). However, issues of cost control and security often severely inhibit the menu, food preparation and meal delivery options in correctional institutions (Stein, 2000; Gater, 2003).

Even in healthcare settings, the norms of acceptable behaviour at mealtimes can be reinforced. It has been noted that elderly patients in care strive to behave at meals according to what they think is acceptable in an institution. Patients with disabilities or handicaps that limit their ability to handle normal crockery and cutlery may be given special equipment to allow them to eat independently with dignity (Sidenvall, 1999).

4.1.3   Medicine-centred meals

This type of meal can be seen in hospitals, nursing homes and to some extent in home-delivered meal services such as Meals on Wheels. From Hippocrates in the 4th century BC to Florence Nightingale in the 19th century, the provision of food suitable for sick patients has been recognised as an important part of their care (see also Chapter 7 by Edwards and Hartwell). In hospital, the food provided to patients is not just another hotel function (like cleaning and laundry), it is part of the treatment, and providing meals that are of high quality and which meet the individuals' specific nutritional needs is an essential goal (Allison, 1999). However, if food is regarded as medicine, often necessary dietary modifications (e.g., liquid or pureed food, low-salt or low-protein diets) can make meals particularly unappealing. It is recognised that in these cases the medical requirements must outweigh the normal culinary expectations.

There are several significant features of institutional meal occasions that differentiate them from meals in commercial foodservices (summarised in Table 4.1). Some of the more important factors are discussed in the following sections.

Table 4.1 Comparison of meals in institutional and commercial foodservices

	Institutional	Commercial
Meals provided	All daily meals to clients	Usually only one meal per day to a client
Meal times	Fixed or narrow limits	Flexible and broader range
Meal location	May be in bed or private room	Usually in dining room or cafeteria
Menu type	Typically cycle menus used	À la carte or daily menu
Menu choice	Sometimes only limited choice at any one meal	Emphasis on providing customer choice
Foodservice system	Cook–chill or convenience systems more common	Usually cook–fresh
Production staff	May have limited professional training	Usually professionally trained
Budget	May be limited by owners	Limited only by customer preparedness to pay
Clients	High proportion may have special dietary needs	Normal population
Legislated or other external standards	Highly regulated – especially nutrition standards	Only food safety standards of concern
Payment	Meals provided as part of a package of service or employment	Meals paid for by client at time of consumption
Accepted level of food waste	High	Low
Feeding assistance	May be provided	Not available

4.2 Types of meals in institutions

In institutional foodservice, the meals may be delivered to the customers or clients in a variety of ways.

4.2.1 Cafeteria style service

In workplace canteens, many military settings, universities and some schools, nursing homes or prisons, meals are usually served in a traditional cafeteria style (either self-service or with serving staff), so that food choices can be made by the customer immediately before consumption. This system has the advantages of allowing the final food choices to be on display for review and assessment by the customer, and also allowing individual client preferences about the portion sizes or combinations of meal ingredients to be met (e.g., asking for sauces to be added or not). This type of service is usually very cost efficient for the meal provider, since there are minimal staff required to deliver or clear meal trays, and it also does not require any special equipment to maintain meal temperatures between service and consumption. It enables last-minute menu changes to be made easily, since there are no printed menus distributed in advance of the meal time. The main disadvantage of this type of service is that the number of choices available will be necessarily limited to those that can be displayed in the available service area. Furthermore, usually food is not made to order and there can be deterioration in the quality of food if hot items are held for long periods of time before service.

4.2.2   Delivered meal trays

In many healthcare institutions, the meals are delivered on individual trays for consumption in bed or in nearby dining areas. The meals may be served in the ward itself from a bulk food trolley, or meals may be served in a central kitchen location, with trolleys of plated meals delivered to the patient areas. This system is particularly suitable for non-ambulatory patients or those who need to be kept isolated for medical reasons from other patients. A significant advantage of the central plating system is that a much greater range of menu choices can be made available – particularly if a cook-chill or cook-freeze meal system is employed. In some hospitals with this system an unchanging à la carte menu with a large number of choices is used (up to 30 entrée choices for example), which allows patients to select from a wide range of foods that are suitable to their current state of health and or appetite. However, while centralised meal service can provide greater menu choice, and perhaps more careful supervision of the accuracy of service (an important factor for many special diets), there are many disadvantages.

The greatest challenge for most tray delivery services is in maintaining a safe and acceptable temperature of the food (both hot and cold). Often there are considerable distances between a central meal plating area near the kitchen and the ward areas. Furthermore, there is also an expectation that all patients will be served a particular meal within prescribed limited time frame (typically a one-hour period).

Three main approaches have been adopted to overcome the problem of maintaining meal temperatures:

(1) Insulated trays or plate covers that passively maintain the temperature of the meals. Such systems can be adequate for periods of up to 30 min.
(2) Heated and cooled meal delivery carts, with the separate hot and cold meal components assembled on to the tray immediately before service.
(3) Reheating chilled meal components and tray service in ward areas. A range of alternative means of reheating can be used including traditional convection ovens, microwave, infra-red and induction heating of plated meals.

Unlike the cafeteria service, with a delivered meal tray, all the components for the meal – including tray cloths, napkins, condiments, cutlery, and beverages – need to be provided at the point of service. This can increase the levels of waste: for example, usually sugar, salt and pepper portions will be routinely provided, even to those clients who do not wish to use them.

Another factor likely to increase waste with a tray service system stems from the fact that normally patients make their menu choices well in advance of the actual meal time (often up to 24 h ahead), without the advantage of being able to see the food beforehand. Patients' appetites can change rapidly, and there is a tendency for many to order all possible items for their trays even though they are unlikely to eat all the food. Furthermore, often serving sizes are standardized and it has been suggested that frail older patients, who may only want small serves, can be put off eating by overly large meals presented to them on their trays (Walton *et al.*, 2006).

In pre-plated tray service systems that use the cook–chill system, a third disadvantage is the general requirement to standardise portion sizes and the amount of food on plates as much as possible; e.g., baked potatoes may have to be cut into smaller pieces to facilitate even reheating (Light and Walker, 1990). Menu choices can also be affected. To prevent drying out of meats, almost always they need to be served covered with a sauce or gravy. Wet entrée dishes that reheat well are usually favoured when cook–chill systems are used over dishes such as grilled meats or eggs, which are more likely to dry out. For these reasons, it has been reported that hospitals using cook–fresh systems are significantly more likely to offer choices of portion size and optional sauces and gravies with meat compared with cook–chill hospitals (McClelland and Williams, 2003).

A last significant disadvantage of the tray service system is the physical challenges for patients trying to eat in bed, particularly if they have problems with mobility or limbs. In a healthcare setting, it is recognised that giving and receiving food is a crucial part of the caring and healing process. Yet concern has been expressed at erosion of the emphasis on this role for nurses in particular, and the devolution of non-nursing duties to other staff. '*Tray meals, with standard serving sizes, plastic containers of butter and jam, and stubborn seals on milk capsules, served by food service staff who sweep in and out, are a far cry from the essentially social occasions of mealtimes in the past*' (Pearson, 1994, p. 325). If patients cannot reach their trays easily, or cannot easily open small portion control packages commonly used for items such as drinks, milk, jams and butter, they may not be able to eat all the food provided. Two alternatives have been trialled to overcome these problems and give patients more assistance to eat: (1) offering mobile patients the option of eating in a dining room setting (Edwards and Hartwell, 2004), and (2) using volunteers to assist patients at meal times (Simmons *et al.*, 2001; Walton *et al.*, 2008).

4.2.3   Ration packs

Whenever possible, a cafeteria-style service is normally used for feeding groups of military personnel. However, considerable research has been invested to develop acceptable individually packaged military ration packs (combat rations) which can be used when mission or tactical operational reasons prevent group feeding (Rock *et al.*, 1998), and to examine the factors influencing food acceptance (Meiselman and Schutz, 2003). Rules of field feeding often forbid consumption of locally procured food, to ensure safety (USARIEM, 2006). The rations provide single meals – ready to heat foil pouches, canned and dehydrated items – which can be consumed hot or cold, with speciality versions designed to meet increased nutritional requirements imposed by exposure to extreme environments (such as extreme cold). They have to be lightweight and stable in a wide range of environments, but providing familiar home-type foods. One of the particular problems for these meals is that over time the monotony of repeated consumption of the same food can lead to inadequate nutritional intakes (Hirsch *et al.*, 2005). Darsch and Moody in Chapter 24 provide more information on US rations.

4.2.4   Supplementary feeding

Mid-meals (i.e., beverage and snacks consumed between the main meals) are an important part of institutional foodservice, especially in healthcare settings. They can be an important occasion at which to increase the nutritional intake of vulnerable clients who may have poor appetites and they can provide more than a quarter of the daily energy intake of patients (Walton *et al.*, 2007). One Australian survey found that most hospitals regularly provided patients with three mid-meals: 98% served morning tea, 99% served afternoon tea, and 95% served supper, and 19% even offered a pre-breakfast early morning hot beverage (Mibey and Williams, 2002).

In addition to normal food items provided at these mid-meal breaks (tea, coffee, milk beverages, biscuits, cake), patients requiring additional nutritional support are often prescribed specially fortified nutritional supplements: typically these are commercially packaged milk-based cold beverages to be drunk from a pack with a straw, although hot versions (soups) are also available. Such supplements may be consumed at normal mid-meal times, or may be delivered in smaller prescribed doses by nurses as part of a drug round to encourage compliance with their consumption. Such supplements would not traditionally be regarded as meals – perhaps because they are not served by foodservice personnel from the kitchen – but they can add substantially to nutritional intakes. One trial of the prescription of a 120 mL sip feed three times daily (providing more than 2200 kJ and 22 g protein) resulted in significantly better energy intakes and weight gain in patients at nutritional risk (Potter *et al.*, 2001).

Another special group of patients are those who, for various reasons, cannot swallow normal food items and are fed either enterally by tube into the digestive system, or nourished parenterally directly into the veins. For such patients, normal meal times are largely irrelevant since infusion is usually continuous throughout the day.

4.3   Menus

In institutions that provide all the daily meals for the clients (e.g. hospitals, boarding schools), it is most common to provide three main meals per day (breakfast, midday, evening), plus a number of mid-meal or snack options. The latter may be served on trays, or from a beverage and snack trolley wheeled around the ward areas. In other institutions, the mid-meals are less likely to be delivered, but supplies may be available for self-service in common dining areas.

An increasing proportion of hospitals are now offering a continental breakfast only. This trend remains a concern because there is evidence that patients may have poorer nutrient intakes when a hot breakfast is not available (Coote and Williams, 1993). Other hospitals have moved away from the traditional pattern of three main meals to one offering four or five meals throughout the day (Puckett, 2004). A typical meal pattern of this kind is shown in Table 4.2, and this can have cost advantages since all hot items can be prepared in one cook's working shift.

In most institutional foodservices, the menus are either an à la carte type (offering a wide range of choices, but remaining the same each day), or a cycle menu (a series of daily menus on a weekly or longer cycle, after which the cycle is repeated). Cycle menus are commonly used in healthcare, prison and school settings to offer variety with some degree of predictability for ordering, budgeting and production scheduling (Spears and Gregoire, 2007). One or two week cycles are common in acute hospitals; 3–4 week cycles are more common in longer-care facilities.

4.4   Nutritional standards for meals

In many settings, the clients are dependent on the institutions to provide all, or a large proportion, of the food they consume and therefore the nutritional content of meals assumes greater importance than in other commercial foodservice operations (Glew, 1980).

Table 4.2   A sample of four meal per day menu pattern for hospitals

8 am Breakfast	11.30 am Brunch	4 pm Main meal	7 pm Supper
Juice	Main hot dish (e.g., lasagne)	Main hot dish (e.g. roast meat)	Soup
Cereal	Salad	Vegetables	Snack (e.g. cheese and crackers)
Bread item (e.g. muffin)	Bread	Dessert	–

Furthermore, the clients are often at higher nutritional risk than the general population. For example, in hospitals or nursing homes a high proportion of patients may be malnourished or have special needs owing to their medical conditions (Thomas *et al.*, 2002) and therefore the nutritional content of the meals needs to be more carefully considered than in a restaurant or cafe. Malnutrition in hospitals is significant problem, and concerted action is needed to ensure meals are not only acceptable to patients but also nutritionally adequate (Beck *et al.*, 2001). Studies suggest more than 40% of hospital patients may have their meals supplemented by food brought in by visitors, but nonetheless meals must be planned to provide at least the recommended dietary allowances for all essential nutrients (Hickson *et al.*, 2007).

In the USA, Military Recommended Dietary Allowances (MRDAs) establish standards for the nutritional content of military rations. The MRDAs are based on the recommendations of the Food and Nutrition Board of the National Research Council. This Board establishes the Recommended Dietary Allowances (RDAs). For some nutrients, the MRDAs have a higher requirement than the RDAs because soldiers are typically more physically active than their civilian counterparts.

A summary of the nutritional standards used for school meal provisions in England and other Western countries has been provided by Harper and Wells (2007). Most countries offer a free school meal to children from poorer backgrounds and some offer a free meal to all pupils regardless of ability to pay. Many countries (including the USA, England, France and Sweden) have nutrient-based standards, typically requiring the school meal to provide around 30% of the estimated daily energy, vitamin and mineral requirements, with no more than 11% energy from saturated fat (Crawley, 2005; CDC, 2006). In the USA, federally funded programmes such as the National School Lunch Program and the School Breakfast Program aim to address the concurrent problems of food insecurity and an increase in the prevalence of overweight among children and adolescents with both nutrient-based standards and requirements for meals to conform with the Dietary Guidelines for Americans (American Dietetic Association, 2006).

4.5   Food waste with institutional meals

Levels of meal waste can commonly be as high as 30–40% in hospitals (Williams *et al.*, 2003), and the health status and type of modified diet that patients are prescribed can influence this significantly. Hirsch *et al.* (1979) reported that the percentage of calories wasted by patients on a regular diet was less than those on a modified diet. Patients ordered high energy/high protein diets are often sent large quantities of food in order to encourage higher energy intakes, but this can result in large quantities wasted (Walton *et al.*, 2007). Patients ordered modified texture meals are another group more likely to have greater levels of plate waste, because of the unappealing nature of the food.

In other institutional settings, the normal levels of meal waste are significantly lower. In a retirement living centre, it was reported to be 20% (Nichols *et al.*, 2002), similar to the levels of 17% in a university dining hall environment, which is probably typical of levels in most cafeteria settings with healthy clients (Norton and Martin, 1991). This compares with levels of 9–11% in some school foodservices (Engström and Carlsson-Kanyama, 2004) and 7% in community-based feeding centres for the elderly (Hayes and Kendrick, 1995).

4.6   Timing of meals in institutions

In healthcare institutions, mealtimes can provide benchmarks that help patients structure their day and give a sense of predictability and security (Holloway *et al.*, 1998). Yet, it is a common experience of patients in hospital that the times the meals are served do not reflect when they would normally be consumed home. In particular, the evening meals are usually served quite early, so there can be a long period of time overnight without food. In two Australian surveys in 1993 and 2001, it was reported most hospitals began serving the breakfast meal between 7 am and 8 am, the midday meal was served between 12.00 noon to 12.30 pm, and the evening meal times were spread over two and a half hours from 3.30 pm to 6.00 pm with most meal service being between 5.00 pm and 5.30 pm (Mibey and Williams, 2002). Similar meal times can be seen in hospitals in the USA and the UK.

The early evening meals lead to long periods of 12–14 hours without access to food, which nurses themselves recognise can be a significant problem (Kowanko *et al.*, 1999). While the reasons for this are partly to enable patients to be ready early for their evening visiting times, another factor is also to reduce the span of foodservice operating hours, as a cost-saving strategy. In prisons meal times can be even more atypical and unwelcome, with evening meals served before 4 pm to allow inmates to locked into cells early (Williams *et al.*, 2006). In other institutions, such as university colleges or workplace canteens, the meal times are more likely to reflect normal experience, and the clients have greater freedom to supplement their food intakes outside the fixed meal service times.

4.7   Methods of counting meals

Counting meals is an important control point for financial and performance management in institutions. Typical performance indicators that might require a census of the number of meals include costs per meal and meals per full-time-equivalent staff.

Most healthcare institutions attempt to keep a daily census or count of the number of meals served to patients and non-patients. One method is to count the number of trays actually prepared for each meal and this has been recommended as the most accurate method to use (Puckett, 2004). However, this method is time consuming and does not include other snacks or supplements that might be provided at the ward level rather than on delivered trays.

A 'meal equivalent' is often used in healthcare settings where patients are provided with many supplementary nourishments in addition to normal main meals. It has been suggested that dividing the number of nourishments by six yields a satisfactory number of meal equivalents (Spears, 2000). Another standardised method is to calculate four standard 'meal unit equivalents' per occupied bed day, assuming that each patient receives three mid-meals (that together may be similar in cost to one main meal) plus three main meals (Institute of Hospital Catering, 1995). Such equivalents do not necessarily equate to the normal concept of a meal, but are used for performance reporting in order to assess trends in efficiency over time and make benchmarking comparisons between institutions possible, without the burden of cumbersome recording of meal numbers.

The methods used to determine the number of non-patient meals served each day vary from one institution to another and depend on the method of payment system employed (e.g. a cash payment system, or whether employees purchase monthly meal vouchers, or are provided free of charge to certain groups). In a cash payment cafeteria, a tray census method may be used, but this does not distinguish between customers who buy a full meal, and those choosing only a beverage or snack item. A common method is to divide the total daily cash sales by a standard price per meal to determine a daily meal equivalent (Sneed and Kresse, 1989). The standard meal price may be based on the average price of each meal component at the midday meal (for example, entrée, vegetable, salad and dessert) or may be based on amounts defined in industrial employment awards, that set prices that a standard meal must be provided to staff.

4.8 Future trends

In healthcare feeding throughout the second half of the 20th century, there was a move away from patient meal service using bulk delivery trolleys in the ward areas (with foodserved by nursing staff) toward centralised meal plating and distribution of individual trays by food service staff. Recently, there has been some reversal of this trend with several recent trials of a return to bulk food trolleys – particularly in nursing home situations. Such systems may result in less waste and greater patient satisfaction but it is unclear how they affect nutritional intake (Kelly, 1999; Shatenstein and Ferland, 2000; Wilson et al., 2001; Hickson et al., 2007).

Other changes that are being trialled in a number of centres are moves away from selective paper-based menus to bedside spoken meal orders, with foodservice staff entering orders directly into hand held electronic devices after interviewing patients (Folio et al., 2002). This allows for meal selections much closer to the time of the meal, without the need for manual tallying of meal orders. A more radical (and more costly) approach is to offer hotel-style room service (enabling patients to order at any time of the day from a restaurant-style menu, with food cooked to order and delivered within 45 min). Such systems are being implemented in order to provide a more client-oriented service and improve patient satisfaction (Sheehan-Smith, 2006).

Greater attention is also being paid to the meal ambience in the dining experience of long stay residents in healthcare settings. Changes in the physical environment (e.g., flowers on tables; background music), meal service (removal of trays and meal covers from dining tables; serving only one course at a time) and nursing practices (e.g., nurses sitting at tables with patients; separating medications from meal times) have been shown to significantly improve client health status and quality of life (Mathey et al., 2001).

In school settings, there is a worldwide trend to impose greater restrictions on the food available for consumption at meal times and between meals. Policies to limit the foods that can be sold in canteens are being implemented in the UK (Golley and Clark, 2007) and Australia (New South Wales Department of Health, 2006), and, in the USA, research is demonstrating the beneficial dietary outcomes when access to high energy snack foods is limited (Cullen et al., 2000; Cullen et al., 2008). At the same time, there is a move to upgrade school foodservices with healthier food options and to introduce more authentically ethnic food that recognises the cultural diversity of the population (Schuster, 2007).

From a menu planning viewpoint, in institutional settings the greater availability of novel functional foods may provide new opportunities for better matching the foods offered to clients with their specific nutritional needs (Williams, 2005). Another trend has been the introduction of familiar branded menu items on institutional menus. Product branding is often used as a sign of quality (Vranesevic and Stancec, 2003) and, in a study conducted on institutionalised stereotyping (Cardello et al., 1996), when individuals were asked to rate their anticipated acceptability of two identical samples of sweetcorn the unbranded sample was rated lower for both the anticipated and actual acceptability. Some US hospitals have even begun offering branded menu choices such as Pizza Hut pizzas.

Lastly, in all foodservice settings there is increasing consumer demand for greater attention to the nutritional quality and environmental impact of the food being offered (Euromonitor International, 2007; Sloan, 2007a). Organic menus have started to appear in the hospital sector and environmental concerns may well have longer term impacts on the technologies employed for meal production and delivery. Recent trends to greater use of cook–chill foodservices, and more portion packaged food and disposable-tray items (in order to reduce dishwashing) have not been made with much awareness of the consequences for energy consumption or environmental impact. These are factors that are likely to have increasing prominence, with a demand for the use of more locally sourced food, recycling and improved energy efficiency (Sloan, 2007b).

4.9   Sources of further information and advice

Aside from the specific references given in this chapter, the following general texts provide good overviews of the structure and management of institutional foodservices and the issues of meal planning and delivery.

Payne-Palacio J and Theis M (2005). *Introduction to foodservice* (10th ed). Upper Saddle River NJ, Prentice Hall – a good general introductory text.

Spears MC and Gregoire MB (2007), *Foodservice organisations: a managerial and systems approach* (6th ed), Upper Saddle River NJ, Prentice Hall – emphasises organisational and management issues.

Puckett RP (2004). *Food service manual for health care institutions* (3rd ed). San Francisco CA, Jossey-Bass – focuses on hospital and nursing home foodservice.

McVety PJ, Ware BJ and Levesque C (2008). *Fundamentals of menu planning* (3rd ed). New York, John Wiley & Sons – contains plenty of detail on the differing menu formats and constraints in an institutional setting.

4.10   References

Allison S (1999), *Hospital food as treatment*, Maidenhead, British Association for Parenteral and Enteral Nutrition.

American Dietetic Association (2006), 'Position of the American Dietetics Association: Child and Adolescent Food and Nutrition Programs', *J Am Diet Assoc*, **106**, 1467–1475.

Batstone A (1983), 'Hierarchy of maintenance and maintenance of hierarchy', in Murcott A, *The sociology of food and eating*, Aldershot, Gower, 45–53.

Beck A, Balknas U, Furst P, Hasunen K, Jones L, Keller U, Melchior J, Mikkelsen B, Schauder P, Sivonen L, Zinck O, Oinen H and Ovesen L (2001), 'Food and nutrition care in hospitals: how to prevent undernutrition – report and guidelines from the Council of Europe', *Clin Nutr*, **20**, 455–460.

Cardello A, Bell R and Kramer M (1996), 'Attitudes of consumers toward military and other institutional foods', *Food Qual Pref*, **7**, 7–20.

Center for Disease Control and Prevention (CDC) (2006), *Guidelines for school health programs to promote lifelong healthy eating: Summary,* Accessed 15 February 2008. Available at: http://www.cdc.gov/healthyyouth/nutrition/pdf/summary.pdf.

Coote and Williams P (1993), 'The nutritional implications of introducing a continental breakfast in a public hospital: a pilot study', *Aust J Nutr Diet*, **50**, 99–103.

Crawley H (2005), *Nutrient-based standards for school food,* London, The Caroline Walker Trust, Retrieved 15 February 2008. Available at: http://www.cwt.org.uk/publications.html

Cullen K, Eagen J, Baranowski T, Owens E and de Moor C (2000), 'Effect of à la carte and snack bar foods at a school on children's lunchtime intake of fruits and vegetables', *J Am Diet Assoc,* **100**, 1482–1486.

Cullen K, Watson K and Zakeri I (2008), 'Improvements in middle school dietary intake after implementation of the Texas public school nutrition policy', *Am J Public Health,* **98**, 111–117.

de Raeve L (1994), 'To feed or to nourish? Thoughts on the moral significance of meals in hospital', *Nursing Ethics,* **1**, 237–241.

Edwards J (2000), 'Food Service/Catering Restaurant and Institutional Perspectives of the Meal', in Meiselman H, *Dimensions of the Meal. The Science, Culture, Business and Art of Eating,* Gaithersburg, Aspen, 223–244.

Edwards J and Hartwell H (2004), 'A comparison of energy intake between eating positions in a NHS hospital – a pilot study', *Appetite,* **43**, 323–325.

Engström R and Carlsson-Kanyama A (2004), 'Food loss in food service institutions – examples from Sweden', *Food Policy,* **29**, 203–313.

Euromonitor International (2007), 'Top ten food trends for 2007', Accessed 17 January 2008. Available at: http://www.euromonitor.com/Top_ten_food_trends_for_2007.

Folio D, O'Sullivan-Maillett J and Touger-Decker R (2002), 'The spoken menu concept of patient food service delivery systems increases overall patient satisfaction, therapeutic and tray accuracy, and is cost neutral for food and labor', *J Am Diet Assoc,* **102**, 546–548.

Gater L (2003), 'Costs and security are top issues in corrections foodservice today', *Corrections Forum,* **12**, 22–25.

Glew G (1980), 'The contribution of large-scale feeding operations to nutrition', *World Rev Nutr Diet,* **34**, 1–45.

Godderis R (2006a), 'Dining in: the symbolic power of food in prison', *Howard J Crim Justice,* **45**, 255–267.

Godderis R (2006b), 'Food for thought: an analysis of power and identity in prison food narratives', *Berkeley J Sociol,* **50**, 61–75.

Golley R and Clark H (2007), 'The transformation of school food in England – the role and activities of the School Food Trust', *Nutr Bull,* **32**, 392–397.

Harper C and Wells L (2007), *'School meal provision in England and other Western countries: a review',* Accessed 18 January 2008. Available at: http://www.schoolfoodtrust.org.uk/UploadDocs/Library/Documents/sft_school_meals_review.pdf.

Hayes J and Kendrick OW (1995), 'Plate waste and perception of quality of food prepared in conventional vs commisary systems in the Nutrition Program for the Elderly', *J Am Diet Assoc,* **95**, 585–586.

Hickson M, Fearnley L, Thomas J and Evans S (2007), 'Does a new steam meal catering system meet patient requirements in hospital?' *J Hum Nutr Diet,* **20**, 476–485.

Hirsch K, Hassanein R, Wutrecht C and Nelson S (1979), 'Factors influencing plate waste by the hospitalised patient', *J Am Diet Assoc,* **75**, 270–273.

Hirsch ES, Kramer FM, Meiselman HL (2005), 'Effects of food attributes and feeding environment on acceptance, consumption and body weight: lessons learned in a twenty year program of military ration research US Army Research (Part 2)', *Appetite,* **44**, 33–45.

Holloway IM, Smith P, Warren J (1998), 'Time in hospital', *J Clin Nurs,* **7**, 460–466.

IGD Food & Grocery Information (2004), *Factsheet: UK Foodservice Market Overview,* Accessed 14 February 2008. Available at: http://www.igd.com/CIR.asp?menuid=67&cirid=110.

Institute of Hospital Catering (1995), *Meal unit methodology,* Sydney, Institute of Hospital Catering.

Kelly L (1999), 'Audit of food wastage: differences between a plated and bulk system of meal provision', *J Hum Nutr Diet*, **12**, 415–424.

Kowanko I, Simon S and Wood J (1999), 'Nutritional care of the patient: nurses' knowledge and attitudes in an acute care setting'. *J Clin Nurs*, **8**, 217–224.

Light N and Walker A (1990), *Cook-Chill Catering: Technology and Management*, London, Elsevier Applied Science.

Mathey M, Venneste V, de Graff C, de Groot L and Van Staveren W (2001), 'Health effect of improved meal ambience in a Dutch nursing home: a 1-year intervention study', *Prev Med*, **32**, 416–423.

McClelland A and Williams P (2003), 'Trend to better nutrition on Australian hospital menus 1986–2001 and the impact of cook–chill food service systems', *J Hum Nutr Diet*, **16**, 245–256.

Meiselman HL and Schutz HG (2003), 'History of food acceptance research in the US Army', *Appetite*, **40**, 199–216.

Mibey R and Williams P (2002), 'Food services trends in New South Wales hospitals, 1993–2001', *Food Serv Technol*, **2**, 95–103.

New South Wales Department of Health (2006), *Fresh Tastes @ School. NSW Healthy School Canteen Strategy*, (2nd ed). Accessed 30 January 2008. Available at: https://www.det.nsw.edu.au/policies/student_serv/student_health/canteen_gu/CMPlanner.pdf.

Nichols PJ, Porter C, Hammond L and Arjmandi BH (2002), 'Food intake determined by plate waste in a retirement living centre', *J Am Diet Assoc*, **102**, 1142–1144.

Norton VP and Martin C (1991), 'Plate waste of selected food items in a university dining hall', *School Food Serv Res Rev,* **15**, 37–39.

Pearson A (1994), 'Eat, drink and be merry', *J Clin Nurs*, **3**, 325–326.

Potter J, Roberts M, McColl J and Reilly J (2001), 'Protein energy supplements in unwell elderly patients: a randomised controlled trial', *JPEN*, **25**, 323–329.

Price CC (1998), 'Sales of meals and snacks away from home continue to increase', *FoodReview*, Sept/Dec, 28–30, Accessed 14 February 2008. Available at http://www.ers.usda.gov/Briefing/FoodMarketingSystem/foodservice.htm.

Price C (2002), 'Food Service', in *US food marketing system*, USDA Economic Research Service Report AER–811, Washington DC, USDA, Accessed 14 February 2008. Available at: http://www.ers.usda.gov/publications/aer811/aer811f.pdf.

Puckett RP (2004), *Food service manual for health care institutions* (3rd ed), San Francisco CA, Jossey-Bass.

Rock KL, Lesher LL, Aylward J and Harrington MS (1998), 'Field acceptance and nutritional intake of the meal, ready-to-eat and heat and serve ration', *Technical Report Natick TR-98/022*, US Army Natick Research, Development and Engineering Center, Natick MA.

Schuster K (2007), 'Making it Authentic, Making it Ethnic', *Food Manag* **42**(4), 42–52.

Shatenstein B and Ferland G (2000), 'Absence of nutritional or clinical consequences of decentralized bulk food portioning in elderly nursing home residents with dementia in Montreal', *J Am Diet Assoc*, **100**, 1354–1360.

Sheehan-Smith L (2006), 'Key facilitators and best practices of hotel-style room service in hospitals', *J Am Diet Assoc*, **106**, 581–586.

Sidenvall B (1999), 'Meal procedures in institutions for elderly people: a theoretical interpretation', *J Adv Nurs*, **30**, 319–328.

Simmons S, Osterweil D and Schnelle J (2001), 'Improving food intake in nursing home residents with feeding assistance: a staffing analysis', *J Gerontol*, **56A**, M790–M794.

Sloan A (2007a), 'New shades of green', *Food Technol*, **61**, 16.

Sloan A (2007b), 'Top 10 food trends', *Food Technol*, **61**, 22–39.

Smith C (2002), 'Punishment and pleasure: women, food and the imprisoned body', *Sociol Rev*, **50**, 197–215.

Sneed J and Kresse K (1989), *Understanding Foodservice Financial Management*, Rockville, Aspen.

Spears M (2000), *Foodservice organizations: a managerial and systems approach* (4th ed), Upper Saddle River NJ, Prentice–Hall.

Spears M and Gregoire M (2007), *Foodservice Organizations. A managerial and systems approach* (6th ed), Upper Saddle River NJ, Prentice Hall.

Stein K (2000), 'Foodservice in correctional facilities', *J Am Diet Assoc*, **100**, 508–509.

Sydner Y and Fjellström C (2005), 'Food provision and the meal situation in elderly care – outcomes in different social contexts', *J Hum Nutr Diet*, **18**, 45–52.

Thomas D, Zdrowski C, Wilson M, Conright K, Lewis C, Tariq S and Morley J (2002), 'Malnutrition in subacute care', *Am J Clin Nutr*, **75**, 308–313.

United States Army Research Institute of Environmental Medicine (USARIEM) (2006). 'Military rations', Accessed 15 January 2008. Available at: http://www.usariem.army.mil/nutri/milrat.htm

USDA Economic Research Service (2007), *Food Marketing System in the US*, Washington DC, USDA, Retrieved 14 February 2008. Available at: http://www.ers.usda.gov/Briefing/FoodMarketingSystem/foodservice.htm.

Valentine G and Longstaff B (1998), 'Doing porridge', *J Mater Cult*, **3**, 131–152.

Vranesevic T and Stancec R. (2003), 'The effect of the brand on perceived quality of food products', *Br Food J*, **105**, 811–825.

Walton K, Williams P and Tapsell L (2006), 'What do stakeholders consider the key issues affecting the quality of food service provision for long stay patients?' *J Foodserv*, **17**, 212–225.

Walton K, Williams P, Tapsell L and Batterham M (2007), 'Rehabilitation inpatients are not meeting their energy and protein needs', *e-SPEN e-J Clin Nutr Metab*, **2**, e120–e126.

Walton K, Williams P, Bracks J, Zhang Q, Pond L, Smoothy R, Tapsell L, Batterham M and Vari L (2008), 'A volunteer feeding assistance program can improve dietary intakes of elderly patients – a pilot study', *Appetite*, **51**(2), 244–248.

Williams P (2005), 'The place of functional foods within hospitality – an opportunity?' *J R Soc Health*, **125**, 108–9.

Williams P and Brand J (1988), 'Food service departments in New South Wales hospitals – a survey', *Aust Health Rev*, **11**, 21–39.

Williams P, Kokkinakos M and Walton K (2003), 'Definitions and causes of hospital food waste', *Food Serv Technol*, **3**, 37–39.

Williams P, Walton K, Ainsworth N and Wirtz C (2006), 'Inmates as consumers: attitudes and food practices of inmates in NSW correctional centres', *Nutr Diet*, **63**(Suppl 1):A11.

Wilson A, Evans S, Frost G and Dore C (2001), 'The effect of changes in meal service systems on macronutrient intake in acute hospitalized patients', *Food Serv Technol*, **1**, 121–122.

Part III

Studying meals

5

The study of Nordic meals: lessons learnt

U. Kjærnes, The National Institute for Consumer Research (SIFO), Norway, L. Holm, Department of Human Nutrition, Denmark, J. Gronow, Uppsala University, Sweden, J. Mäkelä, National Consumer Research Centre, Finland and M.P. Ekström, Örebro University, Sweden

Abstract: This chapter presents a discussion on how eating as an everyday activity is structured in modern societies. Representative telephone surveys conducted in 1997 explored one day of eating in Denmark, Finland, Norway, and Sweden. Eating habits are quite different in the various countries, shaped by national meal conventions and practical coordination, but were at that time relatively structured in terms of what is eaten, when, where and with whom. Most people combine relatively structured, social meals, mainly at home, with less structured events alone. These patterns are first of all influenced by family structure and employment, less by socio-economic distinctions. In the final section, the theoretical and methodological impacts of the approach are discussed.

Key words: Nordic meals, meal patterns, social eating.

5.1 Introduction

Every society has norms and conventions which traditionally regulate eating. There are rules regarding what is considered edible foods, how foods should be combined into dishes and meals, how meals should be ordered with respect to season, daily rhythm, place and social company. In this chapter, we ask how eating as an everyday activity is structured in modern societies. Our approach is comparative and quantitative. We have studied the Nordic countries which, to an outsider, may appear rather homogeneous. But upon closer study we find significant and interesting differences among the countries. Such variations point to important conditions influencing our eating habits in terms of what we eat, when, where and with whom. Based on representative population surveys we have explored a day of eating in Denmark, Finland, Norway, and Sweden. A range of questions can be asked of these data. In this chapter, we will concentrate on our approach for studying contemporary meals, illustrated empirically by data on national variations in eating patterns.

Everyday life has changed considerably during the past few decades, influenced by new organisation of work and new family structures. At the same time supermarkets and food services are offering innumerable varieties of easily prepared and eaten dishes, while mass media are bombarding people with all kinds of messages on healthy, exotic, ethical and modern (or traditional) foods. A growing literature has focused on how these changes influence, and are reflected in, our daily meals. While some contend that modern individuals still eat regular meals in the social company of others, others suggest that eating is more and more going on in a series of random, unstructured events and in social isolation. Many studies present very generalised ideas about what modern life is – or they have assumptions of gradual convergence of life styles. Instead of assuming any processes of similarity or convergence, the Nordic study aimed at exploring this empirically, claiming that eating patterns should be studied with reference to the specific social setting of everyday life in which they happen. A comparative design offers particular advantages in that basic, structuring features taken for granted in one food culture may differ in another.

The next two sections describe in more detail the theoretical underpinnings for our research questions and study design and how eating was operationalized to capture the distinctions mentioned above. An outline of the methodology follows. The section on results concentrates on describing national variations in the daily schedules for various types of eating events and their social context, with some comments attributed to analyses of social differences. This is followed by a discussion of the lessons learnt from the study, its approach and findings. Our study was conducted ten years ago and the final section argues that in the light of changes that may have taken place over the last few years it is necessary to provide relevant data to analyse change over time in eating patterns.

5.2 The debate on meals in contemporary Western societies

The concept of 'a meal' is important if we want to study eating patterns, how they change and vary. According to Douglas and Nicod a *food event* is an occasion when food is eaten, *a structured event* is a social occasion organised by rules concerning time, place and sequence of action (Douglas and Freilich 1983; Douglas and Nicod 1974). Food eaten as part of a structured event is *a meal*. A meal observes the rules of combination and sequence. *A snack* is an unstructured food event without any rules of combination and sequence. Food, and what we do to and with it, is proclaimed to lie at the very core of sociality: it signifies *'togetherness'* (Murcott *et al.* 1992, p. 115). The commensality of eating means that we try to coordinate our actions. To Simmel, the sociability of eating is related to the refinement of social forms of interaction (Gronow 1997; Simmel 1994/1957). Quite a few empirical studies have been added over recent decades that extend and complement these ideas. Many regard eating as a matter of compromise between a number of individually and socially determined concerns (Caplan *et al.* 1998; Ekström 1990; Holm and Kildevang 1996; Kristensen and Holm 2006; Warde and Martens 2000, p. 144).

As an integrated part of everyday life, eating contributes to ordering our days into segments: morning, midday, afternoon and evening. The order and rhythm of eating, the meal pattern, forms intersections between the public sphere of production and the private sphere of reproduction, of afternoon, evening, family and recreation (Aymard *et al.* 1996). In this way, attention is directed towards the organisation of schedules, the particular modes in which food preparation and meals interchange with work and other activities, as part of cyclical calendars as well as throughout the day. These schedules are influenced by societal change. A classical study by Rotenberg shows how eating patterns in Vienna shifted from the early 1900s, throughout the interwar years, and up till the 1980s (Rotenberg 1981). During this period, meals and snacks shifted depending on the organisation of work, not only in terms of time and contents, but even with regard to who ate together and where the meals took place. By the 1930s, industrialisation had substituted a traditional five-meal pattern with a three-meal pattern, in which the men did not go home for the midday meal, and where socialising with friends mainly took place during weekends. The family meal was relocated to the end of the working day. Similar general shifts in meal patterns were identified in a Finnish study (Prättälä and Helminen 1990). The tendency was observed among urban industrial workers already in the 1920s, but a more massive shift in this direction did not take place until the 1960s and 1970s.

A question repeatedly posed is whether traditional meal patterns and meal formats are being disrupted (Murcott 1995). Many use the expression *grazing* (there are also similar versions in other languages) to describe a situation where food is eaten more or less randomly with regard to time, place and contents, according to immediate preferences and concerns. Dissolution of tradition and individualisation is often presented as implying more flexibility and freedom for the individual to choose according to his or her tastes and preferences. However, many contributions have also emphasised negative aspects of complete individualisation, in which each act of choice is to be reflexively considered (Giddens 1994; Sulkunen 1997). A high degree of unpredictability and dissolution is extremely impractical within an everyday context and, for the individual, it may create a basic uncertainty, even anxiety. Fischler has suggested that we are in an era of *gastro-anomie* where regular meals have become increasingly rare and replaced by irregular eating patterns (Fischler 1988). By the term gastro-anomie Fischler refers to a tendency whereby cultural norms for what should be eaten when and together with whom disappear. Following a collapse of traditional and authoritative external rules about what should be eaten, the individual faces a splintered, uncertain and confused situation, where, in the midst of conflicting advice, the individual is left alone, ill-prepared to make decisions about food consumption. Mintz argues that meal patterns are dissolving (Mintz 1996), while Burnett finds indications that the meals that are held to be the very stuff of sociality are in danger of disappearing (Burnett 1989).

These assumptions raise a number of questions. First, to what degree is contemporary everyday eating decoupled from normative regulation and is it the same everywhere? Second, considering how coordination has influenced earlier meal schedules, how is eating organised in modern societies? Gronow and Warde (2001) suggest that increasing individualisation and conventionality need not necessarily be social opposites. With the growing complexity of modern societies we need both flexibility and daily routines. This is supported by Campbell (1996), who contends that life in modern societies can at the same time become de-traditionalised and more habitual. Everyday food-related practices represent a typical example of mundane, routinized practices, closely associated with how we carry out our daily activities (Kjærnes and Holm 2007). These studies indicate that eating patterns develop within strong institutional frames, that common trends can be found, but also that specific conditions must be taken into account. A simple example is the degree and ways in which women have been included in the workforce, which has implications not only for time spent on cooking, but also the daily schedules of eating. The emerging welfare states dealt with this in different ways. In some, mothers stayed at home – and the focus on cooked lunches at home was retained. In other states, women became female wage earners – and elaborate meal services at work places and in schools were established – or a cold lunch became the norm. Yet, eating is not only about workday schedules. Eating patterns have deep historical and cultural roots, while also impacted by political conditions and characteristics of the food market.

Societal organisation and institutional structures will have impacts on the social conventions and standards that regulate food consumption and they will form rather concrete frames for how practices are organised. We must therefore expect to find distinct patterns of similarity and difference according to location, time, and social and institutional context. Patterns emerge in the interplay between practical coordination, institutional restrictions and opportunities, and normative regulation.

Within a given context, for example in a country at a certain point in time, decisive features of everyday eating patterns may be overlooked because what people do when they shop, prepare and eat food is so normalized, trivial and taken for granted. If we look at patterns across social and cultural contexts or over time we find that such 'normalities' may vary considerably and that they change over time (Blake *et al.* 2007; Mestdag 2005; Poulain 2002; Warde and Martens 2000). Events such as workday lunches are typically socially coordinated and strongly linked to shared norms and expectations within specific social contexts. Such 'normal' practices describe not only how things are usually done, but often also how things should be done. At the same time, ideas of normality may prevent us from realising the emergence of new patterns – which at least for a time may appear as 'exceptions' from the daily routine. There may be internal differentiation based on competence and commitment, for example by women adding more vegetables to these lunches than men, and, evidently, there is individual variation and flexibility. Yet, these trivial habits are highly influential when it comes to regulating our eating and therefore are important to study.

5.3 Study design and methodology

Most sociological and anthropological studies of meals are qualitative, in-depth studies of a small group of people (in many cases with female informants). Our research questions about the extensiveness of various types of activities call for an empirical database that can be generalised, i.e., a quantitative and representative design. As we wanted to study how eating is organised and influenced by work and family life, we chose a comparative approach. A challenge was therefore to formulate valid and reliable questions that could reveal the structure of meals within a quantitative and comparative setting. Even though the Nordic countries are similar in many ways, there are significant differences in food habits as well as in the vocabulary for eating. The main meals of the day would not imply the same thing in each country. We soon realised that the Norwegian and Danish 'dinner meal' does not necessarily exist in Sweden and Finland, where a hot lunch is far more common than in Norway or Denmark. Dinner as 'a hot meal one eats after working hours' could not be taken for granted. Moreover, what is eaten may not count as 'dinner' in the respondent's mind if it does not contain the elements that conventions require. This awareness made us realise that we could not use concepts like 'dinner', 'lunch' and 'main meal' implicitly meaning hot or cooked meals at certain hours in our questionnaire.

Starting out with ideas similar to the 24-hour recall method used in dietary surveys, we decided to ask for every occasion of eating something (for practical reasons, just a beverage or only eating chewing gum or sweets, etc, were excluded) during the day before the interview. In our analyses, these occasions are characterised as *eating events*. Thus, we have broadened the concept of a *food event* (Douglas and Nicod 1974) to use as our starting point a very inclusive concept of eating situations. This way we hope to avoid making false generalisations on the basis of cultural, national or ideological prejudices or 'taken-for-grantedness' – i.e., ortodoxography (Bourdieu 1986).

In order to identify social patterns of eating, observations we made of eating events were reconstructed, based on a model of what we call the *eating system*. The eating system includes various types of eating events, the particular composition of foods and dishes of events and their social context, and how these various types of eating events are patterned with regard to chronology and sequence. The model distinguishes between three dimensions: the meal format, the social context of eating, and the eating pattern. Regarding the *meal format*, the composition of eating events is of interest, particularly variations with regard to the degree of complexity (from simple cold snacks/ meals to sophisticated meals with many courses). In describing the *social context* we ask where, with whom, and how (at a table, watching television etc.) eating took place and also who did the cooking. The *eating pattern* is defined by time (the rhythm of eating events), the number of eating events, overall and in terms of various types. These three dimensions do not form a hierarchy. In principle, each of them could be explored separately.

As already mentioned, our approach was to record one day of eating, i.e. eating during the day before the interview. Data were collected in April 1997 using computer-assisted telephone interviews (CATI). We interviewed representative samples of the populations aged 15 years and above (omnibus random sample) (for more details, see Kjærnes 2001). The total sample was 4823 respondents (Denmark 1202, Finland 1200, Sweden 1177 and Norway 1244). Representativity was checked with regard to a range of socio-demographic characteristics and data were weighted with regard to gender and region. Concentrating on the day before, it was important that interviewing was evenly distributed throughout the week. We could not include Sundays since the agencies were closed on that day. This means that the records cover all days of the week except Saturdays. The design of the questionnaire and the data collection procedures in each of the four countries were closely coordinated.

A record of one day of eating allows us to ask for the specific situations of everyday life in which food is taken. But one day will not take account of the variability of an individual's diet and the data therefore do not allow analyses of patterns and routines at the individual level. The questionnaire consisted of loops where questions on the time, the structure, the contents, and the context of eating events, were recorded chronologically: 'When did you first eat something yesterday, what, etc., when did you eat next, what etc.'. A completely 'de-structured' questionnaire would have demanded mainly open questions and highly skilled interviewers. As we already had considerable knowledge about Nordic food cultures from previous studies, a number of concessions were made in order to make the interview more simple to carry out (and thus more reliable) and also to limit work required for coding and recoding. For each eating event, there was a basic distinction between hot and cold food. 'Hot' indicated that cooking had been involved and that the food was eaten hot. Anticipating that hot eating events in many cases are more elaborate and socially structured, cold events may be recorded in a simpler manner. Also for practical reasons, the first eating event of the day had no distinction between hot and cold food – in this text called 'breakfast'.

The questionnaire included few open questions. A detailed description of the study design can be found in Mäkelä (2001).

A considerable part of the analyses was to recode the information on eating events according to the food components and cooking. We distinguished between five types of eating events; 'a snack' event (an ice-cream, an apple or a chocolate bar with/without a beverage), 'a cake event' (a cake, pastry, etc.; may also include a snack), cold food eating events – 'a cold meal' (sandwich, cold cuts, a salad; may, in addition, include a snack and/or a cake), 'breakfast', i.e., the first eating event of the day (often cold foods, like sandwiches or breakfast cereals, but may also involve cooked dishes like porridge), and 'hot meals'. The social context was categorised in similar ways (eating at home or at work, and eating alone or with family members, colleagues etc.). Unfortunately, the social context was recorded only for hot eating events in Finland and Finnish respondents had to be left out of many of the analyses of this aspect of the eating system. Eating patterns were analysed first of all by keeping the focus on timing and daily schedules. Thus, the analytical unit was for the most part the eating event rather than the individual. The statistical analyses involved in this paper include mainly descriptive statistics. Variations between social groups within each country have been tested both with bivariate and multivariate methods, mainly linear and logistic regression. Some of these results are briefly mentioned in the text. A full report can be found in Kjærnes (2001).

5.4   Daily patterns of eating: when and what?

The records show that eating took place from early morning until late at night. However, commonly shared eating hours can easily be identified in all countries. Figure 5.1 shows the daily rhythm of eating in the four countries in terms of frequencies of three types of eating events. The three types, breakfast (first eating event), cold eating events and hot eating events, have been added on top of each other to show the total proportions having had something to eat during one-hour intervals throughout the day.

Figure 5.1 demonstrates that eating in the Nordic countries at the time of the survey was ordered according to distinct rhythms which are relatively uniform within each country, and, for the first part of the day, across the countries. Different national patterns are identifiable in terms of the degrees to which people eat at the same time and the same type of food. Eating hours were most uniform in Denmark (i.e. with the highest peaks), least so in Finland and Sweden. In Denmark more than 50% of the sample had eaten between 12 noon and 1 pm and nearly as many had done so between 6 and 7 pm. In Norway there were clear peaks, but no hours when more than 35% of the respondents had eaten. Finland and Sweden had the highest peaks around lunch time, when between 40 and 50% had eaten, with more dispersion in the afternoon. In all countries, the first eating event, breakfast, is indicated

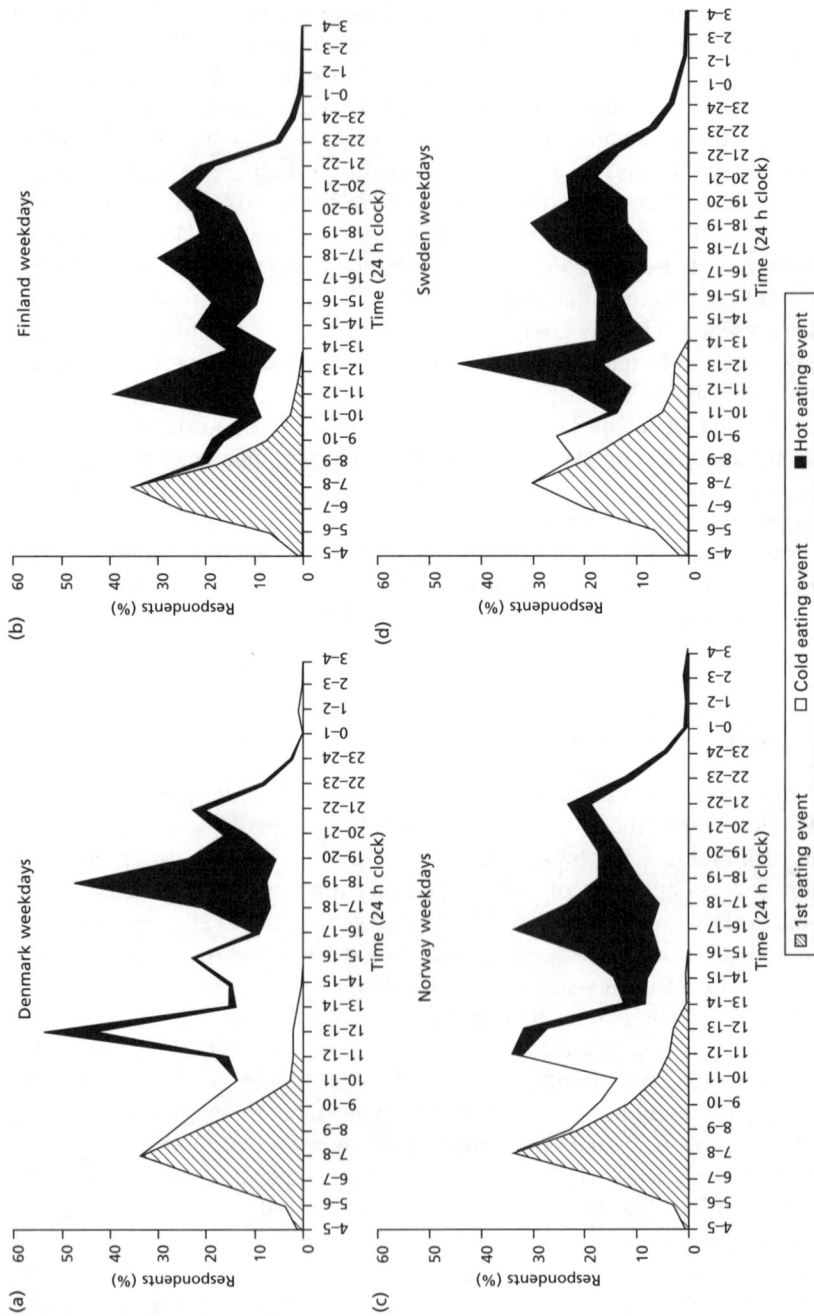

Fig. 5.1 The distribution of eating events during the day on weekdays in (a) Denmark, (b) Finland, (c) Norway and (d) Sweden, accumulated number in 1 h intervals.

by clear peaks. There were also additional, but smaller peaks during mid-morning, mid-afternoon and in the evening. During the first part of the day, the timing of the main peaks is fairly identical between the four countries. After around 3 pm the rhythms start to divert and there are clear differences in the time of the evening meal. Norwegians had an earlier evening meal than in the other countries, whereas the largest 'dinner' peaks are found between 5–6 pm in Finland and between 6–7 pm in Denmark and Sweden. On Sundays, eating hours differed from what is shown in Fig. 5.1 in that eating was generally much more heterogeneous, compared with the weekday patterns (data not shown). Eating events on Sundays were spread out more evenly over the whole day and the day started later.

The alternatives to shared eating hours are either to establish individual rhythms or to skip predictable meal patterns altogether. Individual rhythms would be represented by eating being concentrated in a few, distinct events which do not reflect social conventions about time (or contents) while a lack of orderly patterns would, within our data material, be reflected in a larger number of events (or very few events) taking place at unconventional hours. Table 5.1 shows the number of various types of eating events accumulated over the day. For the majority in all countries, eating was concentrated in relatively few events. Between 80 and 90% of the respondents had 3–5 eating events on the workday before (somewhat less on Sundays). The total number of eating events was slightly lower in Norway than in the other countries. The proportion having eaten more than five times on the day before was marginal and no individual records include more than nine eating events.

We can therefore conclude that at the time of the interviews, there were clear, but also flexible daily rhythms of eating in the Nordic countries. The national differences indicate some variations in flexibility as well as historical legacy, but it may also be due to variations in daily time schedules in schools,

Table 5.1   The daily number of various types of eating events in Denmark, Finland, Norway and Sweden on weekdays. Percentage of respondents (differences between the countries are statistically significant $p<0.05$)

No. of eating events	Denmark	Finland	Norway	Sweden
1	1	1	1	0
2	7	8	8	6
3	29	26	31	31
4	36	36	41	35
5	21	21	16	19
6+	8	7	3	8
Total N	100 (736)	100 (977)	100 (958)	100 (903)
Mean total events	3.94	3.93	3.73	3.93
Standard deviation	0.98	1.15	0.98	1.06
Mean cold events	1.99	1.58	1.73	1.56
Standard deviation	1.11	1.11	0.99	1.10
Mean hot events	0.95	1.44	1.02	1.37
Standard deviation	0.40	0.66	0.49	0.62

workplaces etc. Somewhat surprisingly, however, working hours are significant beyond practical organisation, as people not presently employed tend to follow the same rhythms. This might be due to coordination within the household, but it might also underline the strongly habitual character of eating. Once established, such rhythms are not easily changed. Yet, the considerable proportions of events outside conventional eating hours suggest that these eating events do not represent strict, prescriptive norms but must rather be understood as flexible conventions.

5.5   Hot or cold food?

As indicated by the types of eating events that are distinguished in Fig. 5.1, the peak hours reflect conventional Nordic cultural models of three main meals: breakfast, lunch and dinner. The types of eating events that dominated at different hours were quite uniform *within* the countries, especially during peak hours. Some were also similar *across* the countries. A large majority of the respondents had a hot meal in the afternoon/evening. However, for lunch time eating Fig. 5.1 shows important national differences. Cold 'lunches' dominated completely in Denmark and Norway, whereas 45 and 42% in Finland and Sweden, respectively, had a hot meal between 11 am and 1 pm. Nationally distinct patterns concerning types of eating events are reflected even in the number of various types of eating events of respondents (Table 5.1). In Finland 43% and in Sweden 40% had had two, 2–4% even three, hot meals. In Denmark and Norway, on the other hand, 84 and 78%, respectively, had had only one hot meal on the day before. Ten per cent or less had no hot eating events at all on the day before.

These findings reflect that there are nationally distinct notions of what constitutes 'a meal' as opposed to other types of eating events. As the interviews did not use the term 'meal', our counting is based on eating events according to defined categories that refer to contents as well as degrees of structuration (Douglas and Nicod 1974). We talk about them in terms of 'meals' when these events correspond with (national) meal conventions. In Denmark and Norway eating open sandwiches with a topping of for example cheese, liver paste or ham is generally conceptualised as a meal (Bugge and Døving 2000; Holm 1996). It is the norm (and – as we find – the normal) for lunch-time eating, often for breakfast and a late evening meal as well. This is less the case in Finland and Sweden (Ekström 1991; Jansson 1993; Mäkelä 1995; Prättälä et al. 1993) where meals tend to be associated with cooked, hot meals. In the questionnaire it was not feasible to differentiate between what might be characterised as 'a cold meal' and less significant snacks, even though just a cup of coffee or an apple were generally not recorded. But we tried to account for these distinctions in the reconstruction procedures by differentiating between cold eating events including different categories of food. In Denmark and Norway, 55 and 56%, respectively, had 2–3 daily cold meals with sandwiches, a salad or similar types of dishes/items. The corresponding percentages were 43% in Finland and 35% in Sweden, respectively. Eating events that included only items like a sweet pastry, perhaps accompanied by a drink, were reported more rarely: only 20–30% of the respondents had eaten these types of foods outside the major eating events identified above. These latter figures should be interpreted with caution as people's memory for such types of events may be less reliable. Also, some may not think of crisps or a cake as 'food' or 'eating' at all, instead memorising them as more or less exceptional 'treats' (Døving 2003). Sales statistics show that compared with other Nordic countries Norwegians eat a lot of salty snacks and sweets, while the relationship is opposite in our study.

The dominant daily meal pattern thus consists of 1–2 hot 'meals' plus 1–3 cold eating events (including breakfast), some of which may also be categorised as 'meals'. Minor snacks and cakes represent additions rather than replacements of this pattern; mean numbers varied between 0.4–0.6 per day in the four national samples. There were so few respondents with extreme eating rhythms – i.e. only one or more than five events per day (Table 5.1) – that the proportions seem to represent only what might be expected from the variation and flexibility of ordinary everyday life, like illness and special celebrations. The timing of minor snacks is less strict, but even here we observe clear regularities, probably reflecting the highly organised character of everyday life in the Nordic societies.

In all four countries, people reported eating events that show both complexity and variation. An overview of the contents of eating events is presented in Table 5.2. The table reflects the central position of open sandwiches in the Norwegian and Danish food cultures. Cooked meals take a significant position in the everyday diet, but so also do relatively structured cold eating events. Some, but not many, informants did not eat any hot, cooked food during the day before. Most weekday hot meals had a uniform and relatively simple structure, generally consisting only of one dish, but the complexity of this 'plateful' varied. Starters were very rare (below 10% on any day), while 25–30% of the cooked events included a dessert. In every country, the most typical hot meal plate had two or three components (like spaghetti with meat sauce or potatoes, meat balls and a vegetable on the side). Somewhat different food items dominate in the various countries, like steaks and pork chops in Denmark vs. fish in Norway, 'boiling type' vegetables in Denmark and Norway vs. 'salad type' vegetables in Finland and Sweden, and varying significance of sandwiches, porridge and breakfast cereals at the first eating event of the day. Meals with one component were more common in Norway (such as 'lapskaus' – a beef and potato casserole – or pizza), whereas Swedes had more meals with four or five components. In Norway and Denmark, no meals with five components were reported. One difference that contributes to this is that bread often accompanies hot meals (in addition to potatoes or rice) in Sweden and Finland, while this is much more rare in the western parts of Scandinavia.

Table 5.2 Typical features of various eating events in four Nordic countries. Items used in 15% or more of the respective types of eating events, most dominant items first. Meal hours indicate peaks

	Denmark	Finland	Norway	Sweden
Breakfast	Open sandwiches Breakfast cereals (Sweet pastry Sunday) Coffee or tea Milk (Juice on Sunday)	Open sandwiches Porridge Coffee or tea Juice, milk or water	Open sandwiches (Eggs on Sunday) Coffee or tea Milk, juice or water The simplest one	Open sandwiches Breakfast cereals Yoghurt Coffee or tea Milk or juice The most varied one
Cold eating events	Lunch (12–1 pm)+ snack Sandwiches Sweet pastries Coffee or tea Beer or wine Soft drinks	Snack Sandwiches Sweet pastries, fruits Coffee or tea Water (soft drinks on Sundays)	Lunch (11 am–1 pm) + (evening meal 9–10 pm) + snack Sandwiches (Cake on Sundays) Coffee or tea Milk, (soft drinks on Sundays)	Snack Sandwiches Sweet pastries, fruits Coffee or tea
Hot meals	Dinner (6–7 pm) Meat (steak/ minced, fried), other dishes + potatoes + cooked vegetables (carrots) (+) sauce Water, beer/wine, soft drinks, milk	Lunch (11–12 am) and dinner (5–6 pm) Meat (minced/steak, baked/fried), other dishes + bread, potatoes + uncooked vegetables (cucumber) Milk, water The lowest number of components	Dinner (4–5 pm) Meat (minced, fried), fish (poached – Sunday) + potatoes + cooked vegetables (carrots) (+) sauce water, soft drinks	Lunch (12–1 pm) and dinner (5–8 pm) Meat (minced, fried), fish (fried), other dishes + potatoes, bread + uncooked vegetables (tomatoes) + sauce (+) pickles water, milk, soft drinks, beer/wine The highest number of components

5.6   Social context: where and with whom?

So far, no social criteria have been included in the discussion of the status of meals. Considering the associations made between meals and social life, especially family life, the division between private and public eating is important in debates about contemporary eating habits. The place of eating influences not only social company and social norms but even how food may be provided and prepared and thus the character of the food eaten. Again our results are somewhat surprising. The Nordic populations are relatively well-off, large proportions are employed or go to school, many participate in sports and cultural life, the age for retirement is high, and many live alone. We might therefore expect tendencies of eating shifting from the private to the public sphere. In this study, however, the majority of eating events took place at home, eating at the workplace coming second (Fig. 5.2). Eating in other people's homes and eating out in cafés or restaurants constituted only minor proportions. Frequent eating at restaurants, cafés, bars and fast-food outlets, often seen as a characteristic of modern life, does not show in our data.

At almost all times of the day, more people had eaten at home than in any other place; only 3–4% had not eaten at home on the day before. Eating in the morning generally took place at home. There were also peaks of eating at home in the middle of the day, but they were typically smaller, instead replaced by eating at work. In the evening, there were large and broader peaks of home eating again. In Denmark, more than half of the respondents (52%) had eaten only at home on the day before, while in Norway and Sweden somewhat less than half (41% and 43%, respectively) had done so. Most eating at work seemed to take place as an organised event, usually in a canteen or separate eating room. Eating at a café or restaurant during work hours was slightly more frequent in Sweden compared with the other countries. Eating in other places than at work or at home did occur, but not as a dominant feature; 25% had done so on the day before in Denmark and Norway and 32% in Sweden, including eating in other people's homes (first) and at cafés/restaurants (second). Eating in places like fast-food outlets or in the street was negligible. A frequency question about eating out (in restaurants, cafés, etc.) confirmed this picture, as only about 10% ate out on a weekly basis, the highest proportion found in Sweden, the lowest in Denmark (data not shown). Apart from lunch, eating away from home appeared to be reserved for special occasions.

Figure 5.3 shows that the eating system included a mixture of events taking place alone, with other household members, with colleagues and with friends. A considerable proportion of breakfast eating took place alone, even in non-single person households. Lunch, however, most often took place in the company of family or colleagues. For those who lived with a family, dinner was usually shared with them. This was the case even for those who had cold food in the evening. Eating with friends and others constituted a minor proportion of everyday eating events at all hours. While individual eating took place throughout the day, the peak meal hours were, to a large

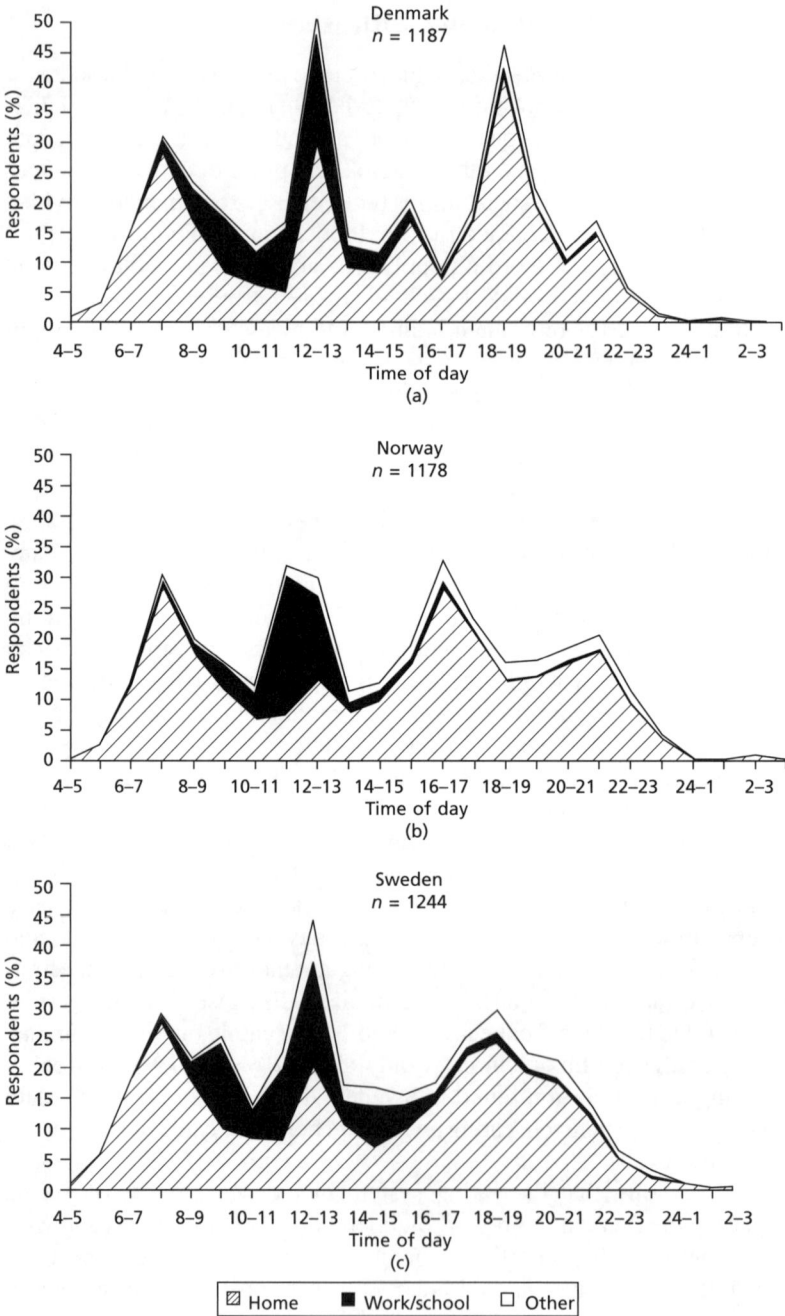

Fig. 5.2   Time and place of eating. Distribution during the day on weekdays in (a) Denmark, (b) Norway and (c) Sweden.

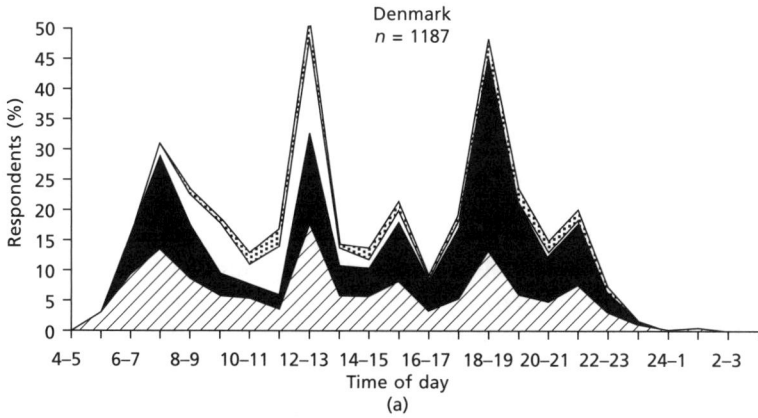

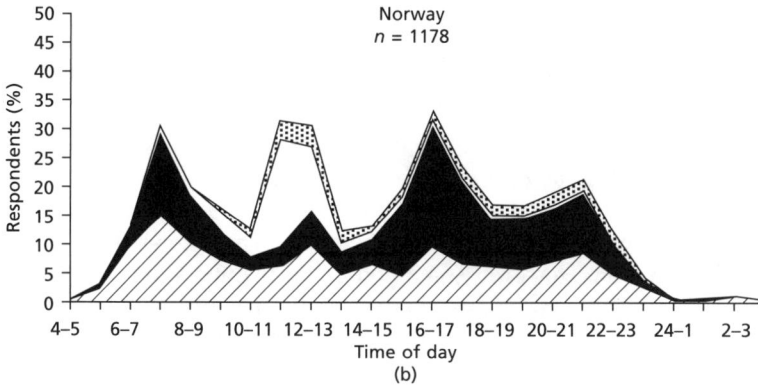

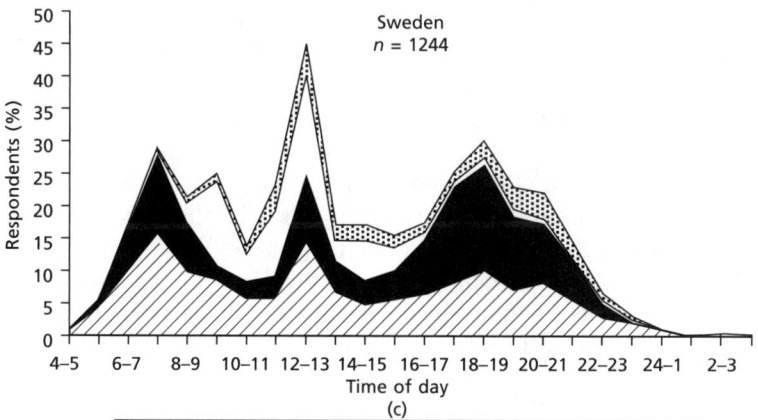

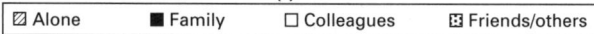

Fig. 5.3 Time and social company of eating. Distribution during the day on weekdays in (a) Denmark, (b) Norway and (c) Sweden.

degree, social events. Moreover, on Sundays, when meals are usually not structured by working hours, family eating dominated, even for a considerable proportion of those who lived alone. About one third had had no eating events with family members, but this proportion was reduced to about 10% in households including at least two adults. The data thus show that even though a considerable part of eating took place alone, eating is also still a social matter in the sense that at least some of the events during the day took place in the company of other people. According to Fig. 5.3, peaks indicating social eating events were generally higher than for events taking place alone. Social eating is also more shared at a national level, to a large degree following conventional eating hours (see above). Many individual eating events took place at the same hours, but the tendency is weaker. With less coordination needed, eating alone tended to be somewhat more evenly distributed throughout the day. These findings suggest that major everyday eating events tend to be socially coordinated and common. The sociability of social eating events was emphasised by such events tending to last considerably longer than the individual ones (data not shown). Individual eating does challenge the pattern of common eating hours, but not to any large extent.

According to qualitative studies, family meals have great significance and meaning as symbols and operators of family cohesion. Our analyses demonstrated that meals at home, especially in the evening, usually take place in company with other household members. This shows that family eating is still common. However, some household members may be absent due to leisure activities or work, so whether family meals in fact do include everybody in the household is another question. The analyses showed that more than half (between 54 and 64%) of those living in multiple person households had at least one hot meal together with the entire assembled household. However, the practice of gathering the whole family, which is often implied in the notion of 'the family meal' did not occur on a regular, daily basis – but rather on an every-other-day basis. This was the case both in households with children and in households without children. Of the eating at home, 65–76% took place at a kitchen or dinner table. A relatively frequent alternative was a coffee table/sofa (including even some events other than coffee or snacks). Television sets are usually placed within view from the sofa in the Nordic countries, and evening snacks seem to be brought to the coffee table, sometimes even the evening meal (most often in Norway, least so in Finland). This was confirmed when inquiring what went on while eating, where about one fifth of the events took place while watching TV. Pizza with the family in front of the TV has been described as a shared family ritual at the end of the week (e.g., Bugge and Døving 2000).

Eating in the company of colleagues occurred on weekdays during the day, but rarely outside regular working hours. Differences between population groups were large, varying primarily according to occupational status. Between 58 and 68% of the occupationally active had eaten with colleagues. Eating in the company of friends or others appeared to play a very small part in the general picture, indicating that such events were not a frequent everyday habit in the general population and first and foremost a leisure-time activity. The tendency to do so was somewhat more common during the weekend than on weekdays (notably our study does not include records for Saturdays).

Eating in passing has been suggested as a term to describe modern eating habits in Denmark, indicating a tendency where people grab something to eat in between or while carrying out other activities (Andersen 1997). It is therefore a version of the English *grazing*. The questionnaire allowed detailed records of such types of eating, operationalised as eating at a fast-food outlet, in the street, etc. and, in the home, eating in other places than the kitchen or at a dinner table. We found that such less structured eating places outside the home constituted a small proportion of daily eating (average 4-6%). At home, very little of the recorded eating took place at other sites (at a work desk, in bed, etc.) or without sitting down at all.

The results indicate that unstructured and highly structured events alternate during the day. This suggests that flexible norms exist regarding eating, allowing for informal, hasty meals which take place while other activities are going on (and there was probably more than recorded here), but that the meal as a ritualised event with social significance and interaction takes a prominent position *as well*. Most individuals seem to enjoy both types of eating on an everyday basis. While we found clear national differences in the daily rhythm and meal formats, the social context does not display the same degrees of nationally distinct patterns. Yet, as the social context of eating will depend on the kind of life we live, such as family type and employment, we might expect national variations along such dimensions to help explain national differences in these respects. We will in the next section point to some issues related to demography and social structure.

5.7   Social differences behind or in addition to national variations?

The Nordic countries are socially highly homogeneous, as reflected even in our data. But the analyses do point to some influences of age, family structure, employment, and social structure, in part via differing daily schedules, in part via other processes. Denmark was the most homogeneous country with regard to social differences in eating patterns, Finland the most diverse. The differences between population groups were often not uniform, but varied from country to country. The different aspects of the eating system did not vary in the same ways. Some background factors, such as educational level and place of residence, showed so diverse and inconsistent patterns that general conclusions are difficult.

Finding that eating patterns are so closely associated with family life and work, we also expected that differences in these respects would matter. In all countries, living together with others seemed to be the one factor that more than anything else increased the probability of following socially shared eating patterns. Thus, those who were living alone tended not surprisingly to have eaten alone to a greater extent than others. People living in households with one or more other adults had eaten more at home, with family members, while single persons had eaten more often at the workplace, with colleagues and with friends (controlled for other factors, such as age). Family status was also the only factor that increased the probability of eating hot food in the afternoon/evening in all four countries. The number of eating events in the home depended more on the presence of adults (and children) than on that of children only. However, it is important to keep in mind that nearly everybody had a quite varied eating pattern, so that even people living alone tended to have had social eating events and a majority of them had had a cooked meal on the day before.

Presuming that older age groups would be more traditional in their behaviour, one would have expected them to eat at more regular hours than younger people. This was, however, generally not the case in our study. But elderly people tend to live alone more often and retirement means they stay more at home – and they eat more at home. They also tended to eat somewhat more often than younger age groups (in Denmark and Sweden). On average, older age groups had one or more extra eating events in the home as compared with the youngest age group. The youngest age group (15–24 years) had eaten with friends to a significantly greater extent than any other age group, approximately half of the people in this age group had done so. But very few did that at the cost of social meals at home.

Being such an important issue in studies of food and eating (Charles and Kerr 1988; Counihan 2004; Ekström 1991; Murcott 1983), gender received considerable attention in our analyses. However, contrary to findings regarding for example food and health, there are only minor gender differences when it comes to eating patterns. Findings from a Norwegian study of drinking patterns (including all sorts of drinks) indicated that it was mainly women living alone who tended to differ – from single men and from households with couples (Bye 1999). The major gender difference that we find in our study is related to who did the cooking in multi-person households. Not surprisingly, across all four countries more women (a female respondent or the mother/wife of a male respondent) had done the cooking of the hot meals. However, an intriguing observation was that while around 80% of the female respondents claimed that they had cooked themselves, considerable proportions among the male respondents (50–60%) said they had done so as well. A cautious explanation might be that the conceptualisation of what constitutes cooking is contested in these welfare states with proclaimed gender equality (Ekström 2006). The exception is Finland, which displays more agreement: women did it (Ekström and Fürst 2001).

Employment status is important with regard to where people eat as well as what. Those who were employed or following a course of education had to a greater extent than others eaten away from home and in the company of colleagues or business contacts, etc. In Norway and Sweden, the tendency for the occupationally active to more often follow the respective lunch conventions is statistically significant, while such distinctions were less clear in Denmark and Finland. In the former countries, we also find some tendencies for people with higher social status, as expressed by educational level, income level and type of occupation, to be more compliant with the dominant national lunch conventions. It can be noticed that these differences in the uniformity of lunch conventions do not depend on whether the meals in question consist of cooked dishes (like in Sweden) or cold sandwiches (like in Norway, where they are generally brought from home).

Many of these variations appear to be straightforward reflections of variations in everyday schedules. However, those who were living alone did not always eat alone, those living with family members did not always eat with their family, and not all of those who were employed had eaten in the company of colleagues or professional contacts. The conclusion is that while clear meal conventions can be identified, they are quite flexible in terms of content, time, place and social company. Few people – and no particular sections of society – live fully up to these conventions throughout a whole day, and very few discard them altogether.

Overall, national distinctions in eating systems are more easily identifiable than social ones, especially with regard to social stratification. Rather than expressing a general tendency, this may reflect the relatively high social homogeneity of the Nordic countries, compared with for example France or the UK. Moreover, while meal patterns seem to be shared within each country, a review found that the selection of food items like fruit and fish was more socially stratified (Roos and Prättälä 1999).

What we do find are influences of family structure and employment, some of which may help to explain some of the national variations. More people live alone in Sweden, while unemployment rates were higher in Finland at the time of the survey. Yet what comes out are first of all nationally distinct ways of dealing with questions of eating and meals. Such national variations are influenced by cultural traditions as well as by current configurations of and relationships between households, the labour market, the state, and the food market. The exact nature of such influences must be subjected to separate analyses. Our data only allow us to suggest this as a line of inquiry for future research.

5.8 Implications of the approach

Our focus is on the structures of contemporary eating. We find that, at the time of our interviews in the late 1990s, socially shared patterns are easily observed. The large majority of the populations across the four Nordic countries include in their daily schedule socially shared meals following dominant, easily recognisable meal conventions. At the same time, eating during the day includes a number of less structured eating events. People eat alone, outside established meal formats, and they eat in many different places. Most likely, there are also minor instances of foods and drinks taken that were not recorded. While small and easily forgettable, such minor snacks may of course add significantly to the overall caloric intake. Yet, it has not been possible to identify particular social groups (for instance young, single people) where such less structured patterns predominate – snacking alone and away from home and only rarely eating conventional meals. 'Gastro-anomie' is a concept indicating a general tendency of dissolution of norms and structures of eating. We do not find support in our data for such an assumption. Less structured eating events seem to represent additions rather than replacements of shared, structured meals. Our study focused on dominant trends and we might find less 'structured' eating among groups which typically drop out of telephone surveys. However, a careful scrutiny of the representativity showed no major bias compared with socio-demographic statistics (Kjærnes 2001). Unrecorded eating of single food items does represent a challenge in our study. We know for instance that sales statistics of typical 'snack type' of foods shows that Norwegians consistently buy more ice-cream, salty snacks, chocolate, and fruit, while their recorded eating of such items in our study is lower than the other countries (ibid). Very little recorded eating might indicate that we missed out a lot. But the proportions of respondents with such patterns are, as already mentioned, very low. The Norwegians' snacking represents indulgence and leisure-time freedom (thus not thought of as 'food' at all) within an otherwise highly puritan and structured food culture (Døving 2003).

The study is based on and formulated within a theoretical understanding of eating as habitual, formed within societal structures and institutions. The findings, some of which have been presented here, give strong empirical support to these assumptions. We find that national differences are not only characterised by distinct cuisines, but also by distinctly different ways in which eating takes place as part of everyday life. The four countries are all welfare states, but they are organised differently, with considerable impact on eating. First of all, this is illustrated by the organisation of lunch. The difference between a hot, cooked meal and cold sandwiches are not only related to taste and nutrition. These differences are closely associated with how lunch is organised at work and in schools. The Swedish and Finnish systems of cooked midday meals require lunch food services from public or commercial institutions whereas the packed lunch with cold sandwiches in Denmark and Norway are generally provided by the households. Thus, they have to do with the societal division of labour and responsibilities with respect to food.

Such habits and influences are, within a given setting, usually taken for granted. This normalised character will in many ways make the impacts stronger and more pervasive, but, at the same time, difficult to study. In recent years, considerable academic interest has been directed towards the situated character of eating events, and how these events form socially contingent but flexible habits. Typically, such studies are based on qualitative methodology. Quantitative and representative approaches seem however more suited to study comparative differences between eating systems – in terms of meal formats, social contexts as well as schedules. But this also requires methodological tools which register what people do, considering the situated character of eating as social events. Only such methods make it possible to sort out large-scale differences and change. So far, there are not many studies of that kind and our Nordic study appears to be unique in adopting a comparative design.

Habitual, socially shared patterns of eating form social institutions which appear to be stable. At the same time, we know that societal developments may influence eating in rather basic ways, sometimes quite rapidly. Our data are cross-sectional and comparable data at other points in time are hard to come by. While brave assumptions are made about change, we can in reality therefore say little about trends. For example, how can we know that unstructured eating events used to be less prevalent in earlier times? And maybe expectations of 'normal' eating patterns are stronger than ever? New products appear in the supermarkets all the time, many have a convenience character, but there is also more emphasis on quality and freshness. Eating out is increasing. At the same time, everyday life in the Nordic countries is characterised by changing family structure and work schedules, while this region has become increasingly wealthy. The public agenda during this period has been characterised by a virtual explosion of interest in food in a number of respects, from cooking to risks and ethics. All of this may amount to a strengthening of tendencies that were hardly observable a decade ago, such as a movement towards more eating away from home and a polarisation between skilled home cooking and everyday, convenience food. A new study is required to answer whether such tendencies are reflected in new meal patterns and social structures of eating. Our study does, however, indicate that it is more relevant to look for social processes of re-structuration rather than de-structuration of eating.

5.9   References

Andersen, J. 1997 *Hverdagens centrifuge – det daglige liv og den moderne livsform [The centrifuge of everyday life – modern daily living]*, Copenhagen: Hovedland.

Aymard, M., Grignon, C. and Sabban, F. 1996 'Food allocation of time and social rhythms. Introduction', *Food and Foodways* **6**(3–4): 161–185.

Blake, C. E., Bisogni, C. A., Sobal, J., Devine, C. M. and Jastran, M. 2007 'Classifying foods in contexts: How adults categorize foods for different eating settings', *Appetite* **49**: 500–510.

Bourdieu, P. 1986 *Distinction. A social critique of the judgement of taste*, London: Routledge & Kegan Paul.

Bugge, A. and Døving, R. 2000 *Det norske måltidsmønsteret – ideal og praksis [The Nordic meal pattern – ideal and practice]. SIFO report no.2*, Oslo: The National Institute for Consumer Research.

Burnett, J. 1989 *Plenty and want. A social history of food in England from 1815 to the present day*, London: Routledge.

Bye, E. K. 1999 *Bruk av drikkevarer – hvem, hva og hvor? (The consumption of beverages – who, what and where?)*. Report No 8, Lysaker: The National Institute for Consumer Research.

Campbell, C. 1996 'Detraditionalization, character and the limits of agency', in P. Heelas, S. Lash and P. Morris (eds) *Detraditionalization. Critical reflections on authority and identity*, Cambridge, Mass.: Blackwell.

Caplan, P., Keane, A., Willetts, A. and Williams, J. 1998 'Studying food choice in its social and cultural contexts: approaches from a social anthropological perspective', in A. Murcott (ed) *The nation's diet. The social science of food choice*, London, New York: Longman.

Charles, N. and Kerr, M. 1988 *Women, food and families*, Manchester: Manchester University Press.

Counihan, C. 2004 *Around the Tuscan table: food, family and gender in twentieth century Florence*, New York: Routledge.

Douglas, M. and Freilich, M. 1983 'Culture and food' The pleasures of anthropology: Merton, New American Library.

Douglas, M. and Nicod, M. 1974 'Taking the biscuit. The structure of British meals', *New Society* 30: 744–747.

Døving, R. 2003 *Rype med lettøl. En antropologi fra Norge [Grouse with non-alcohol beer. An anthropology from Norway]*, Oslo: Pax.

Ekström, M. 1990 *Kost, klass och kön [Food, class and gender]*, Umeå: University of Umeå, Department of Sociology.

Ekström, M. 1991 'Class and gender in the kitchen', in E. L. Fürst, R. Prättälä, M. Ekström, L. Holm and U. Kjærnes (eds) *Palatable worlds. Sociocultural food studies*, Oslo: Solum forlag.

Ekström, M. P. 2006 'Family meals. Competence, cooking and company', Gothenburg: Department for Food, Health and Environment, http://home.edu.helsinki.fi.

Ekström, M. P. and Fürst, E. L. 2001 'The gendered division of cooking', in U. Kjærnes (ed) *Eating patterns. A day in the lives of Nordic peoples. SIFO Report No.7*, Oslo: The National Institute for Consumer Research.

Fischler, C. 1988 'Food, self and identity', *Social Science Information* 27(2): 275–292.

Giddens, A. 1994 'Living in a post-traditional society', in U. Beck, A. Giddens and S. Lash (eds) *Reflexive modernization. Politics, tradition and aesthetics in the modern social order*, Cambridge: Polity Press.

Gronow, J. 1997 *The sociology of taste*, London, New York: Routledge.

Gronow, J. and Warde, A. 2001 *Ordinary Consumption*, London and New York: Routledge.

Holm, L. 1996 'Food and identity among families in Copenhagen – a review of an interview study', in H. J. Teuteberg, G. Neumann and A. Wierlacher (eds) *Essen und Kulturelle Identität – Europäische Perspektiven*, Bonn: Akademie Verlag.

Holm, L. and Kildevang, H. 1996 'Consumer's views on food quality. A qualitative interview study', *Appetite* 27: 1–14.

Jansson, S. 1993 *Maten och det sociala samspelet. Etnologiska perspektiv på matvanor [Food and Social Interaction. Ethnological Perspectives on Food Habits]*: Sveriges Lantbruksuniversitet.

Kjærnes, U. 2001 *Eating patterns. A day in the lives of Nordic peoples. SIFO Report No.7*, Oslo: The National Institute for Consumer Research.

Kjærnes, U. and Holm, L. 2007 'Social factors and food choice: consumption as practice', in L. Frewer and H. C. van Trijp (eds) *Understanding consumers of food products*, Cambridge: Woodhead Publishing.

Kristensen, S. T. and Holm, L. 2006 'Modern meal patterns: tensions between bodily needs and the organization of time and space', *Food and Foodways* **14**(3–4): 151–173.

Mestdag, I. 2005 'Disappearance of the traditional meal: temporal, social and spatial destructuration', *Appetite* **45**: 62–74.

Mintz, S. W. 1996 *Tasting food, tasting freedom. Excursions into eating, culture, and the past*, Boston, Mass: Beacon Press.

Murcott, A. 1983 '"It's a pleasure to cook for him": Food, mealtimes and gender in some South Wales households', in E. Gamarnikow (ed) *The Public and the Private*: Heinemann Educational Books.

Murcott, A. 1995 'Raw, cooked and proper meals at home', in D. Marshall (ed) *Food choice and the consumer*, London: Blackie Academic & Professional.

Murcott, A., Mennell, S. and van Otterloo, A. 1992 *The sociology of food. Eating, diet and culture*: Sage.

Mäkelä, J. 1995 'The structure of Finnish meals', in E. Feichtinger and B. M. Köhler (eds) *Current research into eating practices. Contributions of social sciences. Proceedings of the European Interdisciplinary Meeting, 1–16 Oct 1993, Potsdam*, Frankfurt am Main: Umschau Zeitschriftenverlag.

Mäkelä, J. 2001 'The meal format', in U. Kjærnes (ed) *Eating patterns. A day in the lives of Nordic peoples SIFO Report no. 7*, Oslo: The National Institute for Consumer Research.

Poulain, J. P. 2002 'The contemporary diet in France: "de-structuration" or from commensalism to "vagabond feeding"', *Appetite* **28**, 1–13.

Prättälä, R. and Helminen, P. 1990 'Finnish meal patterns', in J. C. Somogyi and E. H. Koskinen (eds) *Nutritional adaptation to new life-styles. Bibl. Nutritio et Dieata No 45*, Basel: Karger.

Prättälä, R., Pelto, G., Pelto, P., Ahola, M. and Räsänen, L. 1993 'Continuity and change in meal patterns: the case of urban Finland', *Ecology of Food and Nutrition* **31**: 87–199.

Roos, G. and Prättälä, R. 1999 *Disparities in food habits. Review of research in 15 European countries. Report B24*, Helsinki: The National Public Health Institute.

Rotenberg, R. 1981 'The impact of industrialization on meal patterns in Vienna, Austria', *Ecology of Food and Nutrition* **11**(1): 25–35.

Simmel, G. 1994 'The sociology of the meal', *Food and Foodways* **5**(4): 345–351.

Sulkunen, P. 1997 'Introduction: the new consumer society – rethinking the social bond', in P. Sulkunen, J. Holmwood, H. Radner and G. Schulze (eds) *Constructing the new consumer society*, Houndmills, Basingstoke: Macmillan Press.

Warde, A. and Martens, L. 2000 *Eating out. Social differentiation, consumption and pleasure*, Cambridge: Cambridge University Press.

6

Meals and gender

Ø. Ueland, Nofima Food, Norway

Abstract: Gender differences in a meal context are discussed in a historical perspective, and the lines are drawn to perceptions and practices that differ between the genders today. Focus is placed on how differences in food perception and choice between genders may influence behaviour in a meal context.

Key words: gender differences, meals, food choice.

6.1 Introduction

The aim of this chapter is to elaborate on the role of gender in the meal context. Gender refers to the socio-cultural dimension between males and females, while sex, strictly speaking, only describes biological features. Some differences between men and women with regard to food perception can be ascribed to sex as a biological factor, for instance, sensitivity to boar taint, or hormonal impact on perception of bitter taste (Bartoshuk *et al.*, 2007; Bremner *et al.*, 2003). However, in a contextual setting such as a meal, differences observed between males and females will be attributed to socio-cultural factors and should be referred to as gender differences.

Food consumption occasions are mainly ordered in meals and, as a consequence of this, differences between genders will be evident in the study of meals. In the previous version of this book, gender was only briefly mentioned (Mäkelä, 2000) in a labour division context, while gender differences in other relevant areas were not discussed. In this chapter, we will look at how gender differences have appeared in traditional households and how these differences have evolved to the gender differences we observe in Western culture today. The focus on gender differences will mainly be on situations where men and women interact. Gender differences will be discussed in the light of food preferences as well as attitudes and behaviours associated with food in meal settings. While social class has an impact on meal consumption and food choices, gender differences may be more or less influenced by class

– but this is outside the scope of the chapter. Lastly, gender differences will be discussed with a background in Western culture and based on literature from Western Europe and the USA.

6.2   Historical perspective on the role of gender in meal contexts

Historically, food preparation has been the domain of women. Among other things, they were responsible for providing and preparing meals for the rest of the household and, in addition, for rationing and distributing available food so it lasted throughout the year (Bugge, 2005; DeVault, 1991). A direct consequence of this responsibility was the continued health or survival of the members of the household. The food thus provided by women should ensure that men could continue their labour and work to support the household, and that the children could grow up as an insurance for the future. In order of importance, the man would always be served first, receive the best food and the largest portions (Bugge, 2005; Bugge and Døving, 2000; Lupton, 1996). In the food hierarchy, meat was at the top and was served to men first, and to women and children only if there was enough (Bourdieu, 1984; Jensen and Holm, 1999; Lupton, 1996; Roos et al., 2001). Some reports tell of mothers starving themselves so that the rest of the family could eat (Lupton, 1996), while others tell of older children starving themselves to provide for younger siblings (Bugge and Døving, 2000). These historical roles impact today's gendered meal consumption on at least two counts; 1) there is still a strong tendency that a woman living with a man will defer to his food preferences (Bove et al., 2003; Brown and Miller, 2002; Bugge, 2005; Lupton, 1996), and 2) women are still the major drivers for composing and serving healthy meals (Bove et al., 2003; Brown and Miller, 2002; Fagerli and Wandel, 1999; Jensen and Holm, 1999; Øygard and Klepp, 1996).

Even though men historically had first priority when food was served and their wishes and preferences took precedence, women exercised a certain power based on specialised knowledge about food, food preparation, and food stores (DeVault, 1991). This was the woman's domain and men were not to intrude on her territory (Bugge, 2005). Her knowledge was, if not intuitive, at least passed down from mother to daughter or acquired otherwise through formal or informal channels (DeVault, 1991).

In the middle of the 20th century, the housewife cooking for the family was the role model for women in the Western world (McFeely, 2000). However, a trend leading women away from home and into paid work was already in progress at that time. A husband's deep sigh in 1954 'My wife is – unfortunately, a professional woman!' (Author's translation) (Bugge and Døving, 2005, p. 212) was an example of how men would rather have more frugal meals than a working wife. But even taking the female emancipation activities in the

1960s and 1970s into account, sharing of responsibilities in the kitchen has been extremely hard to change. Remnants of this can be seen today in studies showing that many men and women almost unconsciously presume and accept that women know more about meal preparation (Bugge, 2005; DeVault, 1991; Marshall and Anderson, 2002; Roos *et al.*, 2001). Responses from women, in many cases, also imply a reluctance to share the responsibilities of the kitchen (Bugge, 2005). In another study from the Nordic countries, the traditional division of cooking was evident in that more women reported doing the cooking themselves than men (Ekström and Fürst, 2001). However, Ekström and Fürst (2001) reported nuances to this finding in that younger men and men from middle classes reported to do more cooking.

Education opportunities have improved over the years until now, in the Western world, both genders theoretically have equal rights to education at basic as well as more advanced levels. Higher education levels have been associated with healthier eating habits and this is indeed the case for men, whereas women's eating habits are more healthy at the outset and are thus less influenced by education level (Øygard and Klepp, 1996).

As children grow up, new age groups are formed and the habits of previous groups may be phased out or changed. One might expect fewer differences in attitudes towards foods between genders in younger age groups. However, research shows that gender differences in attitudes toward food are evident already at an early age. An Australian study found differences between girls and boys at an age of less than 12 years, where girls were more disgusted with food than boys and, furthermore, mid-teen boys were considerably more happy with food than girls (McNamara *et al.*, 2008). In an American study, young girls reported significantly lower consumption of family meals than did boys (Neumark-Sztainer *et al.*, 2003). As family meals have been shown to be a predictor of healthy eating (Larson *et al.*, 2007), the consequences of this behaviour among girls on future behaviour have yet to be seen.

Traditional gender roles form the basis on which new patterns of gendered behaviour emerge. Although gender differences in meal contexts are not as clear cut today as fifty years ago, since they may be modified by such variables as age and education, differences are still evident in the genders' approach to meals, attitudes, roles in preparation, and in composition and consumption of meals (Bove *et al.*, 2003; Brown and Miller, 2002; Fagerli and Wandel, 1999; Marshall and Anderson, 2002).

6.3   Gender and meals

In the previous section, differences in food preferences between genders were not accorded much attention because external factors such as availability of foods were more important for food provision and meals than having the possibility to choose. Personal food preferences play a larger role in constructing meals today and this needs more attention in the study of how gender differences can influence meal composition (Jensen and Holm, 1999; Ueland, 2007).

In addition, as meals are social as well as eating occasions, differences between genders' attitudes and behaviours are reflected in other factors that influence meals (Bourdieu, 1984; Bove et al., 2003; Brown and Miller, 2002; Fagerli and Wandel, 1999; Lupton, 1996; Marshall and Anderson, 2002). For instance, many studies show that women, to a larger degree than men, choose different foods for their meal depending on type of meal and whether they are eating alone, with a partner, children, or others (Brown and Miller, 2002; Bugge, 2005; Bugge and Døving, 2000; Charles and Kerr, 1988; DeVault, 1991; Fagerli and Wandel, 1999; Lupton, 1996). When a woman eats alone she will snack more and eat more salads. When women prepare food for, or eat with others, they will, more often than not, defer to their tastes, and particularly the man's taste, at the expense of their own preference. Women will try to accommodate both children and men, but one may find that children's preferences, to a certain extent, overrule both men's and women's preferences (Bugge, 2005; Lupton, 1996).

However, some nuances can be found in this picture. A study that looked at what happened to the dietary patterns when men and women changed status from single to cohabitating (Bove et al., 2003), showed that partners will accommodate each other depending more on how rigid their food styles are, rather than gender. In one study (Brown and Miller, 2002), a distinction in meal decision-making could be seen depending on the level of egalitarianism between the partners; a higher level of egalitarianism resulting in fewer gender differences. A study among young Norwegian adults showed that higher education was a major factor in predicting healthy eating habits for both genders, but in promoting more healthy eating behaviours males were more influenced by their partners than females (Øygard and Klepp, 1996). Overall, however, there seems to be a larger dietary change for women than for men, insofar as their diet will converge towards the man's to a larger extent than the other way around (Bove et al., 2003; Brown and Miller, 2002; Bugge and Døving, 2000; Charles and Kerr, 1988; DeVault, 1991; Lupton, 1996).

6.3.1 Gender differences in food choice

Given a choice females will usually choose vegetables, white meat, or fish for a meal, while men will choose meat and potatoes (Bourdieu, 1984; Charles and Kerr, 1988; Jensen and Holm, 1999; Roos et al., 2001; Ueland, 2007). The reasons for these differences are difficult to ascertain, as they are heavily influenced by socio-cultural factors, and a genetic explanation cannot sufficiently cover the differences observed. Health concerns play a larger role in women's food choice than in men's (Fagerli and Wandel, 1999; Jensen and Holm, 1999; Øygard and Klepp, 1996) to the extent that it is impossible to distinguish what is plain hedonic preference and what is learned and socially acceptable preference among women. Health concerns and food choice for women may imply both considerations for acquiring a good-looking body and for providing the best and most nutritious meal for others (DeVault, 1991; Fagerli and Wandel, 1999; Jensen and Holm, 1999; Lupton, 1996). Consequently, women's food choices may be more complicated, because they are more subjected to weighing and consideration of various factors, than men's food choices (Jensen and Holm, 1999). In a study conducted by Kubberød et al. (2002) on the topic of young females' meat consumption, several of the girls said that the best liked dinner dish was a steak. However, they would overrule this preference and several of the girls said they would reduce consumption of meat, and red meat in particular, when they left home. Not eating meat made them feel less full and, by their standards, healthier. Boys on the other hand liked meat best and would also rather eat meat because they thought it was healthy and it tasted good.

Another reason for gender differences in food consumption may be what is considered proper to eat for the different genders. Several studies support this by emphasizing how different foods and eating practices are gendered (Bock and Kanarek, 1995; Bourdieu, 1984; Chaiken and Pliner, 1987; Martins et al., 2004; Vartanian et al., 2007). Food products have been placed in a hierarchy depending on their perceived importance and status (Jensen and Holm, 1999; Roos et al., 2001). Meat is classified as the ultimate masculine food, which is, incidentally, also placed on top of the hierarchy (Barthes, 1975; Bourdieu, 1984; Charles and Kerr, 1988; Fiddes, 1991). Foods classified as female foods are typically vegetables, fruits, low-fat products, and sweets (Jensen and Holm, 1999). Consumption of masculine food would be more acceptable for men than women. Similarly, eating styles are gendered (Bock and Kanarek, 1995; Bourdieu, 1984; Chaiken and Pliner, 1987; Martins et al., 2004; Vartanian et al., 2007). One study (Bourdieu, 1984) commented on how men would eat 'bigger' food, take bigger bites, and drink larger and stronger drinks, while women would eat foods that one can take small pieces of or sip, select foods that men would not eat, and drink smaller and sweeter drinks.

In large parts of the world, vegetarian diets are the rule. In the Western world, vegetarian or low-meat lifestyles are typically a female way of living (Perry et al., 2001; Worsley and Skrzypiec, 1997). There seem to be different motivations for teenage girls and older women, with regard to reasons provided for eating low or no-meat diets. Young girls' motivations are more often associated with the potential for developing eating disorders (Perry et al., 2001; Worsley and Skrzypiec, 1997), whereas in adults vegetarianism can have a health, as well as a moral component.

6.3.2   Breakfast
Breakfast is not the meal where major gender differences are evident in food choice. However, younger females seem to skip breakfast to a greater extent

than males (Bugge and Døving, 2000; Larson *et al.*, 2007; Pliner and Rozin, 2000). Bugge and Døving found that among females aged 15–24 years, 25% reported that they ate breakfast on fewer than five days a week, while 15% of the males reported the same. An explanation for these findings may be that in the youngest age group, more females had weight concerns than boys (Neumark-Sztainer *et al.*, 2004) and that boys in general ate more family meals than girls (Neumark-Sztainer *et al.*, 2003).

In the age group 25–39 years, however, the pattern changed and 17 and 31% of females and males, respectively, reported that they skipped breakfast (Bugge and Døving, 2000). The increase in frequency of breakfast consumption seen for females in the next age group corresponds with typical child-bearing and rearing age, and activities such as making sure that the children eat a proper breakfast so they can 'make it through the day'. Higher breakfast consumption was seen in mothers who reported high health awareness (Boutelle *et al.*, 2007). One study (Bugge and Døving, 2000) did not find any differences in breakfast consumption in the older age groups.

6.3.3   Lunch
One Norwegian study, (Bugge and Døving, 2000) reported that fewer people had lunch on a daily basis than they did breakfast and fewer males reported that they ate lunch than did females. Being unemployed and male was a descriptor for eating lunch less frequently, while this was not a characteristic of unemployed women. One explanation for the differences in consumption frequency between breakfast and lunch, may be that the subjects rise later in weekends and that breakfast and lunch meals are combined in time and registered as the first meal of the day. In another study, (Larson *et al.*, 2007) it was found that young females, and particularly young males, ate lunch more often than breakfast, the frequency of meal consumption increasing in both genders as the day progresses. The differences noted between these two studies may be due to cultural characteristics. Food choices by genders for lunch are not discussed much in the literature, although, as lunch is usually consumed at school or work, one would expect that food choices in canteens will reflect availability and genders' food choices in general. We would hypothesize that women would more often select salads than men, and that women's lunches would to a larger extent be meatless. A Norwegian study on drinking practices showed that women are more likely to choose tea and/ or water to drink, while men drink more coffee and/or milk (Bye, 1999).

6.3.4   Dinner
Dinner is the meal where gender differences are most evident. The context in which dinner is consumed usually involves partner, children, other family members and friends, and the dinner meal consists of a high amount of calories and nutrients and a variety of items or dishes. In the literature, one

will often find that references to unspecified meals evolve into a discussion of dinner practices (Fagerli and Wandel, 1999; Jensen and Holm, 1999; Marshall and Anderson, 2002), and what it all comes down to is that, in family dinners, men will choose meat while women would rather select salads or dishes containing less meat (Bourdieu, 1984; Fagerli and Wandel, 1999; Jensen and Holm, 1999). Dinner foods have been characterized as feminine or masculine, depending how they are perceived by women and men (Bourdieu, 1984; Bove *et al.*, 2003; Jensen and Holm, 1999; Vartanian *et al.*, 2007). Typical feminine dinner foods are salads, pasta, and non-fattening foods, whereas masculine foods are red meat, hamburgers, and high-fat foods. In the same way, beer is perceived to be masculine and is more preferred by males. It has been hypothesised that women serve tempting accessories and many side dishes with the meat so that their lower meat consumption is not visible (Fagerli and Wandel, 1999). If women eat alone, they will choose salads, vegetarian options or soups, or they will just snack. DeVault (1991) commented in her study on a husband, who, when he was served quiche, replied that it was very nice, but he did not want 'breakfast for dinner'.

A dinner can be termed a proper meal when it, in a traditional sense, consists of meat, a starch, and a vegetable, which incidentally, are preferred by many men (Jensen and Holm, 1999; Marshall and Anderson, 2002). In a proper meal, therefore, female preferences would not be equally proper. It seems, however, that this stereotypic view is on the retreat. Many studies show that more men participate actively in food preparation and express preferences for dishes that have been previously associated with female foods (Bove *et al.*, 2003; Fagerli and Wandel, 1999).

Even though most studies cite women as the ones who cook dinner, a closer examination will show that men play a prominent role in the cooking of weekend dinners and on special occasions (Bugge, 2005). Men like to experiment and cook grand meals, while women often prepare everyday dinners. Furthermore, out-door grilling has long been men's domain.

6.3.5   Other eating occasions/meals

Studies show that women are more prone to snacking than men throughout the day (Bye, 1999; Jensen and Holm, 1999). While men will settle for three substantial meals a day, women may skip some of the meals or stem the hunger with snacking between meals. Foods that are often consumed by women between the main meals are biscuits, yogurts, and fruits or cakes, often accompanied by tea or coffee (Jensen and Holm, 1999). Men consume less sweet foods than women, although one study (Charles and Kerr, 1988) found that men preferred heavier, creamier types of cakes.

6.4 Conclusion

Studies show that women choose healthier foods than men, but that they modify their choices when living with a man (Bove *et al.*, 2003; Brown and Miller, 2002; Bugge and Døving, 2000; Charles and Kerr, 1988; DeVault, 1991; Lupton, 1996). One consequence of this is that men living with women will have healthier diets than men living alone as the woman will modify the family's diet in a healthier direction (Anderson *et al.*, 2004; Fagerli and Wandel, 1999; Jensen and Holm, 1999). Eating family meals is also a health benefit for the family members, more so for boys than for girls because the latter consume fewer family meals than boys (Neumark-Sztainer *et al.*, 2003). While females seem to skip more of the major meals than males, they snack more, and eat more light meals in between (Jensen and Holm, 1999). Food choices of men and women are reflected in a masculine and feminine categorization of foods. Typical masculine foods are ranked high in the food hierarchy and examples are red meat, hamburgers, potatoes and high-fat food products, while feminine food products are salads, fruits, pasta, sweets, and biscuits (Jensen and Holm, 1999). Although the meal situation still is very traditional and gender differences are quite noticeable, there are signs of change in that more men profess interest in and liking for more differentiated eating and healthy foods Bove *et al.*, 2003; Fagerli and Wandel, 1999.

6.5 Future trends

In our society where there are changes or something new every day, food habits are proving to be unusually resilient. Despite the fact that more people eat out and on the move, and they eat new foods in new contexts, studies show that our perceptions of meals and what constitutes a proper meal, are surprisingly resistant to change. Traditional gender roles account for some of the differences noted, but it seems that as gender role changes, food behaviours are redefined so that they still uphold some of the old patterns. However, the current focus on the obesity epidemic and increase in health awareness in the population has implications for both genders' attitudes towards foods. In particular, we see that men have become more health conscious in their food choices as well as preferences, and thus the genders converge on perceptions of food. A general increase in education in the population may also contribute to this change in men's perceptions, as education has been shown to influence men's food choices to a larger extent than women's (Fagerli and Wandel, 1999). Furthermore, the increase in general health consciousness will probably influence the concept of masculine and feminine foods and the attitudes about which foods are proper for the different genders to eat. Foods that are traditionally considered to be feminine, such as salads, pasta, chicken, and fish are becoming high status foods, and as such, more acceptable as a full meal for both genders.

6.6   References

Anderson, A. S., Marshall, D. W., and Lea, E. J., 2004, 'Shared lives – an opportunity for obesity prevention?' *Appetite*, **43**, 327–329.

Barthes, R., 1975, 'Towards a psychosociology of contemporary food consumption', *in* E. Forster, and R. Forster, eds., *European Diet from Pre-Industrial to Modern Times.* New York, Harper, p. 47–59.

Bartoshuk, L. M., Marino, S. E., and Snyder, D. J., 2007, 'Age and hormonal effects on sweet taste and preference.' *Appetite*, **49**, 277.

Bock, B., and Kanarek, R., 1995, 'Women and men are what they eat: the effects of gender and reported meal size on perceived characteristics.' *Sex Roles*, **33**, 109–119.

Bourdieu, P., 1984, *Distinction. A social critique of the judgement of taste.* London, Routledge and Kegan Paul Ltd.

Boutelle, K. N., Birkeland, R. W., Hannan, P. J., Story, M., and Neumark-Sztainer, D., 2007, 'Associations between maternal concern for healthful eating and maternal eating behaviors, home food availability, and adolescent eating behaviors.' *Journal of Nutrition Education and Behavior*, **39**, 248–256.

Bove, C. F., Sobal, J., and Rauschenbach, B. S., 2003, 'Food choices among newly married couples: convergence, conflict, individualism, and projects.' *Appetite*, **40**, 25–41.

Bremner, E. A., Mainland, J. D., Khan, R. M., and Sobel, N., 2003, 'The prevalence of androstenone anosmia' *Chemical Senses*, **28**, 423–432.

Brown, J. L., and Miller, D., 2002, 'Couples' Gender Role Preferences and Management of Family Food Preferences.' *Journal of Nutrition Education and Behavior*, **34**, 215–223.

Bugge, A., 2005, 'Dinner – a sociological analysis of Norwegian dinner practice.' [Middag – en sosiologisk analyse av den norske middagspraksis]: Doctoral Thesis thesis, Norwegian University of Science and Technology, Trondheim, 421 pp.

Bugge, A., and Døving, R., 2000, *The Norwegian meal pattern – ideals and practices.* [Det norske måltidsmønsteret – ideal og praksis.], Report nr. 2: Lysaker, National Institute for Consumer Research, p. 1–257.

Bye, E., 1999, *The consumption of beverages – Who, What and Where.* [Bruk av drikkevarer – Hvem, hva og hvor.], Report nr. 8: Lysaker, National Institute for Consumer Research.

Chaiken, S., and Pliner, P., 1987, 'Women, but not men, are what they eat.' *Personality and Social Psychology Bulletin*, **13**, 166–176.

Charles, N., and Kerr, M., 1988, *Women, food and families.* Manchester, UK, Manchester University Press.

DeVault, M., 1991, *Feeding the family: The social organization of caring as gendered work.* Chicago, University of Chicago Press.

Ekström, M., and Fürst, E., 2001, 'The gendered division of cooking', *in* U. Kjærnes, ed., *Eating patterns. A day in the lives of Nordic peoples.* Report no. 7: Lysaker, National institute for Consumer Research, p. 215–235.

Fagerli, R. A., and Wandel, M., 1999, 'Gender differences in opinions and practices with regard to a "Healthy Diet".' *Appetite*, **32**, 171–190.

Fiddes, N., 1991, *Meat: a natural symbol.* London, Routledge & Kegan Paul.

Jensen, K., and Holm, L., 1999, 'Preferences, quantities and concerns: socio-cultural perspectives on the gendered consumption of foods.' *European Journal of Clinical Nutrition*, **53**, 351–359.

Kubberød, E., Ueland, Ø., Tronstad, Å., and Risvik, E., 2002, 'Attitudes towards meat and meat-eating among adolescents in Norway: a qualitative study.' *Appetite*, **38**, 53–62.

Larson, N. I., Neumark-Sztainer, D., Hannan, P. J., and Story, M., 2007, 'Family meals during adolescence are associated with higher diet quality and healthful meal patterns during young adulthood.' *Journal of the American Dietetic Association*, **107**, 1502–1510.

Lupton, D., 1996, *Food, the body and the self*. London, Sage Publications.

Marshall, D. W., and Anderson, A. S., 2002, 'Proper meals in transition: young married couples on the nature of eating together.' *Appetite*, **39**, 193–206.

Martins, Y., Pliner, P., and Lee, C., 2004, 'The effects of meal size and body size on individuals' impressions of males and females.' *Eating Behaviors*, 5, 117–132.

McFeely, M., 2000, *Can she bake a cherry pie? American women and the kitchen in the twentieth century*. Amherst, University of Massachusetts Press.

McNamara, C., Hay, P., Katsikitis, M., and Chur-Hansen, A., 2008, 'Emotional responses to food, body dissatisfaction and other eating disorder features in children, adolescents and young adults.' *Appetite*, 50, 102–109.

Mäkelä, J., 2000, 'Cultural definitions of the meal.', in H. Meiselman, ed., *Dimensions of the meal. The science, culture, business, and art of eating*. Gaithersburg Maryland, Aspen Publishers, Inc., p. 7–18.

Neumark-Sztainer, D., Wall, M., Story, M., and Fulkerson, J. A., 2004, 'Are family meal patterns associated with disordered eating behaviors among adolescents?' *Journal of Adolescent Health*, **35**, 350–359.

Neumark-Sztainer, D., Hannan, P. J., Story, M., Croll, J., and Perry, C., 2003, 'Family meal patterns: associations with sociodemographic characteristics and improved dietary intake among adolescents.' *Journal of the American Dietetic Association*, **103**, 317–322.

Perry, C., McGuire, M., Neumark-Sztainer, D., and Story, M., 2001, 'Characteristics of vegetarian adolescents in a multiethnic urban population.' *Journal of Adolescent Health*, **29**, 406–416.

Pliner, P., and Rozin, P., 2000, 'The psychology of the meal.', in H. Meiselman, ed., *Dimensions of the meal. The science, culture, business, and art of eating*. Gaithersburg, Maryland, Aspen Publishers, Inc.

Roos, G., Prattala, R., and Koski, K., 2001, 'Men, masculinity and food: interviews with Finnish carpenters and engineers.' *Appetite*, 37, 47–56.

Ueland, Ø. 2007, 'Gender differences in food choice.' in L. Frewer, and H. van Trijp, eds., *Understanding consumers of food products*. Cambridge, England, Woodhead Publishing Ltd., p. 316–328.

Vartanian, L. R., Herman, C. P., and Polivy, J., 2007, 'Consumption stereotypes and impression management: how you are what you eat.' *Appetite*, **48**, 265–277.

Worsley, A., and Skrzypiec, G., 1997, 'Teenage vegetarianism: beauty or the beast?' *Nutrition Research*, **17**, 391–404.

Øygard, L., and Klepp, K.-I., 1996, 'Influences of social groups on eating patterns: a study among young adults.' *Journal of Behavioral Medicine*, **19**, 1–15.

7

Institutional meals

J. S. A. Edwards and H. J. Hartwell, Bournemouth University, UK

Abstract: In order to define exactly what constitutes an institutional meal the various ways in which such meals can be categorised are reviewed and their primary characteristics, including perceptions and expectations, contracting out, and alternative production and service styles are described. The serving of institutional meals in a variety of settings, is considered and future trends briefly identified.

Key words: institutional foodservice, institutional catering, institutional meals.

7.1 Introduction

When using the terms 'institutional meals' and 'institutional foodservice', people invariably have an understanding of what is involved; however, defining exactly what constitutes an institutional meal and which parts of the foodservice industry are included and involved is much more problematic. Most basic college and university foodservice textbooks attempt to cover the subject and classify the industry; some (Payne-Palacio and Theis, 2005) simply by listing its component parts; others (Foskett *et al.*, 2003; Davis, *et al.*, 1998) by dividing the industry into two broad segments or categories: Commercial (also called Profit or Private Sector) and Public Sector (also called Not for Profit Sector, Cost Sector, Subsidised or Welfare – but generically referred to as Institutions). The latter category includes universities, schools, armed forces, hospitals, prisons and employee feeding. This broad categorization is shown schematically in Fig. 7.1. Whilst it is difficult to categorise which parts of the industry are included in each sector, as organisations and countries use different selection criteria, a broad indication of the size of each sector, both in the UK and USA, is given in Fig. 7.2, where it can be seen that although the profit sector is larger in both countries, in the UK, for example, something in the order of 37% of the 8.62 million meals served annually are actually in the cost sector (Quest and Needham, 2006).

Cost Sector   Profit Sector
Public Sector   Private Sector
Subsidised   Commercial
Welfare
Institutional

Hospitals, other healthcare
and social services
Schools and universities etc
Prisons etc   Foodservice
Armed forces   Sectors
Public services e.g. police,
fire and ambulance
Employee feeding

Hotels, guest houses etc
Restaurants
Cafes etc
Fast food and franchises
Take-aways
Public houses
Clubs
Transport and travel

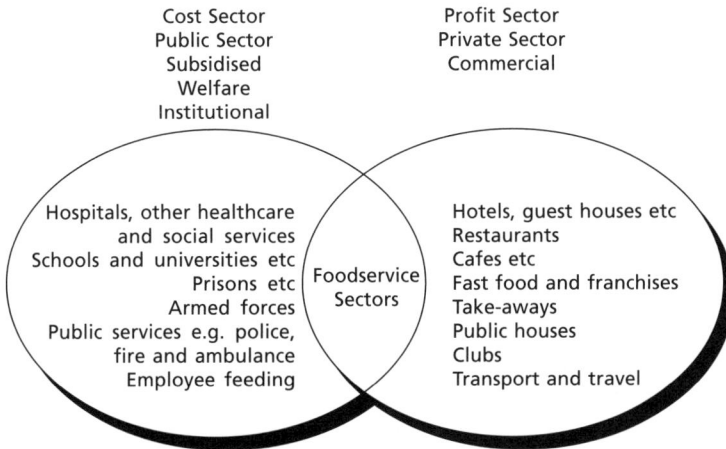

Fig. 7.1 Classifying the foodservice industry.

However, one of the many limitations of categorising the foodservice industry in such a manner (cost and profit sectors) is that it gives the impression that the cost sector is not concerned with profit and by implication has little or no business acumen associated with it. Nothing could be further from the truth, as running a cost or public sector foodservice operation, such as a hospital, is one of the most complex of all foodservice operations (Wilson *et al.*, 1997; Kipps and Middleton, 1990).

An alternative classification could be to group foodservice outlets according to their goal or mission as in Fig. 7.3. Food service outlets exist for different reasons; for some, it is their basic raison d'être or rationale for their existence, for example, a restaurant or café. For others it might be part of the total offering or experience as in a hotel or cruise liner; whilst for others it could be an additional offering or service, which enhances the experience and in doing so generates additional revenue. A final category, which encompasses the basic philosophy of institutional foodservice, is where the meal is part of the business but not the sole purpose. It is there to provide an amenity or facility, which is needed to help and support the business, but is not the main reason that the operation exists. In the United States this category of outlets is often referred to as '*on-site foodservice*' (Reynolds, 1997).

Even so, categorising institutional foodservice in this way presents difficulties. In the workplace, for example, particularly in an industrial or manufacturing setting, the foodservice operation is likely to be a self-service cafeteria, canteen or simply vended beverages and snacks. On the other hand, there might be a separate cafeteria for office staff or more elaborate dining rooms for managers and directors, which is much more up-market and where they can entertain potential business customers. The meals provided are likely to be of a much superior quality, served in surroundings and by staff more commonly found in a public sector restaurant.

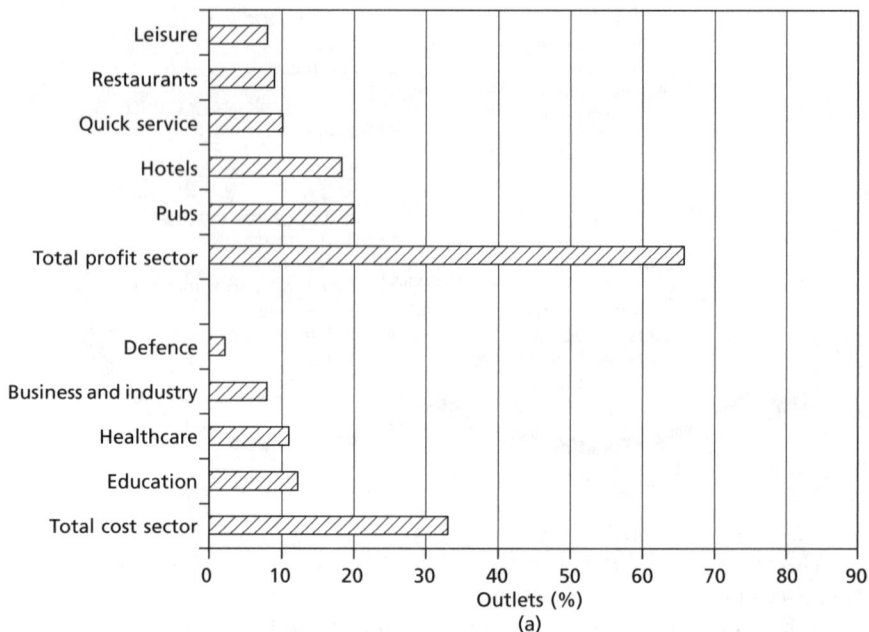

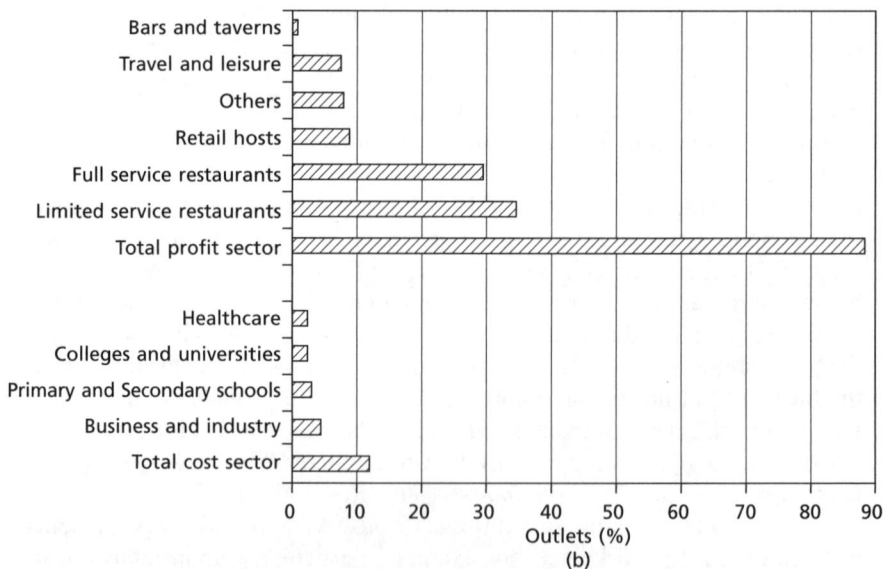

Fig. 7.2 Outlets in each foodservice sector – UK and USA, calculated from data from (a) Quest and Needhan, 2006 (b) Gale, 2007.

As can be seen, therefore, institutional foodservice crosses a broad spectrum of establishments, providing a complete range of meals, using a variety of service styles, and served in various settings. Notwithstanding, the foodservice classification, institutional meals possess a number of broadly similar

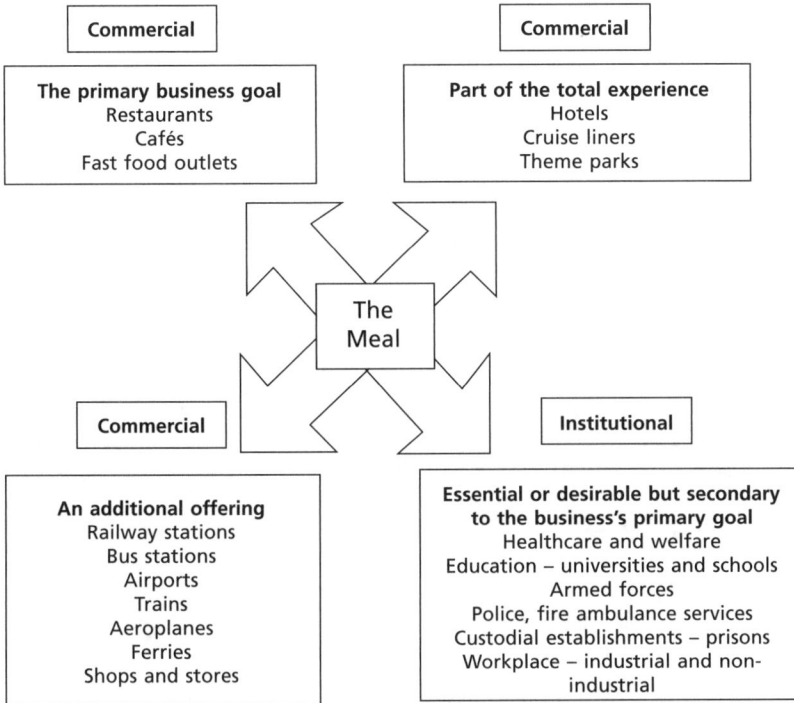

Fig. 7.3   Categorising meals when eating out.

characteristics as outlined in Fig. 7.4; hence when considering institutional meals, we should perhaps consider the meals and their associated components and characteristics, rather than the place where consumption takes place.

7.2   Purpose of this chapter

It is appreciated that institutional meals do not always fit neatly into this categorisation but, using it as the basic structure, in this chapter the institutional meal is considered from the perspective of what is provided rather than from where it is provided.

7.3   Expectations of institutional meals

The threads identified in Fig. 7.4 were reinforced when students were asked what they understood by institutional food and institutional meals, Fig. 7.5. Many people have preconceived ideas of what institutional meals are and have definite expectations, usually negative, of their quality. If an individual

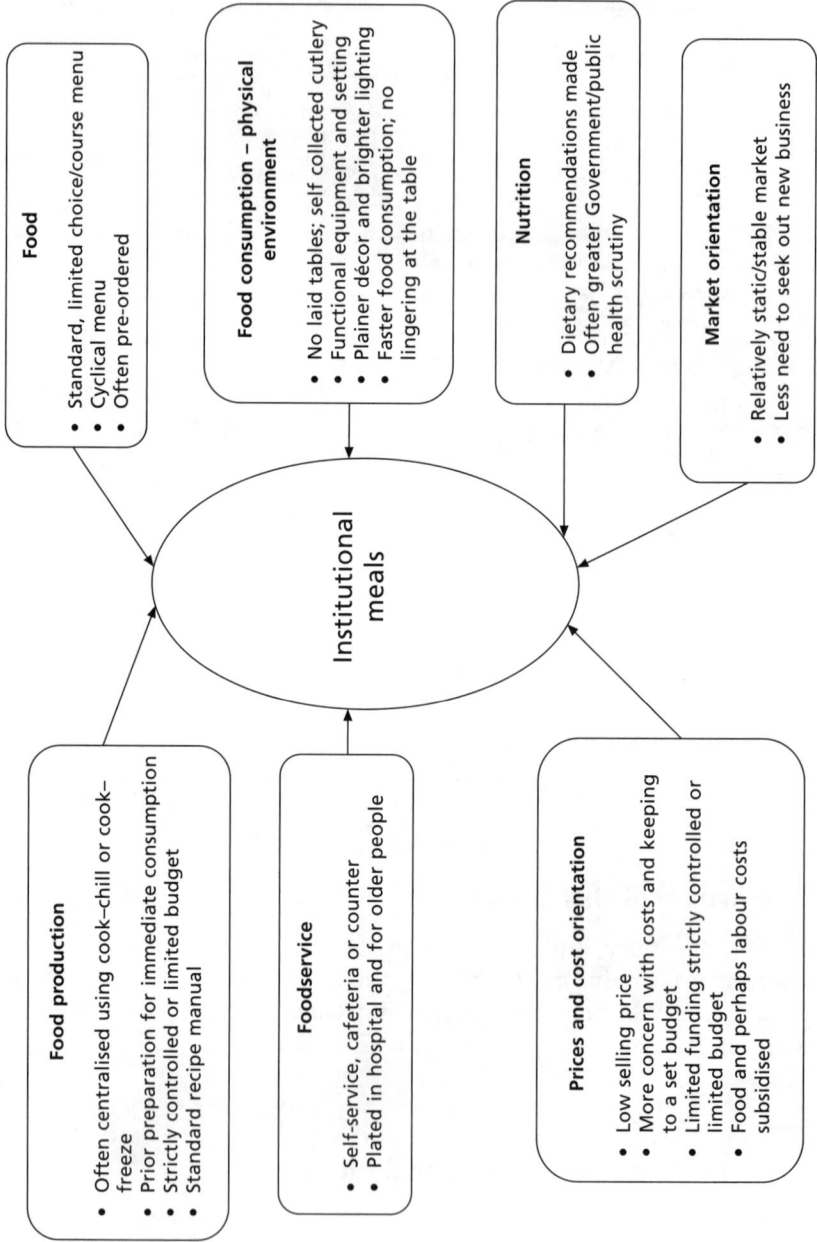

Food
- Standard, limited choice/course menu
- Cyclical menu
- Often pre-ordered

Food consumption – physical environment
- No laid tables; self collected cutlery
- Functional equipment and setting
- Plainer décor and brighter lighting
- Faster food consumption; no lingering at the table

Nutrition
- Dietary recommendations made
- Often greater Government/public health scrutiny

Market orientation
- Relatively static/stable market
- Less need to seek out new business

Food production
- Often centralised using cook–chill or cook–freeze
- Prior preparation for immediate consumption
- Strictly controlled or limited budget
- Standard recipe manual

Foodservice
- Self-service, cafeteria or counter
- Plated in hospital and for older people

Prices and cost orientation
- Low selling price
- More concern with costs and keeping to a set budget
- Limited funding strictly controlled or limited budget
- Food and perhaps labour costs subsidised

Institutional meals

Fig. 7.4 General charactistics of an institutional meal.

What are institutional meals?

First year students (*n* = 55) at the start of their course, and final (4th) year (*n* = 35) students, reading for a honours degree in International Hospitality Management at a University on the South Coast of England were asked what they understood by the term 'institution food' and 'institutional meals'.

Sectors: prisons, schools, hospitals, army, company canteens – secondary to the primary focus of the business – practical (eat to live), all areas standardised – cheap – not marketed

Food: Bland, same colour, unhealthy (fried), lack of flavour, cold, boiled/processed, small portions, rice pudding, chicken, meat and potatoes, mince beef (stereotype 'gruel'), cheap (i.e. subsidised), lack of choice

Food production: Held too long, repetitive menu, mass-produced, packaged.

Foodservice: Sterile, canteen/cafeteria – no sophistication, self-service, fast, plastic trays – functional, no interesting menu description, staff not engaged (female/old) – low skilled

Food consumption – physical environment: No music, colour – white walls – décor basic, noisy, eat at set times, uncomfortable, plastic chairs, eat alone or with people who you would not normally sit with, generally a negative experience – cramped and impersonal.

General comment: Acknowledged that these are stereotypical perceptions based on experience at school and that the image of institutional food is changing as, for example, celebrity chefs become involved in sectors such as schools, hospitals and care homes.

Fig. 7.5 Preconceived ideas of institutional meals.

were to be asked what he or she thought of hospital food, the answer might well be negative, even if they had no recent experience of actually consuming the food.

'Institutional meals, however, are rarely pleasant or social occasions: patients are often left to eat in bed or in isolation, which may inhibit eating.' (Holmes, 2006).

This notion was reinforced in a series of studies by Cardello *et al.* (1996) who termed the phenomenon 'Institutional stereotyping'. They concluded, *inter alia,* that:

- consumers hold strong negative attitudes about both the quality and acceptability of institutional foods;
- compared with commercial food, institutional food has much poorer sensory characteristics;
- associating a food with institutional foodservice decreases anticipated liking; and actual acceptability rarely matches established expectations;
- anticipated acceptability can affect actual liking of the food.

The reasons for this phenomenon, identified by Cardello and co-workers, centred on poor food variety, poor food presentation and a poor physical dining room setting (cf. Fig. 7.4 and 7.5).

7.4 Contracting out

Institutional meals are invariably provided within very tight financial constraints, hence organisations are continuously seeking ways in which costs can be contained and, where necessary, any subsidies either reduced or eliminated. Partly as a result, responsibility for the provision of these meals has been subsumed to outside providers. *Contracting out*, as it was historically referred to in the UK, and *Contract Management Companies* or *Contract Feeders* in the USA (Reynolds, 1997) can be a very emotive issue, for both employees and customers; the former fearing for their jobs, the latter, a fall in quality. Furthermore, as the contract process has increasingly included aspects other than the provision of meals, such as cleaning, *Food and Service Management* is often the preferred term (Quest, 2007). Even so, the notion of a third party providing meals for financial reward, particularly in an institutional setting has become common in many countries. In the UK, in the order of 52% of foodservice outlets in Business and Industry, part of the cost sector, are operated by contractors (Quest, 2007) and worldwide this figure was 53% in 2001 (Euromonitor, 2007) (Fig. 7.6 gives a detailed breakdown).

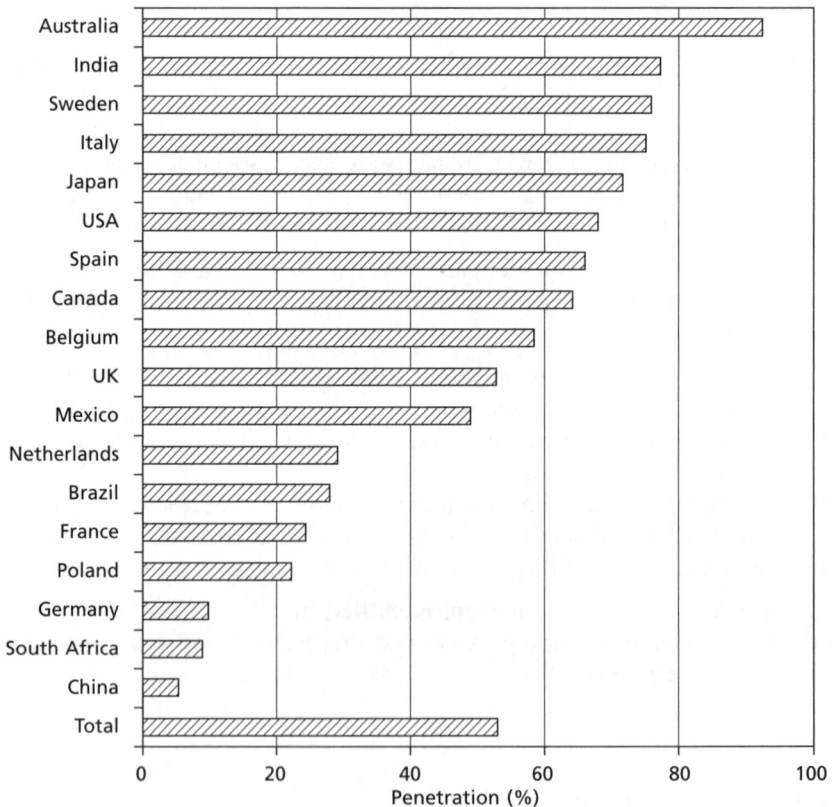

Fig. 7.6 Percentage penetration of contract caterers, 2001 (Source: Euromonitor International, 2002).

In the provision of institutional meals, where they are not the primary function of the business, the use of specialist companies provides a number of advantages, enabling, for example, the organisation to concentrate on its core business. Furthermore, the advantages and disadvantages now provided by contractors apply equally to the profit sector and the distinction between the two sectors is becoming increasingly blurred with many activities in the profit sector being contracted out. Here, it has been argued that the stereotypical image of institutional meals associated with contractors is no longer valid as standards gradually improve (Mintel, 2006) in response to consumer demands. No longer are they prepared to accept 'typical canteen food' but in many cases the competition in an institutional setting are commercial outlets which may be conveniently located in the adjacent high street.

7.5   Institutional meal production systems

In an 'ideal' foodservice operation, fresh, first-class ingredients are prepared by highly skilled chefs immediately before consumption and served by similarly highly skilled staff. While this traditional approach continues in the more expensive restaurants of the profit sector, financial pressures in the foodservice industry and more especially the cost sector, have led to the introduction of food production systems which are divorced or decoupled from service, and to systems which produce large numbers of dishes or components of dishes in advance of their requirement.

These 'alternative' food preparation systems, including cook–freeze, cook–chill, Nacka, A.G.S., Sous-vide and Capkold, some of which are considered in Chapter 4, are particularly pertinent to institutional meals because of the:

- pressures to serve large numbers of customers within very limited time constraints;
- large numbers of customers, often served on fragmented sites such as outlying wards in hospitals, off-site or remote factory/site locations and small staff restaurants;
- requirements for extended hours of service, night shifts and often a 24-hour service;
- rationalisation and reduction of labour, equipment and other overhead costs;
- heightened awareness of the relationship between diet and health, food hygiene and the introduction of new legislation.

Historically though, food quality was not always the highest priority in the production and service of institutional meals, and cost has invariably been the overriding factor. Where meals are not the primary focus of the business, meals are seen as a 'cost centre' rather than a 'profit centre' hence action to contain costs and set very tight budgets has prevailed. More recently, whilst the balance of cost and quality remains a constant challenge, it is now acknowledged that the heightened awareness and demand of consumers, particularly for nutritional quality, healthy eating issues, hygiene and cleanliness and the need to cater for a much more discerning clientele have become an increasingly important focus. This is particularly so where meals are provided to a workforce who regard them as part of their overall pay remuneration.

Furthermore, there is more government scrutiny of 'non-profit' institutional foodservice such as hospitals, prisons and schools as these operations provide either all the meal provision or the main meal provision for the consumer.

7.6   Institutional meal service systems

The service styles for institutional meals are similar to those found in the commercial sector, but typically, a self-service system is the most commonly used. This style ranges from either a self-service or assisted cafeteria, to a counter, buffet or carvery service; the latter being popular in many establishments where the meat (protein) item is served and customers can help themselves to vegetables and salads. In many ways these styles of service are extremely beneficial and are also used in many commercial establishments; for they enable diners to walk up to the service area, see what is actually on offer, rather than have only a menu's description, then choose what they like, and often the actual portion size. The food, or at least many items offered, will have already been cooked and, not taking into account the length of the queue (line) or wait, the time between entering the dining room and sitting at the table with a meal can be drastically reduced. This is particularly important when meal breaks are short.

From a foodservice perspective, self-service outlets are invariably simpler to operate, require fewer staff, both in the kitchen and dining room, and above all are a much more cost effective form of service. In keeping with this service style, the ambience, with effectively part of the kitchen in the dining room requiring additional lighting and so forth, invariably means that the room décor, facilities, tables and seating reflect the need to keep costs to a minimum and so perpetuate the idea of institutional meals.

In some settings, such as hospitals and feeding older people, a plated service is often much more effective as diners are not in a position to go to the service counter and must have their meals brought to them. These meals would normally have been pre-ordered, cooked then covered for transportation, the resulting meal again perpetuating an institutional style. Two categories of plated meals exist: firstly, where the meals are plated centrally in the kitchen and secondly where the food is transported in bulk to the service point (remote dining location, hospital ward) in a trolley where they are plated (Fig. 7.7).

Both systems have relative merits, the main weaknesses being that consumers have to make their meal selection, often up to 24 hours before consumption, without the aid of sensory stimuli; the portion size is standardised; meals, that

Fig. 7.7 An example of a hospital meal service trolley.

is a first course, main course and dessert are often served on a tray together; and, in hospital, new patients receiving a meal ordered by the previous bed occupant or a member of staff. This invariably leads to dissatisfaction, particularly when meals are not as ordered with studies showing error rates as high as 12.9% from 6553 plated meals (Dowling and Cotner 1988).

Sending food to the dining room in a trolley and plating it there offers slightly more flexibility and, in some cases, meal selection and choice of portion size can be made at the point of service with staff interacting with diners, also helping to enhance perceptions and expectations. It does, however, result in higher total food wastage (Hartwell and Edwards, 2003).

In hospitals, there is no unanimous agreement as to which system might be best but in the USA, centralised plated systems tend to be the norm, operated by 81% of hospitals (Silverman *et al.*, 2000) whereas, in the UK, sending food in bulk to the ward is the favoured method (37%) with a centrally plated service being used by 35% or a combination of the two by 28% of operations (Audit Commission, 2001).

In some institutions such as boarding schools and military training facilities, meals are served 'family style'. Two of the largest such operations are the US West Point Military Academy with approximately 4000 cadets (15% female) and the US Air Force Academy. In both of these establishments, cadets enter the dining room together and sit at their appointed table of approximately 10. Meals are then put on the table in dishes and the senior cadet supervises junior cadets who serve meals to the entire table. This style of service is designed, not only to ensure that a large number of individuals can be fed simultaneously, but also forms part of the culturalisation process. The primary disadvantage is that for this number of people to be served together, the food has to be put into containers long before the start of service; hence their sensory and nutritional qualities would have deteriorated.

US Air Force Academy – 'Cadets march into Mitchell Hall nine abreast through doors on the east and west ends. After they take their seats, they are served family style and finish the meal in 20 minutes. To serve the meals quickly (12,000 daily), all cold food items – bread, salad, beverages, etc.– are placed on tables and hot carts are positioned by the tables before the cadets arrive. One waiter is assigned to 10 tables, and the cadets receive their hot food within two and one-half to three minutes after sitting down.

'A first class cadet at each table is designated as the "table commandant". A fourth class cadet sits at the foot of the table and pours beverages and passes food. Even though they have these duties, the fourth-classmen cadets have plenty of time for a good meal.'

Source: Anon (2006)

The following sections consider meals in the larger institutional settings.

7.7   Types of institutions and meals served

17.7.1   Healthcare and welfare establishments (hospitals, older people's homes and day centres)

The provision of hospital meals is a complex foodservice operation, aggravated by various factors, including: the number of stakeholders, varied patient requirements and the logistic complications of siting wards at considerable distances from the kitchen. A schematic representation of the operation is given in Fig. 7.8. Hospital meals are considered to be an essential part of patient care and treatment, and fundamental in aiding recovery. Yet malnutrition continues with up to 40% of patients in the UK being undernourished on admittance; a situation which is not always rectified during their stay (McWhirter and Pennington, 1994). The UK is not alone and countries such as Germany report rates of 27.4% malnutrition (Pirlich et al., 2006) and Australia up to 50% (Banks et al., 2007).

The UK National Health Service (NHS) serves around 300 million meals a year, costing nearly £0.5 billion and is the third largest purchaser of catering services in the UK exceeded only by business and industry and local authority education catering (National Health Service, 1994). Contract caterers serve approximately a third of the 1200 hospitals (Mintel, 2006). The food budget varies but ranges from £1.50 to £8.40 per person per day, for the provision of three meals, seven beverages and snacks if desired although patient satisfaction shows no relationship to costs (Audit Commission, 2001). By comparison, in the USA, the average food cost varies from $3.00 to $8.77 per patient per day (Feldman, 2005).

In Denmark, Sweden, Finland and Norway, recommendations regarding hospital meals have been made at a governmental level with the main emphasis

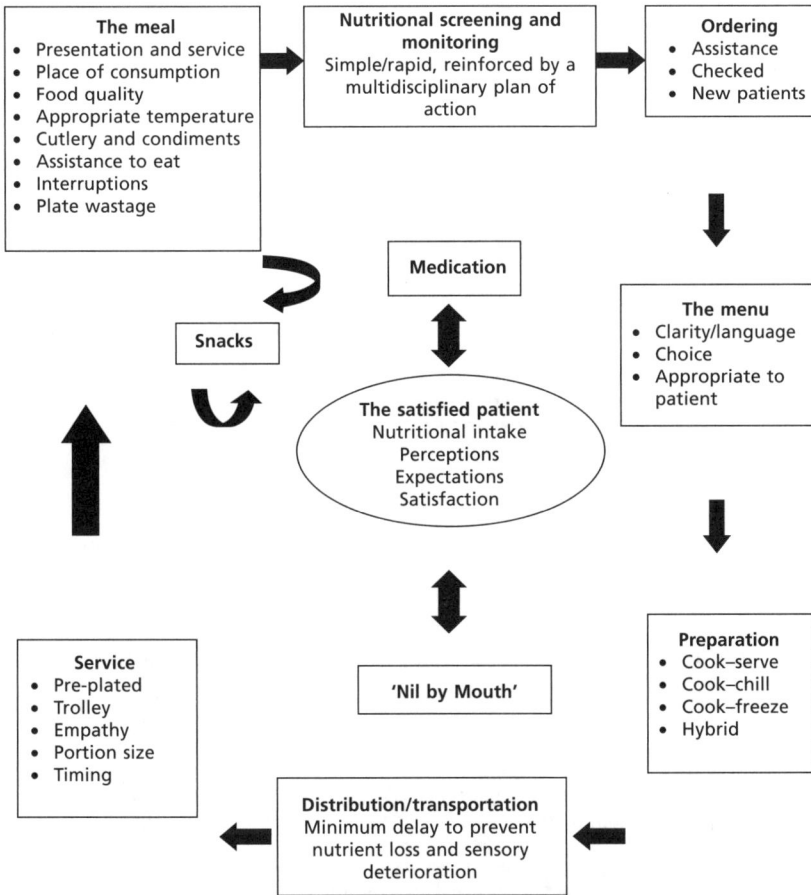

Fig. 7.8 A schematic representation of hospital foodservice.

on the consumption of 'normal' food and meals. Conversely, although France, Germany, Switzerland and the UK have guidelines, these are not always strictly adhered to. The USA has the most stringent control where nutrition is made part of the general requirements for the approval of hospitals (Council of Europe, 2001).

Patients tend to order their meals in advance, often by up to 24 hours, usually from a cyclical menu that changes daily, although some hospitals have developed an à la carte menu similar to those found in a commercial restaurant. (Edwards, 2001). A typical cyclical hospital menu from a UK hospital is given in Fig. 7.9. Once patients have ordered their meals, menu cards are collected from the ward, sent to the kitchen where they are collated. This then provides the basis for the kitchen production and associated activities.

Tuesday	Day Code 3

Name_____

Ward _____

Please put a tick ☑ in the box opposite your choice. If you would like a large portion put a cross ☒ in the box.

Evening Meal
(Please choose one from the following)

1	☐	Chilled Apple Juice	D. (HP)HE
2	☐	Cream of Tomato Soup	D. (HP)

— Main Course —

(Please choose one from the following)

3	☐	Savoury Minced Lamb Pie	D.HP.S
4	☐	Cottage Cheese and Apricot Salad	D.HP
5	☐	Vegetable Nuggets	D.HP
6	☐	Brown Bacon Roll	D.HE.HP.S

7	☐	Creamed Potatoes	D.HE.S
8	☐	Savoury Potatoes	D

9	☐	Mixed Vegetables	D.HE.S
10	☐	Broad Beans	D.HE.HP

— Desserts —

11	☐	Banana Custard	(D)(HP)S
12	☐	Vanilla Ice Cream	D.HP.S
13	☐	Red Leicester Cheese & Biscuits	D.HP
14	☐	Fresh Banana	D.HE.HP.S

HE = *Healthy Eating*, **HP** = *High Protein & High Energy*, **D** = *Diabetic*, **S** = *Soft*, () *Brackets around coded menu indicates products have been made suitable for diets.* Light and Soft menus available on wards. Please ask sister. Condiments on ward.

Fig. 7.9   Example of a hospital menu – UK (Source: National Health Service Hospital, UK).

Feeding the older population in the community

In most Western societies, people are living longer and the preferred cost effective scenario is for them to live in their own homes in the community for as long as possible. At home though, the preparation and cooking of meals may present a challenge, hence a number of countries have developed and use a 'Meals on Wheels' provision which is either government or local authority sponsored, run with the aid of a charity, or as a totally commercial venture.

Whichever way, the older person is able to order meals from a pre-selected menu and have them delivered, either hot and ready for consumption, or chilled for later home regeneration. Many commercial organisations also offer meals which can be ordered for a period (week), delivered frozen in

7-Day Menu Pack

Cost	£23.15	Cost	£23.02
Sliced pork	Apple pie	Sliced lamb in gravy	Apple pie
Shepherds pie	Rice pudding	Fish in parsley sauce	Rice pudding
Sausages in gravy	Chocolate sponge	Lamb & vegetable casserole	Chocolate sponge
Steak & mushroom bake	Bread pudding	Steak & kidney pie	Bread pudding
Lancashire hot pot	Orange sponge	Breaded fish & chips	Jam sponge
Beef hot pot	Bakewell tart	Cumberland sausage	Lemon sponge
Savoury minced beef	Sultana sponge	Corned beef hash	Raspberry & apple crumble

Fig. 7.10   Examples of Meals on Wheels provision – UK (Source: Commercial Supplier, UK).

bulk, held in a freezer, then defrosted and regenerated when required. An example of commercially available meals is given in Fig. 7.10 with internet ordering being an option.

Local authorities in a number of countries also organise local 'day centres' where older people can meet, socialise, share issues of concern, receive advice from various experts such as nurses and social workers and perhaps be provided with a meal. Here meals might be cooked on the premises or brought in from an external supplier, probably hot, and ready for service from a counter.

7.7.2   Education establishments (day and boarding schools, universities)

Historically, the focus in schools was on the provision of adequate nutrients but today, the main issue is that of energy balance and concern with an embedded culture where many children eat a poor quality, high-calorie, high-fat and high-carbohydrate diet both at home and at school (Mintel, 2007). This has led to rising levels of childhood obesity and it is estimated that by 2050, 25% of all children under 16 in the UK could be obese (Foresight, 2007); a similar trend identified globally.

School meals can provide a significant proportion of a child's daily intake especially in the UK where one in nine children go to school without breakfast and one in six go home to no cooked evening meal (School Meals Campaign, 1992). Furthermore, where pupils are resident at the school (boarders), meals provide their primary source of nutrients. This, in part, has led to nutrient-based standards that are to be introduced to all schools by 2009 where the general requirements are for oily fish to be served regularly, bread to be available on a daily basis and a reduction in the use of deep frying. A typical school meal is given in Fig. 7.11.

Internationally, school meals are organised differently but generally fall into two categories: either a meal (hot) is served in canteens and cafeterias in countries such as France, Finland, Sweden, UK and USA, or homemade lunch boxes, fruit breaks and snack outlets predominate in nations such as Denmark and Norway (Council of Europe, 2003). The vending of drinks and

Primary school (Age range 4–11 years old)	Secondary school (Age range 12–16/18 years old)
The Monday Marathon Grilled gladiator beef burger or Vegetable burger in a bread shield Carrot spears or runner beans Potato cannonballs Fighting fit fruit salad	*Salad Days* Baked potato with a choice of Baked beans * tuna * cheese * vegetable curry or Homemade mushroom pizza Help yourself from the sunshine salad bar! Summer pudding or fresh fruit or yoghurt

Fig. 7.11   A 'typical' school meal served in the UK (Source: DFEE, 1998).

snacks are also popular as they provide an additional source of income to the school although recently the focus has been to ensure that only healthy items are available (Mintel 2007).

School foodservice in 2005 was an estimated $8.2 billion industry in the USA serving five billion meals per annum with an allowance of $2.47 per meal (US Department of Agriculture, 2007) while in the UK 3.5 million meals are served daily with an allowance of £1.68 per meal (Mintel, 2007). In many countries, contract caterers again play a dominant role although the trend in UK schools has seen an overall reduction of 7% but an increase in college and university provision (Mintel, 2007). The amount of subsidies varies from all meals being subsidised to the targeting of vulnerable groups, such as the socio-economically disadvantaged. In the USA, The National School Lunch Program, a federally assisted meal programme operating in public and non-profit private schools, and residential child care institutions, provides for nutritionally balanced, low-cost or free lunches which must meet recommendations of the Dietary Guidelines for Americans, which stipulate that no more than 30% of calories should come from fat; less than 10% from saturated fat (US Department of Agriculture, 2007).

Encouraging pupils to consume a healthier diet has led many countries not only to consider the meals themselves but also the meal environment, including the speed of service, length of queues and size of dining area. This focus is seen as part of a 'whole school initiative' where the importance of diet, nutrition, exercise and the environment are considered. Ultimately, a balance must be struck between creating eating patterns that will remain into adulthood and allowing children the freedom to choose meals and dishes with which they are familiar.

However, education provision does not stop at school but continues into both further and higher education. Although some students may continue to live at home, others will move away with many preferring to live independently in private accommodation or in halls of residence. In the latter case, the meal provision can also vary with some having a shared communal style kitchen for approximately eight students where meals can be prepared and consumed. In others, meals will be provided and served either using self-service or

Fig. 7.12   A 'typical' modern university service area.

family style. Many campuses also have a self-service cafeteria style operation (refectory) for the sale of meals to students, staff and visitors.

Many universities today now recognise the opportunities that foodservice operations can provide in that they can help to enhance the overall attractiveness of a university, enhance the student experience and are no longer a 'cost centre' which needs to be subsidised but can actually generate income. As a result, many have responded to this challenge and provide a multi-station 'food court' with a variety of different cuisines, see for example Fig. 7.12.

7.7.3   Prisons and custodial establishments

It is estimated that the world's prison population is in excess of 9.25 million with 2.19 million in the USA, 1.55 million in China and 0.87 million in Russia (Walmsley, 2007). Prisoners are invariably a disadvantaged segment of society and are more likely to have smoked cigarettes, drunk hazardously, taken drugs and practised unsafe sex with a greater number of partners (Condon *et al.*, 2007; Harris *et al.*, 2006): criteria found in many countries and where diseases such as tuberculosis and HIV are also common (Stern, 2001).

Prison meals assume a greater importance than many other situations, firstly because there is an obligation to ensure that all prisoners have the opportunity to choose meals that, when consumed, contribute towards a healthy, nutritionally balanced diet; secondly there is the need to consider personal and religious dietary requirements such as vegetarian and vegan, halal and kosher options, and lastly because meals assume a particular significance to prisoners. Meals play a major role for prisoners conditioning their daily life and symbolising their prison experience (Smith, 2002). Meals relieve the boredom and monotony of a routine existence and a number of authors have suggested that they are a catalyst for aggression (NAO, 1997). An ill-designed menu, inadequate portion sizes, lack of variety and poorly cooked food can contribute to serious complaints and dissension (Blades, 2001). In addition, food is often seen as currency and used to barter for other goods on the underground economy (Godderis, 2006). Even so, nutrient deficiencies such as vitamin A have been reported as being significant health problems in some countries (Mathenge et al., 2007).

There have been few academic studies into the meals provided in prisons, although from the limited literature available it would appear that the standards, quality and type of meals vary considerably around the world. In Russia, for example, the minimal requirements recently signed into law are for: 30 g sugar, 100 g milk, 90 g meat, 100 g fish, 550 g bread and 800 g fruit and vegetables (Anon, 2005); and in the USA, meals are typically very basic featuring 'comfort' foods such as meatloaf, beef stew, mashed potatoes with gravy, macaroni and cheese (Jonsdottir, 1999), with prisoners receiving three meals per day Monday to Friday and two meals per day at the weekend (Elan, 2005). The amount of money spent per person per day on food is difficult to assess. It could be as low as $1.50 per day (Elan, 2005) although a number of prisons also have their own farms where produce is used to supplement inmates' diets. In the USA in 2002, the average spend was in the order of $2.26 (Anon., 2002) and in the UK, the average spend on food ranges from £1.20 to £3.41 (NAO, 2006). What is clear is that the amount of money available is extremely low and, in part, must account for complaints about the monotony and variety of meals available (Smith, 2002).

In the majority of prisons, particularly the larger ones, on-site kitchens, using inmates, supervised by non-inmates, are used to prepare and cook meals (Anon., 2002; NAO, 1997), although larger prisons might also use centralised cook–chill operations or contractors (Jonsdottir, 1999).

The most comprehensive studies of prison meals has been the work of Edwards and co-workers (2001; 2007), who, in two major surveys in the UK, measured the food served and nutritional adequacy of twelve male prisons, two female prisons and two young offenders' institutes. An example of a prison menu is given in Fig. 7.13; inmates having the opportunity to select an entree from a choice of approximately five items. Meals were served from a hotplate/counter where there was an opportunity to freely select potatoes (or other starch items) vegetables, dessert or fruit, bread and condiments. Results show that with the exception of some micronutrients, prisoners are provided with and consume a healthy diet which in the main conforms to current dietary recommendation, although, in all prisons, the intake of salt, and, in one female prison, the percentage of energy derived from fat, exceeded recommendations (Department of Health, 1991).

Day	Lunch	Tea	Pudding
Thursday	Vegetarian Pasta Bake Chicken & Mushroom Pie Halal Jamaican Beef Patti Corned Beef & Pickle Roll Jacket Potato & Coleslaw	Vegetable Supreme Chicken Supreme Halal Chicken Curry Grilled Gammon Pork Pie Salad	Eve's Pudding Fresh Fruit
Friday	Vegetable Pancake Roll Breaded Fish Cheese & Beano grill Cheese & Tomato Roll Jacket Potato & Tuna	Bean & Vegetable Curry Chicken Chasseur Halal Beef Casserole Fish in Parsley Sauce	Sponge Pudding & Custard Fresh Fruit
Saturday	Vegetable Sausage & Fried Egg Bacon, Sausage & Fried Egg Halal Sausage & Fried Egg Turkey Salad Roll Jacket Potato & Curried Beans	Soya Lasagne Minced Beef Lasagne Halal Beef Italienne Rice & Bean Stuffed Pepper Salad Cheese Salad	Sultana Scone Fresh Fruit

Fig. 7.13 Extract of a prison menu in the UK. In addition, potatoes (starch), vegetables, bread and spread are also available (Source: Male Prison, UK).

7.7.4   Armed forces (Army, Navy, Marines and Air Force)

Whilst stationed in barracks, and to a greater extent on a ship, meals are similar to those found in other institutional settings. Soldiers and sailors are served three meals per day, selected from a choice, and invariably served in a self-service style. In an Officers' Mess/Club and often in senior non-commissioned officers' dining rooms, table service similar to that found in commercial restaurants, might be used. Where meals do differ completely is when solders are in the field and under operational conditions.

When soldiers are grouped in close proximity and the tactical situation permits, meals can be prepared centrally in field kitchens and served either in a central location or delivered hot to a remote location. The meals are prepared using a mixture of whatever fresh food might be available; preserved, primarily tinned and dehydrated foods; and other group 'operational' or 'composite' ('compo') ration packs. However, when dispersed, the meals offered would be operational rations, requiring the minimum or no preparation, consumed when the situation permits.

Most countries have operational rations packs, obviously geared to specific country, ethnic and religious needs and tastes and which vary depending on the operational theatre and the nature of the military operation involved. The purpose of these individual rations is to make soldiers independent in their ability to work, fight and survive on their own, or in small groups, but, at the same time, ensure that individuals receive adequate food and nutritional requirements. Countries address this requirement slightly differently using meals and components which are either tinned, retort pouched, or dehydrated. Perhaps the most comprehensive range of operational rations can be found in the US Armed Forces (Darsch and Faso, 2006) where one type of meal is

Menu	Menu	Menu
Beef ravioli in meat sauce	Cheese and vegetable omelette	Chicken breast filet
	Granola w/berries	Corn bread stuffing
Toaster pastry	Toaster pastry	Caramel apple bar
Cookie		
Cracker, vegetable, flavour	Cracker, plain	Wheat snack bread
Cheese spread, jalapeño	Apple butter	Cheese spread, jalapeño
	Cinnamon scone	Candy II
Beef snacks		
CHO electrolyte beverage	Coffee, French vanilla	Coffee, French vanilla
Hot sauce	Salsa sauce	BBQ seasoning
Spoon	Spoon	Spoon
Flameless ration heater	Flameless ration heater	Flameless ration heater
Accessories:	Accessories:	Accessories:
Coffee, cream substitute	Apple cider	Lemon tea
sugar, salt, gum, matches,	Salt, gum, matches, tissue,	Salt, gum, matches,
tissue, towelette	towelette	tissue, towelette

Fig. 7.14   Examples of USA Meal Ready to Eat (MRE, XXVIII) (Source: Darsch and Faso, 2006).

individual packs (Meal Ready to Eat, MRE), each of which contains a 'meal' for one person. Currently, there are 24 meal varieties (Darsch and Faso, 2006), examples of typical meals are given in Fig. 7.14. In the United Kingdom, the approach is slightly different with a pack containing three meals (breakfast, main and snack meals) for a complete 24-hour period; currently with seven varieties, example of meals are given in Fig. 7.15.

Devices to heat the meals range from individual stoves using gas, liquid or solid fuels to a flameless ration heater, which uses a water-activated exothermic chemical reaction to reheat the food pouches.

7.7.5 Business and industry

Business and industry is concerned with the provision of meals in the workplace and it is here where the greatest range of meals is offered. Meals provided for directors and senior managers are likely to be served in a restaurant style setting with table service and so forth. However, the meals served to both shop-floor and office 'workers' are more likely to be served in a cafeteria style setting. In many cases, meals are subsidised and form part of the total remuneration package for employees, and helps to account for the increasing uptake of meals in this sector, where globally the percentage of employees using foodservice facilities has increased from 35.2% in 1996 to 39.1% in 2001 (Euromonitor International, 2002); Fig. 7.16 gives a more detailed breakdown. On the other hand, however, as these meals are subsidised, if not the food, aspects such as overheads and the building might well be, many companies are constantly seeking ways to reduce or eliminate them. In addition, because these meals are not the core business, companies are also interested

Menu	Menu	Menu
Breakfast	**Breakfast**	**Breakfast**
Hamburger and beans	Corned beef hash	Chicken sausage and beans
Snack	**Snack**	**Snack**
Oatmeal block	Oatmeal block	Oatmeal block
Fruit biscuits	Fruit biscuits	Fruit biscuits
Brown biscuits	Brown biscuits	Brown biscuits
Cheese spread	Meat pâté	Cheese spread
Chocolate bar	Chocolate bar	Chocolate bar
Boiled sweets	Boiled sweets	Boiled sweets
Main meal	**Main meal**	**Main meal**
Soup	Soup	Soup
Chicken, mushroom and pasta	Beef stew and dumplings	Lamb stew and potatoes
Treacle pudding	Chocolate pudding and sauce	Fruit dumplings and sauce
Drinks	**Drinks**	**Drinks**
Chocolate	Chocolate	Chocolate
Beverage whitener	Beverage whitener	Beverage whitener
Sugar	Sugar	Sugar
Tea, coffee	Tea, coffee	Tea, coffee
Vegetable stock drink	Vegetable stock drink	Vegetable stock drink
Orange or lemon powder	Orange or lemon powder	Orange or lemon powder
Sundries	**Sundries**	**Sundries**
Chewing gum	Chewing gum	Chewing gum
Waterproof matches	Waterproof matches	Waterproof matches
Paper tissues	Paper tissues	Paper tissues
Water purification tablets	Water purification tablets	Water purification tablets

Fig. 7.15 Examples of UK General Purpose Ration (Source: original packs).

in ways to simplify or reduce their involvement with the provision of meals; hence the growth of contract caterers.

7.7.6 Miscellaneous meal provision

Although spatial considerations have precluded the consideration of all institutional meal settings, it should be borne in mind that a number of other organisations, such as the emergency services, police, fire and ambulance, would also serve institutional type meals. In addition, a number of commercial organisations, such as cafeterias in departmental stores serve meals that have the characteristics of institutional meals (Fig. 7.4) reinforcing the notion that institutional meals are more about the style of meals and their service rather than a sector of the foodservice industry or the place where they are offered.

In each of the scenarios outlined, institutional meals, whilst being secondary to the primary purpose of the business, invariably form an integral component of the business itself being, for example, part of a remuneration package as in business and industry, the primary source of sustenance as in prison or an essential element of treatment as in hospitals. As a result, the provision of meals can be crucial and has led to nutritional recommendations being made in a number of the settings; examples of which are given in Tables 7.1–7.3.

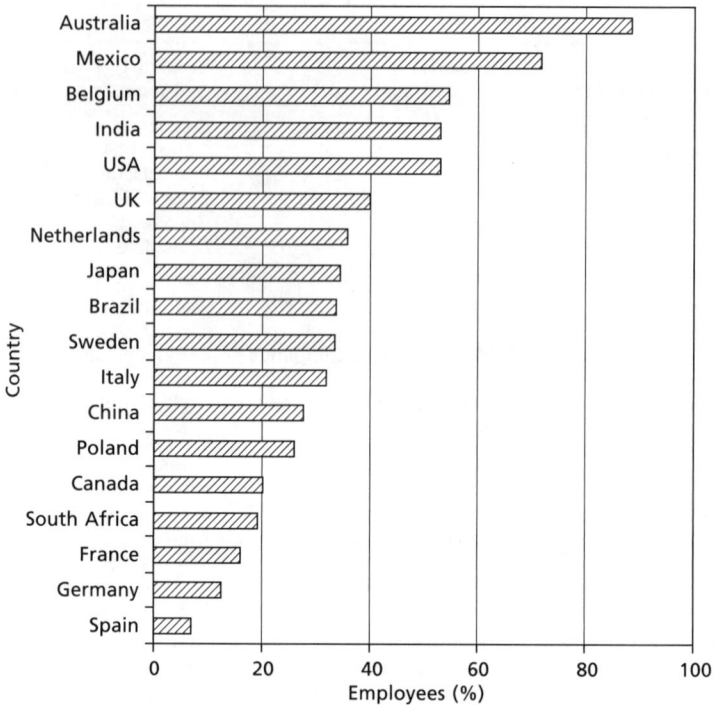

Fig. 7.16   Percentage of employees in 'Business and Industry' using foodservice facilities (Source: adapted from Euromonitor International, 2002).

7.8   Summary, conclusions and future trends

This chapter, in addressing institutional meals, has sought to show that they are characterised by the food and service style rather than where they are served, and being a secondary rather that primary goal of the business, are normally provided to satisfy a need. However, institutional meals are an integral part of the working day and as such have a unique role to play, hence foodservice managers will increasingly need to enhance the experience that these meals bring to their 'customers'.

In many cases, individuals still have negative attitudes, either real or imaginary, towards institutional meals and perceive them to be inferior in quality to those found in the commercial sector. These perceptions are, however, changing as the quality and standard of meals improve, brought about in part, it is argued, by increased competition and the use of contract caterers. No longer are people satisfied with second class food and quality but will increasing expect standards to be improved.

In institutions, the service of meals is often marginalised, although contact personnel can make a significant contribution to the service encounter and satisfaction. It is unfortunate, therefore, that staff appear to be undervalued, paid a lower wage (Pratten, 2003) and perceive their jobs as being for the

Table 7.1 Examples of nutritional recommendations for institutional meals

Nutrient	Unit	Recommendations per day							Per meal
		In childcare[1]	Residential care[2]	Residential care[2]	Residential care[3] Male and female	Hospital[4] Male and female	Prison[3] Male	Prison[3] Female	Community meal Male and female
		1–4 years	5–10 years	11–18 years	>75 years	19–74 years	19–74 years	19–74 years	>75 years
Energy	kcal	903	1780	1845–2755	1955	2225	2515	1930	782
	MJ		7.5	7.7–11.6	8.2	9.4	10.6	8.1	3.3
Fat	g	35.0	70	72–107	76	87	98	75	30
Total	g		237	246–367	260	297	335	257	104
Carbohydrate									
Protein	g	11.0	25	41–55	50	50	55	45	17
Vitamin A	µg	300	500	700	700	700	700	600	260
Vitamin C	mg	21	30	40	40	40	40	40	20
Calcium	mg	260	550	1000	700	700	700	700	280
Iron	mg	5.5	9	15	9	15	9	15	3.6
Salt	g	1.6	4.5	6	6	6	6	6	2.4

Source: Crawley (2007). See also Tables 7.2 and 7.3 on page 124.

1. Examples of menus for each institutional setting are also given in Crawley (2007).
2. Breakdown of nutritional requirements per meal are given in Tables 7.2 and 7.3.
3. Not specifically broken down by meal in Crawley (2007)
4. Not specifically broken down by meal in Crawley (2007), but: minimum recommendations for meals are 300–500 kcal and 18 g protein; sandwich meals 300 kcal and 12 g protein; breakfast, snacks and milk drinks 900 kcal and 30 g protein; the total should provide 1200–2500 kcal per day (Nutrition Task Force 1995).

Table 7.2   Childcare

Nutrient	Meal composition (%)		
	Snack only	Lunch only	Tea only
Energy	10	30	20
Fat	±35	±35	±35
Total carbohydrate	±50	±50	±50
Protein	10	30	20
Vitamins A and C	10	30	20
Calcium	10	30	20
Iron	10	35	25
Salt	10	30	20

Note: See Table 7.1 on page 123.

Table 7.3   Residential care

Nutrient	Meal composition (%)			
	Breakfast	Lunch	Evening	Snacks
Energy	20	30	30	20
Fat	20	30	30	20
Total carbohydrate	20	30	30	20
Protein	20	30	30	20
Vitamins A and C	20	40	40	–
Calcium	20	40	40	–
Iron	20	40	40	–
Salt	20	30	30	20

Note: See Table 7.1 on page 123.

'less intelligent' (Donelan, 2000). The quality of the meal can be diminished by aspects such as poor service, poor physical dining conditions and a lack of empathy by staff; hence attention will need to be paid to the entire experience.

Pressures to reduce the overall costs of supplying meals but at the same time raise quality and standards will continue. This will present a duel challenge to foodservice operators and it is probable that the provision of institutional meals will continue to be contracted out. The divide between the cost and commercial sector is likely to become even more blurred with a greater emphasis on the cost sector to operate and provide facilities and standards traditionally associated with the commercial sector. Facilities which remain idle for much of the time will need to be re-evaluated and put on a sounder financial basis, being used for activities such as private functions.

Increasingly the local sourcing of ingredients is likely to become more prevalent (Mintel, 2006; Foodservice Director, 2007) as local, seasonal produce has both food quality and environmental benefits; although implementation of an organic policy is difficult within the healthcare sector, particularly in the UK, owing to the complex nature of procurement.

7.9 Sources of further information and advice

There are a number of international academic journals which specifically relate to the food service industry, including:

Journal of Foodservice
Journal of Culinary Science and Technology
Journal of Foodservice and Business Research:
In addition, the 'trade press' which in the US would include *Restaurants and Institutions* and in the UK *The Caterer and Hotelkeeper* are useful sources of current information.

7.10   References

Anon. (2002). How prison foodservice is performing. *Foodservice Director* **16**(10), 37.

Anon. (2005). The Russian prison diet. *Russian Life* **48**(4), 11.

Anon, (2006). Mitchell Hall, fact sheet. Available at: http://www.usafa.af.mil/index.cfm?catname=Academy%20Info Accessed 15 February 2008.

Audit Commission (2001). *Acute hospital portfolio: review of national findings*, Wetherby: Audit Commission Publications.

Banks, M., Ash, S., Bauer, J, and Caskill D. (2007). Prevalence of malnutrition in adults in Queensland public hospitals and residential aged care facilities. *Nutrition and Dietetics* **64**, 172–178.

Blades M. (2001). Food and nutrition in the prison service. *Prison Service Journal* **134**, 46–48.

Cardello, A.V., Bell, R. and Kramer, M. (1996). Attitudes of consumers toward military and other institutional foods. *Food Quality and Preference* **7**(1), 7–20.

Condon, L., Hek, G. and Harris, F. (2007). A review of prison health and its implications for primary care nursing in England and Wales: the research evidence. *Journal of Clinical Nursing* **16**, 1201–1209.

Council of Europe (2001). *Food and nutritional care in hospitals: how to prevent undernutrition*, Brussels, Council of Europe.

Council of Europe (2003). *Resolution on food and nutritional care in hospitals*, www.bda.uk.com.

Crawley, H. (2007). *Nutritional guidelines for food served in public institutions*. Report prepared for The Food Standards Agency. St Austell, UK: The Caroline Walker Trust.

Darsch, G.A. and Faso, R.M. (2006). *Operational Rations of the Defence Department*. Natick PAM 30-25 7th ed. US Army Natick Soldier RD&E Center and Defense Supply Centre Philadelphia.

Davis, B., Lockwood, A. and Stone, S. (1998). *Food and Beverage Management*. 3rd ed. Oxford: Butterworth Heinemann.

Department of Health (1991). *Dietary Reference Values for Food Energy and Nutrients for the United Kingdom*. London: HMSO.

DfEE (1998). *Ingredients for Success*. London: HMSO.

Donelan A. (2000). Dietitians and caterers: an uncertain but critical relationship. *Nutrition and Food Science*, **30**, 123–127.

Dowling R.A. and Cotner C.G. (1988). Monitor of tray error rates for quality control. *Journal of The American Dietetic Association*, **4**, 450–453.

Edwards, J.S.A. (2001). Hospital food service – a USA perspective. *The Journal of the Royal Society for the Promotion of Health* **121**(4), 209–212.

Edwards, J.S.A., Edwards, A. and Reeve, W.G. (2001). The nutritional content of male prisoners' diets in the UK. *Food Service Technology* **1**(1), 25–33.

Edwards, J.S.A., Hartwell, H.J., Reeve, W.G. and Schafheitle, J. (2007). The diet of prisoners in England. *British Food Journal* **109**(3), 216–232.

Elan, E. (2005). Prison foodservice officials weigh tight budgets, crew safety. *Nations' Restaurant News*. September. 8 and 26.

Euromonitor International (2002). Contract foodservice – World. Euromonitor International Available at: http://athens.portal.euromonitor.com/portal/server.pt?control= SetCommunity&CommunityID=206&PageID=719&cached=false&space= CommunityPage. Accessed 19 February 2008. Password required.

Feldman C. (2005). Hospital Dietary Policy. *Topics in Clinical Nutrition,* **20**, 146–156.

Foodservice Director (2007). *Hospitals: Beyond the Patient*. June 15, FSDmag.com.

Foresight (2007). *Tackling Obesities: Future Choices*. London: Department of Innovation Universities and Skills.

Foskett, D., Ceserani, V. and Kinton, R. (2003). *The Theory of Catering*. 10th ed. Oxon.: Hodder and Stoughton.

Gale, D. (2007). Pressure points. *Restaurants and Institutions* **117**(1). 77–82.

Godderis, R. (2006). Dining in: the symbolic power of food in prison. *The Howard Journal* **45**(3), 255–267.

Harris, F., Hek, G. and Condon, L. (2006). Health needs of prisoners in England and Wales: the implications of prison healthcare of gender, age and ethnicity. *Health and Social Care in the Community* **15**(1), 56-66.

Hartwell, H.J. and Edwards, J.S.A. (2003). A comparative analysis of 'plated' and 'bulk trolley' hospital food service systems. *Food Service Technology* **3**(3/4), 133–142.

Holmes, S., (2006). Barriers to effective nutritional care of older adults. *Nursing Standard*. **21**(3). 50–54.

Jonsdottir. (1999). Low-cost, high-quality foods are demanded. *The Journal of Foodservice Distribution* **35**(10), 64-66.

Kipps M. and Middleton T.C. (1990). Achieving quality and choice for the customer in hospital catering. *International Journal of Hospitality Management* **9**, 69–83.

Mathenge, W., Kuper, H., Myatt, M., Foster, A. and Gilbert, C. (2007). Vitamin A deficiency in a Kenyan prison. *Tropical Medicine and International Health* **12**(2), 269–273.

McWhirter J.P. and Pennington C.R. (1994) Incidence and Recognition of Malnutrition in Hospitals. *British Medical Journal* **308**, 945–948.

Mintel (2006). *Contract Catering – UK*. London: Mintel International Group Ltd.

Mintel (2007) *School Meals*. London: Mintel International Group Ltd.

National Health Service (1994) *A Report by the Comptroller and Auditor General: Hospital Catering in England*. London: HMSO.

NAO (1997). *HM Prison Service. Prison catering. Report by the Comptroller and Auditor General*. National Audit Office. London: The Stationery Office.

NAO (2006). *HM Prison Service. Serving time: prisoner diet and exercise. Report by the Comptroller and Auditor General*. National Audit Office. London: The Stationery Office.

Nutrition Task Force (2005). *Nutritional guidelines for hospital catering*. London: Department of Health.

Payne-Palacio, J. and Theis, M. (2005). *Introduction to Foodservice*. Upper Saddle River, New Jersey: Pearson.

Pirlich, M., Schütz, T., Norman K. *et al.* (2006). The German hospital malnutrition study. *Clinical Nutrition* **25**(4), 563–572.

Pratten J. (2003) The importance of waiting staff in restaurant service. *British Food Journal*, **105**, 826–834.

Quest, M. (2007). *2007 UK food and service management survey*. London: British Hospitality Association.

Quest, M. and Needham, D. (2006). *Trends and statistics, 2006*. London: British Hospitality Association.

Reynolds, D. (1997). Managed-Services Companies. *Cornell Hotel and Restaurant Administration Quarterly* **38**, 88–95.

School Meals Campaign (1992) School Meals. *Nutrition and Food Science* **23**(5), 22.

Silverman M. R., Gregoire M.B., Lafferty L.J. and Dowling R.A. (2000) Current and future practices in hospital foodservice. *Journal of The American Dietetic Association*, **100**, 76–80.

Smith, C. (2002). Punishment and pleasure: women, food and the imprisoned body. *The Sociological Review* **50**(2), 197–214.

Stern, V. (2001). Problems in prisons worldwide, with a particular focus on Russia. *Annals of the New York Academy of Sciences* 953b, 113–119.

US Department of Agriculture (2007) http://www.usda.gov/wps/portal/!ut/p/_s.7_0_A/7_0_1OB?navtype=SU&navid=FOOD_NUTRITION accessed 3 December 2007.

Walmsley, R. (2007). *World Prison Population List* 7th ed. Summary available at: http://www.kcl.ac.uk/phpnews/wmview.php?ArtID=1635 accessed 15 October 2007.

Wilson M., Murray A.E., Black M.A., McDowell D.A. (1997). The implementation of hazard analysis and critical control points in hospital catering. *Managing Service Quality* **7**, 150–6.

8

Studying meals in the home and in the laboratory

I. Boutrolle, Danone Research, France and J. Delarue,
AgroParisTech, France

Abstract: Hedonic testing methods for assessing food product acceptability, including the more common Central Location Test (CLT), which usually takes place in a standardized location under controlled conditions, and the Home Use Test (HUT), are compared. As CLT conditions limit the possibility of taking into account the meal context, it is necessary to determine whether testing food products under controlled test conditions modifies consumers' perception and liking. Ways in which the tasting conditions may affect food testing results are described and suggestions are made for improving CLT in order to enhance integration of eating/drinking situation variables.

Key words: consumer food testing, Central Location Test, Home Use Test, contextual variables, meals.

8.1 Introduction: current practices for food testing

In the study of the perception and liking of food, it would seem natural to consider meal settings. However, in practice, in both academic and industrial research, food products are usually tested alone and under standardized conditions. The question of whether the testing conditions affect the consumer's perception of food thus arises. From a theoretical point of view, consumption situations and eating/drinking modalities would be expected to exert a strong influence on how food products are perceived and liked. For food companies and scientists who study food choices and liking, study design is thus important.

This section will address declarative hedonic measurements (i.e. relative to pleasure) and the manner in which they are routinely obtained to assist food developers in the issues of developing new products, improving products, reducing costs, or product positioning with respect to the competition. The tests are set up to compare the overall organoleptic performance (i.e. with respect to the sensory properties of foods) of various recipes. Several methodological alternatives are available for the acquisition of data on consumer food perception and liking. Among the methodological alternatives is the choice of the test site: a sensory evaluation laboratory, a room in a public facility or the subject's home. Many other characteristics of the protocol generally necessitate choices on the part of the investigator: the order of product presentation, the quantity of product presented, the number of products presented, the timing of the session, etc. The foregoing examples of the choices the investigator has to make when setting up a consumer test protocol suggests that protocols are highly flexible. However, in practice, the assessment site usually conditions the other characteristics of the protocol. Thus, currently, survey institutes and research from Academic distinguish between two main types of test: the Central Location Test (CLT) in a facility or laboratory and the Home Use Test (HUT). The two types of test will be addressed in order to differentiate them on the basis of numerous methodological aspects in addition to the evaluation site.

8.1.1   Central location tests: tests under controlled conditions

The methodologies set up for central location tests are relatively similar to those set up in the laboratory. Overall, even if the methodologies differ in terms of test site, they are both considered as tests under controlled conditions, unlike home use tests. Questioning a large number of consumers calls for appropriate logistics and hence interviewing several subjects at the same time. Thus, the rooms in which CLT are conducted are generally equipped with several tables independent of each other (or individual booths in the case of a laboratory) in order to prevent subject distraction and communication. The anonymous products are presented in succession in coded identical dishes for all the subjects. The portions presented are frequently smaller than usual portions. It is, however, frequent that investigators require a minimum quantity of product to be consumed ('drink at least half the glass before giving your opinion'). Once tasting has been completed, the interviewer asks the subject to formulate an overall assessment of the product, then may ask the subject a number of 'diagnostic' questions related to the perception and liking of a few specific sensory characteristics or elicit the subject's opinion on aspects of the product. The subject is then required to rinse his/her mouth before tasting the next product(s) proposed and for which the subject answers the same questions. The number of products presented during such tests may be relatively high but it is nonetheless necessary to restrict the number of samples tasted in a given session because of physiological fatigue (adaptation, satiety) and psychological fatigue. However, some studies may be conducted with a pure monadic presentation involving tasting of a single product by subject. This has an impact on the cost of the study.

In contrast to laboratory tests requiring subject pre-recruitment, CLTs are set up in rooms rented in locations frequented by potential purchasers (high streets, malls). The subjects are generally intercepted and selected on the spot throughout the day. Once a consumer has been selected as belonging to the target, he/she is invited to take part in a tasting session that does not exceed 20 min. A longer tasting session would require pre-recruitment and compensating the subjects for taking part in the test. The subjects selected on the spot have less time to spare than pre-recruited subjects. This restricts the number of samples that can be presented and the number of questions relating to the food product. Survey institutes generally propose simple face-to-face interviews rather than self-administered questionnaires enabling the subject to focus on his/her interview and not on his/her neighbor's interview.

8.1.2   Home use tests: tests under natural conditions

The objective of this type of test is to evaluate the liking for foods when consumed at home during a given period of time. The subjects are generally pre-recruited from databases to be representative of the target population of interest and are therefore frequently compensated. In general, the products, masked and coded, are distributed one after the other with a variable time interval between each drop-off (from a few days to a few weeks). This sequential monadic product presentation is time consuming and thus a limited number of different products are usually tested in HUT (at most two or three). The first product is distributed in a large quantity and accompanied by a leaflet containing tasting instructions. The consumer is asked to use the product over a predetermined duration as he/she wishes and often at the frequency the subject chooses. The subject's opinion on the product may be acquired in various manners. An interviewer may interview the subject on the product that has been tested for a few days (overall assessment, then diagnostic questions) during a second visit. The interviewer can also recover a pre-distributed self-administered questionnaire in which the subject has reported his/her assessment after several product intakes. In the case of evaluation of several products, during the second visit, the interviewer distributes the samples of the second product to be tested with the same instructions. Certain home studies also employ a pure monadic product presentation in which each subject only tastes a single product.

8.2   How does meal context affect food preferences?

8.2.1   Role of psychological constructs and attitudes on judgment when eating

The liking scores formulated by the subject during the organoleptic tests is supposed to reflect the subject's opinion on the tested food. The attitude toward the sensory properties of the food is thus what blind hedonic tests attempt to measure. In order to understand why and how the hedonic scores obtained under different conditions may vary, it is appropriate to explore the bases of the concept of attitude which is to be measured.

Attitude: definition and construction process
Attitude may be defined as a cognitive component reflecting the subject's position, from positive to negative, with respect to the object and orienting the subject toward particular actions. Individuals build their attitude periodically on the basis of information that is temporarily pre-eminent or accessible. Attitude is thus a psychological trend that is expressed by evaluation of a particular entity with a certain degree of favor or disfavor (Eagly and Chaiken, 1993). Although attitudes that form in childhood may last throughout life, it may be considered that there is no definitive attitude, that there is not a single attitude toward an object but rather a certain number of attitudes depending on the number of items of information available with respect to the object (Tesser, 1978).

The evaluation judgment is formulated at the time of object encounter. The information underlying the attitude (past feelings, beliefs and behavior) is retrieved from the individual's memory (conscious or unconscious). The construction process largely depends on both internal introspective processes and the external context in which the attitude is expressed. It will thus be understood that attitude is greatly subject to contextual effects.

The mechanisms underlying formation of an evaluation judgment thus naturally apply to formation of the hedonic perception of a food. In fact, the hedonic perception is indispensable to human beings in order to facilitate decision making with regard to the act of eating or rejecting an available food. While eating pleasure may be considered as an unconscious sensory sensation (Chiva, 1992; Pieron, 1974), the sensory message is subsequently interpreted by the subject and may be verbally expressed in a fully conscious manner as 'I like it' or 'I don't like it'. On each consumption occasion, part of the stored information is activated and added to the new information to determine a behavior. Thus, the hedonic perception of a food is a mental construct derived from personal experience, which enables meaning to be attributed to sensory messages and those messages to be associated in a broader perception in which they are meaningful.

The process in food consumption is thus the resultant of at least three stages. Initially, the attention paid to the sensory inputs requires a cognitive effort which activates the knowledge relating to the tested food stored in the memory. Subsequently, an identification stage occurs during which the information generated by the food in its context is interpreted and identified in the light of existing knowledge that has been specifically activated in the memory. The last stage consists of integration of previously perceived information that is identified, stored in the memory and activated. That information may include general beliefs, affective components, visual images, personal experience and moral values.

The evaluations subsequently determine the consumer's attitude toward the food and thus his/her behavior in response to the food in its context. It will thus be observed that the decision to eat a food is the result of a complex psychological process which begins by the experience of the stimulus in its context.

Effect of context on judgment in food-preference studies
Definition of the context
Consumer tests based on sensory evaluation have, as their objective, measurement of the perception and liking of a food solely focused on the intrinsic properties of the latter, i.e. the recipe. However, a large number of studies have shown that almost all human behaviors are influenced by contextual variables and eating behavior is no exception. As previously noted, a large number of items of information related to the context are potentially involved in the formation of the judgment. It thus seems difficult to determine the contribution of the sensory stimuli of the food alone to a hedonic measurement.

Secondly, context is a very broad concept (the term 'situation' is also used). Rozin and Tuorila (1993) define the context as the set of events and experiences which are not included in the reference event itself but are related to it. In the present case, the context may be defined as all the information (conscious and unconscious) not directly related to the food under study and exerting an influence on the perception of that food. Contextual variables may thus be defined as all the variables that influence the hedonic judgment collected other than the variables intrinsic to the tested food, namely its sensory and nutritional properties.

When attitudinal surveys are set up (surveys designed to collect individuals' opinions on a specific theme or object), it is necessary to take into account the influence of the context on the evaluation judgments. It is not rare to observe that people formulate different evaluations of the same object on different occasions, although the object has undergone no change between the two evaluation time points. A probable explanation of the inconsistency between repeated evaluations of the same object that may sometimes be observed resides in the change in the implicit contextual influences operative on the subject at the different evaluation time points. How those influences are possible, the mechanisms of their potential emergence and their consequences will now be briefly addressed.

What mechanisms enable context to influence judgment?
It would be illusory to believe that, in their responses to surveys, the subjects interviewed retrieve a pre-existing response, one 'tailor-made', even though certain subjects may have stored in their memories a pre-existing evaluation of the 'problem'. Attitudes and judgments are, on the contrary, products of the time when the question is asked, but also largely based on information stored in the long-term memory. Contextual effects may mainly be explained by the manner in which the subjects store the information (coding process (Bower *et al.*, 1981)), retrieve it (Tourangeau *et al.*, 1989) and finally integrate the information in the formulation of their judgment (Schuman and Presser, 1981). Tourangeau *et al.* (1989) report that it is mainly the information retrieval component that is influenced (and thus biased) by contextual variables in the generation of an overall judgment.

In fact, the generation of responses to attitudinal questions is not necessarily based on a systematic and rigorous appraisal of all the appropriate information. On the contrary, given the short time the subject has to answer (or more generally to adapt his/her behavior), response generation is rather based on rapid sampling of the information. This sampling may vary from one occasion to the next and the responses may thus change. In certain consumer choices, the resulting responses may not be optimal.

The quantity and type of information retrieved and used to formulate an overall assessment of an object depends (Feldman and Lynch, 1988; Moorthy, 1991) on:

- the availability of the information in the memory,
- the relative diagnosis of the information available,
- the relative accessibility of the information available.

The availability of the information mainly plays a role with respect to the quantity of information retrieved and used. For instance, this may be reflected in terms of the interviewee's 'expertise' with regard to the category of object evaluated: the more the interviewee is an expert, the higher the probability of having information available. For example, wine connoisseurs may have very different judgments than novice wine consumers. More generally, in the case of food evaluation by means of organoleptic tests, there is a clear difference between regular users of the tested food and non-users.

The diagnosis of an item of information has an influence on the selection of that information from other information in that it generally conditions the importance of the information in the construction of the subject's overall judgment. The diagnosis depends, for instance, on the quality of the information. The greater the subject's certainty with respect to the information, the more likely he/she is to call on it. The more pertinent the information seems, the more readily it will be used. An overall assessment may also reflect a comparative opinion vs. another object. In that case, an important item of information may not be used if the subject thinks that the information is not discriminant. Thus, the overall judgment may be constructed while discounting positive information on the tested product if that information is only differentiated in terms of negative information. The manner in which the subject perceives the task requested of him/her may also influence the diagnosis of the information and hence its use. For example, in the case of a hedonic test, if the subject wishes to please the interviewer, the subject may only select positive information, resulting in a more favorable overall judgment than that which the subject really perceives (social desirability bias).

Lastly, the accessibility of an item of information, like the diagnosis, determines whether it will be used in the construction of the overall judgment. Information accessibility depends on both long-term factors (the importance of the information and the frequency of its use) and short-term factors (recent activation, pertinence or relationship to other information that has already been activated).

Moreover, the number of items of information considered (availability) and their selection is a function of motivation and the conditions under which the information was integrated. The greater the perturbation of the integration conditions (short evaluation sequence, distractions, etc.), the less the diagnostic level will be involved in information selection.

In certain cases, information accessibility takes precedence over information diagnosis. Thus, a recently activated but non-pertinent item of information is liable to inhibit the recall and use of pertinent information that is less accessible in the overall judgment. The tendency to base judgment on the most accessible information has been termed 'availability heuristic' by Tversky and Kahneman (1973).

How are contextual effects exerted?

In the field of hedonic testing, and more generally attitudinal surveys, two types of contextual effects have been evidenced: assimilation and contrast. The two types of effects result from the level of interaction between contextual information and the characteristics of the product under evaluation.

The assimilation effect results from diffusion of the contextual value judgment toward target judgment (i.e. product judgment) (Deliza and MacFie, 1996). Numerous studies in the field of experimental psychology have observed that triggering a positive mood may induce more positive attitudes than induction of a neutral or negative mood (Bower, 1981; Deldin and Levin, 1986; Isen et al., 1978). Similarly, the influence of the subject's psychological state on the evaluation of a food has been the subject of a few investigations. Siegel and Risvik (1987), for instance, observed that the hedonic scores assigned to a food are higher when the subjects had previously formulated an opinion on a positive and casual matter than when the opinion was formulated on a more serious and negative topic. The impact of remuneration is also noteworthy. Subject remuneration induces higher scores and decreases the discrimination between tested products (Bell, 1993). The individual's internal state at the time when he/she tastes the tested food is thus of primary importance with respect to the attitude formulated. Contrast effects are the result of the opposite scenario, i.e. when the contextual values and the target do not match. A contrast effect is observed when the judgment does not reflect any relationship or a strictly negative relationship between the value attributed to the target object and the values attributed to the contextual stimuli accompanying the target. The role of expectations in contrast effects is well known. Expectation is defined by Cardello (1994) as the belief that the foods will, to a certain degree, be appreciated or not appreciated before they are eaten. For example, the expectation constructed by a consumer before eating a meal in a famous and luxury restaurant is probably very strong. If the food is finally rated less positively than what the consumer expected, the latter will tend to under-evaluate the food eaten in the restaurant with respect to the evaluation that he/she would have made in a less favorable environment.

Thus, positive contextual information can result in a negative judgment of the target!

8.2.2   Contextual variables that influence food choices and food liking in hedonic tests

As indicated above, the integrative process of hedonic judgment construction largely depends on the contextual information in which it is formulated. When it comes to the evaluation of food and eating pleasure, meal is a central element of context, as the meal itself includes many contextual variables. This section addresses the various studies that have investigated the influence of contextual variables involved in the construction of a hedonic judgment of a food and which may be the cause of differences in the results for tests conducted on the same products under standardized or more natural evaluation conditions.

Quantity of food eaten during tasting

The influence of the quantity proposed and eaten in hedonic tests has been observed in numerous laboratory tests. Thus, several authors (Bellisle *et al.,* 1988; Lucas and Bellisle, 1987; Monneuse *et al.,* 1991; Perez *et al.,* 1994; Zandstra *et al.,* 1999) have reported that high sugar or salt concentrations are more appreciated when the tasting involves a small quantity of samples than when the samples are consumed *ad libitum*. However, other studies have yielded contradictory results (Daillant and Issanchou, 1991; Popper *et al.,* 1989; Shepherd *et al.,* 1991). In addition, Hellemann and Tuorila (1991), Lähteenmäki and Tuorila (1994) and Popper *et al.* (1989) have observed that the hedonic scores obtained after *ad libitum* consumption are higher than the hedonic scores obtained after consumption of smaller quantities.

Hellemann and Tuorila (1991) and Tuorila *et al.* (1994) concluded their work by stating that hedonic judgments constructed on consumption of the usual portion of a food are more pertinent than judgments based on brief exposure. Although the good practices for CLT state that the portion of food presented and ingested is to be equivalent to the usual portion of the product (Köster, 1998), the principle is rarely applied. This is particularly challenging for the evaluation of meals with several component foods. The practice is generally compromised by the fact that several products are frequently tested during the same session and the participants cannot therefore eat realistic portions of all the products. On the contrary, HUT potentially enables the consumption of a realistic quantity of the product at each consumption occasion.

Tested food preparation and food environment

Individual preparation of the food is clearly involved in the formulation of the hedonic judgment and is unfortunately little taken into account in the setup of consumer test protocols under standardized conditions. The question of the value of acquiring a hedonic response on a food prepared in a standardized manner may be raised. The response may be difficult to interpret. The impact of food preparation on its assessment has been little investigated. Most of the studies show an improvement in food perception and liking when the latter are presented at a culturally appropriate temperature (Boulze et al., 1983; Cardello and Maller, 1982; Ryynanen et al., 2001; Zellner et al., 1988). The results obtained for soups or ice creams are not very surprising and the presentation of foods at an appropriate temperature (hot soup and cold ice cream) is currently well incorporated in CLT protocols. In addition, consumption temperature preferences may vary between individuals. This particularly applies to certain beverages, which may be drunk at room temperature or chilled, depending on individual preference. Taking the usual consumption temperature into account can then become more complicated, given that the presentation temperature has to be adjusted to individual preferences. The problem raised here particularly applies to the impact of individual preparation of tested foods on their appreciation. Matuszewska et al. (1997) compared three test procedures to evaluate margarine samples. Initially, the subjects were to spread each sample on slices of bread as they did at home. The second method consisted of tasting slices of bread already spread with 4 g of margarine. The third procedure consisted of a situation in which the margarine samples were tasted alone without bread. The results show that the various samples were better appreciated and discriminated between when consumed in accordance with an individual preparation protocol. Posri et al. (2001) also compared the appreciation of several samples of tea determined by three preparation conditions: an imposed preparation (all the samples were prepared with the same quantities of milk and sugar), an individual controlled preparation (the subjects chose their preferred quantities of milk and sugar and all the presentations were prepared accordingly) and a totally free preparation (the subjects were free to use the quantities they wished). Once again, a strong influence of tea preparation on liking was observed.

In addition to food preparation, the impact of the food environment (other meal items and companions foods) is also frequently neglected in the setup of consumer tests. Obviously, the impact of the assessment of the constituents of a meal on the overall assessment of the meal has been widely observed (Hedderley and Meiselman, 1995; Turner and Collison, 1988; Popper et al., 1989; Tuorila et al., 1990). In addition, in certain cases, although the studies are rare, the appreciation of a food may vary depending on whether it is eaten alone or with other food items, for example in the context of a meal (Eindhoven and Peryam, 1959; King et al., 2004). For instance, King et al. (2004) observed a wider divergence in the assessments of two pizzas when they were eaten in the middle of a meal than when they were eaten individually as in a conventional CLT. These results stress the importance of incorporating the 'meal' variable in the setup of hedonic test protocols for foods generally eaten in meals (King et al., 2007). This is rarely the case in industrial practice. In addition, in practice, a lot of meal research focuses on the combination of a main dish such as pasta (King *et al.*, 2007), pizza (King *et al.*, 2004) or chicken dish (Edwards *et al.*, 2003) and accompanying foods (salad and beverages). It should be noticed, however, that depending on the culture, such dishes may only be one component of a much more complex meal (e.g. antipasti, primo, secondo, contorni and dulci in the traditional Italian meal structure).

Consumption time of day
Taste test implementation at the appropriate times for product consumption is generally recommended (Köster, 1998). However, in contrast to other variables which are well controlled in CLT, tasting time is rarely standardized. The logistic constraint is in fact very strong. For example, it is generally impossible for survey institutes to hire a test room in a public place for just one or two hours. Thus, the CLT sessions are frequently run throughout the day at times when consumers are present. However, consumption time is an important contextual component. Birch *et al.* (1984) observed that the constituents of a breakfast are more greatly appreciated in the morning than in the afternoon, and the constituents of a dinner are more greatly appreciated in the afternoon than in the morning. However, Kramer *et al.* (1992) conducted a similar study without observing any influence of consumption time for the same food categories.

It should also be noted that certain foods, such as meal ingredients, are clearly associated with a consumption time, while other foods, snacks, may more readily be eaten at any time of day. Cardello *et al.* (2000) thus demonstrated a time-of-day effect for pizzas but not for cereals. In addition to the fact that a subject naturally feels more like eating pizza between noon and 2:00 pm than at 10:00 am, the influence of the time of day is also manifested by the participants' physiological state. The subject's physiological state appears to be an important determinant for the pleasure reported on consumption. Cabanac *et al.* (1968), for example, showed that a sweet taste may be particularly appreciated by a fasting subject but not by a subject in a state of satiety. Similarly, Laeng *et al.* (1993) found that subjects score sweet lemon beverages as less pleasant when tasted immediately after a meal. Hill (1974) and Bell (1993) in their respective studies also evidenced a significant effect of the subject's state of hunger on their preferences and hedonic scores, resulting in inferior discrimination between various tested products when the interviewees were hungry. Brunstrom *et al.* (1997) observed that the extent to which the mouth is dry influences the assessment of various beverages.

Social environment
The social environment has been shown to have a marked impact on the quantity of food eaten. One tends to eat more when eating takes place with other people than when eating alone. This effect is termed social facilitation (Berry *et al.*, 1985; De Castro, 1990). However, the phenomenon has not

been systematically verified (Feunekes *et al.*, 1995; Pliner *et al.*, 2003) and the opposite effect may even occur (Cardello *et al.*, 2000). It is probable that the social facilitation effects are due to the time spent eating, which is longer when an individual eats with others than when he/she eats alone. However, Clendenen *et al.* (1994) have shown that the relationship between the protagonists at the time of eating is a factor that significantly contributes to the social facilitation effect. Thus, eating in a familial environment is reported to be more sensitive to the social facilitation effect than eating in a friendly environment or eating with strangers.

Few studies have directly explored the link between the social environment and the degree of food liking. King *et al.* (2004) have shown that the liking of pizza significantly decreases when the pizza is eaten in a group setting, reflecting the opposite of the social facilitation phenomenon. In another study, King *et al.* (2007) did not observe any significant effect of a pleasant and social environment on the scores for various foods/beverages (lasagne, cannelloni, salad and iced tea).

The vast majority of meals are eaten in company. It is therefore legitimate to question the pertinence of tests in which contact between the various interviewees is prevented. However, imposing social contact between people who do not know each other is not necessarily a pertinent solution. In fact, Pliner *et al.* (2003) have shown that social facilitation only has a positive effect in naturally formed groups. In addition, interviewees who may not have any other subjects of conversation may tend to share their opinions on the tested foods in a non-natural manner. This opinion sharing may then influence the judgment of easily influenced people. Cardello and Sawyer (1992) showed that the appreciation of a food may be influenced by information on what other people think. Thus, the setup of a CLT in a social environment would necessitate a specific pre-recruitment phase in order to bring together only people who know each other; this can otherwise be achieved by asking people to bring in friends.

Scope for choosing the food tasted
Choice is present in most natural consumption situations. Thus, it may be supposed that the absence of the choice component in tests under standardized conditions is one of the causes of their poor predictive validity. It is to be noted that with home use tests, the subject can choose to consume the product when he/she wishes, but is nonetheless obliged to consume the test product and not a substitute. Pliner (1992) nonetheless draws attention to the fact that numerous natural situations involve components similar to the situation imposed in the tests. In particular, this is the case for family meals or meals with friends, in which, most of the time, the meal is put on the table without all the participants' prior agreement. Realistic eating situations thus do not necessarily involve a total absence of constraints.

Few studies have specifically addressed the relationship between the degree of freedom in food choice and the appreciation of that food. A few studies addressing monotony or boredom provide some information on the impact of the choice variable on the appreciation of regularly eaten foods. Zandstra *et al.* (2000) observed the time course of the hedonic score for three sauces obtained after several home tastings with different degrees of choice. Repeated consumption of the same sauce for 10 weeks resulted in a decrease in the scores due to the monotony effect. In contrast, the hedonic scores for sauces formulated by subjects who were free to choose the samples to be eaten over that period did not show the same decrease.

Kramer *et al.* (2001) have also observed that allowing military personnel to choose their meals in the field results in an elimination of the monotony effect generated when the same foods are imposed in the laboratory. The scope for choosing the food that one wants to eat is also reported to stimulate the quantity of product eaten (Beatty, 1982; Zandstra *et al.*, 2000). De Graaf *et al.* (2005) studied the impact of food choice on its appreciation. In the laboratory, the hedonic scores were lower for the imposed eating of various foods than when the foods eaten were pre-chosen from a list. Similarly, King *et al.* (2004) tested the impact of adding a choice of menu component in a test procedure conducted in a centralized location and social meal setting. The authors did not observe any influence of the choice variable on the overall assessment of the meal although the hedonic score for certain accompanying foods such as salad increased when those foods had been chosen.

With a view to improving the predictive validity of hedonic tests, De Graaf *et al.* (2005) compared the performances of data generated by a conventional test with those generated by a test providing scope for choice to predict the hedonic scores obtained in natural consumption settings. The authors observed that the laboratory test with scope for choice enabled enhanced prediction from the data obtained in naturalistic settings compared with that generated by laboratory tests with no scope for choice. However, once again, incorporating the choice variable in hedonic test protocols would require a supplementary budget due to the participation of a large number of people in order for all the tested products to be chosen and evaluated an equivalent number of times.

The various studies reviewed above illustrate the investigator's wish to take better account of the influence of tasting conditions on results in the context of consumer tests. Thus, it would now seem clear that consumers are to taste a 'normal' portion of the product before formulating an evaluation. Similarly, foods eaten at very specific times of day are generally to be tested at times appropriate for their consumption. In contrast, making the status of the other three factors (food environment, time of consumption and social environment) more natural in tests is more complex but affords prospects with respect to methodological developments. On the basis of comparisons of the results generated by various increasingly 'naturalistic' test protocols, King *et al.* (2007) point to the urgency of working, as a priority, on the scope for incorporating choice components and the conditions of a complete meal in test protocols before addressing what is probably a less important problem, that of the social environment.

8.3   Food testing under standardized or naturalistic tasting conditions

A number of researchers have addressed the influence of the testing environment overall on the hedonic perception of a food alone or a full meal. The following are to be distinguished:

- studies comparing the hedonic responses obtained in various naturalistic meal environments (different public places, home),
- studies comparing the hedonic responses obtained in a naturalistic tasting environment (public places, home) and in a test environment (laboratory, test room).

8.3.1   A hedonic response sensitive to the consumption site

The change in the appreciation of a given food eaten in different sites has been observed over several decades. Green and Butts (1945), for instance, showed that different constituents of the same meal are scored more severely for a precooked meal eaten in a plane than in a more conventional eating setting. The authors also observed that certain foods such as sandwiches and desserts are particularly popular when served in planes, while other foods such as potatoes are preferred in a more conventional meal setting. In the 1980s, Maller et al. (1980) observed, in the context of a satisfaction survey of hospital food, that the meals eaten by the patients alone in their rooms were more positively rated than the same meals eaten in the hospital refectory. The US Army research center Natick team studied the differences in Armed Forces military ration eating behavior during training maneuvers and in the cafeteria (Meiselman et al., 1988). The role of contextual variables in the interpretation of the difference in behavior depending on the consumption setting seems established. Thus, the monotony of the diet on maneuvers and the soldiers' physical and psychological condition explain why soldiers eat smaller quantities of their rations when on maneuvers than when in the cafeteria. Similarly, the consumption setting consisting of a restricted social environment, limited comfort or the inability to heat the ration may explain the difference in behavior. Subsequently, other studies were set up to compare the hedonic responses to a given food or meal eaten in various more conventional settings. For example, Miller et al. (1995) observed that steaks rated by consumers at home obtained lower appreciation scores than the same steaks rated in a restaurant. Meiselman et al. (2000) and Edwards et al. (2003) also evidenced an influence of the eating environment on the appreciation of a chicken dish (Edwards et al., 2003) or even a meal (Meiselman *et al.*, 2000) served in different public settings such as the company restaurant, a student cafeteria, a refectory in an old people's home, or a 4-star restaurant. The studies showed a ranking of the assessment scores as a function of the prestige of the setting. The studies thus illustrate assimilation effects given the consistency between the valence of the eating environment and the valence of the food judgment. Even though the foregoing studies have the advantage of investigating behavior in natural eating environments, they have the disadvantage of only interviewing populations used to those environments. The influence of the eating site is thus confused with the influence of the subjects' social and demographic characteristics.

The set of studies on the influence of evaluation conditions illustrates the complexity of the issue of generalizing a hedonic response obtained in a given environment to another environment.

8.3.2   A hedonic response sensitive to the artificial character of tasting conditions

Although the value of academic research with regard to elucidating eating behavior requires no further demonstration, the central question for the food industry is to determine whether testing their products under controlled test conditions (centralized location or laboratory) modifies consumers' perception and liking. This question has therefore been the subject of numerous studies whose main results are shown in Table 8.1. Particular attention will be paid to comparing the hedonic data obtained under controlled test conditions (laboratory or centralized location), qualified as artificial, with the hedonic data obtained in more naturalistic tests (home or test restaurant).

How do tasting conditions influence the hedonic responses obtained?
Table 8.1 shows that numerous studies have evidenced changes in results depending on the methodology of the test used. However, various result modification levels are observed:

- a change in the liking score for the foods,
- a change in the degree of discrimination between the foods compared,
- a change in the hedonic ranking of the various foods compared.

Some studies have reported a change in the score depending on the test conditions. Thus, for most of the studies, the scores obtained under naturalistic eating conditions are higher than the scores obtained under artificial conditions, irrespective of the product tested (Boutrolle *et al.*, 2007; De Graaf *et al.*, 2005; Hersleth *et al.*, 2003; King *et al.*, 2004, 2007; Meiselman *et al.*, 2000; Pound *et al.*, 2000; Shepherd and Griffiths, 1987). Even though the opposite phenomenon was observed in a few rare cases (Hellemann *et al.*, 1992), it would appear that the under-evaluation of tested foods under controlled conditions is currently firmly established.

Numerous hypotheses may be formulated with a view to interpreting the

Table 8.1 Results of studies comparing hedonic measurements obtained under artificial and natural consumption conditions

Study	Test products	Method compared	Principal conclusions with regard to method comparison
Miller et al. (1955)	18 soup pairs	Lab vs. Home use Preference	Divergences but results overall the same Products very different: identical results Products very similar: different results (lab more discriminant than home)
Murphy et al. (1958)	6 sardines	Lab vs. In-Home Ranking Degree of acceptance	Same results even though the appreciation deviations were slightly higher at home than in the lab
Calvin and Sather (1959)	Different product categories, 15 pairs	Lab vs. Home use Preference Hedonic score	Hedonic score: same results for 11 out of 15 pairs (but home scores often higher than lab scores) Preference: identical results for 12 out of 15 pairs Different results: home more discriminant than lab (but same ranking)
McDaniel and Sawyer (1981)	6 whiskies tested against a standard 6 pairs	Lab vs. Home use Hedonic score	Different results: home more discriminant than lab (but the same rankings)
Shepherd and Griffiths (1987)	3 eggs	Lab vs. Home use Hedonic score vs. ideal (taste and color of egg yolk)	Taste score: same results Yolk score: different results (home scores higher than lab scores but same ranking)
Hellemann et al. (1992)	1 complete meal	Lab vs. Home use vs. Cafeteria Hedonic score	Different results: lab scores higher than home and cafeteria scores
Daillant-Spinnler and Issanchou (1995)	3 cream cheeses with increasing fat contents	Lab vs. In-home Hedonic score vs. ideal	Low-fat products: same results High-fat products: lab scores higher than home scores Lab test more discriminant than home test.
McEwan, (1997)	10 flavoured crackers	Lab vs. Centralized location vs. Home use Hedonic score	Different results: change in the most appreciated product between-panel score differences but no higher average score

Reference	Study	Results
Meiselman et al. (2000)	Chicken dish	Lab vs. Restaurant vs. Cafeteria	Different results: the restaurant scores were higher the lab scores, which were higher than the cafeteria scores
		Overall and partial hedonic scores	
Pound et al. (2000)	3 bars of chocolate	Lab vs. Centralized location vs. Home use	Same results: the home and lab scores were higher than the centralized location scores but the difference was not significant
		Hedonic score	
Posri et al. (2001)	4 teas	Centralized location vs. Home use (long term)	Different results: change in ranking
		Hedonic score	
Hersleth et al. (2003)	8 wines	Lab vs. Reception room	Different results: the reception room scores were higher than the lab scores
		Hedonic score	
Kozlowska et al. (2003)	5 apple juices	Lab vs. Home use	Different results for one product: the unsweetened fruit juice scored better at home than in the lab (home slightly more discriminant than lab)
		Hedonic score	
King et al. (2004)	Tea, pizza and salad (2 samples per product)	Centralized location vs. Restaurant	Tea: the restaurant scores were higher than the room scores, restaurant more discriminant than centralized location
		Hedonic score	Pizza: same results
			Salad: the restaurant scores were higher than the centralized location scores; centralized location more discriminant than restaurant
De Graaf et al. (2005)	Multiple products (dishes, snacks and meal constituents)	Lab vs. In the field (military maneuver)	Snack products: same results
		Hedonic score	Dishes or meal constituents: different results with change in ranking
Hersleth et al. (2005)	6 cheeses	Lab vs. Home use	Same results: same product ranking but home slightly more discriminant than the lab
		Hedonic score	
Boutrolle (2007)	5 different studies:	Centralized location vs. Home use	Overall, home use scores were higher than centralized location scores (except for iced teas products)
	2 salty cheese crackers	Hedonic score	3 studies with different statistical conclusion:
	2 iced teas		Crackers and iced teas: higher product discrimination in home use test
	3 beverages with fruit juice and milk		

Table 8.1 cont'd

Study	Test products	Method compared	Principal conclusions with regard to method comparison
(King et al. (2007)	2 sparkling waters 2 fermented milks beverages		Beverages with fruit juice and milk: higher product discrimination during the central location test 2 studies with similar conclusion (sparkling water and fermented milk beverages)
(King et al. (2007)	Lasagna, cannelloni, bread, ice tea, salad	Centralized location vs. Restaurant Hedonic score	Different results: the restaurant scores were higher than the centralized location scores
Petit et al. (2007)	2 iced teas	Lab vs. Reception room Hedonic score	Same results: the two products were well discriminated in both tests

change in score depending on eating environment. Among the hypotheses, one relates to the fact that natural evaluation conditions are, *a priori*, more pleasant than the evaluation conditions in a controlled test situation. The tasting rooms are frequently poorly decorated and not always very comfortable. In addition, the interviewees usually have little time to spare given that they are frequently not recruited beforehand but on an on-the-spot basis. In contrast, evaluation in natural settings is supposed to be conducted in a more comfortable environment (or a more familiar environment for home studies) with no evaluation time-related stress. The setting may thus promote the participant's mood and overall well-being. Thus, the high scores assigned to foods under natural eating conditions may only illustrate an assimilation effect due to the participant's well-being. However, few authors have specifically studied the impact of the comfort of the test site on hedonic measurements and those who have done so (Bonin *et al.*, 2001; King *et al.*, 2004) did not really observe any change in the hedonic scores following an improvement in the comfort of the test situation. Very formal evaluation conditions may also induce the subject to perceive the tasting session as an examination and thus be much more demanding with regard to the tested products than the subject would have been in a more natural consumption setting.

Lastly, test conditions that are more or less conducive to consumption of the tested product may also contribute to the difference in scoring. For example, an evaluation conducted at home enables the consumer to eat the product when he/she wants and under appropriate and optimal conditions. Thus, the simple fact of enabling natural conditions of product consumption could promote the participant's pleasure in consuming the product and thus promote a favorable opinion on the tested food.

The phenomenon of under-scoring products under controlled conditions, vs. the scores obtained under natural consumption conditions, is not, at first sight, particularly worrying given that organoleptic tests are set up to compare the performances of several recipes. Thus, if the deviations between the scores remain the same for the products, the statistical conclusions of the tests and, ultimately, the strategic decisions may not be affected even though the score levels are different. However, in the industry, certain development studies on new products or range extensions use action standards based, among other things, on the necessity of obtaining a liking score greater than an imposed limit (frequently 7 on a scale from 1 to 10 in French companies). Obviously, the choice of a score cutoff as an indicator for rejection or acceptance of a product is open to criticism. Whatever the case may be, the scoring shift as a function of evaluation conditions in the context of use of that type of action standard is embarrassing. It would appear obvious that a score of 7 obtained in a CLT does not represent the same degree of liking as a score of 7 obtained, for example, in a HUT. The foregoing thus necessitates revising the action standards or, at the least, differentiating them as a function of test conditions.

In addition to the impact of test conditions on the level of scoring, other studies have clearly evidenced a change in the degree of discrimination between tested products. This may have real consequences on the strategic decisions based on the tests. However, there does not appear to be any consensus with regard to the most discriminant method. While Miller *et al.* (1955) and Daillant-Spinnler and Issanchou (1995) have reported that controlled test conditions may generate more discriminant results than naturalistic conditions, Calvin and Sather (1959), Hersleth *et al.* (2005), Kozlowska *et al.* (2003) and McDaniel and Sawyer (1981) have reported the opposite. King *et al.* (2004) and Boutrolle (2007) observed both types of result, depending on tested product category (for King *et al.* (2004), while teas were better discriminated in a restaurant, salads were better discriminated in a centralized location). The inability to conclude that one methodology is more discriminant than another upsets the preconceived idea of enhanced discriminant potential for tests under controlled conditions related to sequential monadic product presentation in the course of a single session.

Lastly, the results whose consequences are most important in terms of strategic decisions are those of the studies that have demonstrated a change in the hedonic ranking of products depending on the test conditions used (De Graaf *et al.*, 2005; McEwan, 1997; Posri *et al.*, 2001). In the context of screening different recipes of a given product pre-launch, an incorrect ranking would result in incorrect recipe selection. However, in the authors' experience, the methodological comparisons conducted in the context of comparison of the assessment of two or three similar samples have never yielded such a result.

What risk is taken when a test is set up under controlled rather than under naturalistic consumption conditions?
As indicated above and in Table 8.1, the methodological comparisons reported have not all shown evidence of an influence of tasting context on the statistical conclusions of the tests. Focusing on the studies that have conducted several comparisons based on different food categories, it is frequently observed that, in the context of a given study, the methodological comparison results are not the same, although the experimental procedures used were strictly identical. It would therefore appear that the impact or absence of impact of evaluation conditions on the statistical conclusions of the tests are greatly related to the type of food tested. It may, for example, be supposed that evaluation conditions have a very strong influence on the assessment of foods normally eaten in meals and no influence on foods normally eaten as snacks (De Graaf *et al.*, 2005; Boutrolle *et al.*, 2007). In addition, even in the context of foods usually eaten at mealtimes, King *et al.* (2004) observed that the influence of the evaluation conditions on the results seemed more important for accompanying foods (tea and salad) than for the main dish (pizza). The usual mode of eating the various categories of food may therefore be one of the principal reasons for the variability of the influence of evaluation conditions on test conclusions. Finally, the natural consumption mode for snacks is

fairly similar to the consumption conditions imposed in a test under controlled conditions. Consumption is rapid and solitary, with no accompanying foods. In contrast, the differences between the controlled test conditions and natural consumption conditions may be much more marked in the case of, food usually eaten as part of a meals.

Comparative Table 8.2 is intended to help investigators evaluate the risks they incur when they set up a CLT instead of HUT for a specific food category. Table 8.2 may thus be used as an aid in deciding on the advisability of a CLT instead of an HUT. If the usual use of the tested food meets several of the criteria in the left-hand column, the investigator can set up a CLT with less risk of obtaining different results from those generated by an HUT than if tested product use were mainly listed in the right-hand column. It is important to note that the table is not exclusive. A single characteristic in the right-hand column applying to product consumption mode is insufficient reason to definitely exclude a CLT.

Table 8.2 Product consumption mode characteristics that may induce differences in CLT vs. HUT results

Contextual variables	Modalities more often associated to 'snack-like' consumption	Modalities more often associated to 'part of a meal' consumption
Quantity consumed	Standardized (individual packaging), smaller quantities	Adjustable, higher quantities
Mean duration of consumption	Short	Long
Consumption time	Not specific: throughout the day	Specific: morning, noon, night, etc.
Consumption settings	Not specific	Specific: family meal, party, picnic, drinks, etc.
Social environment	Solitary consumption	Consumption with others
Dietary environment	No other food or beverage consumed	Combination of foods and/or beverages
Product preparation	Not personalized	Personalized: food temperature, seasoning, etc.
Product handling	Standardized: direct from the packaging (e.g. can of drink, snack bar)	Personalized: use of silverware or chopsticks for example
Consequences for CLT vs. HUT	Low risk to obtain different results	High risk to obtain different results

8.4   Central location tests versus home use tests

8.4.1   Central Location Text (CLT)

From the foregoing, it will be more readily understood why it is not rare to observe an inconsistency between the judgments formulated by the same subjects during different tasting sessions for the same product. This phenomenon is particularly problematic when one considers that the primary objective of consumer tests is to predict future consumer behavior and hence the fate of the product on the market. But what can be said to a product manager with regard to the probable product positioning on the market when the product has repeatedly generated different assessments and is probably not responsible for the variation in the results? Earthy *et al.* (1997) define the context as all the undesirable variables included in a study which influence its results. The definition in itself reflects the investigators' powerlessness in the face of the influence of contextual variables on evaluation judgments. Thus, the objective of standardizing and controlling test conditions is to control the influence of contextual variables in order to measure only the effect of the product sensory properties on the hedonic response. However, the standardization of food tasting conditions raises the problem of the accuracy of the responses obtained. In the field of behavioral sciences, the concept of accuracy is similar to the criterion of external validity (or predictive validity) of a method, which enables the investigator to generalize from the results obtained on the test population to the population of interest. In the more specific framework of hedonic measurement in consumer tests, the external validity concerns the 'ecological' component and refers to the conditions under which the measurement was obtained. Thus, a high external validity (implying a high ecological validity) enables generalization of the hedonic measurement obtained for a sample of consumers under test conditions to the target population naturally exposed to the product. However, how can one imagine that the hedonic responses obtained under such artificial and hence frequently inappropriate conditions may be predictive of a judgment constructed during real consumption? By setting up tests under standardized and hence artificial conditions, the investigators neglect the determinant contribution of contextual factors in the evaluation of food products, particularly in meal settings. King *et al.* (2004) observed that excluding those variables from research may oversimplify the participants' eating experience, thus providing incomplete and, in some cases, misleading results. It will thus be understood that the external validity of a measurement obtained under consumption conditions extremely remote from reality may be called into question (Drifford *et al.*, 1995).

Despite the external validity problem, the CLT has numerous advantages which underlie its success with manufacturers (in France, 70% of food product consumer tests are CLT). CLT are easy to set up in a relatively short time and with a reasonable budget. Moreover, the controlled evaluation conditions enable generation of more precise data that are easier to interpret than those generated by a HUT. Tuorila and Lähteenmäki (1992) consider that while laboratory situations are perhaps artificial, they constitute a context for studying dietary behavior that enables control over variables that are frequently mutually confounding in studies in natural settings.

8.4.2   Home Use Test (HUT)

Although the standardized and artificial nature of the conditions under which hedonic evaluations are generated in CLT has been called into question, the disadvantages of tests in natural settings are also recognized (Meiselman, 1992; Mela *et al.*, 1992; Pliner, 1992; Rolls and Shide, 1992). As Rolls and Shide (1992) point out, the problem with tests in natural settings is that most of the methods are not precise, do not enable experimental manipulation and are, in addition, very onerous. The budgetary and logistic limitations of HUT thus currently restrict their routine use by manufacturers and, even more so, by academic research. On the basis of estimates for tests conducted on consumer food products, the cost of a HUT in France is generally 50% higher than that of a CLT. The difference in budget mainly derives from the pre-recruitment of subjects, their remuneration and the interviewers' home visits. Moreover, the quantities of product to be supplied for the test are much greater given that each consumer must have sufficient product for a week of consumption. This also contributes to increasing the cost of a HUT relative to a CLT.

With regard to logistics, numerous aspects of the HUT sometimes render implementation very difficult or even impossible. The principal difficulty of the HUT resides in anonymizing the tested products. In CLT, the blind is generally achieved by masking the initial packaging (for individual packagings) or presenting the sample in a neutral receptacle (plate, bowl, glass). In contrast, for HUT, product handling by the subjects frequently necessitates totally repackaging the products in a neutral packaging. Unfortunately, numerous manufactured food products cannot undergo repackaging without their organoleptic properties being impaired.

These difficulties explain why blind HUT is little used for marketed products. The HUT is also associated with an effectiveness problem due to the fact that the HUT cannot be set up with a set of more than two or three products in succession, given the time interval necessary between each pair of products evaluated. The last major disadvantage of HUT is the weak reactivity of the method. Setup and implementation generally require at least one month (for two products) before the initial results can be obtained, while a CLT may be set up so that the initial results are available in a week.

8.4.3   'Conflicting desiderata' for manufacturers

Tables 8.3 and 8.4 show the reasons for preferring CLT and those for preferring HUT. The tables show that the advantages of one method are generally the

Table 8.3 The reasons for preferring CLT

Advantages of CLT	Limitations of HUT
Low cost	Expensive test
Results obtained rapidly	Difficult and long to implement
Tests can be conducted in a mobile home to facilitate changing site and fast contact with different consumers	Consumer pre-recruitment necessary on the basis of databases. Consumers less 'naive' than those recruited on-the-spot
Relative good control conditions (environment, product preparation and presentation, understanding of instructions, etc.): data precise and easy to interpret	No environmental control: large variability in responses due to the variability in consumption conditions
Compliance with the instructions	Requires intense quality control to check compliance with the instructions (Was the product consumed by the right people?)
Few missing data	High probability of missing data
Frequently easier for comparative approaches: several products can be tested during a session	Evaluation of numerous products impossible

Table 8.4 The reasons for preferring HUT

Advantages of HUT	Limitations of CLT
The product is prepared and consumed under normal use conditions	The subject is outside of his/her usual consumption context: preparation, tasting time, social and dietary environment, etc.
The assessment obtained is based on repeated consumption of a large quantity of product	The opinions are collected after a brief contact with the product
Numerous data on product use can be generated and used in the interpretation of the hedonic information	The information on uses and attitudes collected are only declarative and may not be what the consumer experiences at home
No bias due to tasting several products in a short time interval: reliable implicit comparison with the subject's usual product	Disturbance in the subject's references which may lead to non-pertinent judgments

counterpoint of the limitations of the other. Thus, the standardized conditions for CLT make the method practical and inexpensive while generating precise and easily interpreted hedonic measurements even though those measurements may potentially be less predictive of reality. Although more expensive, HUT with the more naturalistic tasting conditions enables hedonic measurements that are potentially more predictive of reality, but little controlled and thus more difficult to interpret with regard to development operations. The contrast between the precision of CLT and the supposed validity of HUT termed 'conflicting desiderata' by Brinberg and McGrath (1985) is thus a problem with which manufacturers are continually confronted when selecting one methodology rather than the other.

8.5   Improving food testing to enhance integration of eating/drinking situation variables?

Currently, the sponsors of consumer tests expect alternative methodological solutions enabling the '*conflicting desiderata*' to be overcome. The last section will therefore address methodological approaches to enhancing both the pertinence and precision of the hedonic data generated by organoleptic tests while considering the budgetary and logistic imperatives of industry. Two modified-CLT solutions are considered with a view to integrating the usual conditions of tasting in the hedonic measurement:

- proposing an environment for consumption that is more appropriate for evaluation of the type of food considered,
- acquiring hedonic information that is a function of the various scenarios for food consumption.

8.5.1   More appropriate tasting conditions

Certain authors have tried to improve the conditions of tasting in CLT so as to make them more appropriate to the usual consumption of the product (Posri *et al.*, 2001; De Graaf *et al.*, 2005; King *et al.*, 2004). But, ultimately, few studies have truly used test protocols that make the CLT tasting situation more naturalistic and appropriate. King *et al.* (2004), tried to modify, in a sequential manner, the status of some contextual variables (meal, choice of foods tasted, social environment, etc.) so as to make the tasting conditions increasingly naturalistic. Accordingly, the present authors chose to simulate two specific eating situations in a CLT setting: a breakfast on the one hand and a more general meal event (lunch/dinner) on the other hand. The two studies were conducted differently. The first compared the hedonic ratings of two recipes of sparkling water with data acquisition in either a classical CLT or modified CLT mode. The modified mode was specially designed to take a lunch/dinner context into account. The second study compared the hedonic ratings of two recipes of crispy bread products with data acquisition in either classical CLT mode or CLT breakfast mode.

Meal context for assessment of liking for sparkling water products
Two groups of women were recruited. One group ($n = 161;$) took part in the conventional CLT design and the other ($n = 160;$) in the experimental meal CLT, which was designed as follows. Each subject was sequentially presented with the two sparkling water products during two different meal sessions,

either at lunchtime or dinnertime. Consumers sat alone at their table to avoid social interaction, which would have been unrealistic because the participants did not know each other. For both meal sessions, a complete meal consisting of tabbouleh, vegetable tart with green salad and fresh fruit salad was served. The participants were served the items sequentially in plain dishes and were instructed to eat as much as they wished and to at least taste all the constituents of the meal including the blinded sparkling water provided on table. After the meal, the participants were asked to score, from 1 to 10 (1 = dislike very much; 10 = like very much), their overall likings for each meal constituent. The participants came back one week later, at the same time of day, for the second meal. They were then told that all the meal constituents had been slightly modified compared with the first session. In reality, however, the same meal was served, but was accompanied by the second sparkling water product.

Figure 8.1 shows the mean hedonic scores for the two groups. The test results clearly depend on the method used. With the conventional CLT, product A was liked significantly more than product B ($p < 0.001$), while there was no significant difference in liking in the meal-CLT ($p = 0.212$). With regard to the impact of the tasting conditions on overall liking, it will be observed that the mean hedonic score obtained for both products was higher for the meal-CLT than for the conventional CLT ($p < 0.001$).

The meal-CLT situation differs in many respects from the conventional CLT situation and a number of those differences may have affected the hedonic scores reported. The lower scores for the products in artificial situation tests than in natural settings is consistent with the observations in previous studies already mentioned. The meal environment may have been especially conducive to water drinking, which may have led to enhanced satisfaction. King *et al.* (2004) have already shown an individual positive influence of the

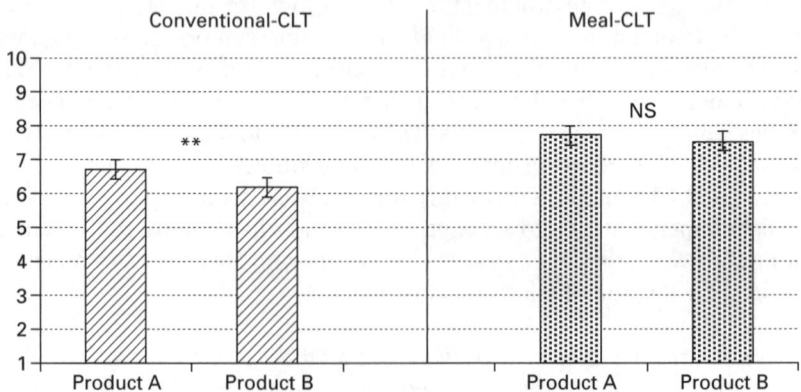

Fig. 8.1 Mean overall hedonic scores for the two sparkling water products in conventional CLT and meal-CLT sessions; *t*-test results: NS, not significant; *, *p*-value <5%; **, *p*-value <1%.

meal environment on the liking scores for beverages. The positive effect on digestive function induced by the sparkling water products was probably experienced during the meal-CLT, but could not have been during the conventional CLT. Naturally, the comfortable environment of the meal situation together with the remuneration and free meals could also have enhanced the subjects' overall well-being and hence scores.

It is noteworthy that the evaluation of the sparkling water products in the context of a full meal not only influenced the hedonic score levels, but also the conclusion inferred from the test, with a non-significant difference between the two products. The conventional CLT conditions, without a meal environment, may have magnified a difference in liking. The difference probably did not exist when the water products were consumed during the meal. Interestingly, diagnostic questions collected after the evaluations tell us that the participants perceived one product as being 'too intense' in terms of sparkle and bubble size, either for the standard CLT or the meal-CLT. Although the 'too intense' aspect of that product could be expected to be less perceived when the water was drunk with food, this finding shows that, even during a meal, the participants still differentiated between the two water products. However, it is striking that in spite of the perception of product B as 'too intense' we observed no difference in overall liking for the two products. Although the hypothesis cannot be verified, it may be that because of the meal environment, there was a change in the predominance of the sensory drivers of liking. Sparkle and bubble size, for example, might not have been so important for consumers while eating, whereas saltiness and bitterness (which were rated as just about right for both products) were more salient. Such information is crucial, given the fact that, in France, sparkling water products are drunk with a meal in 50% of cases (Usage and Attitude survey carried out by Danone marketing research department). Thus, in order to enhance the prediction of liking for such products in real settings in which they are usually consumed as part of a meal, consumer tests under meal conditions are to be conducted.

Breakfast context for assessment of liking for crispy bread products
Two crispy bread recipes (product A *vs.* product B) were evaluated. The two types differed with respect to a number of attributes: shape (round for A and oval for B), texture (A less hard) and taste (A less sweet and more salty). Sixty-one subjects participated in a conventional CLT experiment and 64 subjects in the breakfast-CLT, which was designed as follows.

The subjects were pre-recruited and invited to attend a breakfast session, between 8:00 and 11:00 am, near their workplace. They were asked not to eat before the session. When they arrived, participants were asked to take a tray and help themselves to various breakfast items from a buffet. They were asked to choose their breakfast as if they had been at home and were thus free to choose their usual food items (coffee, hot milk, cold milk, chocolate, tea, orange juice, butter, strawberry jam, apricot jam, honey and chocolate spread). The subjects then sat down at individual tables (in France, the majority of breakfast events occur without social interaction). The interviewers then presented the first crispy bread recipe. Participants were instructed to prepare the bread as they usually did, but to do so in a consistent manner for all the crispy bread samples eaten during the session. Thus, both products were evaluated under the same preparation conditions: type of beverage, type of spread products and dunking (prevalent in France). Participants were asked to score their overall liking for the bread tasted (1 = dislike very much; 10 = like very much). The second crispy bread recipe was then presented with the same instructions. After the overall ratings, participants were asked partial appreciation questions on taste and shape.

Figure 8.2 shows that the difference between products A and B in terms of overall liking was not significant for either protocol (conventional CLT: $p = 0.111$; breakfast-CLT: $p = 0.121$). The consumption conditions during the breakfast-CLT clearly illustrated that eating conditions (100% beverage drinking and spread use) differed quite substantially from those in the conventional CLT (no beverage drinking or spread use). However, the tasting conditions did not influence the differences between the mean overall liking scores. The overall hedonic scores were not influenced either. The products did not obtain a lower score in the standardized situation than in the breakfast situation. This might be explained by the fact that the participants had to wake up earlier than usual to attend the early morning session and were not financially compensated for that. Similarly, the influence of the more naturalistic tasting condition of the restaurant-CLT set up by King *et al.* (2007) was not sufficient to modify the level of hedonic scores vs. the conventional CLT.

Interestingly, further analysis of the breakfast-CLT data shows that dunking behavior influenced the difference between the overall likings for the two products. The data for the 'dunkers' (38% of the subjects) showed a significantly greater liking for product B ($p = 0.004$) while the 'non-dunkers' liked the two products equally well ($p = 0.814$). This was not observed in the conventional CLT for the self-reported 'dunkers'.

The partial hedonic scores also enabled insights as shown in Fig. 8.2. In the conventional CLT, the taste of product B was significantly more liked than that of product A ($p = 0.009$). In contrast, during the breakfast session, the tastes of the two products were liked equally well ($p = 0.962$). Thus, spreading and potential dunking may mask differences in taste and influence taste liking differences. The relevance of focusing on the taste of neutral basic products may be questionable in that, under natural conditions, very few people eat crispy bread without spreads. The responses on shape are also informative. During the conventional CLT, participants did not report a difference of liking for the two shapes ($p = 0.282$), while, under the breakfast conditions, the round shape (product A) was significantly less liked ($p < 0.001$) than the oval shape (product B). Participants may have interpreted the same question in two different ways: in the conventional CLT, the liking for the shape was probably evaluated with regard to visual appearance; in the

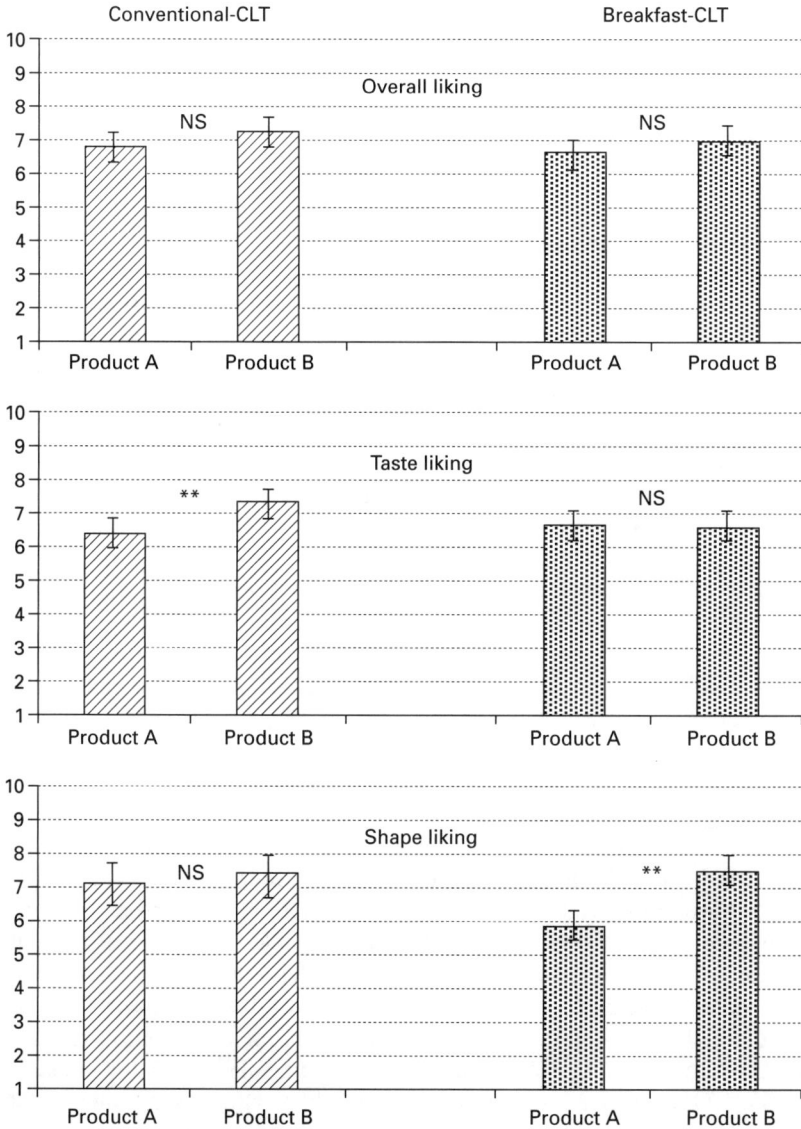

Fig. 8.2 Mean overall and partial hedonic scores for two bread products in conventional CLT and breakfast-CLT sessions; *t*-test results: NS, not significant; *, *p*-value <5%; **, *p*-value <1%.

breakfast situation, participants probably evaluated the convenience of the shapes (e.g. for spreading and dunking). The latter information is important for product developers seeking to improve specific convenience aspects.

In conclusion, the analysis of the diagnostic data revealed that the preparation of the products (spreads) and their use (dunking behavior) had a strong influence on the hedonic perception of product taste and shape. The breakfast-CLT demonstrated the impact of dunking behavior on overall liking data. Given that in France 50% of people dunk bread in hot beverages at breakfast, those consumers' opinions should be taken into account when breakfast bread products are tested.

8.5.2   Hedonic response measured in various imaginary consumption settings

Various attempts to incorporate more realistic tasting conditions in CLT designs have succeeded in integrating specific contextual factors in consumer test designs. However, this reductive approach (singling out contextual factors) presupposes determination of the most natural consumption setting for the tested food product. However, food products are rarely consumed in only one kind of eating situation, which complicates the prediction of 'real life' hedonic responses based on hedonic data obtained from one-situation CLTs. Furthermore, as Köster (2003) has pointed out, perceptual situations are not exclusively defined by objective criteria; they are also defined by the subjects' conscious and subconscious intentions. Everyday life is a world of meanings rather than one of objective facts. Even though a situational CLT would be a more meaningful approach to measuring hedonic responses than a conventional CLT, many of the contextual variables that can be manipulated are objective properties of the context. In real life, they may have different meanings for different people. Thus, Köster (2003) suggests the use of a situation-oriented ('deductive') approach based on evoked rather than physical situations in order not to disturb the subjects' natural perceptions of the situation. The idea is based on measuring the appropriateness of food products to situations, as had already been proposed by Schutz et al. (1977) in the 'item by use appropriateness' method. This method comprised presenting the subjects with a list of foods and list of possible uses (time of day, site, occasion, physiological state, person, etc.) and having the subjects score the appropriateness of each food product for the set of uses proposed. However, few studies have tested the utility of that type of measurement in real food testing (Cardello et al., 2000; Cardello and Schutz, 1996; Lähteenmäki and Tuorila, 1995, 1997). Moreover, the studies address the appropriateness of a product for a situation that is only described in a few words (e.g., 'when I'm eating in front of the television'). That approach to the question is open to criticism since it does not necessarily enable the subject to become involved and really 'project' him/herself into the situation in question. One way of improving situation appropriation by the subject is to make use of his/her autobiographic memory. The aseptic conditions of CLT tasting indeed prevent activation of the autobiographic memory and it is therefore necessary to help the subject recover the memory of consumption events. The approach has already been used by Bonin et al. (2001) to study the influence of the priming of a past positive or negative consumption event with respect to the choice of foods. Thus, Köster (2003) proposes that imaginary situations are evoked and primed with the help of auditory scenarios. The scenarios are defined as short stories about hypothetical characters in specific circumstances. The interviewee is asked to respond to the situation. While that approach has long been used in attitudinal surveys, particularly in the social sciences (Finch, 1987; Nossanchuck, 1972; West, 1982) and marketing (Bitner, 1990; Folkes, 1984; Surprenant and Solomon, 1987), it is only just beginning to be used in the field of food preference measurement (Henry, personal communication; Jaeger and Meiselman, 2004).

The efficacy of the approach was tested on 240 women by collecting consumption intentions in various situations evoked through audio scenarios with a view to comparing the organoleptic performances of two types of salty cheese-flavored crackers. Although very similar, the two products nonetheless differed in terms of appearance, taste and texture. The authors' aim was to verify whether the sensory differences conditioned a different consumption intent in six consumption situations: week-day drinks before the meal, weekend drinks before the meal, picnic, snacking while relaxing, snacking while working, and snacking during transport.

The various scenarios were compiled in the first person singular in a relatively simple style and gave information on a certain number of contextual variables such as the physical environment, but also on the subject's interior condition. For example, the scenario used to illustrate the weekend drinks before the meal situation was as follows: 'It's the long awaited time for drinks before dinner at the weekend. I can at last spend some time with my loved ones. As usual, the conversation is lively and drinks last longer than expected. After a few drinks, it's probably time to have something to eat but nobody seems to want to sit down to the meal. Despite the big meal that's on the way, I take a dish of crackers and offer them around without forgetting to help myself'. Each subject thus listened to six scenarios one after the other and formulated his/her intent to eat the tested cracker in the situation (from 1 'no, certainly not' to 10 'yes, absolutely') together with the frequency of that situation in everyday life.

As observed in Fig. 8.3, the intent to consume recipe B was higher than that for recipe A in almost all situations, but the difference was only significant for the weekend drinks before the meal scenario. Moreover, the intent-to-consume scores for the two products were significantly higher for the weekend drinks than for the other situations ($p < 0.05$). When taking into account all eating situations with respect to the frequency of those situations in the subjects' lifestyles, it is observed that recipe B (mean weighted intent score = 7.16) was significantly favored as compared with recipe A (mean weighted intent score = 6.66). Interestingly, this result could not be anticipated from the sole overall liking scores for the two products that were not significantly different (from a previous conventional CLT study conducted with 240 women). The intent-to-consume by situation data show very high standard deviations reflecting the diversity of the participants with regard to the intent to eat

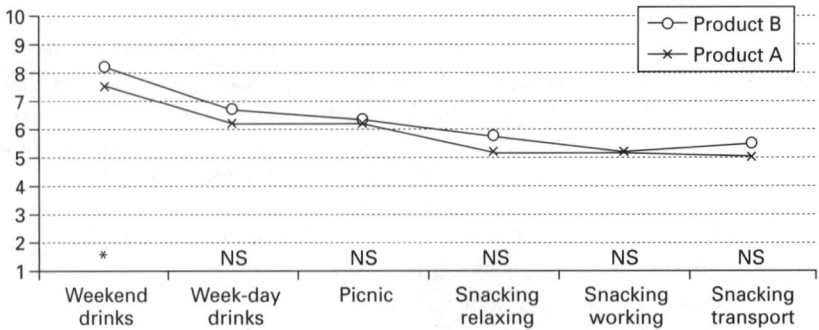

Fig. 8.3 Mean intent-to-eat scores for two cracker products by situation. *T*-test results: NS, not significant; *, *p*-value <5%; **, *p*-value <1%.

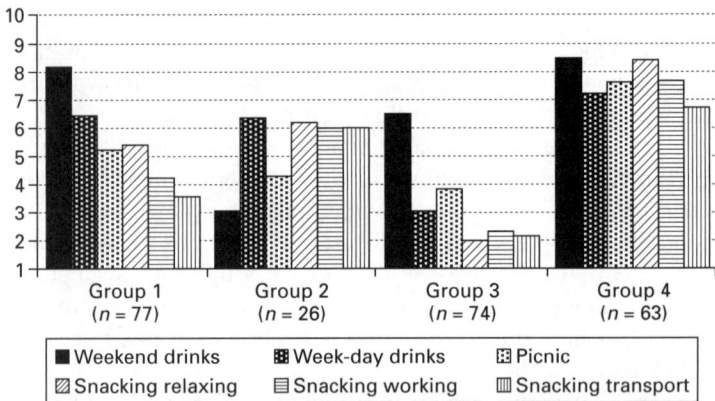

Fig. 8.4 Mean situation frequency profiles for the four groups of subjects identified.

crackers in the various situations. This affords the possibility of investigating the data by subject group on the basis of the responses formulated with the situation occurrence frequencies in the participants' everyday lives. Four different cracker consumption profiles were thus determined and illustrated in Fig. 8.4. Group 1 consisted of subjects who prefer, for the food type in question, consumption situations associated with a social event (drinks before the meal on weekends and picnics) or a solitary but relaxing event (drink before the meal during the week or relaxed snacking). The situations in which consumption is more utilitarian or functional (work, transport) are less associated with the food category. Group 2, which was smaller, consisted of the consumers who do not eat crackers in an environment with marked social interaction (weekend drinks before the meal and picnics) but only when alone (drinks before the meal during the week and snacking). Group 3 consists of the subjects who mainly eat the food type during drinks before the meal at the weekend. In contrast, group 4 consists of participants who eat crackers regularly in all of the proposed situations. Figure 8.5 illustrates the

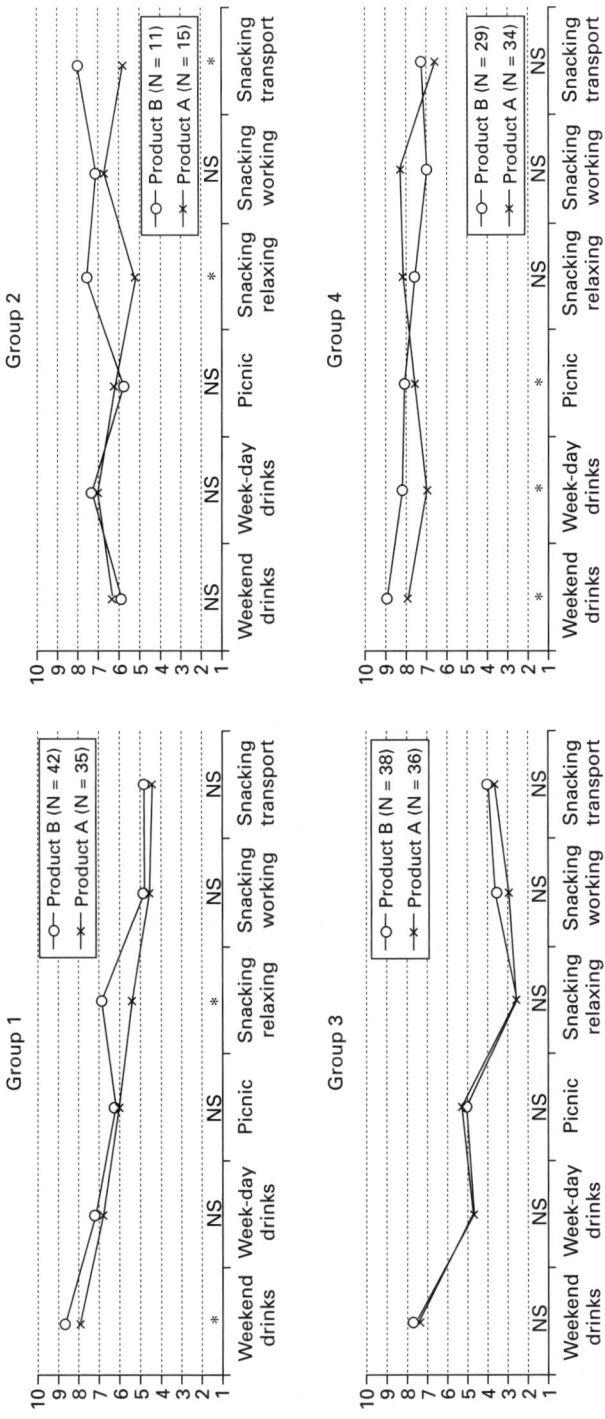

Fig. 8.5 Mean intent-to-eat scores for the two cracker products by situation in the light of the consumer typologies. *T*-test results: NS, not significant; *, *p*-value <5%; **, *p*-value <1%.

analysis of the intent-to-consume data for the two recipes in the light of the consumer typologies and shows that the intent-to-consume profiles differ depending on the use type. It will be observed that the data by typology generate results that are more discriminant than those obtained for the panel overall, except in the case of group 3 for which the two recipes are overall associated with the same intent-to-consume scores, irrespective of situation. In addition, a particularly interesting weak product–situation interaction was observed for group 4, illustrating the fact that a consumer may not want the same sensory properties in a given food category when making different uses of the food.

The variation in tested product consumption intents, depending on the situation, observed in the study points to the value of taking into account food product use in consumer tests. In conclusion, the authors stress the fact that product use is to be taken into account not only in study design but also in the interpretation of the data generated. A real consumer typology based on food eating habits was evidenced both in the present study and in the CLT analyzed above. An understanding of the typology seems essential in order to elucidate consumer preferences.

8.6   Future trends

In line with the numerous criticisms leveled at conventional test methods, in the field of sensory analysis and consumer studies, researchers are almost unanimous with respect to the need to find more reliable hedonic assessment methods. However, while researchers are raising numerous questions with regard to the type of hedonic scale to be proposed or the optimum manner of analyzing consumer data, few proposals for measurement protocols that are radically different from those used conventionally have been retrieved. In addition, research on eating behavior is frequently based on the theory of rational action (Fishbein and Ajzen, 1975), which considers that people make rational choices and are able to explain the reasons underlying their behavior. However, over the last 30 years, psychological research has greatly progressed and raised the issue that behavior is not always under conscious control.

This key component of behavior is currently little taken into account in sensory methodological research which continues to propose attitudinal measurements via direct questioning. We should now turn to the methods used in other disciplines that focus more closely on behavior. That multidisciplinary approach is coming into being as reflected by the increasingly frequent interventions of sociological, ethnological or experimental economics researchers who are invited to present their manner of elucidating consumer behavior at sensory analysis congresses. In addition, research programs relating to the building of experimental cafeterias in order to study food choices under real consumer conditions are underway. The programs are encouraging and have the immense advantage of grouping together research units from different fields of the study of consumer behavior (economics, management, sensory analysis, marketing, etc.).

It would appear that many more doors need to be opened in order to discover new manners of conducting hedonic studies in order to take into account the meal context. However, attention is drawn to the difficulty of adapting the methods reported in the literature to industrial applications. While the proposed methods are frequently tested using caricatural product spaces exploiting products from different categories, consumer test methodology research should more often address the contribution of the methods in the context of the issues encountered in corporate settings, i.e. comparison of products that are more similar at the sensory level. Moreover, rather than being surprised by and criticizing companies that do not exploit the researchers' methodological recommendations, it is also indispensable to remain highly aware of industrial applications with all their budgetary and logistic constraints when developing new methods.

8.7   References

Beatty W W (1982), 'Dietary variety stimulates appetite in females but not in males', *Bulletin of the Psychonomic Society*, **19**, 212–214.

Bell R (1993), 'Some unresolved issues of control in consumer tests: the effects of expected monetary reward and hunger', *Journal of Sensory Studies*, **8**, 329–340.

Bellisle F, Giachetti I and Tournier A (1988), 'Determining the preferred level of saltiness in a food product: a comparison between sensory evaluation and actual consumption tests', *Sciences des Aliments*, **8**, 557–564.

Berry S L, Beatty W W and Klesges R C (1985), 'Sensory and social influences on ice cream consumption by males and females in a laboratory setting', *Appetite*, **6**, 41–45.

Birch L L, Billman J B and Richards S S (1984), 'Time of day influences food acceptability', *Appetite*, **5**, 109–116.

Bitner M J (1990), 'Evaluating service encounters: the effects of physical surroundings and employee responses', *Journal of Marketing*, **54**, 69–82.

Bonin D, Chambres P and Bernard P (2001), 'Influence du contexte sur la perception des produits alimentaires', in Urdapilleta I, Ton Nu C and Saint-Denis C, *Traité d'évaluation sensorielle*, Paris, Dunod, 223–232.

Boulze D, Montastruc P and Cabanac M (1983), 'Water intake, pleasure and water temperature in humans', *Physiology and Behavior*, 97–102.

Boutrolle I, (2007), '*Measurement of consumers' liking for food. Current practices and methodological developments*', Thesis, AgroParisTech, France.

Boutrolle I, Delarue J, Arranz D, Rogeaux M and Köster E P (2007), 'Central location test vs. home use test: contrasting results depending on product type', *Food Quality and Preference*, **18**, 490–499.

Bower G H (1981), 'Mood and memory'. *The American Psychologist*, **36**, 129–148.

Bower G H, Gilligan S G and Monteiro K P (1981), 'Selectivity of learning caused by affective states', *Journal of Experimental Psychology: General*, **110**, 451–473.

Brinberg D and McGrath J E (1985), *Validity and the research process*, Beverley Hills, Sage.

Brunstrom J M, Macrae A W, Qannari E M, Dijksterhuis G B, Hunter E A and MacFie H (1997), 'Mouth state: a nuisance variable in preference tests?', *Food Quality and Preference*, **8**, 349–352.

Cabanac M, Minaire Y and Adair E (1968), 'Influence of internal factors on the pleasantness of a gustative sweet sensation', *Behavioral Biology*, **1**, 77–82.

Calvin L D and Sather L A (1959), 'A comparison of student preferences panels with a household consumer panel', *Food Technology*, **13**, 469–472.

Cardello A V (1994), 'Consumer expectations and their role in food acceptance', in MacFie H and Thomson D M H, *Measurement of food preferences*, London, Blackie Academic and Professional, 253–297.

Cardello A V and Maller O (1982), 'Acceptability of water, selected beverages and foods as a function of serving temperature', *Journal of Food Science*, **47**, 1549–1552.

Cardello A V and Sawyer F M (1992), 'Effects of disconfirmed consumer expectations on food acceptability', *Journal of Sensory Studies*, **7**, 253–277.

Cardello A V and Schutz H G (1996), 'Food appropriateness measures as an adjunct to consumer preference', *Food Quality and Preference*, **7**, 239–249.

Cardello A V, Schutz H G, Snow C and Lesher L L (2000), 'Predictors of food acceptance, consumption and satisfaction in specific eating situations', *Food Quality and Preference*, **11**, 201–216.

Chiva M (1992), 'Les aspects psychologiques des conduites alimentaires', in Dupin H, Cup J L and Malewiak M I, *Alimentation et nutrition humaine*, Paris, ESF, 417–442.

Clendenen V I, Herman C P and Polivy J (1994), 'Social facilitation of eating among friends and strangers', *Appetite*, **23**, 1–13.

Daillant B and Issanchou S (1991), 'Most preferred level of sugar: rapid measure and consumption test', *Journal of Sensory Studies*, **6**, 131–144.

Daillant-Spinnler B and Issanchou S (1995), 'Influence of label and location of testing on acceptability of cream cheese varying in fat content', *Appetite*, **24**, 101–105.

De Castro J M (1990), 'Social facilitation of duration and size but not rate of the spontaneous meal intake of humans', *Physiology and Behavior*, 1129–1135.

De Graaf C, Cardello A V, Kramer F M, Lesher L L, Meiselman H L and Schutz H G (2005), 'A comparison between liking ratings obtained under laboratory and field conditions: the role of choice', *Appetite*, **44**, 15–22.

Deldin P J and Levin I P (1986), 'The effect of mood induction in a risky decision-making task', *Bulletin of the Psychonomic Society*, **24**, 4–6.

Deliza R and MacFie H (1996), 'The generation of sensory expectations by external cues and its effects on sensory perception and hedonic ratings: a review', *Journal of Sensory Studies*, **11**, 103–128.

Drifford V, Lemmet J, Bernard P, Laporte V and Chambres P (1995), 'Acceptance tests and memory of the context of consumption: analysis and experimentation', *Viandes et Produits Carnes*, **16**, 107–111.

Eagly A H and Chaiken S (1993), *The psychology of attitudes*, Orlando: Harcourt Brace & Company.

Earthy P J, MacFie H and Hedderley D I (1997), 'Effect of question order on sensory perception and preference in central location trials', *Journal of Sensory Studies*, **12**, 215–237.

Edwards J, Meiselman H L, Edwards A and Lesher L L (2003), 'The influence of eating location on the acceptability of identically prepared foods', *Food Quality and Preference*, **14**, 647–652.

Eindhoven J and Peryam D R (1959), 'Measurement of preferences for food combinations', *Food Technology*, **13**, 379–382.

Fishbein M and Ajzen I (1975), *Belief, attitude, intention and behavior: an introduction to theory and research*, Reading, Mass: Addison-Wesley.

Feldman J M and Lynch J G (1988), 'Self-generated validity and other effects of measurements on belief, attitude, intention and behavior', *Journal of Applied Psychology*, **73**, 421–435.

Feunekes G I, De Graaf C and Van Staveren W A (1995), 'Social facilitation of food intake is mediated by meal duration', *Physiology and Behavior*, **58**, 551–558.

Finch J (1987), 'The vignette technique in survey research', *Sociology*, **21**, 105–114.

Folkes V S (1984), 'Consumer reactions to product failure: an attributional approach', *Journal of Consumer Research*, **14**, 398–409.

Green D M and Butts J S (1945), 'Factors affecting acceptability of meals served in the air', *Journal of the American Dietetic Association*, **21**, 415–419.

Hedderley D I and Meiselman H L (1995), 'Modelling meal acceptability in a free choice environment', *Food Quality and Preference*, **6**, 15–26.

Hellemann U and Tuorila H M (1991), 'Pleasantness ratings and consumption of open sandwiches with varying NaCl and acid contents', *Appetite*, **17**, 229–238.

Hellemann U, Mela D J, Aaron J I and Evans R E (1992), 'Role of fat in meal acceptance', in *Pangborn Symposium*, Helsinki, Finland.

Hersleth M, Mevik B H, Naes T and Guinard J X (2003), 'Effect of contextual factors on liking for wine – use of robust design methodology', *Food Quality and Preference*, **14**, 615–622.

Hersleth M, Ueland O, Allain H and Naes T (2005), 'Consumer acceptance of cheese, influence of different testing conditions', *Food Quality and Preference*, **16**, 103–110.

Hill S W (1974), 'Eating responses of humans during dinner meals', *Journal of Comparative and Physiological Psychology*, **86**, 652–657.

Isen A M, Shalker T E, Clark M and Karp L (1978), 'Affect, accessibility of material in memory, and behavior: a cognitive loop?', *Journal of Personality and Social Psychology*, **36**, 1–12.

Jaeger S R and Meiselman H L (2004), 'Perceptions of meal convenience: the case of at-home evening meals', *Appetite*, **42**, 317–325.

King S C, Weber A J, Meiselman H L and Lv N (2004), 'The effect of meal situation, social interaction, physical environment and choice on food acceptability', *Food Quality and Preference*, **15**, 645–653.

King S C, Meiselman H L, Hottenstein A W, Work T M and Cronk V (2007), 'The effects of contextual variables on food acceptability: a confirmatory study', *Food Quality and Preference*, **18**, 58–65.

Kozlowska K, Jeruszka M, Matuszewska I, Roszkowski W, Pikielna N and Brzozowska A (2003), 'Hedonic tests in different locations as predictors of apple juice consumption at home in elderly and young subjects', *Food Quality and Preference*, **14**, 653–666.

Köster E P (1998). 'Les épreuves hédoniques', in SSHA. *Evaluation sensorielle, manuel méthodologique*, Paris, Lavoisier, Tec et Doc. 182–203.

Köster E P (2003), 'The psychology of food choice: some often encountered fallacies', *Food Quality and Preference*, **14**, 359–373.

Kramer F M, Rock K and Engell D (1992), 'Effect of time of day and appropriateness on food intake and hedonic ratings at morning and midday', *Appetite*, **18**, 1–13.

Kramer F M, Lesher L L and Meiselman H L (2001), 'Monotony and choice: repeated serving of the same item to soldiers under field conditions', *Appetite*, **36**, 239–240.

Laeng B, Berridge-Kent C and Butter C M (1993), 'Pleasantness of a sweet taste during hunger and satiety: effects of gender and "sweet tooth"', *Appetite*, **21**, 247–254.

Lähteenmäki L and Tuorila H M (1994), 'Liking for ice cream measured with three procedures: side-by-side, after consumption and single samples', *Journal of Sensory Studies*, **9**, 455–465.

Lähteenmäki L and Tuorila H M (1995), 'Consistency of liking and appropriateness ratings and their relation to consumption in a product test of ice cream', *Appetite*, **25**, 189–198.

Lähteenmäki L and Tuorila H M (1997), 'Item-by-use appropriateness of drinks varying in sweetener and fat content', *Food Quality and Preference*, **8**, 85–90.

Lucas F and Bellisle F (1987), 'The measurement of food preferences in humans: do taste and spit tests predict consumption?', *Physiology and Behavior*, **39**, 739–743.

Maller O, Dubose C N and Cardello A V (1980), 'Consumer opinions of hospital food and foodservice', *Journal of the American Dietetic Association*, **76**, 236–242.

Matuszewska I, Barylko-Pikielna N, Szczecinska A and Radzanovska J (1997), 'Comparison of three procedures for consumer assessment of fat spreads: Short report', *Polish Journal of Food and Nutrition Science*, **6**, 139–142.

McDaniel M R and Sawyer F M (1981), 'Preference testing of whiskey sour formulations: magnitude estimation versus the 9-point hedonic', *Journal of Food Science*, **46**, 182–185.

McEwan J A (1997), 'A comparative study of three product acceptability trials', *Food Quality and Preference*, **8**, 183–190.

Meiselman H L (1992), 'Methodology and theory in human eating research', *Appetite*, **19**, 49–55.

Meiselman H L, Hirsch E S and Popper R D (1988), 'Sensory, hedonic and situational factors in food acceptance consumption', in Thomson D M H, *Food acceptability*, London, Elsevier, 77–87.

Meiselman H L, Johnson J L, Reeve W G and Crouch J E (2000), 'Demonstrations of the influence of the eating environment on food acceptance', *Appetite*, **35**, 231–237.

Mela D J, Rogers P J, Shepherd R and MacFie H (1992), 'Commentary. Real people, real foods, real eating situations: real problems and real advantages', *Appetite*, **19**, 69–73.

Miller P G, Nair J H and Harriman A J (1955), 'A household and a laboratory type of panel for testing consumer preference', *Food Technology*, **9**, 445–449.

Miller M F, Hoover L C, Cook K D, Guerra A L, Huffman K L, Tinney K S, Ramsey C B, Briitin H C and Huffman L M (1995), 'Consumer acceptability of beef steak tenderness in the home and restaurant', *Journal of Food Science*, **60**, 963–965.

Monneuse M O, Bellisle F and Louis-Sylvestre J (1991), 'Responses to an intense sweetener in humans: immediate preference and delayed effects on intake', *Physiology and Behavior*, **49**, 325–330.

Moorthy K S (1991), 'Measuring overall judgments and attribute evaluations: overall first vs. attributes first', Technical Working Paper, Marketing Science Institute.

Murphy E F, Clark B S and Berglund R M (1958), 'A consumer survey versus panel testing for acceptance evaluation of Maine sardines', *Food Technology*, **12**, 222–226.

Nossanchuck T A (1972), 'The vignette as an experimental approach to the study of social status; an exploratory study', *Social Science Research*, **1**, 107–120.

Perez C, Dalix A M, Guy-Grand B and Bellisle F (1994), 'Human responses to five concentrations of sucrose in a dairy product: immediate and delayed palatability effects', *Appetite*, **23**, 165–178.

Petit C and Sieffermann J M (2007), 'Testing consumer preferences for iced-coffee: Does the drinking environment have any influence?', *Food Quality and Preference*, **18**, 161–172.

Pieron H (1974), *La sensation*, Paris: Presse Universitaire de France.

Pliner P (1992), 'Let's not throw out the barley with the dishwater: comments on Meiselman's: "Methodology and theory in human eating research"', *Appetite*, **19**, 74–75.

Pliner P, Bell R, Kinchla M and Hirsch E S (2003), 'Time to eat: the impact of time and social facilitation on food intake', in *The 5th Pangborn sensory science symposium: a sensory revolution*, Boston, USA.

Popper R D, Maller O and Cardello A V (1989), 'Hedonic ratings and consumption as measures of preference for a salted entree', *Chemical Senses*, **14**, 739.

Posri W, MacFie H and Henson S (2001), 'Improving the predictability of consumer liking from central location test in tea', in *The 4th Pangborn sensory science symposium: a sense odyssey*, Dijon, France.

Pound C, Duizer L and McDowell K (2000), 'Improved consumer product development. I. Is a laboratory necessary to assess consumer opinion?', *British Food Journal*, **102**, 810–820.

Rolls B J and Shide D J (1992), 'Both naturalistic and laboratory-based studies contribute to the understanding of human eating behavior', *Appetite*, **19**, 76–77.

Rozin P and Tuorila H M (1993), 'Simultaneous and temporal contextual influences on food acceptance', *Food Quality and Preference*, **4**, 11–20.

Ryynanen S, Tuorila H M and Hyvonen L (2001), 'Perceived temperature effects on microwave heated meals and meal components', *Food Service Technology*, **1**, 141–148.

Schuman J and Presser S (1981), *Questions and answers in attitude surveys: experiments in question form, wording and context*, New York: Academic Press.

Schutz H G, Moore S M and Rucker M H (1977), 'Predicting food purchase and use by multivariate attitudinal analysis', *Food Technology*, **29**, 50–64.

Shepherd R (1985), 'Factors influencing food choice', *Nutrition and Food Science*, **96**, 10–11.

Shepherd R and Griffiths N M (1987), 'Preferences for eggs produced under different systems assessed by consumer and laboratory panels', *Lebensmittel Wissenschaft und Technologie*, **20**, 128–132.

Shepherd R, Farleigh C A and Wharf S G (1991), 'Effect of quantity consumed on measures of liking for salt concentrations in soup', *Journal of Sensory Studies*, **6**, 227–238.

Siegel S F and Risvik E (1987), 'Cognitive set and food acceptance', *Journal of Food Science*, **52**, 825–826.

Surprenant C F and Solomon M R (1987), 'Predictability and personalization in the service encounter', *Journal of Marketing*, **51**, 86–96.

Tesser A (1978), 'Self-generated attitude change', in Berkowitz L, *Advances in experimental social psychology*, San Diego, Academic Press, 289–338.

Tourangeau R, Rasinski K A, Bradburn N M and Dandrade R (1989), 'Belief accessibility and context effects in attitude measurement', *Journal of Experimental Social Psychology*, **25**, 401–421.

Tuorila H M, Lehtovaara A and Matuszewska I (1990), 'Sandwiches and milk with varying fat and sodium contents: what is the best combination?' *Food Quality and Preference*, **2**, 223–231.

Tuorila H M and Lähteenmäki L (1992), 'When is eating "real"? Response to Meiselman', *Appetite*, **19**, 80–83.

Tuorila H M, Hyvoenen L and Vainio L (1994), 'Pleasantness of cookies, juice and their combinations rated in brief taste tests and following ad libitum consumption', *Journal of Sensory Studies*, **9**, 209–216.

Turner M and Collison R (1988), 'Consumer acceptance of meals and meal components', *Food Quality and Preference*, **1**, 21–24.

Tversky A and Kahneman D (1973), 'Availability: a heuristic for judging frequency and probabilities', *Cognitive Psychology*, **5**, 207–232.

West P (1982), 'Reproducing naturally occurring stories: vignettes in survey research,' Working Paper, Aberdeen: MRC Medical Sociology Unit.

Zandstra E H, De Graaf C, Trijp H C M and Van Staveren W A (1999), 'Laboratory hedonic ratings as predictors of consumption', *Food Quality and Preference*, **10**, 411–418.

Zandstra E H, De Graaf C and Van Trijp H (2000), 'Effects of variety and repeated in-home consumption on product acceptance', *Appetite*, **35**, 113–119.

Zellner D A, Stewart W F, Rozin P and Brown J M (1988), 'Effect of temperature and expectations on liking for beverages', *Physiology and Behavior*, **44**, 61–68.

Part IV

Eating together and alone

9

A table for one: the pain and pleasure of eating alone

**P. Pliner, University of Toronto Mississauga, Canada, and
R. Bell, Natick Soldier Research Development and Engineering
Center, USA**

Abstract: The prevalence of eating alone and individuals' thoughts, feelings, and behavior when doing so is discussed. Such data as are available indicate that adults do a fair amount of it, while young people eat alone much less frequently. Commensal eating is deeply embedded in cultural consciousness and eating alone is an anomalous behavior. For most people a solitary meal is a highly undesirable situation and often is not considered to be a meal at all. Although individuals typically eat less when they eat alone than they do when they eat with others, for some people, eating alone provides an opportunity to escape public scrutiny and to eat as much as desired. Finally, eating alone is associated with negative nutritional outcomes in both the young and the old.

Key words: eating alone, solitary eating, solo eating, commensal eating, solitary meals, social facilitation.

9.1 Introduction

This chapter is about eating alone. We begin by noting that, in the context of a book on meals, eating alone – a solitary meal – could be considered to be an oxymoron. Sociality is an integral part of meals, and, as we will see, definitions and discussions of meals, both academic and lay, typically include reference to their social nature. An 'eating occasion', to use Douglas' (1972) terminology, is typically not considered to be a meal if it is a solitary occasion. Social eating is the expected state of affairs – the norm. Additionally, eating alone is not a topic about which one finds a large amount of research or theorizing, and we believe this follows from the foregoing point. Indeed, much of the available discussion of and data describing eating alone exist only as they provide a contrast or counterpoint or control for the more usual social eating.

In this chapter, we will address several aspects of eating alone. First, we will examine some of the data on the prevalence of eating alone. It has been suggested that, as a result of changes in demographics and patterns of daily activities, people are eating more meals alone. Some have bemoaned the demise of the family meal, implying that individual family members, no longer eating at the family dinner table, are eating alone in front of the microwave or the television, while others argue that it is alive and well (see chapters on family meals by McIntosh, Dean, Torres, Anding, Kubena, and Nayga, by Fjellström, and by Chrzan in this volume). We will also discuss people's thoughts and feelings about eating alone and some of the behavioral consequences and correlates of doing so. Before beginning, we must acknowledge that eating alone is not a unitary phenomenon. It involves a huge range of situations and many different kinds of people, whose thoughts, feelings, and behavior might be extremely disparate. Thus, among many others, the solitary eater might include: the elderly widow, taking most of her meals alone; the business traveler, eating in her hotel room so as to avoid a solitary restaurant meal (see Jonsson's discussion in this volume of eating alone in restaurants); the office worker, eating lunch at his desk; the child returning from soccer practice after family dinner time; the early riser, stopping for a Starbuck's coffee and muffin to eat in the car on the way to work; the new college graduate, living in his first apartment. Given the relative dearth of published data on eating alone, we will supplement this chapter with some unpublished data of our own on solitary eating.

9.2   How frequently do people eat alone, and is the frequency increasing?

This is not an easy question to answer. As noted above, people eat alone in all kinds of circumstances. In order to be useful, such data would have to be fairly comprehensive, including individuals of many ages and living arrangements and would have to enquire about eating in different circumstances (e.g., at home and in restaurants) and about different meals on different days of the week and in different parts of the world. As it happens, we were unable to find any comprehensive dataset and will report the snippets of data we were able to find, beginning with some of our own. We conducted one study (Pliner and Bell, unpublished data) in which we asked our participants the questions: "How frequently have you eaten alone, in a restaurant, at lunch time?" and "How frequently have you eaten alone, in a restaurant, at dinner time?" Our participants were 168 individuals, mostly in their early 20s, about half of them students. In a second study, we asked about frequency of eating dinner at home alone and dinner in a restaurant alone. In this study, we tested 90 people in the same age group, nearly all students. We note first that among both groups eating dinner alone was relatively rare – even at home. Ten percent of those in the second sample had never eaten dinner home alone, and the modal response was 'once or twice'. Eating dinner alone in a restaurant was even rarer – two thirds of participants had never done so. In the first sample, a third of the participants reported having never eaten lunch alone in a restaurant and 70% of them had never eaten dinner alone in a restaurant. Thus, for these relatively homogeneous groups of young men and women (most of them students living at home with their families), eating alone is a relatively rare event.

Next, we turn to other sources of data. A recent study by Sobal and Nelson (2003) examined meal patterns in individuals aged 18 and above, living in a single county in the USA and consisting of ethnically and socioeconomically diverse rural, suburban, and city dwellers. The participants were asked with whom they 'most often' ate the three major meals of the day – with 'alone' as one of the response options. Eating breakfast alone was reported by 58% of respondents, eating lunch alone by 45%, and eating dinner alone by 19%. Further analyses revealed that 14% reported eating all three meals alone. Using time budget data for a group of Flemish respondents and distinguishing between week days and weekend days, Mestdag (2005) found that on weekdays people ate the day's 'first eating event' alone about 27% of the time, the 'second eating event' alone about 8% of the time, and the 'third eating event' alone about 9% of the time. These figures were lower on the weekend, with the percentages being 11, 6, and 3%, for the three eating events, respectively. There are fairly substantial differences in incidence of eating alone in these two datasets and it is not clear what factor(s) might account for them; however, one very obvious difference is location – the US vs. Europe.

There is an extensive literature (some of which will be described later) on the effects of family meals on various outcomes (see also the three chapters on family meals in this volume); it is relevant for present purposes because many of the studies provide data on the proportion of children and teens eating dinner with their families. To justify including these data in a section on the incidence of eating alone, we cite a survey in which 12–17 year olds, living in the US, were asked what they usually did for dinner when they were *not* dining with their families; the modal response was that they eat at home by themselves (CASA, 2007). Thus, the most common alternative to family dinner, at least for teens, is eating alone. In a study of six European countries, 13 and 15-year olds reported how frequently they ate dinner with their families; nearly 80% reported doing so every day or most days (Zaborskis *et al.*, 2007). Similarly, in a group of American children aged 9–14, 83% reported eating family dinner every day or most days (Gillman *et al.*, 2000); another American survey of 6–17 year olds found that about 75% ate a family meal on all or most days (Child Trends Databank, 2003). Not surprisingly, the data differ according to the age of the child – the older the child, the lower the frequency of participating in family meals (Zaborskis *et al.*, 2007; Gillman *et al.*, 2000; Child Trends Databank, 2003), and boys eat family meals more frequently than girls (Zaborskis *et al.*, 2007; Larson *et al.*, 2007). In summary, although the data on incidence of eating alone that we were able to obtain are quite limited and in many ways not comparable, it appears that, at least for young people living with their families, including children, teens, and young adults, and especially at dinnertime, it is not that common. Sobal and Nelson (2003), questioning a group of individuals who for the most part are older, found a somewhat higher incidence,

Is the incidence of eating alone increasing? It is widely assumed that this is true; however, few data exist to enable us to evaluate this assumption. One reason for eating alone is living alone. At least among elderly Americans, the correlation between living alone and eating alone is high (Torres *et al.*, 1984), and people are increasingly living alone. In the USA over the 35-year period from 1961 to 2006, the number of one-person households has exactly doubled from 13.3 to 26.6% (US Census Bureau, 2006). Similarly, Canadian census data reveal that, in the period 1986–2006, the number of persons living alone increased by 50% (Statistics Canada, 2008). So, people are certainly living alone more frequently than before. What about eating alone? Amato *et al.* (2007), examining changing marital patterns from 1980 to 2000, noted a decline in the frequency with which respondents reported 'almost always' sharing the main meal of the day with their spouse. Of course, we do not know whether those individuals who were no longer eating with their spouses were eating alone. Mestag's (2005) time budget data, described earlier, actually included information from two cohorts, one surveyed in 1988 and the other in 1999. Although there were some meal occasions on which solitary eating had increased significantly, for at least one (Sunday lunch) solitary eating had actually declined and, in most cases, there was no change. In the case of family meals, one survey reports a small increase over the period 1996–2007 (51 to 59%) in the proportion of teens reporting eating with their families at least five nights a week, although this change is not tested for statistical significance (CASA, 2007). In contrast, Mestdag and Vandeweyer (2005) report a sharp and statistically significant decrease in the amount of *time* spent at family meals between 1966 and 1999 from 51 to 27 minutes and the share of time spent with family that occurred at meals declined in the same period from 38 to 25%. Thus, it is difficult to know whether people are eating alone more frequently than they have in the past. Although the debate on the status of the family meal continues (e.g., Murcott, 1997), reports of its death seem greatly exaggerated, given some of the data we have cited.

9.3   What do people think about eating alone?

Nearly any definition, description, or discussion of meals by anthropologists or sociologists involves some aspect of their social nature. As Sobal (2000, p. 119) put it, '[m]eals are social events, as well as food events'. Makela (2000, p. 10) noted that the '…sociability of eating, the fact that a meal is shared with other people, is often considered a necessary feature of meal definition'. Rotenberg (1981, p. 26) described a meal as a 'planned social interaction centered on food'. Much of the academic work focuses on the social functions of meals, sometimes characterizing them as social events that create meaning for the participants or reflect and/or communicate social realities. Douglas (1972), for example, has dealt with meals on this level, distinguishing between the social implications of meals and another type of food event, drinks. At the latter one might entertain strangers, acquaintances, or even workmen, while '...meals are for family, close friends, honored guests' (p. 66). Thus, meals signal the difference between intimacy and distance. Appadurai (1981) has written extensively about the semiotic functions of food and meals in Hindu South Asia, arguing they can serve two diametrically opposed social communicative functions: to homogenize the participants who participate or to heterogenize them. Charles and Kerr's (1988) treatment of family meals describes them as defining family social roles, while other writers describe family meals as a means of forging social bonds (e.g., Pliner and Rozin, 2000), as an important vehicle for the socialization of the child (Blum-Kulka, 1997; Chrzan, this volume; Fjellstrom, this volume; McIntosh, 1996), and as symbols of social coherence (Brown, 1984).

Although there is no dearth of academic theorizing about the social nature of meals, there are few data on layperson's ideas. In order to understand how laypeople understand the social nature of meals, we (Pliner and Bell, 2004) developed a questionnaire in which we presented very brief written 'scenarios' involving eating. One of the scenarios was '...you are eating alone...'. Participants (a group of soldiers (mostly male, mean age = 24) and a group of students (mostly female, mean age = 20)) rated, on a five-point scale, how likely it was that this eating occasion was a snack, breakfast, lunch, and dinner. Their ratings indicated that participants in both groups thought it was very likely that the eating occasion was a snack ($M = 4.2$) and much less likely that it was any of the three meals, particularly dinner ($m = 2.1$). When the scenario indicated that '...you are eating with one other person...' participants were about equally likely to consider it to be a snack or one of the three meals. Finally, when the scenario indicated that '...you are eating with two or more others', participants gave relatively low ratings to the probability of a snack ($m = 2.6$) and much higher ratings of the likelihood that it was a meal, particularly dinner ($m = 4.0$). What these data seem to be saying is that if one is alone, one is unlikely to be eating a meal and, conversely, if one is accompanied by others, one is likely to be eating a meal. Thus, laypeople appear to share the professional academic view that meals are social occasions. Of course, these particular groups of laypeople (students and soldiers) live in communal situations in which meals tend to be eaten commensally. What we do not know is whether other groups, such as those living (and typically eating) alone would respond in the same way.

9.4   How do people feel about eating alone?

Here, academic theorists have little to say, and we are forced to rely on the ideas of writers and philosophers. Interestingly, from these perspectives, eating alone can be either a blessing or a curse. One the one hand, we have sentiments like the ones below, clearly revealing that eating alone is not a good thing – in the first instance because it is uncivilized, in the second because it is unpleasant, and in the third because it is pathetic.

'We should look for someone to eat and drink with before looking for something to eat and drink, for dining alone is leading the life of a lion or wolf.' (Epicurus)

'There is no joy in eating alone.' (The Buddha)

'Sadder than destitution, sadder than a beggar is the man who eats alone in public. Nothing more contradicts the laws of man or beast, for animals always do each other the honor of sharing or disputing each other's food.' (Jean Baudrillard, French philosopher)

Reflecting the view that eating alone is a negative experience is the flood of information available in the popular media; many magazines and newspapers, from the *Wall Street Journal* to *Gourmet* to *Runner's World* have recently featured articles on eating alone. Advice and information about eating alone abound on the internet, where a web-based newsletter devoted to solitary dining (www.SoloDining.com) '…serves up solo dining savvy…' and offers tips and solutions to '…the challenge of eating alone...'. In an interview (Moeller, 1998), the editor of this newsletter expresses her surprise at '…how few people know how to cope with "D.D.S.", or "dread of dining solo"'.

Interestingly, one of her tips is to patronize restaurants that feature communal tables, tables at which diners who arrive individually can eat with others. (See Jonsson, this volume, for a discussion of such communal tables in three countries.) There exist also many more conventionally published guides to solo dining (e.g., Kaminer and Boorstein, 2001; Kaminer and LaBan, 2002) and a recent collection of essays on cooking and eating alone (Ferrari-Adler, 2007). Much of the material available appears to reflect individuals' feelings of discomfort and self-consciousness when eating alone. We should note that we are referring to eating alone in public or in a restaurant – not eating alone in private, which seems to be considered less problematic. Indeed, one alternative to eating alone in public, particularly for business travelers, is to eat alone in the privacy of one's hotel room by ordering room service (see Jonsson, Chapter 12).

Interestingly, it is possible to find quotes in favor of dining alone, as can be seen below.

'Oh, the pleasure of eating my dinner alone!' (Charles Lamb)

'I am inclined] to think that eating is a private thing and should be done alone, like other bodily functions.' (Sylvia Ashton-Warner)

'Before you begin to eat, raise your glass in honor of yourself. The company is the best you'll ever have.' (David Halpern)

The well-known food writer, M. F. K. Fisher (1976, p. 96–99) provides some interesting thoughts in two essays devoted to dining alone. We noted earlier that meals have a social function; they are intended in part to build and strengthen social bonds between individuals. Fisher, whose main interest is in *food*, seems to believe that the social nature of meals can divert individuals' attention away from the food and that an occasional meal alone enables them to 'reconnect' with their food. She recounts an anecdote about Lucullus, the Roman host and epicure, who one day '...grew tired of dining with other men...' and ordered a meal for one. When he complained upon receiving the meal that it was not up to the chef's usual standards, he was told that 'we thought that there was no need to prepare a fine banquet for my lord alone.' His response: 'It is precisely when I am alone that you require to pay special attention to the dinner. At such times, you must remember, Lucullus dines with Lucullus'.

In a second essay, Fisher describes her campaign to '...establish myself as a well-behaved female at one or two good restaurants, where I could dine alone at a pleasant table...' (Fisher, 1976, p. 579-581). Eventually she gave it up because of the '...curious disbelieving impertinence of the people in restaurants...who sniffed at the high wall of my isolation...' and ate her meals at home, '...firmly believing that snug misanthropic solitude is better than hit-or-miss congeniality'. Thus, Fisher, clearly not cowed by eating alone, could not achieve solitude in a public place and was forced to eat alone at home.

Most of the information available about how people feel about eating alone is of the anecdotal type described above. It does not attempt to explain why people feel the way they do or what it is about eating alone that makes it a negative or a positive experience. In order to understand why people feel the way they do about eating alone, we (Pliner and Bell, unpublished data) had male and female undergraduate students complete a questionnaire in which we asked about the extent to which they liked (or believed they would like) eating dinner alone, both at home and in a restaurant. Next were lists of possible reasons why one would and would not want to eat the meal alone in both venues; participants rated their level of endorsement of each on a 7-point scale. Thus, there were four possible sets of reasons, reasons for wanting to (henceforth referred to as 'positive reasons') and not wanting to ('negative reasons') eat dinner alone at home ('home negative/positive reasons') and at a restaurant ('restaurant negative/positive reasons'), respectively. As an example, a home positive reason was 'It would be less work to prepare the food'. A restaurant negative reason 'I would be self-conscious'.

The data from each of the four sets were subjected to factor analysis. The analysis for home positive reasons yielded four factors; people would want to eat dinner at home alone because of: (1) the ability to engage in other activities (e.g., 'I would be able to read while eating'); (2) the ability to have things 'my' way (e.g., 'I could eat whenever I wanted to eat'); (3) the absence of negative aspects of sociality (e.g., 'I wouldn't have to talk while eating'); (4) practical advantages (e.g., 'there would be less to clean up'). For home negative reasons, the analysis yielded five factors: (1) practical disadvantages (e.g., 'it would be more effort than it was worth'); (2) presence of negative emotions (e.g., 'I would be lonely'); (3) experience would not be of a meal (e.g., 'I would feel like I wasn't eating a meal'); (4) absence of positive aspects of sociality (e.g., 'I wouldn't have anyone with whom to share the experience of the food'); (5) absence of norms (e.g., 'I wouldn't know how much I should eat'). The results are presented in Table 9.1.

Many of the factors that emerged when people considered why they would and would not want to eat alone in a restaurant were similar to those for eating alone at home. The analysis for restaurant positive reasons yielded three factors: (1) the ability to have things 'my' way; (2) the absence of negative aspects of sociality; (3) the ability to engage in other activities. Finally, the analysis for restaurant negative reasons yielded four factors: (1) presence of negative emotions; (2) absence of norms; (3) practical disadvantages ('going to the restaurant would be more effort than it was worth'); (4) absence of positive aspects of sociality (see Table 9.1).

Table 9.1   Reasons for wanting to (positive reasons) and not wanting to (negative reasons) eat dinner alone in two venues

Eating dinner alone at home	Rating[a]	Eating dinner alone in a restaurant	Rating
Positive reasons			
Ability to have things 'my' way	4.3	Ability to have things 'my' way	3.0
Practical advantages	3.8	Ability to engage in other activities	2.6
Ability to engage in other activities	3.7	Absence of negative aspects of sociality	2.3
Absence of negative aspects of sociality	3.5		
Negative reasons			
Presence of negative emotions	3.7	Presence of negative emotions	5.3
Absence of positive aspects of sociality	3.1	Practical disadvantages	3.1
Experience would not be of a meal	3.0	Absence of positive aspects of sociality	2.4
Practical disadvantages	2.2	Absence of norms	2.1
Absence of norms for appropriate intake	1.7		

[a]Ratings were on a 7-point scale (1 = 'disagree strongly'; 7 = 'agree strongly'

While participants felt relatively neutral when asked about how much they liked eating dinner at home alone ($m = 3.6$ on a 7-point scale), they clearly did not like (or believed they would not like) eating dinner alone in a restaurant ($m = 1.8$). We then examined the mean ratings, for the various factors, of reasons for wanting to and not wanting to eat dinner alone at home and in a restaurant. In terms of reasons for *not wanting* to eat at home alone, the negative emotions factor was most highly endorsed ($m = 3.7$ on a 7-point scale, as shown in Table 9.1), and the absence of positive aspects of sociality was next ($m = 3.1$). When participants were describing reasons they *would want* to eat dinner at home alone, they endorsed the ability to do things 'my' way ($m = 4.3$), practical advantages ($m = 3.8$), the ability to engage in other activities (3.7), and absence of negative aspects of sociality ($m = 3.5$). When responding in terms of reasons for *not wanting* to eat dinner in a restaurant alone, the negative emotions factor was overwhelmingly endorsed ($m = 5.3$). Finally, in terms of reasons for *wanting* to eat dinner alone in a restaurant, the ability to have things 'my' way was most strongly endorsed (3.0). Thus, in terms of overall ratings, the most strongly endorsed reason for not eating alone in both venues was feelings of loneliness, boredom, and self-consciousness. Participants also considered the ability to have things 'my' way to provide positive reasons to eat alone in both contexts. Thus, some aspects of the presence of others are positive and their absence is associated with decreased reported liking for eating alone while other aspects appear to be more negative and their absence is associated with increased reported liking for eating alone. Given the notion from Jonsson in Chapter 12 that an important category of complaints about eating alone in restaurants involves receiving poor service, it is unfortunate we did not include this as a reason to be rated in our study.

9.5   How does eating alone affect amount consumed?

A great deal of research demonstrates that individuals' social circumstances exert a profound effect on the amounts they eat (Herman *et al.*, 2003). One important, robust, and well known fact about eating alone is that people eat less when they are eating alone than when they eat in company. However, as we will show later, there are some important exceptions to this general rule. Analyzing food diaries generated by members of a community sample, de Castro and colleagues (e.g., de Castro, 1994; de Castro and Brewer, 1992; de Castro and de Castro, 1989) have consistently demonstrated a 'social correlation'; that is, the amounts individuals report eating increase with the number of people present at a meal. It should also be noted that the effect is not linear; rather, it is a power function. Thus, the biggest effect of the presence of others occurs when the meal changes from a solitary to a social one. Indeed, Clendenen, Herman, and Polivy (1994) examined the intake of individuals eating alone, with one companion or with three companions and found that individuals in groups of two and four ate more than solitary diners but there was no significant difference between the former groups. Once again, the big change occurred when the meal changed from a solitary to a social one and the number of companions was of less importance. The effect of the presence of others on eating occurs across a range of meal occasions and contexts; that is, it occurs when individuals are eating breakfasts, lunches, dinners, and snacks; when they are eating on weekdays and weekends; whether or not they are consuming alcohol at the meal; and whether they are eating at home or in a restaurant (de Castro, 1991; de Castro et al., 1990).

Although this finding is typically referred to as a 'social facilitation' effect, in the present context, we can think of it as a 'solitary inhibition' effect; it has been demonstrated in many different populations beyond de Castro's original one, including American soldiers and French and Dutch students (Bellisle et al., 1999; Feunekes et al., 1995; Hirsch and Kramer, 1993), and across a variety of research methodologies, including the food diary approach, direct observation, and the 'gold standard' experimental method (Berry et al., 1985; Clendenen et al., 1994; Edelman et al., 1986; Klesges et al., 1984).

While many researchers have demonstrated that people tend to eat less when alone, few have attempted to explain it and, indeed, most potential explanations approach the phenomenon from the perspective of explaining why people eat more when in groups, not why they eat less when alone. de Castro (1990) has suggested, tested, and rejected several hypothesized mediators of the effect, including the notion that the presence of others might induce arousal, anxiety, or elation, which would facilitate intake, and the idea that the presence of others might increase perceived hunger or the palatability of food. Another possibility is the distraction produced by the presence of others (Hetherington et al., 2006). Probably the most widely-accepted idea is a time-extension hypothesis, proposed explicitly by de Castro (1990): meal duration increases as the number of eaters increases, and this extension of the time spent eating increases the amount eaten. Although they did not formally propose time extension as an explanation for the social facilitation effect, Edelman et al. (1986, p. 81) clearly noticed it, reporting that '...[i]n the social condition, subjects talked with each other and lingered at the table. Many continued eating by nibbling at their leftover food as they sat talking'.

There is substantial correlational evidence for the idea that increasing the number of eaters increases the duration of a meal (e.g., de Castro, 1990; Feunekes et al., 1995; Redd and de Castro, 1991). Bell and Pliner (2003) conducted an observational study of customers in three different lunch settings – a worksite cafeteria, a fast-food restaurant, and a moderately-priced restaurant – in which they assessed the relationship between meal duration and the number of people eating at each table (group size). In all three settings, meal duration increased as group size increased, and the difference in duration as group size increased from one to two eaters was greatest in each venue.

Observing patrons in coffee shops, Sommer and Steele (1997) obtained a similar effect.

Pliner *et al*. (2005) directly tested the time extension hypothesis by having participants eat a meal alone or with one or three other people. Manipulated orthogonally to the number of co-eaters was meal duration; for half the participants the meal lasted 12 min, while for the remainder it lasted 36 min. Their results clearly indicated that meal duration was the critical variable. Participants ate much more in the long meal, but there was no effect of number of co-eaters. Thus, it may be that solitary diners eat less than social ones simply because they spend less time at table.

Of course, these results still do not explain why the solitary eater spends much less time at table than the social one; we suggest that the data presented earlier provide at least a partial explanation. That is, most people consider eating alone, particularly in a restaurant but at home as well, to be an aversive situation; negative emotions as a reason for not eating dinner alone was the most strongly endorsed negative reason in both venues. Thus, eating alone is not a pleasant experience and people attempt to minimize such an experience, terminating the meal long before they would do so if eating with others. However, when circumstances (or experimental requirements) dictate a longer meal, people eat as much when alone as when with others. Next, we turn to a discussion of situations in which the solitary eater eats more than the social eater.

9.6   When does the solitary eater eat more than the social eater?

Modeling

Less well known than the 'solitary inhibition' effect (as we have chosen to reframe the social facilitation effect) is the modeling effect. That is, individuals who are eating with someone who eats heartily eat more than those who are eating with a companion who eats lightly. This effect has been obtained in experimental contexts in which one of the partners is a confederate whose level of eating is predetermined, while the level of eating of the naïve partner is assessed (e.g., Nisbett and Storms, 1974; Pliner and Mann, 2004; Roth *et al*., 2001). It has also been obtained in more naturalistic circumstances in which two freely-eating naïve adults or children are studied (Herman *et al*., 2005; Salvy *et al*., 2007; Salvy *et al*., 2008).

The exquisite sensitivity of individuals to their social circumstances has led one group of researchers to propose a normative framework for eating behavior (Herman *et al*., 2003). Basically, this framework proposes that, in the presence of palatable food and in the absence of other constraints, people are motivated to eat as much as they can. One major constraint is a social norm that prescribes eating 'appropriately', and those who eat excessively may receive the disapprobation of their peers (e.g., Chaiken and Pliner, 1987; Martins *et al.*, 2004; Mooney *et al.*, 1994). What is appropriate (as opposed to excessive) in a given situation can be defined partially in terms of prior experience or cultural knowledge, but appropriateness is also importantly determined by the intake of other eaters in that situation. Thus, an individual's desire to eat a large amount on a particular occasion might be curbed by the presence of a companion whose behavior defines as appropriate a much smaller amount (a low norm). The intake of a solitary eater in an otherwise similar situation might be considerably greater – for two reasons. First, there is no low norm, and second, even if the eater were aware of such a norm, he or she is essentially eating in private. The idea that privacy is important comes from the more general literature on social norms, which suggests that people's adherence to them is motivated at least in part by their impression management concerns. Thus, behaving in accordance with social norms enables people to make a good impression on and earn the approval of others, even if it means going against their desired response (Deutsch and Gerard, 1955). However, if individuals' behavior is concealed from others (i.e., if they can respond in private), they can ignore the norms and respond as they desire without fear of social disapproval (Asch, 1956). So, we might expect that eating alone, essentially in private, will free people from concerns about the impressions that others may form of them and enable them to eat in accordance with their inclination to maximize eating.

This prediction receives support from the experimental modeling literature. A number of experiments have examined participants' behavior in three different circumstances – when eating with a companion who eats a small amount, when alone, and when eating with a companion who eats a large amount. It is typically the case that participants who eat alone eat more than those eating with a companion who eats minimally (Nisbett and Storms, 1974; Conger *et al.*, 1980; Rosenthal and McSweeney, 1979). There is also evidence indicating that individuals eat less when in the presence of a noneating observer (Conger *et al.*, 1980; Roth *et al.*, 2001). For example, in the study by Conger *et al.* (1980) some participants ate alone while others ate in the presence of a noneating confederate described as a participant who had begun the study earlier. Participants ate considerably less in this condition than when alone. Why should individuals' intake be lower in the presence of a noneating observer than when alone? Certainly, no one is modeling a low norm. However, it is likely that individuals' concerns about eating appropriately are strengthened; they become more self-conscious, because eating in the presence of a non-eating observer is the antithesis of eating in privacy. In summary, we are arguing that when social norms might otherwise suppress an individual's intake to less than some desired level, eating alone will allow the individual to eat as much as desired. On such occasions, individuals who are alone should eat more than those who are not.

9.6.1   Are some individuals especially likely to eat more when alone than when with others?

Overweight individuals

Given the emphasis in the previous section on people's self-consciousness in eating situations (more specifically, their consciousness of the impression they are making on others), one might wonder whether there are some individuals who are particularly self-conscious about the amounts they eat. Such individuals should be particularly likely to suppress their eating in the presence of others, eating, therefore, more when alone. One such group might include those who are overweight. In her classic volume, *Fat Is a Feminist Issue*, Orbach (1978, p. 28–30) poignantly describes the overweight woman who wants to eat cake but pretends the cake she buys is for someone else and '...only dare[s] to eat it out in the open when she thinks no one will spot her'. She suggests that such individuals cannot bear other people making the connection between food intake and body size, noting that '...this explains, in part, the public side of the compulsive eater who eats sparingly'. Based on a series of interviews, Zdrodowski (1996) describes eating in public as a painful experience for overweight women, partly because the imagined or actual scrutiny of others makes them self-conscious. Several of her interviewees confessed that their decisions about how much to eat depended on whether others were present or not and that they sometimes sought solitude in order to eat what they pleased. Other researchers, taking an observational approach, have examined actual amounts of food purchased and/or eaten by individuals, obese and nonobese, as a function of their social circumstances. Krantz (1979) observed obese and non-obese customers in a university cafeteria, assessing the amount of food taken as a function of whether they were alone or were accompanied by one or more others. While non-obese individuals showed the usual 'solitary inhibition' effect, taking about 10% less food when alone than when with others, obese individuals showed the reverse pattern, taking about 18% less when accompanied than when eating alone. Similarly, Maykovich (1978), observing diners in a smorgasbord restaurant, found that for those who did not frequent the restaurant regularly, the presence of eating companions had a huge inhibitory effect on amounts consumed for obese and overweight diners, while non-obese diners ate similar amounts whether alone or in company. In accounting for his results, Maykovich argues that obese individuals monitor their diet in public to mimimize critical appraisal. Taking an experimental approach, Salvy *et al.* (2007) had overweight and normal-weight children eat alone and in a group, on two separate occasions. The overweight children ate more when alone than when with a group, while the reverse was true for normal-weight children. Thus, anecdotal, observational, and experimental data converge to suggest that solitary (as opposed to social) eating increases intake for those who are overweight, an effect opposite to that seen in individuals of normal weight.

Binge eaters

Another group that might have particular reason to be concerned about the impression their eating makes on others consists of binge eaters. As noted earlier, people who eat large amounts attract negative attributions from others, and bingers (at least when they are bingeing) are quintessential 'big eaters'. Given all this, one might expect that the presence of others will inhibit bingeing – or to put it the other way, people who binge should be more likely to do so when they are alone than when they are in the presence of others. Indeed, eating alone is one of the topics covered in the Eating Disorder Examination (Fairburn and Wilson, 1993) and is often described as a behavioral characteristic of binge eating disorder on websites providing information to the public (National Eating Disorders Association website; KidsHealth website). According to Heatherton and Baumeister (1991, p. 92), '…[b]inge eaters seem to suffer from an intense sensitivity to other people's opinions and evaluations, and public scrutiny…makes them restrain their eating. Binges are therefore more likely to occur when the individual can escape from the presence of others…'. A considerable number of studies have examined the association between bingeing and social circumstances. An illustrative one is that of Davis *et al.* (1988) who asked participants to keep an hourly diary in which they recorded each instance of their eating behavior, including its classification as a meal, a snack, or a binge, and their social circumstances at the time. While participants reported being alone during less than a quarter of their meals and snacks, they reported being alone during 63% of their binges.

Similarly, Johnson and Larson (1982) found that 88% of binges reported by their participants occurred when they were alone, while Grilo, Shiffman, and Carter-Campbell (1994) found that 86% of binges reported by their participants occurred when they were alone. Although neither of the latter two studies provides data on the prevalence of solitary non-binge eating, given the Davis *et al.* (1988) data, it seems safe to assume that one is considerably less likely to be alone when eating a meal or a snack than when binge eating. Examining bingeing following craving, Waters *et al.* (2001) found that binging was a significantly more common response to craving when the women ate the craved food if they were alone at the time of the craving than if they were with others. A study examining children and adolescents also found that binges (in comparison to other types of eating episodes) were relatively more likely to occur when the youngsters were alone (Tanofsky-Kraff *et al.*, 2007). Finally, in an 18-month prospective study of 12–21 year old females, Martinez-Gonzales *et al.*, (2003) found that the habit of solitary eating (i.e., an affirmative response to the question, 'Do you usually eat alone?') was a significant risk factor for the development of an eating disorder; solitary eaters were more than three times as likely to develop an eating disorder than were those who did not eat solitarily. Thus, the evidence overwhelmingly indicates that bingeing occurs when individuals are alone, and, further, that the habit of solitary eating is predictive of eating disorders.

9.7   What are some of the correlates of solitary eating?

The elderly

Of all age groups, the elderly are most likely to live alone (Palmore, 1984; US Census Bureau, 2006), and both the number and proportion of older adults who live alone have increased dramatically in the United States since 1960 (Davis *et al.*, 2000). Although living alone does not necessarily translate into solitary eating, elderly individuals who live alone are much more likely than those who live with others to eat alone (82% vs. 8%; Torres *et al.*, 1992). The elderly in general have been shown to be at risk for malnutrition and nutritional inadequacy; summarizing their review of aging and nutrition, Martin *et al.* (2006, p. 931) noted that '...it is apparent that weight loss and nutritional risk are common in this population and can be associated with adverse outcomes'. Further, and in the present context, there has been concern that older adults eating and living alone may be particularly at risk for inadequate dietary intake (Davis *et al.*, 2000). When examined carefully, the data are complex and inconsistent. However, given the diversity of populations studied (e.g., in terms of actual age, living arrangements, rural vs. urban residence, socioeconomic status, gender, geographic location, ethnicity), the multiplicity of methods used to assess nutritional status (e.g., analysis of 24-hour recalls, food frequency questionnaires, report of intake of fruits and vegetables, nutritional risk screening instruments), the varying definitions of solitary eating (e.g., living alone, self-report of 'usually eating alone'), the inconsistency is far from surprising. Despite many negative findings (e.g., Lee *et al.*, 1991; Posner *et al.*, 1994), there is a fair amount of evidence showing that those who live alone eat less regularly, consume fewer calories, and eat diets that are more likely to be deficient in major nutrients, although these results are often qualified by gender and age. For example, Davis *et al.* (1990) found that males were more likely than females to suffer from poor nutrition when living alone (vs. with at least one other person) and that the effect increased with age from 55 to 75. McIntosh, Shifflett, and Picou, (1989) found lower nutrient intakes, protein intakes, and caloric intakes in 800 or so elderly participants (mostly women) in those reporting less 'mealtime companionship'. Hendy, *et al.* (1998) found that number of shared meals per day was a significant predictor of nutritional risk (as measured by the Nutritional Screening Index). Donkin *et al.* (1998) looked at fruit and vegetable consumption in elderly men and women, living alone and with a partner and found an interaction between age and gender; single males ate the fewest fruits and vegetables. Thus, although the data are far from consistent, there is a substantial body of literature demonstrating that elderly individuals, particularly men, who eat alone eat less well nutritionally than do those who do not live alone.

What these data do not tell us, of course, is *why* elderly individuals eating and/or living alone might have less nutritionally adequate diets, and, given that the data are correlational, we cannot make any strong causal inferences. In a spirit of speculation, however, we note that one piece of data suggests that the difference in nutritional adequacy is because of lower caloric intake rather than lower nutrient density (Davis *et al.*, 1990). Thus, it is possible that the solitary inhibition effect described above might play a role; that is, people eat less when alone than when in the presence of others. However, the fact that males are more likely than females to suffer from poor nutrition when living alone suggests that other factors may also be responsible. For example, it may be that, as a result of sex role differences in experience with food procurement and preparation (Bell and Marshall, 2003), males are less able to and/or motivated to prepare food for themselves.

Family dinner
There is a burgeoning literature examining the relation between family dinner and various nutrition-related outcomes for adolescents (see Chapters 13, 11 and 10 by Chrzan, Fjellstrom, and McIntosh *et al.*, respectively for more extended discussions). As noted earlier, the most common alternative to family dinner for teens is a solitary meal, so, for the purposes of this section, we will assume that teens who are not eating with their families are usually eating alone. Several studies have shown that adolescents who report less frequent family meals have lower intakes of fruits and vegetables as well as a number of key nutrients, including fiber, calcium, and iron, and have higher intakes of soft drinks and saturated fats (Gillman *et al.*, 2000; Neumark-Sztainer *et al.*, 2003; Videon and Manning, 2003). Other studies have found that the prevalence of overweight is higher for adolescents who report eating dinner with their families never or on some days, in comparison to those who report eating family dinner on most days or every day (Taveras *et al.*, 2005); yet another study found an increased incidence of disordered eating in those adolescents who ate fewer family dinners (Neumark-Sztainer *et al.*, 2004). In a five-year longitudinal study, Larson *et al.* (2007) found that young adult nutritional outcomes (including intake of fruit, vegetables, fiber, and soft drinks) were predicted by frequency of family dinners during adolescence. The correlates for adolescents of eating alone appear to extend beyond even food-related ones; Eisenberg *et al.* (2004) found lower psychosocial well-being among adolescents who ate fewer dinners with their families. Infrequent family dining is associated with higher rates of teen smoking, drinking, illegal drug use and prescription drug abuse (CASA, 2007). We note that all of the data reported are correlational and subject to the usual causal ambiguity. We are not arguing that eating alone is a *cause* of poor nutrition, overweight, eating disorders, well-being, or use of drugs; we are simply describing the association.

9.8 Summary

What do we know about eating alone? To begin, such data as are available indicate that people, adults at least, do a fair amount of it. However, commensal eating is deeply embedded in cultural consciousness. In that context, eating alone is an anomalous behavior. Not surprisingly, for many, if not most, people a solitary meal is a highly undesirable situation and, in some cases, not a meal at all. For others, eating alone provides an opportunity to escape public scrutiny and allows them to eat as desired. Some data reveal that eating alone is associated with negative nutritional outcomes in some groups. Finally, it is clear that there is a great deal that we do not know about the causes, correlates, and consequences of eating alone.

9.9   Dedication

This chapter is dedicated to the memory of Rick Bell, a dear friend and wonderful colleague, who died in February 2007. The chapter, begun before his death, is based on an outline prepared by him and contains many of his ideas. He is truly an author, in every sense of the word, and his collaboration on the final draft was sorely missed.

9.10   References

Asch, S.E. (1956). Studies of independence and conformity: A minority of one against a unanimous majority. *Psychological Monographs*, **70** (9, Whole No. 416).

Amato, P. Booth, A., Johnson, D.R. and Rogers, S.J. (2007). *Alone together: how marriage in America is changing*. Boston, MA: Harvard University Press.

Appadurai, A. (1981). Gastro-politics in Hindu South Asia. *American Ethnologist*, **8**, 491–511.

Bell, R. and Marshall, D.W. (2003). The construct of food involvement in behavioral research: scale development and validation. *Appetite*, **40**, 235–244.

Bell, R. and Pliner, P. (2003). Time to eat: the relationship between the number of people eating and meal duration in three lunch settings. *Appetite*, **41**, 215–218.

Bellisle, F., Dalix, A.-M. and de Castro, J.M. (1999). Eating patterns in French subjects studied by the 'weekly food diary' method. *Appetite*, **32**, 46–52.

Berry, S.L., Beatty, W.W. and Klesges, R.C. (1985). Sensory and social influences on ice cream consumption by males and females in a laboratory setting. *Appetite*, **6**, 41–45.

Blum-Kulka, S. (1997). *Dinner talk: cultural patterns of sociability and socialization in family discourse*. Mahwah, NJ: Erlbaum.

CASA: The National Center on Addiction and Substance Abuse at Columbia University. The Importance of Family Dinners IV. 2007; accessed April 12, 2008. Available at: http://www.casacolumbia.org/ViewProduct.aspx?PRODUCTID= {296A5E1E-B68F-44fa-A64D-95ABC1FB6CA0}.

Chaiken, S. and Pliner, P. (1987). Women, but not men, are what they eat: the effect of meal size and gender on perceived masculinity and femininity. *Personality and Social Psychology Bulletin*, **13**, 166–176.

Charles, N. and Kerr, M. (1988). *Women, food and families*. Manchester, UK: Manchester University Press.

Child Trends Data Bank (2003). Family meals. Accessed April 12, 2008. Available at: www.childtrendsdatabank.org.

Clendenen, V.I., Herman, C.P. and Polivy, J. (1994). Social facilitation of eating among friends and strangers. *Appetite*, **23**, 1–13.

Conger, J.C., Conger, A.J., Constanzo, P.R., Wright, K.L. and Matter, L.A. (1980). The effect of social cues on the eating behavior of obese and normal subjects. *Journal of Personality*, **48**, 258–271.

Davis, M.A., Murphy, S.P., Neuhaus, J.M. and Lein, D. (1990). Living arrangements and dietary quality of older US adults. *Journal of the American Dietetic Association*, **90**, 1667–1672.

Davis, M.A., Murphy, S.P., Neuhaus, J.M., Gee, L. and Quiroga, S.S. (2000). Living arrangements affect dietary quality for US adults aged 50 years and older: NHANES III 1988–19941. *Journal of Nutrition*, *130*, 2256–2264.

Davis, R., Freeman, R.J. and Garner, D.M. (1988). A naturalistic investigation of eating behavior in bulimia nervosa. *Journal of Consulting and Clinical Psychology*, **56**, 273–279.

de Castro, J.M. (1987). Circadian rhythms of the spontaneous meal pattern, macronutrient intake, and mood of humans. *Physiology and Behavior*, **40**, 437–446.

de Castro, J.M. (1990). Social facilitation of duration and size but not rate of the spontaneous meal intake of humans. *Physiology and Behavior*, **47**, 1129–1135.

de Castro, J.M. (1991). Social facilitation of the spontaneous meal size of humans occurs on both weekdays and weekends. *Physiology and Behavior*, **49**, 1289–1291.

de Castro, J.M. (1994). Family and friends produce greater social facilitation of food intake than other companions. *Physiology and Behavior*, **56**, 445–455.

de Castro, J.M. and Brewer, E.M. (1992). The amount eaten in meals by humans is a power function of the number of people present. *Physiology and Behavior*, **51**, 121–125.

de Castro, J.M. and de Castro, E.S. (1989). Spontaneous meal patterns in humans: influence of the presence of other people. *American Journal of Clinical Nutrition*, **50**, 237–247.

de Castro, J.M., Brewer, E.M., Elmore, D.K. and Orozco, S. (1990). Social facilitation of the spontaneous meal size of humans occurs regardless of time, place, alcohol or snacks. *Appetite*, **15**, 89–101.

Deutsch, M. and Gerard, H.G. (1955). A study of normative and informational influence upon individual judgment. *Journal of Abnormal and Social Psychology*, **51**, 629–636.

Donkin, A.J.M., Johnson, A.E., Lilley, J.M., Morgan, K., Neale, R.J., Page, R.M. and Silburn, R.L. (1998). Gender and living alone as determinants of fruit and vegetable consumption among the elderly living at home in urban Nottingham. *Appetite*, **30**, 39–51.

Douglas, M. (1972). Deciphering a meal. *Daedalus*, **101**, 61–82.

Edelman, B., Engell, D., Bronstein, P. and Hirsch, E. (1986). Environmental effects on the intake of overweight and normal-weight men. *Appetite*, **7**, 71–83.

Eisenberg, M.E., Olson, R.E., Neumark-Sztainer, D., Story. M. and Bearinger, L.H. (2004). Correlations between family meals and psychosocial well-being among adolescents. *Archives of Pediatric and Adolescent Medicine*, **158**, 792–796.

Fairburn, C.G. and Wilson, G.T. (1993). *Binge eating: nature, assessment, and treatment*. London, UK: The Guilford Press.

Feunekes, G.I.J., de Graaf, C. and van Staveren, W.A. (1995). Social facilitation of food intake is mediated by meal duration. *Physiology and Behavior*, **58**, 551–558.

Fisher, M.F.K. (1976). *The art of eating*. New York, NY: Vintage Books.

Ferrari-Adler, J. (Ed.). (2007). *Alone in the kitchen with an eggplant: confessions of cooking for one and dining alone*. New York, NY: Riverhead Books.

Gillman, M., Rifas-Shiman, S., Frazier, A., Rockett, H., Camargo, C., Field, A., Berkey, C. and Colditz, G. (2000). Family dinner and diet quality among older children and adolescents. *Archives of Family Medicine*, **9**, 235–240.

Grilo, C.M., Shiffman, S. and Carter-Campbell, J.T. (1994). Binge eating antecedents in normal-weight nonpurging females: is there consistency? *International Journal of Eating Disorders*, **16**, 239–249.

Heatherton, T.F. and Baumeister, R.F. (1991). Binge eating as escape from self-awareness. *Psychological Bulletin*, **110**, 86–108.

Hendy, H., Nelson, G.K. and Greco, M.E. (1998). Social cognitive predictors of nutritional risk in rural elderly adults. *International Journal of Aging and Human Development*, **47**, 299–327.

Herman, C.P., Roth, D.A. and Polivy, J. (2003). Effects of the presence of others on food intake: a normative interpretation. *Psychological Bulletin*, **129**, 873–886.

Herman, C.P., Koenig-Norbert, S., Peterson, J.B. and Polivy, J. (2005). Matching effects on eating: do individual differences make a difference? *Appetite*, **45**, 108–109.

Hetherington, M.M., Anderson, A.S., Norton, G.N.M. and Newson, L. (2006). Situational effects on meal intake: A comparison of eating alone and eating with others. *Physiology and Behavior*, **88**, 498–505.

Hirsch, E.S. and Kramer, F.M. (1993). Situational influences on food intake. In B.M. Marriott, (Ed). *Nutritional needs in hot environments* (pp. 215–243). Washington, DC: National Academy Press.

Johnson, C. and Larson, R. (1982). Bulimia: an analysis of moods and behavior. *Psychosomatic Medicine*, **44**, 341–351.

Kaminer, M. and Boorstein, J. (2001). *Table for one: New York City*. New York, NY: McGraw-Hill.

Kaminer, M. and LaBan, A. *Table for one: Chicago*. New York, NY: McGraw-Hill.

KidsHealth website: http://www.kidshealth.org/teen/food_fitness/ problems/ binge_eating.html

Klesges, R.C., Bartsch, D., Norwood, J.D., Kautzman, D. and Haugrud, S. (1984). The effects of selected social and environmental variables on the eating behavior of adults in the natural environment. *International Journal of Eating Disorders*, **3**, 35–41.

Krantz, D.S. (1979). A naturalistic study of social influences on meal size among moderately obese and nonobese subjects. *Psychosomatic Medicine*, **41**, 19–27.

Larson, N.I., Neumark-Sztainer, D., Hannan, P.J. and Story, M. (2007). Family meals during adolescence are associated with higher diet quality and healthful meal patterns during young adulthood. *Journal of the American Dietetic Association*, **107**, 1502–1510.

Lee, C.J., Tsui, J., Glover, E., Glover, E.B., Kumelachew, M., Warren, A.P., Perry, G., Godwin, S., Hunt, S.K., McCray, M. and Stigger, F.E. (1991). Evaluation of nutrient intakes of rural elders in eleven southern states based on sociodemographic and life style indicators. *Nutrition Research*, **11**, 1383–1396.

Makela, J. (2000). Cultural definitions of the meal. In H.L. Meiselman, (Ed.), *Dimensions of the meal: the science, culture, business, and art of eating*. Gaithersburg, MD: Aspen Publishers, Inc.

Martin, C.T., Kayser-Jones, J., Stotts, N., Porter, C. and Froelicher, E.S. (2006). Nutritional risk and low weight in community-living older adults: a review of the literature (1995–2005). *Journal of Gerontology: Medical Sciences*, **61A**, 927–934.

Martinez-Gonzales, M.A., Gual, P., Lahortiga, F., Alonso, Y., de Irala-Estevez, J. and Cervera, S. (2003). Parental factors, mass media influences, and the onset of eating disorders in a prospective population-based cohort. *Pediatrics*, **111**, 315–320.

Martins, Y., Pliner, P. and Lee, C. (2004). The effects of meal size and body size on individuals' impressions of males and females. *Eating Behaviors*, **5**, 117–132.

Maykovich, M. K. (1978). Social constraints and eating patterns among the obese and overweight. *Social Problems*, **25**, 453–460.

McIntosh, W.A. (1996). *Sociologies of food and nutrition*. New York, NY: Plenum Press.

McIntosh, W.A., Shifflett, P.A. and Picou, J.S. (1989). Social support, stressful events, strain, dietary intake, and the elderly. *Medical Care*, **27**, 140–153.

Mestdag, I. (2005). Disappearance of the traditional meal: temporal, social and spatial destructuration. *Appetite*, **45**, 62–74.

Mestdag, I. and Vandeweyer, J. (2005). Where has family time gone? In search of joint family activities and the role of the family meal in 1966 and 1999. *Journal of Family History*, **30**, 304–323.

Moeller, K. (1998). Table for one: if you hate dining solo, you're not alone. *Entrepreneur Magazine*, April 1998; accessed April 1, 2008. Available at: http://www.entrepreneur.com/magazine/entrepreneur/1998/april/15424.html.

Mooney, K.M., DeTore, J. and Malloy, K.A. (1994). Perceptions of women related to food choice. *Sex Roles*, **31**, 433–442.

Murcott, A. (1997). Family meals: a thing of the past? In P. Caplan (Ed.), *Food, health, and identity* (p. 32–49). London, UK: Routledge.

National Eating Disorders Association website: http://www.edap.org/p.asp?WebPage_ID=286andProfile_ID=41182.

Neumark-Sztainer, D., Wall, M., Story, M. and Fulkerson, J.A. (2004). Are family meal patterns associated with disordered eating behaviors among adolescents? *Journal of Adolescent Health*, **35**, 350–359.

Neumark-Sztainer, D., Hannan, P., Story, M., Croll, J. and Perry, C. (2003). Family meal patterns: associations with sociodemographic characteristics and improved dietary intake among adolescents. *Journal of the American Dietetic Association*, **103**, 317–322.

Nisbett, R.E. and Storms, M.D. (1974). Cognitive and social determinants of food intake. In H. London and R. E. Nisbett (Eds.), *Thought and feeling: cognitive alteration of feeling states* (pp. 190–208). Chicago IL: Aldine.

Orbach, S. (1978). *Fat is a feminist issue*. New York, NY: Berkeley Publishing Corporation.

Palmore, E.B. (1984). *Handbook on the aged in the United States*. Westport, CT: Greenwood.

Pliner, P. and Bell, R. (2004). What to eat: a multidiscipline view of meals. *Food Quality and Preference*, **15**, 901–905.

Pliner, P. and Mann, N. (2004). Influence of social norms and palatability on amount consumed and food choice. *Appetite*, **42**, 227–237.

Pliner, P. and Rozin, P. (2000). In H.L. Meiselman, (Ed.), *Dimensions of the meal: the science, culture, business, and art of eating* (p. 19–46). Gaithersburg, MD: Aspen Publishers, Inc.

Pliner, P., Bell, R., Hirsch, E. and Kinchla, M. (2005). Meal duration mediates the effect of 'social facilitation' on eating in humans. *Appetite*, **46**, 189–198.

Posner, B.M., Jette, A., Smigelski, C., Miller, D. and Mitchell, P. (1994). Nutritional risk in New England elders. *Journal of Gerontology*, **49**, M123–M132.

Redd, M. and de Castro, J. (1991). Social facilitation of eating: effects of social instruction on food intake. *Physiology and Behavior*, **52**, 749–754.

Rosenthal, B. and McSweeney, F.K. (1979). Modeling influences on eating behavior. *Addictive Behaviors*, **4**, 205–214.

Rotenberg, R. (1981). The impact of industrialization on meal patterns in Vienna, Austria. *Ecology of Food and Nutrition*, **11**, 25–35.

Roth, D., Herman, C.P., Polivy, J. and Pliner, P. (2001). Self-presentational conflict in social eating situations: a normative perspective. *Appetite*, **36**, 165–171.

Salvy, S.-J., Coelho, J.S., Kieffer, E. and Epstein, L.H. (2007). Effects of social contexts on overweight and normal-weight children's food intake. *Physiology and Behavior*, **92**, 840–846.

Salvy, S.-J., Jarrin, D., Paluch, R., Irfan, N. and Pliner, P. (2007). Effects of social influence on eating in couples, friends and strangers. *Appetite*, **49**, 92–99.

Salvy, S.-J., Vartanian, L.R., Coelho, J.S., Jarrin, J. and Pliner, P. (2008). The role of familiarity on modeling of eating and food consumption in children. *Appetite*, **50**, 514–518.

Sobal, J. (2000). Sociability and meals: facilitation, commensality, and interaction. In H.L. Meiselman, (Ed.), *Dimensions of the meal: the science, culture, business, and art of eating* (p. 119–133). Gaithersburg, MD: Aspen Publishers, Inc.

Sobal, J. and Nelson, M.K. (2003). Commensal eating patterns: a community study. *Appetite*, **41**, 181–190.

Sommer, R. and Steele, J. (1997). Social effects on duration in restaurants. *Appetite*, **29**, 25–30.
Statistics Canada. E-STAT table. 1986 (2A) basic questionnaire, Provinces to Municipalities; accessed 2 April 2008. Available at: http://estat.statcan.ca/cgi-win/cnsmcgipgm (password required).
Statistics Canada. E-STAT table. 1991 (2A) basic questionnaire, Provinces to Municipalities; accessed 2 April 2008. Available at: http://estat.statcan.ca/cgi-win/cnsmcgipgm (password required).
Statistics Canada. E-STAT table. 1996 Census of Population (Provinces, Census Divisions); accessed 2 April 2008. Available at: http://estat.statcan.ca/cgi-win/cnsmcgipgm (password required).
Statistics Canada. E-STAT table. 2001 Census of Population (Provinces, Census Divisions); accessed 2 April 2008. Available at: http://estat.statcan.ca/cgi-win/cnsmcgipgm (password required).
Statistics Canada. E-STAT table. 2006 Census of Population (Provinces, Census Divisions); accessed 2 April 2008. Available at: http://estat.statcan.ca/cgi-win/cnsmcgipgm.
Tanofsky-Kraff, M., Goossens, L., Eddy, K.T., Ringham, R., Goldschmidt, A., Yanovski, S.Z., Braet, C., Marcus, M.D., Wilfley, D.E. and Olsen, C. (2007). A multisite investigation of binge eating behaviors in children and adolescents. *Journal of Consulting and Clinical Psychology*, **75**, 901–913.
Taveras, E., Rifas-Shiman, S., Berkey, C., Rockett, H., Field, A., Frazier, A., Colditz, G. and Gillman, M. (2005). Family dinner and adolescent overweight. *Obesity Research*, **13**, 900–906.
Torres, C.C., McIntosh, W.A. and Kubena, K.S. (1992). Social network and social background characteristics of elderly who live alone and eat alone. *Journal of Aging and Health*, **4**, 564–578.
US Census Bureau (2006). Current Population Reports; accessed 2 April 2008. Available at: http: www.census.gov/compendia/statab/.
Videon, T. and Manning, C. (2003). Influences on adolescent eating patterns: the importance of family meals. *Journal of Adolescent Health*, **32**, 365–373.
Waters, A., Hill, A. and Waller, G. (2001). Internal and external antecedents of binge eating episodes in a group of women with bulimia nervosa. *International Journal of Eating Disorders*, **29**, 17–22.
Zaborskis, A., Zemaitiene, N., Borup, I., Kuntsche, E. and Moreno, C. (2007). Family joint activities in a cross-national perspective. *BMC Public Health*, **7**; accessed 11 April 2008. Available at http://www.biomedcentral.com/1471-2458/7/94.
Zdrodowski, D. (1996). Eating out: the experience of eating in public for the 'overweight' woman. *Women's Studies International Forum*, **19**, 655–664.

10

The American family meal

W. A. McIntosh, W. Dean, C. C. Torres, J. Anding, K. S. Kubena
and R. Nayga, Texas A&M University, USA

Abstract: In this chapter, the American family meal is examined. Recent trends in
eating by both individuals and families are explored. Frequency of eating various
meals, time spent eating, and which, if any, family members participate in these meals
are studied. Effects of parental employment, income, and other characteristics indicate
the frequency and time spent in some family meals are constrained by resources such
as time and by parental work schedules. Foods currently consumed at family dinner
tables are presented.

Key words: eating patterns, dinner table, family meals, food consumption.

10.1   Introduction

Meals in the United States have exhibited the same general pattern: breakfast,
lunch, and dinner, although the precise names of these meals, their timing,
and the nature of their participants have changed (McIntosh, 1999; Sobal
and Nelson, 2003). In the USA, a longitudinal characterization of the family
meal is a difficult task for a number of reasons; not only have the constituent
members of the family changed, but so has the structure and content of
meals. Furthermore, the historical record provides us with limited detail as
to the relationship among participants, structure, and foods served. For the
purpose of this paper, we define the term 'family meal', both normatively
and descriptively, to mean: (1) all family members are expected to participate;
(2) family members do so when able. Our overview distinguishes the family
meal from the family meal ritual. The former often refers to the material
content of the meal, the items eaten by families, and usual participants
(Douglas and Gross, 1981; Douglas, 1982). The family meal ritual refers to
the patterns of regularity exhibited in the preparation and consumption of
family meals as well as perceived importance by family members.

We also attend to the cultural and normative character of the family meal
ritual; the manner in which repetitive family dining experiences can be described

as moments of social communication whereby social norms are instilled into the participants (Visser, 1992). In conjunction with our description of normative family meals, we will describe the manner in which material circumstances such as income and work schedules or social conflict such as family squabbles may interfere with the communicative character of family meals. We will also examine deviations from the regularity of the family meal ritual. It is clear that individuals often characterize eating events as neither meals nor snacks in the conventional sense (Bisogni *et al.*, 2007). Meals are generally defined as distinct from other eating events such as snacking. Some of these differences are made, at least in the minds of people (and in the minds of some restaurateurs), based on the time of day; others base this on the type of foods served (Marshall, 2000).

Also, we will very briefly examine the role that the social normative meal takes in the debates over eating, and the decline of the family. Gillis (1996) argues the roots of the normative features that Anglo-Americans associate with family meals developed during Victorian times. Finally, we observe that family meals may follow both ethnic and regional lines, providing examples. Throughout the chapter, we will present findings from a study of more than 300 families living in the Houston MSA (Metropolitan Statistical Area). An MSA is defined as a 'densely settled concentration of population of 50 000' and 'comprises the central county or counties containing the core, plus adjacent outlying counties having a high degree of social integration with the central county as measured through commuting' (United States Census Bureau, 2000a). The project was funded through a grant from USDA-ERS-Food Assistance and Nutrition Research Program, 'Parental Time, Role Strain, Coping, and Children's Diet and Nutrition' (43-3AEM-0-80075) (McIntosh *et al.*, 2006). The views expressed in this paper do not necessarily reflect those of USDA. This project will be referred to as the 'Houston project' throughout the chapter.

Participants were 311 children, ages 9–11 or 13–15, and their parents recruited by means of random digit dialing from the Houston MSA during 2001–2002. Each parent completed three instruments:

1   a *telephone survey*: to collect socioeconomic information on the work/home environment such as work hours, work schedule, work flexibility, work to family spillover, and work stress, along with basic demographic information;

2   a *self-administered survey*: to collect information on annual total income, annual earned income, and annual unearned income,

3   a *2-day time diary*: to recording all main or primary activities for 2 days, the length of time of each activity, whether and what other (secondary) activities they engage during the time period in question, where each activity was done, and who they were with during each activity.

The data on secondary activities allowed us to determine how much time parents spent eating while watching TV or time spent watching TV while exercising. For each household one child in the age ranges 9–11 or 13–15 completed:

1   a *personal interview* with questions focusing on eating habits, the home environment, along with basic demographic information and
2   a multi-pass 24-h dietary and activity recall.

These children then maintained a two-day food diary and a two-day activity diary and underwent a brief physical exam. The activity recall and activity diary instruments did not ask the children what else they were doing as they undertook their main activity. This means that, unlike the parents' time diary data, we are unable to present information about the amount of time that children spent eating while watching TV.

10.2   A methodological note

Families meet the definition of Durkheim's social fact; that is such entities are more than the sum of the parts or members (Durkheim, 1962). In one very real sense, the family meal provides a window into a family's life and times. Sociologists approach families and meals through several methodologies, including survey research, which constructs a picture of families via how their members describe family life. However, not all researchers believe that family members' descriptions provide a holistic or intersubjective picture of such a group. Several issues cloud studies of families including the number of family members that ought to be studied and the means of measuring family characteristics via individual family members.

10.2.1   Who should be studied?

In studying families, it is unclear as to how many family members ought to be interviewed in order to capture the family. Many researchers rely on a single respondent to provide information about their families. In studies of family meals, for example, either mothers or the 'principal food preparer' serves as the family reporter (see DeVault, 1991; Charles and Kerr, 1988). Other areas of family research involve an interview of both parents (where present) and two children from samples of families (Conger and Elder, 1994). These have included an observational component which was used to validate questions asked about family functioning during the individual interviews. In addition, researchers such as ourselves have used a combination of interviews with two parents (where present) plus one child with time diary data provided by each participating family member. Time diaries, however, do not capture all that is important about a family meal, such as the amount of interaction or instances of conflict that may arise during a meal. Others attempt to overcome the limitation of individual responses to questionnaires regarding family life and, taking a more qualitative approach, conduct observations of family members, often inside their households (Gilgun, 1999). These offer the opportunity to observe members acting as a part of the family group, often for a significant amount of time. The shortcoming of this approach is that it relies on very small, not necessarily representative samples of families. A variant of this technique is to video-record family activities. Researchers have used video records to create quantitative measures of the frequency of family member interaction and the amount of time family members engage in various activities (Beck 2007; Lewis and Feiring, 1982). Again these are limited by their sample size and their representativeness.

10.2.2   Measuring participation in and other aspects of family meals

Students of the family are long on theory and short on the measurement wherewithal to test those theories (Acock, 1999; Larzelere and Klein, 1987). The full complexity of social life is found in families; families are made up of individuals, dyads, triads and beyond. They can also be described according to structural elements such as size, division of labor, and power structure. Relationships are characterized by frequency of interaction and styles, (e.g., parenting style, consisting of authoritative, authoritarian, indulgent, and neglectful styles which differ based on their degree of demandingness, control, and caring (Baumrind, 1991)). The distinction between nuclear and extended families is also important. Families are also thought of as environments in which individual members operate as well as social systems, characterized by their cohesiveness and adaptability (Copeland and White, 1991). Individuals possess individual goals and opinions which are influenced, to a degree, by their relationships with other family members and by the overall social unit of the family. Quantitative studies tend to use questionnaires, administered to one or more family members. Answers to questions posed to individual family members often serve as a measure of family interaction (e.g., how often does your family eat dinner together?). A more useful variant of this approach is to ask individuals about meals they normally eat and then ask with whom if anyone the meal was consumed; Sobal and Nelson (2003) take this approach, but allow their respondents to define what constitutes an eating event. This garners information about 'non-meal' events such as snacking in the car, which can involve family members. Oddly enough, those researchers who collect dietary recall or dietary record data often ask who the individual eater is during eating episodes, but this data is rarely analyzed. This neglected resource will be discussed in more detail in a later section of this chapter. Other researchers average the scores of family members to estimate a family variable; thus family income is the sum of all incomes made by family members and family dinner as a normative ritual becomes the sum of all family member ratings of this meal. These measures carry more than measurement baggage; for example, what if not all family members answer these questions? Drawing inferences from individual family members about the family as a whole may lead to the individual fallacy (Acock, 1999). Family members may also give conflicting accounts of family conditions such as the level of marital adjustment (Larzelere and Klein, 1987).

The eating patterns of family members are of interest because they probably vary depending on class, ethnicity, and other important social characteristics. Furthermore, they have implications for social relationships and individuals' health. How should we go about measuring these patterns? One way is to use time diaries with the portion of the time diary devoted to 'whom the activity takes place with' made more specific: that is, ask the respondent to record the number and types of individuals he/she was with while doing the activity. This will probably produce some underreporting and of course will exclude other family members' activities (such as breakfast) that did not happen to involve the respondent. However, requesting multiple or all family members to fill out a time diary for the same period of time may overcome this problem. In addition, other problems arise; time diary data shows us, for example, that eating with family means a great variety of things to respondents. It could mean all the family members presently living in the same household, these persons plus some extended family, or some smaller subset of family members such as siblings, parents, one parent alone, one parent plus one or more siblings, cousins, etc. While we believe that most of these combinations count as 'family' in one sense, they may produce very different effects on their participants and it is not clear whether all respondents would refer to a meal with dad, for instance, as a family meal. The Houston study data indicate these combinations of mealtime companions are frequent. Another option is to focus measurement on a key family member such as the mother. However, mothers may not know about secret visits to the refrigerator at home or the vending machine at work. Mothers cannot always smell the hidden snacking on their children's breath. The opposite extreme is to require that all members of the selected family participate. In order to make such an approach feasible, researchers in this area must confront the fact that without incentives, obtaining the participation of multiple family members is very difficult.

10.3   Historical background

The three-meals-a-day pattern appears to have arisen in Western Europe as a result of industry pressure on workers to spend less time eating and more at work. Historically, four meals a day were eaten. In France, for example, the tendency was for these meals to be eaten at home (Aymard et al., 1996; Grignon, 1996). However, eating at home does not guarantee that these meals constituted what we would define as a family meal. At the same time, this does not mean that descriptions of members of 18th century American family members eating whenever they had the opportunity are accurate. The family residence tended to be the place of work and most family members who were not young children or children in school worked and ate there; family members were often joined by persons employed by the family (Gillis, 1996). However, Gillis seems to suggest that meals in the early colonial days were anomic and solitary affairs. Because family members were said to be at home all day every day, this may have encouraged grazing. It would be worth investigating further whether this was indeed practiced by most and considered normative. We could also inquire whether grazing was class-based or more universal. McWilliams (2005), for example, describes a prosperous-sounding family in 1650 as it goes about the business of preparing a large, sumptuous dinner. Each family member was assigned time-consuming tasks in preparing the meal and apparently ate this meal together. McWilliams' example thus may support the claim that family meals were, at least on special occasions, part of upper class life in the 1650s. This report may also be atypical, although it may also represent a normative depiction of ideal family rituals to which family members adhered to a lesser extent on days without ritual significance. Finally, Cinotto (2006) adds that family meals were made more feasible by the spread of the clock in the early 19th century.

10.4   Current meal pattern in the USA

For a variety of methodological reasons discussed above, there are no precise data that fully capture family member participation in meals. Without such data, it is difficult to speak precisely about the present and changing state of the family meal in the United States. In large part this is owing to data collection from individuals rather than from families and from the wording of questionnaires used in the study of meals. We navigate this data sinkhole by presenting a variety of meal-related data generated by a range of methodological approaches.

10.4.1   Frequency of eating meals and time spent eating

The most widely available data regarding meal consumption comes from respondents who report on their own eating. Data generated by dietary recall or dietary record methods general collect data on 'where' and 'with whom' the eating has taken place. However, most software designed to transform these intake data into analyzable form drop the 'where' and 'with whom' from the final dataset. Thus, we have no means of examining family meals with such data. However, we will start with such data because they are so plentiful. These data cover differing decades and base their conclusions on the number of days on which meal participation was measured.

Beginning with breakfast, there is some national-level evidence that the percentage of adults eating this meal declined by about 7% between 1970 and 2000 (Kant and Graubard, 2006) (see Table 10.1). However, most of the available survey data demonstrate a high level of adult participation in this

Table 10.1 Estimates of the frequency Americans eat breakfast

Type of individual	Year(s)	Eating breakfast (%)	Change (%)	Number of days of the week	Population	Study name	Author(s), year
Adults	1980/1994	85	0	1	US	CSFII	Borrud, 1996
Adults	1970/2000	82	−7%	1	US	NHANES	Kant and Graubard, 2006
Adults	1988/1994	74	na*	1	US	NHANES	Kerver et al., 2006
Fathers	2004	62.6	na	2	Houston, TX	Houston Study	McIntosh et al., 2006
Mothers	2004	63.3	na	2	Houston, TX	Houston Study	McIntosh et al., 2006
Children	1970/2001	82	−11%	1	US	NHANES	Kant and Graubard, 2006
Children	1973/1994	60	−26%	1	Louisiana	Bogalusa Heart Study	Nicklas et al., 2004
Children	2004	89.5	na	3	Houston, TX	Houston Study	McIntosh et al., 2006

* not applicable

meal (>80%). At the same time, recent data suggest that mothers and fathers in Houston ate breakfast less frequently. In addition, it appears that younger adults (aged 20–59) were more likely to skip breakfast than older adults (aged 60–90) (Howarth et al., 2007). Slightly larger declines (11%) in children's breakfast participation were found, based on NHANES data from 1993 to 1994, but this same study indicated that over 80% of these children ate breakfast. Two smaller studies provide conflicting evidence; the Bogalusa Heart Study (Louisiana) found that breakfast consumption declined in children by 26% from 1973 to 1994, resulting in a 60% participation rate (Nicklas et al., 2004). The second study of Houston children found a 90% breakfast participation rate (Table 10.1). The frequency of eating breakfast out has risen since the 1990s (Sloan, 2008); however, the contribution of this meal to energy and fat intake remained nearly the same for adults – around 17% (Borrud, 1996). Less than 10% of the Houston parents reported, over the two-day period when each parent maintained a time diary, eating breakfast away from home at least once, but over the three-day period reported on by their children, 20% of these children indicated they ate breakfast away from home at least once (Table 10.2).

Table 10.2 Percentage of fathers, mothers, and children participating in meals and the amount of time spent in those meals over a two-day period. Houston Texas Study

Eating event	People engaging in (%)	Mean time[d] (min)	Standard deviation[d]	Range[d] (min)
Breakfast[a] at home	58.7	18.4	10.8	2.5–52.7
Breakfast away from home	6.6	22.3	19.2	4.9–60.7
Lunch at home	33.1	17.6	8.9	3.7–45.0
Lunch away from home	68.8	35.3	20.4	7.5–135.7
Dinner at home	82.5	27.4	15.5	5.1–90.6
Dinner away from home	20.8	43.1	22.5	7.5–97.4
Breakfast watching TV	45.0	15.8	7.2	2.4–70.8
Dinner watching TV	10.9	36.3	18.5	15.0–106.1
Breakfast[b] at home	60.7	18.7	11.9	1.0–70.1
Breakfast away from home	5.8	16.0	10.6	3.8–45.5
Lunch at home	39.6	21.5	15.2	1.7–105.0
Lunch away from home	47.3	32.6	21.1	7.8–120.1
Dinner at home	84.3	28.2	17.9	9.9–105.3
Dinner away from home	23.4	43.7	29.1	10.0–180.2
Breakfast watching TV	7.6	13.6	6.8	5.1–84.9
Dinner watching TV	17.1	26.7	21.5	10.0–104.9
Breakfast[c] at home	89.5	27.6	17.7	3.4–106.7
Breakfast away from home	20.0	17.6	13.5	3.3–79.6
Lunch at home	54.0	14.3	11.4	1.1–66.7
Lunch away from home	80.0	24.2	16.8	3.6–133.3
Dinner at home	92.9	23.5	13.2	3.3–81.0
Dinner away from home	47.0	21.2	15.1	4.2–75.1

[a]Fathers' time diary, based on a two-day average; [b]mothers' time diary, based on a two-day average; [c]children's time diary, based on a three-day average; [d]excludes those skipping these activities

Adults skipped lunches more frequently than other meals, with 26% not participating in this meal (Kerver *et al.*, 2006). However, Houston data showed lower participation in this meal by mothers and fathers (Table 10.2). In Bogalusa, children's lunch consumption remained steady at near 90% from 1973–1974 to 1993–1994, where a decrease in school lunch participation was offset by an increase in children bringing lunch from home (Nicklas *et al.*, 2004). However, the Houston study found that only 54% of children participated in lunch (Table 10.2), mostly at school. Fewer studies of dinner participation by adults exist than do for children. National-level and more local studies indicate that nearly 90% of adults eat dinner (Howarth *et al.*, 2007). Over 90% of the children in the Houston study ate dinner, taking into account dinners at home and away from home. In summary, the trends in meal participation suggest no extreme declines by US residents. Again, these data tell us little about with whom if anyone these individuals eat.

Turning to the issue of how much time is spent eating, the American Time Use Survey (Bureau of Labor Statistics, 2006, 2008) has found that on average men ate and drank about 1.25 h per day; women, 1.22 h per day (Table 10.3); Black and Hispanic respondents spent less time eating than non-Hispanic whites; and obese individuals spent less time eating than normal and underweight individuals, particularly among females (Economic Research Service, 2008). Comparing these data from the ATUS with the American Use of Time Data (Robinson and Godbey, 1999), we observe that the time adult men and women spent eating decreased by roughly 0.24 h (or about 14 min) per day. Our Houston data show similar averages similar to those from the ATUS survey for mothers and fathers and we found that most of this time was spent eating at home (Table 10.2). However, Houston parents tended to eat longer dinners away from home than at home. The American Time Use Survey presented no data on where and with whom these eating episodes occurred, but others have suggested that meals eaten with children at the table take more time (see also Robinson and Godbey, 1999). Nearly half of the fathers in Houston ate breakfast watching TV, typically spending about 16 min doing so; less than 10% of mothers ate breakfast while watching TV, devoting nearly 14 min to this practice. Children in Houston averaged 73.1 min a day eating meals, 41.9 of which were devoted to meals at home. A

Table 10.3 Time spent in eating activities

Type of individual	Year(s)	Total time spent eating (h)	Study	Author(s), year
Men	2008	1.25	ATUS	Bureau of Labor Statistics, 2006
Women	2008	1.22	ATUS	Bureau of Labor Statistics, 2006
Fathers	2004	1.25	Houston	McIntosh *et al.*, 2006
Mothers	2004	1.18	Houston	McIntosh *et al.*, 2006
Children	2004	1.44	Houston	McIntosh *et al.*, 2006

large portion of this time was taken up by eating dinner; however the largest proportion was taking up by snacking, including the drinking of beverages. These children tended to spend more time eating breakfast and dinner when at home versus away from home, but spent slightly more time eating lunch away from home than at home (Table 10.2). This was probably due to their participation in lunches during school hours.

Siega-Riz *et al.* (2001) found that children do not necessarily follow consistent meal patterns from one day to the next. They classified children in terms of whether these children ate 2–3 meals plus snacks over a three-day period (consistent eaters), 2–3 meals plus snacks 2 of the 3 days (moderately consistent eaters), and one meal plus snacks over the three-day period (inconsistent eaters). Forty-one percent were found to be consistent eaters; only 4% fell into the inconsistent eater category. Males, Whites, school attendees, those who lived in a dual-headed household, and whose mothers had more than 12 years of education were more likely to be consistent eaters. Those who skipped meals tended to have poorer nutrition and be at risk of obesity (Ma *et al.*, 2003; Siega-Riz *et al.*, 2001; Kant and Graubard, 2006). Furthermore, those who skip meals may be skipping family meals, or perhaps more likely, are living in a context which does not encourage nor facilitate such meals. In conclusion, there appears to be no consistent evidence that either adults or children skip meals to a greater extent than in the past nor is there strong evidence that the amount of time spent eating has declined significantly in the past decade. It does appear, however, that participating in meals and the amount of time spent when eating those meals varies by age and other socio-demographic characteristics. It is also clear that a given individual's meal pattern varies over a given week.

10.4.2   Frequency of eating with others

The normative meal pattern includes breakfast, lunch, and dinner; however, there is not an expectation that any or all of the family be present for all of these meals. Breakfast and lunch are generally exempt from this expectation because of the demands of everyday life including work, school, and community involvement. A measurement difficulty arises because events such as coffee breaks and snacking are said to be on the rise, but for our purposes these may not be important sources of family eating as a group. Furthermore, some observers have described the current American meal pattern as 'grazing' and 'continuous eating' (Barer-Stein, 1999), which may interfere with or obviate family meals.

Despite such claims, Sobal and Nelson (2003) found in their survey of adults in New York that 14% ate breakfast, 18% ate lunch, and 67% ate dinner with family. The same percentage of adults ate dinner with family in a recent (2006) Roper Poll, which also noted that two-thirds of these dinners were cooked and eaten at home (Sloan, 2008). As in other studies, respondents' reports of eating with family often leave unclear who this includes and whether this constitutes the entire family, most of the family, or only a few fellow family members. More of Sobal and Nelson's (2003) respondents (21%) reported eating with a spouse rather than family, but again it is not clear whether these couples had children who were not eating with them or whether they were childless. Less than 5% reported skipping any of these meals, but nearly 60% ate breakfast alone, nearly half ate lunch alone, and about 20% ate dinner alone.

In Minnesota 59% of adults have reported that they 'sat down at dinner as a family' four times or more per week (Boutelle *et al.*, 2003, p. 27). Children in Houston indicated that on average they ate breakfast with family twice a week, lunch with family twice a week, and dinner with family less than twice a week. Of those who ate dinner 49% did so with family (Table 10.4). These children assessed that eating dinner with their families was somewhat important and 54% reported that at least one of their parents was at home while they were eating dinner seven days a week. In addition as Table 10.5 demonstrates, children may eat dinner with smaller sets of family members. However, dinner with family tended to involve a greater amount of time spent in this activity than when smaller sets of family members were present. Fathers from this same study reported few instances of eating either breakfast or lunch with their families, but more instances (and time spent) eating dinner with family at home versus away from home. Of interest here is that 52% of these fathers indicated they ate dinner with their families, requiring 35.4 minutes to do so (see Table 10.6). The discrepancy between the fathers' and children's reports of time spent at dinner suggests that children left the

Table 10.4 Meals eaten with family

Type of individual	Year	Type of meal	Individuals reporting family present[a] (%)	Study name; location	Author(s), year
Adults	2002	Breakfast	14	Meal commensality; Ithaca, NY	Sobal and Nelson, 2004
		Lunch	18		
		Dinner	67		
Children	2004	Breakfast	14.7	Parental time; Houston, TX	McIntosh *et al.*, 2006
		Lunch	4.4		
		Dinner	46.7		
Children	2002	All	21.4	EAT; Minnesota	Neumark-Sztainer *et al.*, 2003

[a]The Houston study identified a number of different combinations of other members of the family (e.g., mother, father, brother, sister); for example, 22.5% of Houston children reported eating with their mothers; 7.7% of them reported eating breakfast with parents only; 6% ate alone.

Table 10.5 Patterns of family dinners in which a child is involved. Houston Texas Study Children's time diaries

With whom	Eat dinner (%)	Mean time spent[a] (min)	Standard deviation (min)	Range (min)
Family	46.7	19.7	12.2	3.3–65.0
Parents	13.9	13.2	7.2	3.3–23.0
Mother	16.5	11.9	7.8	1.3–40.1
Father	2.4	9.0	4.2	3.3–20.0
Brother	5.3	9.2	6.3	3.4–23.3
Sister	2.5	8.5	2.4	5.0–10.0
Friends	2.8	15.6	11.5	5.2–30.9
No one	2.5	16.1	10.0	1.5–30.6
Other	6.9	11.4	5.6	1.2–16.2

[a]Children not eating dinner excluded.

Table 10.6 Patterns of family dinners in which fathers are involved. Houston Texas Study Fathers' time diaries

With whom	Eat dinner (%)	Mean time spent (min)[a]	Standard deviation (min)	Range (min)
Family	52.1	35.4	16.5	14.9–105.6
Children only	12.8	24.7	9.1	9.4–45.2
Spouse only	16.6	19.6	13.6	7.5–60.1

[a]Fathers report; excludes those fathers who did not report eating dinner on either of the two diary days.

dinner table much sooner than fathers did. Returning to children's breakfasts, the more time Houston children spent eating breakfast at home and breakfast away from home, the more frequently these meals were eaten with family, suggesting that without the presence of family, these meals would be much shorter in duration or skipped.

Finally, Neumark-Sztainer *et al.* (2003, p. 319) approached this issue by asking adolescents to report on how frequently they ate 'meals with all or most of their family living in their homes.' The adolescents reported an average of 4.5 times per week, which translates into 21.4% of possible meals over a seven-day period. It is likely that the majority of these meals were dinners. Again, while it is difficult to conclude whether family meal participation by children and/or adults has declined, there appears to be a substantial number of adults, parents, and children who continue to eat dinner with members of their families several or more times a week. In addition, some children are eating breakfast and/or lunch with family or parents nearly as often as in the past and children appear to consume these meals and spend more time in the meals when they do if their families are present.

10.4.3   Correlates of eating meals with others

Sobal and Nelson (2003) reported those in larger sized households were more likely to eat dinner with others. For the most part, eating with others was not a function of gender, ethnicity, work status, income, or education. We found in the Houston study that parental income was positively associated with children's time spent eating out for both lunch and dinner, as was mothers' education. However, children with older parents spent less time eating breakfast or lunch out. In such families, children also devoted less time to eating dinner with the whole family. However, parental age, income, and education had no effect on the time children gave to eating other meals at home with family members. Children whose mothers held high prestige jobs spent more time eating breakfast and lunch away from home, but less time eating lunch and dinner at home, but again this variable had no effect on time spent eating with family. In the Washington State study of family dinners, neither maternal marital nor work status made a significant difference in the time families engaged in eating dinner (Ramey and Juliusson, 1998).

The Houston data indicated that children's age was also associated with time involved in eating; older children spent less time eating breakfast, less time eating both breakfast and dinner with their families, and more time eating dinner alone. Male children spent less time eating breakfast at home and female children devoted more time to eating breakfast alone; non-white, non-Anglo children spent more time eating dinner at home and more time eating dinner with their mothers. When compared with children from two-parent families, children of single mothers expended less time eating dinner with family, but more time eating both breakfast and dinner with their mothers. However, children ate breakfast and dinner more frequently if their fathers, mothers, or the children themselves viewed dinner in ritual terms; that is, they viewed dinner as an important family activity. Houston children ate dinner with their families less frequently when their mothers report being 'too tired to cook when they [the mothers] get home at night.' These children ate more frequently with family when mothers reported that they shopped for foods the whole family enjoyed. In Houston families in which dinner time was frequently punctuated by arguments, children went out to eat with their families more often. Perhaps dinner becomes more palatable if the participants are shielded from conflict by the presence of the public.

Family meals may vary by family income. A related issue is poverty and in the USA a frequently used measure is the number of families in or close to poverty. Income thresholds (determined by income and number of persons living in the household) are used to calculate whether families are below, at, or above the poverty level. These calculations are presented in terms of the percent of poverty threshold; e.g., the number of families below the poverty threshold is referred to as families at 'less than 100% of the poverty level.' Using these thresholds, Childtrends Data Bank (2006a) estimated that 27%

of children aged 12–17 whose families were at less than 100% poverty level (below the poverty level) ate between zero and three times a week with their families compared with 32% of such children at the 200% or more than poverty level (above the poverty level) having done so. Of those children at 100% or less than the poverty level, 55% ate six to seven days a week with their family, versus 37% of those whose families were at 200% or more above poverty level. These findings suggest a positive relationship between poverty and eating meals as a family. Whatever difficulties they experience, the poor eat together more often than the better off (Childtrends Data Bank, 2006a). Of interest is that children from low income households, whose mothers had less education, or whose mothers never married were less likely to report mealtimes scheduled at the same time every day.

Older children ate with families less frequently (Childtrends Data Bank, 2006b; see also Milkie *et al.*, 2004). Similar to the income findings, parental education was negatively related to this frequency. While there were only slight rural urban differences, Hispanic children were far more likely (54%) to have eaten meals 6–7 days a week with their families than non-Hispanic black (39.5%) or non-Hispanic whites (38.7%). Among younger children (6–11 years of age) the negative effect of income remained as did the negative effect of parental education. Ethnic differences were similar among 6–11 year olds but the percent of children in each of these ethnic groups who ate more frequently with family was larger than was found among the 12–17 year olds. Children Trends drew these data from the 2003 National Survey of Children's Health. Results from the National Survey of Parents also showed that black parents reported less frequent family meals (Milkie *et al.*, 2004).

In summary, only a few characteristics of children affect their participation in meals. Older children tend to spend less time eating some meals. Gender was less important and evidence for the effects of parental social status (income, education, occupational prestige) on children's time spent eating is mixed. Parental work appeared to reduce time spent in family meals.

10.5   Impact of parental work and television on family meals

Two factors exist whose impact on the family is considerable. The first is employment outside the home and the second is the presence and use of television sets in the home. While only some attention has been given the impact of mothers' work outside the home and family eating patterns (Devine *et al.*, 2003; Devine *et al.*, 2006), researchers have devoted considerable attention to the impact of television on children (Fitzpatrick *et al.*, 2007).

10.5.1 Parental employment and family meals

Some claim that work dampens the propensity for families to eat dinner together. A few studies suggest that it is mother's work outside the home that causes this problem. Using the National Survey of Parents, others have found that the amount of time each parent (when two are present in the household) works reduced the number of meals a family eats together (Milkie *et al.*, 2004). Our Houston research indicated that father's work could have an equally negative effect. This claim has been borne out by others' research (Maddan, 2007) and is illustrated by a recent memoir *Dinner with Dad: How I Found My Way Back to the Dinner Table* (Stracher, 2007).

Children from Houston families in which both parents worked full-time spent less time eating at home and less time eating breakfast with their parents. This may mean that they were eating with their families away from home more frequently or were simply eating less with their parents when eating outside the home. A work-related concern by some has suggested that parents may have non-standard work schedules (involving some weekend days or hours that vary from an 8 am to 5 pm schedule), resulting in family problems (Presser, 2003). If Houston mothers' work schedule was 'standard,' their children spent more time eating at home. In families where fathers had no set work schedule, children tended to devote more time to eating takeout food. However, in families where both fathers and mothers had regular work schedules, children reported more time eating dinner at home. Irregular work schedules also made it less likely that a parent was present in the evening when children were eating dinner. In addition, parents with regular schedules suggested they were more able to participate in meals. Children whose parents had little flexibility in their work schedules spent more time eating food away from home, eating takeout at home, and eating breakfast alone and mothers with little work schedule flexibility had children who spent more time eating snacks.

Work stress is thought to make the routines of daily life away from work all the more difficult. Problems and preoccupation with work frequently spill over into family life, including into the lives of children. Thus, children whose fathers reported more job stress or experienced more work to family spillover spent more time eating lunch away from home. Fathers' job stress and spillover plus mothers' work spillover were all positively related to time spent eating takeout food. Mothers' job stress was also positively related to children's time eating breakfast away from home. The children of mothers who felt their job was important spent less time eating lunch and dinner at home but spent more time eating snacks. One thing that might be concluded from the above is that work matters in the scheduling of family meals. There is some evidence that families with working parents may substitute meals away from home as a means of experiencing food consumption as a family. Others may substitute convenience foods and takeout for home-made meals in order to share meals as a family. However, there is some evidence that the nutrient content of meals eaten in restaurants differs from home cooked meals, and may also affect ritual family meals (see below). Does the same level and type of communication described by observers of family meals take place when the setting is a restaurant?

In the Houston study, we also examined the relationship between parental work and the time children spend eating with others. Parental work variables were not associated with children eating dinner with their families nor did these variables affect the amount of time children spent eating dinner with their families. Mothers' marital strain decreased children's likelihood of eating dinner with their families, but increased the likelihood children ate dinner with their mothers. However, parental work did have an impact on who children ate with. Fathers' job stress increased the likelihood that children either ate dinner with their mothers or ate this meal alone and mothers' job stress and work to family spillover increased the likelihood of children eating dinner with their fathers and less time eating breakfast with their families. Children spent more time eating dinner alone if their mothers picked up takeout food for dinner. Finally, the more mothers reported that breakfast time was disorganized, the less time children gave to eating breakfast and eating breakfast with family. Clearly, parents' experiences at their workplace have effects on their children's participation in meals at home and away from home.

10.5.2 Television and family meals

The frequency with which families and individuals eat their meals while watching television has increased (Saelens *et al.*, 2002). Research on this phenomenon suggests that amounts and types of foods may differ when food is consumed in the presence of a television set and without such presence. Less well understood is the impact television has on family interaction during meals. While we have little direct evidence of this impact, Robert Putnam has suggested a likely causal relationship between the diminishment in social capital and the rise in numbers of hours spent watching television (Putnam, 1995). Putnam's observations lead us to speculate that these family interactions are fewer and more superficial than those that occur when meals are eaten with no television turned on. Of additional interest is the possibility that family meals as rituals are less likely when the television competes for the attention of family members while eating. Children in Houston reported eating more meals watching TV if their mothers were employed and were committed to work, but were less likely to do so if the mothers' work schedules were flexible. The more time children spent either watching TV or using a computer, the more time they devoted to eating breakfast alone. Finally, Houston children in homes where mothers and fathers reported eating meals while watching TV tended to do the same. Time spent watching television had some effects on eating with family, but most noticeably not on family dinners, except under the circumstance that parents chose to consume dinner in front of a television set.

10.6　Family rituals

Bell (1992; 1997, p. 5) noted that features common to all of the work on ritual include 'formality, fixity, and repetition.' However, there is less consensus regarding the basic elements of ritual. The formality and repetition of ritual is tied to its communicative function. Furthermore, rituals serve to create symbols and traditions, which, in turn, are means of social control. For some, rituals cement ties, especially those based on emotion, and thus 'legitimate the traditional nature of established relationships' (Cheal, 1988, p. 637–638). Strict conformity is promoted; however, some observers claim family rituals promote a superficial level of harmony and sociability, with participants masking their true feelings' (Johnson 1988, p. 686).

Family rituals have received specific attention by historians; Gillis (1996), Mestdag and Vandeweyer (2005), and others claimed that the family rituals we are familiar with today originate with the Victorian era. During this period, a common belief arose that children required molding and parents were best suited for this task (Hulbert, 2003). Family ritual was also meant to serve as a 'haven in a heartless world' to borrow from Lasch (1977). The family meal became a ritual thought to support the family institution. These authors were quick to observe that families performing such rituals generally had the means to do so. Those in the middle to upper ranks of the class system possessed the wherewithal to perform family rituals. Furthermore, adherence to the ideology of family ritual likely varied by ethnic group.

Sociologists concerned with child development and family relationships appear to have been among the first to investigate family rituals and their consequences for family members. This work began with Bossard's (1948) analysis of the transcripts of observations made of an unspecified number of family meals. Unfortunately, the author omitted methodological details such as sample size from this study. Bossard concludes from these observations that the family meal serves as 'a clearinghouse for most of the family's information, news, and experiences' (1948, p. 7). Children were said to learn about family roles during these encounters. While the encounters tended to be 'democratic,' they could also re-enforce power differentials in the family. This work was quickly followed by the Bossard and Boll (1950) study of family rituals. They utilized two means in their analysis: the first was a close reading of 100 biographies (73 of which had sufficient detail for analysis) written between 1870 and 1917. They attempted to identify family practices that were repetitive. Using descriptions of these rituals, they classified these events in terms of their perceived importance, required participation, predictability, degree they were scripted, and the degree they were meant to evoke positive emotions and family integration. Also, Bossard and Boll believed it was clear from the biographies that not all families, for example those located in rural areas, had such rituals. Furthermore, not all rituals were pleasant; many evoked negative feelings and memories. Bossard and Boll's second approach involved a survey fielded to approximately 87 undergraduates at a 'prominent East Coast university.'

As a result of these investigations, Bossard and Boll (1950, p. 9) concluded that family rituals involved 'a pattern of proscribed behavior pertaining to some specific event, occasion, or situation which tends to be repeated over and over again…[and] demands rather punctilious observance, admitting no, or at least very few, exceptions or deviations.' In addition, rituals required frequent and exact repetition, adherence to assigned roles, and emotional involvement.

10.6.1   Family meal rituals

Almost all of the 100 biographies studied by Bossard and Boll mentioned family meals. In some families, the meals fit the image of a ritual; in the rest they were simply a case of refueling. This was especially true among lower class and rural biographers. They found that 'in the lower class family, dinner was almost the same as breakfast…the family comes and takes it when and where they want to' (Bossard and Boll, 1950, p. 116). However, they found a tendency for lower class families to have a ritualized special Sunday dinner. The student interviews produced a large number of family rituals, including family meals. Family dinner was the most frequently mentioned of these occasions.

More recent work, while not about the rituals of family meals per se, has focused on the importance of those events involved in 'constructing the family.' In effect, this meant family members sense that they belong to an important social unit as a consequence of integrative activities such as meals eaten together (Charles and Kerr, 1988; DeVault, 1991). It seems clear from these works, furthermore, that family meals are perceived by women as events that should occur regularly and these meals should be carefully planned so that they are considered important by other family members.

Finally, it is clear that certain impediments may stand in the way of family meals. One of these is the television set. Boutelle et al. (2003) hinted that perhaps the same people who ate while watching television were those who claimed they were too busy to eat with their families. This is of particular interest because of the role that television plays in obesity (Caroli et al., 2004; Stroebele and de Castro, 2004; Van den Bulck, 2000). In addition, class differences may be associated with the ability to establish family meals rituals (Fitchen, 1992).

What do we know about family meal rituals in present times? Fiese et al. (2002) provide a useful review of previous research. Data from the Houston study as well as a study of two rural Texas communities indicated that family members perceived that their dinners were family meal rituals (McIntosh et al., 2004). We determined this by asking three members of families to assess the degree to which they perceived (1) these meals as both special and important, (2) attendance was expected, (3) were regular, and (4) most family members participated. More than half of our samples agreed with these perceptions; furthermore, family members' answers were concordant. That is, if one family member perceived dinner in family-ritual terms, other family members were likely to as well. We also found that some families, instead of having a ritualized dinner most nights, had instead a 'special family food night.' Perceptions that dinner was a family ritual were associated with children's perception that it was important to eat dinner with their family and the more frequently they reported doing so. Houston mothers who reported that dinner was a family ritual ate out less frequently, brought home takeout less often, and scheduled meals so the entire family could participate. They also were less likely to report favorable perceptions of convenience foods, and tended to serve foods their family liked. Finally, we found that family meal rituals had consequences for family meals. The amount of time spent by children eating alone was lower when children believed that dinner was a family ritual and children were more likely to eat breakfast with their families if their fathers considered dinner a family ritual. In conclusion, while family meal rituals have little effect on the time spent in meals with other family members, they do appear to increase the frequency of family members eating together and also have an impact on the foods served at these meals.

10.6.2   Parental employment and the family meal ritual

Drawing on data from the Houston study, we found the more flexible and standard parents' work schedules, the more likely family members viewed dinner as a family ritual. Once again work schedules appear to be the deciding factor in family meal practices. Of great interest, however, is that families were able to maintain the perception that their dinners together were an important feature of family life, regardless of parent work hours and experiences of stress and work to family spillover.

10.6.3   Home made food and family ritual

Those who study family meal rituals have neglected the food itself. However, recent work on homecooked meals has suggested a connection to family meal rituals and perhaps the contents of the meals themselves. Moisio et al. (2004) focused on the distinction of homemade as opposed to pre-prepared foods. The idea was that 'home cooked food is considered "authentic" in the creation of the family' (Moisio et al., 2004, p. 363). Informants from this study described 'market made food' as 'sterilized' and 'depersonalized'; in addition, its origins are less well-known. Most importantly homecooked food was seen as the result of an 'act of love' (Moisio et al., 2004, p. 368).

Others connoted family food as unique, fresh, and smelling and tasting good. A Dutch study of family food identified attributes associated with home-cooked meals, and suggested a way that American home-cooked meals might be further investigated (Costa et al., 2007). This study used hierarchical value mapping of homemade meals including a number of 'concrete attributes' such as 'daily task, low cost, shared, made by me, fresh (i.e., prepared from scratch with raw ingredients) and simple (i.e., prepared with basic cooking methods and few ingredients or seasoning)' (Costa *et al.*, 2007, p. 82). Furthermore, values associated with the cooking itself included 'doing my duty, keeping eating habits, enjoyment–pleasure, save money, socializing–belonging, and control' while the meal itself was 'healthy, tasty, and trustworthy' (Costa *et al.*, 2007, p. 82).

Approximately half of US meal preparers reported cooking 'completely from scratch' and another 20% 'mostly from scratch' (Mogelonsky, 1995, p. 15). Others have written that 'the capacity to share homegrown cooking knowledge is critical to our spirit, strength, and happiness as a society' (Zimmerman, 2003, p. 7). However, Sloan (2006) reported that dinners made from scratch have declined in the past few years, with only 32% of evening meals made this way in 2005. Twenty-six percent relied on convenience foods, 17% used restaurant or supermarket take-out, and 23% relied on a restaurant (Sloan, 2006, p.19). Dinners were 'cooked 4.9 times per week' but 20% of adults said they ate 'frozen dinners or store-made, pre-cooked meals' (Sloan, 2006, p. 24).

Others, however, have chosen to take advantage of pre-prepared meals. As one writer noted 'I like cooking, but often run out of time and creativity' (Tugend, 2007). She discovered that 'meal assembly' companies have sprung up, which allow customers to use the facility, food, and recipes for particular food items or dishes. Companies include 'Let's Dish' and 'Super Suppers' (Tugend, 2007). Some have reported on the price and quality of delivered meal services and described their prices, ease of reheating, and eco-friendliness (Gunn, 2006); while others have claimed that the use of convenience foods reflected the 'time famine' experienced by food preparers (Davies and Madran, 1997; Gofton, 1995). Some researchers have measured the time involved in preparing meals of varying complexity. Using videographic data from 32 families in Los Angeles County (CA), Beck (2007, p. 533) reported that commercial foods 'reduced preparation time by 10–12 min on average, and were almost always necessary for hands-on preparation times under 20 min.' Here 'hands-on' time included 'finding cooking utensils, washing or cutting ingredients, unwrapping packages, stirring a pot, and checking food to see if it is fully cooked' (Beck, 2007, p. 534). She also observed that the use of commercially prepared foods in cooking may permit the preparation of more complex meals. Beck's findings may in part be explained by constraints on food preparer time. Mancino and Newman (2007) found that women in higher income households or who worked longer hours outside the home devoted less time to meal preparation, although being married or having a larger family increased this time.

Of interest is the possible effect of the use of convenience and pre-prepared food on family meals. Do families in which these are used spend less time eating meals together? Are they less likely to gather together? Or do they bring benefits that reach the table, motivating participants to eat longer? We found that children's participation in family meals was reduced in households in which mothers report purchasing takeout foods for dinner more frequently.

10.7   Family meal outcomes

A growing number of studies address the impacts of family meals on children. The general theoretical model posits family meals as a normative framework or causative agent which produces some kind of variously measured outcome representative of child well-being. The ritual/normative approach describes family meals as repetitive moments of communicative action whereby children are inculcated with a set of social norms constitutive of healthy eating (Visser, 1992). Compañ-Poveda et al. (2002) argued a connection between adolescent mental health and family rituals such as family meals which reinforce familial cohesion and adolescent identity. Although less conceptually ambitious, a similar argument was put forward by Eisenberg et al. (2004) who identified the causal mechanism as a form of socialization reinforced by consistent routine.

Studies addressing outcomes of family meals on children and adolescents have identified a number of positive outcomes. These have included the reduced use of cigarettes, alcohol and marijuana; improved grades and self-esteem; and a reduction in depression, suicide ideation and suicide attempts (Eisenberg et al., 2004). Compañ-Poveda et al. (2002) identified an improvement in family function, measured by satisfaction with familial support, the family agreement process, perception of family acceptance and support of wishes, feelings of being loved by family, and the amount of time spent with family. Family communication and children's vocabularies have also been said to increase as a consequence of mealtime interaction (Lewis and Feiring, 1982). Furthermore, those who fear for the viability of the American family see the family meal as both a symptom of this demise and as a possible savior. The death of the family supposedly results from family members who live more individuated lives and are more likely to eat alone and at different times of the day than other family members. As a consequence, others in books with titles like *The Surprising Power of Family Meals* (Weinstein, 2006) have attempted to promote the family meal for the benefits it makes to shared time among family members and family solidarity. Some researchers have focused on the negative aspects of meals, particularly when particular family members such as parents make the experience unpleasant (Lupton, 1994). While we do not dispute Lupton's results, we also believe that given the resiliency of the family dinner, if this event is central to family survival, then the American family appears safe.

10.8   Foods found in American meals

Foods commonly consumed at breakfast include ready-to-eat cereals (eaten by 28% of consumers); breads, bagels, rolls and muffins (27%); eggs (16%); fruits (15%); milk (46%); coffee (average intake 2 cups) (33%); 100% fruit juice (19%); soda (average serving $1^3/_4$ cups) (5%) (Moshfegh et al., 2005).

Rates of consumption varied considerably by age group. Ready-to-eat cereal was consumed by 47% of persons between 2 and 11 years of age, but only 22% of those 20 years old or more did so. Milk was consumed by 70% of the 2–11 year olds, compared with only 39% of those 20 or older. In 'What We Eat for "Lunch" in America' Ahuja *et al.* (2005) found that 77% of the US population ate lunch and 'the most popular foods were sandwiches, followed by salads and meat dishes. The most popular beverages were soft drinks, coffee, and juice' (p. A–24). More recently one study has claimed that the sandwich was the most frequently served item at dinner, followed by chicken, beef, and Italian food (Heller, 2006).

Sloan (2006) reported the most popular foods purchased for home or in restaurants were hamburgers and French fries, followed by pizza. 'Stove top cooking and roasting are the two most common forms of home meal preparation.' Danford (2001) observed that cookbooks containing recipes that called for prepackaged foods were popular. 'About half of consumers ate BBQ foods regularly, 31% Southern, and 28% Tex Mex' (Sloan, 2006, p. 24). 'Italian, Mexican, and Chinese remained the most popular ethnic cuisines of food preparers' (Sloan, 2006, p. 26). An indication of the popularity of one of these cuisines is found in the magazine *Chinese Restaurant News*, which documented that there are approximately 43 000 Chinese restaurants in the United States (Chinese Restaurant News, 2007). German and Japanese meals have increased in popularity as well. Take-out food customers tended to prefer 'Asian food, pizza, pasta, Mexican food, and shellfish' (Sloan, 2006, p. 19). Nearly one-third of Americans ate so-called comfort foods such as 'meatloaf, stew, macaroni and cheese, pot roast, sausage...' on a weekly basis (Sloan, 2006, p. 20). Sandwiches remained the most popular food eaten out at lunch time and pizza and sandwiches have become popular restaurant foods (Sloan, 2006). Regional differences in foods eaten developed because of availability of foods in the particular area and the cultural backgrounds of immigrants who populated the region. Margaret McWilliams discussed these origins in her book *Food Around the World: A Cultural Perspective* (2007). In the northeast, American Indians contributed wild turkey, corn and maple syrup to the plain dishes brought by the Pilgrims from England. The results include Boston clam chowder and Indian pudding. Immigrants from the UK, the Caribbean (originally from France and Spain) and slaves from Africa, using greens, pork, cornmeal, and okra, created southern cooking. This was characterized by deep-fried chicken, cornbread, ham, beans with salt pork, and greens. Midwestern immigrants came from Scandinavia, Germany, UK, and Italy with their preference for simple foods like mashed potatoes and gravy, fried chicken, corn on the cob, and apple pie. Preferences for cheese, sausage, beer and pizza remain in this area of the country.

Southwestern cuisine was strongly influenced by the influx of Spaniards from Mexico. Mexican foods with a Spanish influence, such as enchiladas, tacos, guacamole, pinto beans, and tortillas, are common, as is beef. The latter is a reflection of the importance of cattle in the history of this area of the country. The influence from a large number of cultures, particularly those of China, Mexico, Spain, India, and Vietnam, is evident on foods of the Western United States. Combined with a vast array of fruits, vegetables, fish, and other foods produced in the region, the result is what is termed by McWilliams as 'fusion' (2007).

Linda Stradley's (2004) 'What's Cooking America?' website provided additional examples of regional foods. Foods from the Deep South included boiled peanuts, fried catfish, chess pie, chitterlings, Hoppin John, and Southern Fried Chicken (Stradley, 2004). The West Coast featured avocado pie, French dip sandwiches, fish tacos, oyster cocktails, and Hawaiian Spam Musubi. The Mid-Atlantic region touted its Hoagie, Dagwood, Italian, and Beef on Weck sandwiches, Manhattan clam chowder, and the Philadelphia Cheese Steak. The Midwest/Plains region provided Kansas City barbecued ribs, bierocks, St Louis Toasted Ravioli, and gooey butter cake. Barer-Stein (2003, p. 36) noted that much of the East coast was characterized by 'flour mixes such as pancakes, waffles, doughnuts, and even cookies [which] are basically of Dutch origin.' The Midwest, by contrast was settled by Scandinavians and Germans, giving this region 'Scandinavian soups and German beer soups which vie for popularity with Ukrainian Borscht and Finnish cold buttermilk soup' (Barer-Stein, 2003, p. 37). The Northeast provided 'the New England Boiled dinner which contains picklèd beef, onions, turnips, and potatoes' (Barer-Stein, 2003, p. 37). Others described specific foods eaten by those of various ethnic backgrounds. Some examples: foods eaten at breakfast: scrapple, the left over parts of pigs which is of German origin and adapted in Philadelphia; eaten with eggs (Wolf and Smith, 2006). Also scrapple is found in the Shenandoah Valley of Virginia where persons of German ancestry settled. For those of Mexican origin, Huevos Rancheros for breakfast was a popular dish. Julia Child, renowned cook and authority on food preparation, defined American cuisine as a hybrid of many ethnic cooking styles (Algert, 2004).

Regional cuisines are blurring because of changes in demographics and in lifestyle. According to the 2000 Census, 51% of the population of the USA came from Latin America, 26% from Asia, 15% from Europe, and about 8% from other regions (United States Census Bureau, 2000b). This demonstrates a change in immigration with far fewer coming from Europe and many more arriving from Latin America and Asia (Satia-Abouta, 2002). Given the demographic shifts suggested by new migration patterns, other cuisines may either join or displace Italian, Mexican, and Chinese foods as home prepared favorites. Cookbook sales of Vietnamese cuisine, for example, have increased (Dahlin, 2000). In 2002, Tillotson reported that Americans worked more hours per year with less vacation time than any other industrialized country in the world (2002). Consequently, the common complaint about not having enough time to cook may be correct and has spawned a revolution in food preparation. Takeout and ready-to-eat foods have increased to the point that they represent half of the food expenditures by people in the USA (Tillotson 2002). Therefore, food consumption and home cooked meals were changing even more. At the same time, cookbooks remained around 1% of all best sellers, reflecting a continued desire to cook meals at home (Puente, 2003), although some of the current interest in these books has to do with a desire to lose weight (USA Today, 2004).

The above discussion focuses on aggregate eating trends of individuals rather than families, so great care should be taken in drawing inferences about what foods families are consuming. As mentioned earlier, publicly available data sets such as NHANES are based on individual reports of what was eaten in the past 24 h. Food consumption data available from NHANES, because of transformations, comes in the form of nutrient intake. Knowing the number of grams of protein in a person's diet tells us little about the specific foods or the regularity of consumption that resulted in this level of protein intake. In order to capture foods individuals eat and with whom these foods are eaten requires additional codes for these data not a change in data collection methods Typical dietary recalls capture the names of the foods eaten, who the eater was with, and where the eating takes place, but these useful data are usually not coded.

10.9   Future trends

The attention given the American family meal and its perceived importance by both researchers and by the press suggests that families may make a greater effort to eat meals, particularly dinner, together. However, as our own data suggest, family meals may not always involve every family member living in the household. Parental work schedules and children's after school activities will likely continue to preclude full participation. Furthermore, family meal rituals remain important for many families, increasing the likelihood that these families will continue to make an effort to eat at least some meals as a social unit. Television watching during meals may represent a threat to such meals.

10.10   Sources of further information and advice

Lee H G, (1992), *Taste of the States: a food history of America*, Charlottesville, Howell.

Wilson D S and A K Gillespie, (1999), *Rooted in America: food lore of popular fruits and vegetables*, Knoxville, University of Tennessee Press.

http://old.lib.ucdavis.edu/exhibits/food/

http://www.nal.usda.gov/fnic/pubs/bibs/gen/ethnic.pdf

10.11   References

Acock A C, (1999), 'Quantitative methodology for studying families', in Sussman M B and SK Steinmetz, and GW Peterson, *Handbook of marriage and family*, second edition, New York: Plenum Press, 263–290.

Ahuja J K C, Omolewa-Tomobi G and Moshfegh A J, (2005), 'What we eat for "lunch" in America', *J Am Diet Assoc*, **105**(8) Supplement 2, A–24. doi:10.1093/her/cyl110.

Algert S (2004), 'Julia Child at 91 comments on American culinary culture', *Nutr Today*, **39**(4): 154–156.

Aymard M, Grignon C and Sabban F, (1996), 'Introduction', *Food and Foodways*, 12 (3/4), 161–185.

Arredondo E M, Elder J P, Ayala G X, Campbell N, Baquero B and Duerksen S, (2006), 'Is parenting style related to children's healthy eating and physical activity in Latino families?', *Health Education Research*, **21**(6), 862–871.

Barer-Stein T, (1999), *You eat what you are: people, culture, and food traditions*, Buffalo, Firefly.

Baumrind D, (1991), 'The influence of parenting style on adolescent competence and substance use', *J Early Adoles*, **11**(1), 56–95. doi: 10.1177/0272431691111004.

Beck M E, (2007), 'Dinner preparation in the modern United States', *Br Food J*, **109**(7), 531–547. doi: 10.1108/00070700710761527.

Bell C, (1992), *Ritual theory, ritual practice*, New York: Oxford University Press.

Bell C, (1997), *Ritual: Perspectives and dimensions*, New York: Oxford University Press.

Bisogni C A, Falk L W, Madore E, Blake C E, Jastran M, Sobal J and Devine C M, (2007), 'Dimensions of everyday eating and drinking episodes', *Appetite*, **48**(2), 218–231. doi:10.1016/j.appet.2006.09.004.

Borrud L, (1996), 'What we eat in America: USDA surveys food consumption changes', *Food Rev*, *19*(September–December), 4–8.

Bossard J H, (1948), *The sociology of child development*, New York: Harper and Brothers.

Bossard J H and Boll E, (1950), *Ritual in family living: A contemporary study*, Philadelphia, University of Pennsylvania Press.

Boutelle K, Birnbaum A, Lytle L A, Murray D M, and Story M, (2003), 'Associations between perceived family meal environment and parent intake of fruit, vegetables, and fat', *J Nutr Ed Behav*, **35**(1), 24–29.

Bureau of Labor Statistics (BLS), (2006), 'American Time Use Survey. Results', www.bls.gov/newsrelease/atus.t01.htm Accessed 9 February 2008.

Bureau of Labor Statistics, (2008), 'Time Adults Spend in Various Activities', http://www.bls.gov/tus/home.htm#charts Accessed 9 February 2008.

Caroli M, Aggentieri L, Cardone M, and Masi A, (2004) 'Role of television in childhood obesity Prevention,' *Int J Obes* **28** (supplement 3), s104–s108, doi:10.1038/sj.ijo.0802802

Charles N and Kerr M, (1988), *Women, food, and families: power, love, and anger*. Manchester, UK: University of Manchester Press.

Cheal D, (1988), 'The ritualization of family ties,' *Am Behav Scient*, **31**(6), 632–643.

Child Trends Data Bank. (2006a). 'Family meals'. www.childtrendsdatabank.org Accessed 2 March 2008.

Child Trends Data Bank. (2006b). 'Regular bedtime and mealtime', www.childtrendsdatabank.org Accessed 2 March 2008.

Chinese Restaurant News. (2007), 'Cover Page', *Chinese Restaurant News*, **13**(2), 1.

Cinotto S (2006). '"Everyone would be around the table": American family mealtimes in historical perspective, 1850–1960', *New Dir Child Adoles Dev*, *111* (Spring), 17–34.

Compan J, Moreno M T, Ruiz M T, and Pascual R E, (2002), 'Doing things together: adolescent health and family rituals', *J Epidemiol Comm Health*, **56**(2), 89–94.

Conger R D and Elder G H, (1994), *Families in troubled times: adapting to change in rural America*, New York: De Gryuter.

Copeland A P and White K M, (1991), *Studying families*, Newbury Park: Sage.

Costa A I de A, Schoolmeester D, Dekker M and Jongen A, (2007), 'To cook or not: an means-ends study of motives for choice of meal solutions', *Food Qual Pref*, **18**(1), 77–88, doi:10.1016/j.foodqual.2005.08.003.

Dahlin R, (2000), 'Stirring the sales pot', *Publishers Weekly*, **247**(30), 18–19.

Danford N (2001), 'The way to a nation's heart', *Publishers Weekly*, **248**(49), 21–22.

Davies G and Madran C, (1997), 'Time, food shopping, and food preparation: some attitudinal linkages,' *Br Food J*, **99**(3), 80–88, doi: 10.1108/00070709710168914.

DeVault M, (1991), *Feeding the family: the social organization of caring as gendered work*. Chicago, University of Chicago Press.

Devine C M, Jastran M, Jabbs J, Wethington E, Farell, and Bisogni C A, (2006), '"A lot of sacrifices": work-family spillover and the food coping choice strategies of low-wage employed parents', *Soc Sci Med*, **63**(10), 2591–2603. doi: 10.1016./j.socscimed.2006.06.029.

Devine C M, Connors M M, Sobal J, and Bisogni C A, (2003), 'Sandwiching it in: spillover of work onto food choices and family roles in low- to moderate-income urban households', *Soc Sci Med*, **56**(3), 617–630.

Douglas M and Gross, J, (1981), 'Food and culture: measuring the intricacies of rule systems', *Soc Sci Info*, **20**(1), 1–35.

Douglas M, (1982), *In the active voice*, London: Routledge and Kegan Paul.

Durkheim E. (1938) (1962), *The rules of sociological method*, London, Free Press.

Economic Research Service, (2008), 'Eating and health module (ATUS): Table 6', USDA. Accessed May 25, 2008 www.ers.usda.gov.

Eisenberg M E, Olson R E, Neumark-Sztainer D, Story M and Bearinger L M, (2004), 'Correlations between family meals and psychosocial well-being among adolescents', *Arch Pediatr Adoles Med*, **158**(8), 792–796.

Fiese B H, Tomcho T J, Douglas M, Josephs K, Poltrock S, and Baker T, (2002), 'A review of 50 years of research on naturally occurring family routines and rituals: cause for celebration?' *J Fam Psychol*, **16**: 381–390, doi: 10.1037/0893-3200.16.4.381.

Fitchen J, (1992), 'On the edge of homelessness: rural poverty and housing insecurity', *Rural Sociol*, **57**(2), 173–193.

Fitzpatrick E, Edmunds L S, Dennison B A, (2007), 'Positive effects of family meals undone by television viewing', *J Am Diet Assoc*, **107**(4), 666–671. doi: 10.1016/jjada.2007.01.014.

Gilgun J F (1999), 'Methodological pluralism and qualitative research' in Sussman M B, Steinmetz S K, and Peterson G W, *Handbook of marriage and family*, Second edition, New York: Plenum Press, 219–262.

Gillis J, (1996), *A world of their own making: myth, ritual, and the quest for family values*, New York: Basic Books.

Gofton L, (1995), 'Dollar rich and time poor? Some problems in interpreting changing food habits', *Br Food J*, **97**(10), 11–16, doi: 10.1108/00070709510104295.

Grignon C, (1996), 'Rule, fashion, work: the social genesis of the contemporary French meal pattern', *Food and Foodways*, **6**, (3–4), 205–242.

Gunn E, (2006), 'Home and family: outsourcing your family dinner', *Wall Street J*, June 15, p. D.6.

Heller L, (2006), *Convenience trumps health in American eating choices, report*, Food USA Navigator, October, 26. www.foodnavigator-usa.com accessed 16 April, 2008.

Howarth N C, Huang T T-K, Roberts S B, Lin B-H, and McCrory M A, (2007), 'Eating patterns and dietary composition in relation to BMI in younger and older adults', *Int J Obes*, **31**(4), 675–684, doi:10.1038/sj.ijo.0803456.

Hulbert A, (2003), *Raising America: experts, parents, and a century of advice about children*, New York: Alfred A Knopf.

Johnson C L, (1988), 'Socially controlled civility', *Am Behav Scient*, **31**(6), 685–701.

Kant A K and Graubard B I, (2006), 'Secular trends in patterns of self-reported food consumption of adult Americans: NHANES 1971–75 to NHANES 1999-2002', *Am J Clin Nutr*, **84**(5), 1215–1223.

Kerver J M, Yang E J, Obayashi S, Bianchi L, and Song W O, (2006), 'Meal and snack patterns are associated with dietary intake of energy and nutrients in US adults', *J Am Diet Assoc*, **106**(1), 46–53, doi:10.1016/j.jada.2005.09.045.

Lasch C, (1977), *Haven in a heartless world*, New York: Basic Books.

Larzelere R E and Klein D M, (1987), 'Methodology', in Susman M B and Steinmetz S K, *Handbook of marriage and the family*, New York: Plenum Press, 125–155.

Lewis M and Feiring C, (1982), 'Some American families at dinner', in Laosa L M and Segel I E, *Families as learning environments for children*, New York: Plenum, 115–145.

Lupton D, (1994), 'Food, memory, and meaning: The symbolic and social nature of food events', *Sociol Rev*, **94**(4), 664–686.

Ma Y A, Bertone E R, Stanek E J, Reed G W, Herbert J R, Cohne N L, Merriman P A and Ockene I S, (2003), 'Association between eating patterns and obesity in a free-living US adult population', *Am J Epidemiol*, **158**(1), 85–92, doi:10.1093/aje/kwg117.

Mancino L and C Newman, (2007), *Who has time to cook? How family resources influence food preparation*, USDA Economic Research Service Report Number 40, Washington, DC, United States Department of Agriculture.

Marshall D, (2000), 'The British meal', in Meiselman H L, *Dimensions of the meal: the science, culture, business, and art of eating*, Gaithersburg, Aspen, 202–220.

Masddan H, (2007), 'Make room for daddy: long workdays keep many fathers from the dinner table, but some are regaining precious family time', *San Francisco Chron* Sunday, June 17, www.sfgate.com/cgi-bin/article.cgi?file=/c/a/2007, accessed 21 March, 2008.

McIntosh A, (1999), 'The family meal in global times', in Grew R, *Food in global history*, Boulder: Westview Press, 217–239.

McIntosh W A, Torres C C, Nayga R, Anding J, Kubena K S, (2004), 'The family meal ritual: an introduction, description and assessment,' presented at the Annual Meeting of the Association for the Study of Food and Society, June.

McIntosh W A, Davis G, Nayga R, Anding J, Torre C C and Kubena K S, (2006), Parental Time, Income, Role Strain and Children's Nutrition. Final report to USDA ERS FANRP. http://www.ers.usda.gov/Publications/ccr19/.

McWilliams J A, (2005), *A revolution in eating: how the quest for food shaped America*, New York: Columbia University Press

McWilliams M. (2007) *Food around the world: a cultural perspective*. 2nd edn. New Jersey: Pearson Prentice Hall.

Mestdag I and Vandeweyer J, (2005), 'Where has family time gone? In search of joint family activities and the role of the family meal in 1966 and 1999', *J Fam Hist*, **30**(3), 304–323, doi: 10.1177/0363199005275794.

Milkie M, Mattingly M J, Nomaguchi K M, Bianchi S M, and Robinson J P, (2004), 'The time squeeze: parental statuses and feelings about time with children', *J Marriage Fam*, **66**(6), 739–761.

Moisio R, Arnould E J and Price L L, (2004), 'Between mothers and markets: constructing family identity through homemade food', *J Consum Cult*, **4**(3), 361–384, doi: 10.1177/1469540504046523.

Mogelonsky M, (1995) 'Cooking from scratch goes full speed', *Am Demogr*, **17**(3), 15–16.

Moshfegh A, Goldman J and Cleveland L, (2005), 'Breakfast in America, 2001–2002', in *What We Eat in America*, www.ars.usda.gov/ba/bhnrc.fsrg, accessed 5 February 2008.

Moshfegh A and Goldman J, (2006), 'Changes in the dietary patterns and food intakes of children over the past 25 years', *J Am Diet Assoc*, **106**(8) Supplement 1, A–35, doi:10.1016/j.jada.2006.05.126.

Neumark-Sztainer D, Hannan P J, Story M, Croll J and Perry C, (2003), 'Family meal patterns: associations with sociodemographic characteristics and improved dietary intake among adolescents', *J Am Diet Assoc*, **103**(3), 317–322, doi:10.1053/jada.2003.50048.

Nicklas T A, Morales M, Linstres A, Yang S-J and Baranowski T, (2004), 'Children's meal patterns have changed over a 21-year period: the Bogalusa Heart Study', *J Am Diet Assoc*, **104**(5), 753–761, doi:10.1016/j.jada.2004.02.030.

Presser H B, (2003), *Working in a 24/7 economy: challenges for American families*, New York: Russell Sage Foundation.

Puente M, (2003), 'Culinary interest is really cooking', *USA Today*, 28 April, 4d.

Putnam R, (1995), 'Tuning in, tuning out: the strange disappearance of social capital in America', *PS: Polit Sci Polit*, **28**(4), 664–683.

Ramey S L and Juliusson H K, (1998), 'Family dynamics at dinner: a natural context for revealing basic family processes', in Lewis M and Feiring C, *Families, risk, and competence*, Mahwah: Lawrence Erlbaum, 31–52.

Robinson J P and Godbey G, (1999), *Time for Americans: the surprising ways Americans use their time*, second edition, University Park: Penn State University Press.

Saelens B E, Sallis J F, Nader P R, Broyles S L, Berry C C, and Taras H L, (2002), 'Home environmental influences on children's television watching from early to middle childhood', *J Dev Behav Pediatr*, **23**(3): 127–32.

Satia-Abouta J, Patterson R E, Newhouser M L, and Elder J K, (2002), 'Dietary acculturation: applications to nutrition research and dietetics', *J Am Diet Assoc*, **102**, 1105–1118.

Siega-Riz A M, Cavadini C and Popkin B, (2001), 'US Teens and nutrient contribution and differences of their selected meal patterns', *Fam Econ Nutr Rev*, **13**(1), 15–26.

Sloan A E, (2006) 'What, when, and where America eats', *Food Technol*, **60**(1), 19–27.

Sloan A E, (2008), 'What, when, and where America eats', *Food Technol*, **62**(1), 20–29.

Sobal J and Nelson M, (2003), 'Commensal eating patterns: a community study', *Appetite* **41**(2), 181–190, doi:10.1016/S0195-6663(03)00078-3.

Stracher C, (2007), *Dinner with dad: how I found my way back to the family table*, New York: Random House.

Stradley L, (2004), 'What's cooking America?' //http://whatscookingamerica.net/AmericanRegionalFoods, accessed 21 March 2008.

Stroebele N and de Castro J M, (2004) 'Television viewing is associated with meal frequency in humans.' *Appetite* **42**(1), 111–113, doi:10.1016/j.appet.2003.09.001.

Tillotson J E, (2002), 'Our ready-prepared ready-to-eat nation', *Nutr Today*, **37**(1), 36–38.

Tugend A, (2007), 'Assembled off site, the somewhat homemade family dinner', *New York Times*, 29 September, p. C.5. http://proquest.umi.com/pqdweb?index=0andsid=1andsrchmode, accessed 11 March 2008.

USA Today, (2004), 'The nation's best sellers', 28 April, 1d.

United States Census Bureau, (2000a), 'US foreign-born population 2000', Population Division, Ethnic and Hispanic Statistics Branch, http://www.census.gov/population/socdemo/foreign/p20-534/slideshow/tsld002.htm accessed 21 May 2008.

United States Census Bureau, (2000b), 'Standards for defining metropolitan and micrometropolitan statistical areas; Notice', *Fed Reg*, **65**, 249, 82227–82238.

Van den Bulck J, (2000), 'Jane, "TV bad" saeth Tarzan: is television bad for your health….?' *J Youth Adoles*, **29**(3), 273–288, doi: 10.1023/A:1005102523848.

Visser M, (1992), *The rituals of dinner: the origins, evolution, eccentricities, and meaning of table manners*, New York: Penguin.

Visser M, (1988), *Much depends on dinner: the extraordinary history and mythology, allure and obsessions, perils and taboos of an ordinary meal*, New York: Collier Books.

Weinstein M, (2006), *The surprising power of family meals: how eating together makes us smarter, stronger, healthier, and happier*, Hanover: Steerforth Press.

Wolf B and Smith A F, (2006), *Real American food: restaurants, markets, and shops plus favorite hometown recipes*, New York: Rizzoli International Publications.

Zimmerman J, (2003), *Made from scratch: reclaiming the pleasures of the American hearth*, New York: Free Press.

11

The family meal in Europe

C. Fjellström, Uppsala University, Sweden

Abstract: Meal frequency and irregular eating, characterized by skipping traditional regular main meals in favour of discontinuous snacking, has been explored in relation to arguments about 'grazing' becoming more prevalent. In Europe, 'grazing' has not been found to be a major trend at the beginning of the 21st century; on the contrary, a structured meal pattern is still common all over Europe. Sharing meals in a family context seems also to be a lively and present activity all over Europe. It is also an activity that European family members value as important. We can say that those people living together as a family most likely also eat together, and this is true mostly concerning the evening meal or dinner. The family meal seems to be a precursor of both nutritional and mental health. Thus, it is not the question of eating together that is a problem for European families in the beginning of the 21st century, but of what foods they decide to share together in everyday life.

Key words: family meals, Europe, commensality, health, nutrition, grazing.

11.1 Introduction

During recent decades, food researchers have shown an increasing interest in people's meal frequency. It has been pointed out that the three-meal-a-day pattern is common in most cultures around the world, because in most languages there are precisely three distinct meal definitions (Pliner and Rozin, 2000). However, irregular eating – characterized by skipping traditional regular main meals in favour of discontinuous snacking, defined as grazing (Caplan, 1997) – has been recognized as a growing problem in modern society (Kelder *et al.*, 1996). Although grazing has not been found to be a major trend in, for example, the Nordic countries (Prättälä, 2000), Sjöberg and co-workers (2003) found that in-between meals contributed to the major part of Swedish adolescents' energy intake, and not the traditional main meals, as could perhaps be expected. Eating and drinking episodes can be described in many different ways and using diverse definitions. When these eating patterns are placed in varied contexts, they will have different importance to people in everyday life. For instance, dinner is a label mostly associated with sharing food with others, in the late afternoon or evening, and with socializing and conversing (Bisogni *et al.*, 2007). This is how many of us perceive dinner for people around the world, as sharing food in the company of family members at home (Charles, 1995; Mäkelä, 2000; Sidenvall *et al.*, 2000; Wood, 2003; Mattsson Sydner *et al.*, 2007). In the Mediterranean, lunch may be considered the main meal; however, this says nothing about lunch being synonymous with sharing the meal with the family (Tessier and Gerber, 2005).

The connotations of the main meal and of the family meal are not necessary the same. To many, the family meal is usually synonymous with dinner. We may ask ourselves, however, what we define as a family member. Is it the partner we live with, the person we have children with? Is it the nuclear family we associate with the family dinner or are other relatives included? Can an au pair living with a family for longer periods be included and integrated into the family context or does she experience a 'false kin' relationship (Cox and Narula, 2003)? According to Cox and Narula, one third of 144 au pair girls in London households normally ate only with the children in the families they worked for. These au pairs were excluded from a closer relationship with the whole family, of having a kin relationship, and moreover they were treated as children in that they were not served or allowed access to 'adult food'. There are, thus, many different household structures in today's consumer society; however, the present chapter will focus on the nuclear family.

One way of discussing family meals or eating patterns in families would be to look at the eating patterns and daily rhythm of eating, the food itself and the social context of eating (Kjærnes, 2001). Another way would be to examine the family meal from a multidisciplinary perspective, in which three underlying principles can be attached to the phenomenon of family members sharing meals, benefits especially for children and adolescents concerning socialization, mental health and nutrition (Larson *et al.*, 2006).

In this chapter, I will try to capture the family meal from different perspectives based on studies carried out foremost in Europe, but also in North America. In one Nordic study, the definition of a family meal was eating events that fulfil the following criteria: the meal is eaten by a person who lives in a multi-person household, it takes place in the home, it is eaten in the company of other family members, and it is eaten by all household members at the same time (Holm, 2001). This definition did not include eating together in the evening. In another definition of the family meal, discussed by the same Nordic research group, the type of food and how it is cooked was also included in the family meal definition, i.e. the family meal had to be a 'proper meal' (Kjærnes, 2001), using a term from Charles and Kerr's (1988) and Murcott's (1982) studies. The definition of the family meal used here will be much broader: a meal shared with family members in the home. This broad definition is vital when discussing the family meal in Europe, owing to the highly varied data collection methods and definitions found in studies on this phenomenon, conducted by scholars from many different disciplines and scientific paradigms, and in different cultures. The focus will be on adolescents and children. The starting point will be to discuss the existence of the family meal today and in the past.

11.2   The existence of the family meal today

When discussing the existence of the family meal, there may be different associations to this phenomenon regarding frequency and participation, everything from all family members sharing a meal together once a week to eating together every day. Thus, it is difficult to establish when a regular family meal eating pattern is practised. What is the reality of the family meal today in the beginning of the 21st century? In the USA, the problems associated with the declining family meal have been discussed for decades, perhaps more so than in Europe, and lately several studies have examined the family meal in the American culture (Freeman, 2005; Weinstein, 2005). A thorough analyses of the American family meal is given in Chapter 10 by McIntosh *et al.*

In a recent study in Minneapolis, two-thirds of the adolescents reported eating regular family meals, meaning three or more family meals during the past week (Feldman *et al.*, 2007). Because the USA is a country with many cultures, races and ethnic groups, these factors have also been studied in relation to the family meal. More white and Hispanic families eat together regularly than black families do, according to another study conducted in New York (Fitzpatrick *et al.*, 2007). When adolescents and their parents are asked about family meal frequency, their answers can differ (Fulkerson *et al.*, 2006). Older adolescents (school grades 10–12) seem to avoid the family dinner more than do the younger age groups (school grades 7–9). Girls shared more evening meals with their parents than did boys. Almost 60% of younger adolescents and parents reported eating five or more family meals a week. The figures for older teenagers were nearly 40%. Both parents and their children had positive expectations about eating meals with their families, and they also believed that this was important, although parents more strongly agreed with this when responding to a questionnaire item (Fulkerson *et al.*, 2006). Another study by Fulkerson *et al.* (2008) looked specifically at family meal routines among young adolescents (8 to 10-year-olds) and their families. They demonstrated that the family dinner could just as well be eaten at full-service restaurants, purchased from fast-food establishments or picked up as take-away food. Thus, the food did not need to be either home cooked or eaten at home to be defined as a family meal. In their study, 98% of the parents reported that all or almost all family members had shared dinner/supper in the home three or more times during the past week. Eighty per cent said they had eaten dinner together more than five times during the past week (Fulkerson *et al.*, 2008).

In European studies, the family meal has also been targeted as an important issue in modern society (Bugge, 2006; Romani, 2005; Carrigan *et al.*, 2006; Fernández-Aranda *et al.*, 2007). Figures indicating whether the family meal in Europe actually exists as a daily activity have also been illuminated. The eating patterns and the family meal in the Nordic countries (Denmark, Finland, Norway and Sweden) showed similarities and differences (Kjærnes, 2001). Most families had three to five meals a day, but in Sweden and Finland, two of these meals were usually a hot meal, while in Norway and Denmark only one hot meal was consumed daily. One could say that, in Sweden and Finland, it was more common to eat two 'proper meals' each day, compared with Norway and Denmark, where only one such meal was eaten daily. In this context it is interesting to refer to Haastrup's (2003) discussion about the differences on how and why school meals are managed and financed in Denmark and Sweden. In Denmark, school meals consists of food packages (sandwiches) brought from home, while in Sweden almost all school meals, cooked hot meals served at school, are paid by the state. These two different ideologies, the social liberal ideology opposed to the welfare ideology can have an impact on the family meal. In Denmark, where cooked meals are not eaten by children during the school day, the family meals in the evening have become a much more highly valued family activity, than in many Swedish homes. The day before the interview in the study by Kjærnes and colleagues (2001), 71% of the Danish sample, 67% of the Norwegians and 63% of the Swedes had eaten together with family members. Sweden showed the most heterogenic eating patterns related to the social context. However, in all four countries the tendency was clear: for those individuals who lived with a family, dinner in the evening was usually the meal shared by family members. Yet, the most striking result from this Nordic study was that breakfast was the meal most shared with family members. Thus, the Nordic people living in a family context at the time (the data were collected in 1997) usually shared both breakfast and dinner with family members.

In France, 1000 children aged nine to eleven years and their mothers participated in answering a questionnaire in three successive surveys: 1993, 1995, and 1997 (Le Bigot Macaux, 2001). Over the period, the great strength of the traditional French meal pattern was demonstrated, with breakfast, lunch and evening meals being eaten by 97%, 96% and 99% of children, respectively, in 1997. The author does not clearly state that the family members are sharing the meal at dinner time, rather this is indirectly understood from the conclusions: 'In general the traditional French pattern of eating persisted among both the children and mothers surveyed, although there were signs on an increasing trend towards television viewing as a social accompaniment to meals and decreasing family interaction' (p. 145).

It would not be to too bold to claim that many Europeans have a preconceived notion that food and eating are more highly valued in France than in the UK. Pettinger and colleagues (2006) therefore looked at popular stereotypes in Central England and Southern France regarding, for example, eating together as a household, cooking practices and eating out. What they found was that French household members ate together more often (65% compared with 51%), cooked from raw ingredients more often and were more likely to follow a regular meal pattern than were English household members. Of the 2000 questionnaires sent out in each country, equally divided among men and women aged 18–65 years, about 800 were analysed in each country. Also in this study, the researchers did not distinguish which meal was most shared among family members. Over 90% of both the French and the English subjects reported eating dinner (evening meal) every day, but we do not know from these data if the French shared specifically dinner more often than the English did (Pettinger *et al.*, 2006). Whether families in Ireland share meals is not all together clear in Burk and colleagues' (2007) presentation of eating locations among 594 Irish children aged 5–12 years. However, they did find that 89% of all eating occasions occurred at home, which should be a sign of the existence of the family meal in Ireland in the beginning of the 21st century.

The eating pattern in Czechoslovakia, as in most cultures, involves three meals: breakfast, lunch and supper (Haukanes, 2007). Of these three meals, lunch is considered to be the main meal of the day. This is true both for the working week and the weekend. As in many other countries, Czech women are responsible for the food at home, and among the 24 urban and rural living women interviewed in Haukanes' study, two out of three prepared all the meals every day. Most urban living people in Czechoslovakia, men, women and school children, eat their lunch outside home in the company of people other than family members, but at supper time it is common for urban families to share this evening meal. Also breakfast can be shared among families in the city. Among rural living families, however, only a few of them shared meals on a regular basis. At the weekends, all families, urban and rural, would gather for the ritual of eating their Sunday lunch (Haukanes, 2007).

In a Finnish study, using a food frequency questionnaire, over 400 children aged 10–11 years answered, together with their parents, questions on meal patterns during weekdays and weekends (Haapalahti *et al.*, 2003). Almost all children had breakfast (99%) and school lunch (94%) daily. While dinner was prepared daily in 80% of the families, 54% reported eating together daily, and 38% almost daily. Only 8% said they seldom or never shared dinner.

In a European cross-national study, including the Czech Republic, Finland, Greenland, Lithuania, Spain, and the Ukraine, a total of almost 18 000 students, aged 13 and 15, answered a questionnaire on different family activities (Zaborskis *et al.*, 2007). One question concerned eating a meal together with the family every day or almost every day. The 15-year-old students reported less sharing of meals with the family (75% for the girls and 80% for the boys) than did the younger age group (81% for the girls and 84% for the boys). In this more recent study than the previous one mentioned, which also included children from Finland (Haapalahti *et al.*, 2003), there is a slightly different picture of sharing family meals. The Finnish girls (13 years/74%, 15 years/66%), but also the Czech girls (75%/65%, respectively), had the fewest meals together with their family, and the girls in the Ukraine the most (13 years/90%, 15 years/80%). The boys in the Czech Republic reported the least number of shared meals with family members (13 years/79%, 15 years/ 72%), while the boys in the Ukraine reported the most (13 years/92%, 15 years/86%). The figures for the other countries vary between 80 and 86%. There are no data on which meal during the day these figures relate to, however it seems that these adolescents do share family meals on a regular basis. In Lithuania, a total of 369 fifth-grade schoolchildren and 565 parents declared that the most common daily joint family activity was eating a meal together (Garmiene, 2006). The data do not reveal what meal during the day the children and parents shared, yet one can speculate that it is dinner.

11.3   Historical evidence

The notion of the family meal as a fundamental phenomenon of the past has been questioned by scholars (Murcott, 1997; Larson *et al.*, 2006). Do we actually know if the family meal has been a natural part of human everyday life throughout history, and is there any support for the suggestion that it is disappearing today in the beginning of the 21st century? The importance of the family meal and a structured meal pattern has been more or less taken for granted (Murcott, 1997). The arguments about grazing being a late modern phenomenon are based on the notion that a regular eating pattern in the presence of all family members was more or less a natural behaviour in the past and that it may soon be a phenomenon of the past (Fischler, 1980). According to Fischler, people in the past had a fixed number of meals that each had a precise grammar, i.e. constituted specific meal types for specific people. However, according to Murcott, there is little empirical evidence for the assumption that all family members in the past sat down to share their daily food at all meals in a structured way. Before the First World War there are, according to Murcott (1997), no studies that can verify or deny the existence of this phenomenon. She has only found one study from the past, published in 1929, that reveals a concern for the decline in the mealtime as an opportunity for family reunion. The study she refers to was carried out in Middletown, Indiana, from 1924 to 1925, and in some of the interviews from that time people express their concern about not eating together as a family as people did in the past. Because Murcott could find no other empirical evidence that the family meal really existed before this date, she went on to propose two possibilities regarding the family meal in the past. The first possibility is that the family meal had begun its decline somewhere around the period of 1920 as a sign of the modern society, as the interview persons expressed. The other possibility is that the informants in 1924–1925 were expressing an idealized picture of their past. The family meal as they remembered it may not have existed. Thus, concluded Murcott, there is good evidence that the family meal of the past must continue to be followed by a question mark.

There is, however, evidence that around the turn of the century (1900), the family meal was a common and exciting joint activity in Swedish households (Fjellström, unpublished data). This meant that all members of the farmer's household, i.e. the farmer and his family as well as their farm hands and maids, gathered together to share the evening meal, which was often synonymous with the last meal of the day. However, at all other meals during the day, one or several family or household members could be missing. The preliminary analysis of the family meal from a historical perspective is based on ethnological and historical manuscripts and records from a folklore and ethnological archive at the Institute for Dialectology, Onomastics and Folklore Research, Uppsala, Sweden. These manuscripts were collected by female students at the forerunner of today's Department of Food, Nutrition and Dietetics during the period 1897–1916. The students observed daily life in Swedish homes for one or several days during weekdays and focused especially on meal patterns. More than 200 middle-class, working-class and farming households, covering all provinces in Sweden, from north to south, are represented in the data. In more than 90% of the homes/households, all members assembled around the table in the evening. The duration of the meals depended on type of meal and time of day. The family meal in the evening usually lasted for about 30 min, while the cooked meal in the morning and in the middle of the day could last longer. All households had a structured meal pattern including three main meals and two to three in-between meals served at specific hours. Further analyses will be made on these interesting historical data.

An important article in this context was published by the Belgian researchers Mestdag and Vandeweyer (2005). By comparing data from a time-budget, they could analyse the time families in Belgium had spent on different activities in 1966 and 1999. Of all the family activities reported by families in Belgium, the most dominant one during both data collection periods was eating together with family members, and the person most likely to share meals with family members for both years was the mother. However, despite this fact, they could clearly see a decline in joint family activities. The Belgians spent significantly less time with their spouses and children in 1999 than they had done in 1966. Mestdag and Vandeweyer's analyses also demonstrated significant differences between the two years. In 1999, meals shared together with family members had decreased from commonly three per day to only one per day. Breakfast was the least shared meal in 1999, but there was no difference between lunch and dinner, which is surprising. In the late 1990s, unskilled men spent less time sharing meals with their families than did craftsmen and white-collar workers. The time spent eating together, i.e. the duration at table, had decreased substantially from 51 to 27 min (Mestdag and Vandeweyer, 2005). In another study, Mestdag has compared time-budget data from a specific region in Belgium, namely Flanders (Mestdag, 2005). In this study, where the comparison of Flemish time-budget data is made across a much shorter time interval, in 1988 and in 1999, Mestdag concluded that the family meal has changed very little. Eating together was still the common norm in 1999, as was eating three meals a day, confirming that meals are highly structured events. The duration of the family meal was still about half an hour in Flanders in 1999, the same time the families spent eating together in 1988. However, she did find that 'there is a slight increase in solitary eating and work-related interaction on weekdays' (Mestdag, 2005, p. 72).

11.4 Socialization

11.4.1 Commensality at table

Figures illuminating frequency of eating patterns and who eats together with whom can tell us whether or not the family meal as a phenomenon exists in our modern consumer society. What goes on during this meal situation, however, is another matter. What do adolescents think about this family activity in daily life? Margaret Vissers' (1991) definition of commensality is worth noting in today's consumer society: 'togetherness arising out of the fact that we eat at one table' (p. 83). Do young people value togetherness as important and does sharing a meal necessarily involve a table? For the first question, we can find some answers; the other is more difficult. There are no studies that actually give people alternatives when asking whether they share a family meal at a table, or in different chairs without a table, but in the same room. Perhaps some would define a family meal as eating the same food at the same time in the same house, but in different rooms.

Nevertheless, the notion that dinnertime can be an opportunity for joint family activities was expressed by teenagers (9–13 years) in a study comparing family meal conversation in Estonian, Finnish and Swedish families (Tulviste et al., 2002). In a German study examining children's (aged 7–16) eating habits and attitudes, the younger children thought the family was a central part of their daily food life, i.e. the family was an important part of eating and sharing meals at home (Westenhoefer, 2002). In Le Bigot Macaux's study (2001), 82% of the French children valued the dinner as an important event for family time together. The children agreed that it provided an opportunity to have conversations and talk with their parents. Based on her interviews with Czech women, but also based on observations in the Czech countryside, Haukanes (2007) stated that the daily meal is not an important social occasion. Her observations have shown how family members come and go, and only occasionally sit down to share a meal. The women interviewed expressed, on the other hand, the value of sharing meals, but various factors hindered them. Haukanes concluded that 'the ideal of meal-sharing as an important part of family life was not very strongly expressed, nor were there any nostalgic yearnings for shared meals of bygone times' (p. 6). She explained that these ideas among the women are part of the class composition. These ideas are quite different from those we see in other studies from European countries, which shows that it is important not to assume that the family meal is the most ideal, optimal and universal eating occasion.

In the French study by Le Bigot Macaux (2001), 41% of the children's evening meals (dinner) were consumed in front of the television, which caused 20% of the children to agree with the statement: 'We watch television at dinner time too much too often'. The TV is a part of meals also in other countries. With regard to eating meals at home, older children in Germany rated the TV as equal in importance as their parents and siblings (Westenhoefer, 2002). In a Belgian study comparing 1966 and 1999, Mestdag and Vandeweyer (2005) saw that television had lost its importance as a joint family activity and had become a solitary activity. Thus, the family meal could still be intact if the family watched television together, but not when television behaviours are changing towards eating and watching TV alone, perhaps also eating food according to individual preferences. The effect this has is discussed below, under Nutrition.

11.4.2 Conversation at table

In the most idealized picture, we see the family meal as the best place for conversing, learning and sharing intellectual thoughts. During the socialization process at table, language is used as a tool. In different cultures, language is used in varies ways (Tulviste *et al.*, 2002). It was shown, for instance, that Swedish monolingual families talked more at meals than did monolingual and bilingual Finnish and Estonian families living in their own countries, as well as in Sweden. Swedish children were also more active in the family meal talk than were children in the other Finnish–Estonian families. While the latter families commented on table manners, the Swedish families commented more and talked more about moral issues. The Swedes also negotiated more than the others did. The one thing that was unanimous in the 100 families participating in this study, and whose meals were videotaped, was that the mother was the most active commenter and the target for the comments was the children. However, the family meal can also be an arena for conflicts and hostility (Frydenberg *et al.*, 1998). In families where authority has to be established, it is also taught that authority is negotiable. This, on the one hand, can be more challenging for children; on the other hand, in such circumstances there may be a reduced need for rebellion and confrontation. This could be of interest when discussing the negotiation of foods served in families.

11.4.3 The negotiated family meal

When discussing the family meal in Europe, the outcome of sharing a table and food is foremost in focus, but all the activities that precede what may be the outcome at the table should not be forgotten in this context (Charles and Kerr, 1988; de Vault, 1991). Meal planning, shopping, cooking, taking care of wastes and dishes, food anxiety and negotiations and division of labour are all issues involved in the family meal. It is evident that feeding the family has implications both for the foods served at meals shared by family members and for the family members as such. The foods shared at the family meal are part of the way children will learn about food in their culture, perceive food in their everyday life and develop taste preferences.

In a study conducted in Belgium in the mid-1990s, adolescents and young adults were given hypothetical situations in which family members tried to introduce healthy food into the family (De Bourdeaudhuij, 1997). The results showed that the young people aged 12–22 years perceived their own ability to influence family food as rather limited. Their parents were seen as more powerful in introducing health food, i.e. new food habits in the family, and fathers were most powerful. This means that if and when new dishes or foods were introduced into the traditional family food habits, the teenagers perceived the father to be the key person to accept them or not. As Charles and Kerr pointed out 20 years ago; 'It seems that men's preferences have more of an impact on what women cook than those of any other family members' (1988, p. 68). Charles and Kerr, however, also found that the women had the least impact on family food habits. The children's preferences were more important than their own. A Norwegian study among young mothers (children in pre-school) confirms the fact that children in some cases have little impact on the outcome of the family dinner (Bugge and Almås, 2006). The women had a kind of caring ideology for the family as their optimal vision when preparing and severing proper meals, yet there were limited possibilities for individualisation. In reality, conventions and conformity took over food-related work in the everyday life.

In a French study, children who were interviewed demonstrated an overwhelming trust in their mother's food choice and cooking (Le Bigot Macaux, 2001). Ninety per cent agreed with the statement 'mom is a good cook; I like eating what she makes' and 86% with the statement 'I don't worry too much about what I have to eat to stay healthy; my mom takes care of me'. However, 31% also agreed with the statement 'Sometimes I get into fights with my mom about food because we don't have the same tastes' and 71% agreed with 'When I don't like something, I'm often forced to eat it anyway because my mom says that I have to eat all kinds of food' (Le Bigot Macaux, 2001). This dichotomy in the children's answers shows the complexity of food socialisation and the role family meals play in learning about new foods and tastes.

If children and mothers in some studies thought they had little control over food choices, then another picture emerges in other interview studies around Europe, and especially in mothers' narratives about feeding the family. This daily activity is experienced as very complicated and filled with anxieties and considerations of family member's preferences, especially children's food wishes, aversions, and rejections (Romani, 2005; Carrigan et al., 2006). In their qualitative study, Carrigan and colleagues showed how UK women constantly negotiate with themselves, their spouse, and their children on what is the ideal and at the same time the most appreciated food. Homemade food is always seen as the best from the mother's perspective, but not always by the rest of the family members, creating frustration in the women when the family does not appreciate her efforts to serve nutritious, good food at family meals. Also in the French study by Le Bigot Macaux (2001) – in which the children both declared their trust in their mother's cooking and choice of food, and sometimes problems with being forced to eat what was placed on the table – women had concerns about caring for all family members. Almost one third of mothers viewed food as a constraint owing to the burden of planning, shopping for and preparing food. Also Czech women, who are always responsible for the family food, negotiate continuously with their spouses and children about what should end up on the plate in the family meal (Haukanes, 2007).

American adolescents interviewed about food choices and meals in family life revealed different family ideologies (Contento et al., 2006). Some teenagers eat whatever they want, whenever they want to, others have little control over their food choice; they eat what is served at family dinners. The latter group, however, consisted of very few individuals. A recent study in the UK among young and older children (7–11 years) showed that older children have more power over foods served in the home than do younger children (Warren et al., 2008). The authors state: 'At home, most older pupils dictated their choice of meal and parents were, for the most part, presented as amenable and compliant with response preferences for unhealthy food items' (p. 6). Thus, there are diverging results in different studies. Dixon and Banwell stressed the fact that, for parents living in the 21st century, and especially for women, three factors are vital in the process of feeding the family and serving family meals: the shift in children's influences on the family meal, anxiety about the family's well-being, and women's ambivalence about whether or not they should continue to have control over the family meal (Dixon and Banwell, 2004). Thus, children's and especially adolescent's influence on the foods served at family meals are important, and this influence has probably developed and increased over time within the consumer society. These issues should be taken into consideration when discussing socialisation. Who is teaching whom at the table? The impact children's food preferences have on the family meal in our consumer society and the efforts to fulfil every individual's wish related to food can have disturbing effects, such as creating eating disorders (Fernández-Aranda et al., 2007).

11.5 Health

11.5.1 Nutrition

During recent decades, particularly the past few years, the scientific and public interest in the family meal, the frequency of family members sharing meals at home, and its effect on nutrition and health has increased substantially. One reason for this attention towards family meals, which can be detected in the literature, is the rapid growth of obesity, especially childhood obesity. A recent German study showed that an increase in daily meals seems to have a protective effect on child obesity (Toschke *et al.*, 2005). Several recent American studies have pointed out the problems associated with skipping meals, the decreasing family meal and children's dietary intake and nutritional status. Researchers have shown that when American families eat together, it has a positive effect on family members' consumption of fruit and vegetables (Neumark-Sztainer *et al.*, 2003).

Also in Europe similar associations have been made between the family meal and healthy eating. Haapalahti and co-workers (2003) found that children 10–11 years of age who seldom or never shared dinner in the family context ate sweets and fast foods more often than did those who had a regular family meal pattern. However, a study in the UK provided no evidence that eating together with the family had an effect on healthy eating (Sweeting and West, 2005). The author's aim was to study associations between family life and 'less healthy eating' and 'unhealthy snacking'. More than 2200 parents completed a questionnaire about family structure, meals and maternal employment status. Interestingly, an association between daily family meals and less healthy eating was not found, but, among children who had mothers at home all day (full-time home-makers), the likelihood of less healthy eating was higher than among children whose mothers worked part-time. In a recent Spanish study, almost 35 000 people between the ages of 25–64 answered a questionnaire about eating habits and weight (Marín-Guerrero *et al.*, 2008). The results showed that people who skipped breakfast and dinner, and especially women who skipped dinner, had a higher risk of developing obesity, but there was no association between eating away from home and obesity. The authors point out that this is in contrast to what has been found in earlier studies conducted in Anglo-Saxon countries, i.e. the USA and the UK. The reason for this discrepancy is explained by the difference in the types of restaurants in these countries. In the UK and the USA, food eaten away from home is mostly represented by fast food restaurants, while in Spain, the traditional restaurants serve food closer to the model of the Mediterranean diet. Thus, one interpretation could be that the restaurant food culture in this part of Europe is more like traditional home-cooked meals than like the modern fast food culture. However, the important question concerning sharing food or not was not answered in this study. One could speculate that the women and men who had dinner also shared that meal with their family members, whether they had the meal at home or away from home, and that this was the key to not developing obesity. In another Spanish study by psychiatrists dealing with eating disorder issues, it was found that trying to accommodate individual family members' specific preferences in a family setting was linked to eating disorders (Fernández-Aranda *et al.*, 2007). This could be interpreted to mean that when individuals in a family come together to share a meal, they not only need to share the same table, but also the same food if they are to have the optimal meal from a health perspective.

The notion that the television is an especially problematic factor in relation to food habits has long been discussed, and earlier in the present chapter. There are, however, large differences in children's television viewing in the beginning of the 21st century (Vereecken et al., 2006). Comparing countries around the world, the frequency of television viewing varied from an average of 2.0 h in Switzerland to 3.7 h in the Ukraine. The results from this WHO collaborative study covering 33 countries/regions in Europe and North America indicated that those who watched the most TV were boys, 13 years of age and pupils from families of lower socio-economic status. Those who watched TV seemed to consume more sweets and soft drinks and fewer fruits and vegetables (Vereecken et al., 2006). In a study by Fitzpatrick and colleagues (2007), the TV is stressed as being especially problematic in the family meal context. When family members came together and shared a meal at dinnertime, this had a positive effect on servings of fruits and vegetables. However, each night the television was on during dinnertime, despite sharing a meal in the family context, servings of fruit and vegetables decreased. This study comprised nearly 1400 parents living in New York, with children between one and five years of age. Feldman and co-workers also found similar associations in older children (2007). Those girls and boys between 11 and 18 years of age who watched television during dinnertime had a lower intake of fruit, vegetables, and grains, but a higher intake of soft drinks than did those who did not watch TV during this evening meal. Yet, Fitzpatrick et al. (2007) also saw that watching TV during family meals was associated with more healthful dietary intake than was not eating regular family meals at all. Thus, the family meal per se was seen as more important to healthy food habits than was the presence/absence of the TV during the meal.

11.5.2 Mental health

Not only nutritional effects have been seen on adolescents who eat together with the family. In a Spanish study among young people aged 14–23, mental health and meals were studied (Compañ et al., 2002). A comparison was made between a group of adolescents that had their first consultation in a public mental health outpatient clinic (case group) and a comparison group of young people from various educational centres. The young people in the comparison group shared significantly more meals with their parents, both on weekdays, such as dinners, and on other occasions, than did the adolescents in the case group. Compañ et al. concluded that their results show how families of young people with psychological problems 'practice fewer unifying and life cycle rituals' (p. 93).

11.6   Conclusion

In Europe, 'grazing' has not been found to be a major trend; on the contrary, a structured meal pattern is still common all over Europe. (For comparison with the United States see McIntosh *et al.*, Chapter 10, on the American family meal.) Sharing meals in a family context seems also to be a lively and present activity all over Europe. It is also an activity that European family members value as important. We can say that those people living together as a family most likely also eat together, and this is true mostly concerning the evening meal or dinner. Only among Czech farmers, seen in one study, is this tradition uncommon during weekdays, but not during weekends (Haukanes, 2007).

In some cultures, breakfast is also shared among family members, while lunch in the middle of the day is often eaten together with other people, or commensal units, as Sobal and Nelson (2003) would call them. Thus, the concern for the declining family meal in this part of the world does not seem to be that alarming, if there should be any concern at all! Yet, as Mestdag (2005) showed in Belgium, the duration of time spent together at table at the evening family gathering has declined over a period of 30 years. Perhaps it is this behaviour that has caused the fear of a social collapse related to meals in everyday family life. On the other hand, in the archive material from Sweden in the late 19th century, the family meal was rather a quick affair before bedtime (Fjellström, unpublished). It could also be that this specific meal is scrutinized because one group in the family wants to break loose from the bosom of the family, and that group is composed of older adolescents. They either wish to eat in front of the television or together with friends. Joachim Westenhoefer (2002) declared; 'Thus it appears that the family is being replaced by the television as "social company" at meals' (p. 19). This may be too far-reaching a conclusion, but it does show the social changes the family meal has undergone during the past decades. Thus, the balance of the family meal has been disrupted, and those who fear the effect this may have on teenagers' health have been found right; the family meal seems to be a precursor of both nutritional and mental health.

Some researchers have attributed to the family meal the function not only of an instrument that enhances healthy food habits, and thus a protector of obesity, but also of a form of protection against drug use, such as cigarettes, alcohol and marijuana (Eisenberg *et al.*, 2008). The studies referred to above have also shown us the role the shared meal has in the socialisation process, for example in learning the art of conversing or negotiating. Negotiating what food should be served at family meals also seems to be one of the factors concerning most family members. Thus, it is not the question of eating together that is a problem for European families in the beginning of the 21st century, but of what foods they decide to share together in everyday life.

11.7   References

Bisogni A C, Winter Falk L, Madore E, Blake C E, Jastran M, Sobal J, Devine C M (2007), 'Dimensions of everyday eating and drinking episodes', *Appetite*, **48**(2), 218–231.

Bugge A C and Almås R (2006), 'Domestic Dinner. Representations and practices of a proper meal among young suburban mothers', *Journal of Consumer Studies*, **6**(2), 203–228.

Burke S J, McCarthy S N, O'Neill J L, Hannon E M, Kiely M, Flynn A, Gibney M J (2007), 'An examination of the influence of eating location on diets of the Irish children', *Public Health Nutrition*, **10**(6), 601–607.

Caplan P (1997), Approaches to the study of food, health and identity. In Caplan P, *Food, health and identity*, London, Routledge, 1–31.

Carrigan M, Szmigin I, Leek S (2006), 'Managing routine food choices in UK families: the role of convenience consumption', *Appetite*, **47**(3), 372–383.

Charles N (1995), 'Food and family ideology', in Jackson S and Moores S, *The politics of domestic consumption. Critical readings*, London, Pearson Education, 100–115.

Charles N and Kerr M (1988), *Women, food and families*, Manchester, Manchester University Press.

Compañ E, Moreno J, Ruiz M T, Pascual E (2002), 'Doing things together: adolescents health and family rituals', *Journal of Epidemiology and Community Health*, **56**(2), 89–94.

Contento I, Williams S, Michela J, Franklin A (2006), 'Understanding the food choice process of adolescents in the context of family and friends', *Journal of Adolescents Health*, **38**(5), 575–582.

Cox R and Narula R (2003), 'Playing happy families: rules and relationship in au pair employing households in London, England', *Gender, Place and Culture*, **10**(4), 333–344.

De Bourdeaudhuij I (1997), 'Perceived family members' influences on introducing healthy food into the family', *Health Education Research*, **12**(1), 77–90.

DeVault M L (1991), '*Feeding the family. The social organization of caring as gendered work*' Chicago, Chicago Press.

Dixon J and Banwell C (2004), 'Heading the table: parenting and the junior consumer', *British Food Journal*, **106**(3), 181–193.

Eisenberg M E, Neumark-Sztainer D, Fulkerson J A, Story M (2008), 'Family Meals and Substance Use: Is There a Long-Term Protective Association'? *Journal of Adolescent Health*, **43**(2), 151–156.

Feldman S, Eisenberg M E, Neumark-Sztainer D, Story M (2007), 'Associations between watching TV during family meals and dietary intake among adolescents', *Journal of Nutrition Education and Behaviour*, **39**(5), 257–263.

Fernández-Aranda F, Ramon A, Roser D, Solano L, Badoa J M, Krug A, Granero R, Karwautz I, Gliménez C (2007),'Individual and family eating patterns during childhood and early adolescence: an analysis of associated eating disorder factors', *Appetite*, **49**(2), 476–485.

Fischler C (1980),'Food habits, social change and the nature/culture dilemma', *Social Science Information*, **19**(6), 937–953.

Fjellström C (unpublished data), *Family meals in the past*.

Fitzpatrick E, Edmunds L S, Dennison B A (2007), 'Positive effects on family dinner are undone by television viewing', *Journal of the American Dietetic Association*, **107**(4), 666–671.

Freeman M (2005), *Family Meal*, Boston, Fanlight.

Frydenberg E, Säljö R, Anerson P (1998), 'Exploring conflict culture in the family: consistency and continuity of conflict during family meal discourse', *Journal of Psychology and Judaism*, **22**(4), 275–288.

Fulkerson J A, Neumark-Sztainer D, Story M (2006), 'Adolescent and parent views of family meals', *Journal of the American Dietetic Association*, **106**(4), 527–532.

Fulkerson J A, Story M, Neumark-Sztainer D, Rydell S (2008), 'Family Meals: Perceptions of Benefits and Challenges among Parents of 8- to 10-Year-Old Children', *Journal of the American Dietetic Association*, **108**(4), 706–709.

Garmiene A, Zemaitiene N, Zaborskis A (2006), 'Family time, parental behaviour model and the initiation of smoking and alcohol use by ten-year-old children: an epidemiological study in Kaunas, Lithuania', *BMC Public Health*, **6**, 287, 1–9 p.

Haapalahti M, Mykkänen H, Tikkanen S, Kokkonen J (2003), 'Meal patterns and food use in 10- to 11-year-old Finnish children', *Public Health Nutrition*, **6**(4), 365–370.

Haastrup L (2003), 'Mad og måltider I skolen' in Holm L, *Mad, mennesker og måltide - samfundsvidenskaplige perspektiver*, Copenhagen, Munksgaard Danmark, 247–262.

Haukanes H (2007), 'Sharing food, sharing taste? Consumption practices, gender, relations and individuality in Czech families', *Anthropology of Food*, S3, December, Food Chains/ Les chaines alimentaires, on line 21 March 2008, http://aof.revues.org/document1912.html. Downloaded 26 April 2008.

Holm L (2001), 'Family Meals ', In Kjærnes U, *Eating Patterns. A day in the lives of Nordic peoples*, Report 7, Lysaker, Norway, SIFO, 199–212.

Kelder S H, McPherson R S and Montgomery D H (1996), 'Meal Skipping Patterns Among Children and Adolescents', *Journal of the American Dietetic Association*, **96**(Suppl 1), A57.

Kjærnes U (2001), *Eating Patterns. A day in the lives of Nordic peoples*, Report 7, Lysaker, Norway, SIFO.

Larson R W, Branscomb K R, Wiley A R (2006), 'Forms and functions of family mealtimes: multidisciplinary perspectives', *New Directions for Child and Adolescent Development*, **111**(1), 1–15.

Le Bigot Macaux A (2001), 'Eat to live or live to eat? Do parents and children agree?' *Public Health Nutrition*, **4**(1A), 141–146.

Mäkelä J (2000), 'Cultural definitions of the meal', in Meisleman H L, *Dimensions of the Meal. The Science, Culture, Business, and Art of Eating*, Gaithersburg, Aspen Publication, 7–18.

Marín-Guerrero A C, Gutiérrez-Fisac J L, Guallar-Castillón P, Banegas J R, Rodríguez-Artalejo F (2008), 'Eating behaviours and obesity in the adult population of Spain', *British Journal of Nutrition*, on line April , pp. 1–7, doi:10.1017/S0007114508966137.

Mattsson Sydner Y, Sidenvall B, Fjellström C, Raats M, Lumbers M, (2007), 'Food habits and foodwork. The life course perspective of senior Europeans', *Food, Culture and Society*, **10**(3), 368–387.

McIntosh W A, Dean W, Torres C C, Anding J, Kubena K S, Nayga R (2009), 'The American family meal', In Meiselman H, *Meals in science and practice*.

Mestdag I and Vandeweyer J (2005), 'Where has family time gone? In search of joint family activities and the role of the family meal in 1966 and 1999', *Journal of Family History*, **30**(3), 304–323.

Mestdag I (2005), 'Disappearance of the traditional meal: Temporal, social and spatial destructuration', *Appetite*, **45**(1), 62–74.

Murcott A (1982), 'On the social significance of the cooked dinner in South Wales', *Social Science Information*, **21**(4/5), 677–96.

Murcott A (1997),'Family meals, a thing of the past'? In Caplan P, *Food, Health and Identity*, Routledge, London, 32–49.

Neumark-Sztainer D, Hannan P J, Story M, Croll J, Perry M (2003), 'Family meal patterns: Associations with sociodemographic characteristics and improved dietary intake among adolescence', *Journal of Dietetic Association*, **103**(3), 317–322.

Pettinger C, Holdsworth M, Gerber M (2006), 'Meal patterns and cooking practices in Southern France and Central England', *Public Health Nutrition*, **9**(8), 1020–1026.

Pliner P and Rozin P (2000), 'The psychology of the meal', in Meiselman HL, *Dimensions of the Meal. The Science, Culture, Business, and Art of Eating*, Gaithersburg, Aspen Publishers, 19–46.

Prättälä R (2000), 'North European Meals: Observations from Denmark, Finland, Norway and Sweden, in Meiselman HL, *Dimensions of the Meal. The Science, Culture, Business, and Art of Eating*. Gaithersburg, Aspen Publishers, 191–201.

Romani S (2005), 'Feeding post-modern families: Food preparation and consumption practices in new family structures', in *Proceedings of the European Association for Consumer Research Conference*, EACR 2005, Göteborg, Sweden, 15–18 June.

Sidenvall B, Nydahl M, Fjellström C (2000), 'The meal as a gift – the meaning of cooking among retired women', *Journal of Applied Gerontology*, **19**(4), 405–423.

Sjöberg A, Hallberg L, Höglund D, Hulthen L (2003), Meal pattern, food choice, nutrient intake and lifestyle factors in the Göteborg adolescent study, *European Journal of Clinical Nutrition*, **57**(12), 1569–1578.

Sobal J and Nelson K (2003), 'Commensal eating patterns: a community study', *Appetite* **41**(2), 181–190.

Sweeting H and West P (2005), 'Dietary habits and children's family life', *Journal of Human Nutrition and Dietetics*, **18**(2), 93–97.

Tessier S and Gerber M (2005), 'Comparison between Sardinia and Malta: The Mediterranean diet revisited', *Appetite*, **45**(2), 121–126.

Toschke A, Küchenhoff H, Koletzko B, von Kries R (2005), 'Meal frequency and childhood obesity', *Obesity Research*, **13**(11), 1932–1938.

Tulviste T, Mizera L, De Geer B, Tryggvason M T (2002), 'Regulatory comments as tools of family socialization: A comparison of Estonian, Swedish and Finnish mealtime interaction', *Language in Society*, **31**(5), 655–678.

Vereecken C A, Todd J, Roberts C, Mulvihill C, Maes L (2006), 'Television viewing behaviour and associations with food habits in different countries', *Public Health Nutrition*, **9**(2), 244–250.

Visser M (1991), *The Ritual of Dinner*, London, Penguin Books.

Warren E, Parry O, Lynch R, Murphy S (2008), 'If I don't like it then I can choose what I want: Welsh school children's accounts of preference for and control over food choice', *Health Promotion International*, **23**(2), 144–151.

Weinstein M (2005), *The surprising power of family meals*, Hanover, Steerforth.

Westenhoefer J (2002), 'Establishing dietary habits during childhood for long-term weight control', *Annals of Nutrition and Metabolism*, **46**(suppl 1), 18–23.

Wood R (2003), 'Meal', in Katz S H, *Encyclopaedia of Food and Culture, Scribner Library of Daily Life*, Vol 2, New York, Thomson Gale, 461–465.

Zaborskis A, Zemaitiene N, Borup I, Kuntsche E, Carmen M (2007), 'Family joint activities in a cross-national perspectives', *BMC Public Health*, **7**, 94, 1–14 p.

12

Gender perspectives on the solo diner as restaurant customer

I. M. Jonsson and M. Pipping Ekström,
Örebro University, Sweden

Abstract: In this chapter eating out at restaurants, i.e. meal experiences outside home, are discussed. Dining out in upper-class restaurants is a worldwide trend nowadays and includes an extensive range of consumer groups. Internationally, there is a large number of solitary customers, both male and female, for whom lone dining is not as easy as they would like it to be. Furthermore, women, and naturally also men, today wish to take their place in urban settings, for instance by eating in restaurants, even if they are alone. They should be seen as a potentially important clientele group worthy of the attention of the restaurant branch.

Key words: eating out, solo dining, gender, restaurant customer, meal experience.

12.1 Eating in and eating out

'Eating in' (eating at home) constitutes the norm of the individual and collective household or family consumption of food (Valentine 1999). The opposite, 'eating out', is defined as eating outside the home arena (Warde and Martens 2000), and forms a source of pleasure and a favoured leisure pursuit for increasing numbers of people. In other words, 'the exotic other' as compared with eating at home (Ashley *et al.* 2004). Conviviality was to be found, according to Warde and Martens, not only when eating in but also in eating out situations.

12.1.1 The 'meal experience' when dining out

The concept of the 'meal experience' had, according to restaurant marketing, already been formed in the 1960s (Campbell-Smith 1967). Campbell-Smith states that when dining out, the customers' concerns and their experiences were functions of a much wider range of factors than the quality of food and drink alone. The concept of meal experience is alive and needs to be further discussed (Wood 2003). A way of describing and arranging a restaurant meal as a holistic event, including the social and environmental aspects, might be to look at different sides of the customers' experience. Here, it is not only the edible 'product' on the plate that is examined, but also other dimensions, 'the room', 'the encounter' with the restaurant staff and guests, 'the atmosphere/ ambiance' linked to 'the management control system' including business administration, organisational theory, leadership, laws, logistics, food safety, hotel and restaurant security. This is known as the Five Aspect Meal Model (FAMM), and was defined by Gustafsson *et al.* (2006), (see also Chapter 14). Clearly, today's experience-based economy provides opportunities for events of foodservice and chic cuisine (Finkelstein 2004, Morgan and Hemmington 2008, Pine and Gilmore 1999).

'Dining out', from Finkelstein's point of view, is a mannered exercise in how to behave appropriately in a restaurant, as learnt by observation, imitation and practice (Finkelstein 1989). Finkelstein later added to her definition 'that it is not synonymous with eating: it is not simply a biological or even an economic process. Dining out becomes an event that brings the individual, figuratively and literally, into the public arena and exposes him or her to the scrutinizing eye of the other.... When, what, and how we eat becomes a narrative retelling aspects of biography and cultural knowledge ... what we understand and misunderstand to be interesting, urbane, civilized, and pleasurable' (Finkelstein 1998: 214).

A meal experience at a restaurant is therefore in many ways pleasurable in performance terms (Morgan *et al.* 2008). Seeing and being seen is an essential part of the attraction; it is a kind of public eating, challenging because of the feeling of being constantly on display (Finkelstein 1998, Bell and Valentine 1997). In their reading of Finkelstein *et al.* (1997) do not agree with her broad concept of dining out as a metaphor for a total experience of modernity. They feel that the experience of eating is neither as homogeneous nor as passive as she implies, but define 'eating out in a commercial setting is that one does something quintessentially familial – sitting down, for some considerable period of time, at table, to eat – but in the visible presence of strangers' (*Ibid*: 131). They also point to a shift, where what had once been a luxury restricted to elites has now, with the expansion of eating out, come within reach of the whole population and become a kind of democratic process. 'The restaurant', they say, 'provides a location in which the harmonious management of social relations in a public place indicates a degree of mutual tolerance between a wide variety of customers' (*Ibid*: 132). They add, however, that eating out also forms 'a welcome alternative to the privatisation of social life, for the restaurant is a comparatively safe, quasi-public environment in which to enjoy many of the real, if ambivalent attractions of modern urban experience' (*Ibid*: 149).

But eating out it is not always pleasurable. Bell and Valentine discuss the restaurant norm, an aspect that makes the meal eaten out a structured event

(1997). Ashley *et al.* go further, meaning that there is consumer anxiety about the codes surrounding dress and the actual meal event: how to order the different courses, how to use the table utensils set in front of them, and finally, how to tip (Ashley *et al.* 2004, Hansen 2005). Within social sciences, gender often refers to social differences between men and women. For example, there are gendered aspects of rituals surrounding service. If a man is a part of a mixed company it is mostly he who is addressed and, traditionally at least, it is the men who order, taste the wine, make the complaints, and pay the bill (Warde and Martens 2000:126). Furthermore, the norms of social behaviour lead to expectations of a shared restaurant meal experience with more than one customer per table. This is shown by the way the tables are set, with two or more chairs, as well as by the way people watch one another's manners (Heimtun 2007, see also Pliner (Chapter 9)).

Wood suggests that restaurants can be seen as masculine domains, where both the food and the organisation primarily address the requirements and desires of men, with women marginalised and treated as appendages to male clients or just one element of the family (Wood 1992). Cultural geographers maintain that a 'genderising' of towns is in process, meaning that certain areas and large places, i.e. different types of setting, are clearly gendered, some male- and others feminine-codified, public places where the setting is interpreted in gendered terms. It is important not to be in the wrong place at the wrong time. Codifying of place becomes a codifying of people in that environment. A lone woman in an out-of-the-way place late at night might be regarded as bad. Safety aspects curb our urban lifestyle patterns, especially in the evenings and at night. There may also be a 're-genderising' of physical surroundings, where an area frequented during the day by women is a clearly male-codified area at night (Domosh and Seager 2001, Forsberg 2005, Pipping Ekström and Jonsson 2007b, Skeggs 1999).

12.2   Restaurant meals for solo diners

Dining out in upper-class restaurants is a worldwide trend nowadays and includes an extensive range of consumer groups. There is, however, one category, that of the solitary customer, both male and female, for whom dining is not as easy as they would like it to be. A number of attempts have been made to get to grips with the problem internationally, including such concepts as SoloDining – Communal tables (USA); Talking tables (UK); and Tables for sitting together, SAMBORD, (Sweden). A possible conclusion from the phenomenon of eating out alone, according to these three examples, and one of which there is a new awareness, is that a vast number of people make up the singles/lone diner market including all those who are situationally alone. Furthermore, solitary females form a noticeable part of this category. Women today wish to take their place in urban settings, for instance by eating in restaurants, even if they are alone. Many professional women have money and are well able to pay for themselves. They should be seen as a potentially important clientele group worthy of the attention of the restaurant branch. Naturally enough there are also men who need to dine out alone and take pleasure in doing so.

12.2.1   SoloDining – Communal tables (USA)

In America, a website was invented focusing on the needs of the solo diner as a critical part of the restaurant business, since it is obvious that solitary men and women often connect the fear of loneliness with dining out alone. Eating on one's own in a restaurant, café or bar is something all travellers do at some point in their journey and there may be a local need to combat the dread of dining solo, so information is given on solo-dining-friendly restaurants all over the world and the concept has been copied worldwide.

> SoloDining.com serves up 'solo dining savvy' for you – whether you despise or delight in a solitary meal in a restaurant or at home; whether you're fond of fast food or fine dining (or something in between!) and whether you're married or divorced, single or solo, bachelor or bachelorette, widow or widower; business/pleasure traveler or stay-at-home-lover. So pull up a chair and join in. The company's fine. Sooner or later, everyone faces the challenge of eating alone. (SoloDining 2008)

Communal tables might be one way of attracting solo diners – singles, travellers on business or out for pleasure or those whose partners are away – as well as many other customers, including convivial couples who enjoy networking with others. This might be an advantage to the customer, but also to the restaurateur who does not have to take reservations and can be open for as many passing customers as may be practicable. Communal dining revolutionises the idea of 'the biggest wallet, the best table', and in theory at least means that the single diner is not forced to eat alone (SoloDining 2008).

12.2.2   Talking tables (UK)

A British citizen, Brian, aged 44, was keen to be better treated in restaurants when eating out alone and in 2002 started a private UK campaign, which rapidly gained public support (Table for one if you please 2002). His idea was Talking tables, a method of encouraging catering establishments across the country to consider reserving just one table, preferably with room for at least four, as a shared table … a talking table. He suggested that restaurants might have a sign in the window to inform customers of the option to join such a table in order to share conversation with other people. 'It's not a place for heated debates or for selling goods, nor is it a dating service, just a means of exchanging social chit chat over a meal, Brian explained. Nevertheless, the issue is also being taken seriously by the UK catering industry (SoloDining.com/greatbritain-restaurants 2008)

12.2.3 Tables for sitting together (Sweden)

In 2007, the Swedish organisation in support of meals (Låt måltiden blomma – May the Meal Flourish!) introduced the idea of communal tables by means of a special sign to be put up on the entrance, in the first place at a number of leading restaurants in the biggest cities in Sweden (Låt måltiden blomma 2007). The idea came from a man working in a big non-governmental organisation for consumer rights; he had himself been refused a restaurant table because he was single. The idea of SAMBORD (i.e. tables for sitting together and dining out with new friends) was explored at a big Restaurant Expo and generated interest in the media for a short period. It takes time, however, to accomplish such social change. Restaurants need to take their own initiatives to disseminate this idea.

12.3 The essence of a restaurant meal from the perspective of the customer

Three different verifications of customer perspectives on restaurant meals are described below. First comes a description of a Norwegian report on eating out, then 'holiday meal experiences', and finally experiences of the essence of restaurant meals, as expressed by men and women both in a Swedish scientific study and an American internet blog.

12.3.1 The role of eating out (Norway)

The purpose of the report from The Norwegian Institute for Consumer Research (Bahr Bugge and Lavik 2007) was to provide better information on the role of eating out in everyday life. Eating in the home is still a strong factor here but one cannot close one's eyes to the fact that eating out is on the increase when 92% of the respondents prove to have eaten out during the last two months. The most frequented venues in Norway for eating out were pizza restaurants, petrol stations and cafés in shopping centres. Mostly, it took place on a very basic level and was motivated by hunger, e.g. at petrol stations, where the customers were predominantly men, but pizza chains and ethnic restaurants were also reported as being social meeting places and good value for money. Only a small group of well-educated people, mainly in the capital of the country, visited restaurants which were unique and high class. Eating out was a social activity, and the norm was to eat together with friends, one's partner or one's spouse. In the report, numerous tensions in the various consumer orientations were pointed out: health versus hedonism, economy versus extravagance, democratisation versus class distinction, convenience versus enthusiasm (Bahr Bugge and Lavik 2007: summary).

12.3.2 Holiday meal experiences

When travelling or on holiday away from home an important part of the experience consists of the necessity of eating meals, often in different kinds of restaurants. Goffman (1959, 1966) in his classic analysis, discussed the roles of different people acting in varying urban social spheres, including restaurants, while Butler goes further in her suggestion that gender identities constitute performative acts of how to be a man or woman (Butler 1993:42, Butler 1999). From the perspective of theories of tourist mobility, holiday experiences can be seen as temporally and spatially located but at the same time 'situated in the being, becoming or "happening" of the performance of host and guests' (Heimtun 2007, Urry 2000, Bærenholdt *et al*. 2004).

A study on gender and tourism showed eating out to be central. The study was based on 30 Norwegian middle-aged (35–55 years old) single women's oral and written solo holiday experiences (Heimtun 2007). The researcher, who considered herself an experienced solo traveller, made a holiday trip during her research period and realised that she felt constrained, especially by the eating out experience. 'On one occasion I found it hard to select a restaurant and really had to work hard mentally to overcome my fears of public solitude and feelings of loneliness' (ibid: 14). Bourdieu (1989/2003) indicates food as being one of the most important social markers in society, 'relation to food – *the* primary need and pleasure – being only one dimension of the bourgeois relation to the social world' (ibid:196). In the Norwegian study, a feminist interpretation of Bourdieu's phenomenology of social space and the concept of habitus was used in the analysis, showing that middle-aged single women struggle to cope with the holiday experience field by investing economic, cultural and social capital (Heimtun 2007: 230). Tourist brochures show people on holiday sharing experiences, meals and activities and, in this study, the idea is that the symbolic capital bonds with the social capital. For the informants, solo holidays were equated with no social capital or sameness at all. The symbolic power embedded in binding social capital was so strong that these middle-aged single women preferred holidays with friends and did not consider travelling solo or eating out alone at all during the holiday.

On the other hand, for women seeking 'the social identity of the independent traveller' daytime restaurant experiences in cities were very positive. This kind of identity, however, was only embraced by a few of the younger single women, who enjoyed being in charge of their experiences of what to do and when, where to go, and who to talk to. More often it was a matter of expressing the experience of 'the social identity of the social loner'. Few of these women enjoyed eating out alone, and few were able to deal with the feeling of publicly exposed solitude and the tourist gaze without feeling marginalised, especially when visiting a restaurant alone at night. But there were also some women who, over the years, had been transformed into independent travellers not scared of eating out alone at night. The diaries of the single women are filled with descriptions of the joy of sharing a holiday meal and an evening meal especially has a significance that goes beyond the sharing of the food (Heimtun 2007). Warde and Martens point out the atmosphere as being a collective creation, a joint achievement when eating out (Warde and Martens 2000: 208). Crang indicates that the social relationship of a 'holiday meal experience' in a restaurant is materially and discursively organised around three consuming practices: 'gazing, sociality/sensory pleasures and talking' (Crang 1997: 150). These factors provide sources of pleasure and happiness that cannot be brought about by any one individual alone, once again showing the importance of the social dimension of a meal as a place for emotional and symbolic belonging (Simmel 1993).

12.3.3 Restaurant meal experiences as recounted by men and women

A Swedish study on *women dining alone* (Pipping Ekström and Jonsson 2007a, b) was set up with the purpose of investigating the reception of women arriving alone in the evening and ordering dinner, including glasses of wine, in an upper-class restaurant. A further aim was to study how a female solo diner reflects on her situation when in a first class restaurant, and moreover, to discuss with women in working life their feelings about dining alone. Two women A and B individually visited each restaurant at the same time. Ten restaurants were chosen in one of the biggest cities in Sweden.

A typical visit might be described as follows: A table for one at 8.30 pm was ordered in advance by telephone on a Friday afternoon. The restaurant staff replied, 'No problem. You are welcome'. The female guest arrived on time and was met by one of the staff, who escorted her to the table reserved for her. 'Would you like a table where you can look at the other guests or would you prefer to sit a little to one side instead?' asked the female waiter. The guest ordered a glass of wine, and was given the menu and served a roll and butter. She ordered her food and then sat there looking round, slowly nibbling at her bread and sipping the wine in order to fill up the time while waiting for the main course. The service was good, but the staff were a little bit on the reserved side, silent and hesitant, as if wondering why the woman was dining alone. Nothing was wrong, but on the other hand nothing was done to please or give pleasure to this special guest. The other guests in the restaurant did not notice the single female, since they were all busy enjoying their food in company of others. On this occasion the solo woman dined alone at a high-class restaurant with an ambiance of loneliness around her:

Description of the scene: Woman B finished her solo meal fairly quickly. Her five courses were served one after the other without any pauses or conversation to help her stretch out the time. She therefore – as agreed beforehand – sent a short message to the other Woman A saying, 'I have finished my meal. Have you?' Woman A replied, 'My God, I've only just started my main course! Come and have a glass of wine with me'. She also asked the waiter if it was all right for a friend to turn up for a glass of wine. 'Yes, of course', the waiter said. A few minutes later Woman B arrived and was cordially received at the door and escorted to the table. She was immediately given a glass of wine on the house and an extra plate to share the cheese; she also was offered a dessert and coffee. Suddenly, there was a remarkable change in the atmosphere in the small restaurant room, as if everyone – staff and guests alike – was smiling with a sigh of relief. – The Woman A was no longer alone! She did not have to sit on her own any more. She had a companion; once more there was 'order in court'.

Some gender studies try to explain the problem *vis à vis* women taking their place in society, especially in an urban context. Goffman, in his book on *Behaviour in Public Places* from the 1960s, suggests that in many situations certain categories of people may not be authorised to be present, and should they be so, this in itself constitutes an improper act (Goffman 1963/1966). Domosh and Seager continue this discussion in their book *Putting Women in Place. Feminist geographers make sense of the world* (2001). Also, in Sweden, women's place in the city is paid attention to by researchers (Forsberg 2005). Etiquette manuals might provide a guide to the manners expected, says Goffman. We might well ask what has happened since, in an equal society like Sweden. In a well-known Swedish Guide to Etiquette (Ribbing 2005), there is a chapter on eating out, i.e. on 'decent behaviour in restaurants', dealing solely with behaviour in company and an American etiquette book tells us that a woman should think twice before dining alone in a bar, since she may quickly attract unwanted dining companions (Ford and deMontravel 2003). When scrutinising etiquette from a cultural point of view, Ashley *et al.* quote the aphorism often used, that manners make the 'man', and raise questions as to etiquette and gender, seeing manners as a means of gender distinction (Ashley *et al.* 2004).

When discussing with women in working life their experience of solo dining everybody had something to say. Typical comments from the women included, 'How dare you...? I would never...'. Though at the same time, many of them said that they really wanted to be part of the comfortable milieu in a high-class restaurant, with a glass of wine and a good dinner. Some of them also wished to luxuriate in service and attention. On the other hand, there were stories of women's strategies for avoiding eating out of an evening when travelling alone on business. They either ordered something to be brought to their room or had some bread and cheese or a slice of pizza with them. There was one woman who described her car, which was extra equipped with a fridge, where she could keep her evening meal when travelling for work. All these examples, however, dealt with eating to satisfy hunger, rather than for pleasure, while one comment referred to the abundance of eating: 'For a single woman to indulge in a 3 to 5-course dinner including wine is on the verge of indecent extravagance! Vulgar! Most unsuitable!'. There were also questions about the finances of eating out. Some women said they would rather spend the same amount of money on a visit to a spa or buy a new dress or go to the hairdresser.

But men, too, revealed their problems. One said, 'I do not feel well when I have to dine alone. I often spend too much money on the wine because I pity myself in this situation'. Other men confirmed his feelings.

An *American internet* blog (Bruni 2006) relating to *Diner's Journal* in the USA discusses the same theme. The heading, 'All by Themselves', beginning with the following text – describes the idea of eating out socially:

> A few weeks ago, in the span of just a few days, I twice noticed something I don't see all that often in relatively fancy restaurants: someone dining alone … I mean at tables intended for two in the middle of dining rooms filled with groups of three, four or more. And I mean at dinnertime, when a solo diner is more likely to arouse curiosity or even pity, more likely to feel conspicuous. (Bruni 2006)

The blog writer continues by talking to restaurateurs who say that solo dining, i.e. dropping in to a high-end restaurant alone, is not as rare as it used to be and the stigma of eating alone has at least partially gone away. At the same time, however, there are no tables expressly set up for one person and 'the solo diner is going to have an empty seat, which equals lost revenue, across from him or her'. Another way of looking at the solo diner might be, as one restaurateur said, that solo diners bestow the greatest possible compliment on a restaurant because they are there just for the food and service – no business, no romance. In restaurant parlance, 'Here comes tomorrow's six-top' (Bruni, 2006).

Interesting in themselves were the 80 short comments (Bruni 2006) sent in, discussing the blog text. The answers were sorted by us, according to names, with 38 men (48%) and 21 women (26%), but, in addition, there were 21 comments under gender-neutral names. We categorised these comments (see Table 12.1) and saw them as two partly overlapping codes: *Treatments*, which we explained as the behaviour and service the restaurant staff offered the customer and *Feelings*, explained as the customer's experiences of the social dimension of the meal event, including atmosphere. In Table 12.1, some examples of both positive and negative comments are presented. Both men and women expressed themselves positively and negatively but in slightly different ways. One man said decidedly, 'Never had a strange reaction to dine alone...', while one of the women recounted how she safeguarded herself in order to be treated well, 'I have received universally good treatment, I make certain to look like a business traveller and bring a magazine or some work....'

These points of view (Table 12.1) reflect similar ongoing discussions as has been seen in the examples from Norway, Sweden and UK. Both men and women want to be part of the safe, comfortable atmosphere of a high-class restaurant, in company and as solo diners, with the expectation of a pleasurable experience connected not only with the food product but also with the meal

Table 12.1 Some comments from the American blog (Source: Bruni 2006)

Gender	Treatments		Feelings	
	Positive	Negative	Positive	Negative
Men	• I also travel alone for business quite a lot and make it a point to check out the more interesting restaurants in the different cities I visit. Never had a strange reaction to dining alone and as many have said, have received top notch treatment in many establishments	• but the WORST service I have ever gotten is when I've dined solo. … • Granted, when you're alone, poor or slow service can be more noticeable	• Love to dine alone and a great way to treat yourself …but I agree that the days of pitying solo diners should be over • I'm enjoying myself. I order my favourite dish, maybe bring a great book, sip a glass of wine – heaven.	• … why are there so many of us 'foodies' out there who have trouble finding one another? I don't mean business travellers, I mean single folk (like me) who have a passion for food but lack a partner who shares that passion?
Women	• I make certain to go to well-rated restaurants. I have received universally good treatment. I make certain to look like a business traveller and bring a magazine, or some work that I need to study – although often what happens is conversations are struck up with near-by diners. All you need is a smile, friendliness, confidence, and be sure to tip the waiter well so that they don't discriminate against single diners in the future!	• If the 'welcoming team' wants to seat me at the bar or at a table near the restrooms, I leave because I know that they are not welcoming people for dine alone. Who wants to eat at a place where we are considered a hindrance?	• When I was single I did so regularly and now do so (dining alone) mostly when I am out of town on business. I bring a book, but often don't end up reading it as the servers and/or chef often come to chat with me…. • … I might get a few cursory and curious glances upon being seated alone, but heck, I'm never going to see my fellow diners again, nor will they see me so what does it matter?	• …I cannot bring myself to eat out alone without some kind of protection from prying eyes though, be it a book, magazine or when all else fails, I'm ashamed to admit, my BlackBerry. I also cannot bring myself to eat out alone on a Friday or Saturday night.

experience as a whole. There is now a wider range of middle-class people eating out more frequently, the dining-out event having become more relaxed and open for everybody. Some constraints, however, still exist both for men and women as to how to deal with restaurant codes. Both men and women dining alone included in the Swedish study, as well as in the international blog, brought books, magazines and other things to make the restaurant experience easier to handle.

The communal tables offered in an increasing number of restaurants constitute one way of making life easier for those travelling on their own and others too, if they want to have someone to talk to and avoid feeling stigmatised and lonely. It was apparently a little easier for men to take their places in the mainly masculine-coded restaurants of the evening, whilst having lunch was a much less stressful experience for both men and women. There was also a subgroup of restaurant visitors extremely occupied with the sensory experience of food and wine, regarding food as fashion, i.e. 'foodies' (Table 12.1 and Ashley *et al.* 2004: 149). There is an anecdote on the Roman Lucullus (Lucius Licinius Lucullus Ponticus *c.* 110–56 BC), famous for the dinners he gave. He decided on one occasion to dine alone – with himself. To his astonishment he discovered that the dinner he was served did not come up to the standard to which he was accustomed. When he complained his servant answered, 'We thought there was no need to prepare a fine banquet for my lord alone'. 'It is precisely when I am alone', the gourmet icily replied, 'that you are required to pay special attention to the dinner...You must remember, Lucullus dines with Lucullus'. (Fisher 2002, and Pliner's Chapter 9 in this book). In our study, we did not find anyone of either sex with such a strong sense of self-esteem as Lucullus, but we could sympathise with the fact that many wished, in the same way he did, to be especially well treated when they dined with themselves. This is something restaurants need seriously to take on board.

Nowadays, too, the social dimension often appears stronger than the eating experience, to the extent that solo diners have to have something – a book, for instance – with which to protect themselves, since they have no-one to whom they can talk. It might help to train oneself to eat alone, thereby consolidating the diner's sense of self and social status (Crang 1997, Heimtun 2007). Distinctions in food culture in Bourdieu's terms continue to operate, and both Warde and Martens (2000) and Ashley *et al.* (2004) conclude that eating out is still a distinctive field, as we have seen in the examples mentioned in this chapter. Once again, in Finkelstein's words:

> Dining out becomes an event that brings the individual – figuratively and literally – into the public arena and exposes him or her to the scrutinizing eye of the other.... When, what, and how we eat becomes a narrative retelling aspects of biography and cultural knowledge....what we understand and misunderstand to be interesting, urbane, civilized, and pleasurable (Finkelstein 1998: 214).

12.4   References

Ashley, B. Hollows, J., Jones, S. and Taylor, B. (2004) *Food and Cultural Studies. Studies in Consumption and Markets.* London: Routledge.

Bahr Bugge, A. and Lavik, R. (2007) *Spise ute – hvem, hva, hvor, hvordan, hvorfor og når?* [*The role of eating out in everyday eating patterns – how often, with whom, what and how*]. SIFO [National Institute for Consumer Research]. Report no. 6-2007.

Bærentholdt, J. O., Haldrup, M., Larsen, J. and Urry, J. (2004) *Performing tourist places*, Aldershot: Ashgate.

Bell, D. and Valentine, G. (1997) *Consuming geographies. We are what we eat.* London: Routledge.

Bourdieu, P. (1989/2003) *Distinction. A social critique of the judgement of taste.* New York and London: Routledge.

Bruni, F. (2006) www.http://dinersjournal.blogs.nytimes.com/2006/12/15/all-by-themselves, retrieved 2008-04-15.

Butler, J. (1993) *Bodies that matter.* New York: Routledge.

Butler, J. (1999) *Gender trouble. Feminism and the subversion of identity,* New York: Routledge.

Campbell-Smith, G. (1967) *The marketing of the meal experience.* Guildford: University of Surrey Press.

Crang, P. (1997) Performing the tourist product. In: Rojek C. and Urry J. (eds) *Touring cultures. Transformations of travel and theory.* London: Routledge. Pp. 137–154.

Domosh, M. and Seager, J. (2001). *Putting women in place. Feminist geographers make sense of the world.* New York, London: The Guilford Press.

Finkelstein, J. (1989) *Dining out. sociology of modern manners.* Oxford and Cambridge: Polity Press, Blackwell.

Finkelstein, J. (1998) Dining out: The Hyperreality of Appetite. In R. Scapp and B. Seitz (Eds). *Eating culture,* New York Press. Pp. 201–214.

Finkelstein, J. (2004) Chic cuisine: the impact of fashion on food. In Sloan Donald (Ed.) *Culinary Taste. Consumer behaviour in the international restaurant sector.* Hospitality, Leisure and Tourism. Oxford: Elsevier Butterworth, Heinemann.

Fisher, M.F.K. (1937/2002) On Dining Alone. In *Serve it forth.* New York: North Point Press. Pp. 114–118

Ford, C. and deMontravel, J. (2003) *21st-Century etiquette: Charlotte Ford's guide to manners for the modern age.* New York: Penguin Books.

Forsberg, G. (2005). Den genderiserade staden [The gendered city]. In T. Friberg, C Listerborn, B. Andersson and C. Scholten (Eds.), *Speglingar av rum. Om könskodade platser och sammanhang.* [*Spatial Reflections. On gender-coded places and contexts*]. Stockholm: Symposion.

Gustafsson, I.-B., Öström, Å., Johansson, J. and Mossberg, L. (2006) The Five Aspects Meal Model: a tool for developing meal services in restaurants. *Journal of Foodservice,* **17**(2), 84–93.

Goffman, E. (1959) *The Presentation of Self in Everyday Life.* New York: Anchor Books.

Goffman, E. (1966) *Behaviour in Public Places. Notes on the Social Organization of Gatherings.* New York: The Free Press.

Hansen, K.V. (2005) *Restaurant Meal Experiences from Customers' Perspectives. A Grounded Theory Approach.* Örebro Studies in Culinary Arts and Meal Science 4. Örebro Sweden: Örebro University.

Heimtun, B. (2007) *Mobile identities of gender and tourism: The value of social capital.* Thesis, Faculty of the Built Environment, University of West England, Bristol.

Låt måltiden blomma [Let the Meal Flourish!](2007). www.latmaltidenblomma.se/text print.php?id=115 www.latmaltidenblomma/text_print.php?id=209, retrieved 2008-04-15.

Martens L. and Warde A. (1997) Urban pleasure? On the meaning of eating out in a northern city. In Caplan P. (Ed), *Food, health and identity.* London and New York: Routledge.

Morgan, M. and Hemmington, N. (2008) From foodservice to food experience? Introduction to the topical papers. *Journal of Foodservice* **19**, 108–110.

Morgan, M., Watson, P. and Hemmington, N. (2008) Drama in the dining room: theatrical perspectives on the foodservice encounter. *Journal of Foodservice* **19**, 111–118.

Pine, II, B. J. and Gilmore, J. H. (1999) *The experience economy: work is theatre and every business a stage.* Boston, Massachusets: Harvard Business School Press.

Pipping Ekström, M. and Jonsson, I. M. (2007a) *Women dining alone in restaurant rooms* Nordic Consumer Policy Conference 3-5 Oktober 2007, Helsingfors www.consumer2007.

Pipping Ekström, M. and Jonsson, I M (2007b) Genus – i rummet, [Gender – in the room] In Tellström, R, Mossberg L, Jonsson I. (Eds) *Den medvetna måltidskunskapen en vänbok till Inga-Britt Gustafsson [Conscious knowledge of meals: a festschrift in honour of Inga-Britt Gustafsson.]* Måltidskunskap, Culinary Arts and Meal Science 3. Örebro universitet.

Ribbing, M. (2005). *Nya stora etikettboken. Raka råd och enkla regler till vardag och fest [New big book on etiquette. Straight advice and easy rules for everyday and festivities]* Stockholm: Bokförlaget DN.

Simmel G. (1957/1993) Måltidets sosiologi [Soziologie der Mahlzeit/Sociology of the Meal] *Sosiologi idag [Sociology of today],* **23**(1), 3–9.

Skeggs, B. (1999) Matter out of place: visibility and sexualities in leisure spaces. *Leisure Studies* **18**(3) 213–232.

SoloDining (2008), www.SoloDining.com, retrieved 2008-05-02.

SoloDining GreatBritain (2008) www.SoloDining.com/greatbritain-restaurant.html, retrieved 2008-04-15.

Table for one if you please (2002) http://archive. somersetcountygazette.co.uk/2002, 18 November retrieved 2007-01-30.

Urry, J. (2000) *Sociology beyond societies. Mobilities for the twenty-first century.* London: Routledge.

Valentine, G. (1999) Eating In. Home, Consumption and Identity. *Sociological Review,* **47**(3), 491–525.

Wang, N. (2002) The tourist as a peak consumer. In Graham M. Damm (Ed.), *The tourist as a metaphor of the social world.* Oxton: Cabi Publishing, 281–295.

Warde, A. and Martens, L. (2000) *Eating Out. Social differentiation, consumption and pleasure.* Cambridge: University Press.

Wood, R. C. (1992) Gender and Trends in Dining Out. *Nutrition & Food Science,* **5**, 18–21.

Wood, R. C. (2003) How important is the meal experience? Choices, menus and dining environments. In *Strategic questions in food and beverage management.* Oxford: Butterworth-Heinemann. Pp 28–47.

Part V

Teaching through meals

13

The family meal as a culturally relevant nutrition teaching aid

J. Chrzan, University of Pennsylvania, USA

Abstract: From the results of nutritional research amongst adolescents as well as experience in food and nutrition education, it is argued that the family meal, because of its culturally defined structural components, provides an excellent model for teaching healthy eating patterns to children, especially adolescents. The use of meal patterning in established nutrition education programs is reviewed and a case study is given of the process of adapting meal pattern information to a high-school nutrition education program. Possibilities for adapting nutrition education programs to meet educational goals in the future are explored and a plea is made to have more faith in teenagers' relationship with food.

Key words: family meals, nutrition, teaching aid, adolescents' food, healthy eating.

13.1 Introduction

The concept of the family meal has been explored by many scholars in nutrition and food studies, including Mary Douglas (1974 and 1975), Jeff Sobal (1999, 2000 and 2003), Alice Julier (2002 and 2005), Johanna Makela (2000) and Alex Macintosh (1999). Understanding the process of the family meal and how it has shifted in meaning, content and practice over time is integral to food system and nutritional analysis. It has received popular attention lately (see Weinstein, 2005) as an icon of family health and childrearing capacities since, as Weinstein chronicles, many Americans believe there is a link between a perceived decline of the family and a decrease in families eating together. In addition, many people believe that the supposed decline of the family meal has been a contributing cause in the growing obesity levels of children and teen behavioral problems such as drug use and delinquency. As a result, the meal functions as an icon for social action and cultural persistence and carries enormous weight in our national and cultural consciousness.

Because of its near-mythic status as a symbol of family functioning, many citizens and scholars consider the meal the preeminent space for the cultural training of children in manners, social skills and nutrition. However, relatively few nutrition education programs explicitly use the meal as a tool of education and this might be due to the near-sacred status of the meal as a site of social action; while structuralists such as Mary Douglas have deconstructed the meal into its constituent and metonymic parts the practice of nutrition education rarely uses such models as a means to develop or teach healthy eating practices in children. Using published studies, results of nutritional research amongst adolescents as well as experiences in food and nutrition education, I argue that the meal, because of its culturally-defined structural components, provides an excellent model for teaching healthy eating patterns to children, especially adolescents.

The meal, family and otherwise, functions as a facilitator of food intake such that social behaviors often result in dietary engagement because people tend to eat together. Facilitation also smoothes the way past fear of new foods by introducing foods in a social setting, thus modeling risk-free eating to children and socializing youth into eating as adults. Humans learn to eat by watching others and rarely try new foods independently. As Carolyn Korsmeyer (1999) maintains, the family meal exemplifies commensality and creates a shared cultural food background so that food meanings are inter-related and contextualized. In this manner, the aesthetics of the family meal become a meaning-making tool for food education and can teach new dietary habits by embedding dietary goals within a food system which is already accepted and understood. Facilitation can also be used to teach appropriate food groupings since acceptability is reinforced by community approval and enjoyment. Facilitation helps to define aesthetic patterns through the modeling of flavor combinations, as Elizabeth Rozin (1973, 1997 and 2000) demonstrated in her work on culturally determined cuisine principles.

As Jeff Sobal (2000 and 2003) has illustrated, the meal also defines social activity and social functioning through the practice of meal sharing; the structure and process of the meal can be analyzed to unpack social practices since the meal functions as a metaphor for how social activities are organized and accomplished. Those who eat together conduct many other functions of society together (Douglas, 1974) and thus commensality (meal sharing) defines the social networks which encompass the known world of the child and teenager. Teaching an understanding of commensality offers a means to teach the social rules of a society and also the rules for the construction of meals including the timing, contents, and processes of appropriate eating occasions. Just as commensality defines sociality, the structural components of a meal are defined by cultural norms and processes, so most cultural agents learn the constituents of a meal as a result of social maturation. All of these processes can be borrowed when teaching food and nutrition education if the educational paradigm involves group practice and tasting, and healthy practices can be encouraged by emphasizing the cultural elements that cue appropriate eating patterns.

13.2   Overview of the use of meal patterning in established nutrition education programs

Given the seeming applicability of meal modeling as a means to teach good dietary habits, it is surprising that few youth nutrition education programs in the United States utilize the concept of the meal to teach health goals. Indeed, much of the research about meals is embedded within studies about other aspects of food behavior, from power structures and banqueting to validation studies of methods for measuring intakes of specific nutrients. In order to use the meal as a teaching tool, it is necessary to define the meal and its components, describe how they can be used in teaching and to define the educational outcome goals desired. In this case, the meal shall be conflated with the 'family meal' and the focus will be on dinner, since that is the most typically shared meal. Second, the goal of the educational program will be to increase vegetable and whole fruit intake in accordance with the goals of the USDA Team Nutrition curricula (http://www.fns.usda.gov/tn/team.html), which recommends that youth eat a variety of foods including more fruits, vegetables, whole grains and calcium-rich foods, eat lower fat foods more often, and be physically active. The population to be targeted is youth (preteens and teens) from the ages of 10 through 19, since that is the age group with whom I conduct research and for whom I have designed and implemented educational programs. However, these programs could be used with all ages providing that the practical elements of the programs are consistent with the skills and learning capacities of the targeted groups.

In the USA, the family dinner has received much interest of late because its supposed decline is considered emblematic of the decline of the family. *The Surprising Power of Family Meals* (Weinstein, 2005) argues that the family meal been shown to decrease rates of delinquency in teens, increase grades and college acceptance rates, and, as the subtitle asserts, 'make us smarter, stronger, healthier and happier'. However, in this case the meal serves mainly as a metaphor for family practices which increase social interactions and promote hands-on child rearing. The nutritional, categorical or food content of the meal is relatively unimportant and the meal serves as a facilitator of parenting 'best practices'; the primary argument for family adoption of a set dinner schedule is the psychological functioning of children (Boutelle *et al.*, 2001; Compan *et al.*, 2002; Croll *et al.*, 2002; Eisenberg *et al.*, 2004; Fulkerson *et al.*, 2006).

However, meals have cultural content structure, as Mary Douglas has demonstrated (1974 and 1975) and the structural components of the meal are understood to provide a cultural 'grammar' that defines proper eating patterns.

Douglas (1975) provides a tripartite meal scheme which has a stressed food (a main dish or element, such as a meat or meat substitute) and two sides, just as the structure of the day provides for a stressed meal (dinner) and two smaller ones. A 'proper' meal consists of a core item plus two ancillary foods, usually a protein, a starch, and a vegetable or two. Similarly, the stressed meal follows this pattern through a tripartite system of courses including a first or starter (often soup or a salad) then a stressed main course followed by the secondary dessert. A plate must consist of these elements to be a meal and that which has only one or two of the elements is considered a snack. Amongst the British people Douglas studied this translates, roughly, into a meat center and a starch and vegetable accompaniment (meat, potato, vegetable or A +2b). The structure of A + 2b remains consistent throughout additional components of the meal, so that each additional course must metonymically repeat the formula, as must any food combination that resembles a meal. In that capacity, it could be argued that a hamburger becomes a meal for North American eaters because it retains the A + 2b structure (meat, starch, lettuce/tomato) or, with meal analogy, the burger + fries + drink, while a Sunday dinner repeats the A + 2b with a main course and two unstressed courses of a starter and a dessert. As Douglas makes clear, this structure is very similar to the North American meal but is not a French or Chinese meal; each culture defines the structure of the meal and its metonymic analogies using the appropriate cultural categories and meal grammar.

Little research exists about the nutritional consequences of meal structure in the United States, especially among adolescents. Nutritionists tend to focus on the constituent parts of the meal and the resulting overall nutrient intake of the meal. However, even that is uncommon, since the primary method for measuring nutrient intake is either per item (when examining food) or per day, when focusing on clinical or individual intakes. In the United States, diet advice often consists of exhortations to 'choose' more servings of specific nutrients and their attached foods (especially high-fiber grains and vegetables), or to avoid large or frequent servings of other specific but less-favored nutrients, such as sugars, fats, and alcohols. Indeed, the metaphor of choice in food intake is a primary frame used in the conceptualization of food behavior and nutrition education with teens (for examples see Barrar and Brinley, 2001; Croll et al., 2001; Agron et al., 2002; O'Neill and Nicklas, 2002; Hamilton-Ekeke and Thomas, 2007). There is little acknowledgment of how the structure of meals may affect intakes of nutrients or food items perhaps because, as Mary Douglas claims, the meal structure is so powerfully embedded in the cultural consciousness that it is not necessary to elucidate it. There is, however, solid evidence that more frequently eating a family dinner contributes to generally greater intakes of vegetables and their constituent micronutrients among teens (Gillman et al., 2000; Story, et al., 2002; Neumark-Sztainer et al., 2003 a and b, 2004; Granner et al., 2004; Videon and Manning, 2003; Veugelers et al., 2005; Fulkerson et al., 2006; Larsen et al., 2007) and that frequent snacking among youth contributes to greater intakes of fats and calories and fewer vegetables (Nicklas *et al.*, 1997; Serra-Majem *et al.*, 2002; Sjoberg *et al.*, 2003; Stockman *et al.*, 2005). This may result because the unstressed elements of the meal structure drop out of the snack event. Therefore, it is reasonable to assert that educational programs may benefit from combining a choice framework with meal structure education and shifting the goals from choosing specific foods because of their nutritional contents to choosing specific foods because they conceptually complete or make 'culturally real' a meal.

While meal patterning is popular for teaching calorie-reduction or slimming skills to adults (see Gutterson, 2005 and Guiliano, 2006 for good examples), meals are not typically used in teen nutrition education. Since many of the state-approved curricula focus on decreasing obesity levels most emphasize choosing micronutrient-dense foods for health in addition to greater levels of physical exercise. For example, California Project Lean (http://www.californiaprojectlean.org/) and CANFIT (http://www.canfit.org/), the California Adolescent Nutrition and Fitness Program, teach about food use and activity and are designed for low income minority youth while the USDA's excellent Team Nutrition program (http://www.fns.usda.gov/tn/team.html) encourages adherence to the My Pyramid nutrition goals through careful choice of food items and increased physical activity. Michigan State Extension Services 'Jump into Foods and Fitness' Program (JIFF: http://web1.msue.msu.edu/cyf/youth/jiff/) and the Colorado Department of Education's Nutrition Literacy Toolkit for grades K-12 (http://www.cde.state.co.us/nltk/default.asp) demonstrate similar goals, while Utah's Foods and Fitness, Nutrition and Curriculum Guide (http://www.uen.org/ate/family/foods1/) emphasizes an understanding of macronutrients and their function in the body in order to better choose foods and portion sizes. The USDA's various nutrition programs offered through Lifecycle Nutrition on-line resources (available at http://www.nal.usda.gov/) encourage knowledge-based food choices for teens as well as specific lessons in reading nutrition labels. All of these programs are excellent, and provide superb teaching opportunities, but most focus on food items as vehicles for specific nutrients which need to be increased or decreased in the diet rather than as foods or groups of foods to be eaten in a meal or social setting.

It is possible to find meal planning information embedded in many adult-oriented food websites produced by federal or state governments and/or their entities. The CDC's program 'Fruits and Veggies: More Matters' (http://www.fruitsandveggiesmorematters.org/) offers meal-planning tips online although they are not linked to a conceptual notion of a 'good meal' but to nutritional intake of macro and micronutrients. Perhaps one of the best examples of a teen-focused education plan is the online source TeensHealth provided by the Nemours Foundation (www.teenshealth.org/teen) which provides a plate graphic clearly showing proportional servings of a main course flanked by a starch and two vegetables (http://www.teenshealth.org/teen/food_fitness/dieting/portion_size.html). The graphic is accompanied by basic and sensible information about food choices and portion sizes, along with hints on how to avoid foods that provide many calories but few nutrients. This site supposedly is operated by and for teens, and the various stories and examples are written by youth with content vetted by medical professionals. Perhaps teens find graphics of this sort more comprehensible than possibly obscure dietary advice such as 'eat more fiber' or 'decrease fat in the diet'.

In conclusion, the typical nutrition education program for youth applies a choice metaphor to encourage higher intakes of healthy foods, usually considered to be vegetable, fruit, and whole grain in origin. However, the language is often 'nutri-speak', since the lessons consist of encouragement to eat more or less of specific nutrients rather than foods. While many programs, such as the USDA's "Eat Your Colors" and "Five a Day" do emphasize whole food intakes, few encourage teens to conceptualize dietary intake in a meal format rather than as intakes of individual foods, even though most food is eaten as a meal category rather than as a food itself. In other words, the cultural actor doesn't 'have potatoes, some carrots and a piece of meat' or a 'cookie with milk' but rather describes these intakes when speaking to others as well as when planning the day's food as 'dinner' or a 'snack'. Culturally, people eat meals rather than foods. While the components of a meal might be elucidated, the act of eating is most likely going to be described as a particular food event labeled by a meal or snack category. People often think about eating as an event defined by a meal structure, even if they spend the day 'grazing' for individual snacks. Part of teaching youth about appropriate eating consists of enculturating them to the same mental categories used by adults, and teaching them to utilize those categories to appropriately organize their days in harmony with other cultural actors.

13.3 Youth meal patterns

The teenage diet among the students in the United States with whom I have worked is marked by high intakes of foods containing fat, sugars and calories, and lower than needed quantities of vegetables and calcium-containing foods. While teens are generally sufficient in calorie and protein intakes, depending on the group and the age, they may or may not achieve the recommended daily intakes of micronutrients. In addition, the teen diet is shaped by social behaviors and school activities which may encourage disordered eating since many after-school activities limit participation in family meals. Many adolescents prefer to spend time with peers, which may lead to increased involvement in social events that revolve around snacking and fast food restaurants, causing an increase in intakes of sugars and fats. This section will provide a short review of the primary dietary patterns found amongst teens in the United States, followed by a discussion of the meal habits and dietary intake results from a study of African–American teen mothers-to-be from the Philadelphia area. Observations about inner-city teen diet are also culled from my volunteer work with the University of Pennsylvania affiliated organization Urban Nutrition Initiative, a school-based food and nutrition education program that uses active learning to teach healthy diet and activity behaviors.

A number of studies have examined teen dietary patterns in relation to recommended intakes of nutrients and food group categories. Most teens in the United States have adequate intakes of macronutrients but may be deficient in some micronutrients including calcium, folate, Vitamin E and magnesium. High intakes of fortified breakfast cereals often contribute to micronutrient sufficiency (Frary *et al.*, 2004). The food groups most typically at risk in the teen diet are dairy products and vegetables, followed by fruits and proteins (Adams, 1997; Gillman *et al.*, 2000; Rockett *et al.*, 2001; Neumark-Sztainer *et al.*, 2002; Zabinski *et al.*, 2006) and there is a relationship between increased fast food and soda intakes, snacking, high calorie and fat ingestion and decreased intakes of calcium, folate, vitamin C and other micronutrients (Adams, 1997; Harnack *et al.*, 1999; Cavadini *et al.*, 2000; French *et al.*, 2001; Stockman *et al.*, 2005). Frary *et al.* (2004) found an inverse relationship between adolescent intake of sweetened beverages and sugared cereals and vegetables and dairy in general, indicating that snack foods and quick meals (breakfast cereals) were displacing balanced meals in the diets of teens. Few teens eat the recommended number of servings of vegetables and fruits per day (Neumark-Sztainer *et al.*, 2003), with most studies demonstrating between three and four servings only (Rockett *et al.*, 2001) with only 15% of males and 12% of females (11th and 12th grades) eating five or more servings per day (Adams, 1997). If the proposed outcome of a nutrition education program is to increase vegetable and whole fruit intake, then helping teens to reach the recommended number of servings of vegetables and fruits is a simple and practical goal. Unfortunately, frequent snacking and disordered eating (skipping meals, frequent dieting, restricted eating habits, etc.) often results in a decrease of vegetable intake and thus the practicality of encouraging a frequent family or shared dinner is reinforced.

Teens tend to snack more than do adults and younger children (Bigler-Doughtson and Jenkins, 1987; Rockett *et al.*, 2001; Stockman *et al.*, 2005). This occurs in part because they are more independent than younger children, may have more pocket money to buy food and because teen social habits include shared food events. Among the teens I studied, neighborhood food venues such as fast food restaurants, corner stores, and food vending trucks are considered safe places to congregate, especially in inner city areas where residents are aware of neighborhood danger zones. Among the sample of 104 teen mothers-to-be in West Philadelphia (average age 16.9 years) with a combined total of 482 diet recall days, the teens ate 445 breakfasts, 356 lunches, 432 dinners and 886 snacks. Breakfast was the most consistent meal taken, although there was an almost one-to-one correspondence with the number of recalls and the three main meals of the day. Lunch was eaten the least but mid-day intakes were balanced by frequent snacking since two snacks per day were eaten, on average. Furthermore, almost 32% of all food events (which consist of a meal or snack of any sort) are sourced from a local fast food restaurant, lunch truck, corner store, or chain restaurant; take-out meals and foods are far less likely to include vegetables and whole fruits than are home-prepared meals (Guthrie *et al.*, 2005). Most of these meal events were provisioned by the teens themselves; 68% of all meals were prepared or purchased by the teen, with a distant 21% of meals provided by adults (parent or grandparent generations). Altogether 78% of all meals and/ or snacks were provided by either the teen or someone from the same generation, such as a boyfriend or sister. Because of this pronounced meal autonomy the messages promoted by nutrition education programs within a high school setting are of primary importance for teaching the adoption and maintenance of good dietary habits.

Even more compelling is a deeper analysis of meal provisioning in relation to family and friends, which demonstrates how frequently teens in this sample forage for themselves and how the family dinner alters this relationship. For instance, roughly 80% of all snacks and breakfasts are prepared and provisioned by the teens for themselves but they prepare only 30% of dinners; 50% of dinners are cooked and served by adults. Given that more vegetables are eaten during dinner than with other meals in this sample and in other studies of adolescents (Gillman *et al.*, 2000) and that teens eat more vegetables from meals provided by adults (Chrzan, 2008; Gillman *et al.*, 2000; Neumark-Sztainer *et al.*, 2003; Zabinski, 2006) it is reasonable to assume that dinners probably represent the majority of vegetable intake for the sample. Indeed, dietary analysis reveals significantly higher intakes of nutrients found in vegetables and whole fruits (vitamin A, C, and fiber, etc.) in those teens in the sample who eat with more than two adults per day or whose meals are more regularly provided by adults (Chrzan, 2008) which indicates that vegetables and fruits are more likely to be consumed when adults make them available to teens.

An inverse relationship was found when examining teen snacking and take-out meal behaviors in relation to nutrient intakes, adult provisioning and food sharing. Teens who ate with (or were provisioned by) more than two adults per day were significantly less likely to eat corner store snacks (usually snack cakes, candy bars and sweetened beverages) and take-out foods on both weekdays and weekends than were teens who ate with fewer adults. This snack/meal relationship also correlated significantly with higher intakes of nutrients including vitamins A and C, riboflavin and fiber. When the meal components are examined, it is clear that adult presence determines vegetable intakes since almost none of the teens included vegetables when preparing a meal for themselves or their friends. However, dinners prepared by mothers, grandmothers, aunts and great-aunts frequently contained vegetables, especially dark green leafy vegetables such as collard greens, which is one of the iconic foods among African–American families. Adults provided salads more frequently, prepared bean dishes such as baked beans and black-eyed peas, and cooked complex dishes such as chicken and beef stews that contain vegetables in addition to meat. Teens more typically prepared ready-made foods such as frozen pizza, hot dogs, and macaroni and cheese. Overall, the girls who ate with adults had significantly more servings of vegetables and whole fruits per day than did those who more typically provisioned themselves.

Another axis of family food practice that affects vegetable, fruit and nutrient intakes is the Sunday meal. Among the teens I studied, the Sunday meal is the pre-eminent meal of the week and defines the ideal family food event just as surely as it defines the ideal cultural diet. While only 53% of the families eat a family dinner four or more nights a week (and approximately 25% of those meals were provisioned from take-out or fast food sources) fully 72% of families prepare a special home-cooked Sunday dinner, and that is the one meal amongst this sample in which two or more vegetables are likely to be served. It has been my observation that the Sunday meal also reified Douglas' theory, for the proper meal, according to the teens, consisted of a chunk of meat, a side of mashed yams, potatoes or baked macaroni and cheese, as well as a green vegetable, usually string beans, cabbage, or collards. To the teens, this meal is the ideal meal socially, culturally, and nutritionally. It is also cited as the ideal pregnancy diet for optimal mother and child health, a belief I exploited when counseling teens to improve their diets during pregnancy. The important factor is the vegetable, because as meals are replaced by snacks, vegetable intake decreases profoundly. It appears that the metonymic structure of the snack supports the stressed item or the starch, but not the other side, which is the vegetable. So a snack may consist of either a chunk of starch with a side (potato chips and drink) or a meat and a side (a plain hotdog and soda drink) but rarely consists of the starch and a vegetable, or the meat and a vegetable. Because the Sunday meal is so important it provides a model for teaching dietary change since we can reinforce the health lessons with reference to culturally favored food practices.

In summary, teens in West Philadelphia as well as youth in many of the studies previously cited tend to eat fewer vegetables and whole fruits than recommended, and also tend to substitute snacks for full meals. Teens within families that provide meals, especially dinners, eat more vegetables and whole foods, fewer snacks and take-out foods, and have a higher intake of the micronutrients often at risk in the adolescent diet. Furthermore, among the African–American families of West Philadelphia, the Sunday dinner stands as an archetypal meal and symbolizes family stability, ethnicity and nutritional health. The Sunday family dinner provides an emic model of an ideal meal which can be used to reinforce healthy eating habits in education programs designed for teens.

13.4   Case study: adapting meal pattern information to a high school nutrition education program

The Urban Nutrition Initiative (UNI) is a university-community partnership that engages kindergarten through college students in an active problem-solving curriculum to improve community nutrition and teach healthy living goals in the Philadelphia area. UNI uses a variety of teaching techniques to reach students of all ages in classroom, after school and summer programs. The education program integrates food and nutrition lessons with school curricula in order to teach key concepts within a framework of social studies, biology, mathematics, and other courses. The food education curricula use the concept of cuisine and cultural meal patterning to teach nutrition and food education to the same geographic population of teenagers studied for the pregnancy research. By utilizing food beliefs and reported dietary practices and preferences, the education project tailors the nutrition message to the targeted teen cohort and teaches practice-based dietary skills. I volunteered with UNI to create and implement food education curricula and I also teach an ABCS (Academically Based Community Service) course at the University of Pennsylvania in which students work with UNI staff and students to develop and lead various programs for students of all ages in the Philadelphia school system. Many of the ideas for using the meal structure as a teaching tool were developed during school and summer school UNI programs in addition to nutrition education clinical work with pregnant teens from the same population.

Theoretically, to encourage teens to adopt a healthier diet (one with more vegetables and whole fruits, for example) we need to create a conceptual shift with ideational linkages to positive dietary messages. We should retreat from empty vessel metaphors of teaching or assume that students do not know that they should eat more vegetables. Merely providing biological nutritional information in hopes that new knowledge will shift behavior does not work consistently with youth or adults (Greene *et al.*, 1999; Kristal *et al.*, 1999; Barrer and Brinley, 2002). In my experience working with teens and college students, I have observed that high school students already know much about food and they have established tastes and habits, and thus biological knowledge divorced from practical training will never shift food behavior (Neumark-Sztainer *et al.*, 1998; Raynor, 1998; Greene *et al.*, 1999; Kristal *et al.*, 1999; Story *et al.*, Barrer and Brinley, 2002; 2002; Patrick and Nicklas, 2005; Shepard *et al.*, 2006). With food, and especially with the meal metaphor, we can use knowledge already in the students' minds to reinforce positive concepts with interactive lessons involving taste, smell, and sound in hands-on cooking exercises while linking the knowledge content to school educational goals such as history, mathematics, social studies, and geography. This creates multifaceted and internally coherent knowledge blocks that reinforce the dietary lessons. Using models of commensality in lesson plans allows elements of the meal to map onto elements of aesthetics, cultural prescriptions and belief systems so that the meal becomes a complete and contextualized whole according to the dictates of the students' cultures.

Ideally, such lesson plans link vegetable intake and meal patterning to cultural rather than biological knowledge, and build on knowledge substrata rather than assume the need for information replacement. This avoids an empty vessel educational paradigm by allowing students' habits and community values to inform their understanding of meal practice rather than hoping that biological knowledge will translate into practice. UNI uses the concept of 'cuisine' as an organizing and teaching model for the curriculum, where 'cuisine' is defined as all aspects of production, processing, consumption, and disposal that contribute to the construction of food systems, beliefs and cultural understandings that determine appropriate food use in differing human societies. Because UNI creates and supports schoolyard gardens, the educational program is able to link a vegetable grown in the garden to a cultural cuisine, and can promote formal and informal teaching of meal and food behavior during the food production and preparation processes. Students in the high school science classes and summer-school programs participate in knowledge and practice-based curricula designed to alter food habits by encouraging alternative food choices that are validated by social science lessons. As we introduce the vegetable, we are able to talk with the students about the flavor principles – giving them information about how to eat it, what to eat it with, and how other cultural groups form meal patterns and conceptualize 'good food' in relation to meals and personal intake. We talk about how they might use the recipe at home, in a weekday meal or a Sunday dinner, and if the rest of their family would like the new dish. We talk about how to combine vegetables in meals in a manner that tastes good and makes nutritional and social sense and we let them help plan the meals they cook in the kitchen labs.

To provide examples, for one summer school lesson I used the concept of a Chinese meal to illustrate the need to incorporate several vegetables into a dinner menu. I had the class make diagrams of what their plates look like during a typical dinner at home; their American menu had a large chunk of meat ('a big steak!' one young man asserted) flanked by rice or potatoes and a vegetable (most of the children chose broccoli). We then learned a little bit about Chinese menus and how rice is the center of the meal rather than meat, and how several different types of vegetables and meats accompanied the rice according to Chinese aesthetic rules of taste. They had difficulties accepting the replacement of the center stressed meat with rice, especially since the Chinese food take-out places in their neighborhoods offered meals that had stressed (and large portion sizes) of meat mixed with vegetables at the center, accompanied by a small portion of rice. This observation allowed us to talk about the nutritional values of each of the food groups, and the students debated whether rice or meat was more important to health. Then we brought the discussion back to vegetables, and I pointed out that they had fallen into the American meal pattern trap of assuming that the meat or the starch would be the center of the meal. What about making the vegetable the center – would that work for dinner? And what kind of dinner would include two vegetables? They debated this new idea, and decided they liked a balanced plate better, 'one with rice and meat and several vegetables with different colors and textures, so the mouth doesn't get bored and the food tastes interesting', and that having two vegetables was like having a Sunday dinner with the whole family after church: 'meat, and rice, and salad, and boiled greens and cake for dessert!'. We then made a Chinese meal with the vegetables harvested from the school garden so the students would know how to process and cook some of the foods we'd learned about; the students were divided into teams to make each recipe. For lunch (which included a salad, rice and two Chinese-inspired dishes with vegetables and tofu) the students ate as a group at the table, each with a properly set place with plate, utensils, and cups. In this lesson, they learned about specific vegetables, Chinese food aesthetics and menus, the nutritional values of the vegetables we were cooking, how to pick, wash, process and cook several vegetables, and how to set a table correctly. They were also given copies of the recipes so they could cook them for their families at home.

The more formal classroom lessons followed the same pattern although during much of the school year the gardens were bearing little, so food from the store was used. In one social studies class, we learned about dark green leafy vegetables and Ethiopian cuisine and the students cooked several African dishes with collards, potatoes, peppers and peanuts. We talked about African meal planning and how people in Africa eat different kinds of meals than Americans; they learned about millet and cassava and how they might be accompanied by a couple of vegetable dishes (with small amounts of meat) for a 'complete' African meal. Even through the food was utterly unfamiliar to these 10th-grade students, after cooking it they enthusiastically tried and ate greens, yams and tomatoes cooked in an African style and served on enjera bread which had been provided by a local Eritrean restaurant. The lesson covered the geography of Africa (with maps to color and fill out), the history of Eritrea and Ethiopia and why folate, found in dark green leafy vegetables, is important for the body. I even gave them a short lesson on Haille Sallassie, Rastafarianism, and the origin of the term 'jah' popular in modern day rap music – accompanied by music, of course. As with the summer school class, this lesson combined hands-on learning with meal planning, mixed curricula messages, and solid skill building to encourage cooking at home. By introducing new foods in a group setting, the students are more likely to try new vegetables, especially if they know a story about a culture that uses them, because they can dare each other to try the strange stuff (see Adams, 1997 and Jessor et al., 1998 for discussions of how group learning can encourage health behaviors in teens).

This sneaky plan to encourage youth to try unfamiliar foods worked. During the first lesson, the students refused to even try fresh tomatoes from the garden; one young lady took a tiny bite, looked around at the shocked faces of her peers and screamed 'it's nasty!' before spitting it out in a thoroughly dramatic fashion. On the other hand, I have witnessed 11th graders who refused to taste a piece of red pepper watch a peer do so, witness the peer's demonstration of enjoyment, and then try the new food. Over the course of that semester they became more and more adventurous about trying new foods, and asked questions that clearly indicated that they understood the lessons and even cooked the vegetables at home. The idea that 'you had to have at least one vegetable for it to be a meal' was reinforced, and they told me that they were showing their brothers and sisters how to cook vegetables so they could make 'real meals' too. Obviously, this is just one small element of the UNI program; the students also learn about healthy snack choices, exercise, community health and even entrepreneurship by selling vegetables from the gardens at farmers markets. My volunteer work with UNI is merely a small element designed to integrate practice, culture, and vegetable intake via lessons in food anthropology. Querying students about food behaviors in a small group setting allows me to verbally reiterate the need to incorporate vegetables in meals and link the practice of eating vegetables to the much-anticipated and enjoyed Sunday meal. This causes the students to accept that a 'proper meal' always contains one or two vegetables. These lessons allow students to become more thoughtful about food behavior by linking meal or snack foods to items within the Sunday meal. Do chips and soda pop belong in the Sunday meal? NO! Then maybe they are not a good choice as a dinner on Tuesday; even lunch should be like the Sunday meal, if it is going to be a meal – so add a few carrot sticks!

13.5   Adapting meal lessons in nutrition education programs to achieve educational goals

The case studies outlined above provide a tantalizing glimpse into how food education might shift in order to encourage youth to eat more vegetables. However, most state and federal mandated programs provide defined curricula that require an emphasis on nutrition and biology rather than practice. Many of these programs are excellent, and provide good lessons and lesson plans, but in my work with students I have found that teens are very concrete thinkers and require straightforward examples in order to understand many concepts. For most eaters, including teens, a lesson with a message such as 'eat more fiber and vitamin A' isn't applicable to them unless it teaches real and socially appropriate practices that will allow them to understand which foods they should eat, how they should eat them, and why, when and what should accompany the foods. Most teens lack the shopping and cooking skills to adopt new foods with ease; the teens in my study most frequently cook easy, one-step meals such as frozen pizza, macaroni and cheese, or hot dogs with a slice of white bread. In fact, even my University of Pennsylvania students tell me that they hadn't known how to prepare red bell peppers, or string beans, or whatever vegetable was highlighted in that day's lesson, so they rarely ate them at home. They learn just as surely as do the teens in whose schools they are volunteering. My students remind me that there is a general lack of food processing skills among many members of the population – not just teens in inner city areas. Adopting a meal model as a means to teach vegetable intake makes the knowledge and practice accessible in an easily understood, culturally appropriate manner and may encourage compliance, because the educational meal model includes all steps required for planning, food sourcing, preparation, and ingestion and, in addition, provides the conceptual model that drives the planning.

In my experience with working with youth, several elements must be present in order to appropriately integrate vegetable preparation practice lessons and meal planning skills with nutrition education programs. First, youth must be integrated into the design and implementation of the food education messages. UNI uses teens as peer educators and to design and teach lesson plans for younger students. Teens appreciate inclusion, and welcome the opportunity to become a part of the planning of an activity. By designing curricula for younger students, they adopt and integrate the lessons into their own daily food practices. Second, since they are such concrete thinkers the actions must be explicit and direct. If the goal is to increase vitamin A and fiber intake it is not useful to tell a teen to eat more carrots because they have fiber and Vitamin A, and far better to urge them to 'add a handful of carrot sticks to their lunch because the carrots will make them feel better and they taste good'. The actions have to be clearly spelled out and very accessible to the teen. One of the better means I found to encourage pregnant teens to add vegetables and whole fruits to their diets was to urge them to add just one vegetable and/or one whole fruit per meal. This translated into a fruit at breakfast, a vegetable and a fruit at lunch, and a vegetable and a fruit at dinner. Since poor fiber intake often leads to constipation during pregnancy, these were clear and easily followed directions that once absorbed stayed indefinitely and added small 'healthy eating' steps to daily practice. Not long ago I saw one of the teens in my study and the first thing she said to me after a brief hello was 'I'm adding those veggies to my dinner just like you said!' which was ample proof that the lessons, if made concrete and practical, can alter practice to increase healthy dietary habits.

The meal, because of its cultural weight, provides an ideal site for the focus of healthy eating reminders. Just as the students enrolled in the UNI programs learn that a meal has to have specific elements, including vegetables, to be a 'real meal', so do the clinical patients at the pregnancy clinic. The message to add a vegetable to dinner (to make it a 'real meal') even works for adults because it causes them to shift from thinking about dinner as a time-specific event to one that is categorized by content. The students from the University of Pennsylvania who volunteer with UNI not only teach, but adopt, this message of adding a vegetable and a fruit, which they then tell me affects their choice of foods in the dining hall. Simple graphics such as the plate diagram provided by the TeensHealth website allows for a reconceptualization of how to eat because it includes two vegetable portions. For teens used to thinking that a meal might consist of a piece of pizza, soda pop and a brownie such a simple message might reconfigure their categorization of such a 'meal' into a snack. Furthermore, creating a mental meal template that mimics a Sunday dinner and contains a meat or meat substitute, a starch, and two vegetables may also encourage other nutritional goals, such as a decrease in high-fat foods and portion control.

13.6   Strengths and weaknesses of using meals as a means to teach healthy eating behaviors

I would like to argue several key points about this project and about nutrition education in general, because there are a number of holes that can be filled by food studies and nutritional anthropology. Obviously, reordering how we teach about healthy food choices to incorporate a meal model is not a magic panacea – the message must contain specific dietary goals and recommend changes the target population can accomplish with relative ease. Meal-centered curricula require a real and working knowledge of neighborhood food (especially vegetable) availability in addition to an understanding of the limits of the students' practical knowledge about vegetable choice, preparation and intake. We need to know the barriers to creating a complete meal which may include a dearth of cooking skills caused by a reliance on food industry products, a lack of access to healthy food items, lack of kitchen equipment or cooking instructions, and a need to prepare foods for household members with differing dietary needs.

There is too much blind acceptance of the phrase 'the family meal is dead'; research demonstrates that the family meal is not dead, but it is changing, and we need to learn how, why, where, and for whom. We need accurate assessments of skills and problems – or the absence of them – so we can better target practice-based lesson plans. We need to know more about what students do know about food, and how they know it. What are the lines of food communication for differing social groups and cohorts? We cannot assume that everyone gets their information from the same places nutritionists do, nor should they. Eating is universal and it is a universal subject of discourse, practice, and belief. Food studies and research in food anthropology can work to unpack meanings and practices in a manner that a nutritionist cannot or will not. Too often teens' baseline emic dietary understandings are not addressed accurately in nutrition or food education and we need more research on teen food habits in order to develop better food education paradigms.

Above all, we need to believe that teens want to eat with their families as well as their peers, and that they are interested in learning how to choose healthy foods, how to prepare and cook them, and how to appreciate food and eating together. We need to have more faith in teens: it has been my experience that youth in the United States, including high-school-age teens in inner-city neighborhoods as well as university students in the Ivy League do desire a more active relationship with their food, their families, peers and cultures, and they enjoy growing, preparing, cooking, and eating foods that taste good and are good for them. But they also need to be taught how to accomplish these food and dietary goals in a manner that makes sense to them as cultural beings, and the concept of the family meal as a culturally defined model for food consumption provides an ideal means to teach healthy dietary practices.

13.7   References

Adams, Lucy (1997) An overview of adolescent eating behavior: barriers to implementing dietary guidelines. *Annals of the New York Academy of Sciences*, **817**, 36–48.

Agron, Peggy; Takada, Erika and Purcell, Amanda (2002) California Project LEAN's Food on the Run Program. *Journal of the American Dietetic Association*, **102**, S103–S1005.

Barrar, Cindie and Brinley, Colleen (2001) Stages of Change tools to increase fruit and vegetable consumption in high school students. *Journal of Nutrition Education and Behavior*, **33**, 57–64.

Bigler-Doughten, S. and Jenkins, R. M. (1987) Adolescent snacks: nutrient density and nutritional contribution to total intake. *Journal of the American Dietetic Association*, **87**, 1678–1679.

Boutelle, Kerri N.; Lytle, Leslie; Murray, David; Birnbaum, Amanda and Story, Mary (2001) Perceptions of the family mealtime environment and adolescent mealtime behavior: do adults and adolescents agree? *Journal of Nutrition Education*, **33**, 128–133.

Cavadini, Claude; Siega-Riz, Anna Maria and Popkin, Barry M (2000) US adolescent food intake trends from 1965 to 1996. *Archives of Disease in Childhood*, **83**, 18–24.

Chrzan, Janet (2008) Social Support and Nutrition during Adolescent Pregnancy: Effects on Health Outcomes of Mother and Child. *Anthropology*. Philadelphia, University of Pennsylvania.

Compañ, E., Moreno, J., Ruiz, M. T. and Pascual, E. (2002) Doing things together: adolescent health and family rituals. *Journal of Epidemiology and Community Health*, **56**, 89–94.

Croll, Jillian; Neumark-Sztainer, Dianne and Story, Mary (2001) Healthy eating: what does it mean to adolescents? *Journal of Nutrition Education and Behavior*, **33**, 193–202.

Croll, Jillian; Neumark-Sztainer, Dianne; Story, Mary and Ireland, Marjorie (2002) Prevalence and risk and protective factors related to disordered eating behaviors among adolescents: relationship to gender and ethnicity. *Journal of Adolescent Health*, **31**, 166–175.

Douglas, Mary (1975) Deciphering a meal. *Daedalus*, **101**, 61–81.

Douglas, Mary and Nicod, Michael (1974) Taking the biscuit: the structure of the British meal. *New Society*, **30**, 744-751.

Eisenberg, Marla E.; Olson, Rachel E.; Neumark-Sztainer, Dianne; Story, Mary and Bearinger, Linda H. (2004) Correlations between family meals and psychosocial well-being among adolescents. *Archives of Pediatric and Adolescent Medicine*, **158**, 792–796.

Frary, Carol; Johnson, Rachel and Wang, Min (2004) Children and adolescents' choices of foods and beverages high in added sugars are associated with intakes of key nutrients and food groups. *Journal of Adolescent Health*, **34**, 56–63.

French, Simone; Story, Mary; Neumark-Sztainer, Dianne and Hannan, P. (2001) Fast food restaurant use among adolescents: associations with nutrient intake, food choices and behavioral and psychosocial variables. *International Journal of Obesity*, **25**, 1823–1833.

Fulkerson, Jayne; Neumark-Sztainer, Dianne and Story, Mary (2006) Adolescent and parent views of family meals. *Journal of the American Dietetic Association*, **106**, 526–532.

Fulkerson, Jayne; Story, Mary; Mellin, Alison; Leffert, Nancy; Neumark-Sztainer, Dianne and French, Simone (2006) Family dinner meal frequency and adolescent development: relationships with developmental assets and high-risk behaviors. *Journal of Adolescent Health*, **39**, 337–345.

Gillman, Matthew M.; Rifas-Shiman, Sheryl; Frazier, Lindsay A.; Rockett, Helaine; Camargo, Carlos; Field, Alison; Berkey, Catherine and Colditz, Graham (2000) Family dinner and diet quality among older children and adolescents. *Archives of Family Medicine*, **9**, 235–240.

Granner, Michelle L.; Sargent, Roger G.; Calderon, Kristine S.; Hussey, James R.; Evans, Alexandra E. and Watkins, Ken W. (2004) Factors of fruit and vegetable intake by race, gender, and age among young adolescents *Journal of Nutrition Education and Behavior*, **36**, 173–180.

Greene, Geoffrey; Rossi, Susan; Rossi, Joseph; Velicer, Wayne; Fave, Joseph and Prochaska, James (1999) Dietary applications of the Stages of Change model. *Journal of the American Dietetic Association*, **99**, 673–678.

Guiliano, Mireille (2006) *French Women Don't Get Fat*, New York, Alfred A. Knopf.

Guthrie, Joanne F.; Lin, Biing-Hwan; Reed, Jane and Stewart, Hayden (2005) Understanding economic and behavioral influences on fruit and vegetable choices. *Amber Waves*, **3**, 36–41.

Gutterson, Connie (2005) *The Sonoma Diet: Trimmer Waist, Better Health in Just 10 Days!*, New York, Meredith Books.

Hamilton-Ekeke, Joy-Telu and Thomas, Malcolm (2007) Primary children's choice of food and their knowledge of balanced diet and healthy eating. *British Food Journal*, **109**, 457–468.

Harnack, Lisa; Stang, Jamie and Story, Mary (1999) Soft drink consumption among US children and adolescents: nutritonal consequences. *Journal of the American Dietetic Association*, **99**, 436–441.

Jessor, Richard; Turbin, Mark S. and Costa, Frances M. (1998) Protective factors in adolescent health behavior. *Journal of Personality and Social Psychology*, **75**, 788–800.

Julier, Alice (2002) Feeding friends and others: boundaries of intimacy and distance in sociable meals. *Sociology*. Amherst, University of Massachusetts Amherst.

Julier, Alice (2005) Hiding gender and race in the discourse of commercial food consumption. in Avakian, A. and Haber, B. (Eds.) *From Betty Crocker to Feminist Food Studies*. Amherst, University of Massachusetts Press.

Korsmeyer, Carolyn (1999) *Making Sense of Taste: Food and Philosophy* Ithica, Cornell University Press.

Kristal, Alan; Glanz, Karen; Curry, Susan and Patterson, Ruth (1999) How can Stages of Change best be used in dietary interventions? *Journal of the American Dietetic Association*, **99**, 679–684.

Larson, Nicole I.; Neumark-Sztainer, Dianne; Hannan, Peter and Story, Mary (2007) Family meals during adolescence are associated with higher diet quality and healthful meal patterns during young adulthood. *Journal of the American Dietetic Association*, **107**, 1502–1510.

Lytle, Leslie (2002) Nutritional issues for adolescents. *Journal of the American Dietetic Association*, **102**, S8–S12.

McIntosh, Alex (1999) The family meal and its significance in global times. in Grew, R. (Ed.) *Food in Global History.* Boulder, Westview.

Mäkelä, Johanna (2000) Cultural definitions of the meal. in Meiselman, H. (Ed.) *Dimensions of the Meal.* Gaithersburg, Aspen Publishers.

Neumark-Sztainer, Dianne; Hannan, Peter; Story, Mary; Croll, Jillian and Perry, Cheryl (2003) Family meal patterns: Associations with sociodemographic characteristics and improved dietary intake among adolescents. *Journal of the American Dietetic Association*, **103**, 317–322.

Neumark-Sztainer, Dianne; Story, Mary; Hannan, Peter J. and Croll, Jillian (2002) Overweight status and eating patterns among adolescents: where do youth stand in comparison with the Healthy People 2010 objectives? *American Journal of Public Health*, **92**, 844–851.

Neumark-Sztainer, Dianne; Story, Mary; Resnick, Michael D. and Blum, Robert W. (1998) Lessons learned about adolescent nutrition from the Minnesota Adolescent Health survey. *Journal of the American Dietetic Association*, **98**, 1449–1456.

Neumark-Sztainer, Dianne; Wall, Melanie; Perry, Cheryl and Story, Mary (2003) Correlates of fruit and vegetable intake among adolescents: Findings from Project EAT *Preventive Medicine*, **37**, 198–208.

Neumark-Sztainer, Dianne; Wall, Melanie; Story, Mary and Fulkerson, Jayne (2004) Are family meal patterns associated with disordered eating behaviors among adolescents? *Journal of Adolescent Health*, **35**, 350–359.

O'Neill, Carol E. and Nicklas, Theresa (2002) Gimme 5: an innovative, school-based nutrition intervention for high school students. *Journal of the American Dietetic Association*, **102**, S93–S96.

Patrick, Heather and Nicklas, Theresa A. (2005) A review of family and social determinants of children's eating patterns and diet quality. *Journal of the American College of Nutrition*, **24**, 83–92.

Raynor, Mike (1998) Vegetables and fruit are good for us so why don't we eat more? *British Journal of Nutrition*, **80**, 119–120.

Rockett, Helaine R. H., Berkey, Catherine S., Field, Alison E. and Colditz, Graham A. (2001) Cross-sectional measurement of nutrient intake among adolescents in 1996. *Preventive Medicine*, **33**, 27–37.

Rozin, Elizabeth (1973) *The Flavor-Principle Cookbook*, Stroud, Hawthorne Books.

Rozin, Elizabeth (1997) *The Universal Kitchen*, London, Penguin.

Rozin, Elizabeth (2000) The role of flavor in the meal and the culture. in Meiselman, H. (Ed.) *Dimensions of the Meal.* Gaithersburg, Aspen Publushers.

Serra-Majem, Lluys; Ribas, Lourdes; Pérez-Rodrigo, Carmen; Garcya-Closas, Reina; Peña-Quintana, Luis and Arancetac, Javier (2002) Determinants of nutrient intake among children and adolescents: results from the EnKid Study. *Annals of Nutrition and Metabolism*, **46**, 31–38.

Shepherd, J., Harden, A., Rees, R., Brunton, G., Garcia, J., Oliver, S. and Oakley, A. (2006) Young people and healthy eating: a systematic review of research on barriers and facilitators. *Health Education Research*, **21**, 239–257.

Sjöberg, A., Hallberg, L., Höglund, D. and Hulthén, L. (2003) Meal pattern, food choice, nutrient intake and lifestyle factors in The Göteborg Adolescence Study. *European Journal of Clinical Nutrition*, **57**, 1569–1578.

Sobal, Jeff (1999) Food system globalization, eating transformations, and nutrition transitions. in Grew, R. (Ed.) *Food in Global History.* Boulder, Westview.

Sobal, Jeff (2000) Sociability and meals: facilitation, commensality, and interaction. in Meiselman, H. (Ed.) *Dimensions of the Meal.* Gaithersburg, Aspen Publushers.

Sobal, Jeff (2003) Commensal eating patterns: a community study *Appetite*, **41**, 181–190.

Stockman, Nancy K.A.; Schenkel, Tanja C.; Brown, Jessica N. and Duncan, Alison M. (2005) Comparison of energy and nutrient intakes among meals and snacks of adolescent males. *Preventive Medicine*, **41**, 203–210.

Story, Mary; Neumark-Sztainer, Dianne and French, Simone (2002) Individual and environmental influences on adolescent eating behaviors. *Journal of the American Dietetic Association*, **102**, S40–S51.

Veugelers, P. J., Fitzgerald, A. L. and Johnston, E. (2005) Dietary intake and risk factors for poor diet quality among children in Nova Scotia. *Canadian Journal of Public Health*, **96**, 212–216.

Videon, Tami M. and Manning, Carolyn K. (2003) Influences on adolescent eating patterns: The importance of family meals. *Journal of Adolescent Health*, **32**, 365–373.

Weinstein, Miriam (2005) *The Surprising Power of Family Meals*, Hanover, Steerforth Press.

Zabinski, Marion; Daly, Tracy; Norman, Gregory; Rupp, Joan; Calfas, Karen; Sallis, James and Patrick, Kevin (2006) Psychosocial correlates of fruit, vegetable, and dietary fat intake among adolescent boys and girls. *Journal of the American Dietetic Association*, 106.

14

Culinary arts and meal science as an interdisciplinary university curriculum

I.-B. Gustafsson, Å. Öström and J. Annett,
Örebro University, Sweden

Abstract: The philosophical stance adopted in the Culinary Arts and Meal Science (CAMS) education at the Department of Restaurant and Culinary Arts at Örebro University is outlined. This is an interdisciplinary approach, based on the Aristotelian tradition of *episteme*, *techne* and *phronesis* knowledge forms and operationalised through the FAMM (Five Aspects Meal Model) model, bearing in mind the overall biopsychosocial nature of each individual experience. CAMS has moved forward using FAMM as the basis for dealing with the complex, multifaceted nature of producing and studying the meal.

Key words: atmosphere, control management, culinary arts, episteme, five aspects meal model, FAMM, meal science, phronesis, techne.

14.1 Introduction

In 1992, the Swedish Parliament decided that the present Örebro University should provide a university curriculum for chefs and waiters that emphasized 'the aesthetic configuration of the meal' in commercial settings. The aim of this curriculum was to increase the knowledge and the status of the professionals working in restaurants thereby both enhancing restaurant business performance and laying a foundation for research. The first doctoral degree in what is now called Culinary Arts and Meal Science (CAMS) was awarded by Örebro university in 2004 (Nygren, 2004) and there are now six completed doctoral theses in CAMS. The educational approach was distinguished by its emphasis on Aristotle's three forms of knowledge and embraced academic multidisciplinarity, to stimulate scientific thinking and reflection, together with the training of practical skills in handicraft and the ability and creativity to prepare aesthetically pleasing meals. In this chapter the philosophical underpinnings of what is now called Culinary Arts and Meal Science (CAMS) is presented together with a discussion of the progression of the discipline from its multidisciplinary beginnings towards an interdisciplinary future.

A model for CAMS education was derived from experiences of the assessment of restaurants performed by Guide Michelin. The model is called Five Aspects Meal Model (FAMM) (Gustafsson *et al.* 2006) and has formed the basis for CAMS education and research at the Department of Restaurant and Culinary Arts, Örebro for the past 15 years. FAMM assumes *a definition of the meal as 'the eating sphere at a defined occasion'*. Specifically, a visit to a restaurant starts with entering a *room, meeting* a headwaiter, being given a table and thereafter receiving some food and beverages (here called the *product*). In addition, there is a surrounding *atmosphere,* by which we mean the guest's perception of the total situation, and a *control management system,* which encompasses the overall business planning, including controls of economy, logistics in the kitchen and in the dining room, management of personnel resources and laws regarding the handling of food and beverages. *The room, meeting, product and control management system* together create *the atmosphere.* A summary of how these aspects are covered in the curriculum for the CAMS Bachelor degree is presented in Table 14.1. Naturally, PhD education is much more individualised. Nevertheless, students discuss the Five Aspects Model to various degrees during their courses, research projects and in their individual theses. This model simplifies the planning of creative and aesthetic meals, and education based on this model requires the use of the three forms of knowledge, namely scientific, practical and aesthetical knowledge, which correspond to the three Aristotelian principles of knowledge, as discussed below.

14.2   Philosophical underpinnings of the use of different forms of knowledge

14.2.1   Forms of knowledge

In western culture the dominant tradition for defining knowledge has been to consider knowledge as being derived only through science. Implicit in this definition, which has its roots in the writings of Plato 427–347 BC (Bostock 1991; Moravcsik 1992) is that theory is then separated from practice; that there is a distinction between what we think and what we do; between mind and body, brain and hand (Gustavsson 2001, Gustavsson *et al.,* 2008 in press). Plato called this scientifically derived knowledge *episteme.* He said that what we believe to be true must be supported by objective and good arguments to be accepted as *justified true beliefs.* This form of knowledge is about things 'which cannot be otherwise', eternal and universal. Thus, *epistemology* distinguishes between what we believe to be true on the basis of objective evidence and personal opinions or beliefs which do not have

Table 14.1 Curriculum of BA degree courses at CAMS with the content of courses in relation to the different aspects of the FAMM model

Courses	Content	Aspect
Year 1		
Culinary Arts and Hospitality Science	Scientific methods, scientific writing Hospitality – trade of hotel and restaurant Sensory analysis, psychology, nutrition theory of beverages Food safety and responsible beverage handling	Product, meeting, room, management control system and atmosphere
Crafts in Culinary Arts and Sommelier I	Theory and practical training about beverages, guest communication	Product, room, meeting and atmosphere
Crafts in Culinary Arts and Cookery I	Theory and practical training about food, nutrition, chemistry, sensory analysis, microbiology	Product
Meeting, Experiences, Tourism I	Communications systems, theory of interpretation, psychology	Product, meeting and management control system
National studies in the field – restaurant and hotel	Practical work in companies	Product, meeting, room, management control system and atmosphere
Aesthetic creation – restaurant and hotel	Aesthetic methods, design, light, colour, choreography, body language	Product, meeting, room and atmosphere
Year 2		
Crafts in Culinary Arts and Sommelier II	Theory and practical training about beverages Food and beverages in combination, marketing	Product, room, meeting and atmosphere
Crafts in Culinary Arts and Sommelier III	Theory and practical training about beverages Food and beverages in combination, marketing	Product, room, meeting and atmosphere

Course	Description	Category
Crafts in Culinary Arts and Cookery II	Theory and practical training about food, chemistry, microbiology, meal planning	Product
Crafts in Culinary Arts and Cookery III	Theory and practical training about food, chemistry, microbiology, meal planning	Product
Meeting, Experiences, Tourism II	Theory and practical training about the organisation and function of the international meeting and congress industry	Product, meeting, room, management control system and atmosphere
Business Leadership and Planning	Marketing, planning, development and leadership in the business industry	Management control system
National studies in the field – restaurant and hotel	Practical work in companies	Product, meeting, room, management control system and atmosphere
Project Work I, independent	Individual choices of topic within culinary arts and meals science, writing scientific paper	Product, meeting, room, management control system and atmosphere
Year 3 Event	Planning and performing of event according to the five aspects, meal forms training: canteen, à la carte, ceremonial meals	Product, meeting, room, management control system and atmosphere
Service meeting in hospitality industry	Psychology, meeting phenomena and creative culinary art design	Meeting, product
Molecular Gastronomy	The chemistry and physics of cooking and sensory phenomena, creative culinary art design	Product
Project Work II, independent	Individual choices of topic within culinary arts and meals science, writing scientific paper	Product, meeting, room, management control system and atmosphere
Business administration for hospitality industry	Theory of applied business administration	Management control system

objective support. The latter Plato called *doxa*. Plato's conceptualisation of knowledge excluded practical knowledge such as that necessary for survival. Until relatively recently, this Platonic exclusion of practical knowledge has dominated western intellectual thinking and educational systems, with more value being placed on intellectual knowledge than on practical knowledge and skills and was reflected in most university curricula. It is generally agreed that within academia, and within any particular discipline, theoretical knowledge can be transmitted, more or less, within a unitary disciplinary context.

However, the Aristotelian philosophical tradition has described knowledge as being of three forms: *episteme*, which Aristotle saw as the theoretical and scientific knowledge; *techne* he described as the knowledge needed for producing various products and creativity, i.e. practical–productive knowledge; and *phronesis,* knowledge with its focus on the wisdom formed and used in the processes of social interaction within a cultural dimension (Aristotle 384-322 f.Kr. ; Reeve 1992). Phronesis may also be viewed as the ethical dimension of knowledge. The goal of wisdom is to accomplish a good life for people, to increase the human sense of wellbeing and happiness. At a fundamental level, it is this three-dimensional view of knowledge derived from Aristotle which has been adopted as the philosophical framework for the academic discipline of Culinary Arts and Meal Science at Örebro University. To be able to produce a meal which gives the guest a good and satisfactory meal experience one requires scientific knowledge in e.g. chemistry, nutrition, food science, sensory science, psychology, business economics; in other words *episteme*. Also needed are practical skills and aesthetical knowledge (*techne*). Together with *episteme* and *techne* we need the knowledge which will allow us to reflect on why we act as we do (*phronesis*). For example, within the meal science area, *phronesis* can be seen in an ethical approach to food, by using food which has come from well treated animals, ecologically produced products and so on. It can also be regarded as phronesis to offer well balanced healthy menus or to make available ethnic or vegetarian food as required.

At first glance, it might appear that this Aristotelian view of knowledge would result in practice which separates the three knowledge forms across different parallel streams of knowledge/activity combinations. However, as noted by Gustavsson (2001) and Gustavsson and Annett (2008), this is not the case, and it is rather more useful to consider different combinations of the three. Indeed, discussion of the importance of different forms of knowledge has interested many other writers since Aristotle, with relatively recent emphasis being placed on discussion of these issues in relation to educational practice.

14.2.2 Pragmatism

Charles Sanders Peirce is credited with founding the theory of pragmatism, (later called pragmaticism by Peirce) in which knowledge of our habits and our actions are associated (pragma) (Ochs 1998). When we are carrying out a practical activity and encounter a problem we need scientific knowledge or a theory for reflection and solving the problem. John Dewey, like Peirce, was also strongly influenced by science. They both postulated that the truth of knowledge is revealed in the practical outcome of an action when the practical knowledge is combined with science and theory. John Dewey is perhaps best known for his theory of 'Learning by doing' discussed in the book: 'Pragmatism and educational research' by Biesta and Burbules (Biesta and Burbules 2003). Later on, Gilbert Ryle further developed the discussion of practical knowledge in his book *The concept of mind*. He introduced the concepts of *'knowing how'* and *'knowing that'*. 'Knowing how' refers to skills required to carry out a task and 'knowing that' is to understand and have insight when acting and to formulate arguments for the way the action is solved. Ryle was opposed to the Platonic and Cartesian dualistic view of the body and soul (consciousness) being separate, with the body acting mechanistically like a machine separate from the conscious mind. According to Ryle, these processes cannot be separated, and he further defined acting without thinking as 'habitual practice' and acting by reflection as 'intelligent practice' (Ryle 1990). Ryle also supported the idea of combining theory, science and practice to produce true knowledge.

14.2.3   Tacit knowledge

In a similar way, Michael Polanyi regarded all knowledge partly as 'tacit'. His arguments are presented in his book *Personal Knowledge* (Polanyi 1962) where he opposed the positivists' standpoint that all knowledge must be possible to be described scientifically or with arguments based on theory. Polanyi states that 'we know more than we can tell' and meant that knowledge of practical arts can be build up during earlier experiences or passed from generation to generation or also so trained that we do it without reflecting like driving a car, baking bread, playing an instrument, playing golf, etc.

Given this philosophical background to CAMS education of Örebro university, it is clear that a practical consequence is that academic input has been required from a number of contributory academic disciplines, working together towards a common goal. But the key question is to what extent these various contributing disciplines remain essentially separable subparts of the overall general discipline? True interdisciplinarity, the stated aim of CAMS education, is something rather more than multidisciplinary in that it demands integration and synthesis of the contributing disciplines to create something which is more than 'the sum of the parts'. This concept of synthesis is an important one in cognitive psychology, for example, and in that context is regarded as indicative of higher order learning and cognition.

The psychologist Jean Piaget (1896–1980), proposed a theory of cognitive development which became influential in educational training. He suggested that cognitive development followed a hierarchy, sophistication being characterised by the ability to integrate information from multiple sources and to deal with multiple perspectives. Inherent also in Bloom's taxonomy of Educational Objectives (Bloom 1956) is the notion of hierarchical development of intellectual skills, beginning with factual knowledge and leading through comprehension, application, analysis, synthesis and evaluation. As noted by Cargill (2005) synthesis is widely discussed in interdisciplinary theory because it is recognised as one of the major outcomes of an interdisciplinary or liberal education.

14.2.4   Interdisciplinarity

These principles are relevant to consideration of interdisciplinarity as a means of dealing with complex issues which are not amenable to study from just one perspective, and what is more complex than asking 'why, where, when, how and what do we eat'? The main practical problem that arises is achieving a balance between the depth given by a disciplinary focus and the breadth given by input from a range of disciplines. Neither a specific single disciplinary focus nor the pursuit of interdisciplinary relationships should predominate (Davies and Devlin 2007) but from synthesis comes creativity, and it is creativity which should overcome disciplinary limitations.

Psychology, perhaps more than many other disciplines can be viewed as the product of a complex, interdisciplinary synthesis. The various perspectives within the current discipline reflect its origins and history, from its origins in philosophy and introspection through psychophysics, psychoanalysis, behaviourism, the cognitive revolution to name but a few. Perspectives today include neuroscience, evolutionary/biological/physiological, abnormal/clinical, health, cognitive, developmental, personality and individual differences, occupational/organisational, social, gender and cross-cultural. If we look at these different perspectives within the discipline of psychology we can see that they are clearly interrelated to the other disciplines contributing to CAMS education, such as nutrition science, public health, domestic science, sociology, anthropology, ethnology and business economics.

Furthermore, if we compare the various perspectives within psychology with the disciplines contributing to CAMS we can see that the emphasis ranges from biological aspects through health, cognition and social aspects. Indeed, in varying degrees, a similar analysis could also be possible within these other disciplines. Now, if one considers the meal, trying to define it is somewhat tricky, and definitions have varied in emphasis in precisely the same way. As we shall see, where FAMM differs from previous approaches is that it incorporates a broad spectrum of contributory factors. If one tried to define the meal in only its simplest biological sense as an activity performed to consume calories and nutrients, that would be somewhat inadequate. In human evolutionary terms, the subtle social activities and behavioural patterns which are associated with the meal must be thought of as fundamentally related to our very survival and development and evolution as a species.

At a general level, this kind of thinking is certainly not new. For example, Engel (1977) introduced a major shift in the theory of clinical medicine. He made the deceptively simple observation that to understand human beings in health and disease we must understand the three factors: biological, psychological and social; that actions at these three levels are dynamically interrelated; and that these relationships affect both the process and outcome of clinical care. His biopsychosocial model was derived from systems theory with its origins in mathematics and engineering. Adoption of the interdisciplinary biopsychosocial approach to healthcare has led to a much broader general understanding, moving from a biomedical, physician-led environment where patients had little input into the process or opportunity to provide influential feedback and including taking into account the effect on a patient's prognosis of the patient–practitioner relationship. This biopsychosocial approach is analogous to and inherent in FAMM and provides a framework for thinking about the meal and food provision, which moves from traditional food provision represented by a provider led approach to the multifaceted approach advocated by the CAMS education discussed earlier. The main point to be aware of is that no matter which perspective the meal is approached from, be it mainly biological, psychological or sociological, fundamentally all these processes exert their effect at the level of the individual: the way an individual's brain interacts with the world, how it processes the information from the outside world and how it deals with both the autonomic and volitional bodily processes. It is this essence of this multifaceted, biopsychosocial underpinning which has been captured by FAMM in the undergraduate education curricula but possibly even more so within our research field and in the PhD education.

The most explicit delineation of this interaction between the individual and his or her environment is perhaps in consideration of the product aspect, and particularly within the associated field of sensory analysis. Moreover, sensory science is also applied to varying degrees, within all aspects in FAMM since it deals with experiences through the senses. Martens' discussion of the philosophy of sensory science acknowledges that attention must be drawn to understanding of stimulus–response interaction between complex biological material and a complex human sensation–perception–cognition system, resting within a broad, ever changing context (Martens 1999). Furthermore, she discusses various philosophical topics such as the epistemological, linguistic, ethical and metaphysical which will be of relevance to sensory science. Epistemology and ethics are clearly in accordance with Aristotle and the description of the three forms of knowledge discussed earlier. Also relevant is the linguistic approach which deals with language theories concerning the spoken (or not spoken) expression of our sensory impressions and knowledge of the world. The metaphysics approach encompasses theories about 'reality' and limits for scientific knowledge.

14.3   The five aspects meal model (FAMM)

Given this general philosophical and historical background, one can now consider the five aspects of FAMM in more detail (Fig. 14.1). When thinking about models, in general, they tend to serve two related functions; as a tool for trying to explain existing observations and also as a way to guide thinking and to stimulate research. In the early stages perhaps one should aim for scope before becoming embedded in the finer detail of specifying with great precision individual elements of the model (Annett 1996) ultimately to come up with useful tools to enhance not just specific aspects of people's lives but also to help improve overall quality of life. This idea of scope being followed by precision is applied widely in computer science, where the general framework is specified at a general level by systems analysts and the detail of each stage is later provided by the computer programmers. In the case of FAMM, we can now look at each of the five aspects of the model (room, meeting, product, management control system, atmosphere) while at a general level incorporating the three Aristotelian forms of knowledge, and also trying to specify the detail of each aspect, the programming as it were, bearing in mind the overarching biopsychosocial qualities of the human consumer. This philosophical background provides strong support for the various curricula at the Department of Restaurant and Culinary Arts at Örebro University, an interdisciplinary model which has the potential to be developed and utilised at other educational establishments. In Table 14.1 the content of different courses are described and their relation to different aspects of the FAMM model indicated.

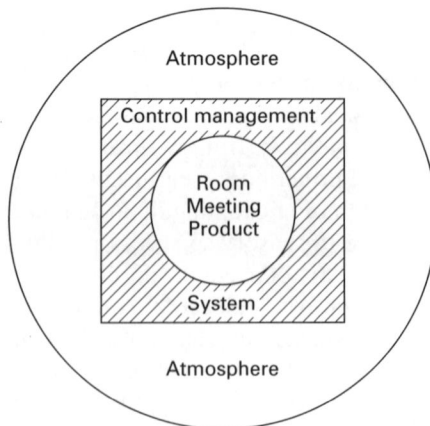

Fig. 14.1   Five Aspects Meal Model (FAMM).

14.4   Types of meals

As noted by Gustafsson (2004), the ultimate aim of all five aspects is the same: to achieve maximum satisfaction in various meal situations for every guest/ customer/ diner. There are of course different expectations depending on the type of eating situation and different ways of grouping these (Edwards 2000). However, the focus at the Department of Restaurant and Culinary Arts, Örebro University is on the production of commercial meals. e.g. à la carte meal, canteen meal, and formal or what we call ceremonial meal (although the same approach can also be applied to more private meal production). *A canteen meal* is a meal which is served to many people at the same time during a short period of time e.g. self-service lunch. An *à la carte meal* is a meal where the guests make a free choice and order from a menu and enjoy this meal during a longer period, e.g. during an evening dinner. A *ceremonial* (formal) meal takes place during a feast of some type either of a private official nature such as the celebration dinners at the Nobel Prizegiving, weddings, anniversaries, government hospitality, academic celebration and so on.

These different meals require different type of Rooms, Meetings, Products, Control Management systems and a different type of atmosphere will be created requiring analytic and synthetic input from the various disciplines and knowledge bases outlined earlier. Indeed, each of these five aspects of the meal do not stand in isolation, but interact with one another to produce the overall meal experience.

14.5   The room

A meal must always take place somewhere loosely designated 'the room'. This could be a dining room at home, a restaurant, at hospital, inside an aircraft, at work, on a beach or indeed just about anywhere. The physical locale can vary enormously and the setting influences the enjoyment of the meal and also contributes to the atmosphere of the meal, as shown in Table 14.1 where both room and atmosphere will appear together within the courses. Therefore, it is somewhat difficult to separate discussion of the room from atmosphere. However, adopting focus on commercial meals, e.g. *à la carte* meals, canteen meals and formal or ceremonial meals means we can limit discussion of the type of room somewhat. Which disciplines have to be involved to create appealing rooms in these different types of meal? As noted below, Campbell-Smith (1967) identified a number of factors relating to the room which affect atmosphere and categorised general internal variables such as temperature, noise levels and cleanliness; general layout variables such as shape and size of room and layout of seating; and table appointments – the crockery, cutlery, etc. More recently Heide and Grønhaug (2006) published a model for atmosphere management which includes roughly the same elements of the room. These authors distinguish between the physical environment 'servicescape' and 'atmosphere', the result of the interaction between the servicescape and the individuals experiencing it. The main physical *antecedent* factors identified by Heide and Grønhaug are rather similar to those of Campbell-Smith. They include what they call ambient factors and design features. Ambient factors include sound, lighting, scent and temperature while design features include all elements such as layout, décor and general sign and symbols which communicate distinctive features of the establishment. The importance of elements such as these have been reviewed (Edwards and Gustafsson 2008a). The antecedent 'room' factors denoted by Heide and Grønhaug are *mediated*, i.e. perceived through the sensory channels, and moderated perhaps by factors such as cultural background, age, gender and expectations. The outcomes or *dependent* variables can be the guest response – cognitive, affective, physiological or behavioural – and also hospitality outcomes, such as guest satisfaction and return visits. It is also clear that to examine these antecedent, mediating and outcome factors in relation to the room also involves each of the biopsychosocial perspectives discussed earlier. The room and atmosphere aspects will be covered in most of the courses within Year 1 studies although they are most focused upon in the 'Aesthetic creation – Restaurant and Hotel' course. They are studied more deeply in Year 2 and 3 within the 'Crafts and Culinary Arts' courses in both the Sommelier and Cookery programs as well as in 'Event' and in 'National Field studies' (Table 14.1).

To understand these elements it is clear that one requires an interdisciplinary approach covering a range of information types corresponding to *episteme*, *techne* and *phronesis*. For example, to study the design features a student would require basic knowledge of architecture, the history of style, epoch, art history and scientific knowledge of textiles, furniture and china. In addition the involvement of designers or artists is also important, with their knowledge of colour, form and design. Some knowledge of psychology, particularly perception and cognition and environmental psychology is required in order to understand how these layout elements of a room are experienced by people. But, furthermore, the students have to combine this basic scientific knowledge with the practical skills of decorating a dining room and to be able to reflect over the choice of style, textiles, colours and china not just in a technical sense but also to reflect at a deeper, ethical level (Schön 2003). In a similar way, study of the ambient factors such as temperature and odour requires knowledge from a varied disciplinary input. Manipulation of these factors is certainly not new in the hospitality industry, although the level of sophistication of their use nowadays requires that basic knowledge of acoustics, lighting and olfaction, for example, needs to be combined with knowledge derived from psychology, sociology and so on, in order to understand the resultant response variables. Heide and Grønhaug (2006) and Edwards and Gustafsson (2008a) provide reviews of some of the literature on the impact of various room variables on the dining experience.

As discussed earlier, synthesis of different forms of knowledge and input from different disciplines should result in a creative approach to 'the room'. As part of the educational process at the department of Restaurant and Culinary Arts at Örebro University, students take responsibility for the production of various large dinner events, supervised by teachers from within both design, cooking, service and management control systems. In such situations, a theme or a story is chosen for the meal occasion and the event created and performed at a practical level in line with that theme. This method has also been applied by the artist and designer Erik Nissen Johansen (Johansen and Blom 2003), who says: 'my main objective is to cater to the guests six senses – the five basic senses, plus their fantasy. If there is a harmony between food, taste, colours, furniture, design, textiles, personnel, culture, attitudes, surface, lightening, guest preferences, smells and historical positioning, the restaurant will leave a clear and easy to communicate memory'. He believes that people nowadays demand more than a delicious meal; they demand an experience. Thereby, Johansen involves different disciplines: art history of the interior, sensory science, psychology, lightning knowledge as well as the practical skills and artistic knowledge to produce the actual meal.

One example of this overall educational approach in practice was a function for which students had responsibility, referred to as the Orrefors project. In collaboration with a designer from Orrefors, Sweden and the Champagne House Veuve Clicquot, students' brief was to produce a lunch and a dinner using FAMM as the guideline. The meals were served at two different locations, and the limitations placed on students were the raw materials: food and drinks and the glassware were the same for both occasions. Guests, who included famous food and wine journalists, chefs and sommeliers evaluated the experience including the room variables, and students carried out observational study of the guests whilst dining. This element was included in the course of 'Crafts in culinary Arts and Sommelier' Year 2 (Table 14.1). Without good quality education and research, the situation may remain as described by Bitner (1992): 'managers continually plan, build and change an organisation's physical surroundings in an attempt to control its influence on patrons, without really knowing the impact of a specific design or atmosphere change on its users'.

14.6   The meeting

The second of the five aspects of FAMM is the meeting. In any meal situation there are several types of meeting, encompassing not only the meeting between customers and staff but also the meeting between one customer and another, and one member of staff and another. These social elements are included in both the Campbell-Smith (1967) and Heide and Grønhaug (2006) models as important contributors to atmosphere. The nature of these social interactions differs according to the type of meal. In a canteen, we mostly meet other guests and very few personnel. The canteen might be fully self-service or be one where staff serve the main dish only. In a more formal evening restaurant, the first meeting might be with the staff that take care of guests' coats, then the head waiter and waiters. These people provide service in different ways and a large discipline of study is service management, usually as part of the general hospitality management discipline. At Örebro University, the meeting aspect is first introduced in the overview course 'Culinary Arts and Hospitality' Year 1, but is focused upon during the course 'Meeting, Experience and Tourism I and II' and in 'Service meeting in the hospitality industry'. Psychology and sociology are important contributory disciplines to the study of these social aspects of the meal experience. From the perspective of the service provider, Martin (1991) describes these social skills as 'convivial skills'. As Heide and Grønhaug note, the attitudes shown by staff towards both the guests and also towards their job, influence every aspect of their behaviour and hence the nature of the experience for the guest.

A scientific understanding of the principles of fundamental aspects of human social behaviour comes mainly from study of psychological and sociological theory covering a wide range of aspects of service. There has been much research on the importance of service for guests covering a wide range of applications, e.g. Andersson and Mossberg (2004), Hansen *et al.* (2004, 2005), Mina (2006), and Tucci and Talaga (2000). Andersson and Mossberg, for example, asked guests how much they would like to pay for different parts of a meal. The guests' answers were that they would like to pay most for the service when dining in the evening while food was more valuable at a lunch meal (Andersson and Mossberg 2004).

More psychological aspects such as the relationship between consumer expectations of service and the perceptions by managers and staff of consumer expectations has been studied and showed a gap between the perception of consumer expectations and the actual expectations (Douglas and Connor 2003). Perhaps it is that service personnel might not realize that they have to act as the host of the restaurant to take care of the guests, almost as though they were their guests at home. This is an aspect of service which seems to be best performed in family restaurants (Mina 2006).

Professional service requires an open minded and communicative personality and an understanding of the role that factors such as impression formation, attribution theory and stereotyping play in social interactions. Furthermore, while the physical layout can determine in some part the social interactions, manipulation of what one might call psychological space is also important. For example, overfamiliarity can result in feeling of invasion of personal space, with consequent negative feelings and experiences (Fisher and Bryne 1975; Marks 1988) A related important factor is knowledge of rules of different aspects of etiquette – not just in the conventional sense but at a general level which encompasses knowledge of different social, religious and political groups and cultures. Of course, one must always remember that social interactions are exactly that – interactions – and the behaviour of the guests themselves can have a profound influence on the behaviour of other guests and on service staff. All of this can be viewed within the biopsychosocial framework discussed earlier and related to the three Aristotelian forms of knowledge. Biological aspects of social interaction include variables such as gender, attraction, dominance and power. Related psychological variables are as mentioned above, things such as attribution, individual/group dynamics, individual differences. Social factors include aspect of culture, perhaps fashion trends and so on. Thus, to be able to handle the complexities of the service meeting requires an education which includes scientific knowledge of human behaviour (*episteme*), practical social skills training (*techne*), combined with elements of *phronesis* – that aspect of knowledge and wisdom formed and used in the processes of social interaction within a cultural dimension.

It is the case, however, that much of the social interaction process usually goes unnoticed at an explicit level. Within the Orrefors project, the students made observations, on how the guests interacted and what they were talking about. In the planning process of such a meal, lunch or dinner, students often put great effort into the timing and the performance process in order to make the meal a great experience. Their expectations are therefore that the guests would talk about the meal within the setting. To their surprise, the results of the observation studies showed that the guests interacted in a normal social way, and did not talk about the meal except when something unusual or unexpected happened (Table 14.1).

14.7   The product

The product aspect of FAMM encompasses our definition food and beverages served as part of a meal. Obviously, this is a large area within our discipline CAMS, which deals with food and beverages from the point of cultivation and production to menus which appear on the table. Even if the students do not actually become involved in actual cultivation of food it is essential that they learn something about the processes e.g. the effect of different growing systems. This involves disciplines such as agriculture, viticulture, enology as well as food science. Haqvin Gyllensköld a famous Swedish cookbook writer, who was interested in food chemistry wrote: 'We prepare food as we like it tasty. We cook the potato as we think it is more tasty than raw. For the same reason we cook and fry vegetables, fish and meat. We mix it and spice it and work a lot with the food. We do that as it gives us pleasure. It seems to me that if we know more of what happens within the food items when we prepare them that could make the cooking both better and more fun' (Gyllensköld 1977).

Gyllensköld was the first person in Sweden to be interested in food chemistry, a subject which later has come to be called molecular gastronomy by Hervé This (This 2006), which is also a course in Year 3 in our curriculum. (Table 14.1). Other writers, such as Harold McGee in his book *On Food and Cooking* also include the history of food, characteristics and growing places besides the food chemistry. That particular book is used as part of the main literature in courses on food technology as it includes both scientific and practical knowledge (McGee 2004) and forms part of the course literature for the course 'Molecular Gastronomy' at Department of Restaurant and Culinary Arts. To be a good chef requires knowledge about the cultivation conditions of the raw materials, storage, ripening, chemical and physical properties and how the different cooking techniques change the molecules of the food. With this level and range of knowledge, it becomes much easier to preserve the good quality of raw ingredients and make use of it in cooking to provide more flavoursome meals. That this requires the involvement of disparate disciplines such as horticulture, food science, chemistry, physics has also been pointed out by Brillat-Savarin in his book *Physiologie du gout* [Brillat-Savarin 1825 (1963)]. Courses in our curriculum are 'Crafts in Culinary Arts and Cookery I, II and III' (Table 14.1).

Another part of the product concerns the sensory qualities also included in the above mentioned courses. Judging sensory qualities requires physiological knowledge about our senses, training the senses and knowledge of sensory scientific methods. Interestingly, the first doctorate degree awarded in CAMS was based on research on food and wine in combination with help of sensory science methodology (Nygren 2004). Sensory education and research requires involvement of disciplines such psychology, physiology, statistics, together with product knowledge from the food technology areas: at a general level, sensory science concern some kind of relation between a product and a person. This relation could be seen as two interfaces, for example the relation between chemical properties of a product and the sensory properties; how the product is perceived. For example, tomatoes with a low pH will probably have a more pronounced acidulous taste. The other interface will most likely be the relation between the descriptive responses to a product and the affective response (Martens 1999). Perhaps the best preferred or liked coffee might be described as earthy, fruity and bitter with a slight acidulous taste. Both interfaces will be useful when applying sensory analysis in projects within the culinary arts field. As Martens (1999) so elegantly captures the essence of the discussion illustrating the different levels of analysis and thus by implication the requirement for a multifaceted approach embracing multiple knowledge forms. 'What is the source of knowledge about an apple? Is it the physical measure of sugars or the perceived sweetness or both? Is it the detected sweetness on the tongue or the feeling of happiness when thinking about the apple?' An ongoing research project at our department within the sensory field is about the sensory language. The aim is to develop a sensory and cognitive language model for description of fruit and vegetables. The model could be used as a tool for communication of food and meals.

Sensory science is also important for the theoretical base within courses which deal with wine and other beverages as well as food, although knowledge of beverages is mainly focused on wine. This knowledge includes aspects of viticulture and enology combined with practical sensory training of different wines together with training in storage, serving and buying wine from different areas, different grapes, etc. Such knowledge is part of examples of areas covered in courses in 'Culinary Arts and Sommelier I, II and III' (Table 14.1). Of course an important part of this sensory training is the description of sensory characteristics of products which could be applied within the restaurant sector. To mention an example: a wine steward or a sommelier could either directly or indirectly influence the sales of wine at restaurants, either directly through their credibility as a salesperson and indirectly through training of other waiting staff, who will be better able to describe the wine to the customers which could increase the sales (Manske and Cordua 2005). Together with the wine comes the wine glass. Obviously, glasses and other utensils used when serving a meal must be included in the room aspect of the meal, as part of the need to reflect over the choice of style, textiles, colours, china (Billing *et al.* 2008). However, glasses are an important instrument for communicating the wine to the human senses, exercising direct effects on sensory experiences such as odour through size and shape. Therefore, it is important to explore experiences and reflections on wineglasses and their contribution to a meal experience in the perspective of craft, art and science within education and research.

One further discipline involved in the product aspect of the meal is nutrition science. Even restaurants have to have knowledge about how to compose menus to support healthy eating as well as be able to serve different diets to people with specific diseases treated with diet, an area covered in the basic course 'Culinary Arts and Hospitality Science as well as in Crafts in Culinary Arts and Cookery I. As the trend towards more and more people eating out continues this will be an important task for the restaurants to support people's desire for healthy eating options – in other words, to put into practice not just *episteme* and *techne* but also *phronesis*. This has been focused upon in the book by Gustafsson and Klein, *Den medvetna kokkonsten* (*The conscious cooking*, Gustafsson and Klein 2006). Within the framework of FAMM, a current doctoral research programme at Örebro university has taken a biopsychosocial approach to examination of several aspects of the meal as product. These include looking at fat content of food items and its related sensory properties and its relation to impact on health (Rapp *et al.* 2007 and Table 14.1).

14.8   Control management system

The control management system aspect of FAMM encompasses the economic aspects of the meal, the overall planning, logistics in kitchen and dining-room, rules and laws regarding both hygienic working practices, authorities' requirements, management of personnel, rules for working hours and conditions, working environment, union business, marketing and so on. Many of these areas can be placed under the general discipline called Business

Administration, with specific themes such as accounting and control, accounting and finance, marketing and change, organisation and management, business development and to some extent legal aspects of the business of restaurants and hotels.

A range of literature from the discipline is used for teaching, e.g. basic literature in accounting, marketing, organization, service management and hospitality management and students are trained to analyse the business situation and capacities of restaurants and hotel companies and to work out a business plan for a company. These elements are included in courses named 'Meeting, Experiences, Tourism I and II' as well as in 'Business Leadership and Planning' year 2 and 'Business Administration for Hospitality Industry' year 3 (Table 14.1). In the specific area of marketing management the book by Kotler (Kotler 2003) can be useful as a teaching tool in this type of course. For service marketing, aspects of the education material from Zeithaml *et al.* (2006) has been used. The law as a discipline is also to some extent involved in this part of the education. Another good example of applied work with the control management systems is a book by Kivela (1994) on *Menu planning* which can be used effectively as course material for Culinary Arts students. Kivela mentions three overall factors, which should be the targets for the business activity: *economic, market related* and *quality and production* targets. These targets have to be well implemented in both leaders and the personnel. The factors are also integrated within each other. If any one factor is deficient, it affects the others. For example, low quality within production or in the raw material might lead to either food having to be discarded thus having economic consequences. Or, if the meal is actually served, the guest will be disappointed and will probably not return. Additionally, the reputation of the restaurant as spread by word of mouth to other people may be damaged, again leading to serious economic consequences.

14.8.1   Economy targets

A chef in a restaurant is responsible for the costs of personnel, raw material and storing. This requires good knowledge of leadership and in coaching as well as knowledge about food and cooking. The economy in the kitchen is based on calculation and estimation based on description of recipes and menus both before and after the production, so that the actual costs and marketing is controlled. The pricing of a menu has to be placed just right for the market position of the restaurant or the special occasion. Both over- and underpricing will lead eventually to economic failure.

14.8.2   The market

A restaurant has to work within a customer-oriented framework to create profitability, which means that the customer/guests must be always in focus and the restaurant owner has to satisfy the guests' needs and demands in competition with other restaurants. The type of guests a restaurant has or wishes to attract determines what menus should be served and offered. This requires knowledge and input of many types, not least an understanding of customer behaviour and how they relate to perceived brand values.

14.8.3 Production and quality

At this level of consideration, whether or not a menu is successful depends obviously on how the raw food material is adapted to the cooking technique for each situation and how the menu is composed at that time, which is the chef's responsibility. An important issue in relation to production and quality is the need for an awareness of all aspects of food hygiene. Lack of knowledge and failure to implement legal requirements and behavioural guidelines will inevitably result in failure. This might be through closure of the restaurant by national regulatory authorities as a preventative measure or, alternatively, in response to actual health-related consequences. There is no more dissatisfied guest than the one who becomes ill as a consequence of eating a meal! CAMS education must therefore include food hygiene as a core part of education. This input can range from education about basic biological processes through an understanding of the interaction between potential infectious agents and human behaviours which may result in less than optimal food quality.

14.8.4 Different meals different logistics and techniques

As mentioned earlier, different menus require different logistics and different cooking and serving techniques. In *canteen* menus often the guests serve themselves and many people are eating at the same time. This will require a special logistic with its focus on the cooking techniques to enable rapid serving, mostly without waiters, of food which can withstand being pre-prepared and kept warm for some time before service. Obviously the best kitchen logistics in this case will minimise standing time for the food. This is an example of where knowledge and implementation of the rules of hygiene are particularly important. Many food-borne contaminants are known to multiply rapidly during food holding time particularly if strict, optimal temperature control is not maintained. Within *à la carte* menus the guests make their choice from '*la carte*' and order their dishes and beverages. In this case the logistical problems are rather different, and require a greater level of 'real-time' cooperation between waiters and the cooks. Both waiters and cooks have to prepare their *mis en place*, which means that all preparation of the food and glasses, laying of the tables has to be done in advance, so that the guests can get their chosen dishes *a la minute*.

A *formal or a ceremonial* meal at banquets, weddings or celebrations requires yet another totally different kind of logistics and leadership. All aspects of the menu and presentation are determined in advance in consultation with the host. In turn, both chefs and waiters must then have the optimal situation to plan, calculate the recipes, prices for profitability and performance of the whole meal. Limitations on various aspects of the occasion are often placed by factors relating to the venue, which may not always have as its primary function meal service. The size and shape may have to be incorporated into planning and optimised by creative design to give the room a special expression. The chef may have to adapt and control kitchen utensils and the number of waiters will have to be adapted to suit the number of guests, the menu and to the room, and so on. Logistical failure at any point in the chain may lead to performance failures such as a dining time which is too long, prolonged waiting time for table service leading to food not arriving on table at the required temperature – perhaps lukewarm or melted. Thus, consideration of the logistical process is very important from before the guests ever appear through to after their departure. Throughout all of this planning process, and in order to deal with the management of the different elements of the meal occasion, it is clear that education must cover a broad spectrum of knowledge, ranging from the biological through to psychological and sociological (Table 14.1).

14.9   Atmosphere

The influence of atmosphere within a room or where food is consumed is often easy to appreciate but difficult to understand or explain. Many factors contribute to atmosphere, and may be categorised according to the features of the FAMM model. By this model, perhaps the three most influential general aspects of the meal contributing to the guests' experience of the atmosphere are the room, the meeting and the product, although the overall management system must not be overlooked (a badly managed restaurant will not be successful, and the effects of the best physical environment and best efforts of individual staff will be diminished by a poor management structure). The atmosphere is highly integrated with the room aspect and a course element is presented in the first overview course 'Culinary Arts and Hospitality Science' Year 1, focused upon in 'Aesthetic Creation – restaurant and Hotel' and also in the 'Event' course and in 'Field studies'. It has to be remembered, though, that the product aspect also affects the feeling of the atmosphere at a restaurant visit. As described earlier, an early model of the elements of atmosphere was that of Campbell-Smith (1967). Heide and Grønhaug (2006) subsume Campbell-Smith's formulation in their description of the *antecedent, mediating and outcome* factors which was mentioned earlier (Section 14.5). This is also in line with the proposal of Wall and Berry (Wall and Berry 2007) who suggest that diners use three classes of variables to judge a restaurant experience:

Functional – the technical quality of the food and service;
Mechanic – the ambience and other design and technical elements;
Humanic – the performance, behaviour and appearance of the employees.

Similarly, Edwards and Gustafsson (2008b) discuss atmosphere of the meal in terms of interior variables, layout and design variables and human variables. These correspond closely with Heide and Grønhaug's antecedent factors. In other words, it is apparent that atmosphere, however one might try to define it, results from an interaction between all the elements of the meal event, perceived uniquely by each participating individual. As such, atmosphere has been identified by the cited researchers above as a key factor for ensuring a successful meal experience.

Given this multifactorial nature of atmosphere, it is obvious that its study must also be multifaceted and proceed at various levels of enquiry, in the same way that the other elements of the FAMM model (Fig. 14.1) imply a multidisciplinary educational approach. Using the Heide and Grønhaug model as a framework for consideration of atmosphere, we can gain some measure of the relative contribution of various disciplines and how knowledge and skills from each might be synthesised. In discussion of the room, for example, it has already been indicated that ambient factors such as sound, lighting, scent and temperature and design factors such as spatial layout, architecture and décor make a significant contribution to atmosphere. Moreover, study of these requires input from many disciplines and differing types of knowledge. Also, as pointed out by Heide and Grønhaug (2007), these are factors which can be controlled by management, who unfortunately may not always have the knowledge and insight to manipulate them successfully. In addition, as Edwards and Gustafson (2008b) conclude, most of the information available currently has been derived from the retail sector, and not specifically the food sector. Interestingly, recent research (Heide *et al.* 2007) suggests that achieving a balance between various sources of input at the practical level of application is not easy – they report that all of the hospitality companies included in their research reported substantial disagreements with their architects and designers. One might speculate that this is indicative of a need for hospitality/restaurant specific interdisciplinary education and research. Or, as Heide *et al.* suggest, perhaps this conflict reflects differences in emphasis on functionality and aesthetics during education – again reflecting the need for an interdisciplinary synthesis of different knowledge forms.

Elements of the meeting, the social factors which also contribute to atmosphere, are again subject to understanding at a number of levels. Edwards and Gustafsson (2008a) review some of the research relating to the psychological and sociological perspectives on density and crowding and on social facilitation, and their effects on atmosphere. They concluded that not only was overall enjoyment of the eating occasion and the food itself – the product – affected by atmosphere, but also amount of food consumed and money spent were also influenced. Overall, it is clear the atmosphere cannot be created or understood by focusing on one factor alone although if one factor has not been achieved suitably, the desired atmosphere can be spoiled even if everything else is in order.

Atmosphere, however, is not a static, one-dimensional thing, perceived by everyone in the same way. Rather, it is a dynamic entity, perceived by individuals in an individually determined way. Each individual can be regarded as presenting a unique set of mediating variables, sensory channels and other biosocial factors derived from personal history. Interpretation and understanding of this mediating process is a clear case for multidisciplinary input – psychology, sociology, biology, sensory studies, ethnology, gender studies, anthropology to name but a few. In a similar way, the outcome measures of atmosphere require an integrated education and interpretation. To understand the guest responses at each of a cognitive, affective, physiological and behavioural level is a complicated business requiring knowledge across a range of disciplines and the ability to integrate that information. Interpreting and relating these possible guest responses in terms of overall hospitality outcomes such as guest satisfaction and being able to produce an adaptive response is the ultimate aim of any good hospitality provider, regardless of the nature of the establishment – restaurant, canteen, hospital or whatever (Table 14.1).

14.10   Conclusions

This chapter began by outlining the philosophical stance adopted in the CAMS education at the Department of Restaurant and Culinary Arts at Örebro University. This is an interdisciplinary approach, based on the Aristotelian tradition of *episteme, techne* and *phronesis* knowledge forms and operationalised through the FAMM model, bearing in mind the overall biopsychosocial nature of each individual experience. Thus, FAMM represents both the educational philosophy and the practical, interdisciplinary educational tool. At the philosophical level, it assumes acceptance that the who, how, when, why, what and where of food consumption can only be understood at multiple levels ranging from the physiological to the societal, and that the individual experience is a product of this complexity. This, in practice, must produce reflective hospitality providers, analogous to the physician who, on adoption of a biopsychosocial model of health, must become capable of maintaining an ongoing self-audit because the performance demands vary from moment to moment (Borrell-Carrio *et al.* (2004) As noted by these researchers, intuition is central to this process, and as Polanyi suggested, professional competence such as this may come from tacit knowledge. Expertise, such as that required to deal with difficult guests or difficult situations which arise unexpectedly, is often apparent in ways that are difficult to explain at a strictly cognitive level – in other words are emergent properties of the process of synthesis and creativity which arises from multiple knowledge sources.

As noted by Davis and Devlin (2007), interdisciplinarity can take a number of different forms that can be thought of as following a continuum. At one end of the continuum is the situation where contributing disciplines have moved beyond the multidisciplinary position. If multidisciplinarity is a simple co-existence without taking into account or even being aware of another discipline's work and contribution to an educational goal, the simplest form of interdisciplinarity is the stage where there is at least recognition and understanding of the contribution made by disciplines other than one's own and how those different perspectives contribute to the common goal.

At its most extreme form, interdisciplinarity has passed through a further stage where disciplinarians take into account their colleagues' contributions and modify their own contribution accordingly, reaching an extreme point where two or more disciplines combine their expertise to jointly address the subject under consideration. The latter results necessarily in modification of disciplinary boundaries and emergence of new synthesised disciplines. Although it is clear that CAMS education has moved from the static multidisciplinary educational position and stepped onto the more dynamic interdisciplinary platform, there is still probably much progress to be made.

As we have seen, the study of food can lead us into contact with the study of diverse areas ranging from basic biological and psychological processes through to consideration of aesthetics and culture. Some years ago Guy Claxton (1980, 1988) compared the multiplicity of areas of study to a group of neighbouring islands between which there was no means of communication – phones, boats and so on. In such a case, he suggested, each island population develops its own language, culture, beliefs and working practices. Every now and again the inhabitants of one island might notice great excitement and activity on a neighbouring island, but, not being able to understand what is being said or to join in, they simply gaze in wonder for a short while and then go back to what they were doing. We believe that CAMS has certainly moved beyond that stage, and using FAMM as the basis for dealing with the complex, multifaceted nature of the study of food, has moved along the interdisciplinary continuum towards producing an evolving but complex, multiply determined understanding of what, where, when, why and how we eat. To use Claxton's analogy, the inhabitants of the numerous culinary islands have begun to communicate through the common language of CAMS and FAMM.

14.11   References

Andersson, T.D. and Mossberg, L. (2004), 'The dining experience: do restaurants satisfy customer needs?', *Food Service and Technology*, **4**, 171–77.

Annett, J.M. (1996), 'Olfactory memory: A case study in cognitive psychology', *Journal of Psychology*, **130**(3), 309–19.

Aristotle (384-322 f.Kr.), Nicomachean ethics (Roger Crisp, 1961-, Trans.). Cambridge, UK: Cambridge University Press, 2000.

Biesta, Gert J.J. and Burbules, Nicholas C. (2003), *Pragmatism and educational research*. Lanham, MD: Rowman & Littlefield.

Billing, Mischa, Åsa Öström, and Erika Lagerbielke (2008), 'The importance of wine glasses for enhancing the meal experience from perspectives of craft, design and science', *Journal of Foodservice*, **19**(1), 69–73.

Bitner, M-J. (1992), 'Servicescapes: the impact of physical surrondings on customers and employees', *Journal of Marketing*, **56**, 51–71.

Bloom, B.S. (ed). (1956), *Taxonomy of educational objectives: The classification of educational goals*. New York: McKay.

Borrell-Carrio, F., Suchman, A.L. and Epstein R.M. (2004), 'The biopsychosocial model 25 years later: principles, practice, and scientific inquiry', *Annals of Family Medicine*, **2**(6), 576–82.

Bostock, David (1991), *Plato's Theaetetus*. Oxford Clarendon.

Brillat-Savarin, A [1825 (1963)], Smakens fysiologi övers. av Physiologie du Goût. Avesta: Wahlström & Widstrand 3:e uppl.

Campbell-Smith, G. (1967), *Marketing of the meal experience : a fundamental approach*. London: University of Surrey.

Cargill, K. (2005), 'Food studies in the curriculum. A model for interdisciplinary pedagogy', *Food, Culture and Society*, **8**, 115–23.

Claxton, G.L. (1980), *Cognitive psychology: new directions*. London: Routledge & Kegan Paul.

Claxton, G.L. (1988), *Growth points in cognition*. London: Routledge.

Davies, M. and Devlin, M. (2007), 'Interdisciplinary higher education: implications for teaching and learning'. Melbourne, Australia: Centre for the Study of Higher Education, University of Melbourne CSHE.

Douglas, L. and Connor, R. (2003), 'Attitudes to service quality – the expectation gap', *Nutrition and Food Science*, **33**(4), 165–72.

Edwards, J. (2000), 'Food Service/Catering Restaurant and Institutional Perspectives of the Meal', in *Dimensions of the Meal. The Science, Culture, Business and Art of Eating*, HL. Meiselman, Ed. Aspen: Aspen Publishers, Inc.

Edwards, J.S.A. and Gustafsson, I.-B. (2008a), 'The five aspect meal model. The Room-Atmosphere', Foodservice.

Edwards, J.S.A. and Gustafsson, I.-B. (2008b), 'The room and atmosphere as aspects of the meal: a review', *Journal of Foodservice*, **19**(1), 22–34.

Engel, G. (1977), 'The need for a new medical model: a challenge for biomedicine', *Science*, **196**, 129–36.

Fisher, J.D. and D. Bryne (1975), 'Too close for comfort: sex differences in response to invasion of personal space', *Journal of Personality and Social Psychology*, **32**, 15–21.

Gustafsson, I.B. (2004), 'Culinary arts and meal science', *Food Service and Technology*, **4**, 9–20.

Gustafsson, Inga-Britt and Örjan Klein (2006), Den medvetna kokkonsten (2 omarb. uppl ed.). Nora: Nya Doxa.

Gustafsson, Inga-Britt, Åsa Öström, Lena Mossberg and Jesper Johansson (2006), 'The five aspects meal model; a tool for developing meal services in restaurants', *Journal of Foodservice*, **17**(2), 84–92.

Gustavsson, B. and J.M. Annett (2008), 'What is knowledge in a knowledge-based society?', *Journal of Foodservice*, in press.

Gustavsson, Bernt (2001), *Kunskapsfilosofi. Tre kunskapsformer i historisk belysning* (2 ed.).

Gyllensköld, Haqvin (1977), *Koka, steka, blanda*. Stockholm: Wahlström & Widstrand.

Hansen, K. Ø. Jensen, V. and Gustafsson, I-B. (2004), 'Payment – an undervalued part of the meal experience?', *Food Service Technology*, **4**, 85–91.

Hansen, K. Ø. Jensen, V. and Gustafsson, I-B (2005), 'The meal experiences of à la carte restaurants customers', *Scandinavian J of Hospitality and Tourism*, **5**(2), 135–51.

Heide, Morten and Kjell Grønhaug (2006), 'Atmosphere: conceptual issues and implications for hospitality management', *Scandinavian Journal of Hospitality and Tourism*, **6**(4), 271–86.

Heide, Morten, K. Laerdal, and Kjell Grønhaug (2007), 'The design and management of ambience – implications for hotel architecture and service', *Tourism Management*, **28**(5), 1315–25.

Johansen, Erik Nissen and Thomas Blom (2003), 'An artistic director of the hotel and restaurant experience through the details, radiance and harmony', in *Culinary Arts and Sciences IV*, pp. 115–124, J.S.A. Edwards and I-B. Gustafsson (Eds.). Örebro University, campus Grythyttan: Worshipful Company of Cooks, Research Centre UK, Örebro University.

Kivela, Jakša (1994), *Menu planning for the hospitality industry*. Melbourne: Hospitality Press.

Kotler, Philip (2003), *Marketing management* (11th ed.). Upper Saddle River, NJ: Prentice Hall.

Manske, Melissa and Glenn Cordua (2005), 'Understanding the sommelier effect', *International Journal of Contemporary Hospitality Management*, **17**(7), 569–76.

Marks, R.B (1988), *Personal selling: an interactive approach*. Boston: Allyn and Bacon.

Martens, M. (1999), 'A philosophy for sensory science', *Food Quality and Preference*, **10**(4–5), 233–44.

Martin, W.B. (1991), *Quality service: the restaurant manager's bible*. New York: Hotel School, Cornell University.

McGee, Harold (2004), *On food and cooking. The science and lore of the kitchen*. New York: Fireside.

Mina, Jo (2006), 'The importance and performance analysis of service encounter quality by types of restaurants', *Journal of the Korean Society of Food Science and Nutrition*, **35**(8), 1076–87.

Moravcsik, Julius M. (1992), *Plato and Platonism: Plato's conception of appearance and reality in ontology, epistemology and ethics and its modern echoes*. Cambridge, MA: Blackwell.

Nygren, I.T. (2004), 'Sensory evaluation and consumer preference of wine and food combinations. Influences of tasting techniques', Sweden: Örebro University.

Ochs, Peter (1998), *Peirce, pragmatism, and the logic of scripture*. Cambridge: Cambridge University Press.

Polanyi, Michael (1962), *Personal knowledge: towards a post-critical philosophy*. Chicago: University of Chicago Press.

Rapp, E., Östrom, Å. Bosander, F. and Gustafsson, I-B. (2007), 'The sensory effect of butter in culinary sauces', *Journal of Foodservice*, **18**(1), 31–42.

Reeve, C.D.C. (1992), *Practices of reason: Aristotle's 'Nicomachean Ethics'*. Oxford: Clarendon.

Ryle, Gilbert (1990), *The concept of mind*. New York: Penguin Books.

Schön, Donald A. (2003), *The reflective practitioner. How professionals think in action*. (8th ed.). London: Arena Ashgate Publishing Limited.

This, Hervé (2006), *Molecular gastronomy: exploring the science of flavor* (M.B. DeBevoise, Trans.). New York: Columbia University.

Tucci, L.A. and Talaga, J.A. (2000), 'Determinants of consumer perceptions of service quality in restaurants', *Journal of Food Products Marketing*, **6**(2), 3–13.

Wall, E.A. and Berry, L.L. (2007), 'The combined effects of the physical environment and employee behaviour on customer perception of restaurant service quality', *Cornell Hotel and Restaurant and Administration Quarterly*, **48**, 59–69.

Zeithaml, Valarie, A., Mary Jo. Bitner, Dwayne, D. Gremler and 2006. 4. ed. (2006), *Services marketing: integrating customer focus across the firm* (4th ed.). Boston: McGraw-Hill, 2006.

Part VI

Meals worldwide

15

The packaged military meal

G. Darsch and S. Moody, Natick Soldier Research Development and Engineering Center, USA

Abstract: An overview is presented of military meals of the Armed Forces of the United States; major changes throughout history from the late nineteenth century to the present day and difficulties encountered in feeding troops are described. The fundamentals of advances in food science are chronicled and a historical perspective is given to the technologies, e.g. in food preservation and packaging, that have played a large role in ration development to achieve safety, shelf stability, utility and variety. The types of rations used in particular wars are described in their historical context and also by the processing technologies employed. There has been significant progress and problem-solving in research and development over the last 200 years to tackle military nutrition.

Key words: military meals, food technology, nutrition, food packaging, war rations.

15.1 Preface and introduction

15.1.1 Preface

Napoleon is often credited with the axiom, 'An army travels on its stomach'. If the Emperor were to take a look at military feeding today, he might well say, 'Plus ça change, plus c'est la même chose…' (the more things change, the more they stay the same). While many changes over the past few centuries have improved the military meal, the fact remains that providing a quality meal to a deployed soldier is still a very complicated and difficult process. This chapter will provide an overview of military meals of the Armed Forces of the USA, discussing the major changes and difficulties encountered throughout history and in the present day.

The basic fundamentals of advancements in food science will be chronicled in this chapter and will also give a historical perspective to the technologies that have played a large role in ration development. The ration, that is, the food provided each day to nourish and sustain soldiers, has changed significantly based on advancements in science and technology. The types of rations used throughout history will be described by their historical context and also by the processing technologies employed and significant progress and problem solving in research and development over the last 200 years.

In the USA, there was little research devoted strictly to military feeding during the nineteenth century, and the army leadership therefore relied on whatever was available through commercial channels to meet the needs of the armed forces. Thus, this chapter will focus the discussion of the military rations of the nineteenth century upon the types of rations used and the problems encountered feeding troops. Meanwhile, obstacles and difficulties in feeding the armed forces in the late 1800s led to the birth and expansion of organized, scientific research for the specific development of military rations in the 1900s. The discussion of rations during the twentieth century will therefore have a much greater emphasis on institutionalized research and development performed through the cooperative efforts of government, industry and academia, which collectively led to higher quality, more nutritious, and more functional rations for America's armed forces.

An additional focus of this chapter will be to discuss advancements in food processing techniques as they relate to the development of military rations. The great discoveries in microbiology and human nutrition of the late nineteenth and early twentieth centuries ushered in a number of new food processing technologies and were vital to the advent of true scientific research and development of military foods and packaging. Each new food preservation technique helped improve the safety, shelf stability, utility and variety of military rations and food in the commercial sector alike.

15.1.2 A historical perspective of military meals

A close look at the cradles of civilization and major military powers throughout history reveals that the most intellectually accomplished and militarily powerful people were geographically established in and around the most fertile flood plains of the world. The fruitful and plenteous crops of these regions provided the necessary nutritional support to sustain advanced societies and large, well-fed armies (Jacob, p. 8–9). Often, though, the extension of an empire's military power was limited by the availability and ability to transport food and water. Inadequate food preservation techniques and restricted transportation resources often limited the option for military campaigns to extend for great distances or lengthy durations. In such instances, armies had to rely on the resources of their enemies. It is no wonder that those warriors who ventured far from their native lands gained a distinction as ruthless barbarians. The Mongols, Huns and Vikings all received reputations as vicious marauders in part because of their reliance on the pillage and plunder of conquered lands as their primary means of subsistence. This is epitomized by Genghis Khan's issue of a straw to each of his men 'so that, in dire need of food, he might draw the blood of his horse' (McDevitt and Samuels, p. 19). From the most ancient civilizations to the modern armies of today, the same dilemma plagues the military leader. The ability to project his force is limited, at least in part, by his ability to provide the soldiers with a sufficient, healthful and acceptable supply of food (US Armed Forces Food & Container Institute, p. 6).

The armed forces of the USA have been deployed all over the world, and no matter where the mission has taken them, they had to be fed. This could be a daunting task, and in almost every armed conflict, there were problems associated with the quality or utility of rations provided to the fighting forces. Often, these problems were identified and led to improvements in operational rations. Unfortunately, the resolution of many ration problems came too late, which meant that each new war was begun with rations tailored to the problems of the previous war. This cycle continued repeatedly, until a focused and funded research program was finally established to continually improve the rations available for military feeding. From the first classes taught to Commissary Officers at the Chicago Quartermaster Depot in 1907, to the current Department of Defense Combat Feeding Program at the Natick Soldier Research, Development and Engineering Center, resources have been devoted to food research and ration development to improve the military readiness of the armed forces. In this chapter, the types of military rations used throughout the history of the United States are discussed, thus identifying the evolution of military rations from the American Revolution to the currently fielded ration. Key problems associated with each war's rations will be discussed to show the catalysts for ration improvement. Historical scientific progress and significant developments in food technology will also be associated with the various advancements in military rations.

15.2   Unique requirements and limitations of military rations

Food for the military has often differed significantly from the normal fare of society. Although the ideal military diet would be the same as that of the civilian community, there are unique military requirements that distinguish a soldier's chow from everyday cuisine. Military rations need a much longer shelf life for both perishable and semi-perishable food, as well as increased packaging protection. Other considerations include achieving nutritional adequacy for the arduous duties of a soldier and attaining universal palatability for the diversity of tastes and cultures that are amalgamated in any given army unit (Risch, p. 175).

15.2.1   Shelf life
A supply of shelf-stable and readily available food has always been required for the mobilization of the army. Military readiness requirements are one of the catalysts for the lengthy shelf life of rations and compel logisticians to

maintain minimum inventory levels that can be quickly available during the initial phases of a military operation. These 'war reserves' are often composed of items that might not normally be consumed by a standing army (US, US Army, NRDEC, *Summary of Operational Rations*, p. 7–8). There are currently over 60 million individual rations in storage in the United States, on strategic supply ships and in prepositioned storage sites throughout the world. These war reserves are meant to provide the first 21 days of subsistence for a major deployment of the armed forces (Viola).

Another major reason for the lengthy shelf life requirement of rations is related to the way that rations are used, namely, garrison or operational. When soldiers were in garrison, that is, stationary at an established camp or fort, the commander was afforded the luxury of purchasing the required commodities almost on an as-needed basis. Herds of livestock could be kept close by and slaughtered when necessary. Other commodities could be purchased from the entrepreneurial tradesmen, called 'sutlers', who could provide almost any legitimate demand. Once a mobilization was required, however, the task of feeding became much more difficult, and operational rations were needed. Soldiers needed a food supply that was mobile and could be carried or transported on a journey of days to weeks during a prolonged campaign. For reasons of both security and logistics, herds could not always be driven alongside the soldiers. Also, bullets, cannons and the proximity of death were enough to send the most profit-minded merchant back to the safety of the rear echelon. Thus, a supply of meat was required that would last at least as long as the army was on the march, hence the reliance on salted and dried meats (and later canned meats) as a staple. The soldiers, therefore, were required to carry as much as five days of rations with them, with the rest of their subsistence following in several supply wagons. Sometimes poor transportation and logistics left soldiers with little or nothing to eat for extended periods (Roosevelt, p. 179–180, 196). Here, we will discuss operational rations, also known as combat rations, in much greater detail than the garrison ration that is fed in fixed dining facilities on military installations. Specialized combat rations are the result of dedicated research and development that provided unique adaptations of existing foods to meet the specific demands of ongoing military operations (US, US Army, NRDEC, *Summary of Operational Rations*, p. 5).

For decades, research and development have been directed at improving the storage potential of various food items. Efforts have concentrated specifically in two main areas: reduction of degradation owing to internal factors, and protection of products from external damage. Scientific principles of food science and new processing techniques have enabled shelf life extension of some very desirable ration components such as fruit and milk.

The shelf life of operational rations remains one of the direct areas of concern for the military field-feeding program. Current requirements for individual rations are for a shelf life of three years at 80 °F (27 °C), and six months at 100 °F (38 °C). This is verified by the use of accelerated storage data correlation (Daley). After storage at the various temperatures for prescribed periods, a sensory panel evaluates the acceptability of the ration components. Data is collected using a 9-point hedonic scale that ranges from extremely like to extremely dislike. For inspection purposes, time–temperature integrators are applied to each case of operational rations. These labels change in appearance over time as the monomers that are incorporated into the printing ink are polymerized. The polymerization reaction is temperature sensitive; therefore, the appearance of the label at any given time reflects a relatively accurate assessment of the aggregate storage temperature history. Thus, the label can be used to obtain a reliable prediction of the remaining shelf life of the products within the case (Taub and Singh, p. 357).

15.2.2   Packing and packaging

Because of the long distances traveled, often over rough terrain, all military supplies absorb an inordinate amount of jarring impacts and exposure to the elements. The packaging materials that would normally protect a product in commercial distribution channels are simply not meant to guard against the outright abuse to which many rations are subjected. In fact, when rations were packaged according to commercial standards, the amount of waste incurred owing to damage far exceeded the increased costs of thicker wraps, stronger cans and more durable boxes (Melson, Clark and Couch, p. 4).

Because of the potential for deployment to all corners of the globe, food items must also be protected from a vast array of storage pests. Packaging materials that are quite sufficient for domestic distribution were often found woefully inadequate for the rigors of the military supply system. Direct exposure to the elements in regions from the tropics to the arctic dictate the need for durable, multifunctional packaging materials (Tauber, p. 86). Even today, the biggest threat to the military food supply is not spoilage microorganisms but the degradation of product quality and stability owing to adverse storage conditions. Refrigeration is still a rarity in many parts of the world, especially tropical and equatorial regions where storage temperatures are often over 100 °F (38 °C).

Even though much of the research and development work on food containers was, and continues to be done by the packaging manufacturing industry, it is the strict military requirements that drive this development (Melson, Clark and Couch, p. 63). Government researchers continue to rely heavily upon industrial research and development to provide new and innovative packaging systems, since many techniques used in the commercial sector can be applied to military use. From time to time, however, unique military requirements necessitate research and development of packaging products for which there is no commercial demand.

Commercial canning in large, pressurized, high-temperature vessels called retorts has been the most ubiquitous method of achieving shelf stability for well over a century. For decades, the most prevalent packaging material used for military rations was the rigid metal container. The advantages of the superior barrier properties and resistance to damage far outweighed the disadvantages of cost, weight and bulk. For thermally processed foods, which comprised a large percentage of the military menu, the only alternative to the metal can was a glass jar – unacceptable for obvious reasons. The canning industry was committed to the use of the cylindrical metal can, which was well established and trusted by the American consumer as the standard for safe, shelf-stable foods. Even though several packaging companies looked into the flexible package because of its acceptability in foreign markets, there was no real commitment to pursue large-scale production. The research conducted to identify alternatives to the 'tin can' had to be initiated and coordinated by military researchers.

Finally, in the late 1950s, testing was performed on flexible packaging materials to replace the rigid metal can. Although there was a potential for acceptance in some foreign markets, there was no commercial application for this type of packaging material in the USA. Still, the promise of substantial savings in weight (29%) and volume (26%) was enough to justify the research. The properties of various barrier films were tested, and eventually, a combination that achieved the desired results was developed. The initial research began in 1959 (Keller). The thinner pouch also took less time to achieve commercial sterility and therefore the food product was not affected as much by the adverse effects of extremely high processing temperatures for retorted products. The US Army Laboratories at Natick, Massachusetts, collaborated with Reynolds Metals Company and Continental Can Company to develop the retort pouch, and assisted greatly in obtaining regulatory approval for the use of flexible pouches as an alternative to rigid metal cans. This flexible packaging, although not readily accepted by the American consumer, has been part of the NASA astronaut feeding program since the late 1960s and was crucial to the transition from the World War II C-ration to the current Meal, Ready-to-Eat, Individual (MRE).

First purchased for fielding in 1979, the MRE consisted of items packaged in flexible pouches. The pouch material for thermally stabilized components was a trilaminate material composed of polypropylene, aluminum foil and polyethylene. The combined barrier properties and protective attributes of these materials were sufficient for the first generation of pouch-packaged meals. Since then, there has been a concerted effort to identify affordable alternatives that provide even better protection. The addition of another layer comprised of nylon fashioned the four-layer, quad-laminate material that is currently being used for the packaging of retorted ration components (Sherman).

15.2.3 Nutritional adequacy

'Beginning nutrition research focused on macronutrients, gastrointestinal digestion and absorption, and nitrogen balance. Research then moved toward the biochemical analysis of the vitamin, the amino acid, and some trace minerals' (Youmans, p. 32). During the American Civil War era, only the most fundamental of nutritional principles were understood. At a time when most military surgeons were just establishing the link between nutritional deficiencies and disease, E.N Horsford studied all aspects of the army ration and recommended several changes based on nutritional considerations. His premise for these recommendations was the assertion that 'the human organism requires, to maintain it in health, both organic and inorganic food. Of the organic, it needs nitrogenous food (protein) for the support of vital tissues, for work; and saccharine or oleaginous (carbohydrates and fats) food for warmth. Of the inorganic, it needs phosphates for the bones, brain, muscles, and blood; and salt for its influence on circulation and secretions' (Davis, p. 9). This foundation of the nutritional knowledge base led to advancements in both military and commercial sectors of food research.

'Available knowledge prior to World War I was used mainly for the prevention or alleviation of dietary deficiency diseases in the individual. Scurvy, rickets and goiter were countered at that level. Meanwhile, just prior to World War II, it became practical to improve common foods with synthetic nutrients as a means of preventing deficiency diseases in large populations. This resulted in the introduction of iodized salt, vitamin D milk, vitamin A fortified margarine, and enriched flour, bread, and corn meal' (Sebrell, p. 3–4). Several ration components are still fortified with nutrients and are key to obtaining the necessary levels of vitamins and minerals established by the military medical community.

The increasing acknowledgment of the importance of nutrition with regard to national defense was well illustrated by the fact that in 1941, the President of the United States established the Nutrition Conference for Defense. This gathering of nearly 1000 experts in the field of food and nutrition provided several recommendations for nutrition training and adopted the NRC's Recommended Dietary Allowances (Sebrell, p. 3). As researchers became more aware of the importance of nutrition, nutritional considerations began to play a larger part in the development of military rations. When evaluating new food processing technologies, for example, scientists now had to consider how those methods would affect the retention of nutrients (Hilbert, p. 90).

Nutrient tolerances and allowances were continually refined as the science of nutrition advanced. New discoveries revealed still more about the necessity and effects of various micronutrients. Even before the dietary needs of women and children were established, the Interdepartmental Committee on Nutrition for National Defense (ICNND) published a 'Suggested Guide to Interpretation of Nutrient Intake Data' (Hegsted, p. 20). These guidelines were for healthy, young men with moderate to heavy activity – exactly the profile of the majority of those in uniform at that time. The suggested amounts of each nutrient were categorized as deficient, low, acceptable and high as shown in Table 15.1.

Because of the distinctive population of the military and their unusually

Table 15.1 Suggested guide to interpretation of nutrient data intake (Hegsted)

	Deficient	Low	Acceptable	High
Niacin, mg/day	<5	5–9	10–15	>15
Riboflavin, mg/day	<0.7	0.7–1.1	1.2–1.5	>1.5
Thiamin, mg/1000 Cal	<0.2	0.20–0.29	0.3–0.5	>0.5
Ascorbic acid, mg/day	<10	10–29	30–50	>50
Vitamin A, IU/day	<2000	2000–3499	3500–5000	>5000
Calcium, g/day	<0.3	0.30–0.39	0.4–0.8	>0.8
Iron, mg/day	<6.0	6–8	9–12	>12
Protein, g/kg	<0.5	0.5–0.9	1.0–1.5	>1.5
Calories	–	–	–	–

arduous duties, the dietary requirements for the armed forces are much different than that of the rest of a nation's citizenry. Developed by the Office of the Surgeon General of the Army, the Military Recommended Daily Intake (MRDI) was established as one of the paramount design parameters for all operational rations. All military rations, whether for garrison or field menus, must be approved by the Surgeon General before fielding.

Originally consisting of proximal analysis for macronutrients (protein, fiber, fat, ash, and moisture), the testing requirements now include detailed nutritional analysis of all operational ration components. Once labeled only with net contents and a very generic and sometimes quite cryptic nomenclature, most current ration components bear the same nutritional labels as commercial food products. The addition of folacin as an essential micronutrient reflects the changing demographics of today's military. Folacin is critical to fetal neurological development during the first trimester of pregnancy. With more women consuming combat rations owing to their proliferation in combat support roles, the Surgeon General not only included folacin as a part of the MRDI, but also established mandatory minimum amounts to be included in operational rations.

A joint service regulation entitled 'Nutritional Allowances, Standards, and Education', gives the Surgeon General, Department of the Army full responsibility for establishing nutritional guidelines for military rations. The regulation provides information in the form of suggested guidelines, such as the MRDI, and also establishes strict nutritional standards for both operational rations and restricted use rations (US Dept of the Army). The current standards, as shown in Table 15.2, are much more comprehensive than the aforementioned 'suggested guide' and reflect the great increase in nutritional knowledge over the past few decades.

Restrictions on the use of certain operational rations are established based on their nutritional inadequacy. Restricted rations are to be used only for periods of up to 10 days for long-range patrol, reconnaissance or assault operations. Survival rations, which are even lower in nutrients than restricted rations, are to be used only under emergency conditions for periods of four days or less (US Dept of the Army 2-1).

Table 15.2 Nutritional standards for rations (as shown in AR-40-25)

Nutrient	Unit	Operational rations	Restricted rations
Energy	kcal	3600	1100–1500
Protein	g	100	50–70
Carbohydrate	g	440	100–200
Fat	g	160 (maximum)	50–70
Vitamin A	μg retinal equivalent (RE)	1000	500
Vitamin D	μg	10	5
Vitamin E	mg tocopherol equivalent (TE)	10	5
Ascorbic acid	mg	60	30
Thiamin	mg	1.8	1.0
Riboflavin	mg niacin equivalent (NE)	2.2	1.2
Niacin	mg	24	13
Vitamin B6	μg	2.2	1.2
Folacin	μg	400	200
Vitamin B12	mg	3	1.5
Calcium	mg	800	400
Phosphorus	mg	800	400
Magnesium	mg	400	200
Iron	mg	18	9
Zinc	mg	15	7.5
Sodium	mg	5000–7000	2500–3500
Potassium	mg	1875–5625	950–2800

15.2.4 Palatability

Perhaps the most overlooked requirement for rations is palatability, or acceptability to the consumer. Indeed, army food has always been stereotyped as tasteless and unappetizing, known by terms such as chow and grub. Since military leaders faced so many challenges just getting any food into the hands of a soldier, providing something that appealed to the troops was a luxury that could rarely be afforded. Therefore, even though it was known that soldiers preferred to eat the types of food familiar to them, they often ended up with hard bread and cold bacon. It was not until an emphasis was placed on nutritional adequacy that palatability became a primary factor for ration development. Nutritionists realized that no matter how balanced the ration was, it would only benefit the soldiers who consumed all of the components. 'In one sense acceptability is the final determinant of the nutritional adequacy of the ration' (Isker, p. 4). During World War II, a great emphasis was placed on palatability as being one of the four objectives of ration development.

Eventually, military leadership acknowledged the need to perform proactive, dedicated subsistence research to improve the acceptance of rations by their soldiers. The Committee on Food Research was formed in 1941 and, through an intensive study of problems with rations, they approved a carefully developed research program in cooperation with 41 universities, five foundations, and five governmental, quasi-governmental and medical laboratories. The functions of the committee, as defined by the Quartermaster General were as follows:

> The Committee will maintain technical liaison with government, quasi-government, and allied government agencies, research institutions, foundations, industrial associations and companies. It will initiate, channel and exchange information with these institutions on subjects relating to fundamental food research and development activities, and ascertain the Army's present and future technical food problems
> (US, Office of the Quartermaster General).

Besides the nutritional considerations, there is another intangible consideration that is greatly influenced by the acceptability of rations. 'The regular serving of palatable food is the greatest single factor in building and maintaining high spirit and morale' (Risch, p. 174). In addressing a group of quartermaster officers, Brigadier General C.O. Thrasher likewise stated, 'No single factor so influences morale of troops as their messes' (Thrasher, p. 4). This increased emphasis on palatability spurred a large number of research projects during the later years of World War II. The research program approved for 1945–46 by the newly formed Committee on Food Research included 21 projects related to food acceptance research (US, Office of the Quartermaster General).

Both the US Armed Forces and their allies conducted several field studies. Many of these, such as the Canadian Army trials of 1943–1944, tested the palatability and acceptability of US rations. Monotony seemed to be one of the biggest problems, and it was recommended that repeat menus be infrequent at best (Johnson and Kark, p. 26). Menu fatigue continues to be a concern with current rations, and increasing menu variety is one of the primary functions of the today's Department of Defense Combat Feeding Program.

15.3   Military meal developments of the 18th century

15.3.1   Types of rations: Revolutionary War

Detailed accounts of the financial and personal devastation of most of the signers of the United States' Declaration of Independence record the price that was paid by early political leaders. Less known, however, is the sacrifice of the farmers and businessmen who became America's first soldiers. These soldiers, known as minutemen, gathered in informal militias well before the establishment of a continental army and had to rely on forage and the generosity of their fellow citizens to meet their requirements for food. However, as the Army became larger and more organized, it was less and less realistic for soldiers to depend on their fellow citizens for subsistence. Their leadership began to work on the daunting task of keeping the soldiers nourished, yet the obstacles to accomplishing this quickly became evident.

In his keynote address to the Symposium on Feeding the Military Man, Emil Mrak cited the 1778 writing of Dr Benjamin Rush stating, 'Fatal experience has taught the people of America that a greater proportion of men have perished with sickness in our armies, than have fallen by the sword – the art of preserving the health of the soldier consists of attending to the following particulars: dress, diet, cleanliness, encampment, and exercise' (Mrak, p. 3). Clearly, feeding soldiers and keeping them adequately nourished was considered the first line of defense in preventing sickness and maintaining the strength of the Armed forces. The first US Army ration, that is, the amount and type of food provided for one soldier for one day, was brought into existence even before the United States declared its independence. 'It was established by Congressional Resolution on November 4, 1775:

Resolved, that a ration consists of the following kind and quantity of provisions: 1 lb. Beef or $^3/_4$ lb. Pork, or 1 lb. salt fish per day; 1 lb. bread or flour, per day; 3 pints of peas or beans; 1 pint of milk per man per day, or at the rate of 1/72 of a dollar; 1 half pint of rice or one pint of Indian meal, per man per day; 1 quart of spruce beer or cider per man per day, or 9 gallons of mollasses per company of 100 men per week; 3 lbs candles to 100 men per week, for guards; 24 lbs. Soft or 8 lbs. Hard soap, for 100 men per week (US Armed Forces Food & Container Institute 1).

15.3.2   Progress and problems: Revolutionary War

Although this was the standard ration that was authorized, many problems were encountered in actually fulfilling the requirements and providing the rations to each soldier. From the very beginning of the United States military, indeed throughout world history, the main problems encountered with military rations have been availability, nutritional adequacy, stability and utility (Risch, p. 175). It does little good to identify the specific diet needed for each soldier if you cannot consistently provide him with that subsistence and in a form which is acceptable.

One of the most infamous military scenarios of the American Revolution is that of Valley Forge. The mere invocation of that name brings to mind the deprivation of food and shelter under the harshest of environments (Marshall and Manuel, p. 320). The irony of the circumstances was that with starving soldiers shivering in the Pennsylvania countryside, there was more than enough livestock to alleviate the hunger of Washington's deprived army just across the Delaware River. Unfortunately, there was not enough transportation support to get the meat to the soldiers who needed it (Meat Serves, p. 169).

When General Washington later requested 40 000 000 lb of meat to feed 30 000 men for one year, the congressional response was somewhat ineffective. Instead of a centralized procurement effort, each state was asked to give their share. The response was slow and ungenerous, which required Washington to personally appeal to the states for compliance. This was just one of many problems encountered by the entangled red tape of government procurement.

As the new government of the USA took shape, mechanisms were established for the purchase and distribution of food to the military. However, it did not take long for another hindrance to limit the availability of rations (Meat Serves, p. 169).

As early as the late 18th century, some rather onerous specifications were established. In order to prevent a few unscrupulous purveyors from taking advantage of the newborn bureaucracy, various rules, regulations and specifications were instituted. One of the more impractical, if not impossible, requirements included the numeric branding of the horns of every animal that was purchased (Meat Serves, p. 169). The bureaucracies of the federal government also required special markings on every container of food destined for military usage. Some purveyors were more willing than others to comply with these new regulations and specifications, and so began the industrial base of military suppliers of subsistence. One of these suppliers was a meat purveyor from New York named Samuel Wilson. Wilson had spent some time as a state regulator and so was familiar with the newly formed red tape. Each case of subsistence was clearly stamped with the initials US and this became the source of an amusing sobriquet. The congenial and popular meat supplier and food inspector became affectionately known as Uncle Sam. At first, the US stamped on each barrel caused soldiers to joke that it was 'Uncle Sam's pork'. The joke soon spread to include all government property, and thus originated the concept of that mythical patriotic icon (Meat for the Multitudes, p. 27).

15.4   Introduction of specialized and standardized rations in the 19th century

15.4.1   Types of rations

American Civil War

Rations remained relatively unchanged until the time of the American Civil War (Fig. 15.1). The significant changes that occurred during this time were the introduction of canned foods as a component of military rations and the beginning of ration classification – different ration components based on the operational circumstances of the soldiers to be fed. It was during this war that an official distinction was made between the type of food provided in garrison and the type of food provided during a mobilized campaign, that is, on the march or in direct combat.

The disparity between the garrison ration and the marching ration can be seen when comparing the two versions authorized for the Union Army.

> Here is just what a single ration comprised, that is, what a soldier was entitled to have in one day. He should have had twelve ounces of pork or bacon, *or* one pound four ounces of salt or fresh beef; one pound six ounces of soft bread or flour, *or* one pound of hard bread, *or* one pound

Fig. 15.1   Union soldier eating Civil War rations. US Army photo (1864).

four ounces of corn meal. With every hundred such rations there should have been distributed one peck of beans or pease; ten pounds of rice or hominy; ten pounds of green coffee, *or* eight pounds of roasted and ground, *or* one pound eight ounces of tea; fifteen pounds of sugar; one pound four ounces of candles; four pounds of soap; two quarts of salt; four quarts of vinegar; four ounces of pepper; a half bushel of potatoes when practicable, and one quart of molasses. Desiccated potatoes or desiccated compressed vegetables might be substituted for the beans, pease, rice, hominy, or fresh potatoes. Vegetables, dried fruits, pickles, and pickled cabbage were occasionally issued to prevent scurvy, but in small quantities.

But the ration thus indicated was the camp ration. Here is the *marching* ration: one pound of hard bread; three-fourths of a pound of salt pork, *or* one and one-fourth pounds of fresh meat; sugar, coffee, and salt (Stern, p. 73).

'No doubt the diet was often deficient. The ration most commonly issued in the field consisted of salt meat (bacon, "sow-belly"), hard bread ("hardtack"), coffee, sugar and occasionally beans' (Steiner, p. 175).

Spanish–American War
The severely downsized military of the late 19th century was ill prepared for the hasty accession and overseas deployment of thousands of troops. The rations of this war gained notoriety as some of the most unpopular in the history of the US Army. This is particularly true of the travel ration that was developed as one of the first self-contained, standardized operational rations.

The travel ration consisted of canned meat, salt pork, hard bread, coffee and sugar. The primary component of the ration was canned meat. The soldiers of the Spanish–American War were unfortunate to be among the first to sample a new item – canned fresh beef. Unlike the well-recognized cured products such as corned beef and 'deviled' meat, this fresh beef was described as stringy and tasteless. Theodore Roosevelt estimated that less than 25% was actually eaten, and spent considerable time deriding the product in his autobiographical novel *The Rough Riders*. 'The travel rations which had been issued to the men for the voyage (to Cuba) were not sufficient, because the meat was very bad indeed; and when a ration consists of only four or five items, which taken together just meet the requirements of a strong and healthy man, the loss of one item is a serious thing' (Roosevelt 61). Once on foreign soil, the soldiers again received the typical combat ration of salt pork and hardtack. This, too, received its share of criticism. Roosevelt wrote, 'Usually, we received full rations of bacon and hardtack. The hardtack, however, was often mouldy, so that parts of cases and even whole cases, could not be used. The bacon was usually good, but bacon and hardtack make poor food for men toiling and fighting in trenches under the midsummer sun of the tropics' (Roosevelt, p. 273). In fact, the tropical climate also played a large role in the spoilage of both canned and fresh meat products during this war. The rations were not designed to withstand the high temperatures and humidity of the Caribbean and Pacific (Meat for the Multitudes, p. 79–80).

By the end of the war, there was a great deal of attention paid to the way soldiers had been fed in their expeditionary excursions to places like Cuba and the Philippines. Because of the increased emphasis on military nutrition during the late 19th century, more importance was placed on designing rations with the soldier's best interest in mind. The pay of military cooks was increased, and they began to receive formal training. 'In fact, around 1902, the US Army established the first training school for baking and cooking at Fort Riley, Kansas' (Darsch and Evangelos, p. 19).

15.4.2 Progress and problems
American Civil War
This was the first time that canned products played a large role in military feeding. Even though the quality and availability of canned foods were unstable, the ability to extend the shelf life of fruit, vegetables and dairy products was a significant accomplishment of great nutritional benefit to

those soldiers who received them. Unfortunately, far too few soldiers were able to benefit from this new technology. Infectious disease and malnutrition devastated both Union and Confederate forces. 'Scurvy was present in the Thirteenth Massachusetts, as they had been exclusively on a salt pork and hardtack diet for over six months' (Steiner, p. 207). The number of casualties related to inadequate food and poor sanitation during the Civil War was staggering. Diarrhea and dysentery, listed as the number one infectious disease in the Federal Army between 1861 and 1866, accounted for 44 558 deaths from the 1 739 135 cases reported. For non-infectious diseases, nutritional deficiencies accounted for over 50 000 reported cases. Scurvy, the second most prevalent non-infectious disease took the lives of 771 soldiers and hospitalized 46 931 (Steiner, p. 10–11). This was probably a huge underestimate, as scurvy was often secondary and unreported among the victims of acute or chronic diarrhea (Steiner, p. 174).

Medical science had still not isolated the role of specific pathogenic bacteria, protozoa and unsanitary conditions of food handling as a cause of diarrhea. 'Medical officers were of the opinion that the inadequate ration, poor cooking, impure water, fatigue, and exposure were prime causes of diarrhea and dysentery, but they were not always in harmony as to which was most important' (Cunningham, p. 186). The large number of troops lost to disease drew the attention of the medical community, but the only indictments of the food supply were spoiled meats (usually not eaten owing to intolerable palatability) and fruits, which had been wrongly implicated as diarrheic for centuries (Causes, p. 623–630). Unlike diarrhea, however, the cause and prevention of scurvy had long been associated with diet. In early 1863, the Army of Northern Virginia was warned to provide more vegetables to their soldiers to prevent the onset of scurvy. 'With the commencement of spring, General Lee ordered a daily detail from each regiment 'to gather sassafras buds, wild onions, garlic, lamb's quarter and poke sprouts' to supplement the ration, but the supply obtained was not sufficient to overcome the deficiency' (Cunningham, p. 206).

Because military medicine had finally made the connection between nutrition and disease, early attempts were made to supplement the military diet with vegetables. Unfortunately, the bulk and weight of them disallowed their use in a fresh state, so a product called desiccated vegetables was developed. A variety of vegetables were 'shredded, mixed, dried and pressed into hard clumps'. Unfortunately, they were extremely difficult to rehydrate and tasted so bad that they earned the nick-name 'desecrated vegetables' (Darsch and Evangelos – Back to the Future, p. 18).

Spanish–American War

In fact, not only did the rations of the Spanish American War receive well deserved criticism for their lack of palatability and utility, but they also were falsely accused of causing the death and illness of soldiers from Cuba to the Philippines. These accusations played a strategic role in bringing about several important changes in the regulation of both military and commercial food.

Like the American Civil War, the Spanish–American War was plagued with an appalling number of casualties owing to infectious disease. This time, however, the nation, and, more importantly, the nation's press, focused on the thousands of soldiers confined to 'fever camps' upon return to the United States. 'Dr. Nicolas Senn, the chief surgeon for the United States Volunteers, reported that not only was typhoid fever rampant among US soldiers, but amebic dysentery and diarrhea were the two greatest enemies of the Spanish army' (Freidel, p. 295).

Once again, far more lives were lost to disease than there were to bullets. This time though, the American people, urged on by a relentless press corps, demanded a reason for such a demoralized Army after a decisive victory. Major General Nelson Miles was in the spotlight and not willing to accept responsibility for these problems. During Congressional testimony and also in press interviews, General Miles blamed an inadequate and tainted food supply as the reason for much of the illness. This 'embalmed beef' scandal was highly publicized, and even though almost every allegation Miles made turned out to be false, there were several actions taken to alleviate the public outcry (Gould, p. 121–122). General Miles made the following remarks, explaining the non-combat casualties to congress. They were also widely reported in the press.

> Refrigerated beef furnished to the troops in Cuba, Puerto Rico, Tampa and elsewhere was 'embalmed' and was a serious cause of illness and distress among the troops…Canned roast beef, furnished to the Army under pretense of an experiment, was a cause of sickness and was unfit for issue as food in any country…What was called canned roast beef was really pulp from which the beef extract of commerce had been boiled out. (Meat for the Multitudes, p. 79)

In subsequent investigations, the meat industry was cleared of wrongdoing, but the Commissary General was court-martialed for the purchases of the unpalatable canned fresh beef that was used in the travel ration (Meat for the Multitudes, p. 80). Another notable result of this scandal was the establishment of the Army Veterinary Service as the food inspection agency for armed forces procurement of meat and dairy products (Miller, p. 3). For the first time, inspectors possessed knowledge of the science of foods, and were not part of the procurement agency. Later developments, such as the uproar over Upton Sinclair's book, *The Jungle*, paved the way for food inspection to become an integral part of America's food industry. Meat inspection laws were passed in 1906 and the Pure Food and Drug Act took effect on January 1, 1907 (Meat for the Multitudes, p. 116).

15.4.3 Processing technologies

American Civil War

Throughout America's history, the military has relied on existing food processing technologies for the production of its rations. This severely limited the choice of items that would meet the particular needs of the armed forces. Early techniques such as smoking, drying and salting provided the only means of preservation of meat and fish items for the military. The assortment of fruits and vegetables available was likewise limited to those tuber and legume crops which either had a long storage life, for instance potatoes, or those that could be preserved by drying such as beans. 'From the Revolutionary War, through the Civil War, and on to World War I the basic military ration was composed of meat, bread and beans' (US, US Army Natick Laboratories 5).

Sometimes a staple, such as bread, could not be adapted to the military's needs in certain situations. To alleviate the demand and provide necessary grain products, unique military items had to be produced. A hard biscuit was developed which was both durable and could be easily transported and issued. This 'hardtack', a military staple from the Civil War through World War I, would never be acceptable as a commercial food product (it was barely acceptable as a military product), yet it provided a vital source of subsistence for hundreds of thousands of American soldiers and sailors (Billings, p. 113). This was just one example of how new military food items were developed that were not necessarily commercially viable. Existing processes were used to meet the military's need of robust and stable foods for an army on the move.

Eventually new food processing technologies, such as canning, enhanced the variety of the soldier's diet by greatly extending the shelf life of what would otherwise be very perishable products. The concept of canning was first established when Napoleon announced a prize of 12 000 francs for a method of preserving food for his armies. Nicolas Appert presented his method of food preservation by the exclusion of air and heating of product in sealed containers. Throughout the 1800s, many ambitious merchants sought to take advantage of this concept, experimenting with canned products such as tomatoes, fish, lobster, meats, and various fruits and vegetables. Peter Durand was granted a British patent in 1810 for his use of tin cans for thermal processing and, in 1817, William Underwood introduced his line of canned meat products, 'deviled' meats that were used by the military. The huge demand for army rations during the American Civil War was a windfall for the canning industry and, in 1854, Gail Borden concocted a canned condensed milk that became a popular supplement to soldiers' diet. Yet, for all this product development, it was not until a half-century after Nicolas Appert first developed the concept of canning that Louis Pasteur provided the first scientific explanation of the principle behind the technology. Pasteur postulated that controlling the growth of microorganisms was the key to food preservation. Finally, in 1895, the Massachusetts Institute of Technology developed the first scheduled process and formally documented the steps necessary to ensure commercial sterility (Food Processors Institute 7).

Spanish–American War
Although operational rations, still in their infancy, left little to be desired, there were enormous strides in garrison feeding during the late 1800s. Even

overseas, in the absence of direct hostilities, there was a more plentiful and diverse supply of food items than ever before. This was a result of two significant developments in food processing: the proliferation and development of the canning industry, and the invention and use of refrigerated transportation on both land and sea.

America's westward expansion and the completion of coast-to-coast rail service created a demand and new markets for canned food products. The canning industry was no longer just a specialized packaging division of the various commodity groups. Brand names began to be used to market specific products. The consistent quality of a particular brand became a selling point and, for better or worse, was the foundation of a company's reputation. This inevitably led producers to invest in the research and development necessary to provide brand name products of consistently high quality. It was during this time that the Massachusetts Institute of Technology published their research on the documented thermal process based on bacterial death rates. This alleviated the need for over processing which, although producing a safe product, was devastating to the quality of canned food items (Food Processors Institute 7).

The increased commercial acceptance of canned foods allowed the military to take advantage of this abundant supply of canned fruits, vegetables, meats and dairy products. Although usually not purchased in quantities large enough to feed the average soldier, many of these items were made available for sale to commissioned officers, who were required to purchase their rations (Roosevelt, p. 196).

15.5   Modernization of military meals in the 20th century

15.5.1   Types of rations

World War I

Even with the scientific advances, and the political pressure for advancement of food safety, military rations had not changed significantly by the beginning of World War I. Efforts had been made to improve the group feeding rations, and improvements in supply lines allowed considerable shipments of food to Europe. Packing plants in Chicago were canning between one and two million pounds of boneless meat daily, most of it destined for Europe. With plants running at or near capacity, packers requested a relaxation of military specifications hoping that they might be able to increase production and eliminate excess costs. The plants supplied not only the American soldier, but allied armies as well. France and Italy purchased almost 30 million cans of meat through the United States Food Administration, a wartime bureau that was headed by Herbert Hoover (Meat for the Multitudes, p. 119).

Because of the great advances of canned products into the commercial marketplace and the improved canning techniques, canned meat finally replaced salted and dried meats in military rations. Although this may seem like a great advancement, the overall ration was strikingly similar to the fare of soldiers since the Civil War. Besides the replacement of salt pork with canned corned beef, the soldier of World War I was still asked to subsist on 16 oz of meat, 16 oz of hard bread, coffee and sugar. The canned meat, of course, was much more palatable than salt pork.

Owing to the trench type, high contact nature of the stationary fronts, it was not possible to feed large groups of soldiers except those that were well behind the front lines. Once again, operational considerations called for operational rations. Unfortunately, the only individual combat ration available was the 'reserve ration' (Fig. 15.2) Strikingly similar to what had been fed to the American soldier for over 100 years, this ration consisted of:

Canned meat	16 oz
Hard bread	16 oz
Sugar	2.4 oz
Coffee, roasted or ground	1.12 oz
Salt	0.16 oz

Although the processing and packaging were better, making the ration more wholesome than some of its predecessors, it still left much to be desired from both a nutritional and a palatability standpoint (Porges 34).

A follow on to the reserve ration was the trench ration. Developed in 1918, this was a group ration designed to feed 25 men. Its main feature was its almost impregnable packaging, designed to withstand chemical gas attacks.

Fig. 15.2 The Reserve Ration, fed to World War I soldiers, was very similar to the rations of the Civil War and Spanish–American War [US Army photo (1959)].

The trench ration had much more variety than the reserve ration, but was still very much lacking the required nutrients for prolonged feeding without adverse effects. Its cost and weight eventually became the reason for its demise before World War II. This ration was assembled at the Quartermaster Depot in Chicago, which had already become the nucleus for military subsistence research, development and procurement. The trench ration was composed of the following components:

50	cans hard bread	8 oz each
10	cans corned beef	16 oz each
5	cans roast beef	16 oz each
4	cans fish	16 oz each
4	cans sardines	4 oz each
25	rations sugar	5 lb in bulk
25	rations soluble coffee	18.75 oz
25	rations salt solidified alcohol	50 oz
25	rations cigarettes	100 cigarettes

All of these components were sealed in a large, watertight metal canister, thus protecting the contents from chemical contamination during gas attacks (Porges, p. 35).

The only other ration of World War I was the emergency ration. Purchased, but not fielded, it consisted of evaporated beef, dried wheat, sweet chocolate, salt and pepper. Because it was packed under a collaborative effort with industry, it is often referred to as the Armour ration (Porges, p. 37).

World War II
The ration development, procurement, production and distribution systems were tested during World War II like never before, nor since. The massive deployment of troops to almost every corner of the globe demanded huge quantities of well protected food suitable to a variety of operational situations. The research and development system was at once overwhelmed by this predicament, but eventually the Subsistence Research Laboratory at Chicago was able to provide a considerable array of rations to feed the soldiers in both Europe and the Pacific. Also established at this time was a new classification system that separated rations into groups, alphabetically, depending on their intended use and type of components. To a great extent, the classification system established during World War II is still in use today.

Type 'A' rations (Fig. 15.3), were the classic garrison-feeding ration. Because they consisted, in part, of fresh or frozen meat, fresh dairy products, and fresh fruits and vegetables, this was the only ration type that required refrigeration. It was normally fed in a well established camp or base, and sometimes brought close to combat lines by the use of small kitchen operations or insulated containers.

Type 'B' rations (Fig. 15.4) required no refrigeration. They were composed of canned, dried, and dehydrated components. B rations could be unitized to

Fig. 15.3 A Rations [US Army photo (1959)].

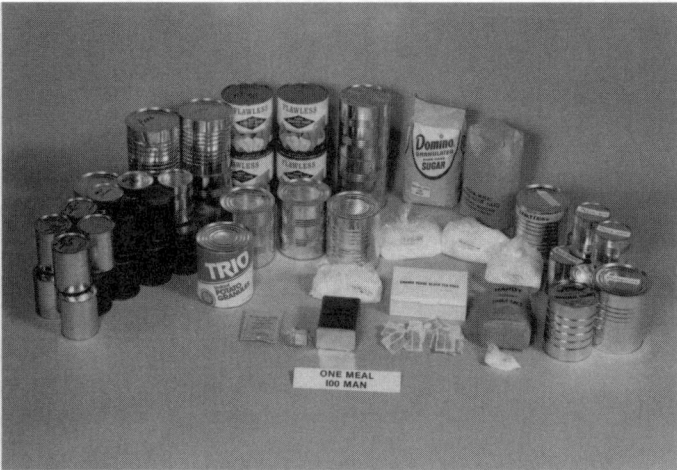

Fig. 15.4 B Rations [US Army photo (1996)].

provide enough of all components (in theory) to feed soldiers in increments of 100. These too required a garrison type cooking and feeding operation, owing to the extensive preparation procedures and the large amounts of water required for rehydration and cooking. B ration components were also used to configure several group feeding rations for small units. The 5-in-1 and 10-in-1 rations were pre-packed modules of all components necessary to feed either five or ten soldiers, respectively (McDevitt and Samuels, p. 107–111).

In 1938, development began on the ration that was to become the most disparaged ration of World War II. The development of the C ration (Fig. 15.5) began well before the war. Type 'C' rations were combat rations. These were true operational rations, in that a soldier would be issued a box that

Fig. 15.5   C Ration components – World War II [US Army photo (1945)].

contained his food for the day. This would be eaten on the march, in the foxhole, or wherever and whenever the opportunity arose. The C ration was composed of two units – a B unit and an M unit. The B unit contained 3–12 oz cans of bread, coffee, and sugar. The M unit contained 3–12 oz cans of meat. C rations, like all operational rations, were not meant for feeding over prolonged periods without supplementation. Like their predecessors, however, these operational rations were sometimes the only food a soldier saw for weeks (McDevitt and Samuels, p. 27–38).

The type 'D' rations (Fig. 15.6) were the survival rations. These were developed to replace the aging and obsolete emergency rations of World War I. The compact bars of the D rations were often used to supplement other operational rations and were popular among the troops because of their light weight and their suitability for eating on the move (Risch, p. 177–178).

Finally, the K ration (Fig. 15.7) was a field ration. Originally developed as a lightweight parachute ration, it ended up being used exclusively as the field ration of choice after it was fielded late in the war. The K ration was the most nutritional operational ration developed, and its lightweight components made its utility better than any of the other rations. Its drawback was the acceptability. In the quest for maximum nutrition, it was decided to pack the ration with nutritionally dense components such as pemmican bars, which were both unfamiliar and unpopular to many of the troops (McDevittt and Samuels, p. 47). The K ration consisted of a Breakfast Unit, Dinner Unit and Supper Unit. The components are listed below (Porges, p. 74–76).

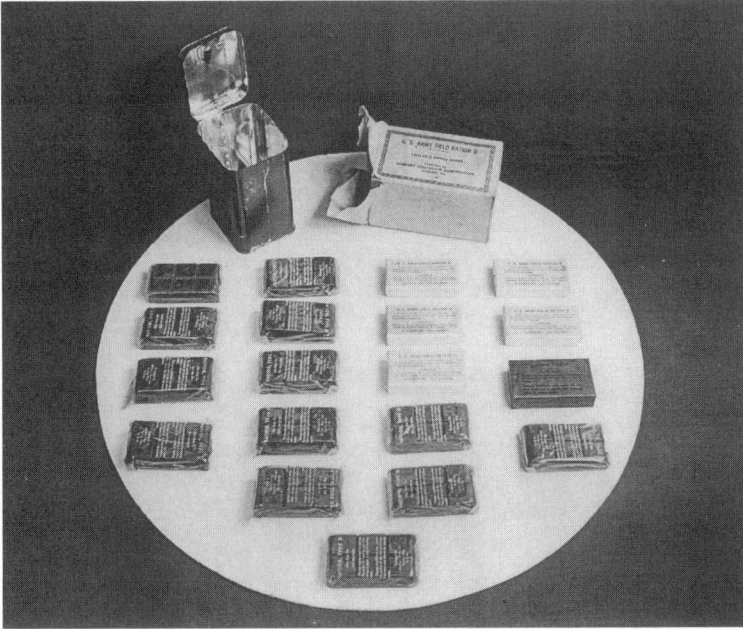

Fig. 15.6   D Rations – World War II [US Army photo (1959)].

Fig. 15.7   K Rations – World War II [US Army photo (1959)].

BREAKFAST UNIT	DINNER UNIT	SUPPER UNIT
K-1 biscuits	K-1 biscuits	K-1 biscuits
K-2 biscuits	K-2 biscuits	K-2 biscuits
Meat	Cheese	Meat
Fruit bar	Hard candy tablets	Chocolate bar (D-rat)
Coffee, soluble	Lemon powder synthetic	Bouillon powder
Sugar cubes	Sugar cubes	Cigarettes
Cigarettes	Cigarettes	Chewing gum
Chewing gum	Chewing gum	Key, can
	Matches	
	Key, can	

Vietnam War

Even as the troops in Vietnam were eating Korean-era C-rations, the next generation of rations was already in the developmental stages. The new mobility and fluidity of the battlefield required smaller, more mobile units. The Army doctrine for field feeding changed drastically in the early 1960s to accommodate this new tactical configuration. Army leadership called for a shift from group feeding to individual feeding, and from issuing individual rations to issuing individual meals (Baker, p. 6).

The first step in this plan was to reconfigure the current menu from rations to meals. The C-ration was replaced with the Meal, Combat, Individual or MCI (Fig. 15.8). Technically it took three MCIs to constitute a ration, but the components were almost identical to the old C rations, and therefore the name C ration stuck to the MCI for its 15-year existence. The MCI was

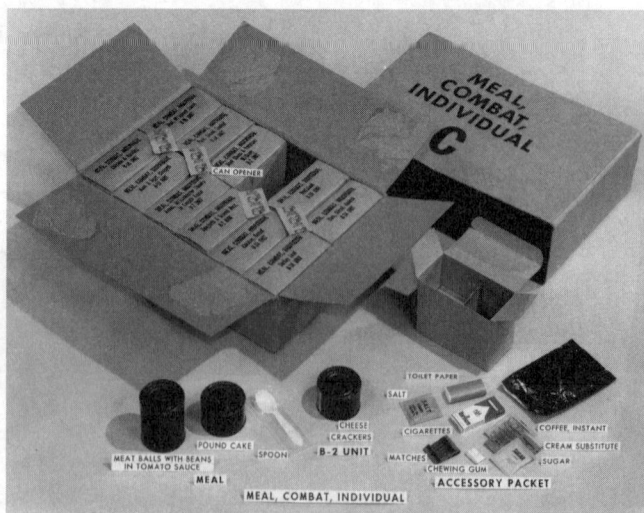

Fig. 15.8 Meal, Combat, Individual – Vietnam [US Army photo (1963)].

certainly an improved version of the C ration. With 12 different meals per case, and a greater variety of canned meats, the acceptability was higher with less monotony and menu fatigue. Like the C-ration each meal consisted of 'a canned meat, a canned fruit, bread or desert, and an accessory packet containing cigarettes, matches, chewing gum, toilet paper, coffee, cream, sugar, salt, and a spoon' (US Armed Forces Food and Container Institute 8).

The MCI, however, was only to be the temporary precursor to the Meal, Ready-to-Eat, Individual (MRE). Established as the individual ration of choice, the MRE was to be a highly acceptable meal that was based on all of the latest research, i.e. flexible packaging, irradiation, freeze-dehydration, and aseptic filling. Targeted for fielding in 1967, the MRE was to be lightweight and as the name implies, ready-to-eat with no preparation required other than the hydration of some components (US Armed Forces Food and Container Institute 30). The ultimate rejection of gamma irradiation as a viable food manufacturing process and delays in obtaining a complete and acceptable menu cycle deferred the actual fielding of this ration until 1979, well after US troops were withdrawn from Vietnam.

Another concept that took longer than anticipated to come to fruition was group feeding using modular, 'factory-assembled' packages that contain all of the necessary components to feed a specified group of soldiers. This future feeding concept, provided in the 1963 publication entitled Operational Rations Current and Future, consisted of the Meal, Uncooked, 25-Man for group feeding when cooking facilities were available. This ration was to contain non-perishable food items sufficient to feed 25 men with minimal supplementation. It would replace the B ration. Likewise, the Meal, Quick-Serve, 6-Man and 25-Man were to contain 'primarily precooked dehydrated components packed in 6 and 25-man modules' for simple preparation and group feeding when kitchens or cooks are not available (US Armed Forces Food and Container Institute 27–28). These rations were never fielded. The same modular concept, however, is being used for current field feeding using Unitized Group Rations.

One type of ration was born of necessity during the Vietnam War. Owing to the guerrilla tactics and jungle warfare in Vietnam, Special Operations Forces were needed to pursue and attack small enemy elements. Many of these small units had to break off pursuit owing to lack of food. They simply could not carry more than a few days' rations with them during hot pursuit of the enemy (Bibber, p. 25). The Food, Packet, Long Range Patrol (LRP) was developed to meet this need. The LRP was a restricted use ration, nutritionally insufficient for extended sustainment, but lightweight and calorically dense enough to fulfill the needs of the Special Forces. The LRP took advantage of freeze-dehydration technology, incorporating a number of dehydrated-compressed components. It also incorporated several compressed cookie bars. Similar to the D ration of World War II, these bars were small in size but packed with calories. The LRP was initially procured in 1968 and purchased through 1983. After a ten-year lull, the item was re-engineered and has been purchased intermittently since 1993 (US, US Army, NRDEC, Natick Pam 30-25 21).

Post Vietnam
As previously discussed, the MRE was the individual ration that finally replaced the tin cans of the C ration with a lightweight, flexible retort pouch. The MRE was the result of almost 15 years of extensive research into the barrier properties of various packaging materials, as well as the acceptability and stability of freeze-dried meats and fruits. Not only did this ration have to be functional, but with the ingress of the all-volunteer Army, developers emphasized improvement of palatability more than ever before. The original MRE consisted, like its predecessor, of 12 different meals per shipping container. The outer package was a durable, brown plastic package that was challenging to open despite the tear notch incorporated into the bag. Inside the meal were a variety of entrées ranging from meat and vegetable stews in retort pouches to freeze-dehydrated meat patties. Each meal also contained a fruit or cake item, crackers, spread, beverage base or cocoa powder, and an accessory packet. This ration has undergone extensive changes, which will be identified in the discussion of the Fielded Individual Ration Improvement Program (US, US Army, NRDEC, Natick Pam 30-2 7-8).

A new concept of group feeding also began to take shape at this time. The module concept of the 1960s was targeted as the preferred method of group feeding. Toward the end of the Vietnam conflict, B ration modules had been configured and fielded for various sized organizational elements, particularly the 100-man unitized B ration (Fig. 15.9). This ration, however, had several drawbacks. First and foremost, it required several food service personnel and

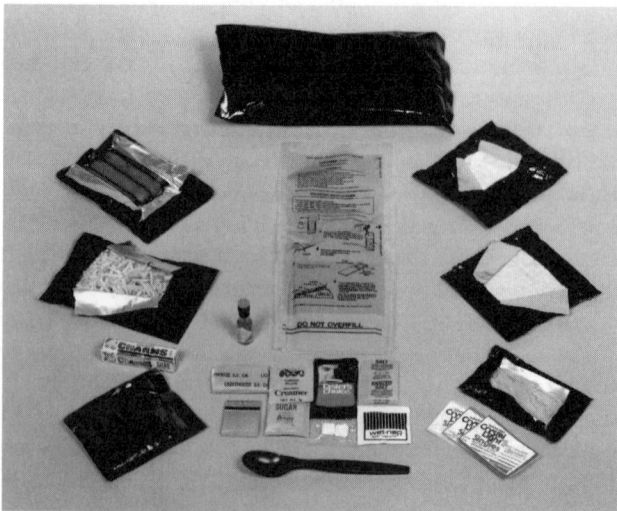

Fig. 15.9   Meal, Ready-to-Eat – Post Vietnam [US Army photo (1992)].

over 75 gal of water for preparation. Plans were made to move toward a more heat-and-serve group feeding ration. Despite the quality and lower weight of dehydrated items, it was decided that retorted meats, starches and vegetables would be used to feed groups of soldiers. A new package, the half-steam-table-tray can was the cornerstone of this new ration. This T ration was ultimately unitized in boxes containing everything necessary to feed a meal to 36 soldiers. After fielding, owing to feedback about excessive waste, the T ration was reconfigured to an 18-man module (US, US Army, NRDEC, Natick Pam 30-2 1-6).

The next development in modular group feeding was the introduction of the Unitized Group Ration, or UGR. Like the T ration, the UGR contains everything needed to provide a group meal, from condiments to cutlery. The configuration, however, is in 50-man modules. Each UGR module consists of three cases, with two modules fitting on a single layer on a standard shipping pallet. Four layers of UGR comprise a palletized unit load and, therefore, provide a unitized ration that will feed 200 troops. The UGR is available in a heat-and-serve (H&S) configuration, the UGR H&S, that uses all semi-perishable products, including tray cans, cylindrical cans and a limited amount of dehydrated products. There is an uncooked UGR configuration, the UGR-A, which includes several perishable components requiring refrigeration and skilled preparation. There is also a cook-prepared shelf-stable option using B ration components, which has been used by the Marine Corps. One of the distinctions of the UGR is its maximum reliance on off-the-shelf commercial components (US, US Army, NRDEC, Natick Pam 30-25).

Other members of the current family of rations include special purpose rations and survival rations. As the name would indicate, special purpose rations are specifically designed for a particular consumer, such as liquid dental rations, or targeted for use in a specific operational environment such as the Ration, Cold Weather and the Ration, Long-Range-Patrol. The latter two rations consist of high-caloric-density, low-weight components. They were scheduled to be combined in 2000 to the Meal, Cold Weather/Food Packet Long Range Patrol. The same components will be packaged into either white or brown menu bags depending on the application (Grabowski). There are currently three types of survival rations procured; General Purpose, Abandon Ship and Aircraft Life Raft. Each of these are restricted use items to be used only under emergency conditions (US, US Army, NRDEC, Natick Pam 30-25).

15.5.2 Progress and problems

World War I

In 1907, the inception of military food research began at the Chicago Quartermaster Depot. The depot commander used this location as an informal school for commissary officers and it eventually became a fully fledged research facility. From 1907 to 1946, this Chicago location was known as the Chicago Quartermaster Depot, the Subsistence Branch of the Supply Division, the Subsistence Research & Development Laboratory, and the Quartermaster Food and Container Institute for the Armed Forces. Regardless of its name, the basic mission was that of coordinating efforts with industry and academia, adapting commercial processes to military applications, and developing more acceptable and functional military rations (Risch, p. 175). Between World Wars, the open lines of communication and established familiarity allowed a rapid assemblage of the research community at the onset of World War II. This link between government, industry and academia continues to be responsible for many recent developments in operational rations.

World War II
Originally designed as a replacement for the 20-year-old reserve ration, the C ration was meant only for temporary feeding during combat operations or at times when A or B rations were unavailable. Many troops, however, sometimes ate C rations for up to 90 days. This, along with a very limited menu selection, contributed greatly to the general dislike for the ration. Often, owing to the unavailability of various meat items, buyers would authorize substitutions with duplicate components. This was exacerbated by the fact that many of the canned meat items were also used in B rations, the only difference being the size of the can. All these factors led to severe menu fatigue. Soldiers not only had to rely on a nutritionally inadequate ration, but they often did not consume all components owing to the monotony of the menu (Risch, p. 182–183).

The unavailability of meat items was simply a matter of the market being overwhelmed with demand. The production surge demanded by the government was astronomical. From 1941 to 1942, government purchases of canned meat items rose from 75.9 million pounds to 920.5 million pounds – a 12 fold increase! The peak of wartime purchasing was between 1944 and 1945. During those two years, 2.09 billion pounds of canned beef were purchased. It is no wonder production could not keep up with requirements (Meat Serves, p. 153).

Evaporated milk was another item in high demand during World War II. Even though evaporated milk had been available long before World War II, it never came close to replacing fresh milk in the diet of the regular American consumer. However, the military demand for suitable dairy products like evaporated milk resulted in the military 'issue of amounts that were seven times that sold on the civilian market' (Remaley, Stolz and Hening, p. 16).

During World War II, the Army Quartermaster Food and Container Institute at its Subsistence Research and Development Laboratory performed extensive research on the improvement of commercially available evaporated milk. The packaging and packing of the product had to be improved to withstand the rough handling associated with the military supply system. The product's sterility had to be further enhanced in order to maintain adequate shelf life in the extreme environmental conditions of locations from the ice fields of Norway to the tropics of the South Pacific. Efforts were also made to improve the palatability of the product. The burnt flavor meant little to the housewife who used the product for cooking, but as a beverage it left much to be desired. The product was improved, but this problem of undesirable flavor was never fully solved. (Remaley, Stolz and Hening 17–26).

After the necessary changes were made, the result was a more expensive, military unique product with little or no commercial application. Unfortunately, this is often the case when the military attempts to use off-the-shelf commercial items. The necessary modifications to suit the military applications create a military–commercial hybrid that inevitably costs more than the original product. The advantages of this dilemma, however, are manifold. For the military, the leveraging of initial product research and development costs allows more concentration on military adaptation. Also, the availability of familiar products, albeit not identical to their commercial counterparts, provides a boost of morale to the soldier. From an industry perspective, many of the enhancements made by the military scientists led to improvements of their commercial products. In many respects, the soldiers and sailors who ate their products provided a diverse test market, and feedback was used to refine and improve their merchandise.

World War II through Vietnam
Thorough historical records recount the problems associated with military rations and field feeding during World War II. The Subsistence Research and Development Laboratory at the Quartermaster Food and Container Institute in Chicago published a 12-volume study entitled, *A Report of Wartime Problems in Subsistence Research and Development*. Subjects ranging from packaging development to fresh fruits and vegetables were discussed exhaustively, and recommendations were provided for future improvements of the entire ration research and development program. Several improvements were made during the war, and many more fell somewhere between the initial planning stages and field testing when the war came to an end. Some wartime problems required immediate attention (Conway and Meyer).

At one point, the initial C ration was completely rejected by soldiers in the Pacific theater, and Army researchers scrambled to redesign the ration with a more varied menu and greater variety of entrées. Researchers at the Subsistence Laboratory had always been aware of the extreme monotony of a menu composed of stew, hash, meat and beans, especially when those same items were duplicated in the B ration menu. The problem was not one of ration development but of procurement and production. The meat industry had to provide huge numbers of canned meat items, and the only items available in quantities large enough to meet the needs of the wartime surge were those few canned items that collectively caused the World War II soldier to utter the word 'Spam' with disdain. The ration, which consisted of 3-M units and 3-B units (all the same) in 1941, was totally re-engineered to include 10 M-units and 6 B-units in 1944. Unfortunately, it was not until the winter of 1945 that the stockpile of old C rations was eaten and the new, improved C ration started to be issued to units in the field (McDevitt and Samuels, p. 27–38).

Another problem that surfaced at the end of World War II was the great loss of expertise from the military subsistence research and development program as many staff members were discharged from service. Colonel Rohland Isker, the Commander of the Quartermaster Food and Container Institute at the end of World War II recognized this problem. In 1947, after he also retired from the military, Colonel Isker formed an organization called the Research and Development Associates for Military Food and Packaging Systems. This organization, comprised of distinguished members of industry, government and academia, was formed to provide collaborative research and development, foster communication, provide expertise on innovative technologies and resolve disputes regarding technical and analytical issues. 'In 1947 a reception was held in Washington, DC to kick off the founding of R&D Associates. In attendance at the reception were the President of the United States, the Chiefs of Staff of the Army and Air Force, the Chief of Naval Operations, other senior military officers and many Senators and Congressmen' (Research and Development Associates 219).

Research continued between World War II and the Korean Conflict, but little changed in the overall composition and classification of operational rations. Field soldiers in Korea were relegated to eating down the surplus of World War II C rations. Improvements in transportation and logistical infrastructure in Japan greatly improved the variety of rations that were available to Corps elements, but line units had to rely on C rations and B rations. Not only did Spam continue as the brunt of jokes, but other items, such as powdered eggs, also gained a reputation as artificial and tasteless. Spray drying of milk and egg products allowed a wider distribution of these staples, but the gross changes in flavor associated with the drying process met with very low acceptability. Improper preparation in many cases also greatly contributed to the poor reputation of these products (Remaley, Stolz and Hening, p. 40).

Post Vietnam
For 30 years, the MRE has been the only individual ration designed for unrestricted field feeding, yet it has still undergone many changes. In fact, there have been more improvements made to the MRE than to any other individual ration in the field feeding system, all of which have been consistently consumer oriented. Field tests are conducted yearly to identify the acceptability of the various ration components, as well as to judge the potential for proposed substitutions. The Fielded Individual Ration Improvement Program is a continuing research project that uses consumer feedback to identify potential changes to the MRE.

Since 1986, the MRE has changed components, packaging or both in every year of production. The first to go from the MRE were the freeze-dehydrated meat items, which were replaced by thermal-stabilized meat products deemed more acceptable during taste panels and consumer testing (US, US Army, NRDEC, Natick Pam 30-2 7-8). Other early improvements included the addition of commercially packaged candies and the eventual replacement of freeze-dried fruits with thermal-stabilized fruits. Also added were four vegetarian entrées and several commercially packaged snack items. Another popular inclusion was the flameless ration heater. This small and inexpensive component produces an exothermic chemical reaction when activated with a small amount of water. The heater allows the solder to have a hot meal, as it raises the temperature of an MRE entrée by 100 °F (38 °C) in 10 min.

Since its inception, the MRE menu has been expanded, first to 18 meals and more recently to 24 different menus. This was driven by an attempt to avoid menu monotony. Eating the same product over and over tends to decrease consumption and detracts from the overall acceptability of the ration. Having a greater selection of entrées and secondary components increases variety and acceptance for those times when the MRE is the only source of food. Even with a greatly expanded menu, it does not take long for Warfighters to identify their favorites. As part of an MRE survey with the 15th Marine Expeditionary Unit, it was discovered that there was an informal 'MRE matrix' which Marines constructed and memorized to identify which snacks were in which menus (Darsch and Evangelos, 'Operational Rations in Combat', p. 31). Individual preferences continue to drive a system of barter and trade among the military consumer. It is said that a pouch of Jalapeño Cheese Spread is a very valuable commodity on the battlefield.

Packaging defects have always plagued the MRE. Inspection data from the in-plant inspectors show that the vast majority of production lots that fail acceptance inspection do so because of holes, abrasions or leaking seals in the pouches. The first several years of commercial production of the MRE were marred by the recall and testing of many retort entrées owing to microscopic leaks in the retort pouch material. An extensive testing program was developed to test all production lots of the flexible retort pouches. Inspectors took samples from each production lot, cut open the pouches, rinsed them and applied fluorescent dye to the interior of the pouch. Pouch exteriors were then examined under a black light to identify any pinhole leaks. This exhaustive process lasted for over two years. Finally, producers of the pouch material, as well as the food processors, identified and corrected most of the causes of the leaks. As mentioned during the discussion of ration packing and packaging, a newer pouch material that includes a nylon layer has alleviated this problem (Boyd).

A trend throughout government contracting that has also extended to rations, and provided its share of challenges, is to do away with detailed government specifications and rely more on commercial, off-the-shelf products. This has been good for the ration program overall, actually lowering the cost of the ration and including familiar and popular brand-name components. Unfortunately, there are very few truly 'commercial' items that meet the three-year military shelf life requirement. Even though there are many ration components that appear to be off-the-shelf items, a recent procurement test found that a large number of the so-called commercial products were unique to the military either with regard to portion size, packaging or formulation.

Persian Gulf War

The first major deployment since Vietnam posed a number of challenges. Between late 1990 and early 1991, over 500 000 troops were deployed to the Persian Gulf region. Transportation assets were strained to near capacity, as was the surge capability of the operational rations industrial base. The Army Surgeon General had approved the MRE as the sole source of military feeding for not more than 21 days, yet many soldiers ate MREs and nothing else for well over a month. Before long, the stored reserves of MREs were depleted. Even though the MRE assembly plants expanded to around-the-clock operations, shortages in components and especially packaging materials prevented the shipment of rations in the quantities needed. This prompted military acquisition personnel to quickly identify an alternative ration to make up this shortfall. The three-year shelf life was not an issue since the rations would be consumed almost immediately, so it was relatively easy to gather commercially available, shelf-stable components for the temporary ration.

The Meal, Ordered, Ready-to-Eat, or MORE as it was called, consisted of a Top Shelf brand entrée that was supplemented with some ration components and some commercial snacks. These were sealed into a clear plastic menu bag, cased, and palletized. One huge problem with this ration, very similar to the problems of early World War II rations, was the lack of variety of entrées. Only a few were available in the quantities necessary. To compound this, one particular entrée was assembled until exhausted, meaning that all cases within a shipping van contained the same entrée. Several horror stories were told of units that would have an entire van load of food arrive, only to find out they would be eating nothing but Top Shelf Lasagna for breakfast, lunch and dinner.

Even the available supply of A ration items was stretched to its limits. Several refrigerated merchant marine cargo vessels were loaded with frozen meats and vegetables and went steaming toward the Gulf. Before long, there was not enough product that met military specifications left in supply channels. Military requirements were waived when items such as frozen ground beef patties had to be obtained from restaurant and hotel distributors.

Another problem was the extremely high temperatures of the region. Several vanloads of rations were lost in port owing to missing documentation. After only a few weeks in the scorching Arabian sun, these vans became ovens that quickly degraded the quality of the rations inside. This prompted an Army Logistics initiative to install radio telemetry on all sea van containers to identify its contents and location at any given time.

Eventually, a commercial contractor was designated to provide the food and operate the foodservice facilities at American bases in the Persian Gulf. This proved to be very effective, and the same type of contract feeding was provided to soldiers in the more recent Balkans conflict. It was very effective in that, after only 30 days of eating MREs, soldiers in Albania were provided hot cooked A ration meals in contractor-operated dining facilities. Although very practical in these cases, contracted field feeding is very dependent on the intensity of the conflict and the duration of hostilities. Like the sutlers of the Civil War, contractors and their employees may not be so quick to enter a volatile and hostile combat situation (Enriquez).

Another lesson learned during the Persian Gulf War was that the diversity of soldiers in the American military included soldiers who required halal and Jewish kosher certified food. These standards require strict oversight of animal slaughter and food processing operations by religious clerics. Attempts were made in 1992 to incorporate dual certified kosher/halal menus into the MRE, but the project was terminated owing to conflicts in different certifying authorities, prohibitive costs, and industry lobbying efforts. Instead, the same items were dubbed 'vegetarian' MRE menus and, although not certified, would be inoffensive and acceptable to all but a few orthodox soldiers. Separate small purchases of certified meals are made available to unit commanders at their discretion. These kosher and halal meals, although shelf-stable, do not meet the stringent packaging and shelf life requirements of the MRE.

15.5.3 Processing technologies

World War II through Vietnam

Even as the Korean conflict was at its height, army researchers began work on a process that they hoped would eventually replace both canned and dried ration components. Ever since the discovery of radioactivity in 1895, scientists had conducted research on its use as a food preservation technique. Preliminary work on food irradiation was performed in 1953 at the Quartermaster Food and Container Institute in Chicago. Initial efforts to use gamma irradiation for food preservation ran into several obstacles. The worst problem to overcome was the appreciable change in flavor and odor that accompanied sterilizing doses for most meat products. Through years of experimentation, this problem was overcome, or at least alleviated, by reducing the dose to the absolute minimum necessary to achieve commercial sterility (Simpson 1). Irradiation, in combination with freeze-dehydration, was targeted as the ideal processing technology for military rations. Even then, however, proponents knew that the stigma of the term 'radiation' was an obstacle to be overcome.

Although scientifically proven to be effective and superior to existing processes, irradiation has still not been used for any military ration. Several irradiation sterilized products have been developed by the Natick facility, used for astronaut feeding and found highly acceptable (Klicka and Smith,

p. 29), but the lack of public acceptance still hinders the exploitation of this food-processing technology. Even with an estimated savings of more than one billion dollars, army leadership continues to wait for the commercial acceptance that they tried so hard to foster over four decades ago (Loveridge). Although some have prematurely predicted the ultimate demise of this technology (Graham, p. 10), public concern over foodborne illness may eventually prompt renewed interest in the use of irradiation as public safety measure (Pehanich, p. 13).

Another food-processing method that the military sought to employ for the next generation of operational rations was freeze-dehydration. Even though the experience of World War II and Korea showed that dehydrated products were of questionable acceptability, and it was well known that the commercial sector had all but rejected dehydration as a viable process, the army still encouraged its proliferation. In September 1960, a government and industry meeting, co-sponsored by the Quartermaster Food and Container Institute and the Research and Development Associates, was held to explain the military's projected need of this technology and to bolster the support of the operational ration industrial base. Only nine months later, an international conference on the freeze-drying of foods was held in Chicago. This highly technical scientific gathering was meant to 'lead to a more attractive economic outlook for freeze-dried foods' (Brockmann v). Although not accepted in the United States as a commercially feasible production method, extensive worldwide research provided a sufficient knowledge base for a small group of defense contractors to venture into this niche market.

Unlike irradiation, freeze-dried meals actually did gain a small foothold in both the commercial and military markets, at least temporarily. The Quartermaster Food and Container Institute developed and field-tested several menus including 'freeze-dried components such as shrimp, hamburgers with mushroom sauce, cottage cheese, and peach slices in syrup'. Meanwhile, both Campbell Soup Company and Armour & Company began test marketing branded freeze-dried products (Simpson). The dehydrated meals were not adopted as an individual meal, but many of those same freeze-dehydrated components were procured to add variety to B rations. Some were even precursors of the short-lived dehydrated entrées in the first versions of the MRE.

15.5.4   Modern challenges for military meal development
Logistical challenges: the Balkans Conflict
Various peacekeeping and peacemaking opportunities arose during the 1990s. The most logistically challenging of these were the two missions to the Balkan region of East–Central Europe (Fig. 15.10). Militarily, the operations were fairly simple owing to the heavy reliance on air power. This new doctrine, a derivative of the Gulf War, allowed the delayed entry of ground forces into a relatively non-hostile environment. The main logistical challenge

Fig. 15.10   Logistical challenges in Bosnia [Dept of Defense jpeg image (1995)].

was geographical, that is, to engineer a large deployment to a mountainous region with relatively little infrastructure. Improvements in transportation, especially airlift capability, eased this burden. There were, however, several difficulties owing to adverse terrain and weather conditions.

Along with peacekeeping operations came the burden of refugee safekeeping. The military ration producers were asked to develop and manufacture a shelf-stable, packaged ration to meet the needs of these malnourished individuals. The ration community met that challenge, producing and fielding the Humanitarian Daily Ration (HDR). This ration, originally produced for the military operation in Somalia, contains no meat products, which ensures widespread cultural acceptability. It is also much less expensive than the MRE. The US Department of State has authorized the fielding of this ration, marked 'A Food Gift from the People of the United States', to refugees and displaced persons all over the world. Somalia, Rwanda, Cambodia, Bosnia, Kosovo and East Timor are some of the places that this non-military ration has been used.

Political challenges: acquisition reform
Another development of the 1990s that greatly affected the way rations were purchased was an initiative called 'Reinventing Government'. A significant element of this initiative was the transition from the traditional detailed specification to performance-based requirements. The former, as the name implies, provided the exact details of how a ration was to be produced including ingredients, formulation, processing parameters and packaging requirements. The new performance-based standards merely gave the end item performance requirements. The contractor was then responsible for all aspects of production, including quality assurance standards, and the government would merely verify that the delivered end item would perform according to the standards.

This process, while sound enough in principle, was somewhat disturbing to the government agencies and the defense contractors involved in the operational rations program. Government acquisition personnel feared the loss of standardization and the inevitable supply failure if defective product infiltrated supply channels. Contractors hesitated to take responsibility for what had always been a burden of the government. Much of the operational rations industrial base lacked sufficient research and development resources to ensure that products would meet the required shelf life, rough-handling, and nutritional requirements (Moody, p. 285). Eventually, however, the operational ration acquisition community was able to adjust to the new performance based requirements, in part owing to the fact that formulations and processing parameters were described in government regulations as 'optional' or 'recommended', but not required.

15.6   Current research and development for military meals

15.6.1   Newest types of rations

The highly acceptable and nutritionally complete rations of today would have truly amazed the soldiers of the previous centuries. Innovation and technology have enabled huge improvements in the military meal. Ration developers of today are taking a visionary approach to identify those future technologies that will enable even more improvement in the care and feeding of the Warfighter of tomorrow (Darsch and Evangelos, 'Care & Feeding', p. 155). The most recent conflicts in Afghanistan and Iraq have generated requirements for new individual and group feeding rations. Fortunately, a robust research and development program was in place to respond quickly and meet the needs that were identified.

First strike ration
The First Strike Ration (FSR) was developed to meet the needs of Warfighters involved in highly mobile, highly intense combat operations (Fig. 15.11). In instances where Warfighters were deployed on two or three day missions, they would be issued the appropriate number of MREs (three per day) as their only supply of food. Challenged by the weight and volume of food along with all of their other gear, they would often open up the MRE packets, removing the components that were ready to consume on the move or easy to carry, and discarding the rest. This process, known as field stripping an MRE, resulted in unacceptably low nutrition at a time when it was needed the most.

The First Strike Ration was developed to alleviate this problem. In fact, the concept of the FSR had been developed long before this need was articulated. Its rapid development and accelerated deployment was a direct result of the

Fig. 15.11   First Strike Ration [US Army photo (2006)].

visionary approach researchers had been taking, trying to anticipate the needs on future battlefields, instead of relying on solutions designed for past conflicts (Darsch and Evangelos, 'Care & Feeding', p. 162).

One FSR contains an entire day's food and consists of eat-out-of-hand, highly acceptable components that are dense in calories and require no preparation. One FSR is about half the weight and volume of the three MREs that it is meant to replace for these short duration missions. It is considered a restricted ration, in that a complete day's ration packet contains approximately 3000 kcal – less than the 3600 required by the Surgeon General. However, compared with field stripping, the FSR provides many more calories because most of its components are much more likely to be consumed.

The FSR relies heavily upon the concept of hurdle technology. Rather than traditional thermal processing, which relies on a high-heat kill step, many FSR components are rendered shelf stable by tightly controlling the product and packaging parameters to preclude the growth of any pathogenic organisms. By controlling the water activity and pH of the product, and excluding oxygen from its package, components such as the shelf stable pocket sandwich are able to attain a shelf life of two years or more.

Unitized Group Ration-Express
The Unitized Group Ration-Express (UGR-E) is another military ration born of necessity (Fig. 15.12). In those instances where small groups are deployed far from field kitchens, their field feeding options are greatly limited. They must subsist solely on MREs, have food brought to them in insulated food containers, or obtain food from the local economy. Each of these options has significant drawbacks. The MRE is not meant to be used as a sole source of food for more than 21 days because monotony will cause a significant drop in consumption. Delivery of meals along often dangerous supply routes

Fig. 15.12 Unitized Group Ration-Express [US Army photo (2005)].

poses great risk to those delivering the food. Consumption of local foods greatly increased the risk of contracting a food-borne illness.

The UGR-E was designed as a 'kitchen-in-a-carton'. This group-feeding ration uses the same polymeric trays of food designed for the UGR-H&S ration, and integrates those trays with a chemical heating system. Each UGR-E contains four trays: an entrée, a vegetable, a starch, and a dessert. The box also contains everything necessary to feed 18 Warfighters, including serving trays, utensils, snacks, and beverage mixes. This one box is engineered so that it can be delivered to remote locations that do not have organic cooks or kitchens. About 40 min before the group is to be fed, a tab is pulled that activates the heaters. This heats the food in the trays, and the unit can serve themselves a hot, group meal.

15.6.2 Processing technologies

Almost fifty years after the army was made executive agent for research into the viability of food irradiation, the technology is finally beginning to gain public acceptance as a safe, alternative food process. One of the biggest barriers to irradiation was the validation of packaging materials to ensure that there is no transfer of substances between the packaging and food during the irradiation process. Army research provided the critical data necessary to approve the packaging materials that are used today in commercial irradiation facilities. Today, the Natick Soldier Research Development and Engineering Center continues to develop irradiated menu items for the NASA space program. The items are formulated and packaged at Natick, then sent to a commercial irradiation plant for processing. Army researchers still perform acceptability testing of these high-dose irradiated products before fielding new menus to NASA. As consumer acceptance grows, the army may once again consider this technology for the processing of operational ration components.

Military researchers today are looking toward other innovative technologies that have potential future application for the production of operational rations. The scientists in the Department of Defense Combat Feeding Program realize that, like irradiation, the only hope of using these new technologies is if they have commercial applications. The processes must be accepted by both the food-processing industry and the consumer as efficient, safe and effective. Currently, research on several innovative processing technologies is being conducted at Natick through cooperative efforts with industry and academia. Current research aims toward non-thermal processes, such as high hydrostatic pressure processing (HPP) and microwave sterilization, as having great potential for use in the operational rations program, as well as other commercial applications (Dunne). The potential for use of HPP for food processing has been studied for well over a decade, and studies have shown that while high pressure has little effect on bacterial spores, it significantly reduces the number of viable non-sporeforming bacteria in a variety of products. Several acidified HHP processed products have been developed and evaluated at Natick through the aforementioned cooperative efforts. Based in part on the military research, several large ready-to-eat meat processors have turned to HPP as a pasteurization step for the elimination of *Listeria monocytogenes*. More recently, the first petition for an HPP commercially sterile low-acid canned food has been submitted to the FDA.

Another focus of current research is to improve packaging materials so that they not only remain durable but have increased ability for atmospheric manipulation inside the package. Until now, oxygen scavenger sachets have been used to extend the shelf life of savory snack items. Recent research, though, has identified packaging materials that have oxygen-absorbing qualities built into them. The polymeric tray that replaced the original metal tray cans are being evaluated to determine whether they can be manufactured with special coatings incorporated into the tray material that scavenge oxygen from the container (Trottier).

Other technologies in the current research program include microwave sterilization, non-invasive shock wave tenderization, radio frequency (RF) processing and osmotic moisture removal. Research into the efficacy of these processes is being conducted in government laboratories through cooperative efforts between government agencies, at universities through partnerships with small business initiatives, and in the industrial setting as a part of cooperative agreements. Scientists at the Natick Soldier Research, Development and Engineering Center will help to develop strategies for the scale up of current pilot-scale systems to commercial processes and develop continuous microwave processing parameters for potential use in the production of military rations (Yang).

15.6.3   Phytonutrients and other performance-enhancing compounds

Another area of operational rations research is the identification of compounds with the potential to increase physical and mental performance. Combat is often characterized by high stress, sleep deprivation and extreme physical exertion that significantly decrease both physical endurance and mental acuity. Research into the identification and delivery of appropriate nutraceutical products could help to counter the degradation of combat capability. This research includes multiple disciplines. Ongoing medical research at the United States Army Institute for Environmental Medicine seeks to identify the most effective compounds. Studies have included the effects of stimulants such as caffeine, metabolic antioxidants, and immune and muscle function promoters (Liebermann).

Meanwhile, food scientists have been engineering potential ration components that could be used for the delivery of performance enhancing ingredients. The 'First Strike Bar', a Natick engineered product similar to commercial sports bars, is currently targeted as one such delivery vehicle. The bar can easily be used to supplement individual rations and may be used in situations that require a particular nutritional fortification. Another Natick engineered product, the Ergo Drink, is a powdered drink mix planned for the provision of supplemental carbohydrates when soldiers are under conditions of continual strenuous activity. Unlike currently available commercial sports drinks, the Ergo drink contains nitric oxide (NO) electrolytes but has a unique combination of polysaccharides, providing a maximum duration of metabolic benefit. Research involving micro-encapsulation is also being performed to provide preservation and isolation of various performance enhancing ingredients (Aylward).

15.6.4   Effect of changing doctrine and operations on military meals

As the military's vision of the future battle field changes, so does the need to feed those who will be fighting the wars of tomorrow. Lighter weight, lower volume rations are a must to maintain the high mobility and agility envisioned for future Warfighters. More technology on the battlefield will require improved cognizance. Cutting edge food and nutrition technologies will be developed to increase mental alertness, enhance cognitive abilities, and reduce combat stress during sustained operations (Manguel).

15.7   Summary and conclusions

Every major military conflict in which the United States military has been involved has been characterized by shortfalls in the adequacy, quality and/or availability of operational rations. Problems such as a limited industrial base, inadequate funding in peacetime, competition for transportation assets and military unique requirements have plagued the army since its inception. Others, like the potential use of biological and chemical weapons, are more recent. Every conflict provided difficult situations to be overcome and problems to be addressed, and for the most part, changes were made to correct those problems. Unfortunately, this resulted in a situation where the rations at the beginning of each war were actually designed to meet the needs of the previous war.

In most cases, the commercial development of new food-processing technologies helped to improve the shelf life, packaging, nutritional adequacy and acceptability of military rations. Usually these technologies had to be adapted or tailored in some way to meet the unique military requirements, but, ultimately, they resulted in the improvement of the soldiers' ration. More recently, as in the case of food irradiation, it was the efforts of military research that provided commercial viability for an innovative technology. Current cooperative efforts attempt to make best use of the research and development resources of government, industry and academia to provide the maximum benefit to both military and commercial sectors. The emphasis on Cooperative Research and Development Agreements (CRADA) and Dual Use Science and Technology (DUST) projects are expected to produce advancements in areas that will be of benefit to both military field feeding and the commercial food industry. The ultimate result of this win-win situation is a safer, less expensive and higher quality food supply.

In the last 50 years, there has been a tremendous amount of foresight in the army's ration research and development program. Even in times of relative peace, the military research community has looked toward the most innovative and practicable technologies to enhance the usability of the family of rations. New rations have been developed to meet the needs of the warfighting community, even in the absence of major military mobilizations. The continual improvement of existing rations ensures an ample inventory of highly acceptable subsistence. Meanwhile, the increase of resources through partnerships with industry and academia generates significant research into key areas to meet the science and technology objectives of the Department of Defense, while simultaneously promoting potentially effective commercial technologies.

Far-sighted and futuristic ideas are also an integral part of the operational rations research and development program. Just as the irradiation research of the 1950s and 1960s are now benefiting the consumer with a safer alternative for pathogen reduction, so the current research may yield significant public health benefits in the future. Although the outcomes are as yet unforeseen, much thought has gone into the targeting of ever decreasing research dollars toward those initiatives with the greatest potential. As the saying goes, if we knew what the outcome would be, we would not call it research.

The establishment of a focused and centrally funded research program in 1907 at the Chicago Quartermaster Depot was the first step in acknowledging that scientific research was imperative to continually improving military rations. Today, the Department of Defense Combat Feeding Program at the Natick Solider Research, Development and Engineering Center is devoted to food research and ration development, ensuring that the United States Armed Forces are supplied with operational rations that are unequaled in quality and utility. The program relies on its vital cooperative relationships with government agencies, industry, and academia in order to provide the military and the nation as a whole with the valuable research necessary to employ new technologies for the safe and efficient production of food.

15.8   References

'Army to Build New Facility for Irradiated Food Study'. *Fort Devens Dispatch*, 27 Sep 1960.
Aylward, Judith. 'Performance Enhancing Ration Component Update'. *Joint Service Operational Rations Forum*. Quality Inn, Petersburg, 21 September 1999.
Baker, Roy. 'Dehydrated Foods: The US Army's Interest'. *Freeze-Dehydration of Foods: A Military-Industry Meeting*, Chicago, 20-21 September 1960 Champaign: Garrard Press, 1961, 2–7.
Bibber, William. 'General says Food Industry Fails US Guerilla Troops'. *Boston Herald*, 20 Nov 1963, 25.
Billings, John. *Hardtack and Coffee*. Boston: G. M. Smith, 1888.
Boyd, Richard. 'USDA Trends and Analysis'. *Fall Meeting. Research and Development Associates for Military Food and Packaging Systems*. Weston William Penn Hotel, Pittsburgh, 1 Nov 1999.
Brockmann, M. C. 'Forward'. *Freeze-Drying of Foods: Proceedings of a Conference*, Shoreland Hotel, Chicago, 12–14 April 1961. Ed. Frank R. Fisher. Washington: National Academy of Sciences – National Research Council, 1962. *v–vi*.
'Causes of Diarrhea and Dysentery'. *Medical and Surgical History of the Civil War*. 12 Vols. Wilmington: Broadfoot Publishing Co., 1990.
'Chronic Dysentery'. *Medical and Surgical History of the Civil War*. 12 Vols. Wilmington: Broadfoot Publishing Co., 1990.
Conway, Hewitt and Alice Meyer, eds. *Operation Studies Number One: A Report of Wartime Problems in Subsistence Research and Development*. 12 Vols. Chicago: Quartermaster Food and Container Institute for the Armed Forces, 1948.
Cunningham, H. H. *Doctors in Gray*. Gloucester: Peter Smith, 1970.
Daley, Barbara. 'Operational Ration Accelerated Storage Data Correlation and Validation (OPRA)'. *Food & Nutrition Research & Engineering Board Program Build*. Natick: DoD Combat Feeding Program, 1999.
Darsch, Gerald. 'Feeding the force: the new millenium'. *Briefing to a Congressional Staffer*. Natick Soldier Center, 9 November 1999.
Darsch, Gerald and Evangelos, Kathy-Lynn. 'The Care & Feeding of Starship Trooper'. *Activities Report of the R & D Associates* **52.1** (1999): 153–168.
Darsch, Gerald and Evangelos, Kathy-Lynn. 'Operational Rations: Back to the Future'. *Activities Report of the R & D Associates* **40.1** (1997): 17–25.
Darsch, Gerald and Evangelos, Kathy-Lynn. 'Operational Rations in Combat'. *Activities Report of the R & D Associates* **55.1** (2002): 31–50.
Davis, Kimberly. 'Soldier Feeding: The Development of Military Nutrition from the Civil War to World War I'. Thesis. Uniformed Services University of Health Sciences, 1993.

Derr, Donald. 'Irradiation Workgroup'. *Fall Meeting. Research & Development Associates for Military Food and Packaging Systems*. Weston William Penn Hotel, Pittsburgh, 2 Nov 1999.

Dunn, Patrick. 'Acting Chairs Message'. *Nonthermal Processing Division (NPD) Newsletter*, 1.1 (1999): 1–6.

Eggers, Russ. 'CORANET Update'. *Joint Service Operational Rations Forum*. Quality Inn, Petersburg, 21 September 1999.

Enriquez, J. 'Kosovo Logistics Review'. *Joint Service Operational Rations Forum*. Quality Inn, Petersburg. 22 September 1999.

Food Processors Institute. *Canned Foods*. 5th ed. Washington: Food Processors Institute, 1988.

Freidel, Frank. *The Splendid Little War*. Boston: Little Brown & Co, 1958.

Garwood, Darrell. 'A-radiated Bacon Sale by Next Year is Sought'. *Washington Post*, 9 Jul 1962.

Gould, Lewis L. *The Spanish–American War and President McKinley*. Lawrence: University Press of Kansas, 1982.

Grabowski, Chuck. 'DSCP Initiatives for MRE XX'. *Fall Meeting. Research & Development Associates for Military Food and Packaging Systems*. Weston William Penn Hotel, Pittsburgh, 1 Nov 1999.

Graham, Karen. 'Food Irradiation; Does it Have a Place in Today's Food Industry'. *The Frozen Food Digest*, Jul 1993, 10–11.

Hallberg, Linnea. 'Radio Frequency Processing of Combat Ration Components'. *Food & Nutrition Research & Engineering Board Program Build*. Natick: DoD Combat Feeding Program, 1999.

Hegsted, D. Mark. 'Establishment of Nutritional Requirements in Man'. *Borden's Review of Nutrition Research*, Mar–Apr 1959: 13–22.

Hilbert, G.E. 'Processing Trends and the Nutritive Value of Foods'. *National Food and Nutrition Institute, Washington, 8–10 December 1952*. Washington: USDA, 1953, 86–92.

Hoover, D. G., Metrick, C., Papineau, A. M., Farkas, D. F. and Knorr, D. 'Biological effects of high hydrostatic pressure on food organisms'. *Food Technology*, Mar 1989, 99–107.

Isker, Roland. 'Introduction; The Problem'. *Ration Development*, **12**. Chicago: Quartermaster Food & Container Institute for the Armed Forces. 1947. 1–6.

Jacob, H. *Six Thousand Years of Bread*. Garden City, NY: Doubleday, Doran and Company, Inc, 1944.

Johnson, Robert and Robert Kark. *Feeding Problems in Man as Related to Environment*. Quartermaster Food & Container Institute for the Armed Forces, 1946.

Kalchayanand, N. 'Interaction of hydrostatic pressure, time and temperature of pressurization and pediocin AcH on inactivation of foodborne bacteria'. *Journal of Food Protection*, Apr 1998, 425–431.

Keller, Robert. *Flexible Packages for Processed Foods*. QMFCIAF Report No. 31–59. Quartermaster Food & Container Institute for the Armed Forces, 1959.

Klicka, Mary and Malcolm Smith. *Food for US Manned Space Flight*. Natick: Food Engineering Laboratory, 1982.

Kluter, R., Nattress, D. T., Dunne, C. P. and Popper, R. D. 'Shelf life evaluation of Bartlett pears in retort pouches'. *The Journal of Food Science*. Nov/Dec 1996: 1297–1302.

Libermann, J. 'Current Rations Studies'. *Joint Service Operational Rations Forum*. Quality Inn, Petersburg, 22 September 1999.

Levin, William. 'Introductory Remarks'. *Symposium on Feeding the Military Man*. Ed. Martin Peterson. Natick: National Academy of Science – National Research Council, 1970.

Loveridge, Vicky. 'Update on Potential Cost Savings for the Use of Irradiation Processing for the Department of Defense'. Memo to recorded. 28 Mar 1996.

Mainwaring, Donald. 'Adam to Oust Refrigeration?' *Christian Science Monitor* [Boston] 26 Sep 1960.

Manguel, Jesus. 'Feeding the Force: The New Millennium'. *Activities Report of the R & D Associates* **52.2** (1999): 203–217.

Marquez, V. O., Mittal G. S. and Griffiths, M. W. 'Destruction and inhibition of bacterial spores by high voltage pulsed electric field'. *Journal of Food Science*, Mar/Apr 1997, 399–401, 409.

Marshall, Peter and David Manuel. *The Light and the Glory*. Old Tappan: Revell, 1977.

McDevitt, Robert, and John Samuels. 'Packaged Operational Rations'. *Ration Development*. Quartermaster Food & Container Institute for the Armed Forces. 1947. Vol. 12 of *Operation Studies Number One: A Report of Wartime Problems in Subsistence Research and Development*. Hewitt Conway Hewitt and Alice Meyer, eds. 12 Vols. 1948.

'Meat for the Multitudes'. *The National Provisioner*, 4 Jul 1981, 27–257.

'Meat Serves in Wartime'. *The National Provisioner*, 26 Jan 1952, 169–183.

Melson, Robert, James Clark, and Robert Couch. *Subsistence Packaging and Packing*. Quartermaster Food & Container Institute for the Armed Forces. 1947. Vol. 5 of *Operation Studies Number One: A Report of Wartime Problems in Subsistence Research and Development*. Hewitt Conway Hewitt and Alice Meyer, eds. 12 Vols. 1948.

Mermelstein, Neil. 'An overview of the retort pouch in the US'. *Food Technology*, Feb 1996, 28–37.

Mermelstein, Neil. 'Interest in pulsed electric field processing increases'. *Food Technology*, Jan 1998. 81–82.

Metrick, C., Hoover, D. and Farkas, D. F. 'Effects of high hydrostatic pressure on heat sensitive strains of *Salmonella*'. *Journal of Food Science*, Nov/Dec 1989, 1547–1549, 1564.

Miller, Everett. *United States Army in Veterinary Service in World War II*. Washington: Office of Surgeon General, Department of the Army. 1961.

Ming, H., Nandram, B. and Ross, E. 'Bayesian prediction of the shelf life of a military ration with sensory data'. *Journal of Agriculture, Biological, and Environmental Statistics*, 1996, 377–392.

Moody, Stephen M. 'Inspection & Acceptance Under Performance Specifications'. *Activities Report of the R & D Associates* **48.1** (1996): 285–401.

Mrak, Emil. 'Food Research and Development for the Armed Forces'. *Symposium on Feeding the Military Man*. Ed. Martin Peterson. Natick: National Academy of Science – National Research Council, 1970.

'Natick-Developed Irradiated Bacon Approved by Food and Drug Agency'. *Suburban Free Press* [Boston] 21 Feb 1963.

Pehanich, Mike. 'Has Irradiation Finally Arrived'. *Prepared Foods*, Sep 1991, 13.

Porges, Walter. *CQMD Historical Studies The Subsistence Research Laboratory*. Chicago: Chicago Quartermaster Depot. 1943.

Potember, Richard. 'Mini Time-of-Flight Mass Spectrometer'. *Johns Hopkins Applied Physics Laboratory Food Safety Initiative*. Natick Soldier Center, 18 Nov 1999.

Qin, B., Zhang, Q., Barbosa-Canovas, G. V., Swanson, B. S. and Pedrow, P. D. 'Pulsed electric field treatment chamber design for liquid food pasteurization using a finite element method'. *Transactions of the ASAE*, Mar/Apr 1995, 557–565.

Research and Development Associates for Military Food and Packaging Systems, Inc. *Proceedings of the Fall 1993 Meeting*. San Antonio: R & D Associates, 1994.

Remaley, Robert, Philip Stolz and James Hening. *Dairy Products*. Quartermaster Food & Container Institute for the Armed Forces. 1949. Vol. 7 of *Operation Studies Number One: A Report of Wartime Problems in Subsistence Research and Development*. Hewitt Conway Hewitt and Alice Meyer, eds. 12 Vols. 1948.

Richert, Walter. *Fruit and Vegetable Products*. Quartermaster Food & Container Institute for the Armed Forces. 1947. Vol. 9 of *Operation Studies Number One: A Report of Wartime Problems in Subsistence Research and Development*. Hewitt Conway Hewitt and Alice Meyer, eds. 12 Vols. 1948.

Risch, Erna. *The Quartermaster Corps: Organization, Supply, and Services*. Washington: GPO, 1953.

Roosevelt, Theodore. *The Rough Riders*. New York: Charles Scribner's Sons, 1926.

Satin, Morton. *Food Irradiation*. Lancaster: Technomic Publishing Co. 1993.

Sebrell, W. H. 'Nutrition – Past and Future'. *National Food and Nutrition Institute*, Washington, 8–10 December 1952. Washington: USDA, 1953, 3–12.

Sherman, Peter. 'Packaging IPT Update'. *Fall Meeting. Research & Development Associates for Military Food and Packaging Systems*. Weston William Penn Hotel, Pittsburgh, 1 Nov 1999.

Simpson, Roger. 'Army Tries Irradiation, Freeze-Drying to give GIs Tastier Rations'. *Wall Street Journal*, 28 Nov 1962.

Steiner, Paul. *Disease in the Civil War*. Springfield: Charles C. Thomas, 1968.

Stern, Philip. *Soldier Life in the Union and Confederate Armies*. Greenwich: Fawcett, 1961.

Taub, Irwin. 'S&T PPET Predicting Pathogen Growth'. *Food & Nutrition Research & Engineering Board Program Build*. Natick: DoD Combat Feeding Program, 1999.

Taub, Irwin and R. Paul Singh, eds. *Food Storage Stability*. Boca Raton: CRC Press, 1998.

Tauber, F. Warren. 'The Contribution of Packaging to the Storage and Delivery of Foods'. *Symposium on Feeding the Military Man*. Ed. Martin Peterson. Natick: National Academy of Science-National Research Council, 1970.

Thrasher, Harold. 'Model Messes for the ETO'. 1943: *Records of the US Army Headquarters, European Theater of Operations, Historical Section*. Dwight D. Eisenhower Library, 1941–1946: microfilm box 7, roll 37.

'Treatment of Diarrhea and Dysentery'. *Medical and Surgical History of the Civil War*. 12 Vols. Wilmington: Broadfoot Publishing Co., 1990.

Trottier, Robert. 'Polymeric Tray Update'. *Joint Service Operational Rations Forum*. Quality Inn, Petersburg. 22 September 1999.

United States. Armed Forces Food and Container Institute. *Operational Rations Current and Future*. Chicago: Food and Container Institute, 1963.

United States Department of the Army. Army *Regulation 40-25 Nutrition Allowances, Standards, and Education*. Washington: DA, 1985.

United States Office of the Quartermaster General. *Quartermaster Corp Manual 17-8 Committee on Food Research, Research Program 1945-1946*. Chicago: Army Service Forces. 1946.

United States Quartermaster Food & Container Institute for the Armed Forces. *A Report of Wartime Problems in Subsistence Research and Development*. 13 vols. Chicago. The Subsistence Research and Development Laboratory. 1946–1947.

United States US Army Natick Laboratories. *Operational Rations Current and Future*. Natick: Natick Labs, 1970.

United States, US Army, NRDEC, *Natick Pam 30-2 Operational Rations*. Natick: Natick Research, Development & Engineering Center. 1992.

United States, US Army, NRDEC, *Natick Pam 30-25 Operational Rations*. Natick: Natick Research, Development & Engineering Center. 1995.

United States, US Army, NRDEC, *Summary of Operational Rations*. Natick: Food Engineering Laboratory. 1982.

Vega-Mercado, H *et al.* 'HACCP and HAZOP for a pulsed electric field operation'. *Dairy, Food and Environmental Sanitarian*. Sep 1996, 554–560.

Viola, Carmen. 'Mobilization Initiatives'. *Fall Meeting. Research & Development Associates for Military Food and Packaging Systems*. Weston William Penn Hotel, Pittsburgh, 2 Nov 1999.

von Loeseke, H. *Outlines of Food Technology*. New York: Reinhold Publishing Corporation, 1949.

Yang, T. 'Microwave Sterilized Meals'. *Food & Nutrition Research & Engineering Board Program Build*. Natick: DoD Combat Feeding Program, 1999.

Youmans, John. 'What Lies Ahead in Nutrition'. *Dedication Symposium on Food and Health*, Geneva, New York, 5-6 May 1960. Geneva, New York: Cornell University, 1960, 32–41.

16

French meals

C. Grignon and C. Grignon, Maison des Sciences de l'Homme
Paris, France

Abstract: The pattern of three structured, regular daily meals in France today and
how it derived historically is described. Its stability is evaluated and future outlined.
The particular case of university students' diet is studied.

Key words: meal pattern, French food, social eating, student diet, history of meals.

16.1   Introduction

According to Goody (1982), French food, reflecting the hierarchical character
of the French society, is highly socially differentiated. Goody had no doubt
in mind the opposition between haute cuisine and everyday cooking but, as
Mintz (1996) pointed out, every hierarchical society does not generate haute
cuisine. The latter appears only in societies where much is made of food,
where consuming food is of paramount importance to the eaters, and where
a decisive general value is placed upon food as a social marker, which is
obviously the French case. Contrary to suggestions about social change, an
increase in the standard of living does not automatically eradicate inequalities
in consumption between social categories, nor does it result in the
homogenization of lifestyles. The standard French food diet, as it appears in
statistics, masks more or less important variations depending on social
categories, income, age, education, and so forth (Grignon and Grignon, 1981,
1999).

However, one of the main features of French food habits is the sharing by
the whole population of usages concerning the daily schedule and the
composition of meals. These implicit rules spontaneously present themselves
to scholarly reflection as traits of culture rather than as social usages. They
can be considered as a pattern or a structure. Their systematic ordering
makes them a subject of choice for anthropologists who worked out the
coherence of this inner logic (Douglas, 1974).

However, an exclusively anthropological approach does not allow us to understand how a cultural pattern manages to form and maintain itself in a complex and conflicted society. For this reason, after describing the contemporary French pattern of meals, we will briefly recount the genesis of this pattern which, once a bourgeois characteristic, becomes a cultural trait common to all classes. Then we will evaluate its stability and outline its future.

16.2   Description of the contemporary French pattern of meals

When speaking of lunch or dinner one refers to implicit rules which in our societies determine the name of the different meals and their composition, the time and the duration of each meal in the schedule of the daytime. A cultural pattern to which each one conforms without being aware of, these rules are a feature of national culture and a set of determined social usages which imposes unequally on different groups and classes.

Such a pattern of meals no doubt exists in societies of the same type as the French one (i.e. western, industrialized and hierarchical). As a result of a particular political, economical and moral history, the French case could be a kind of ideal type. In French contemporary society, the chronological and the non-chronological elements of usages pertaining to meals are closely related to each other. The hour at which one has lunch or that at which one dines, the time one devotes to the various meals, the order in which they occur in the course of the day, their frequency, the daily repetition of this schedule, its variations in the course of the week or year: these chronological elements of the pattern are inseparable from the places where meals are eaten, the company with whom one eats, the activities and social contexts for which meals act as separators and provide a transition, their composition, the importance and meaning one attaches to them, the name one attributes to them, etc.

16.2.1   The daily grid of French meals

To eat normally is, first of all, 'to do one's three meals a day', skipping none. Current usage names them, in chronological order: *petit déjeuner* (breakfast), *déjeuner* (lunch), *dîner* (dinner). In some regions, mostly in the south west of France, one still speaks as during the 19th century about *dîner* instead of *déjeuner* and *souper* (supper) instead of *dîner*.

In a recent survey (INPES, 2004) nine out of ten French people (90.2%) declare to have had their three main meals the day preceding the interview. Less than one out of ten (9.3%) have had only two meals and a minority (0.5%) just one meal. Thus, the traditional daily grid of meals seems to be shared by almost everybody in French society. However, we have to consider that younger people stick a little less to this grid (Fig. 16.1); we have no information about other food intakes during the day such as snacks, nibbling, etc.

The time and the duration of meals are duly scheduled. The breakfast takes place in the early morning, after waking up, before going to work or to school. It is the meal to which one devotes the least time and which is likely to be the most often skipped or at least simplified even if 94.5% of the respondents declare to have it. On average, it lasts 15 min and is taken at home. The workday as well as the school day is divided in two parts by the midday pause, the lunch, which takes place between noon and 2 pm. In Paris, it is the time when cafés and restaurants are packed; in the provinces, it is the time when everything stops or slows down, even the highways' traffic. After this period of time, most restaurants do not serve any more, corporate restaurants and cafeterias are closed and working people who have to lunch late, have to eat sandwiches or go to fast food restaurants. The average duration of the midday meal is 38 min and is taken by 96.9% of the adults aged 18–75. The evening meal or dinner is the one the French care for the most. It takes place at home, at least on weekdays, with the family, from 7 pm in the provinces and later in Paris. It lasts 40 min on average. A fourth meal, the *goûter* or afternoon snack, is traditionally reserved for children and goes with schoolchild status as it takes place at the end of the classes at 4.30 pm However, many university students and some elderly people indulge in it. The total duration of the three main meals amounts to more than one hour and a half – almost the same duration as in a precedent survey (INPES, 1996). If we take into account every food break, the time devoted to food

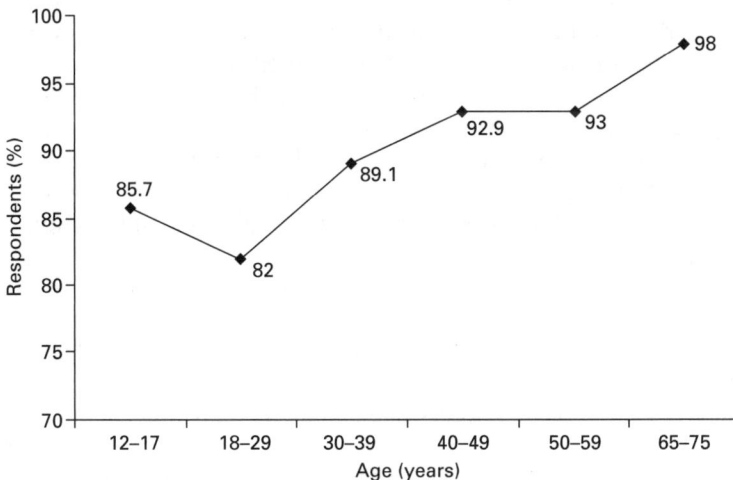

Fig. 16.1 Percentage of respondents having three meals a day, by age (Source: INPES, 2004).

intakes (annual average including weekend and holidays) amounts to 2 hours and 14 minutes and has not changed during the last 25 years (INSEE, 1999).

As well as the finding that one does not eat at just any time, one does not eat at just any place. Meals are taken at appropriate places. In every household, there is a room or at least a part of a room, either in the kitchen or in the living room, which is dedicated to eating meals. Eating at home, even informally, goes with some decorum. A matter of privacy, it is nevertheless a matter of etiquette; a matter of routine, it is also a matter of rituals. To set the table means to put in a certain order the tableware on the table around which the family or the guests sit down. Children are still asked to help setting or clearing the table. Whatever the social disparities in the quality of the tableware or the food, to sit down to table during meals is the minimum of the manners of food, shared by all classes of the society, even if, more and more often and mainly among working classes, the TV set occupies the place of honor.

16.2.2   Composition and structure of the meals

Composition of meals conforms also to well established norms. Common usage distinguishes between small meals, without hot and seasoned cook dishes, and the main meals which in most cases continue to unfold according to a 'plan' organized in successive courses. An ordinary meal, lunch or dinner, is composed of three or four courses: hors-d'oeuvre, main dish, cheese and/or dessert. It is the standardized structure of the meals of the table d'hôte restaurants, where dishes are served one after the other. It is also the structure of the lunch as offered in school or company refectories. It goes with a cold beverage, water, beer, wine or soda. A general and somewhat abstract structure, this sequential organisation of the meals nevertheless determines their compositions, the kind of food that can be eaten, and the kind of dish that can be presented at every step. From this point of view, the menu is quite similar to the plan of a discourse or to the table of contents of a book, which determines what can and will be written at every step of a formal grid, from the introduction to the conclusion. As shown below, though the daily pattern of three meals persists, we witness a simplification of the two main meals.

The present day standard breakfast consists, for about 60% of the adults, most of the time, of a cup of coffee or tea with bread and butter and/or jam or croissant and the like. In spite of nutritional recommendations, the consumption of cheese and fruits along with bread or cereals concerns only 10% of the population and occurs mainly among the young, learned, urban, and well-to-do households. Conversely, 17.5% eat charcuterie, eggs or fish and 14.1% take only one beverage. Children are given hot milk or hot chocolate with bread or, quite recently, cereals which have been strongly marketed by TV advertising.

More than half of the people who eat lunch, have a three- or four-course meal. Women are more likely to have a two-course meal than are men (33.0% versus 27.4%) who prefer three or four courses (61.1% against 55.2%). Older people (60–75 years) are more likely to stick preferably to a three or four course structure than younger ones. The main dish consists mainly of hot cooked meat, vegetables and/or starches, either fresh or processed food. Cold charcuterie or crudités are the usual hors-d'oeuvre. A two-course meal consists of a hors-d'oeuvre and a main dish or, more often, a main dish and a cheese (or a yogurt) or a dessert. When reduced to one dish, which scarcely occurs, the lunch taken at home may consist of a pizza or a quiche with side salad. Cold sandwiches or cakes and pastries are not considered as 'real meals' (in these occasions, one says *'manger un morceau'*).

Contrary to lunch, dinner is more often composed of two courses than three (38.2% versus 34.5%). Women are more numerous than men to have a one- or two-course meal (55.7% versus 50.1%). The three- or four-course structure is preferred by men and people of more than 50 years. Soup, for a long time the staple diet of the French evening meal, is far from being out of fashion; all sorts of processed soups, fresh or canned, fill the place and play the role of the old peasant dish. Two course structures, particularly a main dish and a dessert, are preferred within families with children. Even if one out of two French people (49.8%) watches television while dining, one cannot speak of 'TV dinner' as most of the time the television set is but a reassuring noisy background. As shown in Fig. 16.2, the structure of the main meals tends to simplify over the years. Between 1996 and 2002, the part of the four-course structure at lunch decreased to the gain of the two course one; however, the three-course meal still remains prevalent. On the contrary, at dinner the two-course structure has overtaken the three course one.

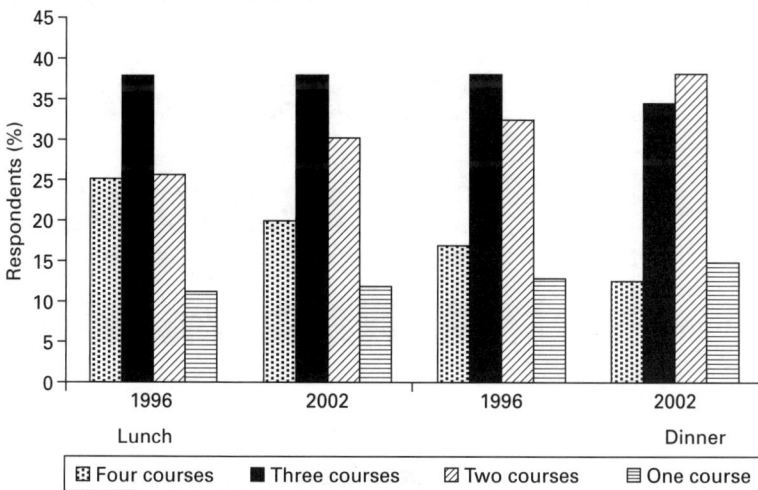

Fig. 16.2 Structure of the main meals in 1996 and 2000 (Source: INPES).

16.2.3   Meals eaten outside the home and family diet

Despite the growing pressure of the economic system that tends to suppress the legal obstacles to the institution of a 'non-stop society' in particular by relaxing the regulations concerning the daily schedule of work, the food of the French is mainly domestic. The share of meals taken outside the home remains small and is increasing very slowly.

Household expenses for meals eaten outside the home which increased up to the 1980s seem to be steady since that time. Their share in the family budget set apart for food went from 22% in 1980 to 23% in 2001 (INSEE, 2005), contradicting the expected scenario of the disappearance of meals taken with the family to the advantage of commercial and institutional feeders. In fact, we can talk of slow erosion of meals taken at home, but certainly not of disruption. More precisely, the increase in meals eaten away from home is attributable entirely to the lunch meal period: the proportion of lunch eaten outside the home went from 26.5% in 1982 to 28.7% in 1991 and 32% in 2002. Despite the drastic growth of the employment rate of women (in 1999 women represent 45.8% of the working population) lunch continues to be an important pause in the workday, and is taken at home when it is possible (68%). In this respect, the provinces differ greatly from Paris and its suburbs where 70% of the inhabitants take at least three lunches a week outside against 31% in provincial towns. In the same way, Parisians spend 31.4% of their food expenditure to eating out. The proportion of dinners taken outside remains stable over the years, around 13%. As shown in Fig. 16.3, lunch and dinner are strong features of family life. To sit down at the

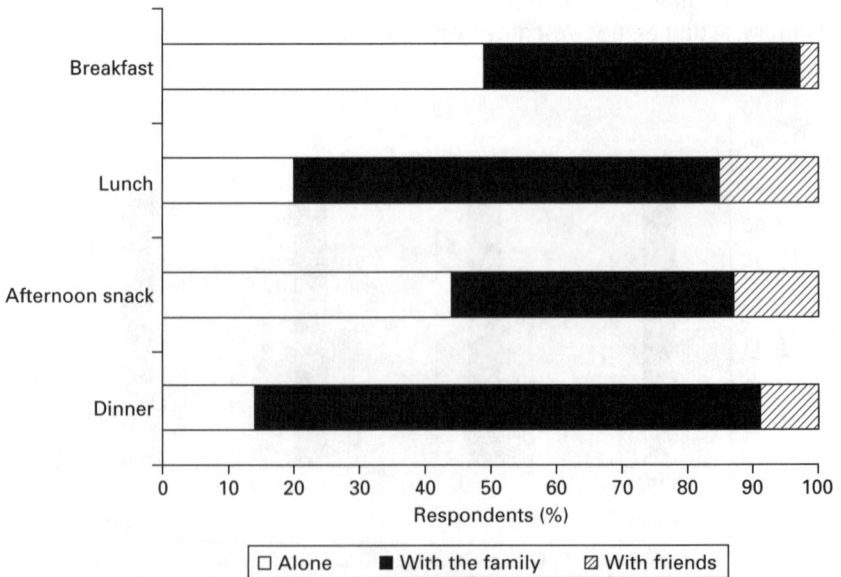

Fig. 16.3 With whom do the adults share their meals? (Source: INCA, 1999).

table, at home, with one's family is still the rule (INCA, 2000). A recent poll shows that, for the French, meals remains a period of time to share, 'a benchmark of the family life' as 88% agreed (SIMM, 2004). At the moment, the progression of meals taken outside is a consequence of the evolution in living and working conditions more than the reflection of major changes in lifestyles.

A detailed account of the social variations of the French family diet would need another whole chapter (Grignon and Grignon, 1999); we will only give here an outline of its main common features. The French do not know how to eat without bread. Bread, and above all baguette, is still present at every meal. It does not stand for a dish but comes necessarily with charcuteries (pâtés, dried sausages, ham) and cheeses and it accompanies crudités (chopped carrots, tomatoes, cucumbers, green salad, pink radishes) and the main dishes. Among main dishes, the 'bifteck frites' (beefsteak with French fries) has lost its supremacy. The consumption of carcass meat (beef, veal, mutton) has decreased during the last 20 years to the advantage of fresh pork meat, cooked ham and, above all, poultry. Meat, whether carcass meat, pork or poultry, is almost always roasted or grilled, less and less boiled. It goes with French fries, mashed potatoes, noodles or rice and vegetables such as French beans, green peas, endives, mushrooms, carrots, leeks, courgettes and ratatouille. Processed sauces (mayonnaise, tomato sauce, ketchup) tend to replace the pat of butter on the table. Complicated and time-consuming recipes are reserved for the weekend and for special occasions (parties with friends or relatives). Fresh fish (mostly fillets of white fish) is eaten on average once a week. Its consumption is decreasing except among older people or educated people. Offal (liver, kidneys, tripe, etc.), of which consumption has drastically decreased, remains a delicacy, especially for men. Whereas pizza has become an ordinary food, other ethnic foods like couscous or paella or Japanese food depend on fashion and are more likely to be eaten outside. Men usually eat cheese (camembert, bleu, etc.) at the end of the meal whereas women and children prefer to choose among a wide range of yogurts and sweetened dairy products. Fewer French end the meal with a seasonal fresh fruit. Fresh pastries, homemade or not, and ice cream are eaten preferably during the week-end, when meals are a little more formal. Non-sparkling mineral water and red wine are the two most common beverages during the main meals, but younger people are more likely to drink fruit juices, beer and sodas.

16.3 Social genesis of the contemporary pattern of meals

The pattern of meals we describe is recent since it developed at the end of the 19th century. It is the product of the more or less fortuitous and often conflictual encounter between social usages peculiar to different groups and cultures: the culture of scholarly institutions, fashionable usages and popular practices.

16.3.1 Rule and fashion

Some of these usages have their source in the serious side of the dominant culture, in the tradition of scholarly institutions. Thus, the norms concerning the schedule and the duration of the meals spread first among the upper class through the medium of schools. The instituting of refectories and scheduled meals, established by the monastic rule during the Middle-Ages, was adapted and secularized by the religious boarding schools (*collèges*). Following the monastic example, the *collège* transformed the taking of food into a pause capable of playing the essential role of temporal marker in the daily schedule, a separator and transitional element between the activities and moments of the day. In subjecting successive generations of the elite to a daily discipline capable of marking and lastingly moulding their attitudes towards the body and time, the boarding schools of the religious colleges managed to insert into the practical culture of the dominant classes a requirement that at the start concerned only a minority of intellectuals. The pattern of four meals a day (breakfast, dinner, snack, and supper) passed without much change from one century and regime to the other, from the boarding schools of religious colleges to those of the *lycées* (high schools). The schedule of meals did not vary very much between 1800 and 1920; despite the changes that occurred in bourgeois and Parisian usages, the name of meals also remain unchanged.

Fashionable usage does not question the principle of the regulated and regulating meal inscribed in the daily schedule; but it distances itself more or less deliberately from the rule imposed by the school. While the ascetic values of study and work order one to gain time and, consequently to rise early, to eat quickly, on time, and early, the aristocratic ideal of leisure, whose diffusion among the bourgeoisie the Revolution accelerates, demanded on the contrary that one be prodigal with one's own time and that of others. This conspicuous laziness found expression in a swing of the bourgeois daytime to the evening and by the social hierachization of the hours of the day, the latest being the most prestigious. By means of being delayed the bourgeois and Parisian dinner occurred, as soon as the first half of the 19th century, at the end of the afternoon, that is to say at the ancient supper's time. The postponement of dinner towards the evening produced hesitations concerning the dejeuner: when one has to take it, and what can one eat? Eventually, at the end of the century, it split into two intakes, the first one becoming the petit dejeuner (breakfast) and the second one the dejeuner (lunch).

16.3.2 The diversity of popular usages

Popular practices also have diverse origins: some come from peasant cultures, local cultures or trade cultures; others are probably once-dominant usages that have gone out of fashion (Grignon, 1993). As it appears in inquiries and testimonies, popular usage in the 19th century seems to have evolved according to the scheme of the one-way diffusion of the dominant patterns from the top to the bottom of the social scale. As a whole, the popular classes remained (or seemed to remain) faithful right up to around the end of the century to the old grid of meals which the ruling classes and the Parisian bourgeoisie had abandoned. It is only toward 1880 that the new grid began to be diffused in the social strata that had the most contact with the dominant style of life, such as Parisian workers. It remains to be seen to what point the popular classes whose family life was dominated and often crushed by work, really conformed to a standard pattern or grid of meals. It is likely that this official grid, convenient reference for informers as well as for respondents, still embraced a wide diversity of usages linked to the very great diversity in the conditions of life and work among the popular classes. The duration of the working hours and the duration of pauses varied strongly as well as the number and the distribution of pauses during the day. The most regulated situation, that of factory workers, was far from being as uniform as one might think. As a general rule, the workday under the Second Empire was 12 h in the provinces and 11 h in Paris (Duveau, 1946). But the length and organisation of the day, the distinction between time put in and actual work, varied according to industry, region, season, and state of the market.

16.3.3 Integration of the popular classes and the spread of the bourgeois way of life

It is thanks to a complete reversal of the dominant ideology that the dominant pattern of the meal in the end emerged and became a cultural trait common to all classes. With the lasting reestablishment of the Republic after 1870, the bourgeois way of life, of which this pattern is an element and a symbol, ceased to be considered exclusively as a distinctive privilege reserved for the elite and became a universal model and appropriate example for all. The popularization of the dominant practices went hand in hand with the intellectual and political triumph of the universalistic ideal inherited from the Enlightenment and the Revolution and transmitted by followers of Saint-Simon and philanthropists. The wish for social integration and assimilation prevailed over the segregationist attitude inherited from the Old Regime, which on the contrary aimed to keep the peasantry and the small folk of the cities in their place – in their order – and to contain the dangerous classes by throwing them outside of 'society' and 'civilization'.

The educative programme of the founders of the Third Republic aimed to include the popular classes in the society, but, by the same token, to assimilate them; cultural assimilation was both the condition and the consequence of social integration. To adopt the bourgeois food ways, and more generally the bourgeois way of life, the popular classes must give up the diversity of their customs and uses, particularly the local ones, as they must abandon the diversity of the local or regional languages (Breton, Provençal, Alsatian, etc.). Imposing prescriptions on food habits and practices was also an opportunity to intervene in a traditional area of religious proscriptions, an attempt to laicize the material life.

The development of a system of schools of meritocratic inspiration was obviously the essential instrument of the politics of popular education that this society brought about. But one also finds traces of this conception in other institutions intended for the 'People', such as the hospital, the prison, or the barracks.

Thus, for example, right up until the last decade of the 19th century, soldiers were subject to a separate diet, which had nothing in common with the bourgeois nourishment reserved for the officers. The soldier's soup lacked the set of divisions that makes up the bourgeois meal and gave it its shape: no menu, a single dish in which the foods are mingled and solid and liquid are mixed together, no individual portions, no place settings. It was only around the end of the century that the troops' diet began to improve. Philanthropy arrived late in the barracks, through the mediation of enlightened military doctors who no longer put up with a diet they considered poorly conceived, irrational, contrary to the law of physiology, and according to them, which could only favour waste and alcoholism. Instead of developing a separate diet for the troops, one offered them the legitimate pattern of the bourgeois meal and, at the same time, the example of the way of life of which this meal is an element. As Schindler said, a first-class doctor-major (1885): 'Distribute to the men one or several dishes per meal; compose each dish of a single kind of food or a simple combination of a meat and a vegetable; vary the nature and the preparation of the food at each meal; in a word, apply to the soldier's feeding the method commonly called bourgeois cuisine; such is the system that we have followed'.

The improvement of the soldiers' conditions of life went hand in hand with an explicit will to educate and to moralize, to take the opportunity of the conscription to give the future head of the family individual habits of hygiene and regularity: 'Let the future head of the family learn in the regiment to make, as he will see done, judicious and economical use of the limited resources available for the daily mess, and he will know how to bring to his table, without straining the most modest budget, substantial dishes that are sufficiently agreeable to his taste to turn him away from the gustatory satisfactions of the cabaret and the inn' (Schindler 1885).

This emergence of an ideal of social integration and assimilation of universalistic inspiration produced at the same time a conversion of worker attitudes and strategies. The industrial revolution did not impose all at once on the popular classes the modern and 'rational' conception of time that was its base. Far from favouring it, industrialization thwarted the diffusion of the dominant pattern of meals in the popular classes. It tended to make the workers, many of whom were cut off from the local and peasant cultures from which they came, a separate category condemned to live according to a particular tempo and a rhythm often off beat. But the 'primitive, rebellious and vociferous' strikes against working conditions were in the long run replaced by organized forms of action (Perrot, 1974). At a time when claims concerning wages were by far the most numerous, demands began to concern the organization of the workday. These demands fitted in the framework of struggles for the reduction of working hours. They concerned mainly the midday break and the possibility of returning home for the noon meal. They bear witness to the wish to lead a 'normal' family life (that is to say one modelled more or less on the bourgeois pattern), the desire to conform to current usage, the right to live 'like every one'.

16.4   The future of the French pattern of meals

16.4.1   Factors of crisis

As shown above, the French continue to refer to the pattern of meals that formed at the end of the 19th century, even if they do not conform to it rigorously (which has probably always been the case: the pattern of meals is an example to be followed, a kind of template, an ideal, too, to which one can approach but one can never wholly realize). This permanency is highly questionable. The present-day pattern, whose stability and durability we no doubt have a tendency to overestimate simply because it is 'our own', is never more than the temporary product of a competitive and often conflictual process that continues to unfold before our eyes. From this point of view, it is obviously exposed to the changes in the 'unique historical constellation' that produced it and, at the same time, poised by the inertia that characterizes long-term phenomena such as cultural traditions.

One has to ask if the contemporary pattern of meals will long survive the elements that played an essential role in its constitution. There is no lack of signs of crisis. Crisis of the dominant family model, to which testifies the drop in fertility and marriage rates and the increase of single mothers. The permanent crisis of employment, which results for a growing number of the unemployed in a loss of temporal references and thus goes with a decrease of the social cohesion in a society where, despite the programmed reduction in the working time, the latter remains the 'dominating social time' (Demazières, 1995). The crisis of the dominant ideology which expresses in revision, indeed abandonment of the ideal of social integration, and return under the cover of cultural relativism to a general conception of an elitist and segregationist society.

It is true that the main changes that have occurred in French society during the 'Trente Glorieuses' (roughly from 1950 to 1980), industrialization, urbanization and increase in standards of living reflected partly in the evolution of consumption patterns (Grignon and Grignon, 1999). Concerning the composition of meals, improvements in standards of living have contributed to the substitution of expensive but quick to prepare foods in place of inexpensive items that require time-consuming preparation. The development of large-scale retailing has no doubt contributed to the increasing consumption of processed foods, frozen or not. The increase of the employment rate of women, when it goes with the persistency of an unequal sharing of domestic tasks between men and women (attested by the greater allocation of time women spend on cooking, housekeeping, or shopping) is likely to encourage the simplification of meals. Moreover, the time devoted to cooking has decreased whereas the time necessary for grocery shopping has increased (INSEE, 1999). The extension of women's education has probably led them to recognize and to respect legitimate norms in health matters or in fitness and contributed to the decline of the three- or four-course meal.

For a very long time, the study of contemporary food consumption in France has been the preserve of economists; whereas numerous and chronological surveys on food purchases are available, it is only recently that the Health Administration, interested in prevention of hazardous behaviours, carried out national surveys on food and health comportments (INPES, 2004; INCA, 2000). The evolution of the ways of living being an important symbolic and economic stake, it is not surprising that the lack of data concerning the schedule and the composition of meals allowed the development of many fantasies and more or less interested self-fulfilling prophecies. Under this theme has flourished c. 1980 a dramatic scenario of change and crisis whose commonplaces are well known: a surge across France of 'American-style' diet; the development, even the hegemony, of fast food; the destructuring of the traditional meal; the generalization of snacking; the disappearance of meals taken with the family to the advantage of commercial and institutional feeders as a result of the crisis in the family and more generally in the society.

Initiated in the field of marketing and then adopted later and popularized by the media and some sociologists, the tenet of meal destructuring served as a scholarly framework that gave rise to the more concrete and clearly more interested notion of snacking, '*grignotage*' (LSA, 1985). If the structured meal was labelled in the name of modernity and individual freedom as constrained, repressive and even harmful, it is perhaps because in today's France it poses an obstacle to the 'extensive consumption' and continuous feeding habits envisioned by some divisions of the food industry (like the candy, pastry, and chocolate industry). By increasing the number of times one eats each day, the destructuring of meals would finally allow the limits of appetite to be overcome. Nevertheless, self-fulfilling prophecies do not always turn out that way; in this case, they encountered the inertia and the capacity of resistance of food habits, and did not succeed to influence common opinion and taste. Nowadays, the thesis of the destructuring of the meal seems to be forsaken. Confronted by the resistance of the facts, professionals of food marketing changed their position and foresee, instead of the generalization of snacking and nibbling, the advent of a complete structured meal processed by the food industry and eaten at home (LSA 2001, Linéaires 2001).

16.4.2   The case of university students

For lack of data covering a sufficiently long period, one cannot precisely evaluate the stability and the chances of continuation of the present-day French pattern of meals. However, surveys carried out among university students give us the possibility of foreseeing the future evolution of ways of living and tastes by observing their present state. University students form a young population whose practices have high life expectancy, and at the same time they are sufficiently extricated from childhood so that their practices are not the pure and simple reproduction of those of the former generations. Moreover, even if unequally, higher education gives access to upper occupations and positions; thus most of the students will belong to prescribing groups whose way of life serves as model and tends to diffuse.

The results of these surveys (Grignon, 1987, 1998, 2000, 2003) do not square with the conventional opinion and folklore about the bohemian life that students supposedly lead, nor do they corroborate the dramatized vision of the food of the future popularized by the media. The last survey, confirming our previous works, shows that far from being unstructured or 'destructured' the students' diet does not diverge much from the established pattern of meals when it comes to the regularity of the main meals. The share of skipped main meals (as a percentage of the total of the meals of the week taken by all the respondents) goes to 3.7% for lunch and 2% for dinner. 85.9% of the students skipped none of the lunches and 91.3% skipped none of the dinners during the week preceding the survey. As for the rest of the French population, the students' food is mainly domestic: two out of three meals are taken at home (theirs or their parents'), 53.2% of the lunches and 81.7% of the dinners. At noon, 57% of the meals taken outside are related to studies or work and organized by institutions or employers: university restaurant and cafeteria (25 and 13.6%), school refectory (9.7%), corporate restaurant (8.7%). Lunch is scarcely eaten outside of places and situations authorized by common use, dinner even more rarely. Only 6.7% of the noon meals and 0.8% of the dinners are quick snacks eaten in the street, in cars, buses, etc. (Table 16.1).

However, the simplification of the main meals is more frequent among students: during the week before the survey, 51.7% of the students ate at least one light lunch and 43.3% at least one light dinner (for example a salad, a soup, slices of topped bread). In the same way, the share of students who skip their breakfast at least once a week reached 47%. The variations with regard to the current model result from various causes: material constraints, health and dietetic concerns, and a lack of social integration. Thus, the omission, even the simplification of the structure of the meals, is a characteristic of poor students and of those for whom poverty prevents them from being full time students such as scholarship holders, students with a regular job, as well as students living far from their university. To simplify or to skip a meal can also be a way of reducing one's food intakes and of conforming to the dominant norms of diet and aesthetic concerns. So the students on a slimming

Table 16.1 University students, type and place of meal (%)
(Source: OVE, 2003)

	Lunch	Dinner
Home	23.9	35.8
Parents and family home	29.3	45.9
Friends home	3.1	6.6
Restaurant and café	6.6	4.9
Corporate restaurant	4.1	1.0
University restaurant and cafeteria	18.1	1.8
School refectory	4.5	1.1
Snacks	6.7	0.8
Skipped	3.7	2.0
Total	100.0	100.0

diet are more likely to simplify their midday meal and girls are more likely to lighten their lunch and to skip the evening meal. In the same way, the more the students loosen their relationship with their family the more they tend to deviate from the established pattern. This is the case for students who do not live with their parents, whether on campuses or in town, alone or with friends. Conversely, the students (whether married or not) who have a family life of their own tend to skip breakfast or dinner less. From this point of view, the more the students approach the standard way of life, the less they tend to deviate from the standard usages. To skip one's meals goes with hazardous behaviour such as smoking, drinking alcohol, taking sleeping tablets or tranquillizers. Contrary to what one should expect, students with the most constrained curriculum are likely to eat more regularly; it is the same for students who stick to a strict schedule of their own.

The results of those inquiries lead one to think that we are far from witnessing a rapid and massive disruption of usage. For all the elements whose combination constitutes the pattern of meals, the majority of students continue to follow the established norms rather closely.

16.5 Conclusion: a paradox

The existence of an elaborate and constraining pattern of meals supposes a high degree of social integration as shown by the programme of education of the popular classes in France at the end of the 19th century or the case of the contemporary university students. This pattern contributes in return to this integration: the interiorizing of the rules and the usages related to food is one of the essential and primordial components of socialization. Insofar as we can rely on the available data, and despite increasing signs of anomie in the French society, its pattern of meals resists far better than predicted. When the structure of the everyday meals tends to simplify, their schedule shows durability. This stability poses a puzzling question to both the sociologist and the anthropologist: how can a cultural feature resist such social changes?

16.6   Sources of further information and advice

The data we rely upon in this chapter are issued by the following institutions:

Institut National de la Statistique et des Etudes Economiques (INSEE), Ministère de l'Economie et des Finances.

Enquête Alimentation, on an annual basis from 1965 to 1983 and on a biennial basis from 1983 to 1991 (the last survey). Average sample of 10 000 households; questionnaire and household diary methods.

From 1992 on, data on food consumption are to be found in family expenditure surveys, less detailed concerning food purchase: Enquête Budget de famille carried out every five year, sample of 10 000 households.

Enquête Emploi du temps: occasional surveys (1976, 1986 and 1999) on the daily time budget of individuals. Sample of 8000 households, 16 000 respondents.

Observatoire de la Vie Etudiante (OVE), Ministère de l'Education.

Questionnaire survey on university students' conditions of living and work, carried out in 1994, 1997, 2000 and 2006, with a national sample stratified according to discipline, academic cycle and year of study, and location of the university ($n = 27\,000$ for each survey). As chairman of the scientific committee of the OVE, Claude Grignon contributed to set up this survey and personally conducted the first four editions.

Institut National de Prévention et d'Education pour la Santé (INPES), Ministère de la Santé, Ministère de l'Education.

Baromètre santé nutrition: survey carried out by phone on a random sample of 3000 persons aged 18–75 in 1996 and 12–75 in 2002. They are interviewed about their food consumptions, the composition and occurrences of their meals, their knowledge and perceptions of food, their physical activities.

Agence Française de Sécurité Sanitaire des Aliments (AFSSA), Ministère de la Santé, Ministère de l'Agriculture.

Enquête Individuelle et Nationale sur les Consommations Alimentaires (INCA) carried out by a private consultant (CREDOC) on a representative sample of 3003 persons aged over three.

Websites of interest:

www.insee.fr

www.ove-national.education.fr

www.paris.inra.fr/aliss

www.inpes.sante.fr

www.afssa.fr

16.7   References

Demazière D (1995), *La sociologie du chômage*, Paris, La découverte, Coll. Repères.

Douglas M (1974), 'Deciphering a meal', in Geertz C, *Myth, Symbol, and Culture*, New York, Norton.

Duveau G (1946), *La vie ouvrière en France sous le second empire*, Paris, Gallimard.

Goody J (1982), *Cooking, cuisine and class: a study in comparative sociology*, Cambridge, Cambridge University Press.

Grignon C (1987), *L'alimentation des étudiants*, Paris, INRA-CNOUS.

Grignon C (1993), 'La règle, la mode et le travail: la genèse sociale du modèle des repas français contemporain', in Aymard M, Grignon C and Sabban F, *Le temps de manger*, Paris, Editions de la Maison des Sciences de l'Homme, 275–323.

Grignon C (1998), *La vie matérielle des étudiants*, Paris, La documentation française.

Grignon C (2000), *Les conditions de vie des étudiants*, Paris, Presses Universitaires de France.

Grignon C (2003), 'Alimentation et santé', *OVE Infos*, 6 (mai).

Grignon C and Grignon C (1981), 'Styles d'alimentation et goûts populaires', *Revue française de sociologie*, **21**(4), 531–569.

Grignon C and Grignon C (1999), 'Long-term trends in food consumption: a French portrait', *Food and Foodways*, **8**(3), 151–174.

INCA (1999) *Enquête individuelle et nationale sur les consommations alimentaires*, Paris, AFSSA.

INPES (1997) *Baromètre santé nutrition 1996*, Vanves, CFES.

INPES (2004) *Baromètre santé nutrition 2002*, Saint-Denis, INPES.

INSEE (1999), 'Enquête emploi du temps', *INSEE Première*, 675 (octobre).

INSEE (2000), *Tableaux de l'économie française en 1999*, Paris, Ministère de l'Economie et des Finances.

INSEE (2005), *Annuaire Statistique de la France*, Paris, Ministère de l'Economie et des Finances.

Libre Service Actualité (1985), 990 (septembre).

Libre Service Actualité (2001), 1709 (février).

Linéaires (2001), 160 (juin).

Mintz S W (1996), *Tasting Food, Tasting Freedom, Excursions into Eating, Culture and the Past,* Boston, Beacon Press.

Perrot M (1974), *Les ouvriers en grève, France 1871–1890*, Paris, Mouton.

Schindler C A (1885), 'L'alimentation variée dans l'armée' *Archives de médecine et de pharmacie militaire*, **5**, 461–491.

SIMM (2004), *Dix ans de pratique alimentaire, la consommation sous surveillance*, Paris, TNS Media Intelligence.

17

Italian meals

E. Monteleone and C. Dinnella, University of Florence, Italy

Abstract: Current Italian meals and meal patterns are described and the structure and composition of breakfast and the main or principal meal (lunch or dinner) is reported. In addition, information about the beverages that accompany everyday meals is provided. Variation in meal patterns, combinations and components owing to social changes is discussed. A picture of the essence of current Italian meals is drawn based on historic, sociological and epidemiological works, official statistics and cooking books and magazines. Regional differences (macro-regions across the country) are described in relation to basic foods, culinary techniques, and flavour.

Key words: Italian food, lunch, dinner, food culture, culinary techniques.

17.1   Introduction

Italy is internationally well known for its food culture. This recognition is based on several factors:

1   the country's historic contribution to culinary arts and gastronomy;
2   the international presence of products which are typical components of Italian meals (i.e. pasta, pizza);
3   the superior preservation of the regional and local gastronomic traditions, compared with many countries which are in the throes of a similar process of modernization (Alexander, 2000);
4   fewer changes over the last 60 years of the meal pattern compared with other European countries (Gracia and Albisu, 2001); the progressive and continuous process of evening out of food habits characterizing Western countries, is proceeding more slowly in Italy, leading to different styles and behaviours compared with other European countries (Albertini and Celenza, 2001, Cipriani, 2003);
5   the social and convivial importance of eating (Alexander, 2000; Gracia and Albisu, 2001; Capatti and Montanari, 1999; Montanari, 1992).

The 'fast food' culture has not penetrated much in Italy compared with other western countries (Gracia and Albisu, 2001). Although there is clear evidence of a shift towards a simplification and modification of the meal structure and meal patterns, the Italian way of eating is still recognisable. This difference in responding to global shifts is owing to both social and cultural reasons: Italy is characterized by a lower crisis of traditional forms of family, and a slower development of new forms of work organization; culinary rules are locally well consolidated; and Italians are well aware of the healthiness of the Mediterranean diet.

A recent survey (CENSIS, 2003) of how foreigners perceive Italians' relationship with food highlights characteristic aspects: the importance attributed to food in general, the ample regional base of the cuisine, the importance given to the main meal, conviviality and the time spent at the table. The importance of some of these aspects in determining Italian eating behaviour are also found in historic, sociological and epidemiological works, in official statistics and cooking books and magazines. From these sources, it is possible to draw a picture of the essence of current Italian meals.

In Italy, as in most western societies, food intakes are typically distributed in breakfast, lunch and dinner. To this division of meals during the day, morning and afternoon snacks have recently been added (Albertini and Celenza, 2001; ISTAT, 2007). On average, 25% of the population, mainly young people, teens and children, consume snacks but not in substitution of lunch or dinner. The following paragraphs describe the structure and composition of breakfast and the main or principal meal (lunch or dinner). In addition, information about the beverages that accompany everyday meals is provided.

17.2   Meal patterns

17.2.1   Breakfast

As pointed out by several authors (Chiva, 1997; Garcia and Albisu, 2001) the Italian breakfast can be defined as a light and cold meal. According to a recent eating habits survey (ISTAT, 2007), 25% of the population does not have breakfast at all, and 46% have milk with 'some' food; typically breakfast is consumed between 6.30 and 8.30 am. The importance of breakfast varies regionally and by age and gender. Breakfast is most regularly consumed in the central–northern part of the country, by adolescents and children. More women than men start their day with this meal. The elderly tend to have a continental breakfast composed of only a drink with bread (de Groot *et al.*, 1998). In this age group, gender differences were not observed. According to a recent study, the breakfast is based on milk (more frequently consumed than tea, fruit juice or hot chocolate) combined with biscuits or snack cakes, and to a lesser degree cereal. Morning coffee is a traditional component of breakfast. In most cases, breakfast is eaten at home although it does not seem to have a particular social meaning, such as bringing family together. The principal reason for skipping breakfast is a lack of time (Vanelli *et al.*, 2005). Furthermore, the majority, including many children, claim to prepare and eat their breakfast alone.

The habit of having breakfast is relatively recent. It began to take hold among Italians after the Second World War and reached its peak in the 1980s. In this period it was the object of important public awareness (Vercelloni, 2001) and dietary education campaigns (Cipriani, 2003) which emphasized the importance of breakfast in terms of daily total energy intake (Vanelli *et al.*, 2005). Throughout the 1990s, domestic consumption of breakfast grew. In the mid-1990s, as a detriment to the importance attributed to breakfast and with no apparent cultural or nutritional support, the mid-morning snack started to become increasingly popular. This habit has become a popular social event for adults and is characterized by an increased consumption of savoury snacks, mostly made up of unbranded products (Vercelloni, 2001). The presence of two light meals during the morning seems to be a return to a widely held Italian habit from the past, rather than a new dietary tendency. Indeed, among the 'rules for health' outlining proper food habits, written by Pellegrino Artusi in *'La scienza in cucina e l'arte di mangiar bene'* (1891), it is suggested to limit the first meal of the day to coffee with toasted bread, milk with coffee or chocolate, to consume a 'solid breakfast' around 11 am and a dinner around 7 pm as the mail meal. While the consumption of snacks continues to grow (ISMEA, 2005), it seems to involve mostly the younger population (ages 11–25). Of this age group, 40 to 60% regularly consume snacks (ISTAT, 2007). This tendency leads to a breakdown of the traditional meal pattern, based on a light breakfast and two cooked meals (lunch and dinner) (Cipriani, 2003; de Groot *et al.*, 1998) in favour of a pattern based on a single main meal and a frequent consumption of light and often cold meals.

17.2.2   Main meals

ISTAT data relative to the 2000–2006 period (ISTAT, 2007) indicate that lunch continues to be the main meal for most Italians: 70.4% of the population identifies lunch as their most important meal of the day, while only 20.9% identify dinner as the most important meal.

The crucial role of lunch among the meals, seems to be notably influenced by life styles. In fact, its importance, compared with dinner, is deeply conditioned by regional and urban contexts. In the northern and central regions of the country the relative importance of lunch is lower than the national average (about 65%), while in the south and in the islands the importance reaches average values of about 80%. Also urban lifestyles exert a great influence: in metropolitan areas, lunch is the most important meal of the day for 64% of the population, while in towns with populations between 2000 to 50 000 inhabitants this figure rises above the national average (74%). Breakdown of these data by age group and gender suggests that one of the determining factors for importance of the main meal is linked to whether the person works outside of home or not. The data for those between 25 and 55 years of age shows a decrease in the importance of lunch as the main meal (65%) and an analogous increase in dinner (30%) with a limited influence of gender.

Whether lunch is consumed, generally reflects the phase of life and social standing. For school children, lunch is most frequently eaten at the school canteen; 59% of children 3–5 years old eat lunch at school; 31.8% of those aged 6–10 (this value is owing to a lower availability of the school meal service). The largest group of people who regularly (more than 90%) consume lunch at home are of those over 60 years of age. Working men are the greatest consumers of away-from-home lunches. In fact, only 55.4% of the male population between the ages of 25 and 55 claim to habitually eat lunch at home, while the remainder regularly have lunch out, for example at a company canteen, at a bar or restaurant or at their work station. Of women in the same age bracket, 75% regularly eat lunch at home. The difference can be plausibly explained by the smaller percentage of women occupied in paying jobs (ISTAT, 2006). Working outside of home seems to be the principal factor determining away-from-home meals. Lunch in Italy is generally eaten between 12 noon and 2.30 pm with a traditional regional variation: earlier in the north and later in the south and in the islands.

A series of considerations point to a decrease in the importance of lunch as the main meal over time, and to radical modifications and simplification of its structure. The increase of food consumption outside the home is considered to be an index of the tendency for the diminishing importance of lunch. In the last 10 years, there has been a strong increase in food consumption outside the home: in 1990, it was 25.4%, compared with the current 30%, and according to estimates (ISMEA, 2007) in the next 20 years it will reach 45 to 50%. Lunch eaten outside of home is also losing its typical structure, as the cost and work routines seem to support the growth of so-called quick meals (i.e. at a bar, self-service, pizzeria) as opposed to more traditional options. These shifts follow similar shifts across Southern Europe (Garcia and Albisu, 2001).

17.3   The traditional structure of the main meal and its evolution

In its most classic form the main meal is, characterized by a sequence of dishes (Cipriani, 2003; Kjaernes, 2006; Montanari, 1992; Rozin, 2000). A meal (staple+meat/fish + salad or cooked vegetable), as often consumed in northern European countries, is different from a meal in Italy where a meal is made up of a sequence of several dishes. Typically, a main Italian meal includes a first course of complex carbohydrates (*primo piatto*, e.g. pasta or rice), a second course that serves as source of protein (*secondo piatto*) and includes a side dish (*contorno*, e.g. salad or cooked vegetable), and as a final course a fruit or dessert (Montanari, 1992). This basic composition of the main meal is found throughout Italy and represents a model, as evidenced in school lunches (De Amicis, 1999; Pagliarini *et al.*, 2005; Policastro, 2005) and in the menus of Italian army (Repubblica Italiana Ministero della Difesa, 2007).

Currently, the traditional menu composed of three dishes is expanded for special meals (i.e. on Sundays and special occasions) to include also a starter course (*antipasto*). Antipasto can be composed of separate cold or hot dishes including seafood, grain and vegetable salads, stuffed vegetables, *omeletes*, slices of salami, *prosciutto*, olives, *crostini,* cheeses, raw fish and meat, and even fruit. In recent history, the starter course has gone through ups and downs. It was very popular in the 1930s and was made up of samples of 'savoury, salty and cold bites'; in the 1950s, its importance in the menu began to diminish as it was seen as excessive and superfluous (Montanari, 1992). According to available statistics (ISTAT, 2007), the first course is essentially pasta, or rice as an alternative consumed on average in a 1 to 5 ratio to pasta. On average, 90% of Italians consume pasta or rice at least once a day, regardless of the region, urbanization or even the age group or gender. The second course is frequently made up of beef or white meat (e.g. poultry and rabbit), rather than of pork, fish or eggs. The only regional difference in this regard involves the frequency of fish in the menu, occurring more often in the central and southern parts and in the islands than in the north. The second course is always accompanied by a side dish of salad or cooked vegetables. Fruit is a daily habit and it is more frequently consumed than dessert. Bread is always a meal component. It is present on every table at every meal. Every region has its local varieties and specialities differing by ingredients, appearance, shape, flavour and texture.

Progressive demographic changes (i.e. slowing down of birth rate, ageing of the population, the decreasing number of members per family) and above all, changes in the general organization of work (i.e. increase in paid work, continuous working hours also for women, the increasing number of dual-career families), along with a tendency to consume meals away from the home, have led to, on the one hand, the crisis in the traditional model of the main meal and, on the other, to new strategies to protect its structure. The composition of the away-from-home meal tends to focus on a single dish, considered an indicator of extra-domestic food consumption (Turrini *et al.*, 2001). Home-prepared meals allow for the various ingredients to be divided into different courses, while this is not possible with meals eaten out. On the other hand, from the 1970s, the presence of convenience foods and ready-to-eat dishes on the market allowed the traditional structure of home-prepared meal to be maintained despite the decreasing time devoted to meal preparation. Recent data (ISMEA, 2007) confirm that this phenomenon continues to increase with regard to both ready-to-eat pasta dishes and fish- and meat-based main dishes, as well as ready salads and vegetables.

The statistical data and scientific works available in the literature are not sufficient to completely describe the transformations that have taken place over time with regard to the structure of the main meal. However, the changes that have occurred in Italian meals can be analyzed through the articles in a popular cooking magazine (La Cucina Italiana, published since 1929), using an approach that has already been successfully applied by culinary art historians (Montanari, 1992) or researchers interested in identifying transformation elements in meal compositions (Mitchell, 2006). For this analysis, four different periods were considered: the 1950s (represented by 1957), the 1970s (represented by 1977), the 1980s (represented by 1987) and the beginning of this century (represented by 2006).

17.3.1 The 1950s

It is only following the economic development in the post-war period that we can speak of an Italian meal pattern and meal composition, because it was only then that the majority of the population benefited from the availability of food. Internationally pasta is no doubt identified as making up an Italian meal, but, in view of the long culinary and gastronomic tradition of the country, its prominent role in Italian meals is a relatively recent phenomenon. From the unification of the country in 1860 to the first decade of the last century, the 'all-Italian way of eating' started to take hold through the institutional distribution of meals in school, military, and factory canteens and in hospital food services. To provide appropriate catering in these contexts, it was necessary to overcome long-standing regional differences, with political and historical connotations, that were reinforced by the diversity of available raw materials. The industrialization of both agricultural and pasta production made it possible to resolve this problem. The ubiquitous availability of pasta and tomato sauce was the basis of unification (Alexander, 2000) and their presence in the Italian diet became a routine. These two products have the advantage of leaving ample room for the use of spices, condiments and preparation methods from region to region. Pasta and tomato, composed and enriched or rendered differently based on the various regional realities, became the pillar of the main meal (Montanari, 1992).

Based on the menus suggested by La Cucina Italiana (1957), the preparation of meals at home for an average family was characterized by two important moments: lunch and dinner. Both were cooked meals served at established times: 1 and 8 pm, respectively. Lunch was indicated as the main meal, with the greatest nutritional input and composed of a sequence of different dishes. The first course was essentially pasta, served in various ways, but almost never with a high protein source such as meat, fish or eggs. In planning a weekly lunch menu, rice or a dish based on potatoes was suggested as possible substitutes for pasta. In any case, the first course was exclusively a source of complex carbohydrates, not of protein or fats.

The second course was made up of meat (alternating between white and red), fish or eggs. The side dish was prevalently composed of a salad or, less frequently, of cooked vegetables. In the 1950s, the main meal was characterized by the presence of a cheese course between the main dish and fruit. Industrial production of typical Italian cheeses such as *parmigiano*, *provolone* and *mozzarella* facilitated this regimen. This was the precursor to the current use of cheese as a component of the meal (lunch or dinner), as substitution of the classic protein source (meat or fish). The presence of fruit at the end of the meal was clearly more frequent than dessert.

Dinner in the 1950s was a hot meal complementing to lunch and differing from it in several ways. For example it tended to be, overall, lighter than lunch; it included frequently a vegetable soup or combination of pasta or rice with vegetables; the second course was more frequently cold including perhaps cured meats or cold cuts; cheese was absent and fruit was consumed cooked rather than fresh. Table 17.1 presents an example of weekly lunches and dinners.

Furthermore, it is evident that the composition of the main meal was characterized by marked seasonality, in particular with regard to fruits and vegetables, but also for some types of meat, such as lamb.

17.3.2   The 1970s and 1980s
The 1970s and 1980s can be considered, and rightly so, the period of modernization when in food habits 'new' became the synonym for opulence (Albertini and Celenza, 2001). A diet based on proteins, especially those from meat, marked access to newly spread well-being and good eating. In fact, the consumption of meat tripled within a ten-year period, passing from 14 kg of meat consumed per year per capita in the mid-1960s to about 62 kg per year per capita in 1975. In terms of meal composition, this meant consolidation of the traditional three dishes, with the second course principally composed of red meat. However, as a symbol of refined habits and modernity, there was a progressive decrease of the first course (generally meaning pasta) in favor of starters and mixed dishes accompanied by rice. During this period, a delocalization took place, both for time and space, as eating became dependent on the food industry (Montanari, 1992). In fact, industrial production led to the reorganization of the transportation and marketing systems of products, as well as to major changes in storage technologies. All this broke down the link between food and territory which until this time had been inevitable. There was also a breakdown of the cyclic aspect of food consumption based on seasonality and weekly alternation of dishes tied to religious holidays such as Carnival and Lent (e.g. Good Friday), the periods of 'abundance' and 'lean'.

Menus suggested by 'La Cucina Italiana' (1977) for the preparation of daily meals indicate the disappearance of the strict distinction and mutual compensation between lunch and dinner that structurally continued to follow the traditional three-course scheme. The composition of the meal, however,

Table 17.1 Example of a weekly menu proposed by 'La Cucina Italiana' in 1957

Monday	Tuesday	Wednesday	Thursday	Friday	Saturday	Sunday
Lunch						
Pasta with tomato sauce	Polenta	Gnocchi	Risotto with pumpkin	Spaghetti with tomato sauce	Onion soup	Tagliatelle with ragout
Beef meat balls	Pork ribbon with cabbage	Spinach omelette	Grilled beef	Buttered eggs	Roasted pork	Roasted duck with potatoes
Salad of celery and carrots	Green salad	Green salad	Salad	Buttered spinach and celery	Green salad	Radicchio salad
Cheese	Cheese		Cheese	Cheese	Cheese	Apple pie
Fresh fruit	Fresh fruit	Fresh fruit	Fresh fruit	Fresh fruit	Cooked fruit	Fresh fruit
Dinner						
Rice in vegetable broth	Vegetable broth with toasted bread	Crumbed mozzarella	Meat broth soup	Polenta	Vegetable broth with rice	Semolina soup
Crumbed beef meat	Pigeon with lemon juice	Liver with marsala wine	Roasted rabbit	Tuna stew	Stuffed cabbage	Sautéed veal chops
Buttered artichokes	Buttered carrots	Potatoes puree	Green salad	Boiled potatoes	Carrot salad	Carrot salad and buttered spinach
Cooked fruit	Fresh fruit	Cooked fruit	Cooked fruit	Fresh fruit	Cooked fruit	Fresh fruit

underwent a drastic transformation. While pasta remained the most frequent first course, it was often substituted by a cold dish (i.e. cured meats or cold cuts, cheeses or eggs) or cooked vegetables to give space to the main dish based on meat, particularly beef or pork.

This shift toward modernity continued throughout the 1980s, but with the addition of three new phenomena (Vercelloni, 2000). The first was the start of the destructurization of meals with a clear simplification of the daily menu accompanied by a progressive decrease in bread and wine consumption, and the beginning of alternative away-from-home consumption of meals. The second phenomenon regards the spread of what can be called 'nutritional conscience', the beginning of the awareness of the relationship between food and health leading to interest in balancing meals from a nutritional perspective. Finally, the speedy rhythm of life, growth in female employment and an increase in single-family households contributed to an increased request for convenience foods.

17.3.3 The 1990s and the current period

The evolution of food habits over the last 15 years has brought about an abrupt halt to processes of 'modernization' and a general rediscovery of neo-traditional values (Vercelloni, 2000). Lifestyles seem to evolve according to the conviction that through one's daily choices it is possible to maintain good health and, in general, guarantee physical well-being. In particular, there has been growth in the awareness of direct relationship between health, well-being and food habits. Moreover, there seems to be a conviction that tasty and satisfying food is not necessarily an unhealthy one; products have appeared that combine these two qualities (Verbeke, 2006). Furthermore, in Italy, as in other Western countries, the culture of the body has taken hold with an equalization of the concepts of physical fitness and health, and behaviours linked to self-moderation rather than to giving up certain foods are generated. This is the modern Italian eating (Albertini and Celenza, 2001) based on the Mediterranean diet model. The strength of this model lies in its ability to combine new demands (nutritional balance and dietary rules to guarantee well-being) with established, traditional values (for natural and wholesome food). The Mediterranean diet seems to respond perfectly to the ideology of interconnected values of 'good', 'natural' and therefore 'healthy'. Indeed, the Mediterranean diet places importance on pasta and olive oil, favours alternative protein sources over red meat, and prefers raw rather than cooked vegetables (Ferro-Luzzi and Branca, 1995). In effect, it is representative of a balanced diet, with less fat and protein and, at least in theory, is more suitable for the reduced energy requirements of modern daily work routines.

The first issue in 1987 of the magazine *La Cucina Italiana* presented an important novelty compared with the past: the traditional everyday menus suggested for the main meal were accompanied by indications by nutritionists to highlight appetizing and perfectly balanced compositions from a nutritional standpoint. For daily menus, this means a return to pasta as the principal first course. Comparing the proposed first course dishes in 1977 and 1987 (Fig. 17.1), pasta holds first place, with a clear decrease of dishes based on rice and especially of those classified as 'other'. In 1977, 'other' consisted of protein sources such as cheese, cold cuts, eggs and canned tuna, while in 1987, there was a greater dependence on sources of complex carbohydrates, such as polenta, thus the traditional division of carbohydrates and proteins between the first and second courses was emphasized. Meanwhile, similarly to the previous period but different from the 1950s, the daily division of meals is not highlighted, nor is the importance of lunch over dinner.

Toward the beginnng of 1990s, there was an important trend with regard to the structure of lunch and dinner: the single course, making it easy to plan the entire meal based on one dish. Only recently has the term 'piatto unico' appeared in Italian language dictionaries and it is defined as 'containing sufficient and balanced nutrition to make up a complete meal alone' (Devoto Oli, 2000). Combining different ingredients, with specific cooking times and condiments, the 'piatto unico' provides the meal as a single portion, with the same items and quantities that could be used for several consecutive dishes. The first Italian cookbook on the topic is by Savina Roggero, 'Piatto unico all'italiana' (1977). Among the compositions, traditional and regional combinations prevail even if European (e.g. quiche lorraine) and non-European (e.g. cous cous, pasta with curry) dishes are included. The single-dish solution finds support among nutritionists for the ease of planning nutritious and well-balanced food. Typical items capable of providing, in a single dish, the nutritional content usually provided by a first and second course are pasta with legumes, pasta with meat and cheese toppings, minestrone soup with grated cheese, and pizza. The success of the single dish is owing to three factors: limited cost, easy-to-find ingredients independent of the region and season and versatility of the dishes (e.g. minestrone soup, pizza and pasta)

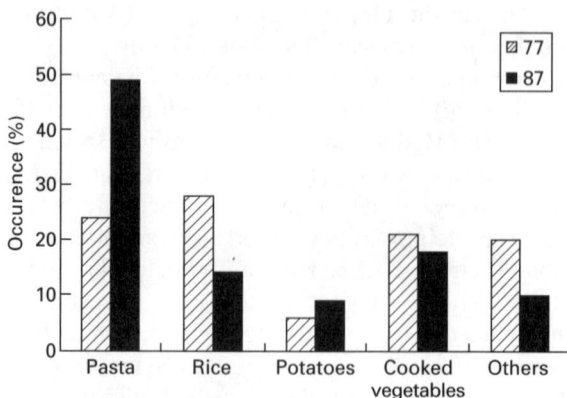

Fig. 17.1 Presence (occurrences %) of different first course in the menus suggested by *La Cucina Italiana* (years 1977 and 1987).

permitting variations in flavour, texture and colour (Montanari, 1998). It is a typical meal for those who eat lunch out and have dinner as their main meal. Different from other 'single' meals (i.e. hamburger), the single dish is seen as nutritionally favourable and, in its Italian form, it requires a table, plate and utensils and therefore includes the ritualistic aspect of a traditional meal. The characteristics of the 'Piatto Unico all'Italiana' may also explain the international success of pasta and pizza all over the world.

Scandals linked to several European food producing companies around the turn of the century had an important impact on Italian food habits (Albertini and Celenza, 2001). Loss of trust in safety of foodstuffs (Vercelloni, 2000) led consumers to search for certainty and increased trust in territorial traditions. This phenomenon is represented by the decrease in beef consumption which, as an important component of second course dishes, gets substituted by other types of meat (i.e. chicken and pork) or fish. The only meat-based products increasingly appearing at the table as a protein source are cold cuts and cured meats. These products, similarly to cheese, have become attractive for their convenience and image linked to typicality, and have acquired a position as a cold dish, as part of daily meals (ISMEA, 2007; Cipriani, 2003 Vercelloni, 2000).

Again, the menus suggested by *La Cucina Italiana* (2006) provide a picture of the food habits of Italians. Among the proposed menus, 6% are vegetarian, a fact not noted in earlier reference years (1957, 1977, 1987). Similar to menus in 1987, the main meal is characterized by an absolute prevalence of pasta as the first course (74%), compared with rice (16%), potato dishes (5%) or other sorts of dishes (5%). As for the main dish, beef is suggested in 32% of the menus, with fish and white meat in 37 and 21%, respectively. The proposal of Sunday brunch menus represents a novelty in *La Cucina Italiana*. This new trend is clearly imported and does not emphasize pasta, but rather vegetable pies and cold meat or fish salads.

17.4   Main meals and regional culinary traditions in Italy

The sequence of three dishes, as well as meal composition, does not differ significantly among Italian regions. Despite the structural uniformity, a thorough description of Italian meals must also consider the great diversification among dishes, which is a result of the varying gastronomic and culinary patrimony found in Italian regions (Capatti and Montanari, 1999; Cipriani, 2003; Piras and Medagliani, 2000). Without going into the historical reasons behind this fact, Tables 17.2–17.4 present well known dishes (also known outside Italy) in their different regional forms. The tables are based on recent sources (Dalla Cia and Pedrazzi, 2006). The intent is not to underline the regional differences, but simply to show the variability among dishes with the same name. The recipes are described according to geographic origin and with the division of distinct flavour principles into three categories (Rozin, 2000).

Table 17.2 Ingredients for lasagne in different Italian regions

Name (region)	Pasta	Basic ingredients	Flavour principles		
			Fat/oil	Liquid component	Aromatic, herbs, spices
Alla bolognese (Emilia Romagna)	Flour Eggs Spinach	Minced beef Minced pork Minced prosciutto Besciamella Parmigiano	Butter	Tomatoes Red wine	Onion Carrot Celery Ground pepper Nutmeg
Alla napoletana (Campania)	Flour Eggs	Pork meat-ball Mozzarella Ricotta Parmigiano Sausages	Extra virgin olive oil	Tomatoes White wine	Onion Carrot Celery Ground pepper
Ai funghi (Liguria)	Flour Eggs	Bacon Porcini Mushrooms Besciamella Parmigiano	Extra virgin olive oil	Tomatoes	Garlic cloves Rosemary Parsley Ground pepper

The first example regards lasagne, that is also largely popular outside of Italy. Table 17.2 outlines the recipes of three different regions. The list of principal ingredients includes typical products from the territories of origin as, for example, prosciutto and parmigiano in Emilia Romagna and mozzarella and ricotta in Campania. The Ligurian recipe, which is simpler and lighter, includes porcini mushrooms, which are distinctively typical of the cuisine of this region. Furthermore, regional differences are evident in all the flavour principle categories.

Fish soup is another traditional dish found along the entire Italian coast, where it is known by various characteristic names (Table 17.3). There are not great differences with regard to the type of fish used, with the exception of *novellame* (fry) which is included only in the Sicilian recipe. The most notable difference involves the type of seasoning used, making each particular recipe unique. For example, *caciucco* in Tuscany is distinguished by the use of hot spicy peppers, while vinegar and saffron are distinctive elements of the recipe along the coastal zone of the Marche. The Ligurian recipe is recognizable for its simplicity and the presence of mushrooms (*zuppa boldrò*). Finally, even if the various recipes may seem to have only marginal differences in composition, the appearance of the dishes is completely different and characteristic.

Territorial peculiarities can also be appreciated considering recipes for pasta in areas that are very near to one another. Recipes from three provinces in Emilia Romagna for stuffed, egg-based pasta cooked and served in broth

Table 17.3   Ingredients for fish soup in different Italian regions

Name	Region	Basic ingredients	Flavour principles		
			Fat/oil	Liquid component	Aromatic, herbs, spices, acidic ingredient
Cacciucco	Toscana	Cuttlefish Baby octopus Rock mullet Angler Scorpion fish Water hen Jumbo shrimp Toasted bread	Extra virgin olive oil	White wine Tomatoes Water	Onion Celery Parsley Red pepper Garlic cloves
Brodetto	Marche	Scorpion fish Water hen Grey mullet Mackerel Hake Sole Rock mullet Cuttlefish Tiger shrimp Toasted bread	Extra virgin olive oil	Water Tomatoes	Onion Parsley Garlic cloves Vinegar Saffron
Zuppa alla siracusana	Sicilia	Scorpion fish Water hen Grey mullet Baby octopus Mussel Toasted bread	Extra virgin olive oil	Water White wine Tomatoes	Onion Bay leaves Carrot Celery Garlic cloves
Zuppa di Boldrò	Liguriae Toscana	Angler Porcini Mushrooms Toasted bread	Extra virgin olive oil	Water Tomatoes	Celery Onion Parsley Carrot Ground pepper
Zuppa di nunnata	Sicilia	Fry Black olives Toasted bread	Extra virgin olive oil	Water	Parsley Garlic cloves
Ciuppin	Liguria	Scorpion fish Water hen Rock mullet Toasted bread	Extra virgin olive oil	White wine Water Tomatoes	Carrot Onion Celery Parsley Garlic cloves

result in different and characteristic shapes and aromas (Table 17.4). In this example, basic ingredients and different flavour principles result in dishes that are similar only in appearance.

The examples selected underline the depth of the roots regarding Italian regional culinary traditions, which may represent a stabilizing aspect for food habits (Fonte, 2002) and thus slow down the globalization of consumption models.

17.5   Beverages with meals

Statistical data relative to the consumption of beverages over the last five years indicate that 90% of the population drink mineral water with meals. It satisfies the daily water needs in more than half the population. Carbonated beverages, wine and beer are consumed daily by about half of the population but, on average, in much more moderate quantities. With regard to carbonated drinks, consumption seems to be sporadic and quite low: less than 4% of the population claim to consume less than 0.5 L per day. Wine and beer are the most popular alcoholic drinks: approximately half of the adult population claim to consume them. Wine is ranked the first among alcoholic beverages, although there are notable differences between the genders. While 72% of men claim to drink wine, the figure drops to 43% for women (ISTAT, 2007). These figures are even more dramatic when the quantity consumed is considered: 237 g/day for men, 69 g/day for women. An analogous difference exists also with regard to the consumption of beer, which is, again, more widespread among men (62%) with mean daily consumption pro capita of 30 g, compared with 31% among women with an average 10 g/day of beer consumed.

While the consumption of moderate quantities of wine appears constant, the consumption of beer is more occasional by nature and conditioned by the season: consumption takes place prevalently during the hottest months of the year. Also, the regional distribution seems to significantly influence the consumption of wine and beer. Apparently men living in the north consume approximately twice as much as men in the south and on the islands. There is the same regional influence on women, as the consumption of wine by women is greater in northern than in southern regions and on the islands.

The consumption of wine is deeply rooted in the food culture of the Mediterranean region, and in Italy in particular. In contrast to the situation in Northern European countries, the consumption of wine does not indicate a special meal or denote attitude for food, but it is considered a natural complement to the meal (Chatenoud et al., 2000). According to the culinary tradition, a properly selected wine accompanies each dish of a meal. Despite the deep changes in meal patterns, this tradition continues to be respected and the consumption of wine is closely linked to daily meals. Indeed, nearly all the adult wine drinkers (96%) claim to consume wine, and alcoholic

Table 17.4 Ingredients for filled fresh egg pasta to be served in broth from various localities in Emilia Romagna (centre north)

Name	Location	Shape	Basic ingredients	Flavour principles		Aromatic, herbs, spices
				Fat/oil	Liquid component	
Tortellini	Emilia, Bologna	Crimped and tied	Pork meat Dry cured ham Parmigiano Mortadella Eggs	Butter	White wine	Rosemary Onion Ground pepper Nutmeg
Cappelletti	Romagna Rimini	Crimped and tied	Capon Pork meat Crescenza Parmigiano Eggs	Olive oil Butter		Ground pepper Sage Rosemary Lemon skin
Anolini	Emilia, Parma	Circular	Beef meat Parmigiano Breadcrumbs Eggs	Olive oil Butter	Tomatoes Red wine Meat broth	Carrot Onion Celery Cloves Red pepper Nutmeg

beverages in general, in association with daily meals. Available data show that the consumption of wine and alcohol in general, is closely related to social behaviour and cultural tradition (DOXA, 2001). It is interesting to note that the daily amount of wine consumed at lunchtime has markedly decreased, further supporting the tendency of this meal to be less important (ECAS, 2001). The consumption of moderate quantities of wine as a natural complement to a meal is even further underlined by its constant consumption during the week, with no substantial difference between the work days and the weekend (Sieri, 2002).

The increased consumption of beverages outside of mealtimes (ISTAT, 2007) in the form of the 'aperitivo' which can be either non-alcoholic (47%) or alcoholic (31%) is a new tendency. Consumption takes place at a bar in late morning or late afternoon and involves the youngest part of the adult population: about 60% of the population between 18 and 44 years of age claim to consume non-alcoholic aperitif. It is interesting to note that the traditional consumption of an aperitif together with a small amount of food is transforming into a true cold light meal which, among young people, even substitutes the regular lunch or dinner.

Finally, a note about the use of coffee, a beverage which has been defined hedonistic and ritual (Diodato, 2001). The current use of the term coffee represents a moment to pause and relax (viz let's have a coffee, coffee break) (Cipriani, 1996). At the same time, its consumption is characterized by a rather rigid order for the different moments of the day and the way of consumption. There is also a sort of code for the way it is prepared based on individual habits and the occasion surrounding its consumption. Espresso, an inevitable breakfast tradition, is also the final element of lunch, whether it is a quick snack or a traditional meal. The consumption of coffee at the end of dinner is less popular.

17.6 Conclusions

Italian meal pattern is characterized by three meals: breakfast, lunch and dinner. In terms of both energy intake and conviviality, breakfast is less relevant than lunch or dinner. The latter two can alternatively represent the main meal of the day, depending on several social factors. However, the consumption of snacks continues to grow involving mostly the younger population. This tendency leads to a breakdown of the traditional meal pattern and to a shift toward more frequent consumption of light and often cold meals during the day, with a single main meal.

The daily main meal is consumed at home and its typical structure is defined by three courses: a first course of complex carbohydrates (primo piatto), a second course source of protein (secondo piatto), a side dish (salad or cooked vegetable), and finally either fruit or dessert. In the last 60 years, this structure has not changed much, despite the 'piatto unico' option for away-from-home meals and the simplification of the daily menu composition.

Changes in meal patterns and composition towards global homogenization of food habits tend to be slow in Italy. Social and cultural aspects account for this evidence. The main daily meal often equates with the concept of family meal that involves the presence of most family members and is structured. On the other hand, recent surveys indicate that family meals are declining. Regional culinary traditions are deeply rooted in the Italian culture and may represent a stabilizing aspect for food habits.

17.7   Acknowledgements

We are greatly indebted to Hely Tuorila for her encouragement to write this chapter and then for her comments on the text. The authors are also grateful to the Culinary Institute 'F. Martini' of Montecatini Terme (PT), Italy for the access to bibliographic sources.

17.8   References

Albertini, A. and Celenza, F. (2001) Report: nuovi trend nei determinanti sociali, culturali e ambientali degli stili alimentari delle famiglie di oggi. *Dossier 2001*. SIAN Az USL Bologna.

Alexander, D. (2000). The geography of Italian pasta. *Professional Geographer*, **52**(3), 553–566.

Artusi, P. (1891). *La scienza in cucina e l'arte di mangiare bene*. Torino: Enaudi. Edition 1995.

Capatti, A., Montanari M. (1999). *La Cucina Italiana. Storia di una cultura*. Roma: Laterza.

CENSIS, (2003). *L'arte del cum vivere: comportamenti, attese, valori condivisi degli Italiani a tavola*. Roma: Centro Studi Investimenti Sociali.

Chatenoud, L., Negri, E., La Vecchia, C., Volpato, O., Franceschi, S. (2000). Wine drinking and diet in Italy. *European Journal of Clinical Nutrition*, **54**, 177–197.

Chiva, M. (1997). Cultural aspects of meals and meal frequency. *British Journal of Nutrition*, **77** (Suppl. 1), S21–S28, doi 10.1079/BJN19970101.

Cipriani, A. (1996). *L'uomo è ciò che mangia*. Firenze: Maschietto & Musolino Editori.

Cipriani A. (2003). *Eating to Live*, Pistoia (Italy). RCP publisher.

Dalla Cia, M.V. and Pedrazzi V. (2006) Le lasagne. *La Cucina Italiana*, Gen., 85–89.

Dalla Cia, M.V. and Pedrazzi V. (2006) I tortellini. *La Cucina Italiana*, Nov., 93–97.

Dalla Cia, M.V. and Pedrazzi V. (2006) Zuppa di pesce. *La Cucina Italiana*, Jul., 83–87.

de Amicis, A. (1999). Valutazione dello stato nutrizionale. In: *Fondamenti di nutrizione umana*. *Mariani* Costantini A., Cannella C., Tomassi G. (eds.). Rome: Il pensiero scientifico Editore. 417–446.

de Groot, C.P.G.M., Schltettwein-Gsell, D., Schroll-Bjørnsbo, K. and van Staveren W.A. (1998). Meal patterns and food selection of elderly people from six European towns. *Food Quality and Preference*, **9**(6), 479–486.

ECAS (2001). *Trends in drinking patterns in fifteen European countries, 1950 to 2000*. STAKES Helsinki.

Devoto, G. and Oli, G.C. (2001). *Il dizionario della lingua italiana*. Firenze: Le Monnier.

Diodato, L. (2001) *Il linguaggio del cibo*. Soneria Mannelli (Italy): Rubbettino Editore.

DOXA Osservatorio permanente sui giovani e l'alcool (2001). *Gli Italiani e l'alcool: consumi, tendenze e atteggiamenti in Italia e nelle regioni. 4th DOXA National survey.* Book no. 14. Roma: DOXA.

Ferro-Luzzi, A., Branca, F. (1995). Mediterranean diet, Italian-style: prototype of a healthy diet. *American Journal of Clinical Nutrition*, **61**(6): 1338S–1345S.

Fonte, M. (2002). Food systems, consumption models and risk perception in late modernity. *International Journal of Sociology of Agriculture and Food*, **10**(1), 13–21.

Gracia, A., Albisu, M. (2001). Food consumption in the European Union: main determinants and country differences. *Agribusiness*, **14**(4), 469–488, doi 10.1002/agr.1030.

Kjaernes, U. (2006) Trust and Distrust: Cognitive Decisions or Social Relations? *Journal of Risk Research*, **9**(8) 911–932.

ISTAT, (2006). *Figli e Famiglie: I tempi delle donne*. Roma: Istituto Nazionale di Statistica.

ISTAT, (2007). *La vita quotidiana nel 2006*. Roma: Istituto Nazionale di Statistica.

ISMEA (2005). *Consumi Extradomestici*. Roma: Istituto di Servizi per il Mercato Agricolo Alimentare.

ISMEA (2007). *Gli acquisti alimentari in Italia: tendenze recenti e nuovi profili di consumo*. Roma: Istituto di Servizi per il Mercato Agricolo Alimentare.

La cucina Italiana (1957). Vol 1–12. Milano: Editrice Quadratum.

La cucina Italiana (1977). Vol 1–12. Milano: Editrice Quadratum.

La cucina Italiana (1987). Vol 1–12. Milano: Editrice Quadratum.

La cucina Italiana (2006). Vol 1–12. Milano: Editrice Quadratum.

Mitchell, J. (2006). Food acceptance and acculturation. *Journal of Food Service*, **17**, 77–83.

Montanari M. (1992). *Convivio oggi: storia e cultura dei piaceri della tavola nell'età contemporanea*. Bari: Editori Laterza.

Pagliarini E., Ratti S., Balzaretti C. and Dragoni I. (2005). Consumer testing with children on food combination for school lunch. *Food Quality and Preference*, **2**(16): 131–138.

Piras, C and Medagliani E. (2000). *Culinaria Italy*. Cologne: Koneman.

Policastro S. (2005). *Children's consumption of, and preference for, school lunch foods.* PhD Thesis, University of Basilicata, Potenza, Italy.

Roggero, S. (1977). *Piatto unico all'italiana*. Milano: Mondadori.

Rozin E., (2000). The role of Flavour in the Meal and the Culture. In H.L. Meiselman (Ed.), *Dimension of the Meal*, (pp. 134–142), Gaithersburg, USA: Aspen Publisher.

Repubblica Italiana, Ministero della Difesa (2007). *Specifiche Tecniche per il Servizio di Ristorazione, Catering Completo e Catering Veicolato.* (published online: www.difesa.it).

Sieri S. *et al.* (2002). Patterns of alcohol consumption in 10 European countries participating in the European prospective investigation into cancer and nutrition (EPIC) project. *Public Health Nutrition*, **5**(6B), 1287–1296, doi:10.1079/PHN2002405.

Turrini, A., Saba, A., Perrone. D., Cialda, E. and D'Amicis (2001). Food consumption patterns in Italy: the INN-CA study 1994–1996. *European Journal of Clinical Nutrition*, (**55**), 571–588.

Vanelli, M., *et al.* (2005) Breakfast habits of 1202 Northern Italian Children admitted to a summer sport school. Breakfast skipping is associated with overweight and obesity. *Acta Biomed*, (**76**) 79–85.

Verbeke, W. (2006). Functional foods: consumers willingness to compromise on taste for health? *Food Quality and Preference*, **17**, 126–131.

Vercelloni, L. (2001). Alimentary habits in Italy from the eighties to the twenty-first century (abstract in English), *Sociologia del Lavoro*, **83**, 141–149.

18

Brazilian meals

R. Deliza, Embrapa Food Technology, Brazil, and L. Casotti, Federal University of Rio de Janeiro (UFRJ), Brazil

Abstract: The meals of daily domestic life in Brazil are described within the context of urban centres, whose miscegenated culture has resulted in a confluence of tastes, aromas and flavours from Africa, Portugal, Germany, Italy and the Orient. The focus is on home cooking in large Brazilian cities, although the growing habit of eating out, associated with many recent changes in family, work and leisure environments, is noted and its effect on family meals is explored. Regional diversities are described as well as the strong 'rice and beans' culture.

Key words: Brazilian meals, rice and beans, miscegenated culture, home cooking, urban food.

18.1 Introduction

The anthropologist Roberto DaMatta (1984, p. 61) describes Brazil as 'a society that struggles with its different self-perspectives'. What perspectives can be chosen to describe everyday meals in a country of such enormous geographical dimensions, enormous socioeconomic differences and in which 185 million inhabitants (IBGE, 2007) live in such an array of subcultures? How can one provide an overview of the quotidian food of a people that can be described in an almost infinite number of categories, combinations and segments?

The authors were forced to make some choices; and they were done based not only on the availability of the information gleaned, but also on what was considered important within the 'meals' category such that Brazilians might be better understood as a society, as a culture and as a people with a history. The codes that determine what one eats and how one eats in Brazil are complex; however, they are rich in information and evidence that is essential to understanding the curious diversity of the Brazilian society.

Owing to the difficulty in finding research studies carried out in smaller cities, the meals of daily domestic life are mainly described within the context of urban centres. The eating habits that are typical of smaller cities or of more rural areas can be found elsewhere in studies on the history of diet and food in Brazil. Our decision to focus on the home cooking of large Brazilian cities brought the additional difficulty of separating home-cooked meals from the growing habit of eating out, a trend that occurs for a multitude of reasons, and which is associated with many recent changes in family, work and leisure environments.

In relation to Brazil, the vast universe of regional cooking makes the very characterization of 'traditional Brazilian food' a challenging one. It is not possible to understand the everyday Brazilian diet without understanding a little about the roots of Brazilian culture. The country's miscegenated culture is based on a confluence of tastes, aromas and flavours. Brazilian food is an amalgam based on the traditions of indigenous people (the first inhabitants of the country), the delicacies of Africa (brought by the slaves who came to work in plantations and in the houses of the wealthy), the luscious cuisine of the Portuguese (the main colonizers), and the food of several other immigrant groups such as Germans, Italians and Orientals who came at different times and to different regions. To this day, Brazil's unique history is reflected in the contrasts to be witnessed at mealtimes (SENAC. DN, 1998, 1999, 1999b, 2000, 2003).

Starting in the 1970s, the literature on dietetics in Brazil began to acquire a more quantitative profile and became more concerned with the concept of food safety and nutrition. In this report on everyday meals served in Brazilian homes, it was decided to describe them in somewhat qualitative terms, in the belief that doing so might elucidate more clearly who Brazilians are, embracing elements of identity, social and economic organization, historical perspectives and ethnographic aspects that go beyond the physiological necessity of eating and nourishing oneself (Fisberg et al., 2002).

Let us begin then, by talking about a past which is still recent – Brazil is, after all, a young country of only a little more than 500 years. We will look at the current situation of Brazilian eating habits, and then at the future by pointing out the recent signs of change in everyday dietary habits.

18.2   Historical perspective

As Brazil was gradually colonized, ingredients, cooking methods and habits gradually became intermingled. Our food is like our blend of races. Multiculturalism is, without a doubt, the most important feature of Brazilian food (Freire, 1987; Câmara Cascudo, 1983).

The various tribes living throughout the Brazilian territory had diverse and complex cultures that influenced the development of the Brazilian people in several aspects, including language, building methods, medicinal and food plants that were utilized and types of foodstuffs (Lima, 1999).

Though the occupation of Brazil began in the Northeast with Dutch, French, Scots and Germans, it was the Portuguese who were the main forgers of the Brazilian genealogical tree. The contact between the Portuguese culture and cultures of different indigenous tribes spread throughout Brazilian territory produced an exchange of values and customs, thereby giving rise to the so-called Luso-Brazilian culture.

What are the principal manifestations of indigenous cultural heritage in the food of Brazil? From indigenous peoples, we inherited the cassava flour [from root that is considered to be the most Brazilian of all plants (Lima, 1999)] in its sweet and savoury variations, chilli, sweet potato, cacao, and heart of palm; countless fruits were also introduced by them as well as a variety of fish and game, each of which required different methods of preparation. It is important to emphasize the role of indigenous women in the kitchen, in the vegetable plantation, in weaving and in utensil-making, and in the ways of domestic hygiene (Lima, 1999). With respect to indigenous foods, the first thing noticed by the Portuguese was the cassava, a root that the Indians transformed using traditional methods into flour that accompanied almost everything that they ate. This flour was gradually incorporated into the Portuguese cooking with its typical meat and fish broths, and later on into the foods of Africa that the slaves had brought with them (Freire, 1987; Lima, 1999).

In relation to the black people that came from Africa, it can be seen that their influence on the development of the Brazilian people has its roots in the many African ethnic groups that had, in turn, been influenced by European invasions of their continent. From black people, we inherited a special kind of fusion: that of religion and food. The food, its special methods of preparation and utensils required to prepare it, take on different meanings within a sophisticated system of power and beliefs in Afro Brazilian religious rituals such as the *candomblé*. To eat and drink is to establish links and communication between man and his Gods or ancestors (D'Oxum, 2008). A live chicken, for example, is given as an offering, along with other foods such as palm oil, extracted from a type of palm tree, beans, maize and cassava. All over the country, there are shops where it is possible to buy foods and objects used in religious cults of African origin. From black people, we also inherited many superstitions that link food with religious rituals (Lima, 1999; SENAC. DN, 1999).

What were the most important things introduced by the Portuguese into the Brazilian diet? From Portugal came cattle, oxen, sheep, pigs and chickens, pigeons and geese. Other foods brought by the colonizer were sugarcane, wheat, rice, butter, garlic, carrots, several fruits such as oranges and lemons and figs, as well as traditional dishes such as salted cod, *cozido*, a traditional Portuguese stew, and sweet dishes such as the popular caramel pudding (Freire, 1987; Lima, 1999).

In the development of the Brazilian society, the importance of the 18th century cannot be overemphasized as a time of ethnic and cultural blending

(Lima, 1999). To the main groups consisting of Portuguese, Indians and Blacks, many other immigrant groups made their mark: French, Spanish, Dutch, German, Polish, Italian, Austrian, Belgian, Jewish, and English, and, a little later on, Japanese.

The sheer range of diversity in Brazil complicates the construction of the country's culinary heritage. Although during the first centuries of colonization, writings on food in Brazil tended toward a fascination with the abundance, the amenable climate and the richness of the soil (in which 'anything grows') in the 19th century, they also mention the inferiority of our foodstuffs relative to those found in Europe. Some criticize the lack of differentiation between the diet of the rich and the poor: the ubiquitous presence first of beans and cassava flour, then later on, of beans and rice (Carneiro, 2003).

18.3   Regional food

Though Brazilian food, from the north of the country to the south, includes the faithful partnership of rice and beans, each region of Brazil has its own style of cooking. Regional differences include the use of herbs, seasonings, sauces and culinary secrets, built upon the miscegenation which has influenced the life of a particular region. In a country that is the size of a continent, travelling from one state to another is tantamount to travelling from one country to another in culinary terms. Regional cookery identifies each place – an added appeal to tourists who are drawn to rich menus and eating traditions that often involve curious methods of preparation, serving and enjoyment.

In addition, the agricultural trends of each region had an impact on regional eating habits as well. The agriculture of Brazil is a testament to the spreading populations that were the result of the economic cycles of great European metropolises, such as the prosperity of the sugarcane and coffee boom, and later on, when faded, those regions were left with a negative legacy.

Regional cooking styles also serve to differentiate the local from the national. The intense internal migratory movements caused by socioeconomic factors, which led to huge contingents of populations moving from the north and northeast to the south or southeast, must also be remembered. These movements also contributed to an interesting exchange and intermingling of regional cuisines. It is not unusual for domestic maids from the northeast to introduce recipes and different ways of preparing regional dishes to families in the Southeast. There are also stories of waiters who, having come from other regions, became the owners of restaurants serving regional food, thousands of kilometres from their cultural roots (Lima, 1999). Studying the distribution of races in the Brazilian population enhances our understanding of the great national diversity in terms of diet. This diversity may be seen by analysing the racial demographics of women, as can be seen in Table 18.1.

Getting to know a little about the roots of regional cookery will help us to understand Brazilian meals. For example, the food of the North, where the

Table 18.1 Race of the female head of household or spouse – Brazil and major regions of Brazil, January 2003 (Percentage)

Race	Brazil	Major region				
		North	Northeast	Southeast	South	Central West
European descent	54.61	25.53	31.82	62.93	81.88	47.29
Miscegenation between European and African descents	37.12	69.06	60.10	26.83	12.67	47.20
African descent	7.30	4.17	7.54	8.97	4.71	4.66
Asian	0.62	0.25	0.26	0.96	0.45	0.54
Indigenous	0.34	0.99	0.28	0.30	0.28	0.35

Source: IBGE (2008).

Amazon is located, is greatly influenced by indigenous culture in its unusual flavours. The use of a countless variety of fruit marks the food of the Amazon region; they include the exciting *guaraná*, *açaí*, *pitanga*, cashew, passion fruit, star fruit, and different kinds of banana, to name but a few (SENAC. DN., 2000).

The various indigenous tribes of the Amazon region contributed to the history of food in Brazil, not only curious types of fish, roots and fruit, but also, different ways of preparing them with fireplaces, ovens built into the earth, utensils made of clay, wood and plant fibre which are used to make equipment to dry, grind, mill, toast, and transport food. More than just a meeting of cultures, the Amazon region is like a veritable confrontation between the cuisines of indigenous people, Europeans and Africans, the latter no less exuberant in their use of oils, seasonings, root vegetables and beans (SENAC. DN, 2000). Table 18.2 shows an example of a typical meal from the Amazon region as well as its ingredients.

In the northeast region, the cuisine of the state of Bahía is well worth noting. Here, there exists a multifaceted cuisine, capable of uniting flavours, aromas and colours that have evolved and changed over time, and with a strong African accent derived from the use of palm oil, coconut milk and pepper. Bahian cuisine can be seen as a vast social and economic laboratory that unites a cuisine using a multitude of ingredients with a religious faith that combines Catholicism with religions brought from Africa. In parallel fashion, Catholic saints are interposed on the religious figures of *candomblé*. Many typical dishes of Bahia are eaten from the hand in the street, where they are purchased from women wearing typical Bahian clothing. An important speciality is the *acarajé*, a bean fritter fried in palm oil – the latter being considered a sign of the permanence of the flavours of Africa on the palate

Table 18.2   Example of a typical meal from the Amazon region

Typical meal	Occasion of consumption	Ingredients
Tambaqui fish stew, rice and *pirão*	Lunch or dinner	Tambaqui fish stew: fish, potatoes, boiled eggs, onion, tomatoes, garlic, peppers, palm oil, coconut milk, salt *Pirão*: dry cassava flour, onion, pepper, salt, fish stock
Dessert: *açaí* pudding		*Açaí* pudding: condensed milk, *açaí* juice, eggs and sugar

Source: SENAC, DN (2000).

Table 18.3   Example of a typical meal from the northeast of Brazil

Typical meal	Occasion	Ingredients
Chicken in blood sauce, *fradinho* beans, rice and yellow *farofa*	Lunch	Chicken in blood sauce: chicken, chicken blood, bacon, garlic, onion, tomato, pepper, vinegar, cooking oil, salt Yellow *farofa*: cassava flour, red palm oil, onion, salt
Dessert: white coconut sweet		White coconut sweet: coconut peeled and grated, white sugar, cloves and cinnamon sticks

Source: SENAC. DN (1999b).

of Brazilians (SENAC, DN., 1999; Lima, 1999). Table 18.3 shows an example of a typical meal from the northeast and its ingredients.

One of the planet's most important ecosystems is found in the middle of Brazil in the *Pantanal* (swamp) region. The constant movement of waters in periods of heavy rains and droughts provides a fertile environment for an enormous variety of fish, game, vegetation and exuberant plants, upon which is based a varied and regional cuisine. Isolated geographically, bringing supplies in was always difficult and this difficulty made way for a cuisine that emphasizes meats and sausages – employing varying and sophisticated preparation methods – as the major standouts (SENAC. DN., 2003). A typical *pantanal* meal and its ingredients are shown in Table 18.4.

The food of the state of Minas Gerais (within the south-eastern region) is characterized by many combinations of foods whose source is purported to be those who came to exploit the region's gold mines. It is said that these miners very much enjoyed snacking which may explain the famous appetizers of Minas Gerais. The lack of food experienced by the region in the colonial period caused Minas Gerais to store a bit of everything and builds the tradition of the 'table of abundance', where concern about how dishes go together is

Table 18.4 A typical *pantanal* meal and its ingredients

Typical meal	Occasion of consumption	Ingredients
Beef with cassava meal, rice, beans, and mashed potatoes Dessert: banana compote	Lunch or dinner	Beef with manioc meal: beef steak, garlic crushed, vinegar, roasted manioc meal, oil for frying Banana compote: banana, sugar, lemon juice and water

Source: SENAC. DN (2003).

Table 18.5 Example of a typical meal from the Minas Gerais

Typical meal	Occasion of consumption	Ingredients
Sautéed pork, trooper beans, shredded kale, white rice, golden fried onions, fried eggs, and crackling	Lunch	Trooper beans: black beans, cassava flour, sausages, garlic, onion, fat bacon for crackling, hard-boiled eggs, salt
Dessert: *Mineiro* guava with typical Minas white cheese		*Mineiro* guava: guava pulp and sugar

Source: SENAC. DN (1998).

scant. An example can be seen in Table 18.5. This region is also characterized by excellent dairies. From Minas Gerais comes the famous *pão-de-queijo* (cheese-bread) which is consumed in café chains all over the country and has been exported to several places around the world (SENAC. DN., 1998; Lima, 1999).

In the south of Brazil where the climate is subtropical, the culinary diversity of the state of Rio Grande do Sul is another outstanding example. The dishes typical of the area were introduced at varying times by the different ethnic groups that inhabited the region. One of the indigenous tribes, the *Guaranis*, is always remembered for the influence of a favoured beverage: no matter where in Brazil they may be, consumption of the beverage serves to identify those from Rio Grande do Sul. *Chimarrão*, as it is called, is a type of bitter *mate* tea that is said to be a stimulant that prevents fatigue; however, its true value lies in the cordiality it effuses since the same drinking spout is shared by several people. The Indians of the region were joined by several colonies of European immigrants, including Spaniards, Germans, Poles, Italians, and Jews, who give rise to a variety and abundance of foods. But within this abundance, it is the variety of meats, typically barbecued, that has a prominent place in the food of the South (SENAC. DN., 1999), as can be seen in Table 18.6.

Table 18.6 Example of a typical meal from the south of Brazil

Typical meal	Occasion	Ingredients and preparation method
Picanha barbecue with rice, beans, *farofa* and salad	Lunch	Cap of rump smothered in coarse salt, cooked over charcoal or in the oven *Farofa*: cassava flour, oil or butter, garlic, onion and salt. Typical salad: lettuce, tomato and onion
Dessert: sago		Sago: red wine, natural grape juice, water, sugar and sago.

Source: SENAC. DN (1999).

18.4 Rice and beans, and other combinations

Ethnic, economic and regional differences all seem to soften with the combination of two foods that predominate in the daily lives of Brazilian families, whether rich or poor, whether residents of inland or coastal areas: rice and beans, a combination considered to be the foundation of Brazilian home cooking (Barbosa, 2007; Câmara Cascudo, 1983; Casotti, 2002; Cheung, 2007; DaMatta, 1984; Fisberg *et al.*, 2002; Maciel, 2004; Mattos and Martins, 2000).

The historian Câmara Cascudo (1983) reminds us that until the 17th century, the mixture of beans with cassava flour was a staple in Brazil and that only later (during the 18th century), when rice production takes off, does the latter replace cassava flour. However, cassava flour was not abandoned; it was, rather, repositioned as one more important component of everyday meals. It is used either plain, or as the main ingredient of *farofa* [cassava flour toasted in a frying pan with olive oil or butter, salt and other seasonings (Lima, 1999)] and seems to provide a better binding element to the foods which were predominant in the daily lives of Brazilian families (Câmara Cascudo, 1983; DaMatta, 1984).

In Brazil, the popularity of 'rice and beans' gave way to an expression used metaphorically to mean 'daily routine'. If somebody quips 'it was just rice and beans', it means that nothing unusual has happened, since all Brazilians recognize this combination to be the routine, the everyday (DaMatta, 1984, p. 56). The anthropologist sees yet further meanings to this sticky mass-forming conglomeration of rice and beans – neither black or brown (like beans) nor white (like rice). This mixture lends itself to new colours, as in our mixing of races. Another expression of Brazilian folklore seems to underline the importance and frequency of the ubiquitous rice. 'Rice of the Party' is the name given to somebody who you always see at social get-togethers; they are never absent – just as rice is never absent at Brazilian mealtimes.

The importance of rice and beans in the everyday Brazilian diet seems to be agreed upon not only by historians and anthropologists who recount the history of our food (Freire, 1987; Câmara Cascudo, 1983) but also by different types of more recent studies (Barbosa, 2007; Casotti, 2002, Cheung, 2007). The wide range of the consensus led Barbosa (2007) to refute (in the case of Brazil), the idea that tastes are becoming homogenized, and concluding that Brazilians have 'homogenized the taste of tradition', since 90% of respondents to her survey buy the basic ingredients for their meals and prepare them at home, demonstrating that they still participate actively in the preparation of meals.

The development, so to speak, of the Brazilian people based on the constant mixing of several races ever since the arrival of the first Europeans seems to be reflected in the food; food which also seems to combine and connect. In France and Italy, dishes are served individually and sequentially; the same is true for Asian countries. In the United States, salads, too, are served as a separate course. In Brazil, however, several types of food are not only served together at the table, but are also placed on the same plate – something which also contributes to the 'relational cookery' of Brazil (DaMatta, 1984).

Many of the dishes which are both typical and important in the Brazilian repertoire are also, in and of themselves, mixtures of ingredients. *Cozido*,[1] *feijoada*[2] and *moqueca*,[3] for example, all seem to typify the Brazilian preference for one-pot meals that are neither liquid nor solid and which combine 'something of everything' (DaMatta, 1984 p. 63). Mention should also be made of highly esteemed secondary ingredients such as scattered-on flours and sauces that help bridge the way between liquid and solid dishes – dishes that value personal relations – relations to be celebrated at bounteous tables (Gonçalves, 2002). In a study that compared Brazilian *feijoada* and American soul food, Fry (2001) noted that in Brazil, *feijoada* carries the idea of integration – a meeting of three races – whereas in the United States, soul food is more associated with black people. Another instance of mixing noted in the study of Barbosa (2007, p. 96) is in the number of cooking techniques that can be found in a single everyday Brazilian home menu: 'rice, beans, steak, french-fried potatoes and stewed green beans' are dishes that involve five different preparation techniques: sautéing, boiling, frying, toasting and raw.

Gilberto Freire, the historian (1997, p. 26), when writing on the sociology of sweets in Brazil, reported that Brazilian sweets have also travelled a path along which ethnic groups, social cultures and classes have intertwined, an intertwining similar to that of our music and football.

As anthropologist Roberto DaMatta (1984) has observed, it is as though

---

[1] *Cozido* is a Luso-Brazilian dish prepared with beef, salted meats, vegetables, potatoes and bananas. It is served with *pirão* – a very thick sauce prepared with cooking broth and cassava flour
[2] *Feijoada* is considered to be the most popular dish in Brazil as well as the most Brazilian. Included with the beans are herbs, cuts of pork, beef, *linguiça* and *paio* (types of sausage), sun-dried beef and different seasonings. There are many regional variations.
[3] *Moqueca* is an Afro-Brazilian dish prepared with fish or shrimp, palm oil, many seasonings and pepper. Regional variations exist.

eating these mixed dishes could attenuate the large economic disparities existing within the Brazilian population or, as in the case of *feijoada*, the best known 'national dish', perhaps even redeem us from the grim period of slavery. This dish has acquired a special place in the hearts of Brazilians and its praises have been sung in verses and poems by important Brazilian musicians and poets: they seem to take pride in the democratic mix of colours and flavours.

18.5   The day-to-day rhythm of meals in Brazil

There is no social system where notions of time and space are absent. In many societies, the two concepts become fused. Time and space is 'constructed' by meals and meals themselves are, likewise, constructed by families. Our different memories of meals in Brazil are related in terms of quality, organization in space and time, the pleasure provided by food, and the exchanges around the table. We shall discuss the main everyday routines of eating at the table and not of the unusual, such as commemorative meals. As such, we agree with the great Brazilian historian, Câmara Cascudo (1983), for whom the study of the history of food should 'proceed on the basis of domestic information'.

The same historian reminds us of the organization of meals in 'Old Brazil' (up to the beginning of the 20th century): the first meal, a 'lunch' at around seven o' clock in the morning; the second, the 'dinner' at around midday; later followed by the 'snack', a short meal around three in the afternoon and the 'supper' at around six o' clock. In colonial Brazil, although mealtimes were described as being without elegance or sophistication, value is always placed on the intimacy of the family reunion; this seems to explain the custom of neither receiving nor making visits at meal times (Lima, 1999) or of considering that a true meal should 'never be had in a hurry' (DaMatta, 1997, p. 197).

This 'traditional Brazil' is substituted by a 'Modern or Contemporary Brazil' with meals that now have the following sequence: 'breakfast', 'lunch', 'dinner', i.e. an eating model based on three main meals is what best characterizes the Brazilian food plan (Canesqui and Garcia, 2005). In Cheung's (2007) study, 84% of the interviewees said they had three (44%) or four (40%) meals per day as well as having a more organized lunch than dinner, the latter being described by many as consisting of a snack of bread, fruit juice and smoothies.

But could it be that the number of daily meals in Brazil has really diminished? Even though doctors and nutritionists recommend three meals be eaten during the day, Brazilians report eating fewer meals than this. Some of the research, however (Casotti, 2002; Barbosa, 2007) suggest that Brazilians have a habit of eating between meals and variously refer to this habit as 'snacking', 'have a little something to eat', 'eating junk' or simply resort to referring to what they are eating using the diminutive form: 'um cafezinho' ('a small coffee'), 'um pãozinho' ('a little bread roll'), um 'lanchinho' ('a little snack'). Among other possible reasons for the decrease in the number of meals consumed or even for the increase in the habit of snacking between meals is the increase in the distances in large urban centres.

With reference to the day's first meal in Brazil, it's interesting to note that, in contrast to Portugal, where the name translates as 'little lunch', in Brazil, the name of the meal translates as 'morning coffee'; however, in other languages, we don't often find instances where the item consumed is the name of the meal. Studies show (Cheung, 2007; Barbosa, 2007; Casotti, 2002) that the main food combination eaten during breakfast in Brazil, across all economic classes, is coffee and bread, although in the upper classes, breakfast includes not only a greater variety of items, but also, items of a higher nutritional value such as fruits (papaya, pineapple, watermelon, melon, banana), yogurt and breakfast cereals. Even though coffee is thought to be a beverage for adults, it is usual to call the children to 'have coffee' even though children and young people evidently prefer milk with chocolate powders.

Which meal is the most important one for Brazilians? The study by Mattos and Martins (2000) commented by Cheung (2007) shows lunch, where rice, beans, meat and vegetables predominate on family plates, to be the most important meal for Brazilians. Being originally the food of slaves, beans are referred to as 'poor man's meat' in some parts of Brazil, owing to their high protein content which can substitute for animal protein. Beans are commonly and traditionally prepared with thick broth and whole beans. Different types and colours of beans can also be found in the daily life of different regions. In Rio de Janeiro for example, black beans are favoured by most, whereas in São Paulo, a nearby city, brown beans are the ones most often seen on family tables. Breakfast in large urban centres is not considered to be a meal which the family eats together; it is frequently consumed individually, led by the coffee–milk–bread trio, by all economic classes. The lunch menu is quite homogeneous in cities throughout Brazil: 94% of interviewees reported eating rice and beans accompanied by red meat (69%), chicken (42%), salads (30%), and spaghetti (24%). What about Brazilian evening meals? Although there is a tendency for the evening meal to become a less important meal in day-to-day family life, and, also, for it to be composed of lighter foods, some studies show that in middle and lower middle classes 'it is common to serve the same menu at dinner as at lunch' (Barbosa, 2007; Casotti, 2002). Examples for the mid-day meal and for the evening meal on week days and Sundays can be seen in Table 18.7.

With respect to the time spent on the three different meals, the study by Cheung (2007) on the consumers of four large Brazilian urban centres indicated that breakfast was the shortest meal: 46% of interviewees said they spent less than 10 min on breakfast, whereas in the other two meals, greater periods of time are more usual (10 to 20 min or 20 to 40 min). Individuals with

Table 18.7   Examples for the Brazilian mid-day and evening meals

Lunch		Evening meal
Mon – Sat	Sundays	Mon – Sat
1 Rice, beans, green vegetable and steak	1 Chicken pie	1 Vegetable soup
2 Rice, beans, fish and mashed potato	2 Lasagne, Russian salad and chicken	2 Chicken soup
3 Rice, beans, pasta, meat	3 Barbecue (beef, sausage, chicken) served with rice and toasted cassava flour	3 Rice, beans, meat and salad
4 Pasta, ground beef and salad	4 *Feijoada* served with rice, collard greens and cassava flour	4 Bread and cold cuts
5 Rice, beans, fried egg and salad	5 Beef or chicken stroganoff	5 Pasta
6 Rice, beans, roast beef and *farofa*	6 Roast chicken with rice, cassava flour and Russian salad	6 Grilled chicken with salad and rice
7 Fish stew with rice and cassava sauce	7 Shrimp *bobó*	
8 Stewed chicken with potato, rice and beans		

Source: Cheung (2007) and Casotti (2002).

smaller incomes who do not have a maid at home or the ease provided by certain appliances, say they eat even more quickly, and, oftentimes, only eat two meals a day (Cheung, 2007).

Regarding the space where meals are normally eaten at home in Brazil, it is important, firstly, to remember that until the middle of the 20th century, kitchens were like appendages to the house, not highly esteemed socially, and characterized as spaces oriented more to the outside yard than to the interior of the house. Kitchens were described as hot, smoky, greasy places, with hanging meats, and not much hygiene. The living room and dining room, however, are described as being spaces more strongly associated with visitors and society – spaces for exhibition and for hierarchy – not so much for those who lived in the house (DaMatta, 1984; Lima, 1999).

More recent studies (Cheung, 2007; Barbosa, 2007), meanwhile, suggest much more informality at mealtimes, both with respect to where meals are eaten, and also regarding the presentation of the family's everyday dishes. The informality of meals includes the TV, for example, which may be on during meals; the microwave, the use of which obviates having to put dishes in serving bowls on the table; and the plating food directly from cooking pans on the stovetop or from plastic containers kept in the fridge. When asked, in Barbosa's (2007) study, about what makes food tasty, 82% said it was flavour, while only 6% of the sample referred to its appearance. Cheung's (2007) study showed that while the majority of interviewees (72%) said they ate meals together as a family and shared the same dishes, 21% said that although they ate the same dishes, the food was consumed at different times of the day by different members of the family.

But while Brazilians report that the appearance of food does not matter, they say that its temperature does. Do Brazilians enjoy hot food despite the predominantly tropical climate? The climate seems not to dampen the preference for hot food. Brazilians have two hot meals per day, not including hot coffee (the national preference at breakfast time), always accompanied by bread, preferably toasted, with butter. It is also interesting to note that the hot climate has not changed another habit brought from Europe a long time ago and which can be witnessed throughout Brazil: the habit of consuming soup (hot) at night-time, and sometimes during lunch too.

18.5.1   Socioeconomic changes affecting food consumption

Another aspect that merits our attention with respect to the Brazilian meal is with regard to the impacts of economic and social changes on food consumption. Over the years, the diet of Brazilians, especially in major cities, has incurred constant transformations, mostly owing to the growth of urban populations and, more recently, to decreased inflation. It is known that urban settlement affects personal lifestyles considerably and, concomitantly, diet standards (Schlindwein, 2006). It is, therefore, important to understand how this process manifests itself in Brazil. Table 18.8 shows the evolution of the process of urban settlement over recent decades. The growing migration of people moving from the countryside to cities over the course of 30 years can be seen in Table 18.8. This demographic change has been observed in all regions in Brazil, and especially so in the Southeast region, which shows the highest rates of urban settlement.

The Brazilian Institute for Geography and Statistics (IBGE) data show that the average number of children per woman decreased from 4.4 in 1980 to 2.3 in 2000, mainly as a consequence of women being incorporated by the workforce. Families without children or whose children are over 30 years of

Table 18.8   Evolution of the population in rural/urban terms, Brazil 1970–2000

	Year				
	1970	1980	1991	1996	2000
Total	**93 134 846**	**119 011 052**	**145 825 475**	**157 070 163**	**169 799 170**
Urban	52 097 260	80 437 327	110 990 980	123 076 831	137 953 959
Rural	41 037 586	38 573 725	35 834 485	33 993 332	31 845 211
Percentage					
Urban	55.94	67.59	75.59	78.36	81.25
Rural	44.06	32.41	24.41	21.64	18.75

Source: IBGE (2008)

age are considered independent and already make up 23% of the population. The reflex of this behaviour is in the tendency of metropolitan populations to combine 'traditional' diet standards (meals consumed at home and which require a certain amount of work before consumption) with 'modern' diet standards (food prepared quickly and easily and food eaten outside the home). In families where women have a high level of education and work outside the home, there is a trend towards the 'modern' diet and, in families where the elderly are present, the 'traditional' standard is upheld (Bertasso, 2000), as evidenced by supermarket sales of ready-to-eat products. A study carried out by Schlindwein (2006) showed that people would prefer to eat frozen food three to four times a week, however, average consumption is only 3.8 times per month. Pizza has a faithful following: class 'A' and 'B' consumers over the age of 20, but the product is also well-liked by children, singles and live-alone divorcees. Sales pick up on Thursdays and normally people buy more than one.

The great variety of ready-to-eat food available in the Brazilian market has stimulated consumers, to an ever greater extent, to choose this type of product, owing to its convenience and the variety available. The population of Brazil includes more than 5 million people who live by themselves, almost 12% of the economically active population. With their sights on these consumers, food manufacturers have brought to market the most practical and innovative products for those with scant time for food preparation. The trend towards urban living in recent decades and the growing role of women in the workforce has drastically reduced the available time for preparing meals. These days, the modern consumer seeks out quality foods which are practical, quick to prepare and tasty. Thus, a new niche has been created in the market: ready-to-eat food. This segment is completely consolidated in large Brazilian cities and the tendency is for the demand for this type of product to increase over the next few years (Schlindwein, 2006).

The food service industry, which includes all eating outside the home and ready-to-eat food purchases (e.g. frozen food and delivered meals), had a turnover of close to R$38 billion in 2005 – 13% more than in 2004. The main factors which have contributed to the growth of this sector have been the advance of women in the workforce (which in the last 30 years jumped from 23% to 43%), and our ever more hectic routines in a globalized world in which time-management impacts lifestyles in all cities.

Another impressive figure is the budget for eating out. According to industry associations, Brazilians spend an average of 26% of their food budget in luncheonettes, restaurants, bars, bakeries and similar places, and this figure is projected to be as much as 40% within 20 years. And it is in this new food sector that the supermarkets begin their exploitation. On the shelves, singles are already beginning to find appropriately sized packages. They are scarce, but they do exist. Nationally, the number of available ready-to-eat meals is also growing. Production has increased almost 700% over the last decade. Today, this amounts to 20 000 tons (18 000 tonnes) per year and a greatly increased range of offerings. There are over 30 lines of ready-to-eat foods including *feijoada*, diced chicken and cream-of-spinach lasagne. Although it is growing rapidly, the prepared food market still has enormous growth potential in Brazil. The trend is that in the future, especially in major cities, numbers of ready-to-eat and frozen meals will overtake home-cooked meals in terms of frequency of consumption. This segment is predicted to grow 410%, from R$73 million annually to $R2.9 billion in turnover, demonstrating that the Brazilian consumer is becoming more cosmopolitan and avid to try out the latest novelties in the world. As such, ever more practical and quick-to-prepare foods are demanded from the industry (Bárbara, 2008).

18.6   Conclusions: relationships surrounding food

The qualitative study performed by Casotti (2002), who interviewed 29 women of different socioeconomic classes, also related examples of the strong 'rice and beans' culture in everyday Brazilian family life. The rice-and-beans tradition brings with it many associations, built into the culture over time, in the routine of Brazilian meal times; the combination is part of a habit that starts in early childhood and is passed on down through the generations. Women interviewed on the subject describe the functionality of the combination of rice and beans as being 'practical' and 'uncomplicated' and also a routine part of meals since 'they have to be there, always'. However, the reports of some interviewees include stories about rice and beans that are associated with feelings about the family: they describe the mixture as being the 'family favourite'; they tell the children to 'eat your beans' and because of this, the combination cannot be absent at the family table; they relate with a certain pride that their husbands insist 'only the beans at home are worth eating' thereby justifying why they do not eat them at work.

We are a 'relational society' where people or their emotional connections overlap with formal contracts or nutritional information. We have a relational cuisine which also expresses our way of being, explains anthropologist Roberto DaMatta (1984, p. 63) who contrasts the social connections, and the combinations of the main Brazilian dishes with those of Nordic or Asian cultures. Regarding the different social relations, DaMatta (1984) reminds us that:

'The world of food in Brazil takes us home, to friends and relatives – to those who share their lives and intimacy with us most intensely. Intimacy occurs at home and at the table.' (p. 63)

'A table that is large, bounteous and happy is also a harmonious one since it allows us to orchestrate over differences in an unequal society – it underscores relationships more than idiosyncrasies.' (p. 62)

Many are the descriptions of aspects that make and unite the Brazilian people: carnival, football and the food from multiple origins. What is the origin of pepper in Brazil? Pepper belongs to all three founding races of the Brazilian people; it accentuates and accelerates the flavours of many delicacies. Pepper was already used in Brazil by the Indians who called it capsicum; they then incorporated *malagueta* pepper brought by the slaves and black pepper planted by the colonizers from the Portuguese kingdom (Lima, 1999).

We are an *'apimentado'* (literally, 'peppery') people i.e. 'lively' and 'mischievous, malicious' in the popular parlance. Malice is also present in the spices of characters such as Gabriela, Clove and Cinnamon, a book by a famous author from Bahia whose popularity increased when the story became a soap opera and a film. Gabriela, a marvellous cook, can be described as 'peppery' or having the colour of the spices 'cloves and cinnamon'. This description would be rapidly understood by the popular imagination: Gabriela has a spiciness which goes beyond her cooking; she is seductive and her skin is what is often referred to as 'the colour of Brazil', a special kind of café-au-lait.

18.7 References

Bárbara, L. 2008. *Pratos prontos*. Available at: http://www.foodservicenews.com.br/materia.php?id=257. Accessed on 23 April 2008.

Barbosa, L. (2007). Feijão com arroz e arroz com feijão: o Brasil no prato dos brasileiros. *Horizontes Antropológicos*, **13**(28), 87–116.

Bertasso, B. F. (2000). *O consumo alimentar em regiões metropolitanas brasileiras: análise das pesquisas de orçamento familiares/IBGE 1995/96*, Master's dissertation in Applied Economics, Escola Superior de Agricultura 'Luiz de Queiroz", Universidade de São Paulo.

Câmara Cascudo, L. (1983). *História da Alimentação no Brasil: pesquisa e notas*. Belo Horizonte: Itatiaia.

Canesqui, A. and Garcia, R. (2005). *Antropologia e nutrição: um diálogo possível*. Rio de Janeiro: Fiocruz.

Carneiro, H. (2003) *Comida e Sociedade: uma história da alimentação*. Rio de Janeiro: Campus.

Casotti, L. (2002). *À mesa com a família – um estudo do comportamento do consumidor de alimentos*. Rio de Janeiro: Mauad.

Cheung, T. L. (2007). *Os comportamentos alimentares de brasileiros urbanos: identificação de tipologia de consumidores e análise das relações dos grupos com os alimentos*. São Carlos: UFSCar.

DaMatta, R. (1984). *O que faz o Brasil, Brasil?* Rio de Janeiro: Rocco.

DaMatta, R. (1997). *A casa e a rua: espaço, cidadania, mulher e morte no Brasil*. Rio de Janeiro: Rocco.

D'Oxum, M. (2008). *A culinária dos Orixás*. Available at: < http://www.mamaafrica.com.br/oferendas.shtm>. Accessed on 10 May 2008.

Fisberg, M., Wehba, J. and Cozzolino, S. M. F. (2002). *Um, Dois, Feijão com Arroz: A Alimentação no Brasil de Norte a Sul*. São Paulo: Atheneu.

Fry, P. (2001). 'Feijoada e soul food 25 anos depois', in Fry P. and Goldemberg M., *Fazendo Antropologia no Brasil*, Rio de Janeiro: Dp&A Editora/Capes.

Freyre, G. (1987). *Casa Grande e Senzala: formação da Família brasileira sob o regime de economia patriarcal*. Rio de Janeiro: José Olympio.

Freyre, G. (1997). *Açúcar: uma sociologia do doce, com receitas de bolos e doces do Nordeste do Brasil.* São Paulo: Companhia das Letras.

Gonçalves, J. R. S. G. (2002). *A fome e o paladar: uma perspectiva antropológica. Encontros e Estudos,* **4**, 7–16.

Instituto Brasileiro de Geografia e Estatìstica – IBGE. *População: indicadores sociais, indicadores sociais mínimos.* Accessible at: < www.ibge.gov.br>. Accessed on February 20, 2008.

Instituto Brasileiro de Geografia e Estatística – IBGE. (2007). *Contagem da população.* Accessible at: <http://www.ibge.gov.br/home/estatistica/populacao/contagem2007/contagem_final/tabela1_1.pdf>. Accessed on May 22, 2008.

Lima, C. (1999). *Tachos e Panelas: historiografia da alimentação brasileira.* Recife: Editora da Autora.

Maciel, E. M. (2004). Uma cozinha à brasileira. *Revista de Estudos Históricos,* **33**, 25–39.

Mattos, L. and Martins, I. (2000). Consumo de fibras alimentares em população adulta. *Revista de Saúde Pública,* **34**(1), 50–55.

Schlindwein, M. M. (2006). *'The influence of the woman's opportunity costs of time on the food consumption of Brazilian family's',* Doctorate in Applied Economics, Escola Superior de Agricultura 'Luiz de Queiroz', Universidade de São Paulo.

SENAC, D. N. (1998). *Sabores e Cores das Minas Gerais: a culinária mineira do Hotel Senac Grogotó.* Basisio A (coord). Christo, M S L and Rocha T, Rio de Janeiro, SENAC Nacional.

Senac, D. N. (1999). *Do Pampa à Serra: os sabores da terra gaúcha.* Basisio A (coord). Lessa B and Medeiros H *et al.* Rio de Janeiro, SENAC Nacional.

Senac, D. N. (1999b). A *culinária baiana no restaurante do SENAC Pelourinho.* Pereira M. V. (coord), Rio de Janeiro, SENAC Nacional.

Senac, D. N. (2000). *Culinária Amazônica: o sabor da natureza.* Basisio A (coord). Lody R. and Medeiros H. *et al.,* Rio de Janeiro, SENAC Nacional.

Senac, D. N. (2003). *Pantanal: sinfonia de sabores e cores.* Basisio A (coord). Sigrist M. and Medeiros H., Rio de Janeiro, SENAC Nacional.

19

Indian meals

C. T. Sen, Chicago, USA

Abstract: The content, preparation, and serving of Indian meals are characterised and the role of religion is discussed. Examples are given of typical meals in middle class urban households in four states; and trends, such as the increased use of convenient foods in urban areas, are outlined. Even though less than a third of Indians are vegetarians, meat plays a minor role in the Indian diet. Meals are based on a grain (usually rice or wheat) and lentils (dal), supplemented by vegetables, meat, fish, and condiments.

Key words: Indian meals, vegetarianism, Hinduism, Islam, caste.

19.1 Introduction

Defining and describing a typical Indian meal is difficult in view of the enormous physical, climatic, ethnic, and religious diversity of a country of more than 1 billion people; 15 official languages and many dialects; eight major religions; and innumerable sects, castes, classes, and other social divisions. Most Indian languages do not even have a word for a meal and use circumlocutions to express the concept. For example, to invite someone to your house for a meal, you would ask them to come and 'eat' at a certain hour, the time indicating whether it is lunch or dinner. While an English speaker might ask 'Have you had lunch (or dinner)?,' a North Indian Hindi speaker would ask 'roti khaya?' ('Have you eaten bread?') and a Bengali speaker 'bhat kheiicho?' ('Have you eaten rice?').

This diversity is evident in the content, structure, location, timing, serving, and organization of meals. Contrary to what many believe, only 31% of Indians are vegetarians, according to the Hindu-CNN-IBN State of the National Survey (Yadav and Kumar, 2006). The proportion ranges from 2–3% in the states of Kerala and West Bengal to 45% in Gujarat and 62% in Haryana and Rajasthan. More surprising, perhaps, are the variations within individual households: only 21% of all households consist only of vegetarian members. Orthodox Hindus, especially Brahmins, and Jains have the greatest restrictions on their diet and eating practices. Nor is there a universal way of eating. Many people sit on the floor and take their food from banana leaves or thalis (round metal plates with a rim) using their right hand, but others sit at western style chairs and tables and eat from china plates using utensils. There are also major differences between regional cuisines.

Still, Indian meals have some basic commonalities. Most Indians (at least those who can afford it) eat four meals a day: two main meals, breakfast or lunch and dinner, and two supplementary meals, breakfast and a light snack in the afternoon. The two core dishes of a meal are a starch (such as wheat, rice, millet, sorghum, or corn) and lentils. They are supplemented by vegetables, meat and fish, and an array of condiments, including sweet chutneys; sweet, sour, hot, and very hot pickles; salads; and yogurt. Alcohol is almost never consumed at meals; the standard beverage is water. Hospitality is highly valued among all groups in India, expressed in the saying 'A guest is god'.

Indian meals do not normally have a sequence of courses. Everything arrives more or less at once, although certain dishes may appear at different points. Even in Bengali cuisine, where dishes are served sequentially, they remain on the table throughout the meal. The Nobel Laureate Octavio Paz, for a decade Mexico's Ambassador to India, writes:

> In European cooking, the order of the dishes is quite precise. It is a diachronic cuisine… A radical difference in India, the various dishes come together on a single large plate. Neither a succession nor a parade, but a conglomeration and superimposition of things and tastes; a synchronic cuisine. A fusion of flavors, a fusion of times. (Ray, 2004).

Relatively little has been written about daily meals in India. Writers of cookbooks and food memoirs focus on the food served at festivals, weddings, and other special occasions and tend to neglect the mundane dishes and routines of daily life. Anthropologists describe social rituals and caste in great detail but do not explain what or how people eat on a day-to-day basis. Some exceptions are Ray (2004), Mahias (1985) and Khare (1976). In writing this chapter, this author's review of the literature has been supplemented by personal experience of meals in the home of Bengali in-laws in India and interviews with approximately thirty people from the subcontinent. They represent a selection of regions and religions, most come from middle-class or upper-middle-class urban backgrounds.

This chapter does not review the meals of the 300 million Indians who live below the poverty line, defined as consumption of 2400 calories a day (World Bank, 2007). For these people, a meal may mean a handful of boiled rice or a couple of pieces of flat bread with chilies or vegetable peels, or roasted chickpea flour mixed with salt and green chilies. The Hindu-CNN-IBN Survey reported that 35% of respondents said that at least once in the past year they or someone in their family could not have two square meals a day; 7% said this happened 'often'. Nor is it intended to be a comprehensive description of all Indian meals. Instead, it should be should be regarded as an initial contribution to an important and neglected area of study.

19.2   Geography and agriculture

India occupies about 80% of the land mass called the Indian subcontinent or South Asia. It is a Federal Republic consisting of 28 states and seven Union Territories (Fig. 19.1). India is the world's seventh largest country in area and second only to China in population. Some Indian states are larger than most countries, and, like countries, have distinctive languages, ethnicities, cultures, and cuisines. Ethnically and gastronomically, India is more diverse than Europe.

Despite rapid urbanization, India remains a predominantly rural country. India is the world's second largest agricultural producing country, and agriculture contributes a third of the country's GDP. However, much of the

Fig. 19.1   Map of India.

farming is subsistence-level so that the ingredients used in the meals of many Indians are produced and consumed locally.

A wide variety of altitudes and weather systems give India an extreme diversity of climates and soils. Geographers divide India into three regions. In the north, the Himalayan range extends 1500 miles from Pakistan and Afghanistan in the northwest to Burma (Myanmar) on the southeast. Melting snows from the Himalayas and seasonal rains feed the great river systems of the subcontinent: the Indus, the Yamuna-Ganga, and the Brahmaputra. Their basins form the 1500-mile long Indo-Gangetic plain, the cradle of India's civilization and its richest agricultural region.

The states of Punjab and Haryana are called the breadbasket of India. In these states the main cereal is wheat, ground into flour and made into flat breads. Dairy farming is an important industry, so that milk, yogurt, and other dairy products are a part of the daily fare. Uttar Pradesh, India's largest state in population, is a major producer of both rice and wheat. In the northeast, West Bengal and Assam produce two, sometimes three, crops of rice each year. Much of the western portion of this, the Indo-Gangetic Plain, including Rajasthan and parts of Gujarat, are arid deserts where farmers grow so-called coarse grains, sorghum and millets.

The Indo-Gangetic plain is separated from the Deccan Peninsula by the Vindyas and other mountain ranges. Over the centuries these mountains have served as a natural barrier between North and South India and enhanced the development of distinctive cultures, languages, and cuisines in the four southern states of Kerala, Karnataka, Tamil Nadu, and Andhra Pradesh. In this region, the main crop is rice.

Famine due to natural and human-made causes was endemic in India from time immemorial; even in the 1950s India relied on food from abroad. Grain and other food production was significantly increased by the Green Revolution of the 1960s, made possible by new farming techniques, massive irrigation projects, and the development of new strains of seeds.

Wheat production more than doubled and by 2003 accounted for 37% of per capita cereal consumption, up from 18.7% in 1963. Rice production also rose, albeit at a slower pace. This growth came at the expense of millet and sorghum, whose role as staples has become progressively less important (Fig. 19.2). According to the United Nations's Food and Agricultural Organization, Indians' average per capita calorie consumption rose from 2000 calories in 1963 to nearly 2500 calories in 2003 (compared with 2900 calories in China and 3760 in the United States) (FAO, 2005).

19.3 Religion

Religion plays a key role in determining what Indians eat. More than 80% of the population is Hindu, but 13.5% – 138 million – are Muslims, making

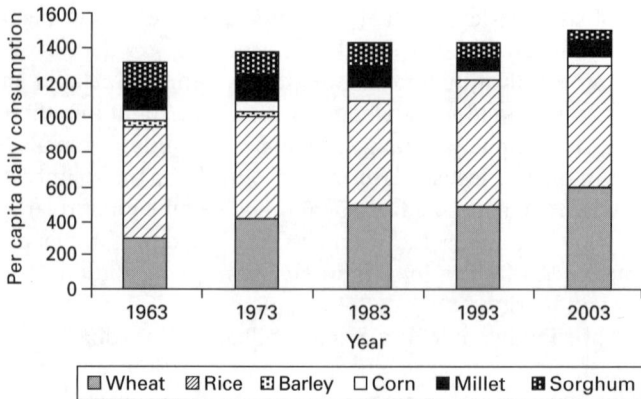

Fig. 19.2   Per capita daily consumption of grains in India, 1963–2003 (Source: Food and Agriculture Organization of the United Nations, http://faostat.fao.org).

India the world's second largest Muslim country after Indonesia. The population also includes 24 million Christians (2.3%); 19 million Sikhs (1.9%); 8 million Buddhists (0.8%), 4 million Jains (0.4%), plus small communities of Parsis, (Zoroastrians), animists, and other religions (India Census 2001).

In traditional India, what and how a person ate was inseparable from his or her religion, life-cycle stage, region, caste and/or social status, gender, health concerns and spiritual beliefs. The subcontinent's diverse religions have their own rules and preferences about what is and what is not acceptable to eat and within each religion there are a myriad of sects and subgroups.

19.3.1   Hinduism

Hinduism is a vast complex of beliefs and practices with infinite variations depending on region, caste, urbanization westernization, etc. It has no sacred texts equivalent to the Koran or the Torah, no prescribed laws, beliefs, or practices that all Hindus follow. All Hindus in theory possess a caste, which traditionally determines two important activities: whom one marries and what and with whom one eats (commensality). Both are related to concerns about purity and its opposite, pollution. Today, the significance of caste is weakening, especially in cities.

In addition to the four major castes (brahmin, kshastriya, vaishya and sudra), there are a multitude of subcastes, called jatis, many of which are rooted in professions and each with its own food restrictions and preferences. More than 55% of the priestly caste, brahmins, are vegetarians, although there are many regional and individual exceptions. Many Bengali brahmins eat fish, for example, while some Kashmiri brahmins eat goat and mutton. Most members of the kshatriya, or warrior caste, eat meat (though not beef), which is believed to promote energy and physical strength. Farmers and merchants (vaishyas) may or may not eat meat depending on their tradition, religious beliefs, and region. Thirty-four per cent of all Indian women are vegetarian, compared with 28% of men, as are 37% of people over the age of 55, compared with 29% of people under the age of 25 (Yadav and Kumar, 2006). Because a vegetarian diet is associated with spiritual serenity, meat is avoided by swamis, yogis and their followers and was traditionally prescribed for widows.

On average, Indians get 92% of their calories from vegetable products and only 8% from animal products (meat, dairy products and eggs). However, growing affluence has led to a rise in the consumption of animal products (Fig. 19.3), although it is still less than in Pakistan, China, or the United States. The reason most people restrict their meat consumption is economic rather than moral or religious: meat is expensive and eaten only on special occasions (something that was true in the Western world until modern times). Even today, some middle-class Indian households eat meat only on weekends.

Even carnivores usually avoid beef, and many states have laws against cow slaughter. The explanation is that the cow is too valuable as a source of milk and dairy products and as a puller of ploughs to kill. Some Hindus, especially in the south and west, and all Jains avoid garlic and onions, which are thought to excite the passions and disturb spiritual serenity. Although pigs are not explicitly proscribed for Hindus, many avoid pork to avoid offending their Muslim neighbors. While not forbidden, alcohol is frowned on by many sectors of society.

Members of the higher Hindu castes were traditionally subject to other taboos and restrictions. They were not supposed to eat items that have stood overnight, leftovers, food touched by an unclean substance or an insect, or food touched by another person. In theory, people could accept food only from members of their own or a higher caste. All lower castes could receive cooked food or water from a brahmin, for example, but a brahmin could not

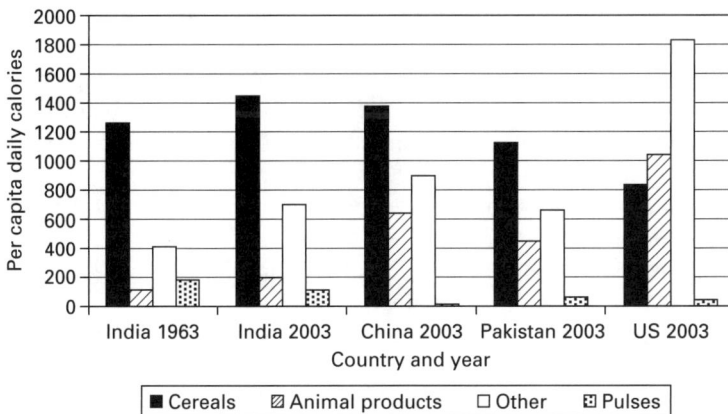

Fig. 19.3 Per capita daily calorie consumption of major food items in India, China, Pakistan, and the United States, 1963–2003.

accept food from them. As a consequence, many cooks are brahmins, especially in South India.

Some practices are rooted in age-old beliefs about the relationship between food and health. A popular division of foods is into hot and cold, depending on their effects on the system after digestion. This has nothing to do with the immediately perceived quality of the food itself, for example, ice is supposed to be hot, and the classifications also vary by region: for example, mangoes are believed to be 'hot' in some regions and 'cold' in others.

In Ayurveda, the traditional Indian system of medicine dating back to 5000 BC, the qualities and attributes of foods are believed to affect one's mental, spiritual, physical health. According to Kautilya, a 3rd century Ayurvedic physician, a 'gentleman's meal' was supposed to consist of one prastha (perhaps 450 g) of pure unbroken rice, $^1/_2$ prastha of lentils, 1/6 prastha of ghee or oil, and 1 64/th of a prastha of salt. (Achaya, 1998). A person was advised to eat twice a day, after morning and evening worship, and to fill half his stomach with solid food, a fourth with liquids, and a fourth should remain empty for the movement of wind.

According to the great (possibly mythical) Ayurvedic physician Charaka, who may have lived in the 5th century AD (Sen, 2004), 'Without proper diet, medicines are of no use; with a proper diet, medicines are unnecessary'. Foods are characterized by six tastes, sweet, salty, sour, astringent, bitter and pungent, that are related in complex ways to a person's physical and mental qualities. A proper diet should contain all these tastes in the correct balance. Even today at least four of the flavors are usually present in Indian meals.

Charaka also laid down some rules for proper eating (Svoboda, 1993):

- Eat properly combined food after digestion of the previous meal to allow a free passage for all substances.
- Eat in a congenial quiet place either alone or with affectionate people so that the mind is not depressed.
- Eat neither hurriedly nor leisurely to appreciate the qualities of the food you are eating.
- Eat without laughing or talking, with concentration, considering your constitution and what is good and not good for you as you eat.
- Do not eat when you are not hungry and do not fail to eat when you are hungry.
- Do not eat when you are angry, depressed, or emotionally distraught or immediately after exercise.
- Keep as large a gap as possible between meals.
- Sit to eat whenever possible facing east.
- Pray, thanking the Creator for the food you are offering your digestive fire.
- Never cook for yourself alone; the gift of food is the best gift at all.
- Feed all five senses: look at the food and savor its appearance and aroma; listen to the sounds it makes, especially when cooking; eat with

your hands to enjoy its texture; chew each morsel many times to extract its flavor.

- Stroll about a hundred steps after a meal to assist the digestive process.
- Do not eat heavy . . . food like yogurt and sesame seeds after sunset and eat nothing within two hours of going to bed.
- Never waste food.

19.3.2 Islam

Although Islam has fewer proscriptions than Hinduism, they must be strictly observed. Islam allows all food except four items: pig meat; blood; any animal that has not been purposely slaughtered as food; and any animal that has been slaughtered in the name of a pagan deity. These last two prohibitions are reflected in Islamic laws about the way in which animals must be slaughtered. Meat killed in this way is called halal, lawful. The opposite, food that is forbidden, is called haram and includes alcohol and any other substance that impairs the senses, memory and judgment. An intermediate category, makrouh, covers dishes that are somewhat frowned upon but not forbidden, such as shellfish and some birds (Davidson, 2003).

Lamb and goat are preferred meats but beef is also eaten. Onions and garlic are used in many Muslim dishes in India. Interestingly, the Hindu-CNN-IBN survey found that 3% of Muslims and 8% of Christians in India are vegetarians. The Unani system of medicine practiced by Muslims also uses diet as a way of restoring balance and health.

19.3.3 Sikhism

Sikhism is a monotheistic religion born in the Punjab in the late 15th century. Many Sikhs are vegetarian, although it is not prescribed by the religion. Their sacred writings contain passages that indicate concern over the morality of killing animals for food, but say that the decision whether to eat meat should be left to the individual.

19.3.4 Jainism and Buddhism

In the 6th century BC, two important religious leaders were born into the ruling families of small North Indian kingdoms. Siddartha Gautama, later known as the Buddha ('enlightened one'), and Vardhaman Mahavira, a wandering ascetic who was a founder of the religion known as Jainism. Both opposed the caste system and the dominance of the brahmins who controlled the ritual sacrifices of animals considered essential for personal salvation. Buddha preached moderation in all things and was not an advocate of strict vegetarianism for lay persons.

Jainism is a much more ascetic religion. Its central doctrine, *ahimsa*, is that everything, including plants and insects, has an eternal soul and should not be slain or injured. The question of eating meat, fish or eggs does not even arise; only 'absolutely innocent' foods are permitted. Prohibited foods include vegetables that grow underground, such as potatoes, carrots, radishes, etc; fruits and vegetables with many seeds (figs, eggplants, tomatoes), fresh ginger and turmeric; foods containing yeast; fruits with green leaves in which insects can hide; buds and sprouts; and onions and garlic (Sen, 2007). Jains cannot eat after sunset in case they accidentally kill an insect. Today, states with a high proportion of Jains (Gujarat, Rajasthan, and Karnatka) have a higher proportion of vegetarians than other states.

19.3.5   Christianity

Christians have few food taboos, although like other groups they may be sensitive to local customs. Pork is a favorite meat among Goans, especially on festive occasions

19.4   Nutritional values of the Indian meal

Indian meals are excellent illustrations of the 'core-fringe-legume' pattern described by Mintz and Schlettwein-Gsell (2001). In large ancient agrarian societies, people consider that a 'real meal' consists of three elements: (a) the core, a complex carbohydrate, such as a tuber or a cereal that is typically bland in taste, homogeneous in texture and eaten at most meals; (b) the fringe, which can be animal, vegetable, or mineral, fresh or preserved, liquid or solid, etc., and has the role of enhancing and enlivening the core; and (c) legumes, which supply proteins that complement and augment synergistically those in the core. Animal protein is relatively scarce in these agrarian societies, where 'the overwhelming majority of the people get hardly any animal protein in the form of meat year-round and eat little or none on a daily basis' (Mintz, 2001).

One of the most nutritious cores is wheat. Half a kilogram provides recommended dietary amounts of protein, iron, Vitamins B1 and B6, and nearly half the daily allowances of calcium and B2; however, it is totally lacking in Vitamins A and C. The consumption of a similar amount of unpolished rice provides two-thirds of recommended protein and all the needed iron, Vitamin B1 and B6, but insufficient amounts of calcium and Vitamins A and C. Foods that supply these missing nutrients (while avoiding overdoses or imbalances) are milk products and dark green vegetables. The quality of a protein is measured by its lysine content, large amounts of which are found in animal products and lentils.

An Indian meal is always centered around a cereal, wheat, rice, millet, sorghum, or corn, which provides 70% of average per capita calorie consumption. The second essential element is lentils (of which there are many varieties) or their cousins, peas, beans, and other pulses. Both lentils and the dish prepared from them are called dal. (As a general rule, thick dals are eaten with bread and thinner, more watery dals with rice.) Relatively small amounts of meat, fish and vegetables, the fringe, are added to enhance the taste and qualities of the main grain.

An Indian meal also includes small amounts of condiments: dairy products, such as yogurt or clarified butter; sweet and sour fruit and vegetable chutneys; and sweet, sour, or pungent pickles. Pickles can be made from fruits (especially green mangoes, and limes), vegetables, nuts, berries, meat or fish and can be sweet, sour, salty, hot, or very hot indeed. Besides adding an accent and a contrasting flavor and texture to a meal, pickles have many health benefits. Spices such as ginger and asafetida are digestives while red chilies are antiseptic. In many parts of India, a meal ends with buttermilk or yogurt, which is believed to aid digestion. Sometimes people end a meal with a paan, a mixture of sliced betel nut, lime paste, and spices wrapped in a betel leaf, which is both a digestive and a breath freshener.

19.5   Description of Indian meals

Traditionally Indians who could afford it ate four meals a day: two main meals, generally breakfast or lunch and dinner, and two supplementary meals: breakfast and a light afternoon meal of snacks, sometimes called tea or tiffin. The timing and nature of these meals reflects such variables as income, location, and the schedule of family members. For rural people, the main meal is a hearty breakfast or early lunch to prepare for the day's labor; dinner is much lighter. In towns and cities in North and Central India, most people start the day with a light breakfast, followed by a large lunch taken either at home or at the office or school, a light afternoon tea when family members return home, and a dinner in the evening. In South India, breakfast is the main meal of the day. (Table 19.1).

Table 19.1   Characteristics of traditional meals in regions of India

Region	Main meal of day	Main starch	Yogurt
East (W. Bengal, Assam)	Lunch	Rice	Not during meal; sometimes sweetened as dessert
West (Gujarat, Rajasthan)	Lunch	Wheat, millet, sorghum	During meal
North (UP, Punjab)	Lunch	Wheat	During meal
South (Andhra Pradesh, Tamil Nadu, Karnataka)	Breakfast	Rice	End of meal

The following section describes the basic meals eaten in four states of India. The meals are typical of those served in a middle to upper-middle class household; that is, one headed by a professional, mid-level official, or affluent businessman or farmer. This class lives mainly in India's cities and very large towns; their lifestyles may incorporate elements of westernization. This overview is not intended to be comprehensive but represents an initial step to describing this important and neglected aspect of Indian foodways.

19.5.1   Uttar Pradesh

With a population of more than 130 million, Uttar Pradesh (UP) is India's most populated state and one of the most prosperous. Around 82% of UP residents are Hindus and about 40% of them are vegetarian. Meat is eaten by a large caste called Kayasthas, the descendants of Hindus who served as administrators for the Muslim rulers.

Among North Indian Hindus, foods were traditionally divided into two categories: *kaccha* and *pukka* (Khare, 1976). *Kaccha* foods are those made at home by boiling or roasting and are served to the family, such as rice, dal (boiled lentils), some breads, and vegetables. This is what everyone, rich or poor, modern or orthodox, eats every day in some form; it is the basic meal of North India. Small quantities of meat and fish are added for non-vegetarians.

The second category, *pukka* foods, are prepared with ghee (clarified butter), an ingredient which is considered ritually pure (statues of deities are often bathed in ghee) and thought to offer some protection against pollution. *Pukka* foods include parathas (round wheat flour breads sautéed in oil); puris (round puffy breads that are deep fried); and many sweets. These foods can be eaten outside the home and offered to guests who are not family members. *Pukka* foods are served at temples and during festivals and feasts.

Although today many no longer think in terms of these categories, and may not even know what they mean, they are still obvious in the content of daily meals. A traditional UP Hindu breakfast, eaten at 8:30 or 9:00, consists of paratha or chapatti (a round flat wheat bread cooked over a flame) served with a vegetable and perhaps some halwa, a sweet dish made from vegetables, nuts, fruits, or grains. Another common breakfast dish is ghugri, a porridge made from steamed lentils, oil and sugar. Western-style breakfasts of toast and jam, eggs and cereal, are becoming more popular, perhaps because they are less labor-intensive. A breakfast eaten by an affluent Muslim family could include keema (spiced minced meat) or nihari, a spicy dark beef stew cooked overnight.

Lunch, traditionally eaten between 11:00 and 12:00, features either rice or bread, dal, and a couple of seasonal vegetable dishes, one wet and one dry – cauliflower in the winter, for example, or bitter gourd in the summer. Some households have bread during the week and rice only on weekends. Non-vegetarians would replace a vegetable dish with a meat in a gravy. Yogurt is standard. Dessert is rare. The dishes are served hot and the bread is always made to order.

A late afternoon snack is eaten everywhere in India when children return from school and workers from their office. This could be a simple meal of fried snacks with a glass of milk or tea, or a more elaborate production with English-style tea sandwiches, pakoras (deep fried vegetables in a chickpea batter), samosas (small triangular-shaped pastries filled with potatoes or vegetables), and western style cakes and pastries and/or Indian sweets purchased from an outside vendor.

Dinner in UP was traditionally eaten as late as 9 pm but today some people dine earlier for convenience. Dinner is basically the same as lunch: bread, vegetable, perhaps salad, and rice. Some families eat yogurt, others do not. Dessert is not an everyday item and may consist of fresh fruit in season or a rice pudding if guests are present. Muslims would add a meat dish such as kabobs or a goat curry to this basic meal.

One informant told me that in her family, dal was not served in the evening because it was considered difficult to digest; another reported that in her family rice was never eaten at dinner because it was believed to thin the blood. (Such idiosyncratic views are not uncommon in India and often belong to the senior male of the family.) Westernized families throughout India may include such dishes as soup, Irish stew, casseroles, roast chicken, cutlets, or chops in their meals. Desserts are rarely served, though a meal might end with seasonal fruit.

Lucknow, the state capital which was one of the Moghul regional capitals, is famous for its refined cuisine that includes rice pulaos; *kormas*, meat cooked in a yogurt gravy; *kalias*, meat cooked in a gravy without yogurt; and various kinds of kabobs.

19.5.2   West Bengal

West Bengal is 75% Hindu and 23% Muslim. Only a very small proportion of Hindu Bengalis are strict vegetarians, and even brahmins enjoy fish. Freshwater and pond fish are preferred to ocean fish, although shrimps are a popular ingredient. The staple is rice.

Bengalis typically start their days with 'bed tea', a cup of tea with a biscuit served in the bedroom by the servant. This is a legacy of the British who introduced tea drinking into India in the 19th century. In West Bengal, tea is served English style with milk and sugar, never with spices as in North India (and the origin of the so-called 'chai' popular in the West.)

Breakfast is served around 10 am or earlier if people have to go to work or school. A 1970 survey of Calcutta households showed that half ate bread with vegetables for breakfast; about 15% had puffed rice or flattened parboiled rice; while the balance had toast (Ray, 2004).

Traditionally, lunch was the main meal of the day, taken between 1:00 and 2:00. Before the days of outside employment, the entire family ate together but today the breadwinner eats outside the home, either at the office canteen or food that is sent from home. The main component of a Bengali lunch is rice; the fringes are vegetables, fish, and yogurt. Dal is always present.

In contrast to other parts of India, a Bengali meal follows a progression of flavor, starting with bitter through salty and sour and ending with sweet. The bitter dish is generally *shukto*, which is made from diced white radish, potatoes, beans, and other vegetables and bitter melon. It is followed by plain boiled white rice, dal made from red lentils, and one or more fried, boiled, and sautéed dishes of seasonal vegetables. If fish or meat are served, they appear at this point. Fish could be fried or served as a *jhol* with a thin watery gravy. The average serving is small: 2 oz per person. Some 40% of all Bengali households eat fish at least three times a week. Meat, on average eaten once a week, is usually goat. Some Bengali Hindus avoid chicken, perhaps because it is a scavenger.

The meat or fish course is accompanied by luchhis, a puffy deep-fried bread made from white flour. The next-to-last course is usually a sweet and sour fruit or tomato chutney. Its function is to remove any heaviness from overeating and, like a sorbet in European cuisine, to clear the palate for the pièce de résistance: the sweet or dessert course, often sweet yogurt.

Afternoon tea is typically an important meal in a Bengali household. Sometimes guests bring sweets or fruits in season. In my husband's family, tea could last several hours and would be a kind of open house during which guests would drop in to visit, sometimes bringing sweets.

Bengalis, like the Spanish, eat dinner very late, starting at 9 or 10 pm, or as late as midnight for some Calcutta families. Dinner is a variation of lunch, although somewhat smaller. The bitter opening course is omitted, and instead of rice, the majority of Bengalis eat wheat-based bread. Fish is served less frequently than at lunch.

19.5.3 Gujarat

In Gujarat, 45% of the population is vegetarian, a reflection of the strong Jain presence. Many Hindus do not eat eggs, onions or garlic; a standard flavoring is a green chili and ginger paste.

Breakfast is traditionally served early, at around 7 am, and includes tea with wheat or millet bread; papri, crisp little squares made from chickpea flour; or puffed rice. Rural families eat lunch between 10 and 11 am before they go to work or school. The core of a Gujarati meal is bread traditionally made from millet or sorghum, grains that are increasingly replaced by wheat. It is served with a sautéed vegetable or a vegetable stew. This is followed by plain boiled rice and dal or kadhi, a spicy yogurt curry thickened with chickpea flour. Gujaratis add a pinch of sugar to dals and vegetable dishes and may serve a piece of sugar on the side. Milk or lentil-based sweets are sometimes served during the meal itself.

Afternoon tea is an occasion for enjoying Gujarat's delicious snacks, called farsans that include a mixture of rice flakes, peanuts, coconut, and spices called *chevda*; *bhelpuri*, a spicy mixture of crisp noodles, puffed rice, tomatoes, onion, potatoes, coriander and tamarind chutney; and *dhokla,* a polenta-like dish made of steamed fermented lentils.

Dinner, typically served at 8 to 8.30 pm, is similar to but smaller than lunch. Sometimes it features a thick bread called bakri and a vegetable or a one-dish meal, such as khichri, a dish of boiled rice, moong dal, and vegetables and nuts.

19.5.4 Andhra Pradesh

Andhra Pradesh (AP) is one of four states in South India, the others being Tamil Nadu, Karnataka and Kerala.The core cereal in this region is rice, and lentils play a much larger role in the diet than elsewhere on the subcontinent.

Breakfast is an important meal in South India and will be familiar to patrons of South Indian restaurants around the world, where it is served as lunch or dinner. A middle-class Hindu family typically eats idlis, soft steamed disk-shaped cakes, or dosas, flat round crispy crepes lightly sautéed in oil. The dough for both is made by grinding rice, black lentils and water into a paste left to ferment, usually overnight. Idlis are served with sambar, a thin spicy lentil soup that may include vegetables, and coconut chutney, a paste made by grinding coconut with spices, and lime juice. The standard breakfast drink is strong filtered coffee mixed with milk that is reminiscent of French café au lait. Another popular breakfast dish is *uppuma*, a semolina porridge with tomatoes and onions.

Lunch is centered around boiled white rice, accompanied by two or three vegetables (such as potatoes, plantains, eggplants, cabbage) sautéed with spices in a little oil or cooked in a gravy; lentils served as rasam (a watery hot and sour dal) or sambar; one or more pickles; perhaps a salad of cucumber or bean sprouts; and mango or other fruit in season. A South Indian meal always ends with yogurt mixed with rice. The vegetables are seasonal and changed every day. Afternoon tea would include tea or coffee, dosas, and uppuma. Dinner is similar to lunch but generally simpler, and the dishes do not repeat those served at lunch. Often dinner includes pappadums, crispy lentil wafers.

Hyderabad, the state capital, was one of the regional capitals of the Moghul dynasty and has a large Muslim population. A breakfast eaten by a Muslim family might include a paratha or chappati served with fried eggs, an omelette, or minced meat and/or sautéed potatoes. Lunch could consist of paratha or puris, rice, a meat dish (beef or mutton) made with a gravy, dal, and yogurt. Dinner at 8 pm would be similar to lunch with the inclusion of a vegetable, such as eggplants. On special occasions, or when there were guests, a pulao or biryani (rice dishes made with spices, vegetables, or meat) might be served.

19.6   Preparation of the meal

The domestic hearth in a Hindu home was an area of high purity and often located next to the area of worship (Achaya, 53). Some households have secondary areas for the preparation of meat. Before entering the cooking area, the person preparing the food is supposed to take a bath and put on freshly washed clothes. Depending on a family's affluence and/or degree of orthodoxy, meals are prepared by a cook, perhaps with assistants, under the supervision of the senior female in the household. Hired cooks were traditionally brahmins, since everyone could eat food prepared by them (hired cooks were always men). In traditional joint families, the daughters-in-law help with the preparation of meals. Before the advent of modern appliances, meal preparation could take hours because of the need to grind spices and slice vegetables. On average women used to spend 3 to 4 h on the morning meal cycle and 4 to 5 h in the evening (Khare, 1976).

19.6.1   Where and how meals are served

Traditionally, Indians of all religions sat on the floor to eat, sometimes on a carpet or raised stool. Many families still follow this custom, but urban and westernized Indians tend to sit on chairs and eat at western-style tables. The food is put into dishes and placed on the table for the diners to serve themselves, or it could be placed directly onto the plate by the mother or a servant. Dishes are prepared for each meal and served hot. Breads are always cooked to order and brought out one by one on a plate and slid onto the diner's plate or tray.

The traditional serving plate is a thali, a circular metal tray with raised edges, shown in Fig. 19.4. Liquid dishes and yogurt are put into little metal bowls, in Hindi called kathoris; rice, bread, chutney, pickles, and fried vegetables are placed directly on the thalis. In some families western-style plates are used. In South India, the traditional serving plate was a banana leaf (Fig. 19.5), sprinkled with water before the meal and later discarded.

The normal and nearly universal way of eating by members of all religions is with the fingers of the right hand (the left hand is considered unclean). Some families use cutlery, mainly forks and spoons, since meat and vegetables are already cut into pieces. Hands and mouths are always washed before sitting down to eat. If bread is part of the meal, the diner breaks off a piece, uses it to scoop up a small portion of the food and pops it into his mouth. For rice or vegetables, the tips of the fingers are used to form a little ball that is scooped into the mouth. Dal and other liquid dishes are poured onto the rice. The skill and neatness with which one eats rice is a mark of good manners; letting the liquid run down one's arm is considered poor form.

An orthodox brahmin householder follows a strict ritual at mealtimes that includes having a bath, changing into freshly washed clothes, and removing shoes or headgear (Achaya, 1994). He is supposed to eat alone and in total

Fig. 19.4   A North Indian thali.

Fig. 19.5   A traditional Kerala meal served on a banana leaf.

silence. In some families, women serve their husbands and children before eating themselves, but this custom is not universally followed.

The standard accompaniment to an Indian meal is water, although my observation is that Indians drink very little with their meals. Orthodox people pour the water into the mouth from the metal tumbler without the vessel touching their lips. A diner NEVER touches the food or plate of another person. Food touched by another person, even a spouse or other family member, is considered polluted and therefore inedible, a concept that still holds sway. In pre-refrigeration days leftover food would be given to servants after the meal or to homeless people or even to cows and birds. Today, there appear to be wide variations in how leftovers are handled. Some families reheat vegetables and rice from the previous night's dinner for breakfast, or take it to the office for lunch.

19.7   Future trends

The booming Indian economy, the entry of more women into the work force, the emergence of dual-income couples, and a decline in the number of household servants have changed to some degree the way urban Indians eat. The role of caste and religion in determining what, where, and with whom one ate continues to weaken, especially among younger people, a process observed more than 40 years ago (Khare, 1976).

One of the most noticeable developments has been a sharp rise in eating out in restaurants, which used to take place only on special occasions, if at all, since restaurants were few and expensive (Sen, 2005). Spending on restaurant meals has doubled in the past decade to $5 billion/year and is expected to double again in five years. Large cities such as Delhi, Mumbai, Chennai and Bangalore, have a proliferation of restaurants ranging from western and Indian fast food outlets to trendy establishments serving Thai, Middle Eastern, Moroccan, Italian, Japanese and other foreign foods. Even in rural areas, where the standard of living is much lower, a quarter of the people have tried soft drinks, a symbol of modernity.

Meal preparation is also changing. Outdoor markets and neighborhood shops with their dingy interiors are being replaced by modern grocery stories that sell once-exotic 'foreign' vegetables, such as broccoli, bell pepper, and asparagus, as well as frozen and dried ingredients and ready-made meals. In 2007 alone, retail sales of packaged food in India rose 15% (Euromonitors, 2008). Butchers in New Delhi sell not only marinated raw meats and pre-cut chicken but cooked kabobs and other meat dishes that can be taken home or eaten on the spot.

The popularity of television cooking shows, such as Sanjeev Kapoor's *Khana Khazana* (the longest running show on Indian TV) together with the publication of cookbooks and lifestyle magazines have made middle-class urban housewives more aware of both the cuisines of other regions of India and other parts of the world (Appadurai 1988). Serving these dishes to one's family and friends has become a mark of sophistication. 'Food boundaries seem to be dissolving much more rapidly than marriage boundaries', writes Appadurai. Preparing these meals is facilitated by the availability of fresh and frozen ingredients from other parts of Indian; for example, ready-made idli batter that is simply poured onto the pan and cooked.

The days when male heads of household came home for leisurely lunches are gone; working men and women generally eat at cafeterias or a fast food stall, and dinner is replacing lunch as the main meal of the day. People in their thirties and forties observe that meals are smaller, simpler, and lighter than those they remember from their childhood. 'We don't have a full lunch or breakfast any more and there aren't as many dishes on the table', commented one person. Health concerns have led many middle class people to reduce their consumption of ghee, oil and coconut milk.

Still, despite modernization and globalization, the fundamental structure of the Indian meal has not changed. The core of an Indian meal remains a carbohydrate, with wheat displacing traditional grains such as millet and sorghum. Although both overall meat consumption and the total number of people who eat meat is growing, meat is still a fringe element in meals, not least because it is expensive relative to other ingredients. Pickles, yogurt, and breads may be store-bought rather than made at home but remain essential components of a meal. A 24-year old office worker said, 'Much to my mother's chagrin, I use store-bought yogurt. And my mother-in-law was upset when she saw that I used Pillsbury flour to make rotis. She still prefers to buy wheat and grind it fresh' (Bhide, 2005).

19.8   References

Achaya, K T (1994), *Indian food: A historical companion*, New Delhi, Oxford.
Achaya, K T (1998), *A historical dictionary of Indian food*, New Delhi, Oxford.
Appadurai, Arjun (1988), 'How to make a national cuisine: Cookbooks in contemporary India', *Comparative studies in sociology and history*, **30**(1), 3–24.
Banerji, Chitrita (1993), *Life and food in Bengal*, New Delhi, Rupa.
Bhide, Monica (2005), 'As cash flows in, India goes out to eat,' *The New York Times*, April 20, 2005.
Census of India, http://www.censusindia.gov.in
Davidson, A (1999), *The Oxford companion to food*, Oxford, Oxford University Press.
Euromonitor, 'Packaged food in India', January 2008, www.euromonitor.com/ Packaged_Food_In_India.
Food and Agriculture Organization of the United Nations, http://faostat.fao.org.
Khare, R S (1976), *The Hindu hearth and home*, New Delhi, Vikas.
Mahais, Marie-Claude (1985), *Délivrance et convivialité: Le système culinaire des Jaina*. Paris, Editions de la maison des science de l'homme.
Mintz, S W and D Schlettwein-Gsell (2001), 'Food patterns in Agrarian societies: the Core-fringe-legume hypothesis,' *Gastronomica*, **1**(3), 40–51.
Ray, K (2004), *The migrant's table*, Philadelphia, Temple University Press.
Sen, C T (2004), *Food culture in India*, Westport, Greenwood.

Sen, C T (2005), 'The North American Indian restaurant menu: the triumph of inauthenticity', *Proceedings of the Oxford Symposium on Food and Cookery*, Totnes, Prospect Books, pp. 391–400.

Sen, C T (2007), 'Jainism: the world's most ethical religion', *Food and Morality: Proceedings of the Oxford Symposium on Food and Cookery 2007*, Totnes, Prospect Books, pp. 230–240.

Svoboda, R E (1993), *Ayurveda: life, health, and longevity*, New Delhi: Penguin.

World Bank, http://devdata.worldbank.org/AAG/ind_aag.pdf.

Yadav, Y and Kumar, S, 'The food habits of a nation', *The Hindu*, August 14, 2006.

20

Thai meals

S.-A. Seubsman and P. Suttinan, Sukhothai Thammathirat Open University, Thailand, J. Dixon and C. Banwell, Australian National University, Australia

Abstract: The chapter begins by providing a brief background to Thailand, highlighting the way that the contemporary culinary culture is the outcome of a unique history of non-colonization, Buddhist religion, the role of the Royal family and national economic plans. The four basic ingredients of Thai meals, fundamental cooking techniques and how ordinary Thai people construct their meals through combinations of dishes with contrasting qualities are then described. These qualities are further elaborated using three concepts: the elements, elegance and energy. Greater detail is given for how meals are sequenced and how people interact at meal time. The continuities in Royal dining are briefly described; in most respects it mirrors ordinary dining. The regional diversity of meals is illustrated by nine case studies showing typical variations in dishes and meal combinations. Current and anticipated trends in Thai meals and meal arrangements are outlined.

Key words: Thailand, culinary culture, cooking, meals, thai food, eating, regional cuisine.

20.1 Introduction

The meals eaten today by more than 60 million 'ordinary' Thai people reflect the country's history, geographic neighbors, centuries-old institutions, regional biodiversity and government-auspiced economic plans. This combination of influences has led to a dominant culinary culture that is easily recognized by foreigners as distinctively Thai as well as to regional cuisines, which vary in their use of ingredients and cooking techniques but which nevertheless share an underlying philosophy: harmony, based on the elements, elegance and energy. One day of Thai meals corresponds to a musical symphony consisting of a harmonious blend of spiritual and metaphysical elements, a simple elegance of presentation, and an energy deriving from a variety of tastes and contrasting incorporation of spicy hot versus non-spicy hot ingredients and dishes.

In the following sections, we explain the variety of influences on Thai meals, paying attention to colonization, religion, the monarchy and national economic directions. The resulting underlying culinary structure is described. We outline the incorporation of four fundamental ingredients to traditional Thai meals being rice, fish, dipping sauce (*Namprik*) and herbs, a variety of cooking techniques, and the accepted meal arrangements and portions. We integrate these practical elements in a typical meal matrix, before revealing something of the regional diversity in meals and because it continues to influence ordinary Thai people's food imaginations, the current royal household's approach to meals.

Based on week-long meal diaries, we describe the meals of nine 'ordinary' men and women drawn from Thailand's four regions and reflecting a range of social status positions within Thailand. Finally, using documentary sources and information obtained in key informant interviews, we report on current trends before providing a section on sources of further information.

20.2   Background to Thailand's culinary culture

'*Tom Yam Kung Disease*' is a well known expression for the Thai economic crash of 1996–1999. The term refers to the spicy soup, enjoyed by Thais and foreigners (*farang*) alike. It reveals the centrality of food to the Thai nation as well as to its economy. As a middle income country, with the highest female labour force participation rates in South-East Asia, Thailand has transformed itself from an agrarian society to a mixed economy and a leading site for global transactions, including agricultural and processed food exports. Since the 1980s, it has also sought to be the premier tourist destination in South-East Asia, and it uses the world-wide popularity of its cuisine to promote culinary tourism.

Approximately one fifth of Thailand's 65 million population live in Bangkok, and the difference in Bangkok's population size compared with provincial cities such as Chiang Mai remain significant. Per capita income differentials between Bangkok and the rest of Thailand also remain wide (USDA, 2004a). Bangkok is where the modern end of the retail spectrum begins and continues to dominate.

Bangkok has become like any other modern Asian metropolis with a variety of retail forms ranging from hawker stalls to side-road cafes, from elegant department stores with food sections to hypermarkets in hypermodern shopping malls, from wet markets to convenience store chains. The hyper/ supermarket phenomenon is a feature of urbanization, and Bangkok has 70% of Thailand's supermarkets and superstores (USDA, 2004a, p. 27). The rest of Thailand maintains its Thai–Chinese shop–houses, hawker or street stalls, and fresh food markets.

Street vendors are very visible on the streets of Bangkok and other major cities. Many people rely on food vendors for cheap and nutritious cooked food and raw goods, and it is a retail form that is easy to establish. Proliferation of street vendors builds on the Thai urban tradition for eating out and taking pre-prepared food home. Rapid urbanization, a result of rural migration, plus long hours for low-paid workers in the informal sector, also makes it harder for the 'urban poor' to cook their own meals (Bhowmik 2005).

Industrialization appears to have brought many Thais the health benefits of greater dietary diversity, as well as the health risks of processed foods, foods of animal origin, more added sugar and fat, and more alcohol (Kosulwat, 2002; Aekplakorn, 2007). This profile of changing food availability is consistent with many Asian countries.

As well as the more common trends of industrialization and urbanization, Thailand has some unique features which influence its culinary culture.

20.2.1   (Non)-colonial influences

Thailand has never been colonized though in the past concessions had to be made to foreign powers and modernity. During King Rama III to King Rama V's reign, from the 1820s–1910s, it used a variety of strategies to resist colonization, including: to be open to, but to play quiet diplomacy with, colonial powers especially England, France and Russia; to minimize conflict domestically through incremental, rather than sweeping, changes; and, to adopt, albeit selectively, those aspects of western modernization deemed appropriate to the Thai cultural context (Lertpanitkul, 2003). These strategies were staged as a cascade, from the royal court down through the hierarchy before being integrated into the lifestyles, including culinary culture, of ordinary Thais (Vespada, 2001).

Thus, although Thai food culture was not subjected to the forces experienced by other colonized nations Thais have readily adopted ingredients and dishes from other countries. Adapting to different culinary influences stretches back many centuries. Early Portuguese missionaries introduced chilies to Thailand while the Chinese brought in a wider range of vegetables and fruit as well as the practice of stir-frying foods. The extensive use of coconut milk, now found in Thai cuisine, is an adaptation of the milk-based curries introduced by Moslem traders. Coconut cream was substituted for Indian butter because butter was not well received in Thai culture (Chulalongkorn, 1971). In addition, spices were reduced and fresh herbs added. Even though many Thai foods and food practices were based on Indian practices, there were not the strict regulations found in the Hindu caste system. Van Esterik (2000) argues that the practice of observing strict Hindu regulations does not fit with Thai values concerning the importance of finding pleasure in food and the ready absorption of new foods and practices.

At present, imported foods do well in Bangkok (especially at locations near higher income families, resident expatriates and tourists) and other major cities such as Chiang Mai (USDA, 2000). Anything Western, especially food, has high status and is considered 'modern' (Walker and Yasmeen, 1996) even though Thais may not like the taste as much as Thai food.

20.2.2   Religious influences

More than 90% of Thais are practicing Buddhists. Like Christianity, Buddhism uses food to create bonds between people and it provides a model for dealing with food: namely the adoption of asceticism to overcome indulgence and lack of control. This principle co-exists with other ideas about food, such as pleasure and choice (Van Esterik, 2004).

Traditionally, Theravada Buddhism stressed moderation and acceptance of any food mixture to maintain life. Some modern forms of Buddhism such as Santi Asoke promote a vegetarian regime and the use of organic produces. Nationwide, most Thais annually adopt the Mahayana Buddhist practice of being vegetarian for about two weeks in October. This generates a demand for vegetarian food variety and vegetable-based food businesses have grown significantly over the last decade (www.kasikornresearch.com, 2007).

20.2.3   Royal family influences

The role of the Royal family cannot be underestimated. The King as the symbol of power in historical Thailand was a natural focus of attention for early visitors to Thailand who were fascinated by royal food and dining practices. Thai rulers were equally open to Western culinary traditions, adapting banqueting styles, employing Western cooks and sending royal children to Europe to learn etiquette and table manners through the 1800s and 1900s. Further, many of the current Thai royal family were born and or educated overseas and employed European food practices to build bridges with European diplomats (Van Esterik, 2000).

The transformation of Western foods into Thai versions was generally undertaken by courtiers and noble class women placed within palaces. They were responsible for inventing new dishes, improving traditional dishes through adding stylistic flourishes, and for adapting foreign dishes.

The current King has supported agricultural development projects designed to improve Thai food production and to increase revenues of small farmers, particularly those from poorer parts of the nation. He has opened an experimental rice cultivation program designed to help raise the standard of living of the Thai people, the majority of whom are farmers. These actions reflect an expectation that Thai royalty would symbolically feed the nation. Royal support for food projects demonstrates trust and openness; and their generosity with food shows status and power.

20.2.4   National economic plans

Since the 1960s, Thailand has shaped its economic and social directions on the basis of five-year national plans. These have encouraged Thailand to become one of the 'leading food producers in the world and the largest food exporter among all SE Asian countries' (USDA, 2004b). More than half of all processed food production is exported, with approximately two thirds of exports being processed seafood products. Thailand's seafood processing sector is the fourth largest in Asia after Japan, China and India. Throughout the 1990s, Thailand was the world's biggest rice exporter (Burch, 1996).

In order to fulfill the Department of Export Promotion's policy of becoming 'Kitchen to the World' (Jamormnam, 2004), Thailand imports significant amounts in dollar terms of soybeans, cotton, soybean meal, crustaceans, ground fish and flatfish and other fishery products: all destined for food processing. Thailand also exports its cuisine, either through Thai restaurants which are among the most popular across the world, or via foreign tourists who pay for cooking classes and other culinary experiences within Thailand.

20.3 Basics of Thai meals

A French Envoy, Monsieur De La Loubère, sent by Louis XIV to the King of Siam in 1687, observed that: *'The table of the Siamese is not sumptuous: As we eat less in summer than in winter, they eat less than we, by reason of the continual summer in which they live; their common food is rice and fish...'*. He continued that *'A Siamese makes a very good Meal with a pound of Rice a day, which amounts not to more than a Farthing; and with a little dry or salt Fish, which costs no more.'*

Geographical factors determine human settlement patterns (Walipodom, 1996). Thai meals have always reflected geophysical conditions which are water-bound: the country is a long peninsula surrounded by the sea, criss-crossed by major rivers and enjoying high tropical rainfall. Thus, rice, aquatic plants and animals especially fish are major aspects of the diet.

20.3.1 Rice

The importance of rice to the Thai people is economic, symbolic and nutritional. On average, a Thai consumes about 130 kg of rice per year. Over 80% of the total population eats rice as a staple food (Piththawatchai, 2005). On the symbolic level, many South East Asians see themselves as 'made up of rice' (Yasmeen, 2000), with the first solid foods for infants being rice (Ho, 1995). Until recently, a mother used to set aside some of her cooked rice, crushed by using the back of a spoon, and feed it to her baby, sometimes replaced by spoon-scraped ripe banana. Rice is important for nurturing the mother's body, and it is assumed that rice is good for her baby as well. A rice-based diet prevails across all groups and classes of Thai society. To eat rice husk or *'kin kleab'*, is used as an expression for abject poverty.

Rice is almost sacred and Thais embody their respect in 'Mae Po Sop – the rice goddess', who is thought to nourish life (Pitathawatchai, 2005). Various stages of rice growing are also associated with motherhood: when rice grains appear on the stalk, they are referred to as 'pregnant'. The white liquid in green rice grain is called *Namnom Khao* or rice milk which can be made into

Kkao Ya koo, a rice juice. It is made by pounding and squeezing young rice grains and leaves, strained, cooked with pandan leaf juice and some sugar. (Sonakul, 1952). It used to be made once a year only because rice growing was then an annual crop, and it was always valued as a nutritious drink. Nowadays *Khao Ya Koo* is very rare because the important issue is a sufficient supply of mature rice to feed a growing population. In addition, other commercialized kinds of fruit juices, soymilk, and corn milk, have taken hold on the drink aisle and shelf.

The association of rice with nurture may also be linked to what has been described as a 'Thai matrifocal culture where women own the fields and its products'. This association is carried through a belief in Mae Po Sop, the rice goddess who protects paddies and rice for the benefit of mankind.

Rice has been the Thai staple food since the beginning. All other plants, vegetables, and fruits are cooked separately and eaten as a complement to rice. The centrality of rice continues: a survey conducted in 1989–90 reporting that most Thais agreed that they did not feel full unless they had eaten rice, that no meal is complete without rice and that rice is the perfect food (Walker and Yasmeen, 1996).

Rice is not only consumed in plain form with *Kap Khao* or add-ons. Roasted ground rice is called for in every *laab* or minced meat salad. Northeastern curries are thickened by *khao buea*, a thickening agent made by grinding soaked glutinous rice grain with pestle and mortar. Many thanksgiving and invocation ceremonies to Thai deities must be accompanied by *Khao kwan*, a decorative bowl of rice. Rice is believed to ward off haunting spirits. Rice is milled and made into rice noodles of different kinds: *Kanom jeen*, spaghetti-like noodles eaten with thick curry; *Kuaytiew sen yai/ lek/mee*, flat wide/flat narrow/vermicelli rice noodle soup; *Kuayjab*, a semi-tubular noodle soup with pork viscera and boiled egg. Many Thai sweets and desserts are also made from regular and glutinous rice. Even more are made from rice flours (Pramote, 2005).

Cultural identity is also associated with the type of rice consumed. Lefferts (2005) draws a distinction between those who eat *Khao neow* or glutinous rice in the Northeast of Thailand and central Thais who prefer regular rice (*Khao chao* which literally means rice of the noble). *Khao chao* was introduced into Thai food through Thai-Indian relationships (Wongted, 1988). It was thus named because it was first consumed in royal households, in contrast to the glutinous rice consumed by commoners. Today, all Central and Southern meals are eaten with regular rice. Northern and Northeastern households eat *Khao chao* more than they did in the past, but they still hold fast to their glutinous rice meal practice.

20.3.2   Fish

Abbé de Choisy observed that 320 years ago, 'Siamese did not like to eat big animal' (Choisy 1644–1724, Smithies 1993). This is because Thailand has always been an alluvial country; it is a perfect place for rice cultivation, and fresh water fish banks. The oft quoted stone inscription 'In water there is fish. In paddy there is rice.' (NFAD, 2004, p. 19) not only tells that Thai food is abundant but also paints a picture of a rice paddy filled with fish.

During the rainy season all rivers, ponds, and canals are flooded with water and fish. This heralds the beginning of the rice season. In preparing the soil for rice growing, farmers form a gigantic rectangular field from mounded earth to capture the water for the entire rice growing period. After the rice yields grains, it is time to drain all the water. This is a joyous time for farmers as fish are easily spotted and caught, and often fermented to make *Pla-ra*.

A Thai meal is always composed of rice and complementary dishes called *kap khao*, or rice add-ons: the most common add-on is associated with fish, eaten along with a wide range of vegetables, spices and herbs. Traditional Thai recipes testify that fish can be incorporated into an unlimited array of dishes. The classic sour curry eaten in all parts of Thailand contains fish in big slices as well as ground fish mixed with curry paste. *Kanomjeen Namya* or rice noodles in curry soup, a popular one-dish meal eaten in all regions, is actually thickened by ground fish in coconut milk, and eaten with various kinds of fresh vegetables.

Fish is also made into desserts: for example, *Pla haeng Tangmo*, a roasted ground dried fish eaten with watermelon; *Khao neow na Pla*, sticky rice and sweet dried ground fish; and sago dumpling with fish stuffing. Some sweets are even shaped after a fish: for instance, *Kanom Ko*, a gritty white large Thai cookie; *Kanom Pla*, a large yellow roasted dough; and *Plakrim Khai tao*, an imitation small fish made with mini rice noodles boiled in sweet coconut milk.

Fish sauce is vital to Thai cuisine and it has a permanent presence much like the salt shaker has a permanent place on the Western table. It is also a common flavoring agent among Southeast Asian cultures. Even ancient Romans used it but called it by a different name, i.e. *garum or liquamen* (Tannahill, 1984). Thai fish sauce is made by fermenting small fish in a heavy amount of salt for one year, after which it is distilled, bottled, and added to virtually all Thai dishes.

There are numerous examples of how fish culture is interwoven into the Thai way of life. One of the five Buddhist Precepts states that one shall avoid killing or harming living beings, so after a weekly congregation at the Buddhist temple, Thais set fish free to compensate for the many lives of caught fish. A bunch of coconut or palm leaf woven fish hangs over the baby's cradle. Every shop sale can be boosted by hanging mini woven bamboo fish and fish catching equipment, for it is believed that clients will flow in like the plentiful fish of Thailand. A person who wants to have two seemingly great things at the same time is said to be 'catching fish in both hands', which risks losing them both. Finally, this earth is believed to be on a giant fish called *Ananda*; earthquakes result when *Ananda* tosses and turns (Thai National Team for Anthology of Asean Literatures, 1985).

The rice–fish system has been described as providing 'a model of near self-sufficiency'. Larger fish are sold and small fish are placed in storing jars and fermented with rice bran and salt to make fish sauce. According to Lefferts (2005), the contrast between the smelly fish sauce and sticky rice of North East Thailand with the mild sea fish based fish sauce and dry rice of Central Thailand is used to differentiate the two regions.

Aquaculture plays an important part in the food security of Thailand and small-scale aquaculture is extremely important for the 'rural poor'. In 2003, aquaculture activities produced 1.046 million tonnes (US1.46 billion value), one quarter of total fish production (FAO website). Aquaculture in Thailand can be divided in two groups: fresh water (developed more than 80 years ago, cultured in ponds, rice fields, cages, ditches) and brackish water (more recently developed along coastal waters where fish are reared in cages and ponds for subsistence).

20.3.3 Namprik

Namprik, or chili paste dip, is a nationwide favorite. It is a dish which reflects Thai food science wisdom that eating spicy hot food encourages the body to perspire profusely, thereby providing a body suitable for the tropical climate.

Namprik's origins are lost in the mists of time past. Its ingredients are chili, shrimp paste, fish, lime, vegetables and so on. *Namprik* accompanies fish, vegetables and other foods. It is made by picking what is available in the kitchen, in the backyard, or in the neighborhood. *Namprik* varieties are created from various seasonal, local sour fruits i.e. green mango, young tamarind that produce different odors and flavors. It is not eaten alone and must be eaten with rice and plenty of vegetables to subdue the resulting tongue-burning. All *Namprik* recipes are hot and spicy.

Namprik is a unique Thai feature food with regional variations. The famous country song and parody written to reflect *Namprik's* importance to the cuisine goes like this:

> First a male sings *'Please tell me lady, just tell me, beauty, what kind of meal do your family have with rice that make you look so pretty and charming, even as an angel'*
> The female replies *'Rice with Namprik of course, makes me look this good...I pick vegetables from nature'* (Thanthong, 1975).

20.3.4 Herbs

Seubsman (2003) has studied the ubiquitous use of herbs in meals, the importance of herbal foods, as well as Thai people's motivations for eating them. She found that Thai people add herbs to all kinds of food and in high volume. Their motivations cover a descending range: (1) health consciousness,

(2) poor health affecting income generation, (3) preventing fragile health/old age, (4) family and friends discouraging unhealthy food habits, (5) the promise of healthy digestive systems, (6) witnessing illness in others and feeling a need to eat better, (7) a liking for spicy hot food with herb content, and (8) feeling that food without herbs is tasteless. The only factor that prevents people from eating herbs is their strong aroma and powerful taste.

Herbs or spices are always visible in Thai cooking (Dahlan, 2003). However, the indispensable part of *kaeng* or curry lies in curry paste made from various herbs such as chilies, galangal, shallots, garlic, and kaffir lime rind. It might be pounded by using mortar and pestle as in the olden days (still common in rural areas), ground in a modern blender, or bought ready-made in a small amount from the fresh food market.

Kaeng can be divided into coconut cream and non-coconut cream base. *Tom* differs from *kaeng* in that different herbs are pounded into paste for seasoning, though some recipes require pounded coriander root, pepper, and garlic, making *tom* less spicy than *kaeng*, unless it is a spicy soup. A note must be added here: there are two types of *tom*, spicy and non-spicy.

20.4   The meal matrix: harmony and contrast

20.4.1   Main cooking techniques

Thai food is categorized according to cooking methods. Essentially due to a limitation of cooking utensils in former days, Thai food originally was cooked without oil, often called *gee* (cooking over medium heat without or with little grease); *laam* (cooking in bamboo stems over a raging fire); and *tom* (boiling in earth pot). *Neung,* or steaming, is a method for cooking sticky rice, many Thai desserts, and northeastern style steamed fish. *Neung* is a method Thais learned from the Chinese. *Ob,* or baking, is mostly for Western food.

Today, the most popular cooking methods include: *tom* (boil for soup), *kaeng* (stir then boil with curry paste), *pad* (stir-fry), *yam* (mixed vegetables, fruit or meat with or without heat), *tam* (spicy salad mixed using mortar and pestle, also used to make *Namprik* by pounding a combination of ingredients to yield a salted, hot, and sour flavor), *plaa* (mixed undercooked/uncooked ingredients, another kind of salad), *tod* (fry; or deep fry), *pao* (put into open fire), *yang* or *ping* (grill), and *lon* (coconut cream base dipping). *Tom* and *kaeng* are made with high water volume. *Tod, pao,* and *yang* are often used for meat or fish.

There are also other regional techniques not explored here.

20.4.2   Meal combinations

Thai meals are an art of combining flavors through the use of mixing and matching different dishes (Uluchata, 1994). Table 20.1 provides a template

Table 20.1 Thai meal matrix: harmony and contrast

Tom/Kaeng (soup/curry) (liquid dish)	Yam/Pad (curry/salad/stir-fried) (pulpy dish)	Tod (fried) (dry dish)	Namprik/Lon (chilli dips)	Nueng (steamed)
Spicy hot, coconut milk base	*Non-spicy hot, non-coconut milk*			
Tay-po curry	stir-fried vegetables w/meat of choice		Shrimp paste chili paste dip	
Red curry	stir-fried mung bean vermicelli			
Green curry	stir-fried mushroom	Omelette	Fermented fish chili paste	
Masman curry	sweet pork	Fried egg	Eggplant chili paste dip	
Kari curry			Catfish chili paste dip	
Bitter gourd/catfish curry			Tamarind chili paste dip	
			Dips made from seasonal fruit/plants	
Sour red curry *Kaeng-kua-som*		Fried dried pork		
Spicy hot, non-coconut milk base		Fried chicken	*Coconut cream base dip*	
Sour curry				
Tom yam soup	Any kind of non-spicy hot stir-fried	Fried dried beef	*Lon* (coconut cream base dip)	
		Fried vegetable	fermented fish lon	
Sour soup (*Tomsom*)		Fried sausage	salted olive lon	Steamed egg
Yellow curry		Fried pork/meat ball	salted crab lon	Steamed vegetables
Forest curry (*Kaeng Pa*)		Fried bird	Shrimp paste in coconut cream	Steamed fish
Clear spicy soup (*keang liang*)		Fried dried fish	(*kapi kua*)	Steamed chicken
Tom klong/tom sap		Fried dried gourami		Steamed clams
Non-spicy, coconut milk base	*Spicy hot, non-coconut milk*	Fried pilchard		Steamed curry cake
Galangal soup	*Pad prik khing* (stir-fried with	Fish cake	Preserved bean curd chilli dip	Curry in banana
Lotus stem and pilchard soup	chili paste &string bean)	Fried frog	Salted crab dip	leaf wrap
Banana flower soup	Stir-fried holy basil		Salted black olive chili dip	
	Spicy minced pork salad		Tomato chili paste dip	
	Shrimp salad (*Pla kung*)		Hell fire dip (extremely hot dip)	

Eggplant salad
Grilled fish or meat salad
Stir-fired ginger with meat of choice
Stir-fired southern style (*Kua-kling*)

Heaven dip (sweet-medium hot)

Spicy hot, coconut milk
Panaeng curry
Chu chee curry

Shrimp oil dip

Mango chili paste dip
Sweet fish sauce (w/neem tree sprig)
Fermented soybean in coconut cream
Fermented fish in coconut cream
Northern style pork and tomato in oil dip

Non-spicy non-coconut milk base
Mung bean vermicelli soup
Five spice soup (*pa lo*)
Ivy gourd leaf soup
Stuffed bitter gourd soup
Sweetsalted soup (*Tomkem*)

for explaining how Thais comprising a meal. Normally, there are three to five dishes comprising a meal, and each should vary in flavor and color. One dish from each column is chosen in constructing a meal. A rough guideline is sour goes with sweet, and hot spicy goes with salted or bland. Whenever there is a chili paste dip or *Namprik*, a curry or salted fish/beef must be served alongside. Sour curry is best eaten with sweet pork, omelet or fried boiled egg with tamarind sauce, for example. Spicy dishes are always served alongside non-spicy ones. All the tastes of sweet, astringent, bland, bitter, hot, and salt must be incorporated to create an appetizing meal. A lover of hot-spicy flavors can pick his/her meal by selecting a couple of very hot spicy dishes but a bland flavored dish is never left out for bland does not mean lacking in taste. More importantly it completes a meal. A steamed dish is picked only occasionally except in the Northeast.

Srisamorn Kongpun, a Thai food expert/consultant suggests some desirable combinations. A pork soup should be served with non-coconut milk base curry. Aside from curry, a fried gourami (*Trichogaster pectoralis*) and a non-coconut milk base dipping sauce is recommended. She comments that two coconut milk base dishes should not be presented in the same set otherwise the meal becomes too rich. A suitable segue is a shrimp paste dip served with fried gourami, and vegetables fresh, boiled or fried.

Any leftover food is often transformed into a new dish for the next meal (Na Songkla, 2002). For example, leftover green curry from dinner can be made into green fried rice the next morning. In the north they have *kaeng hoh* or 'combined curry' which is made from putting all leftovers together with pork loin, lemon grass, fermented bamboo shoots, cellophane noodles, and fried garlic and shallots. *Kaeng samruam*, the Central region's variation on a combined curry is made by putting leftovers such as curry, soup, and stir-fried food in one pot and seasoning anew.

20.5   Harmonious blend of the Thai meal

20.5.1   The elements

For most people, the first thought upon hearing the word 'Thai food' is its hot and spicy flavor which varies in degree according to regional cooking techniques, ingredients, and neighboring country influences. However, behind the energetic flavoring, which will be elaborated later, there are more subtle features of Thai cuisine which derive from Thai people's understanding of their place in nature and the universe. In this section, we focus on how Thais use meals to achieve inner balance and harmonious relations with their biophysical, spiritual and social environments (Chang, 1977).

According to Buddhist influenced Thai medicine, the human body is composed of four *dhâtu* or elements: earth, water, wind, and fire, which correspond to bodily organs such as skin, flesh and bones. 'Earth' is a sustaining part of other elements; 'water' constitutes fluids in the body such as gall, blood, sweat, lymph and mucus; 'fire' provides heat and energy to keep the body warm and help with digestion; 'wind' brings in motions which drive the body such as blood circulation, digestive system, and gastronomic gas (Mahachulalongkornrajavidyalaya, 1996). These elements must be in alignment otherwise an 'element imbalance' occurs, and illness results. One way of balancing is to eat after the fashion of the primary body element, which is influenced by the month of birth.

'Earth' people should eat *faad, waan, mun* and *kem* food: astringent food or *faad* (a combined sour–bitter–brackish taste, found predominantly in Thai herbs, mangosteen, and green guava); *waan* (sweet, as found in milk and palm sugar); *mun* (a potato-like oil composite taste found in peanuts, potato, pumpkin, taro root, and coconut milk); and, *kem* (a salt taste coming from earth and sea salt, fish sauce, shrimp paste, fermented and dried fish). A 'water' person should eat sour or slightly bitter tasting food (found in kaffir lime, lime, orange, pineapple, tomato, bitter gourd, and neem tree sprig). Greasy dishes should be avoided because they counteract water. A 'wind' person should eat spicy hot (found in ginger, galangal, lemon grass, (*Boesenbergia rotunda* (L.) Mansf.), peppers, sweet basil, and holy basil). A 'fire' person should eat cool and non-spicy food (such as water spinach, ivy gourd sprig, Asiatic pennywort, and Siamese cassia) (Kongpun, 1999).

The elemental balance is often affected by seasonal change. Therefore, seasonal and counter-seasonal balance helps to maintain body balance and ward off illnesses caused by external forces (Vespada, 2001). Summer (hot), winter (cold), and rainy season (cool) bring changes to body temperature and if the body cannot aptly adjust itself, illness results. Eating according to climate is an unconscious assimilation between the individual and their environment: it is living in harmony with nature. In summertime, increased body temperature can cripple the 'fire' element, thus affecting the 'water' and 'earth' elements in the body. Constipation, cold sores, feeling tired, thirsty, and feverish result from such an event and can be prevented by eating food containing bitter, cool, astringent and bland flavors. In the rainy season, the wet and cool climate can cripple the 'air' element, causing constipation, fever and cold. This effect can be prevented by eating spicy hot food. In winter, the cold weather will affect the 'water' element, and will cause chapped skin, stupor, runny nose and joint pain. Suitable preventative foods are hot, bitter, spicy and astringent found plentifully in Thailand's tropical plants (Sapcharoen, 1994).

'Elemental' eating does not exist only in theory; it is adapted generally within the Thai way of life. This nature-balance eating practice is particularly evident at the end of the rainy season and the onset of winter, when people get seasonal fever. *Kae* (cork wood tree) flowers which bloom across the country are picked and made into a sour curry for treatment. Another dish called *tomchew* soup is also cooked to prevent a cold since its main ingredient is onion (Kongpun, 1999).

20.5.2   Elegance

Thai food carving and food arrangement on the plate or in the bowl explain the 'ooh-ah' exclamations of foreigners (Fig. 20.1 and 20.2).

Fig. 20.1 'Namprik Platu' (fried fish with vegetable and shrimp pate dip).

Fig. 20.2 Carved papaya by Sumallee Tuntikun.

The effort that goes into presentation and resulting elegance in Thai dishes is well known. On her 1972 visit to Thailand, Queen Elizabeth II, looked at fruits and vegetables carved out of water melon, pumpkin, cucumber, and taro, and asked in surprise, 'Are these carved by hand?'(Promsao *et al.*, 1999).

Thais continue to be innovative in assembling flavors and colors into appetizing bite size food. On special occasions, carved fruits and vegetables are fashioned after flower and animal designs and finished pieces are stored and kept in a cool place or wrapped in damp cloth. Carved eggplant must be soaked in brine or lime water to prevent it from turning brown. Additionally, some Thai desserts are shaped or sculpted into fruits, vegetables, and animal figures. Ordinary Thais present dishes in this way for special occasions.

The food sciences and arts, including elegant presentation, were practiced and perfected during the Chakkri Dynasty (1782 AD – present). Palaces acted as the food culture crossroads where Thai and foreign food cultures merged and evolved. Palace recipes, eating patterns, and food arrangements were distinguished from local ones in their exquisite arrangements and refined flavors. When, in 1932, the absolute monarchy was changed in favor of democratic government, courtiers returned to their former homes often starting up royal cuisine restaurants. They became very successful as the dishes they prepared were dexterously cooked, carved, and arranged with style. Nowadays, food carving is institutionalized in many academic and vocational schools.

In everyday eating, dishes are decorated in some way: either through a simple cut to a vegetable or fruit and the use of colorful foods artfully arranged. A lot of fruit and vegetables are used in this way, and flowers and herbs from the garden act as garnish or an additional focal point of interest. Householders will color foods, like rice, through the addition of pandan leaf for a green effect or carrot juice for an orange hue. Colored foods are seen in fresh food markets and on street vendor stalls as vendors try to entice customers.

20.5.3 Energy

The energy of Thai meals is multi-faceted. In large part, it derives from the use of particularly aromatic herb and spice ingredients in all dishes. Even a mild soup needs coriander root, pepper, and garlic. No stir-fried dish can be made without garlic or pepper sprinkled on top. The incorporation of these ingredients is owing to their availability across the country. Even a fruit-based salad will be dressed with colorful herbs and spices (Fig. 20.3).

The energy also comes from factors already mentioned: the way Thai cooks adapt and improvise foreign food by using local ingredients without compromising the unique spicy hot astringent feature of Thai food. As a result, Thai food is extraordinarily diverse and a typical Thai restaurant meal will contain foods from the traditions of both the royal cuisine and market-based cuisine, with curries derived from Moslem traders, Chinese influenced

Fig. 20.3 Young jackfruit salad.

stir-fried or fried food, regionally influenced foods and European influenced desserts. Energy comes from how different food items are composed into a meal that differs from region to region.

Energy is present in another form: the way some dishes provide a link back to nature. *Tom Yum*, for example, is a clear soup where all the ingredients are visible. Whole leaves and herbs and seafood can be seen, and they remind the eater of the source of the food: in this case, the garden and the sea.

The chatter and interactions that occur during Thai meals is facilitated by the fact that everyone sits in a meal circle and typically shares one central set course. In other words, energy is multi-faceted: it comes from natural ingredients, a taste and group reflections on the dish or taste. A hot dish that causes the sweat to flow, will invariably lead to table commentary on people's capacity for hot food. Diners tease one another, and tears follow with the ensuing laughter.

20.6   Meal sequence

Thai working adults have three meals daily, while monks have two meals a day (breakfast and lunch – always before noon). The statistics show that a higher number of elderly people than working age adults eat three meals a day. People who live in rural areas eat three meals more often than do the urban group (Fig. 20.4).

As breakfast eaters, working adults in urban areas rank bottom, but tie with rural residents as regular lunch and dinner meal participants. In-between

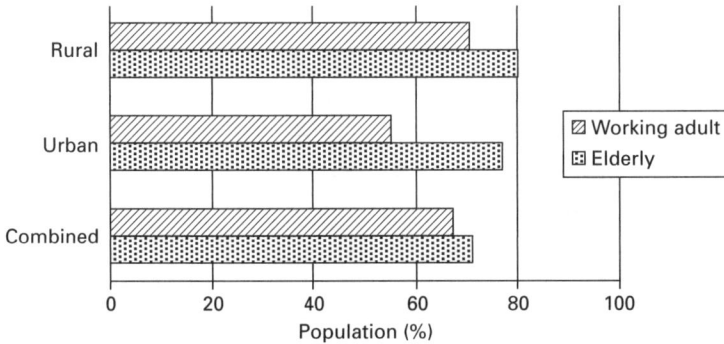

Fig. 20.4 Three meals behavior among urban and rural residents.

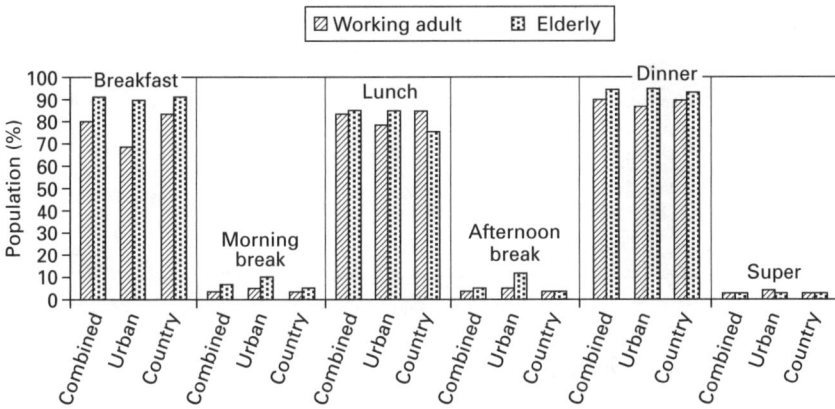

Fig. 20.5 Main meals and in-between meals for urban and rural residents.

meals and snacks (*Khong wang*) are of relatively minor importance compared with main meals. Most people said that they have in-between meals (for example, fruit or a small dessert) on a non-regular basis, although the elderly like to have in-between meals more often than working adults, with the exception of supper which is taken by the working adult group (MOPH, 2006) (Fig. 20.5).

Figure 20.6 shows that working adults (aged 15 and up) generally have in-between meals on a non-daily basis. A high percentage (67%) of the population never have food after dinner while a small group of 5% have morning or afternoon breaks or have food after dinner every day. Most people occasionally have an afternoon break meal.

A popular Thai breakfast is a one-dish meal such as rice and curry, fried rice, stir-fried holy basil and rice, or rice porridge (Fig. 20.7). These are all quick and easy food items. Alternatively, Thais might eat a traditional donut-equivalent, of Chinese origin, called *Pa Tong Go*. It is eaten with hot tea, soy milk or coffee. Western breakfast includes toast, bread, sandwich, coffee, milk, or hot chocolate. It is the least popular breakfast food in Thailand.

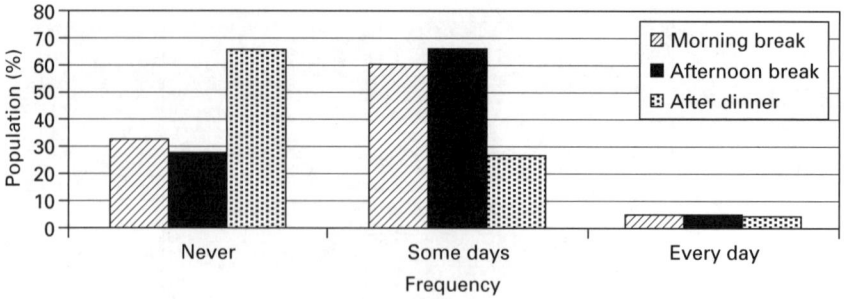

Fig. 20.6   Morning break, afternoon break, and after dinner consumption.

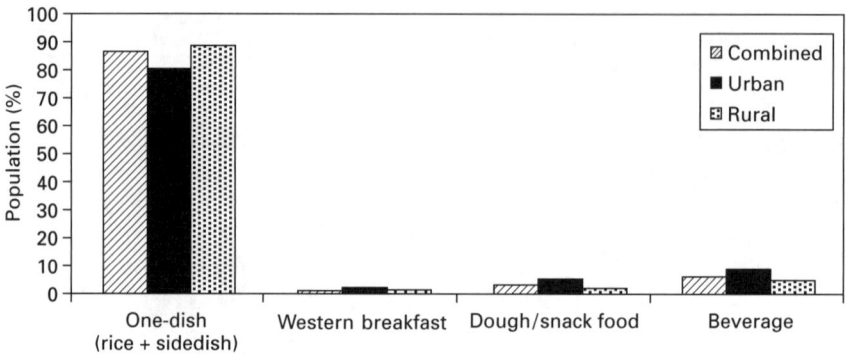

Fig. 20.7   Breakfast meals in urban and rural areas.

20.6.1   Meal arrangements

Commonly, in countryside households or at temple ceremonial events, Thai meal arrangements include sitting around in a circle on the floor or mats. This arrangement is convenient, requiring fewer eating utensils and can accommodate as many people as necessary. The central part is reserved for food and is surrounded by serving plates. Rice plates, which are placed in front of each person, are slightly deeper than flat ones and are more suitable for soggy dishes. There is no knife because each dish is already cut into bite size pieces for convenient consumption. The eating utensils on the right hand side include a spoon for the main dish and dessert spoons and there is a glass of water as well. On the left hand side of the rice plate is a fork. In front of each person and slightly to the left is a serving plate or bowl with its own ceramic or serving spoon (Anupaptripob, 1985).

Customarily, the arrangement is to place complementary foods next to each other. Chili paste or another dip with various seasonal vegetables is usually placed at the middle of the setting or table. Various curries or prepared dishes are placed at opposite ends. Apart from these dishes, a typical set of complementary food will consist of a stir fry dish, *Yam* (various kinds of meat or vegetables) and deep fried dried fish. Vegetable and fruit beautification demonstrates Thais' genuine attention to preparing food with great care and creativity (Anupaptripob, 1985) (Fig. 20.8).

The meal commences with serving rice and water to the honored person or an elder by the person who is sitting nearest. Each person will take the allocate food in front of them by using a communal spoon for each dish, serving their plate first and then passing the dish on to another who sits either on the left or right hand side. Any dish that one does not like or is not familiar with can be skipped and passed on to another. For hygienic reasons, if communal spoons are not provided, people ask for them.

Small portions are valued as part of the harmonious nature of Thai culinary culture. In part small dishes are preferred because of the hot weather, but probably the most important factor is adherence to a Buddhist culture which follows the middle path principle in all things. As Supreme Patriarch His Royal Highness Prince Paramanujitajinorasa (1790–1835) taught:

'Little diet, gets too thin, In appearance,
    Much diet, gets too fat, heavy
Medium diet, no more, no less, thence
    Bestows thee sound vitality.'

Unlike in some cultures, where the spoon and fork are crossed on the plate to indicate feeling full, the Thai way is to place spoon and folk in parallel. The dessert or fruit will be served. To withdraw from the setting

Fig. 20.8

politely, one needs to be considerate of others. If it is necessary to leave earlier one needs to ask first for permission from the main person.

20.7   Continuities in royal class meals

While traveling back from his Europe trip in 1897, King Rama V related his craving for Thai food in a letter to his daughter, Princess Nipanopadol (Chulalongkorn, 1973). He reminisced about fried *Kulao* fish, salted egg, fried mullet, chili paste dip, curry, omelet, shrimp cake, dried fish, *Tay-po* curry, cucumber spicy salad, rice noodles in spicy curry or sweet curry, and shrimp paste cooked in coconut cream dip. Waking up the next morning he made for himself a rice dish with shrimp paste and ham, sprinkled over with lime juice and ground chili.

A century later, King Bhumibholadulyadej's meal does not differ much from his grandfather's. Lady Prasarnsook Tantiwetchakul, the King's royal command-in-chief chef, shares what she prepares for him (Kampa-u, 1988). For breakfast, boiled rice is eaten along with two to three dishes chosen from amongst the following: stir-fried preserved Chinese radish, salted egg, stir-fried Chinese olive, omelet, dried shrimp salad or fermented soybean curd served with chilies and lime wedges, roasted peanuts, fried squid, chop suey, or five spice stew with chicken/pork/egg. Interchangeably, rice porridge with shrimp, pork, chicken or fish is often served. These meats are often fashioned into beautiful designs. His lunch depends on where he undertakes his work. When he switches residential palaces he is served with regional food; for example, young chili paste dip, pork ball and young jackfruit salad are dominant in the north; mixed rice salad (*Khao yam*), stir-fried *sato* (southern bean), and green mango salad are served in the south.

The King's dinner consists of three to four dishes, i.e. a soup, different salads and curries. His favorites include ground dried fish with watermelon and pomfret's egg boiling and eaten with sliced shallots, mangosteen, and spicy fish sauce. He has his meal with hot rice and *namprik* such as tamarind dip, green clustered eggplant dip, Chinese olive dip, salted egg with shrimp paste or *Namprik lung ruer*, shrimp paste dip, with boned fried mackerel, desiccated coconut mixed in mortar with ground peanuts and salt. His *kreung kiang* or side dish consisted of crispy fried and pickled vegetables, ginger, fried gourami, or boiled prawn. Breakfast is a simpler meal than dinner with fewer choices.

20.8   Regional meals

Each regional dish looks and tastes similar throughout the region but when prepared outside the region they become varied. This is because of the absence of local flora and fauna, which when cooked in regional style have unique flavors and aromas. Many regional specialties have become so popular that visitors flock to the region to consume them, and then they become available nationwide. By contrast, many common dishes have become regional dishes owing to the incorporation of local substitutes and cooking styles. For example, fish sauce is a common salt substitute everywhere but fermented fish (*Pla ra*) is preferred in the northeast; in the south the variant is known as *Boo Doo* sauce; and in the north, a crab essence (*Nam pu*) is used.

20.8.1   Central meals
Central people readily adapt food from other cultures and create numerous flavorful spin offs. Their meals however must contain hot, spicy, astringent, salt, sweet, and non-spicy flavor dishes. Many dishes, especially curry with coconut milk are adopted Indian dishes, and then adapted to Thai taste through the use of Thai herbs.

Central region meals are more influenced by the royal kitchen. The food that is made for the royal household is not much different from ordinary households, although the former incorporate greater artistic elaboration (Jamornman, 2004).

20.8.2   Northeastern meals
The Northeastern region is located in a drought prone high plateau with various ethnic groups who share culinary flavors mainly from Cambodia, Vietnam and Laos. Feature dishes include hot chilies, fermented fish, insects, vegetables and plants. Popular dishes are papaya salad and *lab*, a minced or grilled meat salad which has become available and enjoyed nationwide. Most northeasterners prefer cooking with a large amount of herbs and spices. Liquid is used very little as the way people eat in this region is by scooping a handful of sticky rice from a common bamboo container (*Katip*), finger-pressing and forming a round ball which is then dipped into one of the dishes placed in the meal circle. After this same fashion, another bite-size sticky rice ball is dipped into another dish.

20.8.3   Northern meals
The Northern region is influenced by various minority groups including the Tai yai, Tai lue and Karen. Their staple meals consist of glutinous rice, dips and curry normally without coconut milk. However, one popular dish *Khao soi* – noodle in curry' is cooked with coconut milk, and is eaten with complementary items of fresh shallot, pickled cabbage and lemon. They serve their meals on a raised vessel called a '*tok*'.

The sticky rice eating fashion is similar to the Northeastern practice. Nevertheless, Northerners do not spice up their food as much as

Northeasterners; Northerners use herbs and less chili. Northern dishes are influenced by neighboring countries such as Myanmar's curry or *kaeng hung le,* or cooking with pork in a Chinese style. Well known northern ingredients include crab essence and dried fermented bean. Unlike central meals, northern meals use very little coconut milk and cream.

20.8.4 Southern meals

Southerners cook their food with seafood from the engulfing Southern China sea. They cook with aromatic herbs and hot spices. Their choice of ingredients blends well with a tropical climate featuring high humidity. The spiciness of Southern food helps keep body temperature cool and prevent colds. Southern people enjoy shrimp paste chili dip, tamarind chili paste and pickled ginger dips (Seubsman, 2003). A large bowl of fresh vegetables is indispensable for a southern hot meal. They have their meal with white rice, similar to the central region. Southern meals are arranged on a mat and eaten by hand which people say enhances the flavor of the meals (Sithipipat, 1999). However, meals could be arranged on mats or tables depending on preferences and economic status (Setto, 1989).

20.9   Thai meals and meal practices across the regions

Tables 20.2 to 20.10 show samples of food diaries that have been selected to show differences in food consumption according to participants' age, sex, and geographical location. Diaries from the younger people show that they are more likely to eat western style food and to buy more pre-prepared foods.

20.10   Trends in Thai eating behavior

20.10.1   The nutrition transition

The ongoing impact of culinary cultural change is displayed in the nutrition transition. In a 1960 nutrition survey conducted by the Thai government and US Interdepartmental Committee on Nutrition for National Defense (1962), fat energy intake was a mere 9% of total energy. Today, the fat contribution is around 13% to 14% for all age groups over 6 years. Urban youths get more energy from fat than do rural youths and girls receive more fat energy than boys. People in the northeast have the lowest fat energy of all Thais.

At a national level, Thailand will continue to experience trends typical of changing food demand in Asia (see Pingali, 2004, p. 1). These include

1. reduced per capita consumption of rice,
2. increased per capita consumption of wheat and wheat-based products,
3. increased diversity in the food groups consumed,

Table 20.2   A middle-aged woman from the Northeast

Supanee Kampetdee is a 53 year old woman who works for an electric company in Ubon Ratchathani province. The company provides welfare in the form of a special low-cost meal called *sawasdikarn* for workers who pay for them in advance monthly. As a shift worker, her meals at work depend on the shift she is working. For example on the morning shift she will have breakfast and lunch at the company canteen. During the afternoon shift she will have an afternoon break and dinner at work. Having a meal regularly provided by the same cook or food provider is called *fark thong* or 'stomach deposit' meaning her diet largely depends on whatever the company prepares. On the other days at home her food is prepared by her daughter.

Her seven day diary reveals that five days of the week are spent at work. A canteen breakfast always contain western elements i.e. milk, butter, and toast or bread. Lunch at work, in contrast, always has northeastern elements such as spicy minced meat salads, rice noodle curry, and eggplant dip which have been prepared that morning. Whenever she is at home she eats northeastern style dishes such as ant's egg curry, chili paste dip, and fish soup. She often has three main meals a day with an extra snack in between meals.

When at home, her breakfast and lunch is prepared by her daughter and eaten with her husband. The staple home foods foods are rice and fish. On the day she goes out of town, she buys a quick and easy meal such as snacks and stir-fried noodles. Supanee keeps a garden that provides her with fresh ingredients. During dinner at home she likes to sit on the porch and watch TV.

Day 1 (Friday)
Breakfast: rice porridge with pork, bread with sweetened condensed milk
Break: desserts, mango
Lunch: rice, spicy minced pork salad, sour curry, stir-fried vegetable
Break: ice cream
Dinner: rice, fried tilapia, wrapped dumpling, blanched/boiled vegetable, chili paste dip, fruits
Break: -
Day 2 (Saturday)
Breakfast: toast and butter, rice, fried eggs, soup
Break: sour mango with sweet fish sauce
Lunch: rice, spicy eggplant dip, stir-fried vegetables with crispy pork
Break: milk, toast, sweet potato
Dinner: rice, fired fish, sour curry
Break: -
Day 3 (Sunday)
Breakfast: rice, ant's egg curry, chili paste dip, blanched vegetables.
Break: steamed pumpkin
Lunch: rice, chop suey, grilled fish
Break: fruit, milk, snack
Dinner: vegetable salad, soup, fruits
Break:-
Day 4 (Monday)
Breakfast: rice, chili fish dip, fresh vegetables, grilled pork
Break: fruit, ice cream
Lunch: rice, stir-fried vegetables, beef soup
Break: snack, ice cream
Dinner: rice, blanched vegetables, fried fish, fruits
Break:-
Day 5
Breakfast: Ovaltine, snack
Break: sweet drink, snack
Lunch: stir-fried noodles with gravy sauce
Break: -
Dinner: rice, spicy fish soup, fruits
Break: -
Day 6
Breakfast: rice porridge with pork, milk
Break: fruit, snack
Lunch: rice noodle in fish curry
Break: snack, fruit
Dinner: rice, spicy minced beef salad, fresh vegetables
Break: -
Day 7
Breakfast: rice porridge with pork
Break: fruit, milk
Lunch: rice, spicy minced beef salad, fresh vegetables, papaya salad
Break: -
Dinner: rice, chili paste dip, boiled vegetables, sour fish soup, grilled fish, rice noodles
Break: -

Table 20.3   A middle-aged man from the Northeast

Chomchoei Pannoi is a 41 year old laborer who is married with children. His meals are mostly cooked by his wife and eaten at home with the family. Animal protein is featured in his diets, for example, beef, chicken, fish, frog, egg, and pork. The only times he buys food is when he is eating with friends or co-workers. Most of his food is heavily northeastern in style and includes chili paste dips and different kinds of spicy salads which are made by a blending and mixing method, as in a western salad. This food that is eaten unheated rather than like a western-style refrigerated cold dish. All spicy salads require meat of some type, which is first cooked, then removed from the stove and mixed in with many other fresh vegetables and herbs. The dressing is made from lime juice, fish sauce, and ground chili and blended in with the salad before serving. No heat is applied when mixing, and reheating is not appropriate because it ruins the flavour of the fresh herbs and dressing.

Except for noodle soup, which is a stand alone Chinese dish, all his meals are eaten with sticky rice and fresh local plants such as dill, sweet basil, string beans, and *pak paew* (*Polygonum adoratum* Lour.), according to the Northeastern style eating. People use three fingers to roll the rice into a small ball which they dip into a selection of mostly spicy dishes laid out in the middle of the meal circle, followed by eating fresh plants served on the side to balance the hot flavor and enhance the spicy taste in the mouth.

Even though his meals are cooked and eaten in traditional Northeastern style, none of the ingredients are domestically grown or collected from natural sources. They are all bought from the fresh market.

Day 1 (Friday)
Breakfast: chili paste dip with vegetable and fried pilchard, omelette, pan-fried dried beef
Break: orange, banana
Lunch: papaya salad, roasted chicken, rice noodles in curry, sour fish curry
Break: -
Dinner: spicy cucumber salad, omelette, fried drumstick, desserts
Break: instant noodles
Day 2 (Saturday)
Breakfast: omelette, fried beef
Break: iced green syrup, iced red syrup, snack
Lunch: papaya salad, grilled beef
Break: -
Dinner: beef curry, fresh shrimp salad, stir-fried Chinese kale, stir-fried water spinach
Break: -
Day 3 (Sunday)
Breakfast: mushroom curry, fried beef, omelette
Break: snack, mango
Lunch: fried sausage, fried beef, spicy papaya salad
Break: -
Dinner: mushroom curry, fried sausage, spicy cucumber salad
Break: spicy beef soup, spicy beef salad
Day 4 (Monday)
Breakfast: omelette, sour fish curry, coffee
Break: banana, orange
Lunch: roasted catfish, papaya salad
Break: -
Dinner: spicy cucumber salad, fried beef, omelette
Break: -
Day 5 (Tuesday)
Breakfast: stir-fried oyster sauce, Chinese radish soup
Break: -
Lunch: snack, noodle soup, pork ball soup,
Break: -
Dinner: spicy bamboo salad, spicy beef salad
Break: -
Day 6 (Wednesday)
Breakfast: spicy minced pork, spicy beef soup
Break: -
Lunch: papaya salad, roasted chicken, rice noodles
Break: -
Dinner: spicy cucumber salad, omelette, fried drumstick
Break: -
Day 7 (Thursday)
Breakfast: fried mackerel, spicy mung bean vermicelli salad, fried beef
Break: -
Lunch: fried rice with egg
Break: -
Dinner: frog curry N/E style, papaya salad, roasted chicken
Break: -

Table 20.4   A younger Northeastern man living in Bangkok

Tongchai Sapsombat, 37 years old, is a typical Bangkok immigrant from the Northeast who leads a life similar to thousands of other Northeastern working-class laborers. He pays for monthly housing and rents a taxi on a daily basis making a living by transporting passengers through the hectic Bangkok traffic. Living with a small family in a crowded community means he pays a low rent, and low food prices at an occasional market, but he has numerous neighbors.

Every morning, his breakfast is bought from food stalls. Some days he makes it himself. Rice is his staple. It is cooked in the fast food fashion, i.e. fried rice varieties. His daily first meal can be anywhere from 7.30 to 10 am, depending on his luck with his passengers. He is mainly alone for breakfast and lunch, except on the rare occasions when his little family joins him. The second meal of the day is similar to breakfast. Some days he switches to chicken and bitter gourd noodle soup which is a Bangkok favorite noodle dish sold only at street stalls. It is a perfect combination of Chinese noodles served with a bitter tasting Thai plant that has gained in popularity during the past decade.

He only cooks if he is eating with the family. His lunch hour can be anywhere from half-past twelve or as late as half-past two. His children sometimes wait until past 10 o'clock to enjoy rice and a handful of tasty dishes for dinner with him and occasionally with a neighbor. He never eats between meals.

Day 1 (Friday)
Breakfast: rice, stir-fried holy basil with beef, fried egg
Break: -
Lunch: chicken 'n bitter gourd noodle soup
Break: -
Dinner: rice, curry, soup, boiled egg, chili paste dip with vegetables
Break:
Day 2 (Saturday)
Breakfast: rice 'n' chicken w/soybean sauce
Break: -
Lunch: fried rice
Break: -
Dinner: boiled rice, canned pickled Chinese cabbage, salted egg
Break: -
Day 3 (Sunday)
Breakfast: rice, pork omelette
Break: -
Lunch: chicken and bitter gourd noodle soup
Break: -
Dinner: rice, fried fish
Break:-
Day 4 (Monday)
Breakfast: rice, stir-fried holy basil with beef, fried egg
Break: -
Lunch: fried rice with egg
Break: -
Dinner: rice, chili paste dip, fried boiled egg with sweet and sour sauce, vegetable curry (*Liang*), chicken curry
Break: -
Day 5 (Tuesday)
Breakfast: rice, beef soup
Break: -
Lunch: noodle beef soup
Break: -
Dinner: rice, chili paste dip, fried pilchard, vegetable
Break: -
Day 6 (Wednesday)
Breakfast: fried rice with egg
Break: -
Lunch: rice, stir-fried holy basil with chicken, fried egg
Break: -
Dinner: rice, boiled egg, vegetables, galangal chicken soup
Break: -
Day 7 (Thursday)
Breakfast: rice porridge, orange juice
Break: -
Lunch: rice, omelette
Break: -
Dinner: fried pilchard , fish visceral soup, chili paste dip, fresh vegetables
Break: -

Table 20.5   An older woman from the Central Region

Supanni Jiamjurai is a 63 year old retired lady who lives on her own. She usually eats alone except when her children or grandchildren pay a visit. Supanni is typical of a modern older Thai who cooks and buys food by herself, compared with a traditional older Thai who lives in an extended family, is tended by her children and relatives, and who never has to cook or eat alone. Her cooking appliances comprised of a gas stove, pot, pan, rice cooker, and microwave, which is considered a trendy kitchen appliance and is used primarily for reheating. Even though she goes out to buy food every morning, she never eats out nor has more than one side dish with rice. Her meals could reflect the life of city dweller where convenient food of various tastes and selections is at her finger tips and yet she chooses to stay healthy by eating three nutritious meals a day and three snacks.

She does not work so she has ample time for cooking breakfast which she usually eats as late as 9 am. Every day she eats rice which frequently is cooked or boiled with a side dish such as stir-fried vegetables or a soup. Her favorite drink is soymilk which she buys every morning in the neighborhood. She often cooks her food on her gas stove in a pot or a pan but rice is always cooked in an electric rice cooker. Since she has a late breakfast, her lunch is sometimes as late as 2 pm and is always a light meal, i.e. boiled rice or noodle soup, except when she has younger visitors and then she joins the crowd by eating pizza or grilled chicken and papaya salad. Her afternoon snack is fruit. Dinner is always rice with a vegetable stir-fry and at around 8.30 soymilk is her inseparable supper mate.

Day 1 (Friday)
Breakfast: boiled rice, fried fish
Break: soymilk
Lunch: boiled rice, stir-fried pickled radish
Break: orange, banana
Dinner: rice, ash gourd soup, fried halibut
Break: soymilk
Day 2 (Saturday)
Breakfast: rice, cabbage 'n' chicken soup
Break: -
Lunch: pizza, salad
Break: -
Dinner: rice porridge
Break: soymilk
Day 3 (Sunday)
Breakfast: rice, sour mackerel soup
Break: soymilk
Lunch: egg noodle soup
Break: orange, corn
Dinner: rice, stir-fried vegetables
Break: soymilk
Day 4 (Monday)
Breakfast: rice, fermented bean curd dip
Break: -
Lunch: egg noodle soup
Break: orange
Dinner: rice, stir-fried cauliflower
Break: soymilk
Day 5 (Tuesday)
Breakfast: rice, stir-fried French bean
Break: soymilk
Lunch: boiled rice, pickled vegetable, salted egg
Break: corn
Dinner: boiled rice, stir-fried vegetables
Break: soymilk
Day 6 (Wednesday)
Breakfast: rice, pork filled tomato
Break: soymilk
Lunch: pork dumpling, papaya salad, grilled chicken
Break: orange, banana
Dinner: rice, stir-fried squid
Break: soymilk
Day 7 (Thursday)
Breakfast: rice, stir-fried angled gourd
Break: soymilk
Lunch: boiled rice, stir-fried tofu and bean sprouts
Break: orange
Dinner: rice, stir-fried sugar peas
Break: soymilk

Table 20.6 A middle-aged man from the Central Region

Pleng Sanitpon is a 55 year old government employee in Nonthaburi province. He is a Central region man by birth so his meals are not as spicy as a northeastern or southern meal. His age also limits how much hot and spicy food he eats. Fried fish, mung bean vermicelli soup, egg soup, and rice with stewed pig's trotter are all mild and bland in flavour. His meals mostly consist of one-dish types. He does not often have in-between meal snacks either.

On week days he has his breakfast and lunches at the workplace restaurant. On weekends and every evening he eats food that has been cooked fresh by his daughter.

Day 1 (Thursday)
Breakfast: rice, pork blood soup, banana
Break: -
Lunch: noodle soup
Break: -
Dinner: rice, chili paste dip, fried pilchard, egg soup
Break: -
Day 2 (Friday)
Breakfast: rice, soup
Break: -
Lunch: rice, stewed pig's trotter
Break:
Dinner: rice, mixed vegetable sour curry
Break: -
Day 3 (Saturday)
Breakfast: rice porridge with pork
Break: -
Lunch: rice, stewed pig's trotter
Break:
Dinner: rice, stir-fried cuttlefish with curry paste, five spice pork stew, mango
Break: -
Day 4 (Sunday)
Breakfast: rice, fried pork
Break: -
Lunch: rice, curry
Break: iced milk syrup
Dinner: rice, mung bean vermicelli soup, stir-fried pork with curry, mango
Break: -
Day 5 (Monday)
Breakfast: rice, fried tilapia
Break: -
Lunch: rice, fried pork, stir-fried snakehead fish with chilies
Break: -
Dinner: rice, stir-fried chicken with chilies, fried small gourami
Break: -
Day 6 (Tuesday)
Breakfast: boiled rice, salted egg, stir-fried kale with crispy pork
Break: -
Lunch: rice, stir-fried chicken with bamboo shoots, tofu soup
Break: -
Dinner: rice porridge with pork
Break: -
Day 7 (Wednesday)
Breakfast: rice, pork blood soup
Break: hot chocolate
Lunch: rice, ash gourd with chicken, preserved tofu noodle soup
Break: -
Dinner: rice, stir-fried catfish with curry paste
Break: -

Table 20.7   A younger man from the South

Samart Niranam is 31 years old, born and raised in Surattani. He makes a living as a tradesman. At breakfast he never misses rice, often accompanied by sour curry and/or pork, omelette, or soup. His rice porridge and boiled rice contains meat and vegetables so he does not need additional side dishes. His wife is his breakfast companion and they eat together at home. Around 12.30 he sometimes eats lunch alone or with his wife or occasionally with coworkers. He also likes to dine in restaurants. His diet is diverse and includes noodles, fried rice, sticky rice, or sukiyaki. Iced tea seems to be his preferred lunch beverage. In the evening he always has dinner at home with his wife and his parents who are also the family cooks. The standard dishes in the evening are rice and sour curry, similar to breakfast, accompanied by fish, pork and vegetable dishes. Morning break is held regularly around 10 am and afternoon break at 2 pm. His favorite beverages are coffee, water, and orange juice, coupled with snacks and fruits. He does not eat anything after dinner.

Sour curry is Samart's family favorite as it is always present at his dining table, eaten with rice and some fried items such as pork or omelette. This dish is considered part of Thailand's true cultural food heritage and its preference is believed to be ingrained in Thai blood. (Nowadays, a modern variation is to add *Cha-om omelette*, a fried bitter green vegetable resembling dill and beaten egg, into sour curry. It has taken hold in many restaurants and food stalls.) Additionally, the longer the sour curry is kept, the better it tastes and therefore its place is secure on Samart's table.

Day 1
Breakfast: rice, omelette, sour curry
Break: iced syrup, bread
Lunch: noodle soup, iced tea
Break: coffee, snacks
Dinner: rice, soup, pork 'n' eggplant curry
Break: bird's nest beverage
Day 2
Breakfast: rice, fried pork, sour curry
Break: snack, orange juice
Lunch: rice with curry, iced tea
Break: snack, watermelon, coffee
Dinner: rice, fried fish, sour curry
Break: -
Day 3
Breakfast: rice, soup, stir-fried vegetables, sour curry
Break: water, coffee, snack
Lunch: noodles soup, desserts
Break: snack, water
Dinner: rice, soup, fried pork, sour curry
Break: -
Day 4
Breakfast: rice, sour curry
Break: water
Lunch: stir-fried noodles in gravy sauce, Thai desserts
Break: fruits, water
Dinner: rice, stir-fried vegetables, sour curry
Break: -
Day 5
Breakfast: rice porridge
Break: water, coffee
Lunch: mixed meats fried rice, fruits
Break: water
Dinner: rice, fried pork, sour curry, chili paste dip
Break: -
Day 6
Breakfast: rice porridge
Break: fruits
Lunch: sukiyaki soup, rice noodles in curry
Break: coffee
Dinner: rice, sour curry, chili paste dip, blanched vegetables
Break: -
Day 7
Breakfast: boiled rice
Break: fruits, water, snacks
Lunch: papaya salad, sticky rice, roasted chicken
Break: orange juice
Dinner: rice, sour curry, fried pork, stir-fried vegetables
Break: -

Table 20.8 A younger woman from the South

Popeang Yindee is 28 years old and married. She works for the government as a contract worker. Living in the country with her extended family leaves her with disposable income to spend on restaurants. She does not pay rent or a mortgage or security or the monthly maintenance fee required by modern housing developments. From her diary, we can see that her breakfast is always something quick and easy such as rice porridge, boiled rice, or toast and hot chocolate. She rises early enough to prepare food and eats it with her husband (and sometimes her parents) before leaving for work. Around 12 o'clock she has lunch with her coworkers. She prefers a quick meal like noodle soup or rice with curry at a nearby restaurant which costs less than 40 baht (approximately $US1.30) per meal. Sometimes she has a big meal with coworkers who take turns buying food either at KFC or at a Northeastern style restaurant. After work, she usually has dinner around 5 o'clock at home. She eats with her husband and parents. Sometimes she dines out with her husband or she joins friends at a popular sukiyaki restaurant. Eating out can cost a few hundred baht (approximately $US6-8), which is quite high compared with what she pays for market food or at the work restaurant.

Day 1
Breakfast: rice and curry, Ovaltine
Break: cookies
Lunch: stir-fried noodles with gravy sauce
Break: -
Dinner: rice, chili paste dip, stir-fried vegetables, sour curry
Break: -
Day 2
Breakfast: rice porridge with pork
Break: -
Lunch: fried rice with shrimp, assorted sweets in coconut cream syrup
Break: ice cream
Dinner: rice, stir-fried with pork, southern style curry paste, tofu soup
Break: -
Day 3
Breakfast: toast, Ovaltine
Break: -
Lunch: rice noodles–spicy sauce
Break: sweets in iced syrup
Dinner: restaurant sukiyaki
Break: -
Day 4
Breakfast: boiled rice with shrimp
Break: fried banana
Lunch: KFC fried chicken
Break: -
Dinner: rice, omelette, sweet 'n' sour fish, chili paste dip
Break: orange juice
Day 5
Breakfast: rice porridge
Break: -
Lunch: noodle soup
Break: -
Dinner: rice, pork/bamboo shoot curry, stir-fried mixed vegetables
Break: mixed fruit juice
Day 6
Breakfast: toast
Break: pleated Thai dumpling
Lunch: papaya salad, roasted chicken, spicy minced pork salad, sticky rice, fruit smoothie
Break: orange, banana
Dinner: rice, sour curry, fried pork
Break: hot green tea
Day 7
Breakfast: rice porridge with pork
Break: -
Lunch: rice with curry
Break: fried pork balls
Dinner: pan grilled pork buffet
Break: fruit juice

Table 20.9   A young woman from the North

Atinut Duanggat is 24 year-old female working as a psychologist at a government hospital. She lives alone. Every morning she buys bread and milk or sausage for her breakfast, which is bought from a cart vendor or a 7–11 store and is eaten at work. Sometimes she does not get to eat until late. At lunch, she usually eats with a coworker in a restaurant near the hospital. This meal is usually very hot and spicy and consists of curry and spicy salad. On the days when she is dining with her boyfriend her diet is different. No matter where and with whom she eats, all her meals are bought and usually cost around 100 Baht per day (approximately $US3.00).

Day 1 (Thursday)
Breakfast: ham cheese sandwich
Break: fried banana, boiled peanuts
Lunch: pork panang curry, horseradish tree curry, pork entrails curry, rice noodles in curry
Break: papaya salad
Dinner: salad, grilled ribs, shrimp paste dip
Break: orange
Day 2 (Friday)
Breakfast: N/E style sausage
Break: northern rice crispy
Lunch: egg noodles in curry, papaya salad, coconut ice cream
Break: pickled fruits, dessert
Dinner: shrimp wonton
Break: orange
Day 3 (Saturday)
Breakfast: N/E style sausage
Break: -
Lunch: rice, stir-fried string beans with curry paste, northern rice crispy
Break: snack
Dinner: salad, grilled pork, grilled liver
Break: yogurt drink, snack
Day 4 (Sunday)
Breakfast: N/E style sausage, milk tofu
Break: -
Lunch: rice, stir-fried pickled cabbage with egg 'n' mung bean vermicelli
Break: orange
Dinner: fried pork balls, milk tofu
Break: mung bean vermicelli sausage
Day 5 (Monday)
Breakfast: fried dough, Ovaltine
Break: -
Lunch: fried rice, grilled chicken
Break:-
Dinner: boiled rice, salted egg salad, stir-fried celery with tofu
Break: guava, mango
Day 6 (Tuesday)
Breakfast: rice, stir-fried kale with crispy pork
Break: -
Lunch: n/a
Break: potato chips, fish snack
Dinner: sticky rice, pork balls, splash orange juice
Break: prawn chips
Day 7 (Wednesday)
Breakfast: pork bits/mayo bread, chocolate milk
Break: -
Lunch: stir-fried holy basil with minced pork, fried egg
Break: orange
Dinner: boiled rice
Break: orange

Table 20.10   An older man from the North

Pairote Sermma is a 62 year old government official retiree. He runs a flower orchard so he rises very early every morning to get ready for work. His family of three does not eat together often so he is sometimes joined by a neighbor. His breakfast consists mostly of hot beverages in front of TV. He brings his lunch from home or from the market before he enters the orchard so by lunchtime it is not hot. Dinner is fixed at 7:30 pm every day. Every major meal is eaten with rice but he often does not record it because it is understood in Thailand that all food is eaten with rice. Side dishes get recorded because they are the only things that vary in his diary.

Day 1 (Thursday)
Breakfast: coffee, hot chocolate, sesame seeds
Break: -
Lunch: noodle soup
Break: -
Dinner: rice, grilled fish
Break: -
Day 2 (Friday)
Breakfast: hot chocolate, coffee
Break: -
Lunch: five spice stew, stir-fried pork with egg plant and curry paste
Break: -
Dinner: sweet 'n' sour, chili paste dip, boiled vegetables, spicy fish soup
Break: -
Day 3 (Saturday)
Breakfast: hot chocolate, coffee
Break: -
Lunch: sweet 'n' sour, five spice stew
Break: -
Dinner: fried fish, mung bean vermicelli spicy salad
Break: -
Day 4 (Sunday)
Breakfast: hot chocolate, coffee
Break: -
Lunch: noodle soup
Break: -
Dinner: alcohol, fried pork, vegetables, fruits
Break: -
Day 5 (Monday)
Breakfast: hot chocolate, coffee
Break: -
Lunch: noodle soup, rice, crispy pork
Break:-
Dinner: grilled fish, stir-fried vegetables
Break: -
Day 6 (Tuesday)
Breakfast: hot chocolate, coffee
Break:-
Lunch: stir-fried pork with curry paste, five spice stew
Break:-
Dinner: fried fish, soup
Break: -
Day 7 (Wednesday)
Breakfast: hot chocolate, coffee
Break:-
Lunch: fried egg, stir-fried with curry
Break:-
Dinner: soup, fried fish
Break: -

4. rise in high-protein and energy-dense diets,
5. increased consumption of temperate zone products (milk and dairy products, vegetables, meat),
6. rising popularity of convenience foods and beverages.

Thai meals are being affected by changes to biodiversity, the economy, and social and cultural trends. Geographical variation is important for dietary diversity and regional specialities but deforestation has meant that previously plentiful forest fruits, mushrooms, herbs, and many kinds of vegetables and meats are scarce. The food self-sufficiency of the North Eastern people who ate fermented fish has been severely compromised by industrial agriculture and high-yield rice that requires fertilizers and insecticides, which in turn pollute the water in which fish swim. This is the region where Thais have probably had to change their food provisioning strategies most markedly, turning to foods imported from elsewhere in Thailand.

However, there is one upside to the losses of biodiversity. In exchange for thousands of square kilometers of lost forest areas, road construction has improved national food distribution and food diversity, and has made modern living possible for rural as well as urban populations (Dixon *et al.* 2007).

Recent Thailand food consumption statistics compiled by the National Bureau of Agricultural Commodity and Food Standards (2006) reveal the rising popularity of soup noodles, instant noodles, and bread, although rice remains the overwhelming staple (Fig. 20.9).

20.10.2 Incremental changes to meal behaviors

In addition to changing meal preferences, some Thai people are gradually replacing home-grown and home-cooked meals by eating out more often and

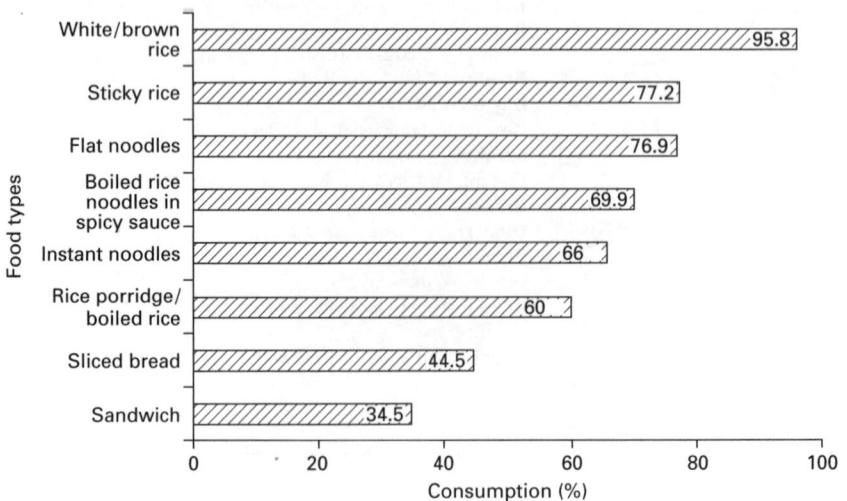

Fig. 20.9 Food consumption of staples.

by obtaining commercially available ingredients and meals. In greeting those walking by their house, Thai people often use the familiar expression, *Ma kin khao kin pla kan* – 'come to eat rice and fish together'. In so doing they refer to sharing a meal which could actually be with or without rice or fish. As modernization permeates through Thai culture, the greeting *Pai kin khao kan* – go to eat rice' is being replaced with the reply *ran nai dee* – 'which restaurant is better'.

Seubsman's (2000) survey of 152 households in the Northeastern provinces shows that an overwhelming number said that they buy their food from markets or food shops but more than half also get some food from natural sources, and close to one half of the sample grow their own food. Exchanging food with neighbors or relatives continues to be a food source for just under one fifth of the sample. There were mixed views towards the changes: only a quarter of respondents thought that the move from food sourced from nature to market sources was favorable. Markets have made their lives easier and more convenient compared with the past, when they had to hand mill their rice, fetch water from the well, and pick firewood. Now they only need to sell their labor and spend the wages on food. Two thirds of the sample considered that scarce natural food and water sources, caused by deforestation and the use of insecticides, had driven them to unnecessary market dependency.

Thai eating behavior is dynamic. It reflects a long tradition of ready-to-eat food which has been made very accessible by the struggle to survive on the part of food vendors. The food landscape of Thailand is now dotted with shops selling Thai-style topping pizza, a holy basil burger, or Korean style food in Thai buffet restaurants. Downtown areas in major cities are filled with international restaurants, and they co-exist with the street food stalls patronized by millions of Thai pedestrians. Both forms of food retail outlets represent a microcosm of interdependency among consumers and the food industry.

Food shops or low-price restaurants are a favorite food provider in every city. From a low-price restaurateur's seven-day diary, analysed as part of the research undertaken for this chapter, a vivid picture emerged of what people eat in Patumthani, an urban area on the periphery of Bangkok. Stir-fried dishes occupy roughly 55% of all made-to-order items. When stir-fried (*Pad*) is combined with other fried dishes (*Tod*), they make up 70% of all sold items. The remaining 30% are soup, curry and dip categories.

Stir-fry dishes are particularly popular at lunch time: this is the quickest way to fix a dish (two minutes maximum). High heat, a wok, and a metal spatula are the only required tools and ingredients are always fresh and crunchy.

The relative popularity of different food retail formats can be inferred from a study involving Sukhothai Thammathirat Open University and data collected from over 87 000 students aged 18 and older living throughout the country (Sleigh 2007) Table 20.11. The highest number eat homemade meals, followed by meal from low-price restaurants (under 100 Baht/meal), and

Table 20.11 Meal sources for a national sample of Open University students

Meal frequency by population attribute	Percentage of population by food sources*						
	Restaurant >200B/meal	Food shop <100B/meal	Western fast food	Work canteen	Street vendor	Home delivery	Prepared at home
Total study population							
Never or 1–2 times a month	95.7	50.4	95.5	42.7	46.0	99.2	16.3
Once a week or more	4.1	25.7	4.2	13.6	31.2	0.7	18.0
Every day or most days	0.2	23.8	0.3	43.7	22.8	0.1	65.7
Rural–rural residents**							
Never or 1–2 times a month	98.0	58.4	97.5	49.2	49.9	99.5	11.0
Once a week or more	1.9	23.5	2.3	13.9	31.3	0.4	13.4
Every day or most days	0.1	18.1	0.2	36.9	18.8	0.1	75.6
Rural–urban residents**							
Never or 1–2 times a month	95.8	46.9	95.6	34.3	43.7	99.2	23.2
Once a week or more	4.0	25.8	4.1	13.0	30.7	0.7	22.6
Every day or most days	0.2	27.3	0.3	52.7	25.6	0.1	54.2
Urban–rural residents**							
Never or 1–2 times a month	94.6	47.5	74.4	46.9	43.8	99.0	12.7
Once a week or more	4.9	23.0	5.2	13.0	31.3	0.9	17.3
Every day or most days	0.5	24.5	0.4	40.0	24.9	0.1	70.0
Urban–urban residents**							
Never or 1–2 times a month	90.7	33.9	91.0	40.6	41.2	88.5	17.7
Once a week or more	8.9	30.3	8.5	14.2	32.1	1.3	21.0
Every day or most days	0.4	33.8	0.5	45.3	26.7	0.2	61.3

* Based on a sample of 87 134 Open University students residing all over Thailand. At the time of survey $US1 was equal to 38 Baht
**Paired categories based on residence at age 12 and at time of reporting in 2005 when sampled students were adults (mean age 29 years)

street vendors. Western food restaurants rank very low as a source of everyday meals, along with more expensive restaurants. Investigated further by residential area, fewer urban residents (61.3% were of urban origin and 54.2% urban immigrants) prepare meals at home most days. Rural residents, both immigrants and those of rural origin, were more likely to cook at home (70% and 75.6%, respectively).

20.11 Future trends

The culinary cultural influences identified at the outset will operate in tension to both energise and impede further changes to meal patterns and meal structures. On the energizing side, is the Thai tradition of ready adoption and openness to new foods and new ways of doing things which facilitates the easy introduction of Western and oriental trends. The Royal Family continues to be highly influential and the King has employed royal resources to intervene in the food system. The Royal Family has also provided a bridge to Western knowledge and people and it mediates between modernity and tradition in many facets of Thai life.

Among the impediments to changing meal practices is Buddhism, which appears to be a conservative but health protective force through its emphasis on asceticism, traditional foods and food practices. The commercially prepared cooked fresh food portions available as alms for monks show that these traditional forces can be co-opted by the market.

International tourism and domestic tourism may influence the Thai culinary culture in three ways: it will reinforce more traditional dishes as tourists demand what they imagine are 'traditional' ingredients and cooking techniques (Henkel et al. 2006). However, it will also create a demand for foreign foods familiar to foreign tourists, offering them a link back to their own country. As local materials (for example, wild mushrooms) become scarce and older cooking methods are considered to be too time intensive then more commercialized ready-to-cook, ready-to-eat variations will become available for everyone.

Economic change plays a vital role in food consumption behaviors. An increase in the cost of natural gas has led to a conversion from food plant agriculture to energy plant agriculture, resulting in higher rice and food prices. When food production costs increase, food manufacturers produce poorer food quality. People are forced to choose between paying more for the same quality food or eating lesser quality food at the same price.

How successive national economic plans will shape the Thai diet is uncertain, but any encouragement of foreign investments will encourage less household food preparation and greater use of ready-to-eat meals. Over two thirds of food retail trade continues to occur through traditional markets, but supermarkets and hypermarkets are rapidly gaining an increasing market share in the major cities (Dixon et al., 2007). The features that have led to an expansion of the supermarket sector in OECD (Organisation for Economic Co-operation and Development) countries apply to Thailand: particularly the entry of women into the workforce and the increased opportunity cost of women's time and their incentive to seek shopping convenience and processed foods to save home preparation time.

Less clear is how the current global shortages in rice availability will play out for Thailand. Given the centrality of rice to so many facets of life and to each and every meal for a majority of Thais we anticipate that the government will continue to regulate this aspect of the food system.

20.12   Sources of further information and advice

Jamornman S (2004) *Thailand: kitchen of the world,* Bangkok, Government Public Relations Department.

Na Songkla W (1988), The Royal favourite dishes: from the hunger diary in the Royal letter No. 42 Written by H.M. King Chulachomklao (Rama V) during his European trip in BE 2440, Bangkok, Samsaen.

Van Esterik P (2000), *Materializing Thailand*, Oxford, Berg.

Vespada Y (2001), (editor), *Thai cuisine: treasure and art of the land,* Bangkok, Amarin Printing.

Walipodom S (1996), *Siam: Thailand's background from the prehistoric period to the Ayudhaya Kingdom*, Bangkok, Matichon (in Thai)

Yasmeen G (2000), 'Not "from scratch": Thai food systems and "public eating", *Journal of Intercultural Studies*, **21**(3), 341–352.

http://www.talad-yingcharoen.com/school/about.php (Srisamorn Kongpun is the Director).

http://www.wandeethaicooking.com

http://thailand.prd.go.th/ebook/kitchen/index.html (Thailand Kitchen of the world – official website)

20.13   References

Aekplakorn W, Hogan M, Chongsuvivatwong P, Tatsanavivat P, Chariyalertsak S, Boonthum A, Tiptaradol S and Lim S (2007), 'Trends in obesity and associations with education and urban or rural residence in Thailand', *Obesity*, **15**(12), 3113–3121.

Anuphabtraipob P (1985), *Royal Kathin ceremony memorial book*, Bangkok, the Secretariat of Thai Cabinet (in Thai).

Bhowmik S (2005), 'Street vendors in Asia: a review', *Economic and Political Weekly*, May 28–June 4, 2256–2264.

Burch D (1996), 'Globalized agriculture and agri-food restructuring in southeast Asia: The Thai experience', in *Globalization and Agri-Food Restructuring. Perspectives from the Australasia Region*, eds. Burch, D., Rickson, R. and Lawrence, G., Aldershot, Avebury, 323–344.

Chulalongkorn, The King (reprinted 1971), *Royal ceremonies in 12 months*, Bangkok, Praepittaya (in Thai).

Chulalongkorn, The King (reprinted 1973), *Far from home*, Bangkok, Amarin Printing (in Thai).

Chang K C (1977) *Food in Chinese culture*, New Haven, Yale University Press.

Choisy, François Timoleon, Abbé de (1644–1724) *Journal of a voyage to Siam, 1685–1686* translated by Michael Smithies (1993), Kuala Lumpur, Oxford University Press.

Dahlan W (2003), *Food and nutrition: follow His Royal Highness King Chulalongkorn's footsteps*, Bangkok, Chulalongkorn University (in Thai).

De La Loubère S (1969). *A new historical relation of the Kingdom of Siam*, Kuala Lumpur, Oxford University Press (facsimile reproduction of 1693 London edition).

Disakul Foundation (1978), *Memorial book distributed on Princess Jongchitrtanom Disakul cremation ceremony* (in Thai).

Dixon J, Banwell C, Seubsman S-A, Kanponai W, Friel S and MacLennan R (2007), 'Dietary diversity in Khon Kaen, Thailand, 1988–2006', *International Journal of Epidemiology*, doi:10.1093/ije/dy1288, 1–4.

Dixon J, Omwega A, Friel S, Burns C, Donati K and Carlisle R (2007), 'The health equity dimensions of urban food systems', *Journal of Urban Health*, **84**(3 Suppl), 118–129.

Health Department, Ministry of Public Health (2006) *The 5th Thailand Food and Nutrition Survey 2000*, Nonthaburi, The Express Transportation of Thailand (in Thai)

Henkel R, Henkel P, Agrusa W, Agrusa J and Tanner J (2006), 'Thailand as a tourist destination: perceptions of international visitors and Thai residents', *Asia Pacific Journal of Tourism Research*, **11**(3), 269–287.

Ho AY (1995), *At the South-East Asian table*, New York, Oxford University Press.

Hutton W (1997), *The food of Thailand: authentic recipes from the Golden Kingdom*, Singapore, Periplus Editions.

ICNND (1962), *The Kingdom of Thailand Nutrition Survey October – December 1960*, Washington DC, Department of Defense.

Jamornman S (2004) *Thailand: kitchen of the world*, Bangkok, Government Public Relations Department. (in Thai)

Kampa-u S (1988), *Royal meal: royal kitchen report*, Bangkok, Siam (in Thai)

Kasikorn Research Center (2007) 'Fluctuating price of rice: global indicators', (http://www.kasikornresearch.com/portal/site/KResearch/menuitem. 458591694986660a9e4e1262658f3fa0/?cid=4&id=14799 (Accessed 30 November, 2007)

Kongpun S (1999), *Eating the 4 element way*, Bangkok, Sangdad (in Thai).

Kosulwat V (2002), The nutrition and health transition in Thailand', *Public Health Nutrition*, **5**(1A), 183–189.

Lefferts L (2005), 'Sticky rice, fermented fish, and the course of a Kingdom: the politics of food in Northeast Thailand', *Asian Studies Review*, **29**, 247–258.

Lertpanitkul S (2003), *Development in Thai context guidelines*, Nonthaburi, Sukhothai Thammathirat Open University (in Thai).

Mahachulalongkornrajavidyalaya (1996), *Thai Tipitakas: Majjihima Nikâya*, Vol. 12 verse 301, Bangkok, Mahachulalongkornrajavidyalaya University press (in Thai).

Ministry of Public Health (2006), *The 5th National food and nutrition survey report*, Bangkok, Ministry of Public Health.

Na Songkla W and Piyawat S (2002), *The Thai Cuisine Book*, Bangkok, Dokya.

National Bureau of Agricultural Commodity and Food Standards (2006), *Food consumption data of Thailand* Bangkok, National Bureau of Agricultural Commodity and Food Standards.

National Fine Arts Department (NFAD) (2004), *First Sukhothai stone inscription: King Ramkamhaeng inscription*, Bangkok, National Fine Arts Department (in Thai).

Natpinit K (1999), *Thai food*, Bangkok, Suan Dusit Rajabhat University (in Thai).

Ninrud N (2007), *Life inside the palace*, Bangkok, Siisara (in Thai).

Panutad (2003), *Composing ancient meals*, Bangkok, Pailin (in Thai).

Pingali P (2004), 'Westernization of Asian Diets and the transformation of food systems:

implications for research and policy'. Rome: Agricultural and Development Economics Division, The Food and Agriculture Organization; September 2004. Report No.: ESA Working Paper No. 4-17.

Pitathawatchai V (2005), *Rice of father*, Bangkok, Direct Media Group Books.

Plainoi S (1998), *Food stories*. Bangkok, Matichon Press (in Thai).

Pramote K (2005), *Four royal reigns*, Bangkok, H.N. Group (in Thai).

Prince Damrongrajanuphab (1987), *Memoirs of Prince Damrongrajanuphab*, Bangkok, Prince Damrongrajanuphab and his daughter Princess Jongchitrtanom Disakul Foundation (in Thai).

Promsao K and Benjasilarak N (1999), *Tracing Thai set menu*, Chiangmai, Wannarak.

Reardon T, and Timmer P C (2005), Transformation of markets for agricultural output in developing countries since 1950: how has thinking changed? In *Handbook 12 Journal of Agricultural Economics*, Vol. 3A, ed. R. Evenson, P. Pingali and T. P. Schultz. Amsterdam: Elsevier.

Sapcharoen P (1994), *Account of traditional Thai medicine theories Vol.1: cause diagnosis text*. Nonthaburi, The Institute of Thai Traditional Medicine (in Thai).

Setto R (1989), *Thai social structure and culture*, Bangkok, Thaiwattanapanich (in Thai).

Seubsman S (2000), *Self sufficient rural society: the Northeast case*. A research report to National Research Council of Thailand, p. 56 (in Thai).

Seubsman S (2003), *Herb content food eating behavior: a research report*, Nonthaburi, Sukhothai Thammathirat Open University, (in Thai).

Sithipipat M (1999), *Eating the Thai way*, Bangkok, Sangdaed.

Sleigh A, Seubsman S, Bain C and the Thai Cohort Study Team (2007), 'Cohort Profile: The Thai Cohort of 87 134 Open University students' *International Journal of Epidemiology*, **37**, 266–272.

Sonakul S (1952), *Everyday Siamese Dishes*, Bangkok, Chantra Press.

Supreme Patriarch Paramanujitajinorasa (1990), *Krisana's teaching poem*, Bangkok, Mahachulalongkornrajavidyalaya University Press (in Thai).

Tannahill R (1984), *Food in History*, New York, Stein and Day.

Thai National Team for Anthology of Asean Literatures (1985), *Traibhumikatha: the story of the three planes of existence: modernized version*, Bangkok, Amarin Printing.

Thantong C (1975) 'What food makes you beautiful?', and 'It's rice with namprik' music and lyric.

The 6th National meetings of traditional Thai medicine doctors and herbs (2004), *Holistic health care: meeting report*, Nonthaburi, The Institute of Traditional Thai medicine (in Thai).

Uluchata P (1994), *Delectable ancient food*, Bangkok, Sangdad (in Thai).

United States Department of Agriculture (2004a), *Thailand exporter guide annual 2004, GAIN Report Number: TH4123*, Bangkok, USDA Foreign Agricultural Service.

United States Department of Agriculture (2004b), *Thailand product brief snack foods industry 2004, GAIN Report Number: TH4082*, Bangkok, USDA Foreign Agricultural Service.

United States Department of Agriculture (2000), *Thailand retail food sector, GAIN Report #TH0116*, Bangkok, USDA Foreign Agricultural Service.

Van Esterik P (1992), 'From Marco Polo to McDonald's: Thai cuisine in transition', *Food and Foodways*, **5**(2), 177–193.

Van Esterik P (2000), *Materializing Thailand*, Oxford, Berg.

Van Esterik P (2004), 'Tasting the other: explorations of power and food in the Thai imaginary', Paper presented at the Conference 'The ambiguous allure of the west? Power, aesthetics and the role of cultures "others" in the making of Thai identities'.

Vespada Y (2001), (editor), *Thai cuisine: treasure and art of the land*, Bangkok, Amarin Printing.

Walipodom S (1996), *Siam: Thailand background since prehistoric period to Ayudhaya Kingdom*, Bangkok, Matichon (in Thai).

Walker M and Yasmeen G (eds) (1996), Contemporary perspectives on Thai foodways, Research Monograph 11, British Columbia, Canada.

Wongted S, Editor (1988) *Commoner rice – royal rice of the Siamese*, Bangkok, Silapawattanatham (in Thai).

Yasmeen G (1996), 'Plastic-bag housewives and postmodern restaurants? Public and private in Bangkok's foodscape', *Urban Geography*, **17**(6), 526–544.

Yasmeen G (2000), 'Not "from scratch": Thai food systems and "public eating" ', *Journal of Intercultural Studies*, **21**(3), 341–352.

Key informant interviews:

Kasemsuwan U (2006), Interviewed by Seubsman S at Lakros restaurant, Bangkok September 5, 2006.

Kongpun S (2008), Interviewed by Seubsman S and Pangsap S on February 2, 2008.

Ninrud N, (2005), Interview by Seubsman S and Somsamai P on February 10, 2005.

21

Chinese meals: diversity and change

J. A. Klein, SOAS, UK

Abstract: This chapter considers the relevance of traditional Chinese food culture for shaping meals in China, in the light of recent changes in food consumption and the introduction of new approaches to food. It highlights diversities and inequalities in Chinese meals, in particular along regional, ethnic, urban–rural, gendered and generational lines. It is argued that, although the last 30 years have seen dramatic changes and growing differences in food, still many of the basic principles associated with traditional food culture are relevant to contemporary Chinese meal practices and discourses on food.

Key words: Chinese meals, traditional food culture, culinary diversity.

21.1 Introduction

In the late 1970s, the archaeologist K. C. Chang (1977a) asserted that there existed a unified Chinese food tradition, which transcended regional and social diversities and which had persisted through periods of momentous change, including that brought about by the communist victory in 1949. Undoubtedly, for most Chinese in the People's Republic, meals at the time of the publication of Chang's piece were similar to what they had been before the revolution, and for centuries before that: barely adequate, monotonous and overwhelmingly based on plant foods, in particular grain staples (Smil, 2004, p. 99). For some, meals were more frugal than they had been before the revolution. The relatively high level of food security enjoyed in China in the late 1970s was impressive, especially considering the famine of 1959–1961, which had claimed the lives of perhaps 30 million people (Yang, 1996). Yet it was achieved through an emphasis on the production and equitable distribution of grain, at the expense of dietary variety. Nor was China producing enough grain – per capita availability of grain and other foods was not much greater than it had been just before the famine (Smil, 2004, pp. 89–95).

Since the late 1970s, mainland China has undergone a period of market-oriented reforms and rapid economic growth, and these developments have had a profound impact on meals. They have brought about substantial increases and diversification in food production, and rising incomes have enabled people to access these foods (Ash, 2006). Overall food consumption, including the consumption of grains and vegetables, rose rapidly in the early 1980s. As the consumption of meat and poultry, fish and seafood, eggs, dairy, fruits, cooking oil and alcohol continued to increase in the 1990s, the proportion of grains and other plant foods in the average diet decreased from 95% in the 1970s to 80% in 2000 (Smil, 2004, pp. 92–103). In the cities, moreover, the daily per capita consumption of dairy products rose from 9.9 g in 1982 to 65.8 g in 2002 (Zhang et al., 2008, p. 40). Since the 1990s, processed foods, packaged snacks, soft drinks and fast foods have become a regular feature of many Chinese diets (Chee, 2000; Smil, 2004, p. 108; Croll, 2006; Zhang et al., 2008). A growing number of people appear to be going through a 'nutrition transition' (Popkin, 1993) from a 'traditional' Chinese diet, dependent on plant foods, to one resembling a 'Western' diet, which relies more heavily on animal foods, fats, sugar and salt (Smil, 2004; Leppman, 2005; Guldan, 2000).

Concerns have been raised over the impact of changing Chinese diets on the environment, food availability and food prices, both domestically and globally (Brown, 1995; Smil, 2004; Ash, 2006). Further, although rising levels of food consumption and greater diversity have led to many improvements in diet-related health (Guldan, 2000, p. 28), the health effects of the 'nutrition transition' have not been unambiguously positive. The shift towards Western-style diets has been linked to rises in obesity, diabetes, cardiovascular illnesses, and certain cancers (Guldan, 2000; Chee, 2000; Smil, 2004, p. 108; Leppman, 2005, p. 46). Zhang et al. (2008) cite a study claiming that the percentage of Mainland Chinese between the ages of 18 and 45 who are overweight rose from 11.9% (men) and 17.7% (women) in 1989 to 27.9% (men) and 28.5% (women) in 2000 (2008, p. 43), and contend on the basis of their own multi-regional consumer study that 'those who broke most with the traditional Chinese food culture are the consumers with the highest Body Mass Index (BMI)' (2008, p. 46). Clearly, not all Chinese are experiencing the 'nutrition transition'. Most, especially in the countryside, continue to subsist largely on plant foods, and under-nutrition is a persistent problem (Guldan, 2000). Like other countries experiencing rapid economic growth and urbanization, China is having to cope with a 'double burden' of under- and over-nutrition (Croll, 2006, p. 178; Gardner and Halweil, 2000).

Knowledge about food is changing, too. Nutritional science is now disseminated through the education system, in government campaigns targeted at young parents (Jing, 2000a; Gottschang, 2000), and in food industry advertising (Zhao, 2000). Advertising has also furthered the idea that the consumption of packaged, branded foods will promote personal happiness and success (Jing, 2000b). These new approaches are challenging indigenous Chinese understandings of food and health. As with food consumption itself, the new approaches are unequally distributed among the population, paving the way for a further diversification of Chinese food culture.

With the rapid changes and growing diversities in food consumption and knowledge, it may be time to reconsider Chang's assertion of a unified and enduring Chinese food tradition. In this chapter, I ask: What role does 'traditional' Chinese food culture play in shaping meals in contemporary China? I address this question through an overview of trends in Chinese meals since 1949, and particularly since the 1980s. The focus is on meals associated with the Han majority. However, I also investigate the boundaries of 'Chinese' meals by complicating the distinction between Han and non-Han meals and by situating Chinese foodways within transnational contexts. All Chinese terms are given in Putonghua (Mandarin), using the pinyin system of transliteration. The chapter draws primarily on published sources, but it includes observations made during ethnographic fieldwork on food culture in Guangzhou, the provincial capital of Guangdong (ten months in 1999–2000), and in Kunming, the capital of Yunnan Province (one month in 2006).

21.2   Traditional Chinese meals

Chang (1977a) defines the traditional Chinese meal in terms of a set of basic principles. Perhaps most important among these is what one might call the elementary structure of the Chinese meal, the 'complimentary dualism' (Thompson, 1994) between cooked grain (*fan*) and side dishes (*cai*) of vegetables, pulses and meat (Chang, 1977a, pp. 7–8). The *fan/cai* distinction serves as the basis for an entire 'grammar' of food (cf. Douglas, 1971). Typically, *fan* are boiled or steamed, unflavoured and served separately from the side dishes. A meal is distinguished from 'snacks' (*xiaochi*) by virtue of containing a proper balance of *fan* and *cai*. The word *fan* can be translated as 'meal' where it appears in compounds such as *wufan* (midday meal, 'lunch') and *wanfan* (evening meal, 'dinner'). The *fan/cai* pair is also used to distinguish different kinds of meals. Thus, everyday, ordinary meals stress *fan* over *cai*. Diners are expected to eat their fill on the grain staple, and meat and vegetable dishes are said to 'down the *fan*' (*xia fan*). Festive meals and banquets are marked by a reversal of the hierarchy between *fan* and *cai*. On such occasions, cooked grain staples are typically served in small quantities at the end of the meal.

A further set of principles has to do with preparation (Anderson, 1988, pp. 182–193; Newman, 2004, pp. 69–85; Chang, 1977a, pp. 8–9). While there are some regional traditions of consuming raw fish and raw vegetables, Han Chinese have defined themselves *vis-à-vis* others as being eaters of cooked foods. A variety of cooking methods, including steaming (*zheng*), boiling (*zhu*), roasting (*kao*) and deep-frying (*zha*), may be employed in the preparation of *cai*. The fact that 'cooking' may colloquially be referred to as 'stir-frying dishes' (*chao cai*) indicates the centrality of that method. In stir-frying the application of heat is intense and brief. This requires that careful attention be paid to chopping and cutting, in order that all ingredients be evenly and adequately cooked. Indeed, the preparation of ingredients in Chinese cooking typically requires much more time than the application of heat, and good knife skills are considered essential (Dunlop, 2008, pp. 76–92). A good cook is also one who is able to balance flavours and textures in a dish and in the meal as a whole. Flavourings vary, but it has been argued that there is a typical 'flavour principle' (Rozin, 1982) achieved through the use of ginger, soy sauce and spring onions, which marks a dish as Chinese (Sabban, 1999).

Recurrent patterns are also evident in the serving and eating of meals (Newman, 2004, pp. 105–128). Typically, diners sit around a table, and eat *fan* from individual bowls with the help of chopsticks. *Cai* are shared between diners and are placed between them on individual serving plates or bowls. Diners pick up pieces of *cai* with their chopsticks, although sometimes serving spoons or serving chopsticks are used for this. Children are taught to show deference to others at the table, for example by eating from all the dishes, not only their own favourite (Cooper, 1986). The same principle of deference towards others holds true for alcoholic beverages and tea, which one should serve to others before serving oneself. More often, however, the mealtime beverage is a soup (*tang* or *geng*), which is indeed 'drunk' (*he*) rather than 'eaten' (*chi*). At ordinary meals, foods tend to be served at roughly the same time, while the sequence of dishes takes on importance at banquets. Seating arrangements are also stressed at banquets to indicate social hierarchies and one's relationship with the host (Kipnis, 1996, pp. 46–57). Fruit may be served before or after a meal. There is no concept equivalent to the Western dessert, although a sweet soup might be served towards the end of a banquet.

The etiquette principles outlined above draw attention to the importance of commensality in creating, maintaining and defining social relationships (Chang, 1977a, p. 16; Kipnis, 1996; Stafford, 2000; Watson, 1987). Virtually all relationships, including those with supernatural beings (Thompson, 1988), are mediated through food. The 'family' or 'household' (*jia*) can be defined as people who eat food cooked from a common stove (Wolf and Huang, 1980, pp. 57–69), and kinship or village solidarity may be regenerated through communal banquets (Watson, 1987). Stafford contends that in China: 'To be "reunited" and "united" is to eat together, whereas the failure to eat together is not merely a symptom of "separation", but is actually constitutive of it' (Stafford, 2000, p. 99).

The composition of a Chinese meal often reflects specific understandings about the relationship between food and bodily health, the basic principles of which have been shared by both laypeople and specialists of traditional Chinese medicine (Anderson, 1988, 229–243; Anderson and Anderson, 1975; Chang, 1977a, pp. 9–10). There is no clear-cut boundary between 'food' and 'medicine', and people have often responded to illness or other conditions by altering their diets. Illness is perceived of in terms of bodily imbalances, which may be caused by and rectified through the consumption of food. Foods channel different *qi* energies into the body, which variously affect the body's own *qi*. Foods and illnesses are classified as 'heating' (*re*) or 'cooling' (*liang*), though some foods, such as boiled rice, are regarded as neutral. Bitter melon (*kugua*), for example, is 'cooling', and may be used to counter the effects of 'heating' conditions such as fevers and acne. Conversely, it should be avoided by those suffering from chills or diarrhoea. Certain foods also have other properties that affect bodily health. For example, 'bolstering' (*bu*) foods are used to rectify conditions of bodily depletion, while 'toxic' or *du* foods should be avoided by those recovering from wounds. People are affected differently by foods, depending on factors such as age and individual constitution. Babies and young children, for example, may be seen as slightly heating and fed accordingly (Wheeler and Tan, 1983), while the proscriptions for pregnant and post-partum women can be staggeringly complex (Martin, 2001). Broadly speaking, everyday meals should achieve a balance between 'heating' and 'cooling', but cooks may also take into account the constitutions of individual diners. They will also consider the effect of external conditions on the body, for example by countering the effects of summer heat with cooling foods and drinks. Cooking methods will also affect the 'heating' or 'cooling' properties of particular foodstuffs, so slowly-simmered stews are favoured in the wintertime and deep-fried, fatty dishes avoided in summer.

Running through these various aspects of Chinese meals are a number of recurrent values. These include balance and harmony – between *fan* and *cai*, flavours and textures, 'heating' and 'cooling', and amongst people sharing meals together. Frugality is important (Chang, 1977a, p. 10). An everyday meal should stress *fan* over *cai* and children should be taught to stop eating when '70% full' (*qifen bao*). Nevertheless, the ethics of frugality has in the Chinese context often not meant a denial of pleasure (Farquhar, 2002). Indeed, the frugality of everyday life can only be understood in relation to the lifecycle events, annual festivals and other extraordinary occasions, during which the pleasures of eating have been paramount, and which have provided much-needed nutrients.

21.3   Regions

For Chang, the basic principles of the Chinese food tradition have not simply endured. Rather, they have shaped change and diversity. They have provided guidelines for Chinese settlers encountering new environments, and offered mechanisms for the incorporation of foreign foodstuffs, such as those reaching China over centuries of exchanges with Central and Western Asia, and those New World crops like potatoes, sweet potatoes, maize and chillies, which began to arrive in the sixteenth century (Chang, 1977b; Anderson, 1988). The question for us is to what extent these principles are still able to guide change and pattern diversity. In this section, I consider regional differences and the ways in which these are shaped by migration, trade and policy. I will suggest that there is evidence of both regional divergence and convergence, and further that the Chinese meal – or aspects of it – is not only still relevant but may be spreading to new places.

21.3.1   Environment

China can be divided into a pastoral region in the West and Northwest and a region of intensive agriculture in the East, South, North and Northeast. On the extensive grasslands of what are today Tibet (Xizang), Xinjiang, Qinghai, Gansu, Ningxia and Inner Mongolia, much food production has focused on the raising of sheep, goats, cattle, horses and camels. Dairy and meat products have been crucial to diets. There are also significant pockets of arable agricultural within the pastoral region, producing grains such as barley, wheat, maize and buckwheat, as well as vegetables and fruits. The densely populated agricultural region, referred to in much of the English-language literature as 'China Proper' in distinction to the largely non-Han pastoral belt, is further divided into a Northern area, where wheat predominates (accompanied by maize, millets, sorghum and rice), and a Southern one, where rice is the main staple crop. Tubers are grown in the North and South, and include potatoes, sweet potatoes and taro. They are often eaten as *cai* but they are also important staples, particularly in poorer regions. Beans and pulses are widely grown, and include soybeans, mung beans and broad beans. In an agricultural system which has favoured labour-intensive plant crops over land-consuming livestock production, soybeans in forms such as tofu (*doufu*) and sauces have been a crucial source of protein and vitamin B12. The rearing of livestock such as pigs, fowl and fish has been complimentary to, and closely integrated into, the growing of crops. Pork is the most important meat, to the extent that the unmodified word 'meat' (*rou*) simply means 'pork', except of course among Muslims. China Proper, especially the South, produces an amazing variety of cabbages, melons, gourds, roots, shoots and other vegetables and fruits. (Anderson 1988, pp. 137–181; Sabban, 1999; Eastman, 1988, pp. 62–71).

The regional boundaries have never been entirely clear-cut, and have become less so in recent decades. Under the grain-first policy of the 1960s and 1970s, many nomadic and semi-nomadic pastoralists were forcibly settled and pastures were converted to croplands, often by Han migrants (Shapiro, 2001; Humphrey and Sneath, 1999). While some of the lost grasslands have been revived, pasture degradation as a result of the continued in-migration of Han farmers and poor livestock mobility is a huge problem in Chinese Inner Asia (Humphrey and Sneath, 1999). Meanwhile, the increasing demand for meat is changing the face of agriculture in China Proper, with farming land increasingly devoted to intensive livestock production and grain increasingly grown – and imported – for livestock consumption (Longworth *et al.*, 2001; Ash, 2006). Moreover, wheat and maize are conquering millets and sorghum in the North, while the North–South divide has diminished as a consequence of increases in rice production in the North and Northeast, following the introduction of new hybrid seeds and wetland reclamation since the 1960s

(Leppman, 2005; Di, 2004). Finally, the move away from the revolutionary era's policy of local self-reliance toward an intensification of transregional and international trade networks in foods has meant that for a growing number of Chinese, especially in the cities, food consumption is now less dependent on local production (Veeck, 2000, 109–110).

21.3.2   Regional cuisines

The food of China Proper tends to be divided into a number of principle regional cuisines, often four, but sometimes five, eight or ten. The widespread four-fold division covers Eastern, Western, Southern styles in the rice region, and a single style in the North. The Eastern style centres on the lower Yangzi River region. Sometimes called Jiangsu cooking (*Sucai*), it is often characterized by a relatively heavy use of cooking oil, sugar, ginger and vinegar. Sweet-and-sour sauces, served with fish, are typical of this region. Sichuan cuisine (*Chuancai*) is sometimes taken to represent the entire Southwest. Chillies in various forms are popular throughout the Southwest, and in Sichuan itself they are often combined with Sichuan pepper (*huajiao*) to create a 'numbing and hot' (*mala*) effect in the mouth. The deep south of China is often represented by Cantonese cuisine (*Yuecai*), with its heartland in the Pearl River Delta region. Cooks here emphasize the importance of freshness and aim to bring out the natural flavours (*yuanwei*) of ingredients – not least fish and seafood – through subtle seasoning and careful steaming or stir-frying. Typical flavourings include the use of fermented black beans and oyster sauce. Soups, especially medicinal broths, are integral to meals. Northern cooking, sometimes referred to as Shandong (*Lucai*) or Beijing (*Jingcai*) cooking, is characterized by a generous use of garlic, leeks, spring onions, salt, vinegar and dark soy sauce, a love of stews, a greater use of mutton and dairy products than in other regions and a wide variety of noodles, dumplings, and pancakes and other savoury cakes (*bing*). (Anderson, 1988, pp. 194–228; Newman, 2004, pp. 87–104; Sabban, 1999; Wang, 1993).

The fourfold division is a useful starting point for thinking about regional culinary differences, but it is broad and incomplete. Like other classifications it reflects the status of particular urban centres and their occupational cooks more than the distinctiveness of ordinary meals (Anderson, 1988, p. 194). The number of distinctive regional styles and 'local flavours' (*difang fengwei*) is virtually infinite, however, a reflection not simply of the diversity of Chinese food but also of the role local cooking styles play in the construction of place-bound identities (Anderson, 1994). Culinary boundaries that may appear arbitrary to outsiders are defended vigorously by locals. Indeed, the number of named and celebrated cuisines seems to have proliferated since the 1990s, as provinces, cities and other administrative units compete with one another to attract tourists and investors and to find markets for their food industries (Ren, 1999, p. 98; Hendrischke and Feng, 1999).

At the same time as new culinary styles are being defined and promoted, the geographical boundaries between regional cuisines are being blurred. With the recent growth of inter-regional trade, migration, domestic tourism and food journalism, cookbooks and other media, a growing number of Chinese people have been able to experience some of their country's culinary diversity (Klein, 2006). The two cuisines that have travelled the most are Cantonese and Sichuanese, but for apparently different reasons. Cantonese cuisine gained in popularity and status in the 1980s and 1990s largely because of the growing economic importance of Hong Kong and the Pearl River Delta region (Gamble, 2003, p. 91). The spread of Sichuan cuisine is closely related to the huge internal diaspora of migrant workers and entrepreneurs from Sichuan Province and Chongqing Municipality. This growing exposure to regional cuisines has enabled people to experiment with previously unfamiliar dishes, ingredients and cooking styles. In Guangzhou, I met people who had begun to incorporate elements of non-Cantonese cuisine into their diets, including Northern-style dumplings and noodles, and local newspapers reported on an increase in the use of chillies among Cantonese following the recent influx of Hunan and Sichuan styles. In the lifestyle pages of local newspapers, journalists took it upon themselves to police the boundaries of Cantonese taste by defining which elements of these cuisines should be considered acceptable to local diners, and which needed to be changed (Klein, 2006).

The movement of culinary styles and ingredients is, of course, nothing new. Nevertheless, the mobility of tastes has been extraordinary since the late 1980s, in particular if considered against the background of the preceding three decades. Between the second half of the 1950s and the early 1980s, the destruction of much of the catering trade, tight regulations on rural-to-urban migration and the encouragement of local self-sufficiency in food meant that people's exposure to outside ingredients and cooking styles was severely limited, even in previous cosmopolitan centres like Shanghai and Guangzhou (Klein, 2008). The renewed mixing of cuisines did not mean that pre-existing cuisines were losing their identities, despite the anxieties of some gastro-journalists. Instead, new foods were shaped by local understandings of meals and flavourings. Northern-style dumplings and noodles may have been popular among many Cantonese in Guangzhou, but these wheat-based foods were typically eaten as snacks, not meals – the latter were defined for Cantonese speakers by the presence of cooked rice. Recipes for *cai* were similarly adapted to local patterns. In the late 1990s, the Sichuanese classic, 'fish-flavoured aubergines' (*yuxiang qiezi*), could be found in popular Cantonese-style eateries throughout Guangzhou, where it was served with pieces of salted fish, an important condiment in Cantonese cooking. The 'fish-flavoured aubergines' served in Sichuan do not contain fish, but rather a sauce made of flavourings typically used with fish.

As repertoires of regional foods become ever more widely disseminated within China, it is possible that we will witness something like the emergence of a national cuisine not only at the level of representation – a project begun in China in the mid-1950s (Klein, 2008) – but also at the level of practice, at least as in post-Independence India (Appadurai, 1988) among high- and middle-income groups in the cities. The development of a national cuisine does not stand in opposition to the flourishing of regional cuisines. Rather, the relationship is a dialectical one (Appadurai, 1988). There is every reason to believe that inter-regional mixing will continue, as will attempts to police culinary boundaries and to define and promote new regional styles.

21.3.3   Ethnicity

Regional cuisines have been used to forge group identities on the basis of native-place origins and as bridgeheads for interactions between groups. These kinds of dynamics, often observed in the context of rural-to-urban migration, have been described by some historians and anthropologists as 'ethnic', regardless of whether the actors involved might also describe themselves as 'Han' (Guldin, 1997a; Honig, 1992). From the point of view of the current Chinese state, however, the most salient 'ethnic' distinctions are those between the officially designated 'nationalities' (*minzu*). In the official classificatory scheme, each citizen belongs to one of the 56 *minzu*, which include the Han majority and 55 'minority nationalities'. Each of these groups is seen to have distinctive cultural characteristics, including its own foodways, and Chinese reference works invariably differentiate culinary practices according to *minzu* (Lu, 1992; Wang and Zhuang, 1994).

Nevertheless, even in multi-*minzu* areas the most salient distinctions for people themselves may be based on criteria such as locality or kinship, and official ethnic categories may tell us little about the content of meals or the relevance of these meals for cultural identities (Wu, 2005). In situations where the official ethnic boundaries converge with or impact directly on local identifications, diets may be broadly similar, but with a small number of foods, often those consumed on festival or ritual occasions, used as markers. Thus, in the 1980s members of the Hui nationality in Fujian in the Southwest, officially classed as pork-abstaining Muslims, in fact ate the same foods as their Han neighbours, but conveyed their separate origins by not using pork in foods prepared as offerings to their ancestors (Gladney, 1991).

In some cases, *minzu* classifications are both locally significant and clearly reflected in everyday meals. Cesàro's (2000; 2002) research on the politics of food among the Muslim, Turkic-speaking Uyghur in Xinjiang provides an apt example. For Uyghurs, daily meals tended to consist of baked breads (*nan* in Uyghur) or noodles, fruit, a limited number of vegetables, tea and, when possible, meat. Although some Chinese influences have long been apparent in Uyghur cuisine, the latter has maintained distinctive meal structures (usually combining the core staple and fringe vegetables and meats in a single dish), ingredients and flavourings. However, Uyghurs in the capital of Ürümchi have recently incorporated several elements of Han cuisine into their cooking including, on certain occasions, meals comprised of boiled rice and stir-fried dishes. This is almost certainly a result of the growing presence of Han in the region, brought about by state-sponsored migration. This same migration is also a key factor in rising ethnic tensions in the region and ethnic nationalism among the Uyghur. Hence, argues Cesàro, Uyghur insist vehemently on the distinctiveness of their foodways not only *vis-à-vis* the non-Muslim, pork-eating Han, but also *vis-à-vis* the Hui, the Chinese-speaking Muslims whom Uyghurs have come to associate with the Han. In a parallel case in South China, an ethnic revival among predominantly pork-eating Uyghurs in Hunan Province may be leading to an increase in pork abstention there (Shih, 2000). To sum up, interaction brought about through migration and trade complicate notions of discrete 'ethnic' foodways, while the growth of 'ethnic consciousness' among some groups has led to a differentiation of meals along ethnic lines, at least in how those meals are represented, if not necessarily in terms of their contents.

21.3.4 Hong Kong, Taiwan and the Chinese diaspora

While national boundaries and policies are clearly crucial factors in shaping Chinese meals, it is nonetheless misleading to define Chinese foodways as being co-extensive with the current borders of China, or even with 'Han' foodways within those borders. Chinese meals in China itself can be viewed in relation to those elsewhere, both for comparative purposes and because they have influenced one another. Hong Kong (now of course a Special Administrative Region of China) and Taiwan are overwhelmingly populated by Chinese, and foodways there have been shaped by waves of migration from the Mainland and the 'ethnic' relations between these groups, by local ecologies and colonial experiences and, particularly from the 1960s, by integration into world markets, which in both places spurred on industrialization, urbanization and the emergence of cosmopolitan middle-classes (Cheung, 2002; Wu, 1997; 2002a). The exposure to non-Chinese foods in Hong Kong and Taiwan – and also Macau (Augustin-Jean, 2002) – has been longer and more sustained than on the Mainland. People have, for example, adopted 'Western' foods such as sandwiches, hamburgers, baked breads and pastries, milk, yoghurt and coffee into everyday consumption patterns, particularly in the mornings (Tsui, 2001), and have created new, Chinese–Western 'hybrid' foods in the process (Wu, 2001). Meanwhile, Chinese in the diaspora have had an ongoing culinary conversation with the non-Chinese world, spreading ingredients, cooking techniques and flavour combinations around the globe, while also developing new, 'improvised' Chinese cooking styles when faced with unfamiliar ingredients and tastes (Mintz, 2007; Roberts, 2002; Wu, 2002b).

These Chinese communities have in turn impacted on the foods of the Mainland. Anderson (2007) argues that in the case of East and Southeast Asia, millennia of food exchanges brought about through migration and trade have in fact created a single, vast and diverse food tradition, a common culinary region with permeable boundaries between 'Chinese' and other cuisines. In recent decades, Hong Kong-, Southeast Asia- and Taiwan-based companies have played a part in the spread of processed snack foods, fast foods and convenience foods in China, providing these industrialized foods in culturally familiar forms. These foods may also (at least in the 1980s and 1990s) have been attractive to Mainland consumers because of associations of Hong Kong, Taiwan and 'overseas' Chinese with an affluent and 'advanced' (*xianjin*) way of life (cf. Gold, 1993). In 1995, the Taiwanese brand Master Kang's (*Kang Shifu*) 'represented more than a quarter of China's instant-noodle market' (Jing, 2000b, p. 153). Hong Kong has had a massive impact on all segments of Guangzhou's catering industry (Klein, 2007) and, in the late 1990s, I witnessed how the ready-made sauces and stocks produced by the Hong Kong-based Lee Kum Kee were making in-roads into both restaurants and homes.

   Tam (2007) has done research in Hong Kong and Guangzhou on what she calls 'convenient-involvement foods', in other words uncooked foods that have been cut, seasoned and combined into a single dish or soup and require only that the consumer empties the contents into a pan and apply heat (presumably adding cooking oil or water) to create something 'home made'. Tam informs us that convenient-involvement foods have spread from Hong Kong to Guangzhou in the 2000s. She argues that, in Hong Kong, these foods had become popular after the economic downturn of the late 1990s, because they were seen as being economical and because they allowed mother–wives to work longer hours outside the home and still live up to expectations that they prepare fresh foods for their families. In Guangzhou, by contrast, convenient-involvement foods were regarded as expensive status symbols, luxury items whose taste was in fact inferior to dishes assembled by women themselves. Tam's work reminds us that even where migration, trade and new food industries may create the conditions for the convergence, even homogenization of Chinese meals, the meanings of these meals may vary depending on socio-economic conditions and local cultural expectations.

21.4   Social diversities and inequalities: the urban–rural divide

The more privileged segments of society have depended less on *fan* and have eaten greater quantities and diversity of *cai*. Social hierarchies in China have been marked not only by the amounts of food people have had access to, however, but also by the consumption of particular ingredients and through styles of preparation and eating (Goody, 1982). There is, for instance, a long-standing hierarchy of starches, which differentiates people by status and wealth. Rice and wheat are perceived to be 'fine grains' (*xiliang*) and are preferred over 'coarse grains' (*culiang*) such as maize, millets and sorghum, and all grains are preferred over tubers such as taro and potatoes (Sabban,

1999). Further, rice is often considered superior to wheat, even in the North and Northeast (Leppman, 2005), and, in general, refined grains are considered superior to less milled ones (Anderson, 1988, pp. 141–145). Unsurprisingly, as incomes have risen in recent decades the consumption of 'coarse' grains and tubers has fallen: between 1982 and 2002, the average daily consumption of tubers dropped from 179.9 g to 49.1 g per person, and the average combined intake of maize, sorghum and millet dropped from 103.5 g to 23.6 g per capita per day (Zhang *et al.* 2008, p. 40).

Certain foodstuffs, such as sea cucumber, abalone, swallow's nest, shark fin and bear paw, have long been used to mark the high status of occasions and diners. The foods are not only delicacies because they are rare (and vice-versa); they are also characteristically prized for their textures more than their flavours – they require the work of a skilled chef to bring out these textures and to enhance their flavours by combining them with other ingredients (Anderson, 1988, pp. 173–174; Dunlop, 2008). Many status-enhancing delicacies are considered to have particular medicinal properties, and their medical potency is enhanced by the fact that they are rare – the animals with the most potent *qi* are those that are 'liminal' in the sense that they fall between established ethno-zoological categories, but also in that they come from 'out-of-the-way' places that are peripheral from the perspective of China Proper (Anderson, 1988, p. 237; Weller, 2006, pp. 34–35).

During the revolutionary years, a degree of success was achieved in levelling culinary hierarchies. However, the gap between country and city was consolidated during this period, and Party cadres and their families often enjoyed better meals than most, owing to higher rations, opportunities for banqueting at public expense, and gifts of food that were received in anticipation of or in gratitude for favours (Yang, 1994; Yan, 1996). Since the economic reforms of the 1980s, social inequalities have intensified. While there has been an overall increase in demand for highly milled rice and wheat and a decrease in demand for 'coarse' grains and tubers, still many people rely on the latter for their daily meals (Smil, 2004, p. 99; Leppman, 2005, pp. 81–82). Meanwhile, the recent banqueting boom among officials and entrepreneurs has taken the historic trade in Chinese delicacies to new heights, exacerbating the threats to several endangered species in China and internationally (Dai, 2002; Donovan, 2004). In this section and the following section, I will show that the growing divisions are apparent not only in the content of people's meals, but also in cultural understandings of food. Table 21.1 provides a few examples of meals from urban and rural settings in different parts of China. It is meant to convey some of the diversity of contemporary Chinese meals, as well as some of their similarities.

21.4.1   The cities
During the revolutionary decades, most urban residents came to enjoy an 'iron rice bowl' of guaranteed employment, housing, pensions and adequate

Table 21.1   Some Chinese meals at the turn of the twenty-first century (listed in order of their elaboration, starting with the most frugal)

1. A morning or afternoon meal in rural Shaanxi Province, North China, 1990s (Liu, 2000, pp. 92–95)	Steamed bread, chillies, pickled vegetables, millet soup or boiled water
2. A family morning meal in rural South Fujian, Southeast China, 1999 (Tan, 2003, p. 183)	Rice porridge, cooked long beans, fried Chinese flowering cabbage, fermented beancurd
3. A family evening meal in Guangzhou, Guangdong Province, South China, 1999 (recorded by the author)	Boiled rice, pork-bone broth with medicinal herbs, minced pork steamed with pickled greens, stir-fried water spinach leaves
4. An informal midday meal in a medium-grade (zhongdang) restaurant in Guangzhou, 1999 (recorded by the author)	Steamed rice, purple laver soup, cold cut chicken (half), fish heads stewed in an earthenware pot, blanched Chinese flowering cabbage, beer
5. A gourmet restaurant meal, Yangzhou, Jiangsu Province, Eastern China, 2000s (Dunlop, 2008, pp. 296–299)	Pre-appetisers: fried peanuts, fermented beancurd, pickled cabbage, slices of ginger Appetisers: salt-water goose, 'vegetarian chicken' made from beancurd skin, 'drunken' river shrimps, sweet–sour cucumber, terrine of shards of pork in aspic served with a vinegar dip; Main courses: hibiscus-flower fish slices, fresh broad beans with straw mushrooms, red-braised carp head with farmhouse beancurd, lion's head meatballs, Wensi Beancurd (beancurd and Jinhua ham) Yangzhou fried rice, broth with mushrooms and seasonal green
6. A banquet in Chengdu, Sichuan Province, Western China, 2000 (Dunlop, 2001, pp. xxvii–xxviii)	Cold dishes: five-spiced 'smoked' fish, chicken with cold rice jelly in a spicy sauce, cucumber in mustard dressing, 'phoenix tail lettuce shoots in sesame sauce, tea-smoked pigeons, dry beancurd with peanuts, Sichuanese cold meats, tripe in hot-and-garlicky sauce Hot dishes: braised sea cucumber, hot-and-fragrant crab with red chillies, fast-fried duck tongues in fermented sauce, steamed pork with rice meal, braised white cabbage with Yunnan ham, traditional bowl-steamed duck with pickled vegetables, braised turtle with potatoes, South Sichuan boiled beef slices in a fiery sauce, fish with pickled vegetables, 'dragon-eye' sweet steamed pork with glutinous rice, soup of green vegetable tips with a chicken-breast coating 'Send-the-rice-down' dish: pickled string beans stir-fried with green chillies and minced pork Snacks: deep-fried sweet potato cakes, boiled dumplings in spicy sauce, leaf-wrapped cones of glutinous rice with Sichuan pepper

food. While households were responsible for purchasing and planning their own meals, prices and distribution were managed by the state. Grain was carefully rationed, as were non-staple foods such as meats, eggs, cooking oil, sugar, soy sauce, and many vegetables. In some cities food rationing was practised until the early 1990s (Leppman, 2005, p. 66). Extra rations were often provided at the Chinese New Year, but the socialist morality dictated that consumption be kept to a minimum and 'waste' avoided at all costs. Few people had their own kitchens, so food storage and cooking were often done in cramped conditions in spaces shared with neighbours (Davis, 2005). The main alternative to eating at home was provided by the canteens, which were established from the 1950s in work units, schools and other institutions. Croll (1983, pp. 231–234) found that around 1980, many urban residents were regularly taking their midday meals in work unit canteens. Some ate there in the mornings and evenings, too, and many others would purchase foods from the canteen and eat them at home in the evening or the following morning. In the anti-consumerism of the Mao era, the majority of commercial dining places were closed. In cities like Guangzhou, the remaining teahouses and restaurants were often integrated into the work unit schedule, with opening times restricted to work breaks, and 'service' styles mimicking those of the canteens (Klein, 2004; Yan, 2000, pp. 209–210).

Shopping for food and preparing and eating meals in urban China in the 2000s are all quite different from the early 1980s. Although the state is still closely involved in food distribution, market forces have been given an ever greater role (Veeck, 2000). 'Free markets' (*ziyou shichang*), which allowed farming households to sell their surplus produce, were reintroduced in the late 1970s. These wet markets proliferated in the 1980s and were joined by supermarkets and, in the 1990s and 2000s, by hypermarkets, including multi-national giants like Carrefour and Wal-Mart. Not only is there now an infinitely greater abundance and variety of local foods available for purchase, but many new foods have entered the urban markets from around China and from abroad. These include processed and packaged foods and drinks, ready-made meals and frozen foods, and also fresh vegetables, fruit, fish and meat, offering greater convenience and allowing urban residents to consume foods that would otherwise be out of season (Veeck, 2000, pp. 109–110; Gamble, 2003, p. 91).

During the course of the 1980s, urban households purchased refrigerators, and with the rapid rise of private housing since the 1990s, they are increasingly likely to have their own kitchens, too (Davis, 2005). At the same time that cooking at home was becoming more convenient, the number of commercial dining establishments also grew exponentially (Klein, 2007). Guangzhou, for example, had 512 registered public eating establishments in 1972 (Gao and Gong, 1999, p. 61). By 1987, the figure had risen to 7851 (Guangzhou, 1988, p. 250). Ten years on it had doubled to over 16 000, not including street vendors (Guangzhou, 1999, p. 273). These establishments ranged from simple shops serving noodles or take-away rice boxes to fast food restaurants to multi-storeyed teahouses to opulent seafood restaurants. Also, many work places continued to provide canteens for their employees. Eating out, while highly differentiated by price, style and occasion, was by the late 1990s in itself no longer remarkable. A study of food purchasing patterns in Nanjing households conducted at that time found that the households' principal shoppers were dining out on average 4.5 times per month, while one cluster identified in the survey as 'convenience shoppers' and comprising 16.7% of the respondents ate out on average 17.3 times per month (Veeck and Veeck, 2000, p. 464). In 1994, 'the Shanghainese already spent 11 percent of their total food expenditures on eating out' (Lu, 2000, 132), and by 2002 that figure had according to some research come to exceed 20% among urban populations in Shanghai, Beijing, Guangdong, Zhejiang and even in some inland provinces like Yunnan (Donald and Benewick, 2005, p. 71). In fact, commercial eating establishments have become focal points in the social life of Chinese towns and cities, providing sustenance during the work day as well as spaces for enjoying good food, for meeting with family and friends, for cultivating networks and for sealing business deals (Yang, 2006).

With the greater availability and variety of foodstuffs and consumption sites, personal choice has become central to the construction of urban Chinese meals (Croll, 2006; Davis, 2005). Food shoppers have become consumers and foods commodities. The commoditization of meals has also posed new challenges. Households have had to learn to cope with food price fluctuations, including periods of rapid food price inflation (Ikels, 1996, p. 59). Such fluctuations are serious matters in a context where people are, in the 2000s, spending on average around 50% of their disposable incomes on food (Croll, 2006, p. 35). Furthermore, the commoditization of food, the expanding geographic scale of food trade and an inadequate regulatory framework have contributed to a proliferation of food safety hazards. These include the over-application of chemical pesticides, food adulteration and contamination, and animal-borne viruses such as SARS and avian flu (Bian, 2004; Dunlop, 2008, pp. 150–171). Conversely, the market itself offers to help consumers to manage food safety risks and other health concerns through the introduction of new value-added products, including certified 'organic foods' (*youji shipin*), chemically-reduced 'green foods' (*lüse shipin*) and 'safe foods' (*wu gonghai shipin*) (the latter are produced 'conventionally' but are guaranteed to be carefully checked for chemical residues). Yet high prices, questionable certification programmes, and the presence of counterfeit goods make these choices problematic, too (Klein, in press; Sanders, 2006; Thiers, 2005; Wang *et al.*, 2008). In short, greater choice in shopping and planning for meals has been accompanied by a more complex management of risks (cf. Thiers, 2003).

New foods and new dining and shopping facilities are not accepted uncritically. They generally need to fit into established patterns and preferences. We have already seen this in the case of the limited success of 'convenience-involvement foods' in Guangzhou. Similarly, despite the ubiquity of supermarkets and household refrigerators, urbanites often have high standards of freshness for fish, meat and vegetables, and in many households food shopping is done daily, as it was during the revolutionary years (Yang, 2006, 70–71; Veeck, 2000, 122). Markets, including hypermarkets, have responded to this preference for freshness, and Wal-Mart in Beijing keeps tanks with live fish, crustaceans and amphibians (Watts, 2006).

Foods choices are also shaped by the *fan/cai* meal structure. As Watson and his colleagues have demonstrated, American fast foods such as Kentucky Fried Chicken and McDonald's hamburgers and fries do not contain sufficient amounts of complex carbohydrates to be considered 'meals' (Watson, 1997; Yan, 1997; 2000; Lozada, 2000). Instead, they are regarded as snacks, and Chinese consumers tend to treat the establishments providing these snacks, not as places to quickly fill one's belly, but as spaces for leisurely sociality. Similarly, Leppman tells us that foods such as soft drinks, jams and jellies, and chocolates are regarded in Shenyang, in the Northeast as 'supplementary foodstuffs', and are thus 'easily adopted because they leave the basic traditional dietary regime intact' (Leppman, 2005, p. 149). In other words, fast foods, soft drinks and sweets do not appear to be becoming a regular part of Chinese meals, but the traditional meal–snack distinction provides a mechanism that allows for their increased consumption – and in some cases 'snacks' may provide more calories than 'meals'.

Across urban China, the main *fan/cai* meals are at midday and in the evenings. Ideally, these meals, especially evening meals, should be eaten in the company of others. By contrast, morning-time food events in Mainland cities, as in Taipei (Tsui, 2001), are less emphasized, more variable and more individualized, something one could eat either as a 'morning meal' (*zaofan*) or a 'morning snack' (*zaodian*). Leppman categorizes urban Shenyang 'breakfasts' in the mid-1990s into four types: 'gruel with steamed or stuffed buns'; 'pancakes or fried breadsticks with some food made from beans'; '*fan* plus *cai*, that is, a menu similar to any other meal of the day'; and 'milk plus yeast bread or cake', which 'most closely resembles the Western breakfast' (2005, 142–143). She found that no one menu dominated Shenyang 'breakfasts'. As Tsui suggests in the case of Taipei, it is in the mornings that many new, 'Western' foods are introduced into Chinese diets. Like afternoon or evening-time snacks, they can be accommodated around the *fan/cai* meal pattern.

Contemporary urban Chinese seem to have rather place-specific understandings of what constitutes a correct balance within each *fan/cai* meal. In urban Shenyang, midday and evening meals in the mid-1990s consisted of cooked grain staple and on average two *cai* of meat or vegetables (Leppman 2005, p. 148). In Guangzhou in the late 1990s, there was widespread agreement that a proper meal at home should consist of cooked rice with 'two dishes and a soup' (Klein, 2004). This was contrasted sharply with restaurant dining. Not only did people in Guangzhou tend to order a greater number of dishes when eating out, but cooking methods were different, too. For meals taken at home, often one dish would be steamed and one stir-fried. Deep-frying was rarely practised at home and roasting was non-existent, as people did not own ovens. In the similarly Cantonese-speaking but wealthier Hong Kong, the 'basic pattern' in the home in the 2000s is 'three dishes and one soup' (Tam, 2007, p. 74). The differences between Shenyang, Guangzhou and Hong Kong suggest not only that meal patterns are localized, but also that they may be affected by levels of affluence.

Within cities, too, social hierarchies impact on meals. Wealth and bureaucratic positions allow certain people access to the expensive restaurants, rare delicacies and imported wines that are well beyond the means of ordinary wage earners, not to mention the now large numbers of laid-off workers and rural migrants. For lower income households in the cities, the focus is on basic sustenance. In 1994, the poorest 10% of urban households were spending over 60% of their incomes on food, in contrast to 40% for households in the highest 10% (Leppman, 2005, pp. 77-79). In addition to economic stratification, lifestyle consumerism has also differentiated urban Chinese diners (Croll, 2006). Thus, educated young people with relatively limited economic means could by the 1990s nevertheless distinguish themselves as being cultivated, through their knowledgeable consumption of Chinese regional cuisines or foreign coffees, whilst deriding conspicuous banqueting as being uncultured and wasteful (Klein, 2006; 2004; Gamble, 2003, p. 156). Taste is often used specifically to distinguish native urbanites from migrant workers, contributing to the reconstitution of the urban–rural divide within the cities themselves (Guang, 2003). At the same time, however, the consumption of food, fashion and other commodities provides a mechanism for upwardly mobile rural migrants to present themselves as urbane (Zhang, 2001).

21.4.2 The countryside

Denied access to urban grain rations, rural residents during the revolutionary years were tied to the countryside. Land was tilled collectively and, in principle, all rural collectives were meant to be self-sufficient in grain. Once the state had taken its share of the harvest, the remaining grain was redistributed by the collectives to each household on the basis of its perceived needs and the labour it had contributed. Goods such as salt, sugar, cooking oil and soy sauce and, in some cases, meat and vegetables, could be purchased from rural shops, but availability was often limited as was villagers' access to cash. For meat, eggs and vegetables, farming households were largely reliant on what they could produce on their small private plots or exchange with other villagers (Croll, 1983; Potter and Potter, 1990).

Since the redistribution of collective land to individual households began in the late 1970s, farmers have become increasingly free to decide what to raise and grow, and to sell their produce directly to markets. Others have sought off-farm employment, including seasonal or long-term labour migration to the towns and cities. Many farmers have benefited from these new opportunities and have been able to improve their diets (Leppman, 2005, pp. 81–82). The daily per capita consumption of meat in rural areas tripled from

22.5 g in 1982 to 68.7 g in 2002 (Zhang *et al.*, 2008, p. 40). However, not all have gained equally. In particular, the frequent banqueting of rural entrepreneurs and officials stands in sharp contrast to the frugal everyday meals of most village folk (Kipnis, 1996; Yan, 1996). Furthermore, the nutritional gap between the cities and the countryside is wide (Guldan, 2000). Rural Chinese were consuming on average twice as much grain in the 1990s as their urban counterparts (Leppman, 2005, p. 82). In Liaoning Province, the difference was threefold (2005, p. 111).

Despite improved access to cash and markets, most farmers' meals are still based on what they can grow themselves (Leppman, 2005) and what they can exchange with their kin and neighbours (Yan, 1996). The villagers of Zhaojiahe in Shaanxi Province, North China, studied in the 1990s by Liu (2000, pp. 82–106) were largely subsistence farmers and 90% of the wheat produced by the village was consumed by villagers themselves. Cooking oil, sugar and salt were purchased in a nearby market town or paid for in kind to itinerant traders. Unlike many parts of rural China, where people ate three meals each day, here meals were taken twice daily, once between nine and ten in the morning and once at around three o'clock in the afternoon. Both these meals consisted of steamed wheat breads (locally called *mo*) downed with some chillies and perhaps pickled vegetables, and accompanied by a 'soup' of boiled water or thin gruel. Any variation of this pattern marked an occasion as extraordinary. If a visitor arrived, the pickles and chillies would be served in two separate bowls, and two additional bowls, one with salt, one with sugar, would be placed on the table. A more important visitor would be served fresh vegetables, eggs or pork; fish was unheard of. On special occasions villagers would prepare and exchange fancy steamed breads (called *huamo*), which were coloured and made into various shapes, and certain days in the ritual year would be marked with special foods, such as dumplings or noodles. The opposition between ordinary meals and special foods structured time and social relationships. Of course, urban Chinese also use food to highlight lifecycle events and annual ritual occasions. But the profound distinction between the 'ordinary' and 'extraordinary' that Zhaojiahe villagers experienced through food was arguably blurred in the cities by the greater diversity and abundance of everyday meals and by the increasing disembeddedness of food consumption from local agricultural cycles and seasonal availability. Differences in meals between country and city are significant not only for their nutritional values, then, but also in terms of their cultural and cognitive values.

However, Zhaojiahe cannot simply be taken as representative of the countryside. Food consumption in villages close to prosperous towns and cities, especially along the Eastern seaboard, are much more varied and abundant than those in the often impoverished inland. Rural industrialization has had a significant effect on the lifestyles of many small towns and villages, especially in the East and Southeast (Guldin, 1997b). Off-farm work and sales of produce have brought cash into villages, while the expansion of markets into the countryside has provided villagers with goods on which to spend that cash. Su (2001) studied the village of Sangcun in Shunde Township in the prosperous Pearl River Delta area. Sangcun villagers had been growing some cash crops for hundreds of years. While the village had long purchased its rice from outside, it had been self-sufficient in pork, eggs, vegetables and fish. Over the course of the 1980s and 1990s, however, Sangcun became more fully integrated into the cash economy of the Pearl River Delta, and farming was becoming increasingly specialized in the areas of vegetable production and aquaculture. At the same time, local farming families were diversifying their occupations and leasing out their valuable land and fish ponds to outsiders. More and more were buying their food, even that which was produced locally, from the village markets. By the 1990s, pigs and fowl were no longer raised in the village at all. In addition to bought fresh foodstuffs, villagers now incorporated ready-made foods into their diets, including so-called 'fish skin dumplings', a local speciality which villagers had once made by hand but which were now produced in factories in nearby Shunde. By the mid-1990s, most households had installed gas stoves and gas cookers, and some had refrigerators and dishwashers. The village now also had a teahouse and two eateries. These establishments introduced wheat-based foods such as steamed bread and fried pancakes into local breakfasts, if not at other mealtimes. Daily meals were becoming increasingly similar to those of nearby towns and cities. However, differences from the cities were still apparent, for example in the relative conservatism of food preferences and in terms of the time spent on preparing and consuming meals during annual festivals and important lifecycle events. For many of the latter, for example, natives of this single-surname village put on elaborate banquets in the ancestral halls, and, while most people in Guangzhou and other cities in the area, now purchased factory-made 'fried sticky rice balls' and 'fried cakes' for the Chinese New Year, in Sangcun these were still made by hand.

In her study of changing diets in and around Shenyang, Leppman (2005) demonstrates how food consumption in different villages varied depending on the extent to which their economies were integrated with that of the urban centres. Partly, this had to do with increases in cash income from selling foodstuffs to the urban markets, and the spread of new food commodities into more affluent rural areas. But urban demands for particular kinds of foods shaped production in villages, and villagers were likely to consume some of these foods. Thus, villages that were able to grow rice for the urban markets also increased their own consumption of this grain relative to other grains, while those who produced milk for the urban market were more likely to consume beef, if not necessarily milk. Nevertheless, Leppman argues that the 'primary distinction' in eating habits was 'that between country and city' (2005, p. 152). Even in the suburbs, variety was much more limited than in the cities, because people consumed mostly the grain and vegetables that they themselves grew. On average, rural households ate only one *cai* at each meal, in contrast to two in Shenyang. Unlike Shenyang, morning meals in all the rural areas showed little variety and followed the same *fan/cai* model as other meals.

To summarize, for many rural Chinese, meals have changed dramatically for the better over the last 30 years. However, often these changes have not kept pace with the cities, or with more affluent rural areas. Even in well-off rural areas with easy access to food markets, many people continue to eat what they and their neighbours produce themselves. The distinction between city and country is not simply a matter of nutritional values, but also of cultural differences to do with the meanings of those meals, in particular the distinction between 'ordinary' and 'extraordinary' eating occasions, which even in 'urbanized' villages such as Sangcun is more elaborated than in the cities. While new food commodities, including packaged snack foods (Jing, 2000b), have made inroads in rural areas, their presence even in places like Sangcun is limited in comparison to the towns and cities.

However, one of the most significant recent transformations, emphasized by the late Elisabeth Croll (2006), has to do with expectations. Throughout rural China, mass media and returning migrants have made people increasingly aware of urban and foreign lifestyles. In the early 1980s, villagers acknowledged that their lot in life as eaters of their own grain was fundamentally different to that of urban residents, who 'ate the state's grain' (Potter and Potter, 1990). By the 2000s, a growing number of farmers felt that they, too, were entitled to enjoy the kinds of meals being consumed in the cities, and many were pursuing livelihood strategies that might allow them to do so.

21.5   Social diversities and inequalities: gender, generation and age

21.5.1   Gender

As several contributors to this book emphasize (see especially Jonsson and Ekström and Ueland, Chapters 12 and 6, respectively), gender is a crucial dimension of meals. In both urban and rural China, revolutionary policies urged women to participate in work outside of the home. However, men were not encouraged to the same degree to contribute to household labour. This pattern has continued since the introduction of reforms, which has seen both an expansion of women's employment opportunities, and an ideological reaffirmation of the 'natural' link between women and the domestic sphere (Wang, 2000; Evans, 2002; Honig and Hershatter, 1988). Women were and are often faced with a 'double burden' of household and outside labour. In rural areas, where agriculture since the 1980s has in many cases been feminized as a consequence of male labour out-migration, both food production and preparation have been left to women and the elderly (cf. Murphy, 2004). In villages and cities, men are often involved in planning meals, and in some cases they shop for foods (Leppman, 2005, pp. 143–145). But women generally continue to be the main cooks in Chinese households. There are regional and urban-rural dimensions to this. Men in Zhaojiahe, Shaanxi did not cook and did not know how (Liu, 2000, pp. 94–95). In Guangzhou, by contrast, most Cantonese men I knew could cook, and in many households they did so from time to time (cf. Yang, 2006, pp. 74–75). Yet, even in Guangzhou, it was often the case that women were ultimately responsible for preparing meals for the family (Tam, 2007).

Meal consumption is also gendered. The preference for sons, in some contexts exacerbated by the strict family planning policy introduced in the late 1970s, has meant that boys' appetites have more often been satisfied than those of girls, especially in the countryside (Croll, 1985, p. 11). Further, despite the symbolic importance of family commensality, men and women have in many cases not eaten together. In homes in late imperial China (defined in this context as 1000–1800 AD), families could eat together on special occasions, but otherwise mealtimes were segregated by gender and age (Bray, 1997, p. 131). Such segregation remains important in many rural villages. In Zhaojiahe in the 1990s, domestic space was strictly divided by gender. Women served the men, who sat at a table placed on the floor, but did not themselves sit down with them to eat (Liu, 2000, p. 46). In the Northeastern village studied by Yan (2003), the principle woman cook often ate alone in the kitchen, after having served her husband, children and (in stem families) father- and mother-in-law, who all sat according to rank at a table placed on the heated bed (kang). However, this began to change in the late 1980s and 1990s, when some villagers introduced large dining tables, at which all family members could be seated together. Nevertheless, 'women still did not dine at the table with important guests' (Yan, 2003, p. 134).

As elsewhere (Jonsson and Ekström, Chapter 12), restaurant culture in China has been male-centred. In Guangzhou, it became acceptable for female customers to visit teahouses in the 1950s – probably not because of communist policies aimed at gender equality in this area, but rather because of the role of teahouses as supplementary workplace canteens. However, even at the turn of the twenty-first century, teahouse sociality tended to be male-centred, and in the city's older neighbourhoods young women rarely frequented teahouses on their own. Those who did had to be prepared to deal with sexual banter and innuendo (Klein, 2004). In Beijing, Yan (2000) found that young women flocked to McDonald's restaurants precisely because they offered a space where, in contrast to Chinese-style restaurants, they could eat alone or socialize with other women without feeling accused of sexual immorality.

21.5.2   Generation and age
The post-Mao era has witnessed an increasing differentiation in food consumption by generation or age group. Even in families that do not live together under one roof or eat daily meals together, adult children, especially

sons, are expected to support their parents out of unconditional filial piety and in reciprocity for the support they received as children. In the People's Republic, this has often worked tolerably well in the cities, where the elderly have also enjoyed other sources of income and support. In the countryside, the elderly are more exclusively dependent on the support of their children. However, with the growing importance of the 'conjugal unit' in rural areas, young married couples are increasingly likely to cook and eat their daily meals separately from their parents. Moreover, the notion of filial piety has arguably weakened in rural China, and reciprocity taken on a more central role. In cases where the parents are not seen to have given adequate support while their children were growing up, children may be less willing to provide reciprocal care to their parents in old age. Indeed, neglect and abuse of elderly parents are growing problems in rural China. Yan describes cases in a Northeast China village of elderly men who had to learn to cook their own meals, as their grown sons were unwilling to take them in, and of an elderly woman who was banned from the dining table and resorted to sneaking into the kitchen to snatch a bowl of rice and any *cai* left in the cooking pot (Yan, 2003, p. 169, p. 175) As the proportion of elderly people in China's population rises rapidly over the next two decades, the quality of their meals is going to be a growing concern. (Whyte, 2003; Yan, 2003; Ikels, 1993; Davis-Friedmann, 1991).

The increasing neglect of the rural elderly stands in sharp contrast to the feeding of urban children and teenagers. The shift in children's diets and food culture is carefully documented in the volume edited by Jing (2000). Following the introduction of the strict family planning policy, parents' hopes for their offspring and for their own security in old age have become pinned on a single child in the cities, or in the countryside more often two children. The health, academic success and affections of the often single child being paramount, parents and grandparents have addressed their concerns and affections through food, and the problem of childhood obesity has already been mentioned. The food industries, often backed by the state and its concerns to build a strong nation through 'scientific' child rearing, have not failed to cater to parents' anxieties, developing a plethora of specific 'children's foods' (*ertong shipin*), and using scientific language to back up the health claims of these new foods. Meanwhile, purveyors of snack foods and fast foods have often specifically targeted children, and the paramount importance of single children has given them the leverage with which to negotiate access to these foods.

The growing power of children to shape their own diets, both directly with pocket money received from their elders and indirectly by influencing family decisions over food, signals a shift in inter-generational hierarchies. If in the recent past, children were expected to eat more or less the same foods as adults and were taught not to question what they were served, today children are knowledgeable consumers of food commodities, and may even instruct parents and grandparents on the correct way to consume new foods

(Guo, 2000; Watson, 2000). Guo (2000) argues that there has emerged a 'generational gap at the table' with respect to 'dietary knowledge'. For grandparents, a proper meal should reflect the 'heating' and 'cooling' properties of foods, and other effects on the body in accordance with the 'folk dietetic' approach outlined earlier. For parents, biomedical nutrition has had a strong impact on their thinking about food. Instead of 'heating' and 'cooling', they are more likely to consider 'proteins' and 'vitamins' when planning meals, and to purchase dietary supplements for their children. For children, however, eating is not about health at all but primarily about 'having fun' and keeping up with the trends in competition with other children. The gap, Guo argues, is more marked in the cities but apparent also in the countryside. It remains to be researched, however, to what extent the approaches to food developed by children in interaction with the marketplace will shape their eating practices as adults and be passed by them on to their own children, or if they will eventually adopt their parents' and grandparents' understandings and practices.

21.6   Conclusion

A dichotomy between 'traditional' agrarian diets and 'industrialized' or 'Western' ones has often been invoked by scholars and food activists (Counihan, 1997; Petrini, 2001; Pollan, 2008). In these studies, traditional diets are presented as being nutritionally balanced and organically embedded in social life and local food production systems. A transition to industrial diets is characterized by these writers, not only by the replacement of plant foods by animal foods and complex carbohydrates by simple ones, but also by a de-socialization of meals, an attenuation of the link between production and consumption, and a confusion about what constitutes a 'proper meal'. While the dichotomy is undeniably a simplification of complex historical processes, it is not without its analytical merits (Mintz, 2006; Wilk, 2006a; 2006b). Certainly, some aspects of the transition are apparent in the dietary changes I have described in this chapter. China's traditionally fibre-rich, plant-based diet has been challenged by an increase in the consumption of animal foods, sugars and fats. Moreover, in many places daily meals are now disembedded from local production systems and reliant on the purchase of food commodities, including fresh and processed foods that may have travelled long distances. This process is likely to continue: In 2002, 39% of China's population lived in urban areas, compared with 18% in 1978 (Donald and Benewick, 2005, pp. 26–27). Fifty per cent of the population was employed in agriculture in 2002, compared with 68% in 1982 and 84% in 1952 (2005, pp. 38–39). Yet, despite ongoing urbanization and industrialization, at present at least half of China's people retain direct ties to the land, and many people hold on to ideas about food that have been challenged by new culinary languages. There is a case for arguing that Chinese meals have become fragmented along the urban–rural divide, as well as along the lines of age, income and social status. In the past, such divisions may as Chang (1977a) argues have been about the quantity or quality of foods consumed within a shared framework of meaning. Today, they are also about cultural understandings of food and the body and about the roles meals play in shaping social life.

Should we then conclude that although China's 'traditional' food culture continues to shape the daily meals of rural folk and the elderly, it has not endured in the cities and is likely to become irrelevant in the future? I doubt it is so simple. To begin with, there are arguments against the notion of culinary 'fragmentation'. The expansion of markets, migration, new agricultural technologies and mass media have actually created the conditions for certain convergences in Chinese meals. An example would be the spread of wheat and, especially, rice across the country. Even in Xinjiang, a region where rice has played only a minor role in the diet, we have seen the growing popularity of meals constructed along the cooked rice/stir-fried dishes model. Another example of convergence would be the blending of regional styles into a national cuisine, however limited and contested that national cuisine may be.

Further, people's food preferences in both the cities and the countryside continue to be guided, more or less self-consciously, by many of Chang's culinary principles. Some of these principles can be thought of in terms of 'habitus', a set of enduring, embodied predispositions, which structure people's practical engagement with the world (Bourdieu, 1977; cf. Appadurai, 1991). Culinary principles such as habitus may change over time in reaction to changing external conditions, but usually only slowly and without actors' conscious awareness. The *fan*/*cai* dualism, the basic structure of the Chinese meal, is usefully approached in this light. Although it is, of course, named and therefore consciously upheld in some sense, its power arguably stems from a taken-for-grantedness, embodied through everyday (childhood) experience. For over half of Yan's (1997, p. 47) Beijing informants, a McDonald's meal was not a 'proper meal', not simply because it did not satisfy them intellectually, but because it did not satisfy them corporeally – it was not filling. As we have seen, this enduring predisposition did not prevent them from eating the burger, but it did mean that it was more likely to be successfully incorporated into eating patterns as a snack food. Indeed, the increase in the consumption of non-staple foods in relation to staple foods would suggest not only the growing significance of snacks in Chinese diets, but also that the embodied balance between *fan* and *cai* is actually quite flexible. Surely, the flexibility of the *fan–cai* meal will contribute to its endurance, even as the meal may itself come to represent a diminishing proportion of people's total caloric intake.

Nevertheless, culinary principles are not only significant as embodied habitus. In certain contexts they may be the objects of explicit deliberation and debate, and of attempts by certain actors to change them in particular directions. Jing (2000b) argues that in the Gansu village he studied, dietary understandings were backed up by different 'cultural authorities'. It was left to villagers themselves to decide whether to approach food in terms of the

476 Meals in science and practice

government-backed 'scientific' language of nutrition; in terms of its 'heating' and 'cooling' properties, supported by the authorities of traditional medicine and popular religion; or as a commodity whose purchase would enhance personal happiness, as was suggested by television advertisements. The new multiplicity of dietary messages and cultural authorities could, as Jing demonstrates, cause anxiety. But Jing also makes it clear that decisions around food were shaped by villagers' practical engagement in the world rather than by an exclusionary logic. As numerous studies have shown, Chinese have often sought healing through 'traditional' dietary therapies, Chinese medicine and biomedicine, without necessarily excluding either one (e.g., Anderson, 1988, p. 230; Ahern, 1975). There is no reason to assume *a priori* that a nutritional approach to meal construction, or for that matter a commercialized view of food, would either exclude or confuse an understanding based around the notion of *qi* (cf. Bradby, 1997).

The ability to analyse one's own foodways in relation to new foods and ideas about food is important. As Mintz (1996) and Sutton (2001) have argued, a crucial part of maintaining a cuisine is the ongoing conversation that people sharing that cuisine have about it. While the power of changing social structures and of Jing's 'cultural authorities' to shape people's meals should not be denied, that power is always mediated by a myriad of everyday dietary discussions, deliberations and decisions. In recent years, for example, there has been a re-appreciation of 'coarse grains' and other 'simple' foods among some people in urban areas, who appreciate them for flavour, nutritional value and, perhaps above all, nostalgic value (Leppman, 2005, p. 148; Hubbert, 2007). Further, any argument that 'nutritional science' or commercial knowledge are marginalizing folk dietetics needs to take into account the massive popularity of traditional-style dietary therapies witnessed since the 1990s, not least in the cities (Farquhar, 2002).

At certain moments such conversations and re-evaluations may coalesce into popular social movements, aimed at influencing food consumption in a particular direction, and often involving the reinvention of 'traditional' food values. These social movements should be added to Jing's cultural authorities, certainly in urban China. Vegetarian cuisine has a long history in China, where it has been associated above all with Buddhism (Kieschnick, 2005). In the last decade, a new vegetarian movement has been emerging among educated young people in the cities. The movement draws inspiration both from the Chinese Buddhist tradition and from recent transnational debates on environmental protection, animal welfare, human health and food safety (Shi, 2004). Even more recently, environmental activists have been promoting the consumption of 'organic' and similar foodstuffs (Klein, in press). These movements appear to be creating new consumerist lifestyles groups (*ibid.*), adding yet further layers to China's multifaceted food culture. Yet it is also possible that they will have a more far-reaching and enduring impact on the way people construct their meals, on food policy and on the food industries. The two developments are not necessarily mutually exclusive (cf. Belasco, 1993).

21.7   References

Ahern E (1975) 'Chinese-style and Western-style doctors in northern Taiwan', in Kleinman A *et al.* (eds) *Medicine in Chinese cultures*, Washington, DC, US Government Printing Office, 209–216.

Anderson E N (1988) *The food of China*, New Haven and London, Yale University Press.

Anderson E N (1994) 'Food', in Wu D and Murphy P D (eds), *Handbook of Chinese popular culture*, Westport, CT, Greenwood Press, 35–53.

Anderson E N (2007) 'Malaysian foodways: confluence and separation', *Ecology of Food and Nutrition*, **46**, 205–219.

Anderson E N and Anderson M L (1975) 'Folk dietetics in two Chinese communities, and its implications for the study of Chinese medicine', in Kleinman A *et al.* (eds) *Medicine in Chinese cultures*, Washington, DC, US Government Printing Office, 143–175.

Appadurai A (1988) 'How to make a national cuisine: cookbooks in contemporary India', *Comparative Studies in Society and History*, **30**(1), 3–24.

Appadurai A (1991) 'Dietary improvisation in an agricultural economy', in Sharman A, Theophano J, Curtis K and Messer E (eds), *Diet and domestic life in society*, Philadelphia, Temple University Press, 205–232.

Ash R F (2006) 'Population change and food security in China' in Tubilewicz C (ed.), *Critical issues in contemporary China*, New York and London, Routledge, 143–166.

Augustin-Jean L (2002) 'Food consumption, food perception and the search for a Macanese identity' in Wu D Y H and Cheung S C H (eds), *The globalization of Chinese food*, Richmond, Surrey, Curzon Press, 113–127.

Belasco W (1993) *Appetite for change: how the counter-culture took on the food industry*. Ithaca, Cornell University Press.

Bian Y (2004) 'The challenges for food safety in China', *China Perspectives*, **53**, 4–11. Online: http://www.cefc.com.hk/uk/pc/articles/art_ligne.php?num_art_ligne=5301 (accessed 6 November 2005).

Bourdieu P (1977) *Outline of a theory of practice*, trans. Nice R, Cambridge, Cambridge University Press.

Bradby H (1997) 'Health, eating and heart attacks: Glaswegian Punjabi women's thinking about everyday food', in Caplan P (ed.), *Food, health and identity*, London and New York, Routledge, 213–233.

Bray F (1997) *Technology and gender: fabrics of power in late imperial China*, Berkeley, University of California Press.

Brown L (1995) *Who will feed China? Wake-up call for a small planet*, New York, Norton.

Cesàro M C (2000) 'Consuming identities: food and resistance among the Uyghur in contemporary Xinjiang', *Inner Asia* **2**(2), 225–238.

Cesàro M C (2002) *Consuming identities: the culture and politics of food among the Uyghur in contemporary Xinjiang*, unpublished PhD thesis, University of Kent at Canterbury.

Chang K C (1977a) 'Introduction', in Chang K C (ed.), *Food in Chinese culture: anthropological and historical approaches*, New Haven and London, Yale University Press, 1–22.

Chang K C (ed.) (1977b) *Food in Chinese culture: anthropological and historical approaches*, New Haven and London, Yale University Press.

Chee B W L (2000) 'Eating snacks and biting pressure: only children in Beijing', in Jing J (ed.), *Feeding China's little emperors: food, children, and social change*, Stanford, Stanford University Press, 48–70.

Cheung S C H (2002) 'Food and cuisine in a changing society: Hong Kong' in Wu D Y H and Cheung S C H (eds), *The globalization of Chinese food*, Richmond, Surrey, Curzon Press, 100–112.

Cooper E (1986) 'Chinese table manners: you are *how* you eat', *Human Organization*, **45**(2), 179–184.

Counihan C (1997) 'Bread as world: food habits and social relations in modernizing Sardinia', in Counihan C and Van Esterik P (eds), *Food and culture: a reader*, New York and London, Routledge, 283–295.

Croll E (1983) *The family rice bowl: food and the domestic economy in China*, London, Zed Books.

Croll E (1985) 'Introduction: fertility norms and family size in China', in Croll E *et al.* (eds), *China's one-child family policy,* London, Macmillan, 1–36.

Croll E (2006) *China's new consumers: social development and domestic demand*, London and New York, Routledge.

Dai Y (2002) 'Food culture and overseas trade: the trepang trade between China and Southeast Asia during the Qing Dynasty' in Wu D Y H and Cheung S C H (eds), *The globalization of Chinese food*, Richmond, Surrey, Curzon Press, 21–42.

Davis D (2005) 'Urban consumer culture', *The China Quarterly*, **183**, 692–709.

Davis-Friedmann D (1991) *Long lives: Chinese elderly and the communist revolution*, 2nd revised edition, Stanford, Stanford University Press.

Di L (2004) 'Land-use patterns and land-use change', in Hsieh C-m and Lu M (eds), *Changing China: a geographical appraisal*, Boulder, Westview Press, 17–31.

Donald S H and Benewick R (2005) *The state of China atlas*, Berkeley, University of California Press.

Donovan D G (2004) 'Cultural underpinnings of the wildlife trade in Southeast Asia', in Knight J (ed.), *Wildlife in Asia: cultural perspectives*, London and New York, Routledge Curzon, 88–111.

Douglas M (1971) 'Deciphering a meal', in Geertz C (ed.), *Myth, symbol, and culture*, New York, Norton, 61–82.

Dunlop F (2001) *Sichuan cookery*, London, Michael Joseph.

Dunlop F (2008) *Shark's fin and Sichuan pepper: a sweet-sour memoir of eating in China*, London, Ebury Press.

Eastman L E (1988) *Family, fields, and ancestors: constancy and change in China's social and economic history, 1550–1949*, New York and Oxford, Oxford University Press.

Evans H (2002) 'Past, perfect or imperfect: changing images of the ideal wife', in Brownell S and Wasserstrom J N (eds), *Chinese femininities/Chinese masculinities: a reader*, Berkeley, University of California Press, 335–360.

Farquhar J (2002) *Appetites: food and sex in post-socialist China*, Durham, NC, Duke University Press.

Gamble J (2003) *Shanghai in transition: changing perspectives and social contours of a Chinese metropolis*, London and New York, RoutledgeCurzon.

Gao X and Gong B (1999) *Guangzhou meishi* (Guangzhou delicacies), Guangzhou, Guangdong Sheng Ditu Chubanshe.

Gardner G and Halweil B (2000) *Underfed and overfed: the global epidemic of malnutrition*, Washington, DC, Worldwatch Institute.

Gladney D C (1991) *Muslim Chinese: ethnic nationalism in the People's Republic*, Cambridge, MA, Council on East Asian Studies, Harvard University (Harvard East Asian Monographs 149).

Gold T B (1993) 'Go with your feelings: Hong Kong and Taiwan popular culture in Greater China', *The China Quarterly*, **136**, 907–925.

Goody J (1982) *Cooking, cuisine and class: a study in comparative sociology*, Cambridge, Cambridge University Press.

Gottschang S K (2000) 'A baby-friendly hospital and the science of infant feeding', in Jing J (ed.), *Feeding China's little emperors: food, children, and social change*, Stanford, Stanford University Press, 160–184.

Guang L (2003) 'Rural taste, urban fashions: the cultural politics of rural/urban difference in contemporary China', *Positions East Asia Cultures Critique*, **11**(3), 614–646.

Guangzhou Nianjian Bianzuan Weiyuanhui (ed.) (1988) *Guangzhou nianjian* (Guangzhou yearbook), Guangzhou, Guangzhou Nianjian Chubanshe.

Guangzhou Nianjian Bianzuan Weiyuanhui (ed.) (1999) *Guangzhou nianjian* (Guangzhou yearbook), Guangzhou, Guangzhou Nianjian Chubanshe.

Guldan G S (2000) 'Paradoxes of plenty: China's infant- and child-feeding transition', in Jing J (ed.), *Feeding China's little emperors: food, children, and social change*, Stanford, Stanford University Press, 27–47.

Guldin G E (1997a) 'Hong Kong ethnicity: of folk models and change', in Evans G and Tam M (eds), *Hong Kong: the anthropology of a Chinese metropolis*, Richmond, Surrey, Curzon, 25–50.

Guldin G E (1997b) *Farewell to peasant China: rural urbanization and social change in the late twentieth century*, Armonk, NY, M.E. Sharpe.

Guo Y (2000) 'Family relations: the generation gap at the table', in Jing J (ed.), *Feeding China's little emperors: food, children, and social change*, Stanford, Stanford University Press, 94–113.

Hendrischke H and Feng C (eds) (1999) *The political economy of China's provinces: comparative and competitive advantage*, London and New York, Routledge.

Honig E (1992) *Creating Chinese ethnicity: Subei people in Shanghai, 1850–1980*, New Haven and London, Yale University Press.

Honig E and Hershatter G (1988) *Personal voices: Chinese women in the 1980s*, Stanford, Stanford University Press.

Hsu C (2007) *Creating market socialism: how ordinary people are shaping class and status in China*, Durham, NC and London, Duke University Press.

Hubbert J (2007) 'Serving the past on a platter: Cultural Revolution restaurants in contemporary China', in Beriss D and Sutton D (eds), *The restaurants book: ethnographies of where we eat*, Oxford and New York, Berg, 79–96.

Humphrey C and Sneath D (1999) *The end of nomadism? Society, state and the environment in Inner Asia*, Cambridge, The White Horse Press.

Ikels C (1993) 'Settling accounts: the intergenerational contract in an age of reform', in Davis D and Harrell S (eds), *Chinese families in the post-Mao era*, Berkeley, University of California Press, 307–333.

Ikels C (1996) *The return of the God of Wealth: the transition to a market economy in urban China*, Stanford, Stanford University Press.

Jing J (ed.) (2000a) *Feeding China's little emperors: food, children, and social change*, Stanford, Stanford University Press.

Jing J (2000b) 'Food, nutrition and cultural authority in a Gansu village', in Jing J (ed.), *Feeding China's little emperors: food, children, and social change*, Stanford, Stanford University Press, 135–159.

Kieschnick J (2005) 'Buddhist vegetarianism in China', in Sterckx R (ed.), *Of tripod and palate: food, politics, and religion in traditional China*, New York, Palgrave Macmillan, 186–212.

Kipnis A (1996) *Producing guanxi: sentiment, self, and subculture in a North China village*, Durham, NC, Duke University Press.

Klein J A (2004) 'Reinventing the traditional Guangzhou teahouse: caterers, customers and cooks in post-socialist urban south China', unpublished PhD thesis, University of London.

Klein J (2006) 'Changing tastes in Guangzhou: restaurant writings in the late 1990s', in Latham K, Thompson S and Klein J (eds), *Consuming China: approaches to cultural change in contemporary China*, London and New York, Routledge, 104–120.

Klein J A (2007) 'Redefining Cantonese cuisine in post-Mao Guangzhou', *Bulletin of the School of Oriental and African Studies*, **70**(3), 511–537, doi: 10.1017/S0041977X07000821

Klein J A (2008) '"For eating, it's Guangzhou": regional culinary traditions and Chinese socialism', in West H and Raman P (eds), *Enduring socialism: explorations of revolution and transformation, restoration and continuation*, New York and Oxford, Berghahn Books, pp. 44–76.

Klein J A (in press) 'Creating ethical food consumers? Promoting organic foods in urban Southwest China', in Pieke F (ed.), The anthropology of contemporary China, special issue of *Social Anthropology*, **19**(1) (February, 2009).

Leppman E J (2005) *Changing rice bowl: economic development and diet in China*, Hong Kong, Hong Kong University Press.

Liu X (2000) *In one's own shadow: an ethnographic account of the condition of post-reform rural China*, Berkeley, University of California Press.

Longworth J W, Brown C G and Waldron S A (2001) *Beef in China: agribusiness opportunities and challenges*, St Lucia, Queensland, University of Queensland Press.

Lozada E P, Jr. (2000) 'Globalized childhood?: Kentucky Fried Chicken in Beijing', in Jing J (ed.), *Feeding China's little emperors: food, children, and social change*, Stanford, Stanford University Press, 114–134.

Lu H (2000) 'To be relatively comfortable in an egalitarian society', in Davis D S (ed.), *The consumer revolution in urban China*, Berkeley, University of California Press, 124–141.

Lu K (ed.) (1992) *Zhonghua minzu yinshi fengsu daguan* (Grand spectacle of the national culinary customs of the Chinese), Beijing, Shijie Zhishi Chubanshe.

Martin D (2001) 'Food restrictions in pregnancy among Hong Kong mothers', in Wu D Y H and Tan C B (eds), *Changing Chinese foodways in Asia*, Hong Kong, The Chinese University Press, 97–122.

Mintz S W (1996) *Tasting food, tasting freedom: excursions into eating, culture and the past*, Boston, Beacon Press.

Mintz S W (2006) 'Food at moderate speeds', in Wilk R (ed.), *Fast food/slow food: the cultural economy of the global food system*, Lanham, MD, Altamira Press, 3–11.

Mintz S W (2007) 'Asia's contributions to world cuisine: a beginning enquiry', in Cheung S C H and Tan C-B (eds), *Food and foodways in Asia: resource, tradition and cooking*, London and New York, Routledge, 201–210.

Murphy R (2004) 'The impact of labor migration on the well-being and agency of rural Chinese women: cultural and economic contexts and the life course', in Gaetano A M and Jacka T (eds), *On the move: women in rural-to-urban migration in contemporary China*, New York, Columbia University Press, 243–276.

Newman J M (2004) *Food culture in China*, Westport, CT, Greenwood Press.

Petrini C (2001) *Slow Food: the case for taste*, trans. McCuaig W, New York, Columbia University Press.

Pollan M (2008) *In defence of food: the myth of nutrition and the pleasures of eating*, London, Allen Lane.

Popkin B M (1993) 'Nutritional patterns and transitions', *Population and Development Review*, **19**(1), 138–157.

Potter S H and Potter J M (1990) *China's peasants: the anthropology of a revolution*, Cambridge, Cambridge University Press.

Ren B (ed.) (1999) *Zhongguo shjing* (The Chinese classic of food), Shanghai, Shanghai Wenhua Chubanshe.

Roberts J A G (2002) *China to Chinatown: Chinese food in the West*, London, Reaktion Books.

Rozin E (1982) 'The structure of cuisine', in Barker L M (ed.), *The psychobiology of human food selection*, Westport, CT, AVI Publishing, 189–203.

Sabban F (1999) 'China: eating and cooking', in Davidson A (ed.), *The Oxford companion to food*, Oxford, Oxford University Press, 171–173.

Sanders R (2006) 'A market road to sustainable agriculture? Ecological agriculture, green food and organic agriculture in China', *Development and Change*, **37**(1), 201–26, doi: 10.1111/j.0012–155X.2006.00475.x.

Shapiro J (2001) *Mao's war against nature: politics and the environment in revolutionary China*, Cambridge, Cambridge University Press.

Shi Y (2004) *Sushizhuyi (Vegetarianism)*, Beijing, Beijing Tushuguan Chubanshe.

Shih C-Y (2000) 'Uygurs that eat pork: a note on ethnic consciousness in Changde's Uygur-Hui townships', *Issues and Studies*, **36**(3), 180–198.

Smil V (2004) *China's past, China's future: energy, food, environment*, New York and London, RoutledgeCurzon.

Stafford C (2000) *Separation and reunion in modern China*, Cambridge, Cambridge University Press.

Su J (2001) 'The changing foodways of a village in the Pearl River Delta area', in Wu D Y H and Tan C B (eds), *Changing Chinese foodways in Asia*, Hong Kong, The Chinese University Press, 35–45.

Sutton D E (2001) *Remembrance of repasts: an anthropology of food and memory*, Oxford and New York, Berg.

Tam S M (2007) 'Convenient-involvement foods and production of the family meal in South China', in Cheung S C H and Tan C-B (eds), *Food and foodways in Asia: resource, tradition and cooking*, London and New York, Routledge, 67–82.

Tan C-b (2003) 'Family meals in rural Fujian: aspects of Yongchun village life', *Taiwan Journal of Anthropology*, **1**(1), 179–195.

Thiers P (2003) 'Risk society comes to China: SARS, transparency and public accountability', *Asian Perspective*, **27**(2), 241–251.

Thiers P (2005) 'Using global organic markets to pay for ecologically based agriculture development in China', *Agriculture and Human Values*, **22**, 3–15, doi: 10.1007/s10460-004-7226-z.

Thompson S E (1988) 'Death, food, and fertility', in Watson J L and Rawski E S (eds), *Death ritual in late imperial and modern China*, Berkeley, University of California Press, 71–108.

Thompson S E (1994) 'Riz de passage? Life bytes and alimentary practices', *China Now*, **149**, 10–12.

Tsui E Y-l (2001) 'Breakfasting in Taipei: changes in Chinese food consumption', in Wu D Y H and Tan C B (eds), *Changing Chinese foodways in Asia*, Hong Kong, The Chinese University Press, 237–255.

Veeck A (2000) 'The revitalization of the market place: food markets of Nanjing', in Davis D S (ed.), *The consumer revolution in urban China*, Berkeley, University of California Press, 107–123.

Veeck A and Veeck G (2000) 'Consumer segmentation and changing food purchase patterns in Nanjing, PRC', *World Development*, **28**(3), 457–471.

Wang X (1993) *Hua-Xia yinshi wenhua* (The food and drink culture of China), Beijing, Zhonghua Shuju.

Wang F and Zhuang H (eds) (1994) *Zhongguo yinshi wenhua cidian* (Dictionary of Chinese food and drink culture), Hefei, Anhui Renmin Chubanshe.

Wang Z (2000) 'Gender, employment and women's resistance', in Perry E J and Selden M (eds), *Chinese society: change, conflict and resistance*, London and New York, Routledge, 62–82.

Wang Z, Mao Y and Gale F (2008) 'Chinese consumer demand for food safety attributes in milk products', *Food Policy*, **33**, 27–36, doi: 10.1016/j.foodpol.2007.05.006

Watson J L (1987) 'From the common pot: feasting with equals in Chinese society', *Anthropos*, **82**, 389–401.

Watson J L (ed.) (1997) *Golden arches East: McDonald's in East Asia*, Stanford, Stanford University Press.

Watson J L (2000) 'Food as lens: the past, present, and future of family life in China', in Jing J (ed.), *Feeding China's little emperors: food, children, and social change*, Stanford, Stanford University Press, 199–212.

Watts J (2006) 'Wal-Mart leads charge in race to grab a slice of China', *The Guardian*, 25 March: 20–21.

Weller R P (2006) *Discovering nature: globalization and environmental culture in China and Taiwan*, Cambridge, Cambridge University Press.

Wheeler E and Tan S P (1983) 'From concept to practice: food behaviour of Chinese immigrants in London', *Ecology of Food and Nutrition*, **13**, 51–57.

Whyte M K (ed.) (2003) *China's revolutions and intergenerational relations*, Ann Arbor, Center for Chinese Studies, The University of Michigan.

Wilk R (ed.) (2006a) 'From wild weeds to artisanal cheese', in Wilk R (ed.), *Fast food/ slow food: the cultural economy of the global food system*, Lanham, MD, Altamira Press, pp. 13–27.

Wilk R (2006b) *Home cooking in the global village: Caribbean food from buccaneers to ecotourists*, Oxford and New York, Berg.

Wolf A and Huang C S (1980) *Marriage and adoption in China, 1845–1945*, Stanford, Stanford University Press.

Wu D Y H (1997) 'McDonald's in Taipei: hamburgers, betel nuts, and national identity', in Watson J L (ed.), *Golden arches East: McDonald's in East Asia*, Stanford, Stanford University Press, 110–135.

Wu D Y H (2001) 'Chinese cafe in Hong Kong', in Wu D Y H and Tan C B (eds), *Changing Chinese foodways in Asia*, Hong Kong, The Chinese University Press, 71–80.

Wu D Y H (2002a) 'Cantonese cuisine (*Yue-cai*) in Taiwan and Taiwanese cuisine (*Tai-cai*) in Hong Kong' in Wu D Y H and Cheung S C H (eds), *The globalization of Chinese food*, Richmond, Surrey, Curzon Press, 86–99.

Wu D Y H (2002b) 'Improvising Chinese cuisine overseas', in Wu D Y H and Cheung S C H (eds), *The globalization of Chinese food*, Richmond, Surrey, Curzon Press, 56–66.

Wu X (2005) 'The New Year's Eve dinner and wormwood meal: festival foodways as ethnic markers in Enshi Prefecture', *Modern China*, **31**(3), 353–380, doi: 10.1177/0097700405276353.

Yan Y (1996) *The flow of gifts: reciprocity and social networks in a Chinese village*, Stanford, Stanford University Press.

Yan Y (1997) 'McDonald's in Beijing: the localization of Americana', in Watson J L (ed.), *Golden arches East: McDonald's in East Asia*, Stanford, Stanford University Press, pp. 39–76.

Yan Y (2000) 'Of hamburger and social space: consuming McDonald's in Beijing', in Davis D S (ed.), *The consumer revolution in urban China*, Berkeley, University of California Press, 201–225.

Yan Y (2003) *Private life under socialism: love, intimacy, and family change in a Chinese village, 1949–1999*, Stanford, Stanford University Press.

Yang D (1996) *Calamity and reform in China*, Stanford, Stanford University Press.

Yang M (1994) *Gifts, favors, and banquets: the art of social relationships in China*, Ithaca, Cornell University Press.

Yang X (2006) *La fonction sociale des restaurants en Chine*, Paris, L'Harmattan.

Zhang L (2001) *Strangers in the city: reconfigurations of space, power, and social networks within China's floating population*, Stanford, Stanford University Press.

Zhang X, Dagevos H, He Y, van der Lans I and Zhai F (2008) 'Consumption and corpulence in China: a consumer segmentation study based on the food perspective', *Food Policy*, **33**, 37–47, doi: 10.1016/j.foodpol.2007.06.002.

Zhao Y (2000) 'State, children, and the Wahaha Group of Hangzhou', in Jing J (ed.), *Feeding China's little emperors: food, children, and social change*, Stanford, Stanford University Press, 185–212.

22

Australian meals: as reported by Australian adult women

D. N. Cox, CSIRO, Australia

Abstract: The role of Australia's complex history of immigration in shaping current eating patterns and food choices at meals is described. In the absence of recent national survey data, descriptions of Australian meals are based upon 8-day (2 × 4 days) weighed food records from 62 adult women. Eating occasions were estimated by peaks in diurnal dietary energy intakes revealing distinct three-meals-a-day patterns, less distinct breakfast times at weekends and some evidence for large Sunday lunches. Cold breakfast and lunches dominate weekdays and Saturdays. Food choices were diverse reflecting both historical (popularity of red meat) and recent influences (Asian style meals).

Key words: weighed food records, energy, women, food choice, Australia, meals.

22.1 Introduction

22.1.1 Scope

Australians generally have access to abundant, safe and affordable food that has, because of recent cultural diversity become a large variety of experiences and tastes. The notable exception is the radically different circumstances of Aboriginal and Torres Straits Islanders (ATSI). In general, literacy levels are high and Australians have good access to health and nutrition information (though they might not always have a good understanding, Hendrie *et al.*, 2008) and education and so have the potential opportunity to make informed food choices. With the notable exception of ATSI, levels of infant mortality are generally low, life expectancy is generally high, and Australia ranks among the healthiest countries in the world with some of the highest living standards (as measured by physical quality of life indices). However, Australia also has one of the highest rates of overweight (40.5% of males and 24.9% of females) and obesity (17.8% of males and 15.1% of females) in the world (based upon self report from 2004–5 National Health Survey, ABS, 2006) with high prevalence of diet and lifestyle related chronic diseases (ABS, 2006).

22.1.2 A brief history

The first migrants to Australia arrived more than 40 000 years ago and the population of Australia just before European contact has been estimated to be between 250 000 and 750 000 possibly comprising of more than 500 different cultural groups, most of which were hunter–gatherer societies. Permanent European settlements were first established about 220 years ago and many Aboriginal and Torres Strait Islander cultures were devastated as European settlement spread.

Migrants in the nineteenth century were primarily of British and Irish origin with notable exceptions being German Lutherans to South Australia, Chinese to the goldfields and Pacific Islanders to the Queensland. The European population reached 3 million in the first 100 years and 16.5 million in the next 100 years with the current population around 21 million. The ATSI population now comprises only about 1.5% of the total Australian population.

The ethnic diversity of migrants to Australia increased after World War II, although British migrants remained the most numerous. The numbers of other European settlers increased particularly between 1947 and the early 1970s, many of these settlers were persons displaced by the war in Europe followed by southern European migrants, particularly from Greece and Italy. Since the mid-1970s immigration from Europe has slowed, although this has been offset partly by a rise in immigration from New Zealand.

Immigration from the Middle East and Asia has increased markedly as a proportion of the total migrant intake since the 1970s, and there has also been significant immigration from South and Central America. The origins of immigrants in recent years have tended to be areas of political unrest or war, such as the former Soviet Union, the former Yugoslavia, the horn of Africa and Iraq. Temporary residents, particularly skilled labour and students of south-east and eastern Asian origin, are another recent part of the Australian population and their food preferences are being met by Asian grocery and food service providers thus increasing the availability of Asian foods.

In summary, the present population of Australia is multicultural in nature with varying influences upon food choice. Nevertheless, without specific targeting of cultural minorities, what is known about 'Australian' food habits are largely based upon those that classify themselves European Australians. The same is true for the data presented below.

Australia has been known as 'the lucky country' based on a general abundance of living and recreational space, good (but highly variable) climate and, in respect to food, an abundance (compared with the migrants' origins) of cheap high quality meats (owing to major sheep and cattle production) and, partly due to climate and more recently partly due to southern European market garden habits, fruits and vegetables. The associated 'clean and green' image associated with the lucky country is not necessarily reality particularly when per capita carbon emissions are the highest in the world. This, in turn, has been attributed to high dependence upon cheap fossil fuel energy and high car usage in low density suburban settlement.

Since the UK's incorporation into the European Union (EU) and Australia's loss of preferential trading privileges, trade and general greater integration with Asia has increased. Japan has become Australia's largest trading partner and food commodity production (particularly beef and cereal grains) has been oriented towards Asian export markets with some influence on domestic consumption (for example, it is possible to buy Australian wagyu beef).

Also in recent decades there has been a large increase in wine production with associated reduction in beer and increase in table wine consumption whilst alcohol consumption in Australia has remained stable (1997–2004) at around 9.8 l per adult. Of 45 'established market economies' surveyed for alcohol consumption in 2005, Australia ranked 22nd in terms of per capita consumption (McCarthy, 2007).

22.2   Sources of data on the Australian meal: nutrition surveys

Unlike most developed nations, national nutrition surveys, which may provide useful information on food intakes, are not regularly undertaken in Australia. The last national survey (1994/5) used a 24 h recall method which is both limited and dated. Indeed, most dietary survey methods do not lend themselves well to understanding which foods are eaten, when and by what proportion of people because those methods' main purpose are to elicit nutrient composition and hence estimates of nutrient intake. Nevertheless, one type of nutrition survey method, the weighed dietary intake diary (Cameron and Van Staveren, 1988), does have the potential to describe foods eaten across time, across a period of days. The limitation of this approach is discussed below.

22.3   Definition of 'a meal'

Many reports (including other chapters in this volume) describe some of the difficulties of defining what is a meal and what is a snack (for example, IUNS Committee II/2 Symposium on 'Methodology to Identify and to Assess Eating Patterns'). There is an emerging consensus that people themselves define a meal or a snack and therefore this subjectivity negates definitions by *a priori* methods, for example, number of foods, food type, or minimum amount of dietary energy intake in a given period of time. To avoid these problems we took an approach that simply looked at patterns of energy intake over hours of a day over the week (see Fig. 22.1 and 22.2). The peaks of energy intake were taken to be defining meals and the lower energy intakes between those peaks labeled as snacking occasions. The latter are not discussed. Chapter 5 by Kjaernes *et al.* takes a similar approach.

22.4   Methods

The primary source of information to describe Australian meals were two sets of four-day weighed intakes (each of three week days and one weekend day) for 62 women from South Australia who had volunteered for a study that sought to validate a food frequency questionnaire (Lassale *et al.*, in preparation). Most, but not all, women provided two sets of data, eight days in total per individual. Approximately half (32) the women completed the study during the spring and summer of 2007 and the other half during the autumn and winter of 2008. As such we combine a range of seasons but do not distinguish between seasons. The limitations of this approach are discussed further below. Weights and heights were measured and socio-demographic data, including perceived cultural identity (with options for multiple responses; ABS, 2001) was also collected

22.5   Food categorization

Individual food items were originally coded using dietary intake nutrient composition software codes (Foodworks, http://www.xyris.com.au/). Previous classifications of food categories are either commodity based or nutrient based and therefore not amenable to eliciting foods 'as eaten'. Furthermore such data tend to break down foods as eaten into the component parts in order to determine nutrient composition and intakes. Therefore, foods were reclassified manually 'as eaten' (i.e. components were combined) and common combinations e.g. breakfast cereals and milk relabeled simply as 'breakfast cereals' (Appendix, p. 502). Considerable reference was made to the raw data (food diaries) to refine the classifications. There were numerous challenges to this process, including the recombining of bread and various fillings to create 'sandwiches' and at best these data remain as estimates of foods eaten at differing times of the day.

Foods were further classified as to 'cold and savory'; 'hot or cooked and savory'; 'cold and sweet and/or dessert' or 'hot or cooked and sweet and/or dessert'. Because of the diversity of foods eaten particularly at lunch and dinner reference to the raw data (food diaries) was made so as to combine foods as eaten. Because of the nature of the original data (which did not record whether food was eaten hot or cold) interpretations were made of the food items in the context of the recorded eating occasion. Lunch items were the most challenging to characterize for several reasons.

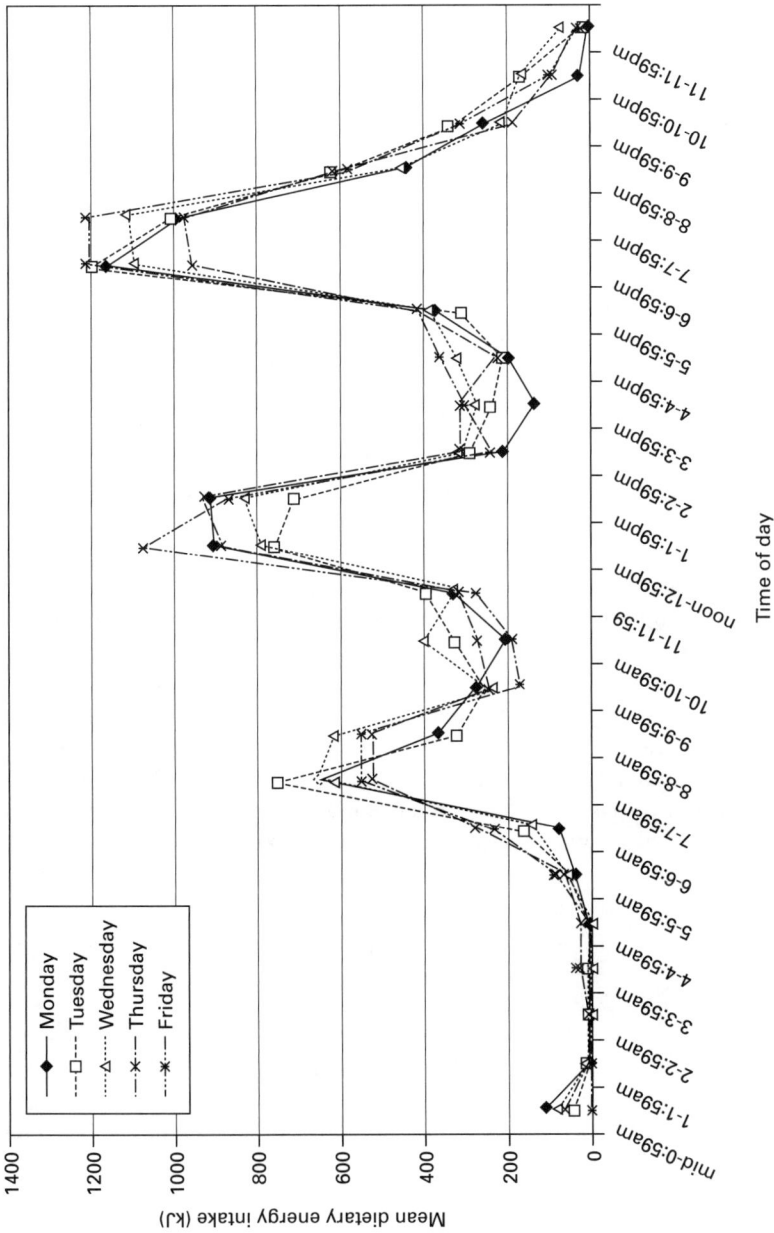

Fig. 22.1 Mean dietary energy intake per hour for weekdays.

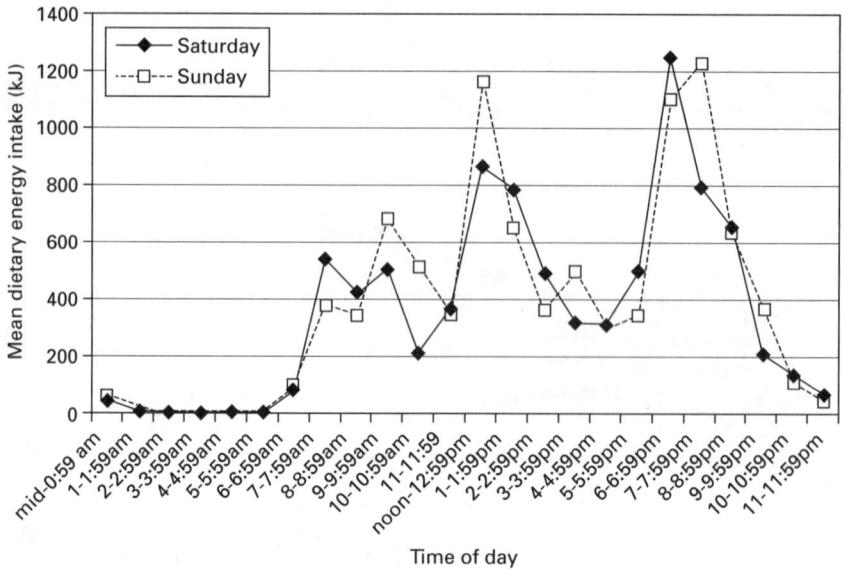

Fig. 22.2   Mean dietary energy intake per hour for Saturday and Sunday.

Firstly, observation of the data (supported by other observation and anecdote) revealed Australian women were in the habit of eating 'leftovers' from the night before (often re-heated in a microwave); secondly, meats (preserved, processed, red, poultry and other) could be eaten hot or cold and the reliability of this classification is questionable. Fish was assumed to be canned and cold by its description (e.g. tuna, canned) unless it was otherwise stated (e.g. fried fish or included with other cooked items). Nevertheless the reliability of these classifications is limited by the original data set and the interpretation of the author and co-workers. It is also worth noting that it is likely that water was not comprehensively recorded and therefore results on beverage consumption exclude water.

22.5.1   Day of week and hours of the day

Others, notably de Castro (2004) have separated weekdays from weekends when investigating energy intake patterns but we rejected de Castro's classification, which takes *a priori* time blocks and includes Fridays as part of the weekend. In contrast we simply divide the day into hours and, on the basis of the results, treated Saturday and Sunday separately.

22.5.2   Frequency

In the main, results are reported as frequencies of participants reporting eating items at a specific eating occasion time in order to estimate the popularity of foods within an eating occasion. The frequencies are averages across the weekdays (5 days) and frequencies for the separate weekend days of Saturday and Sunday. Hence, they represent the popularity of foods eaten at particular times of the day, weekdays combined and Saturdays and Sundays.

22.5.3 Analysis
Mean difference tests and non-parametric (Chi^2) category difference tests were undertaken and reported below as significantly different when $p < 0.05$.

22.6   Results and discussion

22.6.1 Participants
The 65 women ranged in aged from 38 to 65 years with a mean age of 51 (standard deviation 6.7). All were permanent Australian residents and most (80%) described themselves as Australian. However, approximately 10% identified themselves with a dual cultural background of Australian and British, or English, or Scottish with only one identifying herself as Aboriginal Australian. The remainder identified with various cultural identities including New Zealanders and other Europeans.

On average these women were overweight [mean body mass index (kg m^{-2}) 28.9, ranging from normal (19.1) to obese (44.6)] with 38.5% normal; 29% overweight and 32% obese. Hence, we have a 5% higher prevalence of overweight and twice as many classified as obese as national estimates of weight status suggest (ABS, 2006).

The majority (65%) was university educated, a far greater proportion than national estimates of 19.1 (ABS, 2001). Almost half reported taking dietary supplements and only one reported that she was a vegetarian.

22.6.2 Eating occasions
Defined eating occasions (i.e. meals) are based upon the total dietary energy estimates as recorded by hour of day for weekdays combined (Fig. 22.1) and for Saturday and Sunday (Fig. 22.2). For clarity, errors (standard errors or deviations) are not shown but these data are available from the author upon request. There were no significant differences in mean energy consumed across the weekdays overall. Intakes across the hourly time points did not differ by day of week with the exception of Friday lunch which has significantly higher energy intakes, compared with Tuesdays and Wednesdays. Both Saturdays and Sundays differ from weekdays. Saturday and Sunday do not differ by total mean energy intake but the patterns of consumption differs and by simple observation (Fig. 22.2) and testing revealed mean intakes at time point 12–1 pm is significantly higher for Sundays than Saturdays; hence,

we report these two days separately. Similarly, eating on a Sunday morning differs from Saturdays and other days of the week.

22.7 Weekday meals

Observation of these data suggest that in terms of eating patterns there are three distinct peak energy intake occasions on weekdays and these could, therefore, be considered three traditional meals, breakfast, lunch and dinner (the latter also commonly referred to as 'tea' in Australia).

It is important to note that the weekday data reported (descriptively) below are reported consumption over five weekdays hence it does not follow that people eat all the possible foods everyday i.e. the percentages are proxy measures of popularity of foods eaten at particular times and meals.

22.7.1 Weekday breakfast

Only a few individuals (17%) reported eating before 6 am but observation of the raw data (not shown) reveal some individuals consume pre-breakfast followed by a larger breakfast or eat on two early morning occasions. All of our participants ate between 6 and 10 am with a distinct peak between 7 and 8 am; we, therefore, consider these eating occasions to be the Australian breakfast.

Early, pre-breakfast or two-part breakfasts mirror what was classified as breakfast (below), with breakfast cereals with milk, bread with sweet or savory spreads, raw fruit and yoghurt characterizing intakes. No cooked food was reported consumed before 6 am. In terms of beverages, half reported drinking tea, whilst a third reported drinking fruit juice. No coffee was reported consumed before 6 am

Breakfast comprised of 24 food types and seven beverage types (Appendix, p. 502). The majority of foods (63% by category) were consumed cold, comprising breakfast cereals (with milk), raw fruit, bread (or rarely crackers) with or without butter or margarine with or without sweet or savory spreads. About 80% of participants reported consuming breakfast cereals (with milk). Almost half of the participants reported consuming raw fruit. Similarly, half the participants reported consuming bread (or, rarely, crackers) with about two-thirds of those using accompanying butter or margarine and half using sweet and half using savory spreads. Yoghurt was reported consumed by about a third of participants. Of the hot cooked foods, eggs were the most popular (17% of participants) closely followed by cooked fruit. Muffins or cakes were reported eaten by about 15% of participants. Raw vegetables, processed meat and cheese were reported consumed by about 10–12% of participants each. The remainder comprised lone consumption of foods (including fish, rice, and confectionery). In summary, cold foods dominate weekday breakfasts.

Beverage consumption at breakfast was divided in popularity almost equally between coffee and black tea (usually consumed with milk), a little more than half and half respectively, with a few people drinking both. About one in six reported drinking herbal teas. Less than one third reported consuming fruit juice.

22.7.2 Weekday lunches

All participants reported consuming foods and/or beverages between the hours of 11am and 3 pm. This 'lunch' comprised of 53 food types and nine beverage types (Appendix, p. 502).

The proportion of total food items consumed was 43% 'eaten cold' with 8% that could have been eaten 'hot or cold' (meats and fish) and 49% eaten 'hot'. However, foods eaten during this period have to be considered in terms of numbers of people eating these foods in order to gauge their popularity in meals. This is explored below.

Cold and savory
Bread (or, rarely, crackers) were reported consumed by 80% of participants throughout the weekdays. Interestingly, only about half the participants reported adding butter or margarine to their bread or crackers. Raw vegetables (salads) were reported eaten by two-thirds of participants and sandwiches reported eaten by half the participants. Cheese (40% of participants); savory spreads; (22% of participants); processed meats (15% of participants) were popular accompaniments to bread. Over one third (36%) reported eating fish, with the vast majority eating (cold) tinned tuna at this time. With a recent large increase in availability, 'sushi' was reported eaten by 17% of participants.

Hot and/or cooked and savory
Over one third (36%) reported eating cooked vegetables at lunch and poultry, red meat and eggs was consumed similar proportions of participants (17, 17 and 18%, respectively). Soups (Asian noodle and vegetable and meat types combined) and so-called 'fast food' (pizza, burgers, fried fish, hot dogs and fried chicken) were reported consumed by about a quarter of participants each (28 and 25%, respectively). Those reporting eating cooked starchy foods for lunch were similar (~15%) across pasta and pasta dishes; rice and rice dishes and potatoes (including 'hot chips' or fries). With over one third of participants reporting eating Asian foods (soups, Asian sauces, Asian dishes, sushi and noodles), there is some evidence, supporting popular anecdote, that Asian foods are influencing Australian food choice even amongst these participants who, predominantly, consider themselves 'Australian'. Pies and pasties (often accompanied with tomato sauce), whilst anecdotally considered a common part of the 'traditional' Australian lunch or snack menu, were only rarely eaten, possibly because our sample comprised of only high socio-economic status females.

Cold and sweet and/or dessert
Raw fruit was reported consumed by 80% of participants at lunches. Over half the participants reporting consuming sweet baked goods (muffins, cakes, biscuits) or confectionery at this time. Nuts and seeds were surprisingly popular (over one third of participants) and yoghurt (one quarter of participants) had more consumers than ice cream/gelati (10% of participants)

Hot and/or cooked sweet and/or dessert
Hot desserts were only rarely reported consumed at lunch (cooked fruits and miscellaneous desserts) amounting to a popularity of 10% of participants each.

Beverages
Coffee was the most popular beverage reported consumed (about two thirds of participants); followed by black tea (about half the participants) and about one third reporting drinking 'soft' drinks (cordial, or carbonated sodas). About one sixth reported drinking herbal teas and about 10% hot chocolate or fruit juice. Alcohol (mostly wine) was rarely reported consumed (only four participants).

22.7.3 Weekday dinners (evening meal)
All participants consumed food during the period 6–10 pm. It is widely reported, anecdotally, that Australians eat early in the evening. Based on the evidence from this sample of women, there is evidence that only some Australians eat early in the evening and that peak eating varies between 6 pm and 8 pm.

Of the 53 food types reported eaten, only 38% could be classified as cold food types suggesting that hot foods are more important for dinner.

Cold and savory
Bread (or, rarely, crackers) was the most frequently reported cold food consumed (by 90% of participants), with only about one third using butter or margarine and only 10% using savory spread. This suggests bread is an important accompaniment to other foods (below). Salads (raw vegetables) were the next most popularly cold savory foods and reported eaten by almost three quarters of participants with about one third of these reporting using salad dressing. Over 40% reported using ready-made sauces (as flavoring to hot dishes, below) such as tomato (ketchup) or 'brown' sauce confirming the popular belief that ready-made sauces are important in Australian food choice. Cheese was reported consumed by about half the participants (however, some of this was as ingredients in hot dishes and we cannot be certain of its role as a cold food). Preserved vegetables (e.g. pickles or olives) were reported eaten by about one in five. A few reported eating sandwiches (10%) or sushi (5%) during this time.

Hot and/or cooked and savory
Cooked vegetables were the most frequently reported hot food consumed (by 90% of participants) followed by potatoes (72%). Other foods rich in carbohydrates that were popular were rice (46%) and pasta (25%) and pasta dishes (20%). Asian dishes were reported eaten by almost half the participants (46%) with noodles eaten separately by only 10%. In terms of protein rich foods, red meat (using the Australian definition: beef, veal and lamb) was the most popular and reported eaten by over two-thirds of participants (68%) providing evidence that red meat is still a very important part of Australian evening meals. Asian dishes (for example, beef or chicken curry, in (simmer) curry sauce (with/without rice); Pork stew/stir fry, in sweet sour sauce; rice fried, with vegetables (and meat); dhal, lentils; chicken, with noodles, in satay/Asian sauce) were reported consumed by 45% of participants lending further weight to the assertion that Australian diets are becoming 'Asian' in character. Other protein foods included poultry (mostly chicken) reported by 45% of participants and 12% as fried chicken. Fish was reported eaten over one third (38%) and legumes and eggs were reported consumed by about a quarter of participants. So-called 'fast foods' (pizza, fried chicken, fried fish, burgers) were only reported consumed by 12% for the first three and 6% for the latter. Soup was rarely reported eaten (15%) suggesting that it is rarely eaten as an entrée (starter course). Pies and pasties were reported eaten more frequently than lunch but still relatively rarely (15%).

Cold and sweet and/or dessert
Raw fruit was reported consumed by 70% of participants at dinner (significantly less than lunchtime). Yoghurt was equally as popular as ice cream/gelati with one third of participants reporting consuming these cold 'desserts' at dinnertime hence more popular than at lunch items. Confectionery was reported consumed by more than half the participants during this time.

Hot and/or cooked sweet and/or dessert
We cannot be certain of the hot or cold nature of foods classified as desserts but less than 20% reported consuming miscellaneous desserts. Far more, almost one half of participants, consumed cakes and biscuits during this time.

Beverages
The most popular beverage consumed was wine (by over half the participants). However, tea and coffee were almost as frequently consumed (by slightly less than half the participants). About one third reporting drinking 'soft' drinks (cordial, sodas) or fruit juice. About 15% reported drinking herbal teas and hot chocolate. Beer or other alcohol was only rarely reported.

22.8   Saturday meals

The most notable contrast between weekend eating occasions and weekday eating occasions is the lack of a distinct peak in morning mean energy intakes for this group of Australian women as breakfast times vary considerably. It would be reasonable to hypothesize that this was owing to an absence of work commitments. Secondly, for 90% of participants breaking fast occurs later than weekdays (after 7 am and usually before 10 am), but it is much more difficult to distinguish between 'distinct' meals, i.e breakfast blends into mid-morning snacks

22.8.1   Saturday breakfast

Breakfast comprised 20 food types and seven beverage types (Appendix, p. 503). The majority of foods (70% by category) were consumed cold, comprising breakfast cereals (with milk), raw fruit, bread (or rarely crackers) with or without butter or margarine with or without sweet or savory spreads. About 50% of participants reported consuming breakfast cereals (with milk) compared with 80% across weekdays. Less than a quarter (22%) of the participants reported consuming raw fruit. About one third of the participants reported consuming bread (or, rarely, crackers) with about two-thirds of those using accompanying butter or margarine and half using sweet and half using savory spreads. Yoghurt was reported consumed by about 17% participants. Of the hot cooked foods, eggs were the most popular (14% of participants) but cooked fruit was rarely reported. Muffins or cakes, raw vegetables, processed meat and cheese were rarely consumed. In summary, cold foods dominate Saturday breakfasts with the suggestion of less (compared to weekdays) cereal consumption and greater reliance on bread.

Beverage consumption at breakfast was again divided in popularity almost equally between coffee and black tea (usually consumed with milk), with 80% of the sample consuming one of these. No one reported drinking herbal teas. Less than a quarter reported consuming fruit juice.

22.8.2   Saturday lunch

About 90% of participants reported consuming foods and /or beverages between the hours of 12 am and 3 pm. This 'lunch' comprised of 42 food types (significantly less than total weekdays) and nine beverage types (Appendix, p. 503).

The proportion of total food items consumed was 31% 'eaten cold' with 8% that could have been eaten 'hot or cold' (meats and fish) and 61% eaten 'hot'; hence more hot foods were eaten for Saturday lunch than weekday lunches. The suggestion is that participants consumed a simpler breakfast and a more complex lunch.

Cold and savory
Bread (or, rarely, crackers) were reported consumed by 46% of participants. Interestingly, only about half the participants reported adding butter or margarine to their bread or crackers. Raw vegetables (salads) and sandwiches were reported eaten by only a quarter of participants (less than weekdays): cheese (21% of participants); savory spreads (8% of participants); and processed meats (14% of participants) were popular accompaniments to bread. In terms of other food types, there are almost as many food types reported as participants with no one food being more popular than any other. In other words, on any one weekend day people choose widely from the variety available.

Hot and/or cooked and savory
About one quarter (24%) reported eating cooked vegetables at lunch. Hence, these data suggest that about one third of participants ate no cooked or raw vegetables at all on the Saturday on which they recorded their intakes. Less than 10% reported eating potatoes. As in the Saturday breakfast, frequencies of reported consumption were almost as numerous as the food items (Appendix, p. 503).

Cold and sweet and/or dessert
Raw fruit was reported consumed by about a quarter of participants at lunch. About one in five participants reported consuming sweet baked goods (muffins, cakes, biscuits) and confectionery at this time. Yoghurt and ice cream/gelati were reported by only 5% of participants. Again there was wide diversity of reported consumption (Appendix)

Hot and/or cooked sweet and/or dessert
No consumption of hot desserts (e.g. cooked fruits and miscellaneous desserts) could be identified.

Beverages
Coffee was the most popular beverage reported consumed (about one third of participants); followed by black tea (about one quarter of the participants) and about 14% reporting drinking 'soft' drinks (cordial, sodas).

22.8.3 Saturday dinner (evening meal)
Approximately 95% of all participants consumed food during the period 6–9 pm. There was a distinct peak (Fig. 22.2) between 6 and 7 pm, which one could hypothesize is related to subsequent social and/or leisure activities on the Saturday evening. Hence, when eating at home (the majority of responses), generally, most participants eat early on a Saturday in contrast to weekdays (above).

Of the 46 food types reported eaten, 43% could be classified as cold food types, with 8% that could have been hot or cold suggesting that hot foods are only marginally more important for dinner in terms of foods eaten overall.

Cold and savory
About one third reported eating raw vegetables (salads) at this time and dressing was rarely reported (6%). Bread was reported consumed by over a quarter of participants; savory snacks by 10%; margarine, butter or savory spread and ready made sauces used by 6%. Sandwiches, sushi or seafood were rarely eaten.

Hot and savory
Cooked vegetables were the most frequently reported hot food consumed (by 45% of participants) followed by potatoes (26%), Other foods rich in carbohydrates that were reported eaten were pasta (16%), rice (8%) and pasta dishes (5%). Asian sauces were reported eaten by almost 13% of participants with noodles eaten rarely. In terms of protein rich foods, poultry (mostly chicken) was the most popular and reported eaten by 13%) in contrast to more red meat eaten throughout weekdays. Other protein foods reported eaten were diverse included red meat reported by 11% of participants and 12% as processed meat and other meat. Fish or seafood was rarely reported eaten. Of the so-called 'fast foods', pizza was the most popular, with pies, fried fish, fried chicken, and burgers (all 3% each) Again, soup was reported eaten by 15% suggesting that it is rarely eaten as an entrée (starter course) or main course. Hot savory main items are generally diverse, but there is evidence of meat, carbohydrate and some form of vegetables consumed as a Saturday dinner.

Cold and sweet and/or dessert
Raw fruit was reported consumed by about a third of participants at dinner. Cakes or biscuits were reported consumed by about a quarter of participants. About one in six participants reported consuming confectionery at this time. Ice cream/gelati (15%) was more popular than yoghurt (5% of participants).

Hot and/or cooked sweet and/or dessert
Consumption of hot desserts (e.g. cooked fruits and miscellaneous desserts) was rarely reported.

Beverages
The most popular beverage consumed was wine (by about one third of the participants). Beer or other alcohol was reported about 10%. Tea and coffee were equally frequently consumed (by slightly less than quarter of the participants). Few reported drinking 'soft' drinks (cordial, sodas), fruit juice, drinking herbal teas or hot chocolate.

22.9   Sunday meals

Similarly to Saturday (and again in contrast to weekday eating occasions), there was the lack of a distinct peak in morning mean energy intakes for this group of Australian women (Fig. 22.2) as breakfast times vary considerably. Secondly, for 96% of participants breaking fast occurs later than weekdays (after 7 am and extends to 11 am) however, in contrast to Saturday, there is some evidence of cooked Sunday lunch, a tradition that is often, anecdotally, described as a phenomena of the past.

22.9.1 Sunday breakfast

Again breakfast (like Saturdays) comprised 20 food types and seven beverage types (Appendix, p. 504). The majority of foods (75% by category) were consumed cold, comprising of breakfast cereals (with milk), raw fruit, bread (or rarely crackers) with or without butter or margarine with or without sweet or savory spreads. Similarly to Saturdays, about 50% of participants reported consuming breakfast cereals (with milk). About a quarter (24%) of the participants reported consuming raw fruit. About one third of the participants reported consuming bread (or, rarely, crackers) with about two-thirds of those using accompanying butter or margarine and half using sweet and half using savory spreads. Yoghurt was reported consumed by about 15% of participants. Of the hot cooked foods, eggs were the most popular (15% of participants) but cooked fruit was rarely reported. Muffins or cakes, raw vegetables, processed meat and cheese were rarely consumed. In summary, cold foods also dominate Sunday breakfasts.

Coffee was the most popular beverage consumed at breakfast by about half the participants in contrast to other days of the week. Black tea (usually consumed with milk) was notably less popular, with a third of the sample consuming. Herbal teas and hot chocolate were rarely reported. Less than a quarter reported consuming fruit juice.

22.9.2 Sunday lunch

About 83% of participants reported consuming foods and/or beverages between the hours of 12 am and 2 pm. This 'lunch' comprised 35 food types and seven beverage types (Appendix, p. 504).

Of total food types, 51% were 'cold' and 41% 'hot' with 8% that could have been eaten 'hot or cold' (some meats and fish).

Cold and savory

Bread (or, rarely, crackers) were reported consumed by 48% of participants. Interestingly, only about 15% of the participants reported adding butter or margarine to their bread or crackers. Raw vegetables (salads) were reported eaten by a third of participants. Sandwiches, cheese and savory spreads; (17% of participants) were similarly reported. Again, in terms, of other food types, there are almost as many food types reported as participants with no one food being more popular than any other. In other words, on any one day people choose widely from the variety available.

Hot and/or cooked and savory
Only (12%) reported eating cooked vegetables at lunch in contrast to salads (above). Hence, these data suggest that half the participants ate no cooked or raw vegetables at all for Sunday lunch. About 15% reported eating potatoes. Meat consumption was very varied (for example, poultry, processed meat 7%; red meat only 5%; other meat 3%. Fish, eggs, and Asian dishes were eaten by 5% of participants each. Other frequencies of reported consumption were almost as numerous as the food items (Appendix, p. 504). Notably, less 'fast food' was reported consumed (no pies, fried fish or pizza) and only rarely fried chicken or burgers. However, overall 43% reported eating a cooked savory (usually meat based) 'main' dish at this time.

Cold and sweet and/or dessert
Raw fruit was reported consumed by about a third of participants at lunch. Yoghurt and ice cream/gelati were reported by only 5% of participants. Again there was wide diversity of reported consumption (Appendix, p. 504)

Hot and/or cooked sweet and/or dessert
Cooked fruits and cakes were the only foods reported (5%).

Beverages
Coffee and soft drinks were the most popular beverages reported consumed (one in five participants); followed by black tea and wine (10% of the participants). Of the other beverages there was diversity of reports.

Late lunch
About 21% of participants reported eating a hot savory dish (see lunch above, for example, meats, pasta dishes and fish) and cooked vegetables or salad between the hours of 2 and 6 pm. Observation of each individual's eating pattern suggests that in the majority of cases these were late lunches. Therefore, there is some evidence for an 'English' late Sunday lunch but only for a minority of participants. Note that the percentage reporting eating Sunday lunch exceeds 100% because some individuals ate both between 12 noon and 2 pm (peak lunchtime eating) and after 2 pm (late lunch).

22.9.3 Sunday dinner (evening meal)

All participants consumed food during the period 6–9 pm. There was a distinct peak (Fig. 22.2) between 7 and 8 pm, suggesting that people tend to eat later on a Sunday than on a Saturday. One could hypothesize that subsequent social and/or leisure activities are less common the evening before the start of the working week and consequently influence Sunday evening mealtimes.

Of the 41 food types reported eaten, 56% could be classified as cold food types, with 8% that could have been hot or cold suggesting that cold foods are only marginally more important for dinner in terms of foods eaten overall.

Cold and savory
Similarly to Saturday meals, about one third reported eating raw vegetables (salads) at this time but dressing was rarely reported (6%). Bread was reported consumed by half the participants, savory snacks by 7%; margarine or butter, savory spread and ready made sauces used by 6%. Sandwiches and sushi were more popular than Saturday choices. This suggests that, for some, a hot lunch is followed by a cold dinner.

Hot and savory
Potatoes were the most frequently reported hot food consumed (by 15% of participants) followed by cooked vegetables (13%). Other foods rich in carbohydrates that were reported were rarely eaten [rice (5%) and pasta dishes (3%)]. Asian dishes sauces were reported eaten by only 5% of participants. In terms of protein-rich foods, poultry (mostly chicken) was the most popular and reported eaten by 7%. Other protein foods reported eaten were diverse, including fish (5%), red meat (3%), processed meat (7%) and other meat (3%). Seafood was rarely reported eaten. So-called 'fast foods' were rarely reported eaten. Again, soup was reported eaten by 15% suggesting that it is rarely eaten as an entrée (starter course) or main course. Hot savory main items are generally diverse, but there is evidence of low consumption of hot meals on a Sunday evening.

Cold and sweet and/or dessert
Similarly to Saturday, raw fruit was reported consumed by about a third of participants at dinner. Cakes or biscuits were reported consumed by about 5% of participants. About one in six participants reported consuming confectionery at this time. Ice cream/gelati (15%) was more popular than yoghurt (5% of participants).

Hot and/or cooked sweet and/or dessert
Consumption of hot desserts (e.g. cooked fruits and miscellaneous desserts) was rarely reported.

Beverages
The most popular beverages reported consumed were soft drinks and coffee (by about one fifth of the participants). Wine or other alcohol was reported by about 10%.

22.10   Limitations of the data

We only have good quality data on well-educated Australian women there being a general dearth of data on the dietary habits of men. Whilst we had access to data on younger women, the irregularity of their lifestyles influenced the poor quality of the food intake data. Hence, whilst weighed intakes are a rich source of data the nature of the task determines that only a certain type of person (with a high degree of structure in their lives), who may have regular eating habits, is likely to provide such data. Nevertheless, we have captured data on regular meals for adult women.

22.11   Conclusions

Within the limitations of the data, we have shown some of the eating patterns that can be described as Australian meals, at least for well-educated Australian women. Breakfasts are dominated by cold cereals and are consumed in regular time slots probably as a result of work commitments. Without work commitments, breakfast consumption, whilst similar in composition, is much more varied in time of consumption on weekend days. Sandwiches and bread accompanying savory foods dominate weekday and Saturday lunches. In contrast, there is evidence that for many people (over half our participants) the Sunday cooked lunch still exists, with cold foods eaten in the evening. Whilst red meat is a very popular source of protein food, there is some evidence that Asian foods are an important part of Australian meals (and may also contain small quantities of meat). However, on any one day, choice is very diverse. Interestingly, these data suggest that vegetable consumption is below that of current recommendations for health (Australian Government). There is a wide range of food intake across the weekday evening lending only partial support to the widely held belief that Australians tend to eat early in the evening. It would appear to be true for Saturday evening possibly owing to subsequent leisure activities. In contrast, Sunday food intake is later in the evening which one could hypothesize could be a result of the satiety of Sunday lunch. Alternatively, it may be socially determined (daytime leisure activities). These should be further explored.

22.12   Acknowledgements

Colleague Dr Jennifer Keogh assisted with the design of the original study from which these data were derived. Student interns Charlotte Guilbert (INSFA, Rennes) and Camille Lassale (AgroParisTech) collected the data and Camille and colleague, Adam Harrison helped with analysis.

22.13   References

ABS (2006) National Health Survey 2004–5: Summary of Results 4364.0.
Australian Bureau of Statistics. (2001). Census of population and housing: 2001 Census basic community profiles: Main areas – by location name: South Australia: Basic

Community Profile: B12 Highest level of schooling completed (data table online) (accessed 2005 June). Available from URL http://www.abs.gov.au/websitedbs/ D3110124.NSF/24e5997b9bf2ef35ca2567fb00299c59/034b261536480e03 ca256c3a0000-d6a8!OpenDocument.

Australian Government. Go for 2 & 5® Campaign. Canberra: Australian Government, 2005. (Cited 25 June 2008) Available from URL: http://www.gofor2and5.com.au.

Cameron M E and Van Staveren WA (1988), Manual on methodology for food consumption studies, Oxford Medical publications.

de Castro JM (2004) The time of day of food intake influences overall intake in humans. *Journal of Nutrition* **134**: 104–111.

Hendrie GA, Cox DN, Coveney J (2008) Exploring nutrition knowledge and the demographic variation in knowledge levels in an Australian community sample, *Public Health Nutrition* DOI: 10.1017/S1368980008003042.

IUNS Committee II/2 Symposium on 'Methodology to Identify and to Assess Eating Patterns'.

Lassale C, Guilbert C, Keogh J, Syrette J, Lange K and Cox DN, (in preparation) Validation of the CSIRO food frequency questionnaire (C-FFQ) against weighed dietary intakes.

McCarthy P, (2007) Drinkwise Australia Ltd http://www.drinkwise.com.au/Common/files/ Alcohol_consumption_in_Australia.pdf.

22.14   Appendix

Table 22.A.1 starting on page 502 lists food and beverage groups by day and meal.

Table 22.A.1 Food and beverage groups by day and meal

Weekday

Breakfast

Food: Bread & crackers, Breakfast cereals, Cake & biscuit, Cheese, Confectionery, Eggs, Fish, Fruit cooked, Fruit dried, Fruit raw, Meat processed, Muffin English, Muffin other, Nuts & seeds, Rice, Sauce ready-made, Savoury snack, Spread fat, Spread sweet

Beverage: Coffee, Cordial/soda, Fruit juice, Herbal tea, Hot chocolate, Vegetable juice

Lunch

Food: Bread & crackers, Breakfast cereals, Burger, Cake & biscuit, Cheese, Crustaceans, Dessert, Dressing, Eggs, Falafel, Fish, Fried chicken, Fried fish, Fruit cooked, Fruit dried, Fruit raw, Hot dog, Ice cream/gelati, Legumes, Noodle, Nuts & seeds, Other Asian, Other cereals, Pasta dish, Pies & pasties, Pizza, Potato, Rice, Rice dish, Sandwich, Sauce Asian, Sauce European, Sauce readymade, Savoury snack, Soup Asian/noodle, Soup vegetable/meat, Spread fat, Spread savoury

Beverage: Coffee, Cordial/soda, Dairy drink, Fruit juice, Hot chocolate, Other alcohol, Tea, Wine

Dinner

Food: Bread & crackers, Breakfast cereals, Burger, Cake & biscuit, Confectionery, Crustaceans, Dessert, Dressing, Eggs, Fish, Fried chicken, Fried fish, Fruit cooked, Fruit dried, Fruit raw, Hot dog, Ice cream/gelati, Legumes, Main Asian dish, Nuts & seeds, Other Asian, Other cereals, Pasta, Pies & pasties, Pizza, Potato, Rice, Rice dish, Sandwich, Sauce Asian, Sauce European, Sauce ready made, Savoury snack, Shellfish soup, Asian/noodle, Soup vegetable/meat, Spread fat, Spread savoury

Beverage: Beer, Coffee, Cordial/soda, Dairy drink, Herbal tea, Hot chocolate, Other alcohol, Tea, Wine

Spread sweet
Topping
Vegetable cooked

Vegetable raw

Yoghurt

Saturday

Bread & crackers
Breakfast cereals
Cake & biscuit
Cheese
Confectionery
Dressing
Eggs

Fruit cooked

Fruit raw
Meat procesed
Muffin English

Coffee
Cordial/soda
Dairy drink
Fruit juice
Hot chocolate
Tea
Vegetable juice

Main Asian dish
Meat other

Meat poultry
Meat processed

Meat red
Muffin English
Muffin other

Spread sweet
Sushi

Topping
Vegetable preserved

Vegetable cooked
Vegetable raw
Yoghurt

Other Asian

Pies & pasties

Pizza
Potato

Rice
Rice dish
Sandwich

Sauce Asian
Sauce ready-made

Savoury snack
Shellfish
Soup

Coffee

Cordial/soda

Dairy drink
Fruit juice

Herbal tea
Hot chocolate
Tea
Vegetable juice

Wine

Bread & crackers
Breakfast cereals
Burger
Cake & biscuit
Cheese
Confectionery
Crustaceans

Dessert

Dressing
Eggs
Fish

Spread sweet
Sushi

Topping
Vegetable preserved
Vegetable cooked

Vegetable raw
Yoghurt

Noodle
Seeds
Nuts & seeds

Other Asian
Pasta

Pasta dish
Pies & pasties
Pizza

Potato

Rice
Sandwich
Sauce Asian
Sauce ready-made

Meat other
Meat poultry

Meat processed
Meat red

Muffin English
Muffin other
Noodle

Bread & crackers
Breakfast cereals

Burger
Cake & biscuit

Cheese
Confectionery
Crustaceans

Dessert

Dressing
Eggs
Fish

Beer

Coffee

Cordial/soda
Fruit juice
Hot chocolate

Other alcohol
Tea

Wine

Breakfast		Lunch		Dinner	
Food	Beverage	Food	Beverage	Food	Beverage
Muffin other	Coffee	Fruit dried	Asian/noodle	Fried chicken	Savoury snack
Noodle	Fruit juice	Fruit raw	Soup vegetable/meat	Fried fish	Soup vegetable/meat
Nuts & seeds		Hot dog	Spread fat	Fruit cooked	Spread fat
Spread fat		Ice cream/gelati	Spread savoury	Fruit dried	Spread savoury
Spread savoury		Legumes	Spread sweet	Fruit raw	Spread sweet
Spread sweet		Main Asian dish	Topping	Ice cream/gelati	Sushi
Topping		Meat poultry	Vegetable preserved	Legumes	Topping
Vegetable cooked		Meat processed	Vegetable cooked	Main Asian dish	Vegetable preserved
Vegetable raw		Meat red	Vegetable raw	Meat other	Vegetable cooked
Yoghurt		Muffin English	Yoghurt	Meat poultry	Vegetable raw
		Nuts & seeds		Meat processed	Yoghurt
		Yoghurt		Meat red	
				Muffin English	
Sunday					
Bread & crackers	Coffee	Bread & crackers	Coffee	Bread & crackers	Beer
Breakfast cereals	Fruit juice	Burger	Cordial/soad	Breakfast cereals	Coffee
		Meat red		Noodle	
		Muffin English		Seeds	
				Nuts & seeds	

Cake & biscuit	Herbal tea	Cake & biscuit	Noodle	Dairy drink	Cake & biscuit	Pasta	Cordial/soda
Cheese	Hot chocolate	Cheese	Nuts & seeds	Fruit juice	Cheese	Pasta dish	Fruit juice
Confectionery	Tea	Confectionery	Pasta dish	Other alcohol	Confectionery	Pies & pasties	Herbal tea
							Hot chocolate
Eggs	Veg juice	Dessert	Potato	Tea	Dessert	Pizza	Tea
Fish		Dressing	Rice	Wine	Dressing	Potato	Wine
Fruit raw		Eggs	Sandwich		Eggs	Rice	
			Sauce ready-made				
Meat processed		Fish	Savoury snack		Fish	Sauce Asian	
Muffin English		Fried fish			Fried chicken	Sauce European	
						Sauce ready-made	
Nuts & seeds		Fruit cooked	Soup vegetable/		Fruit cooked		
			meat				
Potato		Fruit dried	Spread fat		Fruit dried	Savoury snack	
Rice		Fruit raw	Spread savoury		Fruit raw	Shellfish	
Sauce European		Ice cream/gelati	Spread sweet		Ice cream/gelati	Soup	
						Asian/noodle	
Sauce readymade							
		Legumes	Vegetable cooked		Legumes	Soup vegetable/	
						meat	
Spread fat		Main Asian dish	Vegetable		Main Asian	Spread fat	
			preserved		dish		
Spread savoury		Meat poultry	Vegetable raw		Meat other	Spread savoury	
Spread sweet		Meat processed	Yoghurt		Meat poultry	Vegetable	
						preserved	
						Vegetable cooked	
Vegetable					Meat processed		
preserved							
Vegetable raw					Meat red	Vegetable raw	
Yoghurt					Muffin other	Yoghurt	

Part VII

Meals in practice/meals as art

23

Chefs designing flavor for meals

K. Vetter, McCormick and Company, USA

Abstract: Creative and well thought out flavor, food and menu design, the lifeblood of the food industry, are discussed. The basic elements and considerations when designing flavors and food for various segments of the food industry, including independent operators and corporate operations in both commercial and non-commercial segments, are outlined. The creative processes used by chefs to assist in the development of new flavors and food for meals are described.

Key words: flavor, menu design, chefs, meals.

23.1 Introduction: the art and process of flavor design by chefs

The essential job of chefs is to please their patrons by providing the best food and dining experience that they can. Menus are the chefs' canvas for expressing themselves to the guests. The food and flavors on the menus are a reflection of their style, personality, training, passion and love for the craft they have chosen as a profession. A perfectly executed sandwich can be as intriguing and satisfying as the most avant-garde chef's tasting menu. By understanding some basic principles, a chef can transform simple, pedestrian ingredients into a masterpiece that seems to validate the saying 'the whole is greater than the sum of its parts'. The chef should consider the five basic tastes, the visual elements of the dish, aromas, textures, complementary and contrasting elements of the dish and the overall context of the dish. By ignoring or not understanding these elements one may be designing food and flavors that fail, but, by becoming a master at manipulating these variables, the chef will be able to design food and flavors that are sure to satisfy.

Many chefs, restaurants and large companies have a process for developing new foods and flavors. These processes can be driven by factors that are routine and everyday such as: identifying items that need to be moved out of inventory because of over ordering, or the chef might have too much of an ingredient that has not sold as well as expected. Or, it can be as simple as a chef, cook or sous chef deciding what he wants to run that night based on an intriguing dish or ingredient he saw in a magazine or on a cooking show. In the corporate world, the development process can be so complex as to require in-depth consumer research, product or category immersions, ideation and proto-cept development, concept testing and validation, product development, consumer validation of the finished dish and operations testing that takes nine to twelve months. The famous Spanish chef, Ferran Adria of El Buli, shuts down his restaurant for six months to enable a team of creative chefs to do research and development on what the next season's menu will be – all this to repeatedly impress diners through his complex cooking methods and never-before-seen flavor combinations. Whether the process is simple or complex, it is an important element in the design of new foods and flavors.

23.2   Elements of flavor

In order to better understand flavor, one should understand the link between taste, aroma, sight and texture. Simply put, flavor is the interaction of the visual appeal, aroma, basic tastes, and tactile sensations. The visual cues or presentation of the dish is typically the first impression and interaction that the diner has with the dish. This first impression can make or break the execution of the dish. Right or wrong, we judge people, places, things, and food quickly based upon first visual impression. While not directly connected to the physical consumption of the dish, the visual nature of the dish is paramount for creating the right first impression. The balance of how people perceive the basic tastes and ultimately flavor is as individual as the human finger print and is a great challenge for the chef trying to develop exciting new flavors. Aroma is typically, but not always as we will explore later, the second sense that is engaged in the eating experience. It can be an indication of something that is ripe and at its peak for consumption and pleasing to the palate, or aroma can be a warning sign of something that has spoiled and not fit for consumption. A large part of what is experienced as flavor has to do with the aromas that are coming from the ingredients that compose a finished dish. The tongue can only distinguish the basic sensations of sweet, salty, bitter, sour and umami, along with heat, astringency and texture. It is the interaction of the basic tastes, together with the aromatic or olfactory characteristics driven by the ingredients or the preparation method, and the visual cues of the dish, as well as the stimulation by textural attributes, that allow people to really experience and enjoy food as flavor.

One way of thinking about the effects of aroma on flavor is by considering how the selection of highly fragrant, aromatic ingredients provides a much more fulfilling eating experience: a Thai red curry is a perfect example of a dish that relies on its ingredients to provide the first hint of deliciousness even before you have had the first bite. As the ingredients are prepared and simmered together, a marriage of harmonious flavors and aromatics is forming. While enjoying a red curry, the first thing a diner experiences are the visual cues of the finished dish. Next, the diner is engaged by the tantalizing aromatics of lemon grass, galangal, kafir lime, perhaps a whiff of coconut milk and the background notes of chili and fish sauce. This aromatic prelude sets the expectations for what is to come and has the diner anticipating the delicious flavor even before it has entered the mouth. As the first bite is taken, they might notice the light sense of richness and sweetness from the coconut milk playing off of the contrasting acidity from the lime juice and complimenting the saltiness of the fish sauce. The second bite may find the diner pondering how well the assertive nature and fragrance of the galangal, kafir lime leaf and lemon grass contrasts with the subtle sweetness of the shrimp and earthy undertones of cilantro.

A great example of how aroma can be the first sense engaged is the simple act of caramelizing onions for a burger. This process not only affects the experience of the person eating that burger by providing a richer, deeper more satisfying and complementary flavor to the meat, but also affects the people seated in the restaurant who can smell the rich and delicious aroma of the onions and butter caramelizing on the flat top grill wafting through the air. Another great example would be to imagine walking towards a Cinnabon® store at an airport. The first sensation that hits the senses even before the stand is in sight is the intense aroma of the cinnamon filling. As you get closer, you begin to smell the rich aroma of the freshly baked dough. So far you have not seen or tasted anything, but you have begun to enjoy the experience nonetheless. Finally, you reach the counter and see the sweet gooey icing dripping down the sides of the warm, lightly golden brown bun. As you prepare to enjoy this treat the first thing you do is deeply breathe in all of the enticing aromas that make up the cinnamon bun. Cinnamon, butter, vanilla and the sweet brown notes of the dough fill your nose. Your mouth begins to water. As you begin to enjoy the warm bun, you taste the first bite of the flaky outside ring of dough slathered with the aromatic and intensely flavored cinnamon butter smear and sweet vanilla icing. As you get closer to the center you notice how the texture of the dough changes and becomes softer and slightly chewy. The intense cinnamon smear is on both sides of the dough now and provides a bit of a mouth tingling sensation from its high cinnamic aldehyde content. Finally, you take that last perfect bite of warm, soft, sweet cinnamon and icing drenched goodness. That was truly a complete sensorial eating experience. Try that same experience with a stuffy nose and you will be sure to be left flat and unsatiated.

As effective as aromas can be to please, they can also detract from the dining experience by overwhelming the dish or by being out of balance. Consider the pungent, powerful aroma of raw garlic in a pasta sauce or the acrid smell of a burnt steak. These aromas signal that something is wrong and again have an impact on the expected flavor of a dish, yet this time in a very unpleasant way.

Humans have always used their sense of taste as an indicator of food quality and food safety. Bitterness can be associated with things that are poisonous or harmful to the body. Many people can be sensitive to the effects of bitterness in food and unless this basic taste is predominant and familiar in one's own cuisine, can be a polarizing attribute and difficult to balance.

Balancing of all of these elements is one of the most challenging aspects for the chef, who must not only be skilled at understanding the sensory characteristics of the individual ingredients that make up the culinary pantry but how the physical properties of these ingredients change and interact with each other in the context of preparation method. Chefs across all segments of the food industry are gaining a better understanding of how combining and manipulating all of the sensorial aspects of the finished meal can provide something beyond a simple eating occasion; a true sensory-loaded dining or eating experience. All of these elements can and should be considered when designing food and flavors for any area of the industry. Striving for balance and harmony in every dish is the ultimate goal for the chef or culinarian.

23.3   Elements of the chef

As mentioned, a chef's task when designing a great dish or meal is to ensure that the dish delivers on many levels to the consumer. First, the dish must sound intriguing and impact the consumer on an emotional level to appeal to their sense of culinary adventure, their sense of comfort and home or quite simply fit the mood they are in. The dish must be visually appealing, have an enticing aroma, an appealing mouth-feel and texture and above all, have great flavor. Almost as important is to be able to define what about that dish will compel the guest to order that same dish again or to tell others about how great it is. Dishes that only succeed once or for a very short period of time (one hit wonders) certainly have their place, but sustained and repeatable success is key.

As a chef sets out to create a successful new menu item, a subconscious process begins to take place. If one were to ask the average restaurant chef the process of how they go about designing flavor for their menus, they might respond with a puzzled look, a shrug of the shoulders and a response such as 'I honestly don't know. I've never really thought about how I do it. I just do it!' For many, this craft is truly a process of developing new flavors and menu items through feeling and intuition. Culinary skills are honed through not only years of formal education at universities and culinary schools but mostly through hands-on mentoring by experienced chefs, constant collaboration with fellow chefs and most importantly first-hand experiences with ingredients, cuisines and cultures. The process of designing flavors for meals may come naturally to those who choose the profession of food as a career but few have formal training as to why they are making the decisions they do when putting together the elements of a single dish or an entire meal. In order for chefs to fully realize the potential that lie within themselves for designing great flavors for menus they must learn to rely on their own instincts and sense of what works and what does not. Experience and field immersion are the best ways for a chef to gain a solid foundation in the fundamentals of a cuisine or set of ingredients. This experiential trial-and-error method helps the chef build an understanding of the key elements of the cuisine or ingredients that will allow them to balance the elements of the dish perfectly.

Conversely, some chefs have taken the exact opposite approach and have an extremely dynamic and detailed approach when considering the elements that each ingredient or cooking technique may contribute. Some of these chefs break it down to almost a molecular level when evaluating the interaction of ingredients. This meticulous attention to detail is necessary when executing dishes that are very complex not only in flavor but also in how the meal is experienced at the table. One example that comes to mind includes serving a fall-inspired dish accompanied by smoldering oak leaves and twigs whose aroma and visible wisps of smoke mentally transport the guest to relive the experience and emotional connection of a fall day in their mind even before taking the first bite. By setting the stage with external stimulus that is not food, the chef is adding another level to the experience and enjoyment of the dish. A second example might be how a chef has chosen to deconstruct a glass of wine and recreate it through food with the flavor elements that make up the wine. Depending on the varietal of grape or style of wine, the chef may choose to represent various small individual tastes including the elements of oak, butter, apple, minerals, sweetness, acidity or specific citrus The guest will be instructed to taste each element separately first then combine all of the flavors into a single bite. The second, all encompassing bite should remind the guest of a sip of the chosen style of wine the chef is recreating. These chefs are attempting to deliver not only a great food and flavor experience but a truly over-the-top, interactive, extraordinary dining experience.

'Molecular Gastronomy', an in-vogue practice whereby chefs apply ingredients and techniques that have traditionally been only used in science labs or industrial food manufacturing to restaurant dishes, points to a breed of chefs who consider many non-traditional factors when developing new menu items. A group of world-renowned chefs authored a statement on 'new cookery' that reinforces their position that there are three basic principles that guide them: excellence, openness and integrity. The chefs also feel that one should embrace culinary tradition and build upon it, as well as embrace innovation including techniques, ingredients, appliances, information and ideas. Finally, the chefs agree that a spirit of collaboration and sharing is essential to developing this potential (Adria *et al.*, 2006).

These chefs have taken the elements of presentation, flavor, texture, and aroma to an entirely different level by manipulating them in very unexpected ways through the science of physics and chemistry, leaving the guest wondering how they were able to achieve such new and novel textures, combinations and food and flavor experiences. A great example of how chefs are engaging the guests in the experience is at Wylie Dufresne's New York Restaurant WD-50. The restaurant serves a bowl of steaming hot aromatic broth with a bottle of unlabeled white liquid. The chef or server instructs the guest to take the bottle and squirt it into the steaming bowl of broth in a constant stream. To the guest's amazement the liquid immediately and magically transforms into al dente shrimp noodles as it hits the hot broth. Not only does this interactive activity add another layer of flavor and texture to the dish but adds an all important WOW factor to the dining experience.

A chef's personal style can be influenced by many factors including where they were born and raised, their background, education, life experiences, life style and previous occupations. A chef's style is apt to change and become more refined and develop into their own signature style over the course of their career. Many factors along the course of a budding chef's life will determine his culinary perspective and style. All of their experiences will either complement or contrast their style of cooking by leading them towards one style or philosophy or perhaps away from another. No chef looks at any palette of ingredients with the same eye. The same set of ingredients given to any ten chefs would probably lead to a vast array of very different dishes from those chefs based on their training, experience with the ingredients, knowledge of regional and global cuisines and cooking techniques. The diversity of chefs' perspectives on food and technique is what drives change on the menu and makes this such as exciting and ever-evolving industry.

23.4   Inspiration and authenticity

A chef's inspiration for flavors can come from many sources and at any time. Culinary inspiration may come from something as simple as a walk through a local farmers' market to see what is in season and available at its peak, or by something as exciting as eating their way through the food stalls of a food market in Thailand or Vietnam and experiencing new ingredients or preparations they have never tasted before. Inspiration may come by looking through resources such as cookbooks, food photography, recipes, food magazines, and menus. As chefs are doing research for potential new ideas and menu items they may not look to these resources to duplicate what they see or read, but to consider how they may adapt ingredients, dishes or techniques for their own operation. While recipes may serve as a detailed road map to success for some, most chefs are more interested in evaluating the combination of ingredients, the interaction of those ingredients in the recipe and how they are being prepared. A chef's own unique perspective and cooking style enables them to adapt a recipe to fit their own personal or business need. This ability to cook through intuition is something that sets a professional chef apart from many home or novice cooks. A chef is used to experimenting in the kitchen to adjust all of the elements that make the dish or meal perfect. The ability to make rapid adjustments (cooking on the fly) ensures that each part of the meal that is served exceeds the expectations of the diner.

In addition to recipes, chefs may also look to tasting menus for inspiration when designing meals. The returning trend of 5–20 course tasting menus in high-end restaurants has allowed the chef to truly design the meal experience for their patrons. While chefs have long studied the ingredients of a singular dish, the opportunity to study the overall progression of a lengthy meal has re-emerged in the past 5–10 years. A chef must pay close attention to each course as it is being planned for such an extravagant dining experience. These ultimate culinary dining experiences can range in price from $50.00 to $500.00 and should deliver to the guest the perfect meal.

A tasting menu could be compared with a great piece of music. As the music starts the listener is pulled in and engaged by the opening notes, the harmonious collaboration of the instruments slowly building in intensity and complexity, sounds become richer, more vibrant and more exciting to listen to. Finally, the music reaches its crescendo providing the perfect emotional and auditory experience for the music lover. Slowly the music gently brings the listener back down from their musical euphoria leaving them completely satiated and longing for their next perfect musical experience. Just as the musician, so must the chef orchestrate his food to deliver to the guest an experience that entices and excites but never overwhelms at any point in the progression of the meal. The chef should at first slowly set the stage and introduce the diner to the culinary journey they are about to embark on. With each passing course, the chef should continue to build upon the previous course by introducing bolder and more intriguing flavors, textures and other culinary or interactive elements of surprise. The crescendo course (or courses) should embody the soul of what the chef's intention is for this magnificent dining experience. This should be the best of the best and should delight the guest with a culinary experience never before enjoyed. Finally, the chef should begin to wind down the culinary excitement with their final dishes allowing the guest to savor the entire meal experience and begin to anticipate what the next adventurous culinary dining experience may be.

Inspiration for a chef may also be sparked from something as abstract as art or architecture. Similarly to food, both art and architecture are made up of many layers of color, texture, patterns, materials and dimensions. As most chefs could certainly be classified as creative, the ability to appreciate and find inspiration from these non-food forms is easily achieved. A chef may also look for inspiration not only from what is new and happening in the food world today but also from looking to the past to see how the classic foundational cuisines and dishes can be adapted to today's taste. A lot can be learned from looking at the history and evolution of food through time and culture. It is important to fully understand not only the culture and cuisine of a certain region of the world, but what were the circumstances and events that helped shape and evolve that cuisine. Only then will you have a true appreciation for how closely many cuisines are connected. This is particularly important when developing flavors for menus that need to convey authenticity. Whether the menu idea is to be 100% authentic or inspired by authenticity, the chef should have a firm grounding in the ingredients, cooking techniques, and flavor principles for that specific cuisine or region. Within a region or cuisine there can be huge variation of ingredients, techniques and preferences so, although learning through experience and immersion may seem like a daunting and never-ending task, it is welcomed and cherished by the chef who is driven to learn. By understanding why the ingredients exist in the cuisine and the circumstances of how they came about and how they are used or combined the chef will be better prepared to develop appropriately.

Understanding the methods of preparation and cooking techniques are equally as important because they drive how the flavors are developed. There is a huge difference in how flavors come to life through quick cooking methods such as stir fry and sauté or lower and slower methods as braising and pot roasting. Once a chef has immersed themselves in the basic understanding of ingredients and techniques they can move onto exploring the most critical aspect: the flavor principles of the cuisine. As Elizabeth Rozin points out in her book *Ethnic Cuisine* (Rozin, 1992) the flavor principles are the most crucial element of a cuisine and drive the direction of the majority of the flavor in the dishes. The flavor principles not only include identifying ingredients and techniques but also how the dishes are balanced by the basic tastes of sweet, salty, bitter, sour and umami that those ingredients bring. A great example from Rozin's book is the flavor principles of Chinese cuisine. The basic flavor principles would include ingredients such as soy sauce, rice wine and ginger root. To extend this flavor principle to Peking cuisine you would add miso and/or garlic and/or sesame. To extend the basic Chinese flavor principle for Szechuan, you would add sweet, sour and hot. For Cantonese, you would add to the basic flavor principle black bean and garlic. The flavor principles also consider what other tastes or sensations including hot or piquant are important to consider. The flavor principles of a cuisine will address the foundational flavors of the cuisine and help guide the chef in building truly authentic flavors. As mentioned, even if a chef is not trying to develop a 100% authentic dish, having this important background information will help them develop flavors and dishes that still reflect the essence of the cuisine or region. This will lead to dishes that make sense and are not viewed as just being a random medley of ingredients and techniques that leave the consumer confused and disappointed.

23.5   Considerations for menu flavor design

When putting together a menu item there are many components that should be considered. How will this dish taste is the obvious first consideration, but equally as important from a sensorial perspective is how will this dish be presented? What are the aromas that the guest should experience even before their first bite? What should this dish sound like as it approaches the table; while it is being eaten? What sensations might the guest experience while eating this dish? What memories or emotions or feelings should this dish elicit while being consumed? Before a chef even begins to start thinking about food they also need to consider a host of other important factors when developing a single dish or an entire new menu. No matter if a chef works for a single-unit independent restaurant, a regional or national chain, or serves guests in a non-commercial foodservice establishment, one of the key questions that should be asked is, does this dish fit my concept? The chef should clearly understand if the concept's vision is being driven by a specific theme or cuisine and what that theme or cuisine represents. A restaurant's identity is the one thing that sets it apart from its competition in the highly competitive foodservice industry. The menu can offer significant advantage in drawing guests to an establishment. This is a universal consideration from the top-end $500.00 per person, trend-setting restaurants down to the most back-to-basics quick service restaurants.

Another important factor to consider is what the menu philosophy of the operation is. Is it being driven by the chef or by other factors such as an owner, a management team or the brand identity? Just as important as understanding the overall restaurant concept and identity, is the full comprehension of what the menu itself should be conveying. Is the cuisine the cutting edge brainchild of a super-chef, who wishes to amaze and surprise with over-the-top creations to quickly become a Mecca for foodies the world over; should it have a family-friendly feel for everyday dining by offering classic comfort foods; is it a destination restaurant for special celebrations that highlight classical French cuisine; or does it perhaps convey seasonal simplicity that is good for both the guest and for the planet? While the décor and surroundings may convey the overall theme and concept, the menu is the true reflection of the spirit of the establishment. Without a complete comprehension of what that spirit of the menu embodies, the flavor design efforts may fail.

As most chefs and their teams who are responsible for developing new menu items tend to like to push the limits on culinary creativity, it is important to understand what the boundaries of the concept or the menu are. Foodservice innovation relies on the chef or menu development team to be able to push their concept's ability to deliver what the guests expect as well as to offer them items that they may not expect but which will still surprise and delight them. Chefs need to consider if their guests really will be able to understand what they are trying to convey with this dish or will there be a disconnect with what they have designed and what their guests have come to expect. By having a good understanding of the answers to these questions, a chef will have a far better chance of developing dishes that not only satisfy the chef's own passion for culinary creativity but for developing dishes that resonate with the customers and quickly become favorites. In this sense, a chef has to be a bit of a consumer psychologist as well as a culinarian; to be able to predict how much unexpected sensation is a delightful surprise rather than a disconnected disaster.

There are several organizations that can assist chefs with developing the needed skills for menu and flavor development. The American Culinary Federation® (ACF®) is an organization that has traditionally provided skills and business training and certification opportunities for chefs in the United States. This organization has primarily been geared towards the traditional foodservice chef working in restaurants or institutions. Another organization, the Research Chefs Association® (RCA®), is geared towards chefs and other technical professionals that are working in research and development for the food industry. The RCA® promotes a discipline called Culinology®, which combines culinary arts with technology and science. A Certified Research Chef will have training in not only the classical and foundational aspects of cooking but also will have an understanding of how science and technology are used to create meals, menu items or menu components that can be produced with not only optimal flavor in mind but also produced efficiently and profitably from a manufacturing and operations standpoint.

23.6   Menu flavor design by segment

While there are clear differences in how different segments of the foodservice industry go about designing flavor for new menu items, there are many similarities as well. There is no right way or wrong protocol, there are opportunities for all to learn about the different approaches that are taken to develop great food and flavors. Competition for 'share of stomach' is fierce in the foodservice industry across all of the segments so chefs and menu designers should always be on the lookout for ways to modify their approach to food and flavor design. The foodservice industry is generally classified in several different ways. The first slice is independent vs. corporately owned. Independents are typically controlled by one or a few owner/operators who establish the policies that the restaurant or establishment will be governed by. Corporate foodservice establishments are controlled by a parent company and have a much more formal structure as to how things are done.

The second way to split the industry is by commercial and non-commercial businesses (Table 23.1). The commercial segment encompasses the traditional restaurant sector. The non-commercial segment is comprised of dining establishments which are on-site in an office building, school campus, healthcare facility or other form of business or entertainment venue with large numbers of employees, residents or visitors to feed.

On the *commercial side* there are five basic categories. The first category is white table cloth or fine dining restaurants; very high-end dining where food and service reign supreme. There is no room for compromise in quality of ingredients or ingenuity of ideas. Guests will pay big money and will

Table 23.1 Non-commercial foodservice versus commercial foodservice

Commercial foodservice includes:	Non-commercial foodservice includes:
Fine-dining restaurant	Business and industry
Casual-dining restaurant	Healthcare and assisted living
Fast-casual	School K–12
Quick-service restaurant	College and university
Convenience store	Sports and entertainment
	Prison and corrections

expect big results from the kitchen. The second category is the casual dining segment. These restaurants can be likened to the neighborhood favorite spots where you may go with friends and family for good conversation and fare. The menus tend to be a blend of familiar favorites with a nod towards culinary adventure. A concept's identity plays a major role in defining the menu in this category. The third group is the 'fast casual' restaurants, a hybrid between casual and quick service, offering more casual upscale menu items but in a faster format than a casual dining full service restaurant. Fourth, the 'quick service' or 'fast food' restaurants are the workhorse restaurants that millions of people rely on everyday for something fast and satisfying for breakfast, lunch, dinner or something in between. The fifth segment, 'convenience store dining and foodservice operations' have begun to successfully branch out from just the roller-grill experience to a more satisfying and interesting dining option that is both fast and filling, and often fresh as well.

The non-commercial side is comprised of essentially six segments. As several of these segments rely on the daily feeding of the same group of people through multiple day-parts (breakfast, lunch, dinner, snacks), the non-commercial chef needs to pay particularly close attention to the dynamics of their meal and menu design. The non-commercial operator needs to take into consideration the type and style of foods being served, the frequency of the dishes being served, the popularity of the menu items and the ease of portability or the ability to eat the meal at another location such as their office desk or on the go. From a business perspective the non-commercial operator should also consider the ease of preparation and service for the kitchen staff and the food cost and profit margin for the items. Having a good menu variety that includes selections of hot and cold, healthy and indulgent, comfort food and new or unique dining options, seasonal specialties, local and exotic fare will help keep their diners interested, satisfied and most importantly, repeat customers.

The 'business and industry' segment of non-commercial foodservice caters to those working in offices and the industrial sector. The goal of these operations is to keep the workers on site for their dining needs by offering a wide range of fare that meets an even wider range of demographic segments, nutrition needs and flavor preferences. 'College and university' dining has evolved successfully to meet the changing needs and lifestyle of today's college-bound generation. Gone is the typical tray line with the three-sectioned plate. This has been replaced by a more upscale station style format that offers students a wide range of selections from familiar favorites to authentic ethnic food experiences. The power of choice is a key factor in this segment. When you enter the foodservice facilities of some colleges and universities you will now find a very wide selection of choices including sushi, made-to-order stir fry stations, organic and vegan selections, ethnic food stations, grill and deli stations for made to order burgers, sandwiches and wraps and so much more. The diversity of the population in colleges and university has essentially forced this segment to rethink and adjust how they not only design their meal selections but how they deliver them as well. By providing access to a wide range of food types, styles and cuisines and by changing the service format to a much more customizable approach, this segment has truly been able to meet the needs of its patrons.

The 'K-12 school foodservice' program, while still driven by regulations and commodities, is evolving to have a much more focused approach to nutrition and the power and benefit of making better choices. Some school districts have gone so far as to hire professional chefs to help in the design and implementation of new meal programs that are better tasting, more nutritious and more appealing to the students. The introduction of fresh vegetables and seasonal foods that are properly prepared and the updating of kids' favorites are becoming more commonplace. Favorites such as pizza and macaroni and cheese have been revised to include more healthy ingredients such as whole grain or have been modified to reduce fat and sodium per serving. These changes are not only teaching kids how to make better choices but are teaching them to enjoy new foods.

'Hospitals, healthcare and assisted living' facilities are also in an evolutionary phase as patients and residents have higher expectations for quality of life and quality of food. Residents are no longer complacent with just accepting what is offered at meal time. As the Baby Boomer generation moves into the next phase of their lives they are not willing to accept mediocrity as it relates to food. Higher standards and expectations from the residents and patients have put increasing pressure on these facilities to improve and upgrade menu selections to more contemporary meals and service.

'Sports and entertainment' is driven by stadium and large entertainment venue foodservice operations. While still largely fueled by beer, hot dog and nacho sales, there is much more emphasis on offering more upscale items and items that are healthier. Showing up amongst the typical fare are grilled chicken, salads, fresh fruit, sushi and even vegetarian options. Suites and skyboxes also offer the sports and entertainment segment a venue to offer ultra-high end options for VIPs. Custom menu options can be developed per client based on criteria such as food preferences of the attendees, party theme, or seasonality. It is likely that this higher level of service and options will continue to trickle down into the larger scale foodservice offerings within this segment.

'Prison feeding', while heavily affected by cost limitations still offers an opportunity for innovation and flavor exploration. Operators have frequent opportunities to create satisfying meals within budget, using unexpected surplus meal ingredients.

Within and across all segments of the foodservice industry, from the most expensive white table cloth restaurant to the roller-grilled items at the convenience store, there is clear passion for providing guests with high quality, innovative menu items that are flavorful, satisfying and anything but everyday.

23.6.1   Independent operator flavor and menu design

The advantage that an independent restaurant may have over a corporate foodservice establishment is flexibility in both process and in structure of introducing new flavors and menu items. Typically, an independent operator has fewer layers of management that need to be satisfied before a new item can be introduced onto the menu. This freedom is enviable to those who have been tied by the constraints of a corporate approval process to roll-out a new item. However, there is far less risk to an independent operator who has a dish that does not do well. A flexible and nimble independent foodservice operator may be able to change the flavor of their menu for a number of reasons. First, they may make a change based on a chef's whim. Because the chef may have the autonomy to make decisions based on the direction he sees fit for the restaurant's menu there needs to be little or no discussion necessary. A change may be based on a number of ingredients in inventory that need to be moved quickly. All foodservice operators rely on tightly controlled costs making it crucial to assure that there is no excess inventory that could spoil or go to waste. Independent operators may not have the means to track historical data on how menu items, be it permanent menu items or daily specials sell from day to day or month to month. Not having this data may hinder a chef's ability to accurately predict how much to order. As a restaurant's bottom line is already razor thin, eliminating waste is a business objective that is taken seriously.

Spontaneous changes to menu may also be based on a favorable deal from a vendor or on availability of a seasonal ingredient at its peak. Independent operators rely on their network of vendors to provide them with a wide array of information including what is coming into season, fluctuations in commodity and major markets, as well as availability and pricing. Vendors can also supply information on market shortages, new products or ingredients that are entering the market and even current and emerging flavor and consumer trends that can help give them a competitive advantage.

Chefs have a good sense of which seasonal ingredients come into the market but also rely on their sources to provide them with details of how the supply is developing including how weather conditions are affecting the season, product availability, product quality and potential price fluctuations based on market conditions. For those establishments that rely on seasonal bounty to design the flavors for their menus, the relationship with their vendors help them plan what flavors and dishes are placed on the menu at any given time of the year. Chefs also rely on their vendors to provide them with information on how the major commodities such as wheat, dairy, protein and oil are performing. The price of these commodities can dramatically affect how a business operates, so the chef or menu maker needs to be sensitive to market movement so that appropriate action can be taken. Consider for example a casual dining pizza and pasta restaurant. As the price of wheat and dairy continue to rise, profit margins begin to fall. A decision will need to be made on how the menu will need to change to compensate for these dramatic cost increases. Given that increases in wheat cost drives the cost of both pizza dough and pasta and increases in dairy drive the cost of cheese, the establishment needs to plan for additional new higher profit menu items to off-set the declining margins on their core business. By understanding the consumer and flavor trends in the market place, the chef can create menu items that make sense for his business and weather the storm through the tougher times.

Even though there is far less mandated process for new menu development, most independent establishments will have a process in place to research and test most new potentially permanent menu items. In most typical independent high end white table cloth restaurants, the menu vision and development process is clearly set on the shoulders of the executive chef. The chef is responsible for communicating his vision for the menu to his support culinary staff as well as the front of the house staff whose job it is to represent the kitchen's creations. The chef may take sole ownership for designing and creating new flavors for the menu but the process will most probably include collaboration with his other chefs and cooks. Certainly a singular process does not exist for menu design and development within the independent category but rather a large number of processes that are adapted to fit the style of the restaurant and those involved in the design. High-end restaurants may have an advantage over other segments of the food industry. Cost, while always a consideration, is less of a hurdle for fine dining restaurants. Higher end restaurants may have the monetary resources and staff to support executing a complex dish that uses costly ingredients.

One process that can be found in the white table cloth segment entails the chef creating the new menu item on his own to ensure that the balance and execution of the dish is exactly the way he wants it to be. The chef will then train his staff on how to properly execute the dish to his high standards then expect his staff to be able to prepare the dish time after time. Another process is that the chef can delegate the process of flavor and menu design to his culinary staff with the expectation that the staff will need to present their creations to the chef for approval. The chef would expect that the dish would make sense and fall within the concept and menu philosophy of the establishment. The culinary team member would be expected to have an understanding and ability to articulate what his design vision was, how it fits into the structure of the menu, what the basic elements and flavor principles are and how these all work together to form a cohesive dish. If the staff member cannot defend his rationale on why he did what he did, the dish may not make it to the menu.

Another, more complex, process that can be found in an independent restaurant is the collaboration of the culinary team on the redesign or invention of the total menu or a singular dish. This process tends to focus more on drawing from the individual creative perspectives of each culinary member. While each member is charged with creating from within their own realm of new and exciting ideas, the collaboration will ultimately make the final product stronger. The entire culinary team will meet to discuss the overall boundaries or the scope of the project, discuss potential ingredients, what the basic elements of the menu or dish may be and how the elements of the dish might be experienced and presented to the guest. A series of working sessions will follow to evaluate and understand the individual components of the dish, evaluate how the basic tastes and elements can be manipulated and optimized for maximum impact, and determine what might be all the ways to make this dish perfect. This intensive collaboration of culinary talent not only serves to develop a fantastic flavor or dish but also serves to train and develop the group into more knowledgeable chefs and cooks as well as allows everyone to begin to develop synergies for future creative session.

In an independent casual dining or quick service establishment, the vision and development process may also fall to the chef but is also more likely to include influence from the owner or manager as well. Unlike quick service dining, the fine dining guest may be more willing to put themselves completely in the hands of the chef to go on a guided culinary adventure. As people use casual and quick service restaurants as an almost everyday adjunct to supplement their meal requirements, the approach needs to be different. This style of dining depends far more on what the consumers' expectations are for that particular foodservice establishment. Menu design needs to reflect this consideration. Casual and quick service restaurant menu designers need to have an in-depth understanding of what is happening in their category. They need to be mindful of who their consumer is and what they need to do to keep them coming back for more. This understanding can only be gained through constant communication and dialog with their regular clientele. The team that is responsible for new flavor and menu development needs to spend significant time talking to and gauging the needs of their core customer.

In addition to understanding who comes through their door regularly, and why, they need to understand the other potential consumer segments, or the people that are not coming through their door and why not. Independent operators generally do not have the luxury of market research departments to find the answers to these questions and must rely on their vendor partners who can provide them with this type on information. Even though the day-to-day rigor of running a busy restaurant leaves little time to investigate some of these details, understanding consumer differences and behaviors around age, gender, ethnicity, generation, income segments and the overall state of the economy greatly improves the chance of developing successfully.

Similarly to the white table cloth segment, the independent casual and quick service restaurant chefs will do menu and flavor development on their own using adaptations of some of the same techniques. Additionally, these mid-scale establishments may also rely on their vendors to bring them complete menu solutions. A restaurant's vendors or other foodservice food manufacturers frequently develop turnkey menu programs that help a restaurant design and launch new menu items that are a fit for their operation. Typically, these programs will include information on food and consumer trends to help the chef operator get a feel for what is on trend now and what may be becoming a trend in the next year or two. These programs will also include recipes and tips for specific menu items that support these trends. It then becomes very easy for the chef to adapt a recipe to best fit their operation. As part of the program promotion, the company that has supplied the menu solution will also make available menu inserts or point of sale promotional material for the guests to review at the dining table. This level of assistance from an outside resource makes the transition from concept to menu item much safer, simpler and successful for the small independent operator.

As the skill level of the staff can vary greatly, these restaurants need to develop products that are not only satisfying to their customers but can be executed consistently by the staff. Food manufacturers and distributors have done a fine job in providing information and menu solutions for restaurants that may not have the resources available for in-depth menu development. By providing operators with a variety of options such as scratch-cooking recipe and menu ideas, speed-scratch cooking components and ideas, and ready-to-use products, the casual and quick service operators have the ability to offer on-trend, exciting menus that can be easily and consistently executed without over-committing resources to a lengthy development phase. Vendor-provided scratch cooking recipes allow the chef to immediately implement a specific pre-tested recipe, or to make modifications to create a new dish inspired by the provided recipe. Speed-scratch cooking involves combining pre-prepared products that a chef is likely to have on hand with or without other home-made products to create another finished meal or meal component. Some examples of speed-scratch cooking would include a chef taking a prepared ranch dressing and combining it with a southwest seasoning to create a southwest ranch dressing or perhaps taking a pre-marinated Italian chicken breast, cooking it and serving it over purchased black pepper pasta with a scratch-made sauce. A ready-to-use product is a product that can be served as-is with no additional preparation needed other than simply heating the product to the appropriate temperature. Some examples of ready-to-use products could include frozen heat-and-serve Alfredo sauce, pre-marinated and pre-cooked beef fajita strips, or aseptically packaged hollandaise sauce.

23.6.2   Corporate segment flavor and menu design

Where as the independent operator has the ability to be flexible and nimble, most corporate flavor and menu design initiatives seem to be driven by process. This contrast is not necessarily a bad thing. Over the years, the typical cycle time for new product development has decreased so that new products can be launched faster. This speed to market for new products has been in response to competitive pressures, the ever-changing consumer palate and the need to differentiate restaurant concepts through the menu, as well as by advances in the creative process itself.

A critical point of difference in the menu design for corporately owned foodservice operations is the management of risk. Independent restaurants are able to respond to their personal inspiration and create and launch new menu items almost on a whim, or for a single evening. The corporate organization, being significantly larger, has much more momentum and risk in sales and profit to manage and control. This consideration has resulted in highly defined and proprietary stage-gate processes for development of new menu items. National restaurant chains have evolved new menu development workflows to manage the complexity of a variety of vendors, supply chain logistics, nutritional content, operator training, and costing. Each corporation has a unique pathway that is designed to work for their operations, their corporate culture and their comfort with risk vs. speed to market.

In these cases, the senior chef has shared responsibility for menu development as well as responsibility for ensuring that the flavor of the finished product remains true to the original flavor profile that was developed in the kitchen. As there can be some changes in flavor as the product moves from a culinary created gold standard to a commercially viable menu item, the chef, food scientist and restaurant operations personnel must work together to keep the flavor and texture integrity of the dish, but make sure it can be profitably produced and executed by the restaurant.

The final step in most stage gate processes is to run an operations test in the restaurant. This may be a small four-restaurant test or a 100-unit test. The purpose of this is to gain guest feedback on the menu items as well as prove that the new concept is operationally sound and can be consistently executed. At the close of the market test, a go or no-go decision is made by a management team, not the chef alone.

While there are many steps to launching a new product on a regional or national scale, the thoroughness and discipline of a rigorous process provides a better chance of success in the competitive marketplace and provides a platform for continuous success in introducing new flavors and menu items, even globally. Corporations pride themselves on their systematic process for creating and commercializing new menu items, and many consider their go-to-market process an important point of differentiation and competency.

23.7   Conclusions

Chefs use a combination of artistic creativity, field experience and systematic process to design and develop successful new flavors and menu items. Since there is no clear singular best practice that serves all segments, corporations, restaurants or chefs, each chef must determine the process that will best fit their business. A chef's passion for building his knowledge of ingredients, cooking methods, and cuisine styles will largely determine how creative he will be at designing foods and flavors that work.

Understanding basic elements of taste, aroma, texture, visual aesthetics and a touch of human psychology will also play a large part in the success the chef will have in designing great flavor. When combined expertly, these elements provide the chef with the necessary fundamentals to create within an operation's given boundaries or to push beyond the boundaries into truly new food innovation.

23.8   References

Adria, F.; Blumenthal, H.; Keller, T.; McGee, H. (2006) *Statement on the new cookery*
Rozin, E. (1992) *Flavor principles*. Penguin Press.
American Culinary Federation (ACF). www.acf.com.
Research Chefs Association (RCA). www.culinology.org.

24

Creating concepts for meals: perspectives from systematic research and from business practice

H. R. Moskowitz and M. Reisner, Moskowitz Jacobs Inc., USA, G. Ishmael, Decision Analyst, Inc., USA

Abstract: This chapter presents a systematic approach to understanding the meal, based upon experimental design of ideas and concept research, first focusing on creating and evaluating ideas behind the mind (promise testing or benefits testing), and then moving on to two different approaches to testing full concepts. The first is concept screening, which tests full concepts and requires the respondent to identify 'how well does this concept perform'. The second is experimental design, where the components of concepts are systematically varied, an approach to identify 'what features work, what features do not work'. Possibilities are explored for creating a database of ideas for meals, using linked studies, each of which is set up by a 'master experimental design' (the so-called Innovaid™ approach).

Key words: conjoint analysis, idea generation, concept development, screening, innovation, optimization.

24.1 Introduction: meals as collections of products versus meals in the experience economy

When a business practitioner thinks of meals in terms of commercial aspects, two thoughts occur. The first revolves around the meal as a product in which the objective is to identify the product features, as well as rules about what goes together, what does not, what type of segments exist, and so forth. The objective is to optimize the particular offering(s) on the shelves of supermarkets. The second dimension to the concept of meals is the meal as an experience. In their seminal work *The Experience Economy*, Joseph Pine and James Gilmore identified Customer Experience as the emerging key differentiator in today's services economy (Pine and Gilmore, 1999). As they point out,

80% of the work force in the United States is employed in the Services Sector, with 17% in manufacturing and only 3% in agriculture. Given these numbers, the next major paradigm shift in the service economy probably will involve customer experience. The issue becomes what can be done in terms of systematized knowledge development to make the experience of the meal optimal, in light of changes in society, and changes in the way people eat meals, including situations, foods, focus on nutrition, focus on social aspects with others sharing the meal, etc? Those businesses that can understand, create and then deliver an exceptional customer experience will succeed more frequently in this experience economy. Such understanding of the importance of customer experience is beginning to become a popular topic among scientists, a definite signal of the emerging importance of the 'science of experience' (see Schifferstein, 2007 for one of the first comprehensive books in this area).

We will spend this chapter exploring the different ways in which business researchers develop and explore ideas for meals. In contrast to science, whose goal is to create a body of knowledge about the meal, and, by doing so, 'understand' meals, the objective of business is to understand the meal occasion and its products in order either to create the proper products, or the proper experience. The goal of business is to create and to sell, hopefully guided by knowledge that itself adds to the sum of human understanding while it guides commercial application.

With that said, we now move on to the first part of the chapter, dealing with the creation and evaluation of ideas. The second part of the chapter will deal with the application of these principles in a case history about a bagel restaurant, showing how one applies ideas about meal situations and products to create a restaurant environment and its offerings.

24.2   Concepts as blueprints

Those who work in industry know that the genesis of a product is an idea. Ideas, however, need to be fleshed out, to be tested and then to be optimized in order to reach their fullest potential. Indeed, if one looks at books about new product development, such as Robert Cooper's *Winning at New Products* (Cooper, 1993), one will see at least one step devoted to concept development and concept testing.

This chapter begins with basic methods used in companies (benefits or promise testing), and moves on to screening ideas (ConScreen or concept screening), and finishes up with systematic development of concepts by experimental design of ideas (also called conjoint analysis). We will present the theory, the practice, a discussion of how the study is run, and finally some new approaches to better understand the mind of the customer.

It is important to note at the outset of this chapter that there is very little archival literature on the creation of concepts in the world of business, where concept research is primarily used. Certainly business schools teach concept testing, and there is literature in market research on methods to test concepts, such as the types of scales and attributes to use. Of course, advertising agencies teach copywriters how to write ads, as part of the creative process. However, in the main, concept research is done in a time-limited, often journeyman-like fashion, to answer specific problems, for a specific company that needs answers about the viability of an idea. This focus on problem-solution means that the world of concept research does not produce archival literature in the way other sectors of consumer science might do. There are no 'rules' of concepts, no general laws that are waiting to be discovered. Anyone delving into the world of concepts will find that for the most part the archival information lies both in the company files itself, dealing with specific products, or lies in the hands of market research companies which do testing of concepts, and thus need a normative base from which to base their predictions of success or failure.

In recent years, the authors have tried to remedy some of these problems through a number of books and papers (Moskowitz *et al.*, 2005), as well as creating a database for ideas called the It! Databases (e.g., Moskowitz *et al.*, 2005). These are worth a more detailed look, because they will augment significantly the information and points of view presented in this chapter. Other authors have also reviewed the array of methods, both at the global or macro level and at the specific or micro level having to do with a specific product. Van Kleef and van Trijp (2007) reviewed the topic, and provide an excellent summary of alternative approaches to this increasingly important area of commercial practice.

The reader should note that when the topic of concepts and ideation is 'meals', it is important to put the meal in context of an eating occasion that is provided by a third party, such as a restaurant. Unless there is such a nugget of 'action to be taken' rather than mere description, the notion of ideation and indeed the notion of concept have no meaning. Concept research in the context of meals has meaning only when the meal itself is something whose components can be varied, whether those components are the ambience, the food, the day-part, etc.

24.3   Ideation: the act of creating the raw materials for ideas, products and experiences

The easiest approach to ideation requires that the participants come up with small parts of ideas, rather than complete ideas. Realistically, this task is easy. People have a hard time coming up with finished ideas, but in a properly structured environment people have little problem coming up with the pieces of the ideas. The pieces need not be coherent, and no one is expected to be judged on a phrase. In contrast, a person is often quite concerned about being judged for a full concept, no matter how reassured one may be by a third party that the results of the ideation session will not be immediately judged nor blame assigned.

Let one think that the area of ideation is barren of points of view, however, we should mention the various methods that researchers have suggested for ideation, or the discipline of coming up with idea. Ideation is the name given to the wide-ranging practice(s) of coming up with ideas. For the most part there is no fixed 'best practice' for coming up with ideas, and thus the notion of practices, rather than prescription. Rather, ideation has developed as a practice from the efforts of thousands of practitioners over decades, all faced with the task of new product and service development.

Practitioners have developed a wide array of methods by which to develop these ideas that underlie new products and services. The meal combines both, and so ideas created for the meal will span a wide variety of such ideas. Some of these ideation methods involve brainstorming, where a moderator leads a group of respondents through some exercises in an attempt to create the correct environment for creative thinking. The output of this exercise is presumed to be better in terms of the quality of data. Within the world of brainstorming one can find many different methods. Practitioners apply specific 'twists' to ideation, whether they require participants to create collages of pictures taken from the Internet to visually represent how they would describe the new product or service (so-called Zaltman metaphor elicitation technique; Zaltman, 1997), or read a set of prescribed articles so that everyone is 'on the same wavelength' (planned invention sessions).

Although creativity flourishes in a setting that is free of judgment and promotes expression of new ideas, that is not to say effective ideation is unfocused and without parameters. The results of traditional, 'anything goes' ideation can be disappointing and counterproductive. 'It is as if during a brainstorming session each participant is trying to make the other participant laugh at the craziness of an idea' (de Bono, 1995). There is evidence that disciplined, guided exercises that provide participants with a relevant frame of reference in which to ideate produces a larger number of actionable ideas, in contrast to the often-seen 'anything goes' approach to ideation (Callahan et al., 2005).

A key to successful ideation provides tools which allow the participants in an ideation session to expand and diversify their thinking. 'It is not enough to be innocent and uninhibited and to have a creative attitude. The normal behavior of the brain in perception is to set up routine patterns and to follow these. In order to cut across patterns we can use deliberate techniques ... These techniques can be learned, practiced, and used deliberately' (de Bono, 1995).

The world of market research practice abounds with different methods to pull out the creative ideas from individuals. Some practitioners prescribe homework at the start of the process, with the respondents being told about the general problem (e.g., concepts for a new restaurant), and then instructed to bring in pictures, etc. to address the problem. That effort presumably stimulates a respondent's creative juices. Once in the invention session, or whatever name is given, the respondents share their collages and then proceed to the invention of ideas.

SCAMPER, used by Decision Analyst, Inc., is an effective tool to generate new ideas by deliberately changing an existing product or service. Participants are assigned to think about a specific product or service, and then follow the disciplined steps prescribed by SCAMPER (Table 24.1). For example, participants in an ideation session for sportswear might be told to think of a water bottle and then instructed to use come up with a new idea using R (reverse). One resulting idea might be clothing that releases moisture onto the wearer's skin to cool the body.

Another expansion and diversification technique used by the market research community is known as 'Clean Slate'. The participants acknowledge the rules under which they currently operate, and then *consciously* break each rule to develop a new idea. For example, participants in a session to generate ideas for new paper products that help people manage their time would write down what they associate with paper-based time management products. Such rules might include 'paper products contain static information'. The rules are shared with the group, and then participants generate ideas that do not conform to the rules they wrote. The acts of acknowledging the current state of being and then deciding as a group to ignore that state, give participants explicit permission to think of new ideas that otherwise would not have been acceptable.

Projective techniques, often used in focus group discussions, are useful for ideation in that they enable participants to subconsciously incorporate their own needs and desires into the idea generation process. This approach works particularly well when the participants' expressed and actual beliefs or desires are not aligned. Projective techniques attempt to dim participants'

Table 24.1 What snippets of ideas look like from ideation

1	How about using a Bagel restaurant to cater a brunch or may be your kids' swim meet?
2	When I think of a neighborhood Bagel store, I expect to be hit with the aroma of freshly baked bagels when I walk through the door. I envision an old-fashioned deli with bins of fresh bagels in a variety of flavors…and sizes. My kids can never finish a whole bagel so a smaller sized option wouldn't go to waste.
3	What would really be nice is if a Bagel store would use fresh local produce like lettuce and tomatoes.
4	Breakfast is always my favorite meal so I'd like the atmosphere to be very relaxing. Newpapers available, maybe some local artwork on the walls. Big, comfy chairs. I want it to feel like it's my local hangout. You know, kind of like where everyone knows your name.
5	I love Bagels but I have to watch my weight. I'd really like a Bagel that has a full taste, but is lower in calories and carbs.
6	I never know what to get my coworkers for their birthdays. How about a Bagel of the month club like they have those Wine and Fruit of the Month clubs?

awareness of self and encourage a more unconstrained state of mind, which minimizes personal judgments and barriers to creativity and thinking. One such approach, called 'Zoom Out', blends visualization and projective technique by instructing participants to imagine themselves in a specific setting, such as a cafe. Using their five senses and their imaginations, participants explore the café, and select an individual he or she would consider interesting. Next, participants are directed to focus on the selected individual's hand, and notice everything about it – skin texture, jewelry, nails, etc. Then participants 'zoom out' until they can see the individual's entire arm, and once again note all its detail. This process continues until participants can view the entire individual in his or her immediate surroundings. After briefly discussing who and what they saw, participants begin generating ideas for the individuals in the café.

Ideation does not have to focus on generating ideas for new products or on new enhancements to existing products. In keeping with the desire to optimize the customer experience, there are approaches which are particularly good for generating such ideas. The feelings and emotions customers experience while preparing and/or consuming meals are key influencers of how they perceive satisfaction. Therefore, if the goal is to generate ideas for enhancing the customer experience, ideation could shift away from creating products and features to ways in which to associate desirable emotions with a product. 'A Picture Is Worth… Lots!' is a technique in which participants are directed to find images and objects they associate with specific feelings. Then, using those images and objects as starting points, participants generate ideas they believe would allow them to experience the specified emotions. Such emotions might manifest themselves in the form of ideas for flavors, packaging, scents, and textures, or perhaps ideas for messaging and positioning.

Technology-based methods are now becoming popular for ideation. Thus, there is the extended online ideation session, where participants log into an online message board and take part in idea generation for several days, which often enhances creative output. (Silverthorne 2002) In a typical extended online session, participants log into a protected message board environment where a facilitator leads them through a series of creativity exercises and activities which change daily. Participants can see each other's postings in the board, and are encouraged by the facilitator to interact and build on each other's ideas, much as they would during an in-person ideation session. Because the idea generation takes place over a period of days, participants have the opportunity to think about the exercises, leave the session and ideate on their own, and return to the session with a fresh round of innovative ideas to stimulate the group. On average, a 6-h in-person idea generation session can yield hundreds of ideas and idea fragments, whereas a time-extended online ideation session can produce thousands.

Even more technologically based are methods known by the rubric of collaborative filtering (Dennis and Valacich, 1993). Participants in this method participate in a web-based procedure. One variation, called brand Delphi™

consists of three sections. The first section requires participants to select ideas from a list, based on their relevancy. The second section requires the participants to provide new ideas. The third section requires the participants to rate some older ideas. The outcome of this exercise is a bank of ideas, in which the ideas have been mulled over, polished and grown or not by the interest of respondents. Those ideas that look promising will continue to be selected when they are presented. Those ideas that are simply 'cost of entry' will have high ratings, but not necessarily be chosen frequently because they 'don't excite' the participant.

At the time of this writing (early 2008), it is fair to say that most researchers and marketers are familiar with one or another method of ideation. Ideation is the creation of ideas. These ideas can be snippets, later to be combined into full concepts, or they can be the concepts themselves. Whatever the method for creating these ideas, snippets or fully completed concepts, it is necessary for the marketer to have a stream of possible concepts in the market. Consumer tastes change. It is important to keep up with these changes, and have a method for creating new ideas to fill new opportunities.

24.4   Systematized ideation for experimental design (conjoint analysis)

Another method that works well creates these ideas in a systematized manner. The authors use this second approach to create ideas that will be later used in their experimental design of ideas (see below). For now, we concentrate on the approach only. The moderator leading the sessions sits with a group of participants. The immediate task is to create a set of lists of 'silos' (i.e. categories or buckets). These silos are organizing phrases that encompass a number of related elements. The term silo is easy to understand, and ensures that there is no confusion between the term 'category' as a set of related elements in the exercise and 'category' as a set of products in the way the marketer uses it.

The moderator goes around the table, instructing each participant to provide a silo relevant to the problem. Each person provides one silo and then the floor is given to the next person. After a few minutes, the silos have been identified. Then the process repeats; this time with the moderator instructing each person in turn to select one silo that has been previously identified, and contribute one element to that silo. The order soon breaks down (e.g., after 10–15 min), as the respondents become increasingly involved in the task. Within 2–3 h, the full set of silos and elements has been pulled out from a now excited group of participants. We see an example of this ideation in Table 24.2. Note that this is just a small part of the set of ideas. The task was to create ideas for a bagel store, and the type of food and meals that the store would provide.

Table 24.2 Attributes, percentage accepting concept (top 3 box), and average ratings from a ConScreen study for the two concepts shown in Fig 24.1a and 24.1b, respectively

Question	Anchors	Concept 1 (Text)	Concept 2 (Text + pictures)
Top 3 Box % rating purchase as 7,8,9	1–6 coded as 0% 7–9 coded as 100%	39%	54%
Purchase interest	1 = Definitely would not purchase 9 = Definitely would purchase	4.5	6.4
Uniqueness	1 = Not at all unique 9 = Very unique	3.4	4.5
Appetizing	1 = Not at all appetizing 9 = Very appetizing	4.9	6.8
Meal occasion	1 = I would eat alone 9 = I would eat with my family	7.9	2.3
Frequency of eating	1 = Would eat infrequently 9 = Would eat frequently	6.5	3.8
Gender	1 = More appropriate for males 9 = More appropriate for females	5.9	6.4

24.5   Testing full concepts – ConScreen, norms

Simple ideas by themselves constitute only the building blocks of concepts. The concept itself is more complex, comprising reasons for buying the product, a description of the product, pricing, packaging, etc., and perhaps even graphics. We are not talking here about an advertisement, nor are we talking about a simple idea, but something in between. Another word for this in-between is a vignette, where we describe the product or the situation.

We see two concepts about meals in Fig. 24.1. One concept provides simple ideas about a meal enjoyed in a bagel restaurant (Fig. 24.1a). The concept simply describes in bare-bones terms what is happening. The second concept comprises a set of paragraphs, written as if an ethnographer or author were sitting in the restaurant observing what is happening and trying to add meaning and feeling to the bare-bones description (Fig. 24.1b). Both types of stimuli have been tested thousands of times. There is no one right format for the concept, but the rule of thumb for concept screening is that there should be enough information for the reader to rate the concept on rating scales that may be of interest (e.g., interest in this type of meal, this type of restaurant, or degree to which the concept describes a serious meal versus a fun meal, etc.).

What does one do with these concepts? The researcher presents these concepts to the consumer respondent, either by mail, occasionally but less frequently by phone, in a mall, or increasingly frequently over the internet.

Concept as a series of 'bullet points'

- Bagels baked from scratch in many varieties and flavors
- The perfect breakfast pairing – freshly brewed coffee and a homemade bagel
- A relaxing place to read the paper and enjoy a delicious meal
- An assortment of delectable deli meats, cheeses, fish and spreads for your every bagel topping desire
- A wide selection of cookies, cakes and desserts to satisfy your sweet tooth

(a)

Welcome to your neighborhood Bagel Restaurant, where you can enjoy Bagels baked from scratch in an assortment of variations and flavors. Visit us anytime of the day for all your meal occasions. We offer something for everyone's needs. Stop by before work where you will find a relaxing place to sit back and read the morning paper and enjoy a delicious meal, or perhaps your personal favorite ... a perfect breakfast pairing of freshly brewed coffee and a homemade bagel.

We don't stop at Bagels. We also offer an assortment of delectable deli meats, cheeses, fish and spreads for your every bagel topping desire. And when you're finished with your meal, be sure to stop by our bakery section to see our wide selection of cookies, cakes and desserts to satisfy your sweet tooth.

We're more than Bagels; we're your neighborhood restaurant, where good food means good times.

(b)

Fig. 24.1 (a) A concept made from 'bullet point ideas', or short statements, unconnected. (b) A concept comprising a connected set of paragraphs and a picture.

The concept may be a stand alone, or the concept may appear in a collection among other concepts, either of the same type (restaurant) or different (e.g., restaurants and telephones). Whatever the venue of testing, whatever the surrounding concepts may be, the fundamental question is whether or not 'this is a good idea'. Afterwards, the questions are 'among which respondents does this concept do well or poorly', and 'why is this concept a winning idea or a losing idea' (so-called diagnostics). When the focus is on selling the 'restaurant', the respondent rates the restaurant, based upon the concept. When the focus is on the meal in the restaurant, the researcher would first direct the respondent to a specific aspect of the meal, and then instruct the respondent to rate the meal on that aspect, based upon reading the concept.

The bottom line for concept tests is a diagnostic report, similar to the schematic report shown in Table 24.3. If the company doing the concept research has a bank of concepts and knows how well the products or services embodied in these concepts subsequently performed in the market, then there is always the possibility of estimating total sales or shipments of the product, or total patronage of the restaurant. These estimates are left to specialist companies. The business community has spawned these specialist companies, who focus on meals and restaurants as 'products' that are consumed in the 'experience economy'.

Careful attention must be paid when selecting the correct rating attributes for our attributes. They must be specific to the category and address the objectives of the research. For the two concepts for the bagel restaurant as seen in Fig. 24.1a and 24.1b, we chose the relevant attributes shown in Table 24.3.

The aforementioned tests use the consumer as a measuring instrument, to rate the acceptance and communication of single and compound ideas. Certainly, respondents can generate the communication profile of the concept. At the same time, there is very little true learning about what 'drives' the concept. We do not really learn rules from concept tests. We simply measure, and if there are a sufficient number of test concepts we might see patterns that lead to hypotheses, which, in turn, may lead to confirmatory experiments. It is with experimental design of ideas that we truly go beyond simply measuring what exists to understanding the drivers, and why good concepts or poor concepts perform the way they do.

24.6   Experimental design of ideas: databases of what works

Experimental design or so-called design of experiments works by systematically varying the factors under the experimenter's control, testing these among the target population, and developing rules or equations that relate the independent variables to the dependent variables. All this said, the goal of experimental

Table 24.3 A sample of some of the strong and weak performing attributes tested by respondents. Strongly performing elements are in bold.

		Total	Male	Female
	Base size	152	35	117
	Constant	31	17	36
	Silo A = objective			
A1	When only the freshest bagels will do	**9**	**14**	7
A6	When you are looking for a fast healthy lunch	**9**	**11**	9
A2	When you are looking for a quick, satisfying lunch	7	**8**	7
	Silo B = ambience			
B1	A clean, bright restaurant for a quick bite or a relaxed meal	**11**	7	12
B4	A light, casual place where you can get a really satisfying meal	**10**	**14**	9
B2	Conveniently located right near your home	7	**15**	5
	Silo C = selections			
C1	A deli counter filled with your favorite sandwich fillings	**10**	**17**	8
C6	Bagel and a cup of fresh soup... nourishing and filling	**9**	**10**	8
C5	Try our healthy lunch alternatives... bagel, fresh fruit and yogurt	7	3	8
	Silo D = bagels			
D1	All bagels made fresh each day on premises	**9**	**16**	6
D6	Baked fresh every day from scratch... just the way you like them	**9**	**17**	7
D2	Lower carb bagels...for your special needs	5	2	6
	Silo E = bagel fillings			
E1	A wide selection of flavored cream cheeses...see our special of the day	6	**8**	5
E2	50% reduced fat cream cheeses...comes in a variety of flavors	4	**10**	2
	Silo F = additional services			
F5	Have a party platter delivered right to your home – no hassles for you	**9**	**12**	8
F6	A gourmet's selection of coffees and teas	4	**8**	3
F2	One stop shopping for your all your breakfast and brunch needs	1	**11**	–2

design is to understand how nature works. If we assume that the stimulus being tested comprises a number of variables, then the best way to understand what each variable contributes creates alternative combinations of these variables, measures the response to the combinations, and then estimates the contribution of each variable. It is impossible to do this estimation on one test combination alone, despite the attempts of many researchers to the contrary, e.g., by analysis of the distribution of responses to a single concept across individuals. You need stimulus variation, and responses to that variation. Only then is it possible to uncover the rules underlying the performance of the elements of the concept (Box *et al.*, 1978).

When it comes to the application of experimental design of ideas to a meal, the reader should again keep in mind that the focus of this chapter in on the business application of such ideas and concepts, rather than on the description of 'what exists'. Thus, the application would deal with the specific 'experimenter-controllable' features of the meal occasion, and the responses to those features.

The original design of experiments worked with physical stimuli which were varied, and then assessed by, e.g., instruments or general performance. By the middle 1960s, however, it became clear that psychologically based stimuli could be used instead of physical stimuli, and that the response could be a psychologically measured response rather than a physical response. In the end, psychologists recognized the value of design of experiments. Indeed, one psychologist, Norman Anderson (1970), went so far as to label the approach 'functional measurement'. Using regression and analysis of variance, Anderson was able to deduce the way stimulus variables combined to generate a response.

About the same time as Anderson was experimenting and publishing, two mathematical psychologists, Luce and Tukey (1964), were working on the mathematical foundations of measurement. They developed a procedure called conjoint analysis, heralded in their ground-breaking paper as a 'new form of fundamental measurement'. Although the early work involved the theory of measurement, the fundamental idea was similar to Anderson's, namely, to understand the workings or impact of components of a mixture by measuring the response to the mixture and deconstructing that response.

Conjoint measurement as a research tool enjoyed some vogue in mathematical psychology, but reached its apex when it was applied to marketing issues (Wittink *et al.*, 1994). It was now possible through experimental design to create test concepts (mixtures of ideas), and use powerful procedures to understand what components of these ideas really worked.

In conjoint measurement, the researcher follows some fairly straightforward steps, with some remarkable results. We need not present the entire approach here. It has been presented extensively, and in a variety of alternative implementations by various researchers (Green and Srinivasan, 1980; Moskowitz *et al.*, 2005). Suffice it to say, the following steps represent what is probably close to an irreducible minimum, and will suffice as the basis for our case history regarding eat-in and take-out meals in a bagel restaurant:

1. Identify what can be said (general silos), and options or elements within those silos. This is probably the most important step. When we talk about 'what can be said', we mean short, easy to assimilate ideas.
2. Create an experimental design in which these individual elements are systematically combined together in small, easy to read combinations,

called test concepts. Mix and match these elements by experimental design, to create small, easy-to-read concepts comprising 2–4 elements each. The experimental design ensures that the elements are statistically independent of each other, that each element appears several times, and that the appearance of the element is against different background. Essentially, this step is a 'mix and match', using statistical criteria. It is vital that the elements be statistically independent of each other because the analysis can not really work if the elements co-vary. Co-variation or correlation means that the individual contributions of the elements cannot be easily determined. Every respondent evaluates a unique set of these combinations, avoiding the possibility of systematic bias owing to one particular choice of combinations of elements.

3. Invite the respondents to participate in the study; Fig. 24.2 shows an example of the invitation by e-mail). The interview in which the respondents participate should be relatively short, preferably 15 min or so, in order to avoid drop-outs (MacElroy, 2000).

4. Orient the respondents in the task by means of an introductory screen at the beginning of the study (Fig. 24.3).

5. The respondent reads the test concept, and evaluates it as a totality on one or more rating scales. Each respondent is presented with 48 test concepts, comprising the elements. Each element appears three times in the set of 48 concepts (Fig. 24.4).

6. Transform the ratings to a binary scale. A rating of 1–6 is transformed to a 0, denoting 'not for the particular occasion' (depending upon the scale). A rating of 7–9 is transformed to a 100, denoting 'for the particular occasion' (depending on the scale). The transformation follows the

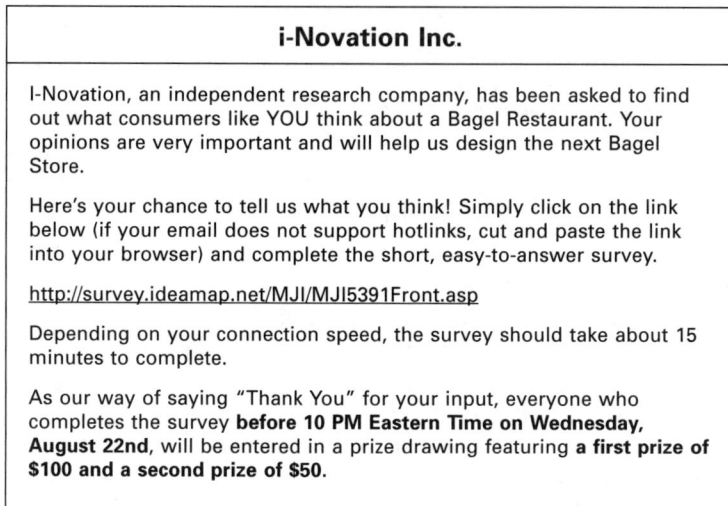

---

i-Novation Inc.

I-Novation, an independent research company, has been asked to find out what consumers like YOU think about a Bagel Restaurant. Your opinions are very important and will help us design the next Bagel Store.

Here's your chance to tell us what you think! Simply click on the link below (if your email does not support hotlinks, cut and paste the link into your browser) and complete the short, easy-to-answer survey.

http://survey.ideamap.net/MJI/MJI5391Front.asp

Depending on your connection speed, the survey should take about 15 minutes to complete.

As our way of saying "Thank You" for your input, everyone who completes the survey **before 10 PM Eastern Time on Wednesday, August 22nd,** will be entered in a prize drawing featuring **a first prize of $100 and a second prize of $50.**

Fig. 24.2 E-mail invitation.

Welcome to our bagel store and meal study.

The whole world of eating meals is changing. You're probably aware of your own meals, what you eat, where you eat and when you eat

We are trying to discover what people think about bagel stores, and dining there.

In the next section you'll be reading small 'vignettes' or descriptions about a BAGEL STORE.

Read the ENTIRE vignette, and rate it on the following question and scale.

'How interested are you in having a meal in this BAGEL RESTAURANT?'

1 = Not at all interested ... 9 = Extremely interested

Please click ">>" to continue

Fig. 24.3   Orientation page.

The taste of fresh New York deli bagels

A light, casual place where you can get a really satisfying meal

A deli counter filled with your favourite sandwich fillings

An assortment of cholesterol-free tofu toppings

How interested are you in having a meal in this BAGEL RESTAURANT?

1 2 3 4 5 6 7 8 9

1 = Not at all interested ... 9 = Extremely interested

Fig. 24.4   Example of a test concept.

conventions of market research, which deals with the proportion of respondents falling into a specific category, rather than the intensity of the feeling. In turn, this convention comes from the intellectual heritage of market research, which traces its origins to applied sociology. In contrast, experimental psychologists would be more likely to use the ratings themselves, which show the intensity of feeling.

7. Relate the presence/absence of the concept elements to the rating. The most common method is regression analysis, with the independent variables being the different elements (present or absent in the concept), and with

the dependent variable being either the rating or some transform thereof, such as the binary transform. For the data reported here the dependent variable will be the binary values, 0 or 100, respectively.

8. The outcome of this analysis is a simple model showing the marginal or part-worth contribution of each element to the rating or to the transform of the rating. To the degree that the analysis is straightforward (e.g., ordinary least-squares regression), the interpretation will be correspondingly easy. The model is expressed by the simple equation:

$$\text{Binary rating} = k_0 + k_1(\text{Element } A_1) + k_2(\text{Element } A_2)...k_{36}$$
$$(\text{Element } F_6)$$

9. Researchers have created a number of variations of Steps 1–6, such as the nature of the design (one design for everyone versus multiple, isomorphic designs), the type of rating scale, the creation of individual level models versus group models, etc. In the end, however, the output is a set of numbers, one per element per subgroup, showing the impact or driving power of that element.

24.7   Learning from a case history

The remainder of this chapter deals with a case history involving a meal at a bagel store. The objectives of the case history are to instruct on research procedure, and to provide actual data pertaining to the bagel store and the meal that might be eaten there, or taken home. It is important to note that these data come from a national study, rather than a study based in one market. The Internet allows studies of concepts to be executed worldwide, if desired, transcending the traditional limits found in central location tests. Keep in mind that the approach can be used for almost any type of meal or meal situation, as long as the components can be identified.

The typical design and field implementation have already been described in part above. The data appear in Table 24.4, which provides most of the information that we need to understand experimental design, and the contribution to meal concepts.

24.8   Interpreting the results: steps in the systematic analysis of results

1. The study ran for 48 h, during which time 152 respondents completed the interview, with 117 females and 35 males. It is typical to see a higher proportion of females participating in online studies. When the research requires specifics numbers of respondents (e.g., by gender or age) respondents who fit the criteria are specifically recruited to participate,

Table 24.4 Results from the study for a meal at a bagel store. The numbers in the body of the table come from the regression model, with the dependent variable being the binary transform (0 = not interested in the store, 100 = interested). The elements are rank ordered by performance among the total panel. Strongly performing elements are in bold.

		Total	Male	Female
	Base size	152	35	117
	Constant	31	17	36
	Silo A = objective			
A1	When only the freshest bagels will do	**9**	**14**	7
A6	When you are looking for a fast healthy lunch	**9**	**11**	**9**
A2	When you are looking for a quick, satisfying lunch	7	8	7
A3	When you crave the delicious taste of a soft bagel	7	**15**	5
A4	A casual, friendly place for a meal with your friends	7	4	**8**
A5	The taste of fresh New York deli bagels	7	*8*	6
	Silo B = ambience			
B1	A clean, bright restaurant for a quick bite or a relaxed meal	**11**	7	**12**
B4	A light, casual place where you can get a really satisfying meal	**10**	**14**	**9**
B2	Conveniently located right near your home	7	**15**	5
B6	Interesting art from local artists available for sale on the walls	1	4	1
B3	Located in your favorite mall...a perfect place for a quick bite	**−3**	**−6**	−3
B5	Tables with backgammon boards and checkers for your entertainment	**−5**	1	−7
	Silo C = selections			
C1	A deli counter filled with your favorite sandwich fillings	**10**	**17**	**8**
C6	Bagel and a cup of fresh soup...nourishing and filling	**9**	**10**	**8**
C5	Try our healthy lunch alternatives...bagel, fresh fruit and yogurt	7	3	**8**
C3	A wide selection of homemade cookies and desserts	6	2	7
C2	A special section with fruit or vegetable smoothies as a nutritious beverage option	4	6	3
C4	When you want a light, healthy lunch...bagel and a yogurt	1	1	1
	Silo D = bagels			
D1	All bagels made fresh each day on premises	**9**	**16**	6
D6	Baked fresh every day from scratch...just the way you like them	**9**	**17**	7
D2	Lower carb bagels...for your special needs	5	2	6
D4	Bagels are low in fat and only 170 calories	5	8	4
D3	We offer a variety of bagels...for every taste bud and every mood	2	3	2

Table 24.4   Cont'd

		Total	Male	Female
	Base size	152	35	117
	Constant	31	17	36
D5	Our scrumptious bagels will give you the energy you need	1	–3	2
	Silo E = bagel fillings			
E1	A wide selection of flavored cream cheeses… see our special of the day	6	**8**	5
E2	50% reduced fat cream cheeses…comes in a variety of flavors	4	**10**	2
E3	Reduced fat whipped cream cheeses…light and luscious	4	3	4
E4	Healthy fish choices like smoked salmon, whitefish, tuna and more for your bagel toppings	–3	5	**–6**
E6	An assortment of cholesterol-free tofu toppings	**–16**	–3	**–20**
E5	Tofu toppings…a great source of Isoflavins	**–18**	**–6**	**–22**
	Silo F = additional services			
F5	Have a party platter delivered right to your home – no hassles for you	**9**	**12**	**8**
F6	A gourmet's selection of coffees and teas	4	**8**	3
F2	One stop shopping for your all your breakfast and brunch needs	1	**11**	–2
F3	We ship bagels anywhere throughout the United States	0	2	–1
F4	Make your next office meeting a success with a bagel buffet	0	5	–1
F1	Plan a party with catering from your local bagel store	–1	9	–4

rather than allowing any qualified respondent to participate, as was done here.

2. By creating individual-level models the researcher need not worry about the disproportionately large numbers of females. The results from the total panel can either be 'reweighted' to make men and women each contribute 50% of the results, or the genders can be analyzed separately. In this chapter, we have chosen to analyze the genders separately.

3. Next turn to the constant, which is a measure of basic interest. The additive constant in the regression model, i.e., the intercept, is an estimate of the percent of respondents who would rate the concept as 7–9 on the scale 'How interested are you in having a meal in this Bagel Restaurant?' The constant is 31 for the total panel. This 31 means that 31% of the respondents are interested in a Bagel Restaurant if there were to be no elements present in the concept. Clearly, all concepts had elements, so the additive constant is an estimated parameter. Still, the constant is an important number to keep in mind. The constant is a baseline number.

One can compare this number across different groups of respondents in the set of 152 individuals, or across studies. Just for comparison purposes, credit cards show an additive constant of only about 15 or 20, meaning that only one person in about 5 or 6 is interested to begin with in credit cards, whereas for the bagel store we already start with about one person in three interested in having a meal there.

4.   Looking at the total sample, we see strong motivating power of the elements to drive interest in having a meal in the bagel restaurant. The only exception is the silo that deals with ' Bagel Filling'. The *Objective* silo drives interest when talking about 'Fresh Bagels' and 'A fast, healthy lunch'. In *Ambience*, respondents were motivated by mentions of 'A clean, bright restaurant for a quick bite or a relaxed meal', and 'A light, casual place where you can get a really satisfying meal'. *Service* is also very motivating as seen in 'A deli counter filled with your favorite sandwich fillings' and 'Bagel and a cup of fresh soup…nourishing and filling'. The *Bagel* silo has winning elements that speak of 'Made Fresh every day and Made from scratch'. Respondents were interested in *Additional Services* that 'will deliver a party platter to your home'. The one topic that turned off the respondent was the mention of Tofu fillings and toppings. These participants were interested in having a meal when the restaurant was described in more traditional terms, such as a nice clean, quiet ambience, fresh meals and baking done on the premises daily.

5.   Turn now to gender, and look for differences, which appear quite readily in Table 24.4. Males look at the bagel store as providing a snack, whereas we will see below that females look at the bagel store as providing a meal. Males are highly motivated by mentions of fresh New York bagels, quick satisfying lunches, and the delicious taste of soft bagels. They want to hear about their bagel, imagine it before eating it. They also want something that is near home. Convenience is important to them with a deli counter where they can grab a bagel and a cup of soup. They want to know their bagels are made fresh on the premises each day, with a wide variety of cream cheeses. Men also like elements that mention catering and one stop shopping, again the convenience factor. So what do males look for in a bagel store that acts as a restaurant? It looks like they want bagels that are fresh and remind them of a New York deli with all the convenience of a neighborhood shop.

6.   Now turn to females. Females are more interested in a place they can have a meal with their friends, rather than a quick bagel to eat on the run as a snack. Ambience is important to females. They are looking for a place to meet for a light, healthy meal with friends, whether it is a quick bite or a relaxed meal. They are not overly motivated by mentions of bagels and fillings. It is more about the meal for the ladies. Perhaps a bagel is part of it if it is accompanied by fresh fruit and yogurt or a cup of soup.

24.9   What goes together: does a light healthy meal differ from a healthy meal alternative?

The elements in the concept vary in a systematic way. Furthermore, each respondent saw a different set of concepts. Thus, for any silo it becomes straightforward to sort all of the concepts across the full set of respondents into the seven strata. These seven strata include all those concepts with no element from the silo, all the concepts with one specific element from the silo, and so on. Since a silo has six elements there are six such strata, and as mentioned, the seventh stratum wherein the concept is absent all elements from the silo.

Thus, we have the opportunity to sort the data set of concepts into different strata. One stratum comprises all concepts which have the element 'light healthy alternative to a meal'. For the remaining five silos, six elements per silo, we know which elements were present in each test concept. How strongly does each element perform in the presence of this common element? Now, do the same analysis, this time looking at the stratum with the element 'light healthy meal'. How do the same remaining elements perform? The only difference between the two analyses is the specific element, whether this element is the light healthy alternative or the light healthy meal, respectively. The name for this specific analysis is 'scenario analysis' (Gofman, 2006).

We see the results of the scenario analyses in Table 24.5. We deal with two sets of test concepts. The first set comprises all concepts having the element 'light healthy alternative to a meal'. The second set comprises all concepts having the element 'light healthy meal'. These two sets of concepts are mutually exclusive, and constitute only a fraction of the total set of concept. By creating a model that shows the performance of the 30 elements in the remaining five silos, we can compare the performance of specific elements in the presence of each of these two elements (light healthy alternative versus light meal, respectively).

In the presence of the meal alternatives the strong performing elements are about the food, whereas in the presence of the light lunch the strong performing elements are about location and ambience. Thus, the elements in the concept interact. The presence of one element (e.g., light healthy meal) affects how other elements in that concept will perform. Through the scenario analysis, one identifies how a statement about the nature of what is being eaten and why drive responses to other elements in the same concept. We have, therefore, a glimpse into the mind of the respondent as the respondent integrates information about a restaurant and meal occasion.

24.10   Creating optimal concepts using the database

A key benefit of experimental design is the ability to recreate new concepts, not before seen, by combining elements (Moskowitz *et al.*, 2002). If, in fact,

Table 24.5 Scenario analysis showing those elements which perform well in the presence of two specific elements defining strata. One stratum is defined by the element that talks about the light meal, whereas the other stratum is defined by the element talks about an alternative to the meal.

	Try our healthy lunch alternatives… bagel, fresh fruit and yogurt	When you want a light, healthy lunch… bagel and a yogurt
	Lunch alternative	Light, healthy lunch
Additive constant	33	38
Strata of concepts featuring 'lunch alternative' **Winning elements (strong performers)**		
Our scrumptious bagels will give you the energy you need	22	−2
Bagels are low in fat and only 170 calories	17	−7
All bagels made fresh each day on premises	13	1
Baked fresh every day from scratch…just the way you like them	13	0
Reduced-fat whipped cream cheeses…light and luscious	11	6
A light, casual place where you can get a really satisfying meal	9	−1
We offer a variety of bagels…for every taste bud and every mood	9	−1
When you crave the delicious taste of a soft bagel	9	−7
Lower carb bagels…for your special needs	9	0
Strata of concepts featuring 'Light and healthy lunch' **Winning elements (strong performers)**		
Conveniently located right near your home	3	21
A casual, friendly place for a meal with your friends	0	17
A clean, bright restaurant for a quick bite or a relaxed meal	7	15
The taste of fresh New York deli bagels	−1	10
When you are looking for a fast healthy lunch	8	8

a great deal of innovation is recombining components into new mixtures, hitherto unknown, then the database we have developed in Tables 24.4 and 24.5 is perfect for innovating. The exercise we followed tested the elements in combination, with each element appearing against many different backgrounds. Therefore, those elements that perform well should do so because they perform well in general, and not because they performed well in a specific, and limited set of combinations. We need not be concerned about taking elements that were tested in isolation, and somehow putting them together for the first time in a new concept. In that latter case, there is always the possibility that the elements in isolation will interact with each other in unknown ways when combined in a concept. Our permutation strategy, which presents each person with a unique set of combinations, guards against this unhappy possibility. Furthermore, the scenario analysis in Table 24.5 permits us to uncover any special positive or negative interaction among elements, should such interaction be of interest.

24.11   Future trends

This chapter began with a discussion of ideas as foundations for concepts, then moved to the evaluation of complete concepts through ConScreen (concept screening), and finished with an example of systematic design of concepts. Throughout, we have seen the value of a structured approach, whether the structure be in the elements or in the assemblage of elements into test concepts. It is the authors' opinion that simply screening concepts as fixed combinations without any systematic design may at times and perhaps fortuitously uncover winning ideas and even foreshadow new trends. However, such concept screening does not really create a bank of knowledge about what ideas drive concepts, nor does it create a science. Rather, conventional concept screening creates a database of ideas, from which one might, if sufficiently disciplined in data collection and analysis, estimate the likely success of future ideas. Such success is measured by putting the 'metrics' of the test concept into a model, which in turn estimates the dependent variable, such as sales.

What is missing, of course, is the knowledge of how to create these winning concepts, or indeed why they are winning. One might instruct the respondents to 'circle the ideas' that appeal to them, as some practitioners do in focus groups or in the Web interview. Yet, this is again a post-hoc approach, which provides diagnostics of the 'patient', but not a science of 'patient health'. From what is circled we may learn, or at least think we learn, why the concept 'works' or 'does not work'. Yet it is not quite clear that the method really uncovers things that are not obvious. We do not know how to create new concepts, nor do we have a database of ideas from which to begin. We do have the ability to generate ideas as needed, through focus groups and ideations, but this 'just in time' development may not be the best way to found a true science of concepts.

Perhaps the first thing that is needed is the wider use of experimental design for concepts, to understand the different aspects of the meal. We see some evidence for the increasing acceptance of designed experiments in business schools, where the ideas of conjoint analysis have been accepted for 35 years, and where the students find conjoint analysis to be quite useful in their business practice. We may expect to see more of these studies as experimental design of ideas becomes increasingly popular in the universities.

We may expect advances in *concept science* (italics ours) from the adventurous few consumer-research companies that write so-called 'white papers' and present them at conferences. Sadly, we probably will see fewer advances from client companies in the food and hospitality businesses because practitioners in the companies are prohibited from publishing proprietary data. Nor will we see many advances from consumer research companies, which are accustomed to executing studies having 'normative databases' that are used to interpret the results. Reliance on norms means focus on a single performance score, not the search for underlying patterns to understand 'how nature works'.

24.12   Databasing the mind and developing this database for meals

Beyond the acceptance of conjoint analysis, there needs to be more effort put into creating a database of knowledge for meals. The data presented in Tables 24.4 and 24.5 provide just the beginning. There is much more to do. Some similar work was reported by Moskowitz *et al.* (2005), where they dealt with a fast food restaurant. Other work was reported by Moskowitz and Reisner (2008), dealing with health issues and a deconstruction of information about restaurants taken from the Internet. Finally, still other work was reported by Moskowitz *et al.* (2002) who deconstructed information about quick serve restaurants, and tested these snippets of competitive information by experimental design. These studies are just initial forays into the creation of a science that would be later called 'mind genomics (Moskowitz *et al.*, 2005; Moskowitz *et al.*, 2006).

In order to create this database it is important to develop a way of thinking about connected databases, using experimental design and a common structure of concepts. The first such approach was the It! databases begun in 2001 with a food database called Crave It!™ (Beckley and Moskowitz, 2002; Moskowitz and Gofman, 2007). The ultimate objective of the approach was to provide the marketer and the product developer with a data set comprising elements and utilities for different products. With such a dataset it would become easier to innovate by first understanding how a product or service category 'works' with empirical data, and then subsequently providing a tool for the creation of newer ideas (Moskowitz *et al.*, 2002).

A second approach, Innovaid™ (Moskowitz, 2007), is more appropriate for our treatment of meals as combinations of ideas. The approach deals primarily with actionable elements, in contrast to brands and emotions, which the Crave It!™ database was designed to do. Innovaid™ was designed to spur innovation by presenting a set of elements to be used in concept creation. The Innovaid database comprises three sections, one for food, one for beverage, and one for lifestyle. The elements in Table 24.4 above come from the lifestyle section, specifically the elements dealing with a bagel restaurant. The lifestyle section of the Innovaid™ database deals with restaurants, and other dimensions of the meal, such as the nature of a store that sells prepared foods for immediate home consumption as a meal.

Each of the Innovaid studies in the database comprises six silos, and each silo in turn comprises six elements. The database of elements is available for anyone to use. When consumers participate in the different studies, the result will be a database showing how each of the 36 elements for each eating situation 'drives' the response. In addition, each of the studies in the database comprises an extensive classification questionnaire, to define who the respondent is, and the attitudes towards meal situations. Table 24.6 shows the elements for three of these Innovaid studies.

24.13   Conclusions

We have looked at the development of concepts for the meal in two ways. First, we have dealt with the issue of concepts as stimuli to be evaluated. We have seen how concepts can be created, either as gestalts, or systematically varied, and how they can be evaluated by consumers to guide the business. This first part belongs properly in the world of business.

The second part of the chapter deals with the systematic variation, testing, and analysis of concepts. Traditionally, the primary goal of such systematic exploration is to learn about what makes a good concept in the hopes of increasing concept performance in a consumer test through this knowledge. We introduce the notion of experimental design of concepts, however, for another purpose as well, namely to create a science of the consumer/customer mind, based upon patterns of responses to systematically varied stimuli. Such systematic exploration opens up the possibility of archival databases of concept elements, as well as information about what works, and what does not in the mind of the consumer. With such information, concept research may greatly contribute to the science of food, to the study of eating, and to understanding the myriad subjective factors of hospitality during the meal occasion.

Table 24.6 Silos and elements for three Innovaid studies relevant for meals (beachside restaurant, food magazine, and a supermarket with a prepared food section)

	Beachside restaurant	Food magazine	Store – prepared food section
	Silo 1 – Objectives		
O1	More than hot dogs and French fries… beachside dining with a twist	A magazine where healthy food is the star…give up the fat, not the flavor	For the gourmet in a rush, who still wants to have a good dinner
O2	Boxed lunches that offer healthy eating alternatives…salads, grilled vegetables, fresh fruit	A magazine focusing on those who want to eat right…tasty recipes that are good for you	New choices for dinner – for the sophisticated side of you
O3	Vegetarian menu available…so you can eat well and feel healthy	For today's health-conscious consumer… give up nothing except the fat and calories	Healthy choices for your family, ready for you to take home
O4	Menu selections with healthy alternatives…feel good eating well	The magazine that shows you how to eat well…healthy recipes and healthy tips	For the shopper who wants something exciting for dinner
O5	Healthy yet delicious menu items	The magazine of great food, good health and smart living	Quickly heat in the microwave and dinner is ready
O6	Grab a healthy meal or snack for the beach	For those who like to eat right…recipes that not only taste good but are good for you	For the person always on the run…pick up a quick, healthy lunch or dinner
	Silo 2 – Ambience		
A1	Air-conditioned interior	The photography displays the food to advantage…colorful "good-enough-to-eat" photos	A beautiful display of good food, all ready to eat
A2	Comfortable tables and chairs outside	Written with high-quality journalism and impeccable nutrition credentials	Conveniently located right in your supermarket, no need for extra stops
A3	Tropical surroundings	Recipes include preparation time, ease of preparation and complete nutritional information	Everything looks so good, you want to try it all!
A4	Reggae music in the background	Easy access index on the front cover…see what's inside and know where to find it.	One quick stop and you have a healthy meal that your family will love
A5	Relaxed dress code…come as you are	Tear-out recipe cards in every issue…add them to your collection	An entertaining and scintillating food department

A6	Bare feet always welcome	Recipes are printed on special paper…easy to wipe clean	Try something new…no need to worry about how to prepare it…it's ready to eat

Silo 3 – Services

S1	An ice cream parlor with low fat/no fat/sugar free ice creams	Contains healthy recipes that you can incorporate into your daily cooking	A large variety of vegetarian dishes
S2	Call ahead to place your order and its ready when you are to hit the beach	Each issue contains a series of healthy recipes that take 30 minutes or less	Soups, salads, main dishes…we have it all
S3	Amount of protein, carbohydrates, fat and calories are listed next to each selection	Features creative, easy-to-follow recipes with the best nutrition information available	Prepared with only the freshest, high quality ingredients
S4	Light snack menu available…whole wheat pretzels, veggie chips, low fat popcorn	Every issues features exciting tips from dieticians and other health experts	Soups…four types made fresh every day
S5	Uses only fresh, local produce purchased daily	You'll find tips on how to cook with fewer calories and less fat	A salad bar where everything is crisp and fresh
S6	Atkins, South Beach and Weight Watchers menu items served	Features exciting culinary innovations… keep up with the latest happenings	Bread bar…a wide selection of individual loaves and rolls

Silo 4 – Special features

SF1	Limbo Lane…. Beat the Height of the Day for a free dessert	When you want to find good recipes using healthy ingredients	Immaculately clean…just the way you want it
SF2	Waitered cabana tables on the beach… because you deserve some extra pampering	If you're looking for a healthy food magazine to cook with, this is the one	Freshly prepared daily
SF3	Sunset smoothies….hot drink specials for those cold summer nights	When you want recipes that are healthy, without sacrificing the taste	All natural ingredients make a delicious meal healthy, too
SF4	Saturday Night Luaus….Every Saturday during the summer	When you're concerned about what you eat and how to prepare it	Heart healthy dishes for those on special diets
SF5	Summer Loving….Singles leave a message for a fellow beachgoer on our Maui Meeting Wall	Great ideas for using leftovers … and still make a good meal	Organic meals are our specialty
SF6	Family Fridays featuring Castle Kahuna sandcastle contest	A fun way to learn about healthy new products that are on the market	Buy just what you need for each of your family members…please everybody

Table 24.6 Cont'd

	Beachside restaurant	Food magazine	Store – prepared food section
	Silo 5 – Benefits		
B1	You can still have reliable, healthy selections for your special diet when at the beach	Special monthly feature 'Cooking for a Crowd'…recipes that will delight your guests	Freshly cooked vegetables ensure a vitamin enriched meal
B2	Beachside take out that fits your healthy lifestyle	Special monthly feature 'Cooking for One'…eating alone doesn't have to be boring	A beautiful salad bar – a healthy accompaniment to any meal
B3	Treat your body with fresh, healthy ingredients	Special monthly feature 'Recipes from the Readers'…share your favorites	Homestyle food for a nutritious meal…just like you would prepare for yourself
B4	Take out with confidence knowing your nutritional needs will be met	Special monthly feature 'Cooking with Children'…fun recipes to share with your kids	Low in fat…so its good for you
B5	Knowing you have nutritional choices	Special monthly feature 'Cooking with the Family'…a meal the whole family can prepare and enjoy	No artificial ingredients or preservatives…made from scratch
B6	To know that you can still eat healthy even at the beach	Special monthly feature 'Meal of the Month'…the recipes and shopping list for a complete meal	A list of all ingredients is posted with every dish
	Silo 6 – Additional services		
AS1	Fresh tropical drinks offered… alcoholic and non alcoholic	Buy it at your local newsstand	Let us cater your next party…your friends will love how good our food tastes
AS2	Meatless meals available for a healthier heart	Look forward to getting your copy every month in the mail	Satisfaction guaranteed…let us know how you feel about our food
AS3	Menu items can be altered to your diet specifications	Check out the magazine online…anytime you want, anywhere you want	Tuesday night specials…save on all salad bar items
AS4	Fruit and vegetable smoothies as a beverage option	Available at your favorite bookstore	Friday night specials…pick any three items for a complete meal
AS5	A menu which allows you to mix and match your food selections	Pick up a copy at your favorite health food store	Kids section…special foods your children will love
AS6	Newspapers, magazines and paper-backs sold for beachside reading	Pick up a copy near the check-out register at your local supermarket	Call in your order and we will have it ready for you to pick it up and run home

24.14   References

Anderson N H (1970), 'Functional measurement and psychophysical judgment', *Psychological Review*, **77**, 153–170.
Beckley J and Moskowitz H R (2002), Databasing the consumer mind: The Crave It!, Drink It!, Buy It!, Protect It! and the Healthy You! Databases. Paper presented at the *Institute of Food Technologists Annual Meeting*, Anaheim.
de Bono E (1995), 'Serious creativity', *The Journal for Quality and Participation*, **18**(5), 12–18.
Box G E P, Hunter J and Hunter S (1978), *Statistics for Experimenters*, New York, John Wiley.
Callahan R, Ishmael G, and Namiranian, L (2005), *The Case For In-The-Box Innovation*, ESOMAR Innovate!, Paris
Cooper R G (1993), *Winning at new products: accelerating the process from idea to launch*. 2nd ed., Reading, MA, Addison Wesley.
Dennis, Alan R and Joseph S. Valacich (1993), 'Computer brainstorms: more heads are better than one', *Journal of Applied Psychology*, **78**, 531–537.
Flores L (2005), 'What can research learn from biology?' *Admap*, (Special Issue on Innovation)' November, 45–48.
Gofman A. (2006) 'Emergent scenarios, synergies and suppressions uncovered within conjoint analysis'. *Journal of Sensory Studies*, **21**, 373–414.
Green P E and Srinivasan V (1980), 'A general approach to product design optimization via conjoint measurement', *Journal of Marketing*, **45**, 17–37.
Luce R D and Tukey J W (1964), 'Conjoint analysis: A new form of fundamental measurement', *Journal of Mathematical Psychology*, **1**, 1–36.
MacElroy B (2000), 'Variables influencing dropout rates in Web-based surveys', *Quirks Marketing Research Review*, www.quirks.com, Paper 0605.
Moskowitz H R (2007), 'Consumer-driven concept development and innovation in food product development', in *Consumer-led Product Development* (ed. H. MacFie), Cambridge, Woodhead, 142–182.
Moskowitz H R, Flores L, Beckley J, Mascuch T, Cleveland C and Ewald J (2002), 'Crossing the knowledge and corporate to systematize invention and innovation', *Esomar Congress*, Barcelona.
Moskowitz H R, German J B and Saguy I S (2005), 'Unveiling health attitudes and creating good-for-you foods: the genomics metaphor, consumer innovative web based technologies', *CRC Critical Reviews in Food Science and Nutrition*, **45**, 165–191.
Moskowitz H R and Gofman A (2007), *Selling Blue Elephants: How to Make Great Products That People Want Before They Even Know They Want Them*, Upper Saddle River, NJ, Wharton School Publishing.
Moskowitz H R, Gofman A, Beckley J and Ashman H (2006) Founding A New Science: Mind Genomics. *Journal of Sensory Studies*, **21**, 266–307.
Moskowitz H R, Itty B, Manchaiah M and Ma Z (2002), 'Learning from the competition II: A case history dissecting in-market quick-serve-restaurants communications through conjoint analysis', *Food Service Technology*, **2**, 19–33.
Moskowitz H R, Porretta S and Silcher M (2005), *Concept Research in Food Product Design and Development*, Ames, IA, Blackwell Publishing.
Moskowitz H R and Reisner M (2008), 'Using high-level consumer research to create low-caloric and pleasurable food concepts and products', in *Energy is Delight: A Brain-to-Society Approach to the Prevention of Childhood and Adult Obesity. Volume 2: Behavioral, Socio-Cultural, and Market Adaptation for Pleasurable Survival in the Modern World of Plenty* (ed. L. Dube, A. Bechara, A. Dagher, A. Drewnowski, R.F. Yada). In press.
Pine J and Gilmore J (1999), *The Experience Economy*, Boston, Harvard Business School Press.

Schifferstein H N J (ed) (2007), *Product Experience*, Elsevier.
Silverthorne S (2002), 'Time pressure and creativity: why time is not on your side, Q&A with Teresa M. Amabile and Leslie A. Perlow', *Harvard Business School Working Knowledge,* Item 3030.
Van Kleef E and van Trijp H C M (2007). 'Opportunity identification in new product development and innovation in food product development'. In: *Consumer-led Food Product Development* (ed. H. MacFie), Chapter 14, pp. 321–341, Cambridge, England, Woodhead Publishers.
Wittink D R, Vriens M and Burhenne W (1994), 'Commercial use of conjoint analysis in Europe: results and critical reflections', *International Journal of Research in Marketing*, **11**, 41–52.
Zaltman, G. (1997), 'Rethinking market research: Putting people back in'. *Journal of Marketing Research*, **34**, 424–437.

Part VIII

Further perspectives on meals

25

Meals, behavior and brain function

K. E. D'Anci and R. B. Kanarek, Tufts University, USA

Abstract: The role of meals in mood and cognitive performance is discussed. An overview is given of the nature of mood and cognition in psychological research followed by a description of how individual macro- and micronutrients affect brain functioning. The chapter concludes with a discussion of individual meals and how mental function is affected by meals, and what types of meal composition have beneficial effects on mood and cognition.

Key words: mood, cognitive performance, meals, nutrients, brain function.

25.1 Introduction

Curiosity about the ways in which foods affect brain functioning, emotional well-being and behavior is long-standing. Throughout history, foods have been consumed not only for their nutritional value, but also for their proposed ability to improve disposition and mental functioning. For example, foods as diverse as figs, honey, chocolate, oysters, and onions have been relished by those who believe that these foods stimulate sexual prowess, while for several thousand years, coffee and tea have been consumed for their capacity to decrease fatigue, increase alertness, and stimulate mental performance (D'Anci and Kanarek, 2006). Moreover, across cultures and eras, herbs and spices such as turmeric, garlic, ginseng and ginger have been prescribed for their value in enhancing a number of mental functions, including memory and mood, while more recently, fish has been promoted as a 'brain food'.

Within the last 50 years, scientific investigations have confirmed many of these beliefs about the roles of foods in modulating brain activity, and ultimately observable behavior. The present chapter will explore the empirical evidence for relationships between foods and mental functioning. More specifically as this is a volume dedicated to meals, rather than individual nutrients, the chapter will concentrate on the impact of meals in determining feelings and behavior.

25.2   Mood and cognition

Before assessing the role of meals in moderating brain functioning and its consequences, it is important to provide definitions of mood and cognition. Mood and cognition are the reportable or observable facets of brain function. Mood, which is defined as 'a temporary, but relatively sustained and pervasive affective state' (Colman, 2001) covers an array of affective states ranging from calm to rage, alertness to fatigue, and elation to depression. Most research exploring the role of meal intake on mood has used self-reported questionnaires as a measure of mood state. One of the most commonly used questionnaires, the Profile of Mood States (POMS; McNair *et al.*, 1971), asks individuals to rate their feelings of fatigue, vigor, confusion, anger, depression, and tension. In contrast, other questionnaires measure specific attributes of mood (e.g. alertness, depression or anxiety). Other standardized mood scales include the Stanford Sleepiness Scale (Hoddes *et al.*, 1973), and the Activation/Deactivation Adjective Check List (Thayer, 1989), which rates energy, tiredness, calmness, and tension. To gain a more complete picture of the effects of food on mood, it is recommended that a combination of scales be used, and that the specific scales and mood variables that are assessed be considered when interpreting findings.

Cognition refers to the processes that we typically think of when we use the term 'higher mental functions'. These processes can be clustered into several domains including learning, memory, language skills, reasoning, strategic thinking, and attention. Cognition is frequently assessed by observable performance. For example, learning and memory may be determined by an individual's ability to learn, and recall a list of words, or the locations of countries on a map, while verbal comprehension may be measured by evaluating an individual's understanding of a complex reading passage.

While mood and cognition are often studied individually, they do not function independently, but rather work in tandem with one affecting the other. Mood, for example, can be altered by cognitive performance, e.g., feeling 'down' after doing poorly on a test. Cognition can, in turn, be affected by mood; an individual experiencing mild anxiety may perform well on a test, but high anxiety in the same individual may result in poorer cognitive performance.

It is important, in any discussion of the interaction between cognition and mood to distinguish performance from ability or aptitude. For example, individuals suffering from fatigue, considered a mood state in psychological research, may show impaired performance on an arithmetic problem solving task, a measure of cognitive performance. Overall IQ or intelligence, however, is not expected to fluctuate with mood states.

The relationship between food and mental functioning is a two-way street. As discussed in this chapter, intake of different foods can alter mood and cognitive performance. However, it should also be recognized that mood and cognition can have a significant impact on food intake and food choices.

People eat, or don't eat, for a variety of reasons that are not limited to hunger. People may eat in response to a wide array of feelings including happiness, fatigue, boredom, or anger. On a meal-by-meal basis, the effects of mood on food choices may be modest. After a bad morning, a lunch of ice-cream sundaes may seem fine, but then afternoon snacks and dinner tend to follow normal patterns. However, in some populations, mood can have long-lasting effects on food and meal intake. Individuals with depression related to body image, may become anorexic or bulimic (Salbach-Andrae et al., 2008; Strauman et al., 1991), while elders living alone may limit nutrient intake as a result of loneliness (Ferry et al., 2005; Pliner, Chapter 9 in this volume).

Cognition also plays a major role in food intake. Our memory for a previous meal affects our choices for the next meal. One's overall diet may be somewhat monotonous: eggs and toast for breakfast, soup and sandwich for lunch, a hot meal for dinner. However, memory of what was eaten before will affect subsequent food choices: Memory of an egg and toast breakfast would probably lead to choices *other* than egg salad for lunch or an omelet for dinner. Cognition can also proactively affect food intake. Beliefs about the nutritional value or caloric content of a meal will affect food choices within that meal. For example, women who were provided food labelled as low-fat consumed more calories in a subsequent meal than those told that the foods were higher in fat (Shide, 1995). In daily life, people apply many cognitive heuristics in selecting, preparing, and eating foods including memory for foods liked or disliked, beliefs about the nutritional value of foods, and learning new cooking techniques.

25.3   Nutrients, mood, and cognition

Examination of the role of meals in moderating mood and cognitive behavior is challenging as most meals are comprised of a variety of foods which themselves contain a number of components which could affect mental functioning and behavior. Additionally, the specific effects of any one food component can be influenced by other food components consumed within a meal. For these reasons, information on the effects of individual nutrients on brain function and behavior will be reviewed briefly before we assess the role of meals in moderating mood and behavior. Nutrients eaten in a meal include macronutrients, such as carbohydrate and protein, micronutrients, such as vitamins and minerals, and other compounds found in foods including caffeine and phytochemicals.

25.3.1   Macronutrients
Carbohydrates
Carbohydrates, found in fruits, vegetables, grains, and honey, as well as processed foods, are broken down during the digestive process to glucose,

which is the primary fuel for the brain. After absorption from the gastrointestinal tract, glucose is carried in the blood stream to the liver, brain and other tissues. As the brain does not store glucose and lacks the enzymes necessary to convert amino acids and fats into glucose, the brain is dependent upon circulating blood glucose levels for fuel, and experiences consequences related to fluctuations in blood glucose levels (Benton and Nabb, 2003; McCall, 2002; Morris and Saril, 2001; Sieber and Trastman, 1992; Wenk, 1989).

There is a growing body of evidence to support the proposition that alterations in glucose intake, and consequently blood glucose levels translate into changes in mood and mental performance. An acute reduction in blood glucose levels (hypoglycemia), as might result from failure to consume sufficient amounts of foods, can lead to feelings of anxiety, dizziness and confusion, and a decline in the performance of mental tasks (Hoyland et al., 2008; D'Anci and Kanarek, 2006). The effects of acute hypoglycemia are typically reversed soon after eating. On a longer term basis, impairments in peripheral and central glucose utilization, such as occur in diabetes, are often accompanied by negative mood states and impairments in cognitive performance (Elias et al., 2005; Messier et al., 2004).

While reductions in the availability of glucose leads to decrements in behavior, acute intakes of glucose, other sugars and foods containing significant amounts of carbohydrates which increase blood glucose levels and the supply of glucose to the brain are associated with improvements in mood and cognitive behavior. In particular, intake of high-carbohydrate foods enhances subjective reports of energy and facilitates performance on tests requiring sustained attention (Benton and Nabb, 2003; Hoyland et al., 2008; Mahoney et al., 2007; Markus, 2007; Messier 2004). These results have contributed to the rapid rise in the intake of sports drinks to improve both physical and mental performance (Winnick et al., 2005).

Protein

Protein is critical for health and development. Failure to consume sufficient amounts of protein is associated with poor cognitive development, impairments in behavior, and deficits in motor skills (Grantham-McGregor and Baker-Henningham, 2005). Moreover, protein-malnourishment in children is almost always accompanied by negative mood states including fatigue, depression and a lack of motivation. In many children, one of the first signs of recovery from malnutrition is a smile (Worobey, 2006).

Intake of dietary protein may influence brain function and behavior via the roles of amino acids in neurotransmitter synthesis. For example, it has been proposed that intake of high-protein foods containing the amino acid, tyrosine, the dietary precursor of the neurotransmitters, dopamine, norepinephrine and epinephrine could improve performance in working memory and information processing tasks in individuals subjected to environmental stressors (Worobey and Kanarek, 2006). Other research has demonstrated that varying dietary levels of tryptophan, the amino acid precursor for the neurotransmitter, serotonin, alters behavior, particularly in individuals experiencing negative mood states such as depression or anxiety. For example, consumption of α-lactalbumin, a whey protein containing relatively large amounts of tryptophan, increased the plasma ratio of tryptophan to the other large neutral amino acids, enhanced performance on a memory-scanning task, and decreased feelings of depression in stress-vulnerable individuals (Markus *et al.*, 2002). Additionally, intake of α-lactalbumin produced positive effects on abstract visual memory in individuals with and without a history of depression (Booij *et al.*, 2006). It has been proposed that the effects of α-lactalbumin on mood and cognitive behavior are the result of diet-induced increases in brain serotonin, a neurotransmitter important in maintaining mood and in learning and memory.

Fats
Dietary fats, or lipids, contribute to brain function in a variety of ways. Approximately 2/3 of the dry weight of the brain is comprised of lipids. Fatty acids are important in neuronal membrane structure, neurotransmission, and ion channel function – all of which contribute to brain functioning (Worobey, 2006). Recent work, however, has demonstrated that not all fats are equal with respect to their role in brain functioning. Diets rich in monounsaturated and polyunsaturated fats are associated with improved brain functioning when compared with diets high in saturated and trans-fats. As an example of the positive aspects of fat intake, docosahexaeonic acid (DHA) and eicosapentaenoic acid (EPA) omega-3 fatty acids, found in fatty cold-water fish, are particularly important for brain development and function (Cohen *et al.*, 2005; Fleith *et al.*, 2005; Worobey, 2006). Low-intake or deficiencies of these compounds have been associated with cognitive decline, particularly Alzheimer's disease (Freemantle *et al.*, 2006). In comparison, consumption of DHA and arachidonic acid (AA: an omega-6 fatty acid) improves developmental scores in infants (Fleith *et al.*, 2005), while maternal intake of fish during pregnancy has been associated with higher visual attention, language, and psychomotor development scores in the offspring. On the other end of the age spectrum, intake of fish high in omega-3 fatty acids may slow the development of dementia and mild cognitive impairment frequently observed in the elderly (Morris *et al.*, 2005; Lim *et al.*, 2006).

25.3.2   Micronutrients and phytochemicals
The beginning of the 20th century saw the discovery of vitamins, and the use of vitamins to treat deficiency diseases. Good nutrition started to be understood in terms of providing balanced combinations of nutrients and led to the promotion of balanced meals. Nutrition policy began to focus on the quality of the diet, rather than quantity. The USDA developed its first food guide in 1916 and provided basic food groups from which meals should be made to maintain and promote health. Meals began to be considered in terms of more

than physical health, and the importance of good nutrition for mental performance also began to be recognized and studied (Brozek, 1947; Brozek, 1955; Sherman, 1950; Peraza, 1946). Populations particularly at risk for cognitive impairments in connection with deficiencies in micronutrient intake include the elderly and alcoholics, both of which groups may not consume sufficient amounts of food or suffer from problems that decrease nutrient absorption, and very young children who have specific nutrient requirements for brain development.

B vitamins particularly vitamins B1, B3, B6 and B12, are important players in the maintenance of cognitive ability and mood (D'Anci and Rosenberg, 2004, 2005; Kaplan *et al.*, 2007; Tepper and Kanarek, 2006). Less than adequate intakes of these vitamins are frequently accompanied by neurological problems, impairments in cognitive performance, and mood disturbances. For example, clinical manifestations of vitamin B1 deficiency (beri-beri) include decreased initiative, increased irritability, fatigue, depression and confusion, and problems with both short- and long-term memory. Inadequate intakes of vitamin B3 lead to the development of pellagra which is associated with sleeplessness, signs of emotional instability, loss of memory, and severe depression (Tepper and Kanarek, 2006). Vitamin B12 deficiency is particularly prevalent in older people and may contribute to the neurological deficits and cognitive impairments often seen in this population (Morretti *et al.*, 2004; Lewis *et al.*, 2005). Further support for a role of these vitamins in maintaining optimal brain functioning comes from the observation that dietary intake of vitamins B6 and B12 are positively correlated with better memory performance in adults (Bryan and Calvaresi, 2004).

A deficiency of another B vitamin, folate, is often accompanied by neurological problems and mood disorders such as depression. It has been established that mothers consuming insufficient amounts of folate during pregnancy have a higher risk of giving birth to infants with neural tube defects than adequately nourished mothers. Moreover, owing to decreased absorption, dietary inadequacies, and drug interactions, the elderly are at particular risk for folate deficiency. Although severe folate deficiency is rare in the US, particularly after flour fortification with folic acid, subclinical deficiencies may prove to have significant consequences in health status across the life span. It is recognized that people with 'low-normal' levels of folate may be at risk for some of the negative consequences of folate deficiency (e.g. Lindeman *et al.*, 2000; Bottiglieri *et al.*, 2000). Importantly, such marginal deficiencies and related negative sequelae may be reversible with dietary improvement and/or supplementation.

Increased oxidative stress is hypothesized to impair cognitive functioning, particularly with aging. Dietary intake of antioxidants such as vitamins C and/or E may offset the cognitive decrements which have been associated with putative oxidative stress (Martin *et al.*, 2002a). Vitamin E, found in vegetable oils, nuts, and seeds, contributes to membrane stability and protects against free radical damage in cellular membranes. Vitamin E is also important in synthesis of neurotransmitters and in neurotransmission. A deficiency of vitamin E may result in brain cell damage and impaired neural communication (Martin *et al.*, 2002a; Martin *et al.*, 2002b). Vitamin C, found in citrus fruits and a variety of other fruits and vegetables, functions as a cofactor in several enzymatic reactions. Low levels of both C and E have been associated with mild cognitive impairment, dementia, and Alzheimer's disease (Rinaldi *et al.*, 2003), while diets rich in these antioxidants may contribute to the maintenance of cognitive functioning (Berr, 2002; Scarmeas *et al.*, 2006).

In their comprehensive review on vitamin and mineral influences on mood, Kaplan and colleagues (2007) describe several possible models for nutrient-mood interactions. Mood disturbances may be related to inborn errors of metabolism; in such an individual, there may be adequate intake of a nutrient, but decreased absorption or altered metabolism affects nutrient efficacy. Deficient methylation reactions can influence mood through alterations in neurotransmitter synthesis and function. Frank nutrient deficiency can alter gene production and can produce diseases related directly to deficiency, such as beriberi. Absorption of nutrients can also be influenced by age, illness, and medication, all of which can, in turn, impact mental functioning.

Ingestion of whole fruits and vegetables is positively correlated to cognitive performance (Kang *et al.*, 2005; Martin *et al.*, 2002a) and reduced risk of depression (Sánchez-Villegas *et al.*, 2006). It is hypothesized that a diet rich in fruits and vegetables contributes to brain health not only as a result of improved vitamin status, but also as a result of the phytochemicals and antioxidants found in these foods. The role of phenolic phytochemicals in health is becoming a highly popular avenue of research. Phenolic compounds are found in plant foods and extracts such as wine and grape juice (resveratrol), chocolate (epichatechin), blueberries (anthocyanins), herbs and spices (oregano, turmeric, fenugreek polyphenols), and green tea (green tea polyphenols). Recent experimental studies suggest that chocolate may benefit mental performance on verbal and visual memory tasks, and improve impulse control (Raudenbush *et al.*, 2008), while epidemiological studies indicate that diets rich in green tea (Kuriyama *et al.*, 2006), and fruits and vegetables (Kang *et al.*, 2005) and red wine (Panza *et al.*, 2004) may reduce the risk for dementia and have protective effects on executive brain function throughout aging. Although much remains to be learned about phytochemicals and cognition, the prevailing hypothesis is that phenolics exert their benefits on health via their antioxidant capacity (Shetty and Wahlqvist, 2004).

In summary, the nutrients above are a shortlist of the many compounds found in our food that are known to affect brain function, neural communication, and brain structure. Individual foods with a myriad of macro- and micronutrients comprise each meal and every individual follows a unique meal pattern and diet. The richness of our diets contributes to differences in our physical and mental health, and contributes to cognitive function, and may protect against cognitive decline in aging.

25.4   Meal intake, cognition, and mood

While the effects of individual nutrients on brain function and behavior have been extensively studied, it must be recognized that people do not consume nutrients in isolation. Most people consume the majority of their food as meals which contain a variety of different nutrients. For example, even a humble snack of an apple and a handful of nuts is comprised of carbohydrates, fiber, vitamins and minerals, fats and protein.

For most individuals, meals provide the bulk of calories eaten during the day, with snacks adding additional, needed energy. Missing a meal increases hunger, irritability and distractibility, and decreases alertness and motivation (Neely et al., 2004). Hunger leads to cognitive decrements not only because of short-term changes in nutrient status, such as hypoglycemia, but also because attention shifts from tasks at hand to the physical signs of hunger and the drive to eat. In general, eating a meal reduces hunger, and attention then shifts back to the tasks at hand. Different meal compositions, however, produce different effects on alertness and sleepiness (Wells et al., 1998). For example, eating meals high in carbohydrates and relatively low in protein can result in sleepiness or fatigue (Fischer et al., 2002; Nabb and Benton, 2006), whereas the addition of protein improves alertness and cognitive performance (Fischer et al., 2002; Paz and Berry, 1997). Interestingly, meal influences on mood and cognition are not universal, but can vary according to age or gender. For example, women may be more susceptible to sleepiness following a high carbohydrate meal whereas men may be more likely to show calmness after the same meal (Spring et al., 1982–1983).

When considering research on meals and brain function, it is important to distinguish laboratory meals from meals eaten in real-world settings. By definition, experimental meals are artificial constructs of nutrients of interest, and experimental meals can have limitations with respect to palatability and acceptability to individual participants (this is similar to limitations described in Chapter 8 by Boutrolle and Delaruse). In the laboratory, meals are chosen and designed with control, consistency, and replication in mind. Food is carefully weighed and measured, and, in some cases, is developed especially for a given study. Nutrient content is balanced in different conditions, and meals are made to be as similar in terms of weight, caloric content, and palatability as possible for each participant. In some cases, test meals consist of suspensions of nutrients, akin to liquid meal replacement drinks. The benefits of these liquid suspensions include the ability to finely control nutrient composition, and to conceal from participants the true nutrient composition of the meal. Known as 'blinding', this concealment helps prevent experimental bias when participants and/or researchers may inadvertently affect study results through expectation. Liquid suspension meals have been criticized as being 'artificial' and far-removed from real-world meals, but they have utility and validity for meal-based research.

The majority of meal-based research in has focused on breakfast, although snacks and lunch have also received attention. The evening meal, dinner, or supper, has received relatively little attention regarding its effects on brain function and behavior. This pattern is, in part, driven by the considerable focus on school meal programs, and academic performance. Although the field has grown to include study of mental performance at all ages following all meals, breakfast remains the most studied area

25.4.1   Breakfast

Breakfast is considered by many people to be the most important meal of the day. After an overnight fast, the morning meal supplies energy needed for physical and mental activities. In seminal research on the effects of breakfast on physical and mental health, 'backward children', children who would now be classified as children with delayed mental development, were fed a balanced breakfast daily (Seymour and Whitaker, 1938). After two months, the children showed improvements in academic tasks such as arithmetic and English class. In a subsequent study, undernourished children were fed a daily breakfast for two months (Peraza, 1946). The breakfast, a dubiously appetizing meal of milk, an orange, dry yeast tablets and cod liver oil, improved many health markers in the children, including body weight and 'improved disposition for work'. Since these original studies, many experiments have confirmed the importance of eating breakfast in both children and adults, although this meal is commonly skipped (Gross et al., 2004).

Breakfast contributes significantly to the maintenance of adequate nutritional intake. Individuals who eat breakfast on a regular basis, on average, consume a more nutritious diet, have micronutrient intakes which better match the recommended daily allowances and are less likely to be overweight than those who forgo the morning meal (Rampersaud et al., 2005; Vermorel et al., 2003). Breakfast also provides nutrients that are essential for mental functioning and cognitive performance. Children and adults who consume breakfast perform better on tasks measuring problem solving abilities, logical reasoning, short- and long-term memory, and attention than those who skip breakfast (e.g. Kanarek, 1997; Mahoney et al., 2005; Nabb and Benton, 2006; Pollitt, 1995; Rampersaud et al., 2005; Wesnes et al., 2003; Wyon et al., 1997). Eating breakfast is also associated with increased vigor and reductions in fatigue (e.g. D'Anci and Kanarek, unpublished data; Nabb and Benton, 2006; Pasman et al., 2003). In older people, eating a complex carbohydrate for breakfast improved cognitive performance in people with poor memories (Kaplan et al., 2000), and eating fat, protein or carbohydrate after an overnight fast improved memory performance and executive performance in healthy older people (Kaplan et al., 2001). Both the energy content and nutrient content of a morning meal can alter the effects of breakfast on cognitive behavior. For example, research has shown that children who eat a high-calorie breakfast perform better on tests of creativity, physical endurance, mathematical reasoning, than those who consume a low-energy breakfast (Michaud et al., 1991; Wyon et al., 1997).

In recent years, the concept of the glycemic index of foods or glycemic load of a meal, the rate at which a food or meal causes blood glucose to rise and remain elevated, has become an important topic in research on the behavioral effects of breakfast. Generally speaking, foods with a lower glycemic index or with greater proportions of complex carbohydrates or protein may be more beneficial in improving cognitive performance and alertness than foods with a greater proportion of simple sugars (Benton *et al.*, 2003; Fischer *et al.*, 2002; Fischer *et al.*, 2004; Mahoney *et al.*, 2005). Boys and girls given oatmeal (low glycemic load) for breakfast performed better on spatial memory and short-term memory tasks and had higher attention than when given a ready-to-eat cereal (high glycemic load) or nothing (Mahoney *et al.*, 2005). In college-aged students, breakfast intake produced positive effects on spatial memory, short-term memory and attention, and memory performance was significantly better in those with better glucose tolerance or after eating a meal that produced slower, sustained release of glucose into the bloodstream (Nabb and Benton, 2006). Attention to the glycemic load of meals is of particular concern to individuals with type 2 diabetes. Diabetics must pay strict attention to blood glucose levels to maintain health. After meals with high glycemic loads, diabetics experience high levels of blood glucose, which has negative effects on cognitive performance, although mood remains stable (Greenwood *et al.*, 2003). Individuals with type 2 diabetes given a low glycemic breakfast have improved cognitive performance compared with when they were given a high glycemic load breakfast (Papanikolaou *et al.*, 2006).

There is growing evidence that the protein content of a breakfast meal plays a role in determining subsequent cognitive performance (Fischer *et al.*, 2002; Mahoney *et al.*, 2005). For example, using a laboratory liquid meal of isoenergetic suspensions of protein and carbohydrate with similar sensory properties, Fischer and colleagues (2002) found that short-term memory, attention, and reaction time performance was better following intake of a protein meal than after intake of a carbohydrate meal with a high to medium glycemic index. They concluded that the protein-rich meal was associated with less variation in glucose metabolism and/or higher metabolic activation and modulation of neurotransmitter synthesis.

Although numerous studies have demonstrated that breakfast intake can improve cognitive performance, not all studies have shown beneficial effects of breakfast on cognitive behavior. Variables such as the nutritional status of the individuals, the time between breakfast and mood assessment, cognitive testing, the type of tests employed, and the quantity and quality of the morning meal, must be taken into account when evaluating how breakfast affects mental functioning (Kanarek, 1997). With respect to nutritional status, for example, although breakfast intake benefits both well nourished and more poorly nourished individuals, breakfast's positive effects on cognitive performance are more pronounced in those suffering from less than adequate nutrient intake (Pollitt, 1995; Simeon and Grantham-McGregor, 1989). The time interval between breakfast and cognitive testing must also be considered when evaluating breakfast's importance for performing mental tasks. Two hours after a high-protein breakfast, older individuals reported being more tense and less calm than after a high-carbohydrate breakfast. Two-hours following a high-carbohydrate lunch these same individuals demonstrated impairments in cognitive performance and concentration relative to when they consumed a high-protein lunch (Spring et al., 1982–1983). In studies in Israel, students who were tested within an hour of eating breakfast did better on tasks measuring learning and memory than those tested after a two-hour delay (Vaisman et al., 1996). Conversely, in other studies, for children tested multiple times throughout the morning, the positive effects of breakfast on cognitive performance became more pronounced as the morning progressed (Ingwersen et al., 2006; Wesnes et al., 2003). In studies looking at protein to carbohydrate ratios, high carbohydrate breakfasts were associated with improved attention and reaction time for the first hour after meal consumption, but then performance dropped off at later time points. Following intake of balanced protein to carbohydrate ratio or high protein meals, performance improved after the first hour of ingestion. Finally, high protein intake was associated with the best performance in a short-term memory task over the full 3.5 h of testing (Fischer et al., 2002).

The majority of studies assessing the effects of breakfast on mental performance have only compared the effects of single meals. However, the few long-term studies which have measured the effects of school breakfast programs on academic performance and behavior have provided further evidence of the importance of breakfast. Children enrolled in school breakfast programs have lower rates of tardiness and increased rates of school attendance and improved attention and academic performance than children who either consume breakfast at home, or skip the meal altogether (Meyers et al., 1989; Murphy et al., 1998; Reynolds et al., 2001; Richter et al., 1997; Sampson et al., 1995).

25.4.2 Lunch

At mid-day, most people eat lunch. Lunch size can vary based on individual preference, culture, and available time. For some people, lunch is the larger of the two main meals of the day with a smaller evening meal, although the majority of people in the US eat a moderately sized lunch followed by a larger evening meal. While the 'post-lunch slump' is addressed in the following section on snacks, it is important to consider that food choices at lunch can either offset or exacerbate a mid-afternoon dip in performance.

Intake of lunch, in general is associated with decreased energy, and fatigue, particularly if the meal is high in carbohydrate (Smith et al., 1988), but is also seen with other meal compositions (Christensen and Redig, 1993; Smith et al., 1988; Wells and Read, 1996). This effect may be because of normal circadian rhythms in humans, as decreased mood and energy are frequently seen in the afternoon (Stone *et al.*, 2006). Men and women given a high-carbohydrate meal or high-protein mid-day meal showed reduced vigor and increased fatigue, regardless of meal composition (Christensen and Redig, 1993). Greater increases in fatigue were seen 2 h after the high-carbohydrate meal relative to no meal, further supporting the notion that high-carbohydrate meals promote sleepiness throughout the day. Eating a lunch with balanced proportions of carbohydrate and fat was associated with better performance on a simple reaction time task, less fatigue or drowsiness, and being more cheerful up to three hours after meal consumption compared with meals high in either fat or carbohydrate (Lloyd *et al.*, 1994). People fed a high-fat lunch, performed more slowly, but also more accurately on a selective attention task than those fed a low-fat lunch. Moreover, individuals reported feeling calmer following the high-fat lunch than after the low-fat lunch (Smith *et al.*, 1994 b). Although lunch-time meals are followed by drowsiness or sleepiness, attention to the nutrient composition of the meal can benefit mental performance. A brief post-lunch nap may counter the dip in mental acuity commonly seen after lunch (Takahashi *et al.*, 1998; Waterhouse *et al.*, 2007).

25.4.3   Snacks

Food is not just consumed at main mealtimes. For many individuals, particularly young people, snacking accounts for a substantial proportion of their daily energy intake (Jahns *et al.*, 2001; Nielsen *et al.*, 2002; Zizza *et al.*, 2001), particularly when another meal has been missed (Savige *et al.*, 2007). Mid-morning and mid-afternoon are common times for snack food consumption, and these times are linked with peaks in both negative emotion and tiredness (Stone *et al.*, 2006). Morning coffee breaks and the 'post-lunch slump' are times when people briefly stop the work (or school) day and re-energize with a quick snack. Work in our laboratory and others shows that such snacking can have direct benefits on mental performance in children and young adults (Busch *et al.*, 2002; Kanarek and Swinney, 1990; Mahoney *et al.*, 2007).

On a day-to-day basis, morning or afternoon snacks can produce positive effects on mental performance. In children who had eaten small breakfasts, a mid-morning snack improved concentration on school-related activities (Benton and Jarvis, 2007). Feeding a mid-morning confectionery snack to children who had already eaten breakfast improved mid-morning memory tests relative to having no snack (Muthayya *et al.*, 2007), but did not improve attention in these children. After an overnight fast, children given a confectionery snack in the morning showed improved attention compared with the placebo condition (Busch *et al.*, 2002). Additionally, children given the confectionery snack, made fewer errors on a vigilance task, and made significantly more correct responses on that task in comparison to those given a placebo.

After lunchtime, it is natural to experience a dip in energy, in fact many cultures, particularly in a warm climate, still observe a post-lunch rest period such as a siesta. During this post-lunch period, people take short naps and return to work refreshed in the middle of the afternoon (Monk, 2005). In societies that do not observe such traditions, the post-lunch slump is countered with vending machine snacks such as candy bars or with cups of coffee in an attempt to maintain mental alertness and functioning. Although confection-based (sugary) snacks may be perceived as unhealthy, they do have positive effects on mental performance. In college-aged students, intake of either a confectionery snack or fruit-flavored yogurt significantly improved short-term memory and attention relative to the placebo condition (Kanarek and Swinney, 1990). In related research, consumption of a confectionery snack improved spatial learning and memory performance in college-aged men, although they had better sustained attention in the placebo condition. In the same study, school-aged boys also performed better on spatial learning and memory tasks following a confectionery snack, with no improvements in attention (Mahoney et al., 2007). As with breakfast intake, factors such as the time interval between meals, the type of food eaten, and the types of mood scales and cognitive tests employed can significantly affect the outcome of snacks on mental function.

In the long-term, intake of certain types of snack foods can have a large, cumulative impact on mental function in adults (Kuroda et al., 2007) and physical and mental development in at-risk children (Neumann et al., 2007). In Japan, where light broths are consumed as part of breakfast or as a snack, long-term daily intake of a savory bonito (fish) broth significantly improved vigor and overall mood in fatigued individuals (Kuroda et al., 2007). In this study, cognitive performance was similarly increased relative to fatigued individuals not receiving the broth. Bonito broth, while minimally caloric, is hypothesized to exert its actions on mood and brain function either through antioxidant pathways or neurotransmitter release. In a recent longitudinal study based in Kenya, school children were given mid-morning snacks at school and were followed for up to 2.5 years. Children who were provided milk or meat protein in addition to a plant-based morning snack had better growth rates and muscle development than children given no additional protein, or those who received no snack (Neumann et al., 2007). Children who were given meat with their snack showed better academic and cognitive performance, and also showed more initiative and leadership during playtime than the other groups. The presence of protein, vitamins and minerals in a meat-based snack may all contribute to physical and mental development seen in these children. In this study, snacks were only given when school was in session, not during breaks or holidays, suggesting that supplementation need not be continuous to see a benefit.

25.4.4   Evening meal

The evening meal has received the least attention in meal and mental function research. This may be due to the general notion that, for the majority of people, the evening meal represents the end of the productive portion of the day. As with other meals, individuals given an evening meal perform better on cognitive tasks involving reaction time and reasoning compared with those given no meal (Smith *et al.*, 1994a). However, evening meals may not affect attention in people following a normal circadian pattern. There are, however, a number of studies examining meal type and mental functioning in shift workers, individuals who start the work day at different times to the general work force, and who also face changing work times. Thus, an individual could work a 4 pm to midnight shift one week, an work an 8 pm to 4 am shift another week. Such shifts in work patterns disrupt eating and sleeping patterns. Maintaining mental acuity of people engaged in shift work is important (LaDou, 1982; Murphy *et al.*, 1979).

In shift workers, it is essential to address decrements in alertness, attention, and mental acuity. However, alertness and cognitive performance may be independent of each other, and not necessarily affected by similar meal compositions. Shift workers given a high-fat, moderate-carbohydrate, low-protein test meal showed improved cognitive performance and attention than when given a high-carbohydrate, low-fat meal (Love *et al.*, 2005). Sleepiness and alertness were not affected by meal type in this study. In other studies, sleepiness ratings increased following a high-carbohydrate meal relative to a high-protein meal, and cognitive performance and alertness was improved following a high-protein meal relative to a high-carbohydrate meal (Paz and Berry, 1997). In a study examining meal composition and performance in people kept awake for 24 h, relative to high-protein meals, high-carbohydrate meals were again associated with sleepiness over the entire day and with slower performance on a reaction time task (Lowden *et al.*, 2004). A high-fat diet, particularly consumed as the evening meal, was associated with less sleepiness and better reaction time performance relative to the high-carbohydrate condition. The researchers in this study suggest that, based on the available evidence, shift workers should consider meals with smaller proportions of carbohydrate to maintain alertness.

25.5   Conclusions

Meal and mental performance research is complex and must address variables not only related to the meal itself, but to other meals eaten throughout the day, the general nutritional status of subjects, and time of day. The intent of meal research is not to provide a prescription for optimal performance, but rather to expand our understanding of how food combinations can interact to affect performance. By definition, studies examining meals and food intake are artificial constructs and do not mirror real-life eating behavior. As evidenced above, improving one's mood or mental ability is not as simple as increasing carbohydrates or reducing fat in the diet. Each component of the diet contributes to mood and cognitive performance, and, once combined in a meal, can have vastly different effects than when its components are eaten in isolation. Meals that are high in any one macronutrient have variable effects on performance; meals high in carbohydrates increase subjective ratings of sleepiness, meals high in protein reduce fatigue, and meals high in fat can improve accuracy. When looked at in a broader context, we can generalize that balanced meals and meals that have a low glycemic load tend to produce improvements in mental performance.

25.6   References

Appleton KM, Peters TJ, Hayward RC, Heatherley SV, McNaughton SA, Rogers PJ, Gunnell D, Ness AR, Kessler D (2007), 'Depressed mood and *n*-3 polyunsaturated fatty acid intake from fish, non-linear or confounded association?' *Soc Psychiatry Psychiatr Epidemiol*, **42**,100–104.

Benton D, Jarvis M (2007), 'The role of breakfast and a mid-morning snack on the ability of children to concentrate at school', *Physiol Behav*, **90**, 382–385.

Benton D, Nabb S (2003), 'Carbohydrate, memory and mood', *Nutr Rev* **61**, S61–S67.

Brozek J (1947), 'Review of the psychology of diet and nutrition', *J Appl Psychol*, **31**, 348–350.

Brozek J (1955), 'Nutrition and psyche with special reference to experimental psychodietetics', *Am J Clin Nutr*, **3**, 101–113.

Busch CR, Taylor HA, Kanarek RB, Holcomb PJ (2002), 'The effects of a confectionery snack on attention in young boys', *Physiol Behav*, **77**, 333–340.

Buydens-Branchey L, Branchey M, Hibbeln JR (2008), 'Associations between increases in plasma *n*-3 polyunsaturated fatty acids following supplementation and decreases in anger and anxiety in substance abusers', *Prog Neuropsychopharmacol Biol Psychiatry*, **32**, 568–575.

Christensen L, Redig C (1993), 'Effect of meal composition on mood', *Behav Neurosci*, **107**, 346–353.

D'Anci KE, Kanarek RB (2006), 'Dietary sugar and behavior', in *Nutrition and behavior, a multidisciplinary approach,* Oxfordshire, UK, CABI Publishing.

D'Anci KE, Kanarek RB (2006), 'Caffeine, the methylxanthines and behavior', in *Nutrition and Behavior, a multidisciplinary approach,* Oxfordshire, UK, CABI Publishing.

D'Anci KE, Rosenberg IH (2004), 'Folate and brain function in the elderly', *Curr Opin Clinical Nutr Metabol Care*, **7**, 659–664.

D'Anci KE, Rosenberg IH (2005), 'B vitamins and the brain, depression', *Nutr Clinical Care*, **8**, 143–148.

Elias MF, Elias PK, Sullivan LM, Wolf PA, D'Agostino RB (2005), 'Obesity, diabetes and cognitive deficit, The Framingham Heart Study', *Neurobiol Aging* **26S**, S11–S16.

Ferry M, Sidobre B, Lambertin A, Barberger-Gateau P (2005), 'The SOLINUT study, analysis of the interaction between nutrition and loneliness in persons aged over 70 years', *J Nutr Health Aging*, **9**, 261–268.

Fischer K, Colombani PC, Langhans W, Wenk C (2002), 'Carbohydrate to protein ratio in food and cognitive performance in the morning', *Physiol Behav*, **75**, 411–423.

Fischer K, Colombani PC, Wenk C (2004), 'Metabolic and cognitive coefficients in the development of hunger sensations after pure macronutrient ingestion in the morning', *Appetite*, **42**, 49–61.

Freeman MP, Hibbeln JR, Wisner KL, Brumbach BH, Watchman M, Gelenberg AJ (2006), 'Supplementation with omega-3 fatty acids may help reduce postpartum depression', *Acta Psychiatr Scand* **113**, 31–35.

Green P, Hermesh H, Monselise A, Marom S, Presburger G, Weizman A (2006), 'Red cell membrane omega-3 fatty acids are decreased in nondepressed patients with social anxiety disorder', *Eur Neuropsychopharmacol* **16**, 107–113.

Greenwood CE, Kaplan RJ, Hebblethwaite S, Jenkins DJ (2003), 'Carbohydrate-induced memory impairment in adults with type 2 diabetes', *Diabetes Care*, **26**, 1961–1966.

Gross SM, Bronner Y, Welch C, Dewberry-Moore N, Paige DM (2004), 'Breakfast and lunch meal skipping patterns among fourth-grade children from selected public schools in urban, suburban, and rural Maryland', *J Am Diet Assoc*, **104**, 420–423.

Hoddes E, Zarcone V, Smythe H, Phillips R, Dement WC (1973), 'Quantification of sleepiness, a new approach', *Psychophysiology*, **10**, 431–436.

Howland A, Lawton CL, Dye L (2008), 'Acute effects of macronutrient manipulations on cognitive test performance in healthy young adults; a systematic research review', *Neurosci Biobehav Rev*, **32**, 72–85.

Iribarren C, Markovitz JH, Jacobs DR Jr, Schreiner PJ, Daviglus M, Hibbeln JR (2004), 'Dietary intake of *n*-3, *n*-6 fatty acids and fish, relationship with hostility in young adults – the CARDIA study', *Eur J Clin Nutr*, **58**, 24–31.

Jahns L, Siega-Riz AM, Popkin BM (2001), 'The increasing prevalence of snacking among US children from 1977 to 1996', *J Pediatr*, **138**, 493–498.

Kanarek RB, Swinney D (1990), 'Effects of food snacks on cognitive performance in male college students', *Appetite*, **14**, 15–27.

Kaplan RJ, Greenwood CE, Winocur G, Wolever TM (2001), 'Dietary protein, carbohydrate, and fat enhance memory performance in the healthy elderly', *Am J Clin Nutr*, **74**, 687–693.

Kaplan RJ, Greenwood CE, Winocur G, Wolever TM (2000), 'Cognitive performance is associated with glucose regulation in healthy elderly persons and can be enhanced with glucose and dietary carbohydrates', *Am J Clin Nutr*, **72**, 825–836.

LaDou J (1982), 'Health effects of shift work', *West J Med*, **137**, 525–530.

Lloyd HM, Green MW, Rogers PJ (1994), 'Mood and cognitive performance effects of isocaloric lunches differing in fat and carbohydrate content', *Physiol Behav*, **56**, 51–57.

Love HL, Watters CA, Chang WC (2005), 'Meal composition and shift work performance', *Can J Diet Pract Res*, **66**, 38–40.

Lowden A, Holmbäck U, Akerstedt T, Forslund J, Lennernäs M, Forslund A (2004), 'Performance and sleepiness during a 24 h wake in constant conditions are affected by diet', *Biol Psychol*, **65**, 251–263.

Mahoney CR, Taylor HA, Kanarek RB (2007), 'Effect of an afternoon confectionery snack on cognitive processes critical to learning', *Physiol Behav*, **90**, 344–352.

Mahoney CR, Taylor HA, Kanarek RB, Samuel P (2005), 'Effect of breakfast composition on cognitive processes in elementary school children', *Physiol Behav*, **85**, 635–645.

McNair D, Lorr M, Droppleman L (1971), *Profile of mood states manual*, San Diego, Educational and Industrial Testing Service.

Messier C (2004), 'Glucose improvement of memory, a review', *Eur J Pharmacol*, **490**, 33–57.

Messier C, Awad N, Gagnon M (2004), 'The relationship between atherosclerosis, heart disease, and type 2 diabetes and dementia', *Neurol Res*, **26**, 567–572.

Meyers AF, Sampson AE, Weitzman M, Rogers BL, Kayne H (1989), 'School Breakfast Program and school performance', *Am J Dis Child*, **143**, 1234–1239.

Monk TH (2005), 'The post-lunch dip in performance', *Clin Sports Med*, **24**, e15–e23.

Murphy JM, Pagano ME, Nachmani J, Sperling P, Kane S, Kleinman RE (1998), 'The relationship of school breakfast to psychosocial and academic functioning', *Arch Pediat Adoles Psychiat*, **152**, 899–906.

Murphy TJ, Winget CM, LaDou J (1979), 'Fixed vs. rapid rotation shift work', *J Occup Med*, **21**, 318–326.

Muthayya S, Thomas T, Srinivasan K, Rao K, Kurpad AV, van Klinken JW, Owen G, de Bruin EA (2007), 'Consumption of a mid-morning snack improves memory but not attention in school children', *Physiol Behav*, **90**, 142–150.

Nabb SL, Benton D (2006), 'The effect of the interaction between glucose tolerance and breakfasts varying in carbohydrate and fibre on mood and cognition', *Nutr Neurosci*, **9**, 161–168.

Nabb S, Benton D (2006), 'The influence on cognition of the interaction between the macro-nutrient content of breakfast and glucose tolerance' *Physiol Behav*, **87**, 16–23.

Neely G, Landstrom U, Bystrom M, Junberger ML (2004), 'Missing a meal, effects on alertness during sedentary work', *Nutr Health*, **18**, 37–47.

Nielsen SJ, Siega-Riz AM, Popkin BM (2002), 'Trends in food locations and sources among adolescents and young adults', *Prev Med*, **35**, 107–113.

Nemets B, Stahl Z, Belmaker RH (2002), 'Addition of omega-3 fatty acid to maintenance medication treatment for recurrent unipolar depressive disorder', *Am J Psychiatry*, **159**, 477–479.

Nemets H, Nemets B, Apter A, Bracha Z, Belmaker RH (2006), 'Omega-3 treatment of childhood depression, a controlled, double-blind pilot study', *Am J Psychiatry*, **163**, 1098–1100.

Ness AR, Gallacher JE, Bennett PD, Gunnell DJ, Rogers PJ, Kessler D, Burr ML (2003), 'Advice to eat fish and mood, a randomised controlled trial in men with angina', *Nutr Neurosci*, **6**, 63–65.

Neumann CG, Murphy SP, Gewa C, Grillenberger M, Bwibo NO (2007), 'Meat supplementation improves growth, cognitive, and behavioral outcomes in Kenyan children', *J Nutr*, **137**, 1119–1123.

Papanikolaou Y, Palmer H, Binns MA, Jenkins DJ, Greenwood CE (2006), 'Better cognitive performance following a low-glycaemic-index compared with a high-glycaemic-index carbohydrate meal in adults with type 2 diabetes', *Diabetologia*, **49**, 855–862.

Pasman WJ, Blokdijk VM, Bertina FM, Hopman WP, Hendriks HF (2003), 'Effect of two breakfasts, different in carbohydrate composition, on hunger and satiety and mood in healthy men', *Int J Obes Relat Metab Disord*, **27**, 663–668.

Paz A, Berry EM (1997), 'Effect of meal composition on alertness and performance of hospital night-shift workers. Do mood and performance have different determinants?', *Ann Nutr Metab*, **41**, 291–298.

Peet M, Horrobin DF (2002), 'A dose-ranging study of the effects of ethyl eicosapentaenoate in patients with ongoing depression despite apparently adequate treatment with standard drugs', *Arch Gen Psychiatry*, **59**, 913–919.

Peet M, Murphy B, Shay J, Horrobin D (1998), 'Depletion of omega-3 fatty acid levels in red blood cell membranes of depressive patients', *Biol Psychiatry*, **43**, 315–319.

Rampersaud G C, Pereira M A, Girard B L, Adams J, Metzl J D (2005), 'Breakfast habits, nutritional status, body weight, and academic performance in children and adolescents', *J Am Diet Assoc*, **105**, 743–760.

Reynolds AJ, Temple JA, Robertson DL, Mann EA (2001), 'Long-term effects of an early childhood intervention on educational achievement and juvenile arrest: a 15-year follow-up of low-income children in public schools', *JAMA*, **285**, 2339–2346.

Richter LM, Rose C, Griesel RD (1997), 'Cognitive and behavioural effects of a school breakfast', *S Afr Med J*, **87**(1 Suppl), 93–100.

Rogers PJ, Appleton KM, Kessler D, Peters TJ, Gunnell D, Hayward RC, Heatherley SV, Christian LM, McNaughton SA, Ness AR (2008), 'No effect of *n*-3 long-chain polyunsaturated fatty acid (EPA and DHA), supplementation on depressed mood and cognitive function, a randomised controlled trial', *Br J Nutr*, **99**, 421–431.

Salbach-Andrae H, Lenz, K, Simmendinger N, Klinkoski N, Lehmkahl U, Pfeiffer E (2008), 'Psychiatric comorbidities among female adolescents with anorexia nervosa', *Child Psychiatry Hum Devel*, **39**, 261–272.

Sampson AE, Dixit S, Meyers AF, Houser R (1995), 'The nutritional impact of breakfast on the diets of inner-city African-American elementary school children', *J Nat Med Assoc*, **87**, 195–202.

Sánchez-Villegas A, Henríquez P, Bes-Rastrollo M, Doreste J (2006), 'Mediterranean diet and depression', *Public Health Nutr*, **9**, 1104–1109.

Sánchez-Villegas A, Henríquez P, Figueiras A, Ortuño F, Lahortiga F, Martínez-González MA (2007), 'Long chain omega-3 fatty acids intake, fish consumption and mental disorders in the SUN cohort study', *Eur J Nutr*, **46**, 337–346.

Savige G, Macfarlane A, Ball K, Worsley A, Crawford D (2007), 'Snacking behaviours of adolescents and their association with skipping meals', *Int J Behav Nutr Phys Act*, **17**(4), 36.

Sherman HC (1950), *The nutritional improvement of life*, New York, Columbia University Press.

Shide D (1995), 'Information about the fat content of preloads influences energy intake in healthy women', *J Am Diet Assoc*, **95**, 993–998.

Smith A, Maben A, Brockman P (1994a), 'Effects of evening meals and caffeine on cognitive performance, mood and cardiovascular functioning', *Appetite*, **22**, 57–65.

Smith A, Kendrick A, Maben A, Salmon J (1994b), 'Effects of fat content, weight, and acceptability of the meal on postlunch changes in mood, performance, and cardiovascular function', *Physiol Behav*, **55**, 417–22.

Smith A, Leekam S, Ralph A, McNeill G (1988), 'The influence of meal composition on post-lunch changes in performance efficiency and mood', *Appetite*, **10**, 195–203.

Spring B, Maller O, Wurtman J, Digman L, Cozolino L (1982–1983), 'Effects of protein and carbohydrate meals on mood and performance, interactions with sex and age', *J Psychiatr Res*, **17**, 155–167.

Strauman TJ, Vookles J, Berenstein V, Chaiken S, Higgins ET (1991), 'Self-discrepancies and vulnerability to body dissatisfaction and disordered eating', *J Pers Soc Psychol*, **61**, 946–956.

Stone AA, Schwartz JE, Schkade D, Schwarz N, Krueger A, Kahneman D (2006), 'A population approach to the study of emotion, diurnal rhythms of a working day examined with the Day Reconstruction Method', *Emotion*, **6**, 139–149.

Takahashi M, Fukuda H, Arito H (1998), 'Brief naps during post-lunch rest, effects on alertness, performance, and autonomic balance', *Eur J Appl Physiol Occup Physiol*, **78**, 93–98.

van Gelder BM, Tijhuis M, Kalmijn S, Kromhout D (2007), 'Fish consumption, *n*-3 fatty acids, and subsequent 5-y cognitive decline in elderly men, the Zutphen Elderly Study', *Am J Clin Nutr*, **85**, 1142–1147.

Waterhouse J, Atkinson G, Edwards B, Reilly T (2007), 'The role of a short post-lunch nap in improving cognitive, motor, and sprint performance in participants with partial sleep deprivation', *J Sports Sci*, **25**, 1557–1566.

Wells AS, Read NW (1996), 'Influences of fat, energy, and time of day on mood and performance', *Physiol Behav*, **59**, 1069–1076.

Wells AS, Read NW, Idzikowski C, Jones J (1998), 'Effects of meals on objective and subjective measures of daytime sleepiness', *J Appl Physiol*, **84**, 507–515.

Winnick JJ, Davis JM, Welsh RS, Carmichael MD, Murphy EA, Blackmon JA (2005), 'Carbohydrate feeding during team sport exercise preserve physical and CNS function', *Med Sci Sports Exerc*, **37**, 306–315.

Worobey J (2006), 'Direct effects of nutrition on behavior, brain–behavior connections', in *Nutrition and behavior, a multidisciplinary approach*, Oxfordshire, UK, CABI Publishing, pp. 25–42.

Worobey J (2006), 'Effects of chronic and acute forms of undernutrition', in *Nutrition and Behavior, a multidisciplinary approach*, Oxfordshire, UK, CABI Publishing, pp. 63–80.

Worobey J, Kanarek RB (2006), 'Short-term effects of nutrition on behavior, neurotransmitters', in *Nutrition and behavior, a multidisciplinary approach*, Oxfordshire, UK, CABI Publishing, pp. 43–62.

Zizza C, Siega-Riz AM, Popkin BM (2001), 'Significant increase in young adults' snacking between 1977–1978 and 1994–1996 represents a cause for concern!', *Prev Med*, **32**, 303–310.

26

Designing meal environments for 'mindful eating'

J. L. Le Bel and R. Richman Kenneally, Concordia University, Canada

Abstract: Young people's representations of a meal are examined as well as their memories of their childhood foodscapes in order to better understand the features and aspects of the spaces within the home where food is prepared and consumed that can promote 'mindful eating'. Using quantitative survey methodology, the pleasure young consumers derive from different types of meals is examined and the features or characteristics that define or make a meal for them are explored. Drawings of childhood's foodscapes submitted by participants were analyzed. These surprisingly rich renditions reveal a number of important themes and dimensions of their childhood foodscapes. These include the centrality of the kitchen as a prime domestic foodscape; the repeated articulation of the significance of the main meal as a family-based activity, often completed without distractions such as television; and the importance of the kitchen table as the most concentrated component of the domestic foodscapes.

Key words: meals, pleasure, design, kitchen, children, kitchen table.

26.1 Introduction

In this chapter, we investigate mental representations of meals and memories of childhood foodscapes as a means to explore the role and impact of the physical environment and objects therein on eating styles and relationship to food. Marketing professionals and environmental psychologists have long known the importance of a number of environmental variables on individuals' attitudes and behaviors. Environmental psychologists (e.g., Mehrabian and Russell, 1974), for instance, describe environments in terms of the pleasure and arousal they instill in people and how those dimensions of environments shape behaviors within them. In marketing, consumer researchers have examined the effects of so-called 'atmospheric' variables, such as music, color and crowding, on consumers' shopping behavior (for a review, see Turley and Milliman, 2000). Today, many brands and companies, especially in the food retailing and foodservice industries, employ sophisticated 'experience design' whereby specific elements (e.g., wall texture, color, music) are carefully assembled to create desired emotional responses and well-being amongst consumers (e.g., Smith and Wheeler, 2002). By contrast, the impact on individual behavior of design and features of domestic food areas has not been examined as thoroughly as those in the commercial domain.

We aim to identify general dimensions and specific aspects of domestic foodscapes that have the potential to shape one's relationship to food and eating. To do so, we focus on young consumers having recently left the family home. The loss of cooking skills by this group has been lamented and these consumers are very much the target of restaurant marketers' efforts, making them an appropriate group to study. We draw from various theoretical frameworks and research methodologies to explore their mental representations of meals and memories of childhood foodscapes. This is a hitherto insufficiently explored aspect of the socialization process by which children acquire important habits, preferences and knowledge about food, cooking and eating that will, in many cases, last a lifetime with positive and/or negative consequences on individual health. Our investigation is rooted in the belief that the physical environment in which we prepare and eat food in the home, along with the material and visual culture associated with these activities and their wider performative implications, directly influence the acquisition of attitudes and behaviors towards food. Surprisingly, while foodservice professionals, for example, have long appreciated the influence of décor on their customers' behavior, the thesis that the home's physical environment can have equally profound impact on its inhabitants' food attitudes and eating behavior has not been systematically explored, and may come as a surprise to many.

We first discuss changes to today's domestic foodscapes. Within that context, we then explore young consumers' representations of a meal in terms of the features or characteristics they deem defining of a meal. We then examine the architecture of domestic foodscapes. We report on a project designed to reconstruct the narrative of consumers' childhood engagement with food using personal drawings of childhood foodscapes. Together, these two sources of data point to insightful avenues from which to rethink the design of meal environments.

26.2   Today's domestic foodscapes

The word 'foodscape' has thus far been reserved for macro-level examination of the food supply system (e.g., Winson, 2004) or of the variability and availability of food in specific rural and urban area (e.g., Cummins and Mcintyre, 2002). At the micro level, the immediate built environment has a more direct impact on the choices and eating behavior of individuals within a household. The physical space where food is eaten is one of the basic dimensions underlying memories and conceptualizations of everyday eating and drinking episodes and includes (1) the general location (e.g., home, work, car) and a specific place (e.g., in front of the television), (2) access to food storage and preparation facilities (e.g., shared kitchen in a dorm), and (3) condition of the physical environment (e.g., temperature) (Bisogni *et al.*, 2007). To date, micro-level analyses have examined specific areas within the home and focused on issues such as pantry management (Baranowski *et al.*, 2007) and food availability and visibility within the kitchen, table, and plate (Sobal and Wansink, 2007). We use the word 'foodscapes' to denote those areas where food is prepared and consumed, and the objects therein. By 'domestic' we mean those foodscapes inside the home (this important distinction between the domestic sphere and the commercial marketplace is also explored is this book in the context of home food testing versus commercial food testing by Boutrolle).

26.2.1   The battle for your food dollar

Changes taking place in food distribution and retailing are infiltrating the household and having profound consequences for domestic foodscapes and the food preparation activities and consumption experiences taking place there. Perhaps most significant is the impact of new foods and food delivery and distribution practices. In 2008, the American restaurant industry is expected to generate 558 billion dollars in sales and serve over 133 million meals per day (NRA 2008). This represents roughly 49% of a household's food dollar going to 'food-away-from-home' (FAFH) and 15% of meals. Equally significant for our modern consumption modes is the fact that the majority of restaurant meals are no longer consumed in restaurants: 58% of restaurant meals are eaten off-premises and the home is the most popular location where 'take out' restaurant meals are consumed (followed by the car and the workplace). For instance, fully 54% of all fast-food takeout food is eaten at home (Mintel, 2007).

At the moment, it is nothing less than a war that is being waged for the consumer's food dollar. On one side are the 'food-away-from-home' preparers (i.e., foodservice providers) and on the other side are the 'food-at-home' providers (i.e., grocery stores and food manufacturers) and each is increasingly crossing over the other's territory by developing new foods, new cooking techniques and new delivery modes. For instance, food manufacturers are now offering a variety of alternative preparation methods like ready-to-heat (e.g., pre-cooked frozen vegetables now packaged in microwavable bags), ready-to-cook (e.g., marinated chicken breasts), and 'speed scratch' (e.g., sauces, two-step cake mixes, ingredients like chopped garlic) all designed to save precious food preparation time. Restaurants, such as Applebee's and Outback Steakhouse, are now offering 'curbside delivery' where customers need only to drive into specially marked parking bays (after phoning in their order) and a staff member delivers the food to their car.

Whether these tactics by food and foodservice marketers are in response to consumer demand or proactively shaping it, they are nonetheless changing our eating behavior and the very definition of what constitutes a meal. While some niche segments care about taste and health-value and are willing to pay the related premium for these benefits, the vast majority of consumers put convenience and ease of cooking as their top concerns when deciding what to cook and eat at home, according to market analysts at the NPD Group (NPD, 2006). As a result of the many enticements to spend our food dollar, today's food environment, both at home and outside of it, often results in eating behavior that has been characterized as 'mindless' (Wansink, 2007) with negative health outcomes, such as higher rates of obesity.

26.3   What makes a meal *'a meal'*?

Given the growing commercialization of prepared foods and the increasing frequency of eating in alternative venues outside the home (such as one's car), the very meaning of a 'meal' may be changing in significant ways. In this modern context, what actually constitutes or makes a meal? And are meals still a source of pleasure? The answer to these questions will vary across cultures (e.g., Mestdag, 2005; Rozin, 2005) but the powerful influence of the physical environment is likely to be a universal feature.

Categorization theory offers a potentially insightful perspective to look at some of these questions. Categorization theory (e.g., Rosch *et al.*, 1976) holds that our knowledge system is structured into categories that have three levels: 'furniture' for example is a concept at the superordinate level (often pan-cultural), at the intermediate levels 'chairs' may be a sub-category belonging to furniture and at the basic or exemplar level an 'armchair' would fit in the sub-category 'chair' (with intermediate and basic levels often varying across cultures). Within a category, some objects are more representative than others; this is referred to as the hierarchical or graded structure of categories. For example, 'office chair' might be deemed more characteristic of the category 'chair' than say a 'bar stool' even though these two pieces of furniture share some common features. Membership by an object to a category depends on the object possessing the defining features or characteristics associated with that category. Not surprisingly, marketing researchers, for instance, expend considerable effort to uncover what these features are, since possession of the desired features by a product can significantly increase the likelihood that consumers will like it and purchase it.

If we think of a meal as a broad, superordinate concept (such as furniture in the example above), then what are the features that define a meal? To answer this question, we surveyed 150 undergraduate students (62 men, 88 women, average age 20.3 years). First, participants were asked to indicate the intensity of the pleasure they derived from their last meal (on a 10-point scale where 1 = very little pleasure; 10 = a great deal of pleasure). Next, they indicated how representative each of 35 different features might be of a meal (on an 8-point scale where 1 = not representative all, to 8 = very representative). Finally, participants were asked to indicate the intensity of pleasure they typically derive from 11 different types or forms of meals (on the same 10-point scale used for the first question).

26.3.1   Features of a meal

Table 26.1 lists the 35 features of a meal evaluated by our participants in descending order of representativeness. The defining features are revealing

Table 26.1   Features of a meal

Based on your own personal definition of a 'meal,' please indicate to what extent you believe a meal possesses the following features or characteristic. A meal is/has…

1.	An important source of pleasure	6.7*
2.	Includes a main course	6.7
3.	Includes vegetables	6.3
4.	Fun	6.3
5.	Includes protein	6.1
6.	Tastes better with conversation	5.9
7.	Eaten out in restaurants	5.7
8.	Eaten with family members	5.6
9.	Shared	5.5
10.	Homemade	5.4
11.	Requires a table and chairs	5.0
12.	Eaten at home	5.0
13.	Eaten at regular times	4.7
14.	Requires silverware	4.4
15.	Eaten in dining room	4.4
16.	Many courses	4.3
17.	Can be 'grab-and-go'	4.2
18.	Time consuming to prepare	4.2
19.	Includes dessert	4.0
20.	Requires good manners	4.0
21.	Starts with appetizers	3.9
22.	Eaten in the living room	3.8
23.	Eaten out of necessity	3.8
24.	Ends with a dessert	3.8
25.	Eaten in the kitchen	3.7
26.	Eaten while watching television	3.7
27.	Storebought	3.4
28.	Eaten from disposable dishes	3.3
29.	Eaten while doing other things	3.1
30.	Eaten from takeout containers	3.1
31.	Eaten alone	3.0
32.	Requires to be dressed up	2.5
33.	Eaten in silence	2.1
34.	Eaten while standing up	1.8
35.	Waste of time	1.3

*1 = Disagree totally, 8 = Agree totally, n = 150

of these young consumers' representation of a meal: the meal is a source of pleasure, includes a main course, vegetables, and protein; it tastes better with conversation, can be shared, especially with family members. Respondents, true to the fact that their generation are more frequent users of restaurants than their parents (Mintel, 2007), associated meals with 'eaten out in restaurants'. 'Fun' may be surprising to see as a meal attribute but is an important food dimension for younger consumers, so much so in fact that 'fun foods' targeted to younger consumers are now a fast-growing $16 billion industry (Elliott, 2008). At the same time, and somewhat paradoxically perhaps, a meal was also seen as 'homemade'.

Equally instructive are the features deemed less representative and those in the middle of the graded structure described in Table 26.1. Our respondents did not see meals as eaten standing up, out of disposable dishes or takeout container, alone or in silence. Nor they did associate meals with 'storebought' (which suggests that for them 'eaten in restaurants' is not the same as 'storebought'). This bodes well for the future of mindful eating. Gender differences were detected in the data presented in Table 26.1. For instance, women believed that meals include vegetables (6.82 out of 8) more so than men (5.7 out of 10). Women also believed that meals are shared (5.8) more than men (5.1) and can be eaten while doing other things (3.3 versus 2.7 for men). Men on the other hand, believed that meals start with appetizers (4.4) more strongly than women (3.6), that meals require people to be dressed up (2.8 versus 2.1 for women), and that meals can be eaten while standing up (2.1 versus 1.6 for women).

26.3.2   Pleasure from meals: company and location matter

Food, in theory, is a chief source of pleasure in life (Rozin, 2005). In reality, how do different meal formats or types of meal stack up in terms of their ability to produce pleasure? Table 26.2 presents the pleasure associated with eleven different types of meals by men and women. Differences between men and women ($p < 0.05$) were uncovered for eight of the eleven meals and are highlighted in bold in Table 26.2. For both men and women, the top three meal categories are enjoyed in the company of family or a sweetheart with women experiencing more pleasure than men from each of those meals. Men experienced more pleasure than women from eating at malls food courts and at a school cafeteria, and breakfast at Starbucks is more pleasurable for women than for men. Interestingly, a self-made 'brown bag' lunch is more pleasurable for women than for men and only mildly more pleasurable than eating fast-food in a car. Recall that participants rated the pleasure from their most recent meal at 7.6, which places it fairly high within this nomenclature of different meals. It is interesting that meals with greater social and emotional components (with sweetheart, with family, with friends) are deemed more pleasurable. It also appears that meals eaten in contexts that include social cues can lead to lesser subsequent consumption and snacking than when

Table 26.2   Reported pleasure from different meals, by gender

In general, how much pleasure might each of the following types of meal bring you?

	Men	Women
Restaurant dinner with your sweetheart	**8.7**	**9.1**
Sunday night dinner with family members at home	**8.2**	**9.1**
Weekday dinner with family members at home	**7.3**	**8.2**
Fast-food meal eaten with friends	5.8	5.2
Meal eaten alone while watching television	5.3	4.8
Meal eaten at a food court in a mall	0.4	3.7
Meal at a school cafeteria	**4.0**	3.3
Breakfast at Starbucks (or competitor)	**3.2**	**4.2**
On-the-go food at an airport	3.2	2.7
'Brown bag' lunch at work	**3.1**	**3.7**
Fast-food meal eaten in your car	**2.6**	**2.0**

*1 = very little pleasure, 10 = a great deal of pleasure, $n = 150$

eaten without such cues (Pliner and Zec, 2007). Therefore, socials cues and social pleasure should be designed into meal environments.

The list of 35 features in Table 26.1 was assembled after a brainstorming session with research assistants. The goal was to cover various aspects of a meal, including components or elements of a meal (appetizers, vegetables, desserts, etc.) as well as the physical and social environments of a meal (dining room versus kitchen, eaten alone versus shared, etc.). Curiously, the most representative feature ('an important source of pleasure') only ranked 6.7 out of the 8 point scale, suggesting that perhaps the most representative features of a meal for this group was not elicited or part of the original list we presented to them. Interviews, focus groups, and open-ended questions may be necessary to uncover the more representative features and to achieve a more complete inventory of the key features of the 'meal' mental category. Finally, it is interesting that besides disposable and take-out dishware, the features relating to the physical environment and objects of meals (e.g., kitchen versus dining room, silverware, distractions like television) did not fall at either end of the graded structure but in the middle, perhaps because the props and objects and the milieu where we eat on a daily basis are taken for granted and not attended to consciously. We might for instance, appreciate and remember the fine linen or dishware at a restaurant we visited on a special occasion but daily props and environmental features of domestic foodscapes may not enjoy the same top-of-mind awareness. We might, for instance, develop a particular affection for a specific dishware, linen or other table embellishment without realizing that this object has entered our mental representation and that it is shaping our relationship to the foods we eat. Direct observation or in-depth interviews would seem warranted here to further explore not only the props and physical objects of importance in people's representations of a meal but also their more complex relationship to what constitutes a meal and the pleasure individuals derive from using such props. Nonetheless, taken together these results begin to document the structure of the category 'meals' for younger people.

26.4   Architecture of domestic foodscapes

The architecture of domestic foodscapes, and, in particular, although not exclusively, the kitchen, has been interrogated by a number of scholars, although much remains to be done. For example, historical material on the kitchen as an architectural space, especially during the twentieth century, has been analyzed in a number of ways: as a backdrop for examining preferences of individuals regarding the design of their kitchens (Freeman, 2004); as a site of gender narration or gendered landscape (Watkins, 2006; Llewellyn, 2004; Domosh, 1998); or as a domain central in the integration of new technology in the home (Lupton and Miller, 1992; Parr, 1999; Hardyment, 1991; Cowan, 1985; Hand and Shove, 2007).

Extremely significant to this chapter are studies of the kitchen as a physical landscape of interaction, enculturation and socialization, as stated by Buckley: 'the kitchen is far more than architecture, it is a concept which defies material limits to become a space of domestic fantasies, both homely and unhomely, of the family and the nation-state' (1996: 441). The kitchen as a concentrated emotional space, central in the construction of personal identity, has also been explored: Supski, for example, refers to the 'kitchen as home' metaphor in her analysis of the kitchens of immigrant Australian women after World War II (2006: 137; see also Pascali, 2006; Rolshoven, 2005). It has also been interrogated as a domain reflecting and activating self-expression or the projection of lifestyle (Dorfman, 1991; Plante 2002, Attfield, 1999; Floyd, 2004). Much more than simply a room in which to prepare and serve food, the kitchen is where 'chemistry and passion intersect, where conflicting sensibilities coexist.... [It is] all about the possibility of transformation...egg whites...beaten into *soufflé praline*...the kitchen is the place in the house where the ordinary become extraordinary' (Busch, 1999: 50).

If the kitchen serves, then, as a major zone in the construction of identity and in the construction of the 'meal' category, that is, at least in part, because the design of the space itself and its contents generate certain performances and behaviors. For example, meal preparation was the subject of substantial analysis, especially during the first half of the twentieth century, by observers who studied the steps women took as they prepared a meal, i.e., from stove, to sink, to refrigerator, to storage area and so on, and then attempted to interfere with the patterns of navigation through that space in the interest of instilling practices of scientific management (Sparke, 1995; Lupton and Miller, 1992; Bullock, 1988). On the other hand, researchers have pointed out that these and other inducements to modify domestic practices in the kitchen (not to mention the other spaces in the home), are often met with resistance and modification as these spaces are occupied through everyday use. Llewellyn cites the behavior of some families living in experimental working-class housing in England during the 1930s, whose response to the small kitchens they were obliged to live with, was to eat their meals there even if it meant 'perching children on top of work surfaces and the cooker [stove]' (2004: 48; see also Corrodi, 2005; Cromley, 1996; Hollows, 2000; Johnson and Lloyd, 2004).

26.5   Memories of childhood foodscapes

26.5.1   Domestic foodscapes as a socialization mechanism

Lessons learned around the family dining table often have lasting influences. The socialization process by which children acquire dispositions towards food and eating habits has been richly documented (and is addressed in this book in the context of family meals in Europe by Fjellström and in America by McIntosh). This process is believed to evolve in stages roughly defined by age (John, 1999) and permeable to a variety of influences. Different perspectives have been brought to bear to investigate the complex antecedents and outcomes of this socialization process. Parental and sibling influences have emerged as a key factor in the process. Parents' own food repertoire (Guidetti and Cavazza, 2008), mealtime communications (Orrell-Valente *et al.*, 2006) and interactions with children (Hays *et al.*, 2001) have been shown to influence children's attitudes and behaviors. One of the lesser understood aspects of this socialization process is the role played by the physical environment where food is prepared and eaten (along with the objects therein). Bell and Valentine (1997) identified the dinner table (whether in the kitchen or elsewhere) as particularly important for the socialization or 'civilization' of children; disciplining children, teaching them table manners, and reproducing 'the "family" in more positive ways' (63–64) are examples they give of enculturation processes that take place there. Except for the distracting influence of watching television at mealtime (Davison *et al.*, 2006; Buijen *et al.*, 2007; Fiates *et al.*, 2008) we know rather little of the physical environment and objects that capture children's attention and shape their relationship to food.

26.5.2   What did your childhood kitchen look like?

To explore the elements of the domestic foodscapes that may focus eaters' attention and promote mindful appreciation of food and its many pleasures, we invited 53 undergraduate students (24 men, 29 women; mean age 18.9 yr) to tell us what their childhood foodscapes looked like. These students were not part of the previous group but were drawn from the same school. Participants provided basic personal information (sex, age, height, weight, cultural background) and then were invited to describe the kitchen and dining areas in the home where they grew up. Specific questions guided them through the process (e.g., was the meal generally a formal occasion, or a casual one and how so? Did you eat in the same room in which the food was prepared? Who prepared your meals? How were cooking/serving/food acquisitions functions allocated?). Finally, participants were asked to 'sketch or briefly describe the layout of the kitchen/dining spaces' where they grew up.

The level of detail was quite extraordinary in certain cases. One participant actually drew food on the dining room table noting, as well, the chair of the 'head of family', which was the only one of nine that had arms. Several identified a spice rack, and one noted 'chotchkys [kitschy ornaments] on window sill'. 'Inedible fish' on a plate on the kitchen counter was pointed out by another participant, along with a pair of candlesticks (with the mention 'one night per year') and a television with waves radiating from it. Another participant added an additional page to his questionnaire so he could illustrate the kitchen and dining room separately, and drew the 'tooth pick [sic] container (shape of rabbit)' on the table. The same participant explained that if his grandparents came to visit, his parents and sisters 'move one seat each'. Two participants drew flowers on the kitchen table, and one of these added drawings of little animal ornaments on the shelf above the kitchen sink.

The analytical approach to make sense of these drawings and text draws from the growing body of researchers of architecture and design who evaluate material and visual culture from the perspective of the user (Kostof, 1995; Upton, 1998; Mellin, 2003). Material culture studies devote primary attention to artifacts: architectural spaces; objects including food, clothing, tools and instruments; articles used for work and for recreation; photographs; drawings; as reflections, activations, and performances of cultural ideologies, rituals, and behaviors (Prown, 1982; Glassie, 2000). The significance of these artifacts of everyday life is acknowledged by constructing ways of looking at a particular social group by studying its objects first, and using them as a basis through which to derive certain assumptions that are then tested by the subsequent exploration of additional artifacts, as well as texts, modes of performance, etc. Hence, an analysis of drawings such as these does not evaluate on the basis of whether they adhere to some kind of design or visual aesthetic, or even of whether they accurately portray their subject, but are valued because they can be interpreted as visual articulations of, in this case, childhood domestic foodscapes. First, each drawing was studied in order to determine qualitatively any patterns, consistencies and differences that arose across the cohort with regard to the responses. Next, in order to gain a more finely-grained perspective on the patterns that emerged, the ten most detailed, descriptive drawings were selected and studied as one concentrated group.

26.5.3   The daily family meal

What soon became evident was that roughly three-quarters of the participants reported that they habitually shared the principle meal of the day with their family. While some specified that the television was kept on during the meal, a surprising number reported that the focus was on the table, and this seems in most cases at the behest of the parents. A few indicated that prayers were part of the ritual; that proper table manners were expected; that the son was expected to wear a shirt to the table; that, once seated, no one could leave the table until everyone was finished. In short, participants either over-reported the domestic foodscape as more focused and family-oriented than the norm, or seem anomalous *vis-à-vis* the popular but anecdotal assumption that turn-of-the-twenty-first century homes are characterized by a less structured domestic foodscape, with individuals routinely eating in bedrooms, in living rooms in front of the television, and everywhere but at the kitchen table. To be sure, on the one hand, the growth of convenience and single-portion prepared foods may well promote 'grazing' and solitary eating rather than communal meals and the acquisition of mindful eating habits. On the other hand, our respondents' own reports match Mestdag's (2005) results and suggest that whatever grazing might be taking place has not jeopardized the existence of the typical or traditional meal. The predominance of structured family meals among our respondents does not imply that grazing and unstructured eating episodes are inherently negative, but, rather, supports the hypothesis that structured family meals contribute to positive mindful eating experiences. Further study, perhaps combining market-level sales data with family-level observational or diary data, is needed to explore the circumstances when grazing versus traditional meals take place along with the content, structure and pleasure derived from these two types of eating episodes.

The ten submissions having particularly expressive drawings showed an even higher incidence of family meals without distractions. Remarkable is the number of these families who reportedly ate their main meal without outside intrusion: five specified that the television was turned off; two additional students did not specify whether there were distractions, but reported that their family (in one case) discussed the day's activities and (in the other case) spent two hours eating dinner, and a third eating dessert. One participant indicated that the television was always on during the meal, another, that it was normally on but turned off during more formal gatherings, and the tenth did not eat with her family.

26.5.4   The centrality of the kitchen table

Equally important was the discovery that over half the participants reported eating their principal daily meal either at the kitchen table, or at the dining room table in the four instances when there was no table in the kitchen. That is, 24 routinely ate at a table in the kitchen; seven in the dining room, and ten in an open-plan configuration. In all of these cases, the participant drew a table, and also specified his/her designated seat in roughly two-thirds of the questionnaires submitted. Only five usually ate in rooms designated as other than kitchen or dining areas (in the remaining cases, the location was either unknown, or not consistently in any given room). Hence, in only a few cases did participants report that after the food was prepared, 'we would serve our food and go where we felt comfortable. I ate in my room on my bed in front of the TV'.

Focusing on the ten most detailed drawings, all but two of these participants ate at the kitchen or dining room table, plus one at 'a' table (there are four in the drawing: in the kitchen, dining room, close-in porch, and outside deck, and 'chairs and table' were even drawn in the 'hot tub room'). Only one student, of Northern European cultural origin, whose drawing stood out, usually ate on her bed or sofa. In fact, her drawings and story were particularly insightful: she routinely ate breakfast in the car, admitted that she seldom ate in the kitchen, and she would eat about half her dinner as she would prepare it. She routinely brought a book to read at the table and reported that her family's tendency to argue at the table was a strong disincentive to a communal dining experience. Still, her responses to other parts of the questionnaire suggests that she enjoys cooking and is otherwise involved with various aspects of food, even caring significantly about a nicely laid table.

Given these two general themes, there seems to be a consistent family intimacy operating around the kitchen or dining room table that suggests its major role in the food experience: the table is an important micro-foodscape in itself. This echoes the views of Christopher Alexander, a very influential architectural theorist during the 1970s, who wrote that the kitchen table 'will be the first and most important centre. It will be where you share meals, talk, work, and relax with a cup of coffee. . . . The table is the source of pleasure and of practical work together' (quoted in Kähler, 2005: 77; see also Rolshoven 2005). The table is also a means of giving some relative order to the context of the space in which it is located: Leatherbarrow's reading of the table in a restaurant pertains to that of the kitchen or dining room as well as concentrating the surrounding visual field 'much like a building site or even a city' (2004: 219) so that the individual elements of the vista around, above, or below the table 'constitute something like an atmosphere, a disposition, or mood that is not easy to describe but is never unclear' (2004: 219). The collective visual and, indeed, sensorial experience, is what he calls the 'character' that 'is often what is memorable about settings' (2004: 219). The kitchen table, it can thus be suggested, is a critical element of what makes the kitchen setting memorable and conducive to mindful eating.

26.5.5   Domestic foodscapes and mindful consumption

To shed further insights into the dynamics and themes uncovered by our examination of participants' textual and graphic descriptions, we had asked participants to complete Bell and Marshall's (2003) Food Involvement Scale (FIS) and to indicate the amount of pleasure they derived from 23 lifestyle activities or objects (e.g., shopping, reading, movies, sports, going out to bars, etc.). Although statistical comparisons are not meaningful with such a small sample, it is nonetheless noteworthy that our ten focal participants who drew detailed drawings enjoyed cooking, food shopping and a nicely laid table more than the overall sample. As for lifestyle activities, while statistical comparisons are not our objective, it appears that both our ten focal participants and the others appreciate the pleasure of food equally. Both groups derive equal pleasure from food but for the latter group memories of domestic foodscapes do not appear to affect subsequent food behaviour to any perceivable extent. Other differences suggest that lessons learned around the table may spill over into domains other than food: for instance, our ten focal participants got more pleasure from school and reading than the overall sample but less from going out to clubs, shopping and watching television. These, obviously, are not statistically significant differences but still warrant to be further explored with larger samples and alternative methodologies.

26.6   Implications for designing meal environments

What are some implications that arise from these findings? For architects and designers, for example, it might suggest that the kitchen as physical space, whether as a room by itself, or in a more open-plan configuration, deserves consideration as a foodscape that can facilitate the interaction of the household. While this seems obvious, it needs to be understood in the context of well-intended twentieth-century experiments with the kitchen as a segregated food-preparation space, or a galley-style kitchen, neither of which was designed to accommodate a table at which diverse domestic activities, including eating, could take place. The findings of this study might also induce designers to think about matters of comfort and convenience in the design of eating spaces, for example the use of non-precious materials that facilitate relaxation and are easy to keep clean.

The surprisingly traditional behavior of the participants seems to point to previously-dismissed potential opportunities for mindful eating in the household. This draws attention to the possible intrusions that may prevent the development of mindful eating practices. For instance, although many participants noted the absence or silencing of the television set during dinner, only one mentioned background music while eating, which has been shown to lead to longer meals and larger amounts eaten (Stroebele and de Castro, 2006). As regards current trends towards foods brought in from stores and restaurants in the household, and branded foods on the kitchen or dining room table, one question that arises is the effect that these might have on the quality of the meal experience. In France, for instance, children's socialization process focuses on the acquisition of skills and knowledge to appreciate food and to make distinctions about the taste and preparation of food and meals (Leynse, 2006) and to develop a sense of place (Leynse, 2007). Is the acquisition of such skills still possible when the table includes cellophane- and vacuum-packed industrial foods?

With regard to the importance of the table, as both a physical space for congregating but also as a socialization milieu, its role and the mechanisms of its influence warrant further inquiry. Why does the table have such power of attraction and transformation? The setting does have an impact on the pleasure we derive from food. For instance, Meiselman *et al.* (2000) found that the same meal served in a restaurant was deemed more acceptable than when served in a laboratory and the meal served in a laboratory was deemed better than when served in a cafeteria. Is it the case that a meal eaten at the kitchen table simply tastes better? By which magical property does this happen? Does the table top setting and design have any bearing on the taste of the food? Interestingly, over 35 participants admitted, on the questionnaire (the last item of the Food Involvement Scale), that table settings mattered to them and this was even more so for our concentrated group that submitted the more detailed drawings: is this the residue of childhood eating habits centered on the table, or does it speak to the power of the tabletop design and table objects in triggering mindful eating? Just as different eating locations create different expectations that, in turn, influence the acceptability of the food served in these locations (Edwards *et al.*, 2003), it is probably more likely that the table top setting and its objects create expectations that focus one's attention on mealtime proceedings and thus promote the mindful appreciation of food.

Finally, many participants noted that they took part, albeit in modest measure, in meal preparation (often in the form of washing/peeling vegetables). This is in line with the findings from industry experts that close to three quarters of today's teens are involved with meal preparation (Lempert, 2004). To what extent do domestic foodscapes facilitate the acquisition of food preparation skills? Our concentrated group of ten participants who submitted more detailed drawings also expressed more involvement in cooking and derived more pleasure from it than other participants. Was this due to parental coaching, or might it be attributed to some feature of the domestic foodscapes? For instance, observations of children at play in a toy kitchen revealed marked differences between boys' and girls' behaviors (Matheson *et al.*, 2002: 202). Girls involved the doll (even carrying it while cooking) in their activities in the toy kitchen and prepared complicated recipes or multi-course meals. Boys were more likely to engage in repairs and use the microwave, would serve snacks (rather than prepare a meal) or served excessive amounts of food. The potential of the domestic foodscapes to invite and promote exploration and development of cooking skills warrants further inquiry.

26.7   Limitations and future research directions

Naturally, our findings must be interpreted within the limitations of our chosen sampling frame and methodology. These students were not versed in making architectural drawings, and so their drawings have to be read more as a product of the imagination than an accurate portrayal. This circumstance has a certain advantage, in that students would have been unaware of the need to adhere to the conventions [i.e., straight lines (only one student used a ruler); drawing to scale; line thicknesses] that architects depend on to communicate precisely. Hence, what was on the page was, essentially, what the student wanted on the page, in response to the questions being asked.

The decision of what to include, then, and the prominence it received in the drawing consequently become, in and of themselves, opportunities to evaluate the emphasis given by the student to a particular part of the drawing's content, and became a variable for understanding the significance of the domestic foodscape of each participant. Admittedly, this cohort of students is already pre-selected as interested in food, is university-educated, middle class, and can afford an Ivy League education (participants were recruited at Cornell University where the first author taught). Future research should extend this investigation to households of different socio-economic backgrounds, that are equally if not more at risk of developing negative health consequences related to mindless eating.

We have implicitly defined mindful eating in relationship to its opposite: mindless eating or a mode of consumption characterized by inattentive and often excessive ingestion of high-calorie density foods and snacks (Wansink, 2007). And yet, 'mindful' eating begs to be defined more thoroughly. Could it mean different things to different individuals? Mindful of food's origins? true costs? cultural traditions? family culinary traditions? And how does mindful eating reveal itself behaviorally? Can an individual eating a burger on the run still mindfully (as in consciously) savor it? Must mindful eating require more time and effort than its alternative? Can one realistically eat mindfully all the time? Advancing our comprehension of mindfulness in this context appears imperative to devise promotional strategies (and not merely in a marketing context but also from a health promotion perspective) effectively tailored to our evolving modern obesogenic environments.

Future research could also rely on different methodologies to document not just the physical features of the domestic foodscapes but also what gets eaten (and how) in relation to the domestic foodscapes. For instance, longitudinal studies have examined changes in dietary behavior (e.g., Lake *et al.*, 2004) and a recently-released report by the NPD Group reveals that the top ten breakfast items served to children of new parents have changed very little in the last 20 years: the only recorded change is that fruit juices were part of the top ten in 1987 but are no longer, and waffles have now made it to the top ten list (NPD, 2007). Yet, the longitudinal impact of domestic foodscapes along with the eating behaviors therein begs to be studied further. What will our participants' eating behavior look like in 20 years? What will the eating behavior of today's toddlers be when they reach adulthood, especially for those now living in tight quarters with open-space floor plans and no 'kitchen' table *per se*?

The growing popularity of prepared foods and restaurant 'take outs' has been shown to impact the definition of 'homemade' and matters of family identity (Moisio *et al.*, 2004). What impact will it have on the domestic foodscapes? What impact will technology have? For instance, many manufacturers of commercial cooking equipment are now offering residential versions or product lines that alter the very meaning of cooking. For example, the 'Turbo Chef', used by industry giants like Subway and Papa John's Pizza, is now available for the home (at a price tag of $7,500). The oven will roast a chicken in 20 minutes and bake a frozen pizza in two and a half minutes. Will mindful eating prove to be compromised by this further distancing of the food-preparation experience, the feel, smell, and look of ingredients being washed, chopped, seasoned, etc., from the eating experience? Will this distancing devalue the foodstuff itself because the eater loses the ability to see how much of an ingredient was wasted, for example? On the other hand, hopefully the technology of the future will offer means to enhance the quality of the meal at the kitchen table. Notwithstanding Cowan's argument (1983) that microwaves and similar innovations simply raised expectations of food-on-demand and actually increased the workload of the home-maker, it is easier to buy a pre-cooked lasagna and heat it up than to make one. And if that lasagna is served at the kitchen table where family members have the time to relax and share the day's events, what are the net repercussions of such an outcome? Isn't that mindful eating? Inevitably, mindful eating in the domestic foodscape, is a concept of complexity: acknowledging its myriad constituent elements is in itself a formidable task.

26.8   References

Attfield, J. 1999. Bringing modernity home: open plan in the British domestic interior. In *At Home: An Anthology of Domestic Space*, ed. I. Cieraad, 73-82. Syracuse: Syracuse University Press.

Bell, D. and Valentine, G. 1997. *Consuming Geographies: We Are Where We Eat*. London: Routledge.

Bell, R. and Marshall, D. W. 2003. The construct of food involvement in behavioral research: scale development and validation. *Appetite* **40**: 235–244.

Bisogni, C. A., Falk, L. W., Madore, E., Blake, C. E., Jastran, Sobal, M. J. and Devine, C. M. 2007. Dimensions of everyday eating and drinking episodes. *Appetite* **48**: 218–31.

Buckley, S. 1996. A Guided Tour of the Kitchen: Seven Japanese Domestic Tales. *Environment and Planning D: Society and Space* **14**: 441–61.

Bugge, A. B. and Almås, R. 2006. Domestic Dinner: Representations and practices of a proper meal among young suburban mothers. *Journal of Consumer Culture* **6**(2): 203–28.

Bullock, N. 1988. First the kitchen – then the façade. *Journal of Design History* **1**(3–4): 177–92.

Buijzen, M., Schuurman, J. and Bomhof, E. 2007. Associations between children's television advertising exposure and their food consumption patterns: A household diary–survey study. *Appetite* **50**: 231–39.

Busch, A. 1999. *Geography of Home: Writings on Where We Live*. New York: Princeton Architectural Press.

Corrodi, M. 2005. On the kitchen and vulgar odours: the path to a new domestic architecture between the mid nineteenth century and the Second World War. In *The Kitchen*, ed. K. Spechtenhauser, 21–24. Basel: Birkhauser.

Cowan, R. S. 1983. *More Work For Mother: The Ironies Of Household Technology From The Open Hearth To The Microwave*. New York: Basic Books.

Cromley, E. 1996. Transforming the Food Axis. *Material History Review* **44**: 8–22.

Cummins, S. and Macintyre, S. 2002. A systematic study of an urban foodscape: the price and availability of food in Greater Glasgow. *Urban Studies* **39**(11): 2115–30.

Davidson, K. K., Maschall, S. J. and Birch, L. L. 2003. Cross-sectional and longitudinal associations between TV viewing and girl's body mass index, overweight status, and percentage of body fat. *Journal of Pediatrics* (July): 32–37.

Domosh, M. 1998. Geography and gender: home, again? *Progress in Human Geography* **22**(2): 276–82.

Dorfman, C. J. 1992. The garden of eating: the carnal kitchen in contemporary American culture. *Feminist Issues* **12**(1): 21–38.

Edwards, J. S. A., Meiselman, H. L., Edwards, A. and Lesher, L. 2003. The influence of eating location on the acceptability of identically prepared foods. *Food Quality and Preference*, **14**(8), 643–652.

Elliott, C. 2008. Entertaining eats: children's 'fun food' and the transformation of the domestic foodscape. Presented at 'Domestic Foodscapes: Towards Mindful Eating' conference, J. LeBel, R. Richman Kenneally (Eds.), Montreal, Canada, March 21–22.

Fiates, G. M. R., Amboni, R. D. M. C. and Teixeira, E. 2008. Television use and food choices of children: qualitative approach. *Appetite* **50**: 12–18.

Floyd, J. 2004. Coming out of the kitchen: texts, contexts and debates. *Cultural Geographies* **11**(1): 61–73.

Freeman, J. 2004. *The Making of the Modern Kitchen: A Cultural History*. Oxford: Berg.

Glassie, H. 2000. *Vernacular Architecture*. Bloomington: Indiana University Press.

Glassie, H. 1982. *Passing the Time in Ballymenone: Culture and History of an Ulster Community*. Philadelphia: University of Pennsylvania Press.

Guidetti, M. and Cavazza, N. 2008. Structure of the relationship between parents' and children's food preferences and avoidances: An explorative study. *Appetite* **50**: 83–90.

Hand, M. and Shove, E. 2007. Condensing practices: ways of living with a freezer. *Journal of Consumer Culture* **7**(1): 79–104.

Hardyment, C. 1991. *From Mangle to Microwave*. London: Blackwell.

Hays, J., Power, T. G. and Olvera, N. 2001. Effects of maternal socialization strategies on children's nutrition knowledge and behavior. *Applied Development Psychology* **22**: 421–37.

Hollows, J. 2000. *Feminism, Femininity and Popular Culture*. Manchester University Press.

John, D. R. 1999. Consumer socialization of children: a retrospective look at twenty-five years of research. *Journal of Consumer Research* **26**: 183–213.

Johnson, L. and Lloyd, J. 2004. *Sentenced to Everyday Life: Feminism and the Housewife*. Oxford: Berg.

Kähler, G. 2005. The kitchen today: and a little bit yesterday: and tomorrow, too, of course: on kitchen styles and lifestyles. In *The Kitchen*, ed. K. Spechtenhauser, 75–92. Basel: Birkhauser.

Kostof, S. 1995. *A History of Architecture: Settings and Rituals*. New York: Oxford University Press.

Lake, A. A., Rugg-Gunn, A. J., Hyland, R. M., Wood, C. E., Mathers, J. C. and Adamson, A. J. 2004. Longitudinal dietary change from adolescence to adulthood: perceptions, attributions and evidence. *Appetite* **42**: 255–63.

Leatherbarrow, David. 2004. Table talk. In *Eating Architecture*, ed. J. Horwitz and P. Singley, 211–28. Cambridge, MA: MIT Press.

Lempert, P. 2004. Youth must be served, *Progressive Grocer* **83**(6), 18.

Leynse, W. 2007. Journeys through 'ingestible topography': socializing the 'situated eater' in France. In *European Studies: Food, Drink and Identity in Europe*, ed. T. M. Wilson, 129–158. Amsterdam: Rodopi.

Leynse, W. 2006. Learning to taste: Child socialization and food habits in France. *Appetite* (Abstracts) **47**: 384–401.

Llewellyn, M. 2004. Designed by women and designing women: gender, planning and the geographies of the kitchen in Britain 1917–1946. *Cultural Geographies* **11**(1): 42–60.

Lupton, E. and Miller, J. A. 1992. *The Bathroom, the Kitchen, and the Aesthetics of Waste: A Process of Elimination*. Cambridge, MA: MIT Visual Arts Center.

Matheson, D., Spranger, K. and Saxe, A. 2002. Preschool children's perceptions of food and their food experiences. *Journal of Nutrition Education and Behavior* **34**: 85–102.

Mehrabian, A. and Russell, J. A. 1974. *An Approach to Environmental Psychology*, Cambridge, MA: MIT Press.

Meiselman, H. L., Johnson, J. L. Reeve, W. and Crouch, J. E. 2000. Demonstrations of the influence of the eating environment on food acceptance. *Appetite* **35**: 231–37.

Mellin, Robert. 2003. *Tilting: House Launching, Slide Hauling, Potato Trenching, and other Tales from a Newfoundland Fishing Village*. New York: Princeton Architectural Press.

Mestdag, I. 2005. Disappearance of the traditional meal: Temporal, social and spatial destructuration. *Appetite*, **45**, 62–74.

Mintel. 2007. *Off-premises Eating, US*, London: Mintel International Group Limited.

Moisio, R., Arnould, E. J. and Price, L. L. 2004. Between Mothers and Markets. *Journal of Consumer Culture* **4**: 361–84.

NRA. 2008. *2008 Restaurant Industry Forecast*. Washington, DC: National Restaurant Association.

NPD. 2007. When it comes to feeding young children breakfast, new moms today take cues from their own childhood. Press release, October 24.

NPD. 2006. Convenience trumps health as the driving force behind how America eats. Press release, October 24.

Orrell-Valente, J. K., Hill, L. G., Brechwald, W. A., Dodge, K. A., Pettit, G. S. and Bates, J. E. 2007. 'Just three more bites': An observational analysis of parents' socialization of children's eating at mealtime. *Appetite* **48**: 37–45.

Parr, J. 1999. *Domestic Goods: The Material, the Moral, and the Economic in the Postwar Years*. Toronto: University of Toronto Press.

Pascali, L. 2006. Two stoves, two refrigerators, due cucine: the italian immigrant home with two kitchens. *Gender, Place and Culture* **13**(6): 685–95.

Plante, E. 2002. *The American Kitchen 1700 to the Present: From Hearth to Highrise*. Markham, Ontario: Fitzhenry and Whiteside.

Pliner, P. and Zec, D. 2007. Meals schema during a preload decrease subsequent eating. *Appetite*, **28**, 278–288.

Pocius, G. 1991. *A Place to Belong: Community Order and Everyday Space in Calvert, Newfoundland*. Montreal: McGill–Queen's University Press.

Prown, J. 1982. Mind in Matter: An Introduction to Material Culture Theory and Method. *Winterthur Portfolio* **17**(1): 1–19.

Rosch, E., Mervis, C. B., Gray, W., Johnson, D. and Boyes-Braem, P. (1976). Basic objects in natural categories. *Cognitive Psychology*, **7**, 573–605.

Rolshoven, J. 2005. The kitchen: terra incognita: an introduction. In *The Kitchen*, ed. K. Spechtenhauser, 9–15. Basel: Birkhauser.

Rozin, P. 2005. The meaning of food in our lives: a cross-cultural perspective on eating and well-being. *Journal of Nutrition Education and Behavior*, **37**, S107–S112.

Rozin, P., Fischler, C., Imada, S., Sarubin, A. and Wrzesniewski, A. (1999). Attitudes to food and the role of food in life in the USA, Japan, Flemish Belgium and France: possible implications for the diet–health debate. *Appetite*, **33**(2), 163–180.

Sobal, J. and Wansink, B. 2007. Kitchenscapes, tablescapes, platescapes, and foodscapes: influences of microscale built environments on food intake. *Environment and Behavior* **39**(1): 124–42.

Smith, S. and Wheeler, J. 2002. *Managing the Customer Experience*. New York: Prentice Hall/Financial Times.

Sparke, P. 1995. *As Long as it's Pink: The Sexual Politics of Taste*. Kitchener, Ontario: Pandora Press.

Stroebele, N. and de Castro, J. M. 2006. Listening to music while eating is related to increases in people's food intake and meal duration. *Appetite* **47**: 285–89.

Supski, S. 2006. 'It was another skin': the kitchen as home for Australian Post-War immigrant women. *Gender, Place and Culture* **13**(2): 133–41.

Terrenghi, L. 2008. Design for sharing cooking experiences: potential and challenges of computing technologies. Presented at *Domestic Foodscapes: Towards Mindful Eating*, Montreal, March 20–22, 2008.

Turley, L. W. and Milliman, R. E. 2000. Atmospheric effects on shopping behavior: a review of the experimental evidence. *Journal of Business Research*, **49**(2), 193–211.

Upton, D. 1998. *Architecture in the United States*. Oxford and New York: Oxford University Press.

Wansink, B. 2007. *Mindless Eating. Why We Eat More Than We Think*. New York: Bantam.

Watkins, H. 2006. Beauty Queen, Bulletin Board and Browser: Rescripting the Refrigerator. *Gender, Place & Culture* **13**(2): 143–52.

Winson, A. 2004. Bringing political economy into the debate on the obesity epidemic. *Agriculture and Human Values* **21**: 299–312.

27

Kosher and halal meals

E. Clay, Shared Journeys, USA, G. Marks, USA, M. M. Chaudry, Islamic Food and Nutrition Council of America, USA, M. Riaz, Texas A&M University, USA, H. Siddiqui, White Jasmine LLC, USA, J. M. Regenstein, Cornell University, USA

Abstract: This chapter presents the information necessary for a professional to develop minimal proficiency for appreciating and providing kosher and halal meals, as well as other religiously defined meals. The aim is to sensitize professionals to the cultural, religious, political, economic, and public policy barriers to accessing 'good' meals, which for Muslims and Jews, and most persons of any religious faith, go a long way to making one holy and whole.

Key words: Jews, Muslims, religious food practices, kosher halal.

27.1 Introduction

The central focus of this chapter is what constitutes a good meal for Muslims and Jews and how to address the challenges of securing those meals. The primary audience for this chapter is food professionals, or professionals in training, many of whom will be unfamiliar and quite possibly uncomfortable with religious topics discussed openly in professional settings.

Addressing food issues compels us to see religious practices as deeply personal, but not private, with profound public implications that may surface many closely held assumptions about religious faith and make conversation difficult. This audience will span secularists and Christians, as well as persons of diverse religious faith who may know very little about Judaism or Islam in general, let alone their food practices. As the markets and locales for sacred foods expand, so does the need for education across differences.

This chapter may test the patience of devout Muslims and Jews, as it offers a very basic background on the practices, laws and beliefs of Islam and Judaism and as it attempts to translate those concerns and concepts into less familiar terms accessible to outsiders. Some statements may, on the other hand, challenge the preconceptions of non-Muslims and non-Jews about a subject they may know little about. The authors hope this chapter will promote a greater understanding within both the wider religious and secular communities alike.

While there is one Torah and one Quran, there are diverse, practical interpretations of many of the laws. But the diversity is not limitless, so this chapter will outline major variations, and points of conflict, within each tradition. What is most vexing, however, from a descriptive point of view, is the almost infinite variation in the ways individuals choose to be observant. While this chapter cannot address questions regarding individual practices, it should provide the information to proficiently engage in conversations with individuals that result in our being able to honor their needs for good meals.

This chapter addresses how a meal is perceived. For Jews and Muslims, a meal needs to conform to the formal religious rules for foods. So foods and the recipes that turn foods into meal components are subject to detailed rules affecting processing and preparation. The meal is framed and structured by religious considerations, e.g., the role of a prayer at the beginning and end of a meal and other aspects of the meal itself. The meal is part of a religious cycle of ordinary and special meals. The latter are holiday fasts and feasts. Religious considerations may define who shares the meal.

27.2   What is a 'good' meal for Muslims and Jews following the food laws?

In Judaism and Islam, all the good that is around us and all that is possible comes from and through G-d. [The custom of not spelling out G-d is one that belongs to some within the Jewish faith including the senior author, who thanks his co-writers for indulging him.] The fact of something is neither good nor bad; it just is. Yet the meaning of something is either in keeping with G-d, and good, or it is at odds with G-d and not good. To become one with G-d is to become sanctified, transfigured and holy. Becoming holy or good is not just emotionally pleasant, or physically satisfying, but profoundly whole. In the Hebrew Bible, the very first commandment given to humans is a dietary law, 'From the Tree of Knowledge of Good and Evil do not eat'. Much later, G-d promises to write such knowledge upon human hearts. We live in an intermediate time.

For Jews and Muslims, the Torah and the Quran are G-d's communication to human beings of the laws and teachings that guide human beings in assessing what is good or not good. The Law is not G-d, but rather the communication from G-d. Obedience brings one closer to G-d. For Christians, Hindus and Sikhs, this resonates with their theistic teachings. For Buddhists, the Dharma fulfills a similar role as the law and teachings.

What is meaningful is not neutral. In philosophical or religious approaches that are atheistic or agnostic, matters of meaning predispose adherents to endorse and accept, disapprove and reject, or merely watch carefully what they experience. Confucians and Taoists have accepted texts that serve as reference points for their judgments. Secularists and atheists are part of less centrally organized and oriented cultural and social movements that have arisen in opposition to the misuse or excessive control of religious (or in some cases, cultural) conventions. Secular and atheistic approaches have produced specific public policies and social norms that have from time to time received popular support, but they have not produced any singular, enduring texts, outlining coherent organizations or teachings. In the West secular and atheistic approaches are mostly focused on policies that maintain and define the role of government to shape national and international order, especially through regulating human behavior and enforcing educational standards in public schools. In an analytical sense, governance, from this perspective, has become religion.

Discovering the reality of what is good, not good or not yet known is an ongoing process for all persons, whether they live life with reference to a text-based religion, a non-text-based religion (such as animism or some forms of paganism) or within the evolving ideas of secularism or atheism. Where there are no scriptures, there are practices and policies that function in very similar ways. Human beings need a point of departure for organizing and understanding themselves in relation to each other and this larger reality, often called G-d, and for limiting bad behavior.

This may seem ironic, or obvious, but no scriptures, or policies or practices exist because G-d or reality needs them. But rather human beings need some structures to understand and act in a world of ambiguity and complexity that boggles the human mind and heart. In some traditions, the teachings are described as being from G-d, from reason, from the regularities of nature, or simply how we have always done it, because humans need some orienting. What *is* will endure and change, whether or not human beings understand it, in spite of all efforts to understand. For all religious and secular traditions, there is a mystery in the relationship between the extremes of what is lawful and not lawful, what is permitted and not permitted, what is good and evil. For all traditions, there is a mystery in the relationship between the extremes of what is lawful and not lawful, what is permitted and not permitted, what is good and evil. Within a religious tradition, the devout can appreciate the balancing and integration of the extremes into a continuous, subtle fabric of experience, but to outsiders this dichotomy may appear very black and white.

Food is a major thread in the biblical epic and this same epic is again revealed in the Quran. Adam's moral decay was accompanied by a transformation of eating habits, from easily obtained fruit in the Garden of Eden to difficult to secure grains and vegetables outside of the garden. Later, humanity's moral decline at the time of Noah led to another dietary change, to the permission to eat meat. Yet, human history is not simply a matter of continual decline, but marked as well with periods of spiritual growth, such as Abraham and Moses, incorporating other dietary changes, the various laws of kashrut. Food serves both as a sign of degeneration as well as a means of amending it. The observance of kashrut, in addition to aiming for holiness, has proved a significant contributor to the Jewish national survival. In the words of Jean Anthelme Brillat-Savarin (1825/1972), 'The destiny of nations depends upon what and how they eat'. By our food, we declare and affirm who we are and who we want to be. To paraphrase Ahad Ha'am_(1856–1927), 'It is not so much that the Jews have kept kosher, but keeping kosher has kept the Jews'. Like many of these quotes that come from the Jewish or Muslim perspective, each would easily be accurate for the other community. Thus, also for Muslims, food provides an organizing framework.

The boundaries between what is permitted or not permitted, lawful or unlawful, good or evil forces us to become human subjects of a reality greater than ourselves. We accept and receive a world we did not create; this is humility. We are governed by our choices, not deterministic instinct; this is freedom. We become freer as we make more and more choices about what we have received; this is responsibility. The dualism forces us to examine the margins, where our judgments make a difference. What we may have determined to be good or evil may turn out to be the opposite. This possibility forces upon human beings an even more profound humility and a paradoxical imagination, minimizing the damage that might be done in the name of religious fervor. The great goodness that is possible to achieve with some personal and intellectual discipline, invokes awe and wonder.

27.3   A good meal makes one holy and whole

Both Judaism and Islam give constant attention to what enters the mouth throughout the day. This is an on-going part of life and serves to constantly remind adherents of their relationship to the Supreme Being. Participants need to constantly strive to meet the religious strictures that are subject to divine evaluation on the Day of Judgment.

The word halal means lawful for Muslims, while the word kosher means 'fit', referring to the fitness of food for eating. Thus, in both of these religious communities a meal is more than just a way to provide sustenance but also imbued with significant religious meaning. Obedience to G-d as defined by the food laws contained in scripture and interpreted in communities over generations makes people holy.

Many religious laws were actually easier to carry out or trivial when Jewish or Muslim communities lived in isolation from others. Nowadays, the food industry, in trying to serve these communities' needs, must deal with issues of operating in ways acceptable to multiple religious and secular groups, as well as to the demands of science and government policy, which often present some interesting practical challenges. (Note: All Islamic quotes are from Husseini, 1993.)

> Eat of the good things We have provided for your sustenance, but commit no excess therein. (*Al-Qur'an* 20: 81)
> O people! Eat of what is on earth lawful and good... (*Al-Qur'an* 2: 168)
> Eat of the good (tayyab) things we have provided you. (*Al-Qur'an* 7: 160)
> ...And prohibit for them only the foul. (*Al-Qur'an* 7: 157)

These quotes establish the routine eating of meals as a moral matter, including a moral judgment on overeating and waste.

By way of contrast, because of early conflicts over authority with Judaism, the moral status of Christians is more dependent upon what emanates from individual actions within one's community and that community's response, than what enters an individual's body. Most Christians, as well as secularists, do not have elaborate food laws, though many Roman Catholic and Anglican Christians will, for example, choose to eat fish rather than meat on Fridays and restrict their choice and intake of food during the season of Lent and refrain from eating in the period immediately before Holy Communion. Dieting as the careful selection of what one chooses for meals has, for many, assumed the role of a spiritual discipline or an act of religious devotion. This is not to single out Christians or secularists but to emphasize that what we put in our mouth has a special if not sacred status, regardless of 'religious' orientation.

For many Christians, the highest form of worship focuses on food. It is in the mystery of the Eucharist or Communion, where food and drink serve as the symbol for humanity's basic relationship of dependency upon G-d, the fruits of biological life and upon one another. Death is central to appreciation of food, both for animal and vegetable products. There are strong differences of opinion across Christian institutions, and no consensus, about how this ritual should be interpreted. For some, the serving and eating of ordinary meals becomes highly ritualized with prayers and seating patterns for those dining that reflects gratitude and obedience to G-d and one another. For others, only the formal ritual, conducted in a religious sanctuary, with specialized serving equipment, captures the meaning of the experience.

Not surprisingly, more often than not food functions as a metaphor in Christianity for teaching about justice and the consequences of injustice. Direct teachings about food in Christianity have to do with feeding the hungry and offering hospitality to others, but not necessarily all others. The Apostle Paul specifically suggests that Christians should not eat meat sacrificed to idols, because it may confuse immature persons of faith who may become drawn to or offended by the idol, because it may falsely imply an endorsement of the idol or appear to compromise the integrity of the community. The issue of food for idols is also a major theme in both Islam and Judaism.

The issue of 'idolatry' may sound quaint to the non-religious person. An idol is a misrepresentation of the power and nature of 'reality' or G-d. To try to live according to something that is likely untrue or unreal would be to miss the entire purpose of the life of religious faith and careen off into something that would be misguided and damaging to self and others. For the devout it is a core issue of only endorsing representations that are true to the integrity and ambiguity of 'reality' or 'G-d', properly understood.

In terms of science, for example, we accept the representations of Newtonian physics, even if we know it has misrepresentations of the actual workings of reality. Those representations are useful, and are accurate at a certain level of generality, even if they do not conform to the more rigorous, complex standards of quantum mechanics. So there is no problem of idolatry there. Likewise, within a specific religious tradition, G-d or 'reality' has many names and representations, for example, wisdom, decisive action, and fecundity are all dimensions of G-d or reality, but none, by themselves, are acceptable or adequate for describing the whole of 'reality' or G-d. Reality or G-d embraces what human beings experience as complexity, ambiguity and paradox.

The issue of idolatry is particularly vexing for both religious and non-religious persons who are faced with unfamiliar religious or secular practices. Two kinds of misunderstandings are most dangerous and undermine achieving respect and building relationships across different religious and secular practices. The first involves honoring a fractured view of reality that fosters invoking one god in competition and opportunistic conflict with another god. The second involves rendering divine justification for the pursuit of human vanity in order that opportunists may prey on others or seek unfair advantages for themselves.

First, those who adhere to Western monotheism, or its secular equivalent of the rational unity of reality, often fail to understand the nature or images of G-d and polytheism within Hinduism. When these different gods are named or revered, there is a tendency to recoil in judgment against dividing the divine, or dividing reality, against itself. But a fluent understanding of Hinduism would recognize that each individual god represents a dimension of the divine, often in animal or human form. Eating food offered to a fractured sense of the divine could be a problem for a devout Christian, Jew or Muslim, who holds to the inherent unity of G-d. Being seen eating such food might render suspect one's religious integrity. If one were viewed by a fellow religionist who was ignorant of the reality of Hindu practice, one might be subject to attack within the community, or one might cause the ignorant and insecure religionist to doubt the integrity of the discipline of religious faith.

Second, while the extreme growth in material consumption and the knowledge that has supported it has reduced the impact of illness and the harshness of life on many, consumerism has come to be seen to be in close association with sexual and gluttonous visions of power, virility and control that are at odds with the complexity of achieving human happiness. The images of advertising, dress and status are regarded by many as idolatrous in that they reduce what is the complex pursuit of wholeness in a larger context to the hedonistic pursuit of individual pleasure at the expense of others. From a religious or secular approach, the experience of abundance available to all is distorted to an experience of abundance for the self that can only be available to the few.

As a matter of differentiation among religions, it has been proposed that pork became increasingly associated with Christian holidays and diets in an effort to distance Christianity from both Judaism and Islam in Europe.

It bears mentioning that non-Muslims and non-Jews in the developed world often fail to appreciate the freedom that a strong structure of rules provides Jews and Muslims and other highly ritualized religions, in part because of the absence of simple, clear rules outlining spiritual disciplines and rituals within much of Christian theology. The rules and rituals of law may limit obsessive impulses, creating a context for greater freedom.

Meals are an opportunity to gain nourishment, but if that were their sole purpose, then we would be developing a set of optimized foods like we do for the animals. That would allow us to optimize what we learn from scientific reductionism but would make eating into a pretty boring event. So humans, clearly have lots of other factors that determine what they eat and why they eat it. Some of our readers will be food professionals and already know this, but the profession is so caught up in reductionism that even we miss the bigger picture.

The first and foremost constraint on diet throughout most of history was the limiting factor of what food was available. You can only include in the meal that which is available, given the land and growing conditions, given that which is within the span of trade at the time or given seasonality and the ability to store and preserve food through the off-seasons. Thus, food technology was often driven by the need to preserve food and what was possible evolved with the technology.

So the traditional foods that we eat today were originally very locally based. As commerce began to expand, new food items were added to the mix. But even within a relatively closed community, where meals have a lot of commonality, the individual homemakers will develop their own 'take' on these items. This may even evoke a competitive framework when people talk about the person who makes the best version of the local specialty. As options for different foods emerge, early adopters will use these new items. Others will stay closer to the traditional foods. Thus, personal/family decisions mitigate community norms.

As communities diverge and become less 'homogeneous' along various sociological measures, e.g., travel, education and employment, the range of foods found within the community increases and the pressure for conformity also may decrease. Obviously, issues of class and economic status will further impact the food choices even within a closed community. Food choices then become a reflection, in part, of where one stands in the community and one's relationship to that community. Purchasing and eating are not often done in isolation, and thus extraneous issues are part of the picture, i.e., what do you want your neighbors to see you buy and see you eat?

What is often fascinating is that despite the many ethnic, religious and cultural differences of various communities, the solution to practical problems of providing a functioning food supply often takes the same form. So the issue of minimizing expensive ingredients by 'dilution' with lower value materials, often carbohydrates, has been solved in many societies by almost identical products such as the Polish pierogi, Ashkenazic Jewish kreplach, Ukrainian vareniki, Italian ravioli, Korean mandu and Chinese wonton.

On the other hand, human biology, e.g., lactose intolerance, the resources available, e.g., does keeping dairy animals make sense in a particular environment, have led to interesting contrasts. Thus, in Europe, dairy and its preserved form, cheese, plays an important role as a 'storage' food, the Asian communities have processed soybeans (a legume that has an excess of the amino acid lysine) into similar materials and created a source of calcium along with a protein that can be dried and stored and then complimented with grains (with their excess of the amino acid methionine) so that good nutrition is obtained.

A 'good' meal reflects G-d's blessing of food and nutrition for our moral sanctification and physical well being. Jews and Muslims understand a good meal by the way in which it conforms to the kosher and halal food laws.

27.4   Obedience to G-d and the law makes one free

For Muslims and Jews, the Law is the intellectual framework for daily life. While the quote below happens to be from a Jewish source, the basic concepts frame the issue for those Muslims and Jews who practice these laws.

> And ye shall be men of a holy calling unto Me, and ye shall not eat any meat that is torn in the field (Exodus XXII:30). Holiness or self-sanctification is a moral term; it is identical with...moral freedom or moral autonomy. Its aim is the complete self-mastery of man.
>
> To the superficial observer it seems that men who do not obey the law are freer than law-abiding men, because they can follow their own inclinations. In reality, however, such men are subject to the most cruel bondage; they are slaves of their own instincts, impulses and desires. The first step towards emancipation from the tyranny of animal inclinations in man is, therefore, a voluntary submission to the moral law. The constraint of law is the beginning of human freedom.... Thus the fundamental idea of Jewish ethics, holiness, is inseparably connected with the idea of Law; and the dietary laws occupy a central position in that system of moral discipline which is the basis of all Jewish laws.
>
> The three strongest natural instincts in man are the impulses of food, sex, and acquisition. Judaism does not aim at the destruction of these

impulses, but at their control and indeed their sanctification. It is the law which spiritualizes these instincts and transfigures them into legitimate joys of life. (Grunfeld 1972)

There is a paradox that both Judaism and Islam must deal with – religious law is absolute and eternal, but its interpretation is evolving. Both Judaism and Islam have created legal systems with judges and courts that shape the on-going reality of their communities, providing continuity and change simultaneously. Thus, within Judaism and Islam, 'The Law' is not just the original text, but the body of interpretation that has developed over time. Change happens within the body of interpretation.

In both communities, because of the vicissitudes of history, the religious community has fragmented so that the process of interpreting the law today may take place within different sub-communities. Within Judaism there are many different 'centers of law' although some relative internal hierarchy seems to exist within each tradition – Hasidic, Orthodox, Conservative, Reform and Reconstructionist. The Sunni/Shiite split early on within Islam has led to a legal divergence between these two groups. Within the Sunni community, with its wide geographic and cultural diversity, four major schools of legal thought – known as the Hanafi, Maliki, Hanbali, and Shafi'i schools, in honor of their founders – have developed. Some legal differences exist among these groups.

An important consequence of this history, is that no one Jew or one Muslim or even one 'group' can speak as the definitive voice of either community. The actual legal religious 'authority' for any one person, whether Jewish or Muslim, is very much dependent on a number of factors of both fate and choice. Both communities function with a great deal of diversity and there is no ultimate authority that can define a set of monolithic religious values. There is no single governing body that can speak for all Jews or Muslims and expect its view to prevail. There is a tension between unity and diversity, integration and differentiation that is typical of any large, historic institution.

In secular terms, the Law provides the primary framework for the formation of a human being in a process referred to as socialization. The framework of Law should not be confused with the reality of the good. It is a discipline for achieving the good that is possible. Socialization processes are neither good nor bad. They just are the way individuals become persons formed within community. Socialization processes can be rigid, deterministic and inherently fragile, whether secular or religious. They can also be more robust and prepare the person for flexibility, strength, and clear personal and intellectual judgment. In the absence of socialization, each person does the good as seen in one's own eyes, without regard to the impact upon others or on the larger environment. The Law and other socialization processes curb individualistic opportunism.

While we may never be free of stereotypes, a few critical stereotypes, across religious and secular differences, hinder effective communication. As an example, the Christian Gospels thoroughly condemn legalism, rote adherence to rules for the purpose of self-justification or self-righteousness within the community. This has sometimes been used to criticize the core value of Jewish and Muslim rituals. However, both Jews and Muslims condemn such behavior, too. From a slightly different perspective, secularists often dismiss the felt need for religious order as willful ignorance and partisanship, rather than as an expression of a necessary discipline for daily life. All of us need to learn to be more tolerant and appreciative of the significantly different perspectives different groups have on a life well lived.

Indeed, all religious systems risk becoming venues for partisanship and sanctimoniousness, losing sight of the more modest goal of small steps toward the sanctification of self, others and the wider world; or the simple, graceful acceptance that what is may be blessed by G-d, without any vain human striving. All religions have teachings about acting in both complete transparency and complete secrecy as distinct, but opposite, ways to make known what is important and to hinder vainglory.

27.5   Kosher and halal food laws

The following is a general discussion of the key components of the food laws. These laws define what is and what is not to be eaten. To understand a kosher or halal meal, one needs to fully understand what cannot be eaten and how foods must be prepared to meet these laws. This section tries to frame this discussion by sharing with readers the key elements of these laws. For a more detailed paper, please see Regenstein et al. (2003). A more complex exposition of kosher commercially is given by Blech (2004) and of halal commercially by Riaz and Chaudry (2004).

27.5.1   Kosher and halal: overview of core practices

Auditing and certifying foods is highly developed within Judaism and increasingly so within Islam. While Jews are only 0.2% of the world's population and Muslims 20%, their food practices and labels are present around the globe, setting a standard for product certification processes that may be emulated by other religious and secular groups. Among Muslims, most will keep some part of the law, although many in America no longer keep the complete set. In Judaism, most do not observe the laws in their traditional form, but may still keep symbolic parts, especially around holidays. For both groups, the avoidance of pork is the most commonly retained practice. Secondarily, Jews may avoid shellfish and Muslims alcohol.

The role of religious supervision in the food production facility, which traditionally was provided by the local religious authorities and members of the community, is to simply validate (audit) the process. This was and is a matter of community governance, enforcing religious policy within a particular community. Kosher and halal food rules are not about the religious leaders' blessing of food, even though in most cases members of both communities say blessings of thanks before and after meals. Authorizing kosher does not even require a rabbi nor an imam for halal.

The kosher laws cover four major topics: (1) permitted animals; (2) prohibition of blood; (3) prohibition of mixing of milk and meat; and, (4) the special rules for Passover. In addition other minor rules can have an impact in interesting ways on how Jews deal with life in modern times and how the food industry must adjust its practices. The Muslim community has laws covering the first two areas, i.e., allowed animals and the prohibition of blood, along with rules prohibiting alcohol. The first two topics, as they do overlap, are taken up together below to better permit some comparison of the two sets of rules.

In addition, inanimate objects also require supervision. Indeed, the first item ever to receive rabbinical supervision, was invented by Israel Rokeach in Lithuania in 1870. It was a pareve soap, i.e., a soap that was made with no animal products.

27.5.2   Permitted animals

With respect to kosher mammals, these are only ruminants with a full split hoof and which also chew their cud, i.e., all of the allowed mammals are ruminants. This is a fairly small category of animals, making the choice of meats highly restrictive. Muslims have a wider range of acceptable animals, but eliminate all of the carnivorous animals and specifically ban the pig. Practically, the most important ban is that for the pig – for Jews because it is the only animal that has split hooves but does not chew its cud and for Muslims because it is Quranically defined as unhealthy. Note that the ideas of unhealthy and unclean are religious designations and need not meet a modern medical or sanitary definition – so the evaluation of the scientific validity of these claims as responses to religious mandates is not an appropriate scientific topic.

For Jews, the rules regarding permitted animals are mainly an issue of their food use. The use of slaughtered animal products for purposes other than food is not an issue. On the other hand, for many Muslims, the strong prohibitions against the pig in the Quran have led Muslims to prefer an almost total avoidance of the pig in all shapes and forms. This can be a challenge for a multi-cultural society, where some parts of society use the pig as an important part of their food culture.

Traditional domestic birds are acceptable as kosher, while many other birds, particularly birds of prey, are not. For any fowl to be considered kosher, it must have a *mesorah* (oral tradition), which may vary among communities. The acceptable Muslim list also prohibits any of the birds of prey but permits ostrich, emu and the like, which are not kosher. Kosher fish must have fins and removable scales, which is scales that can be removed without tearing the skin. Often practically, kosher fish will be filleted and most of the skin removed while leaving a small tab with the skin and scales still attached. For Muslims, they may eat any creature that lives solely in the water. But there exist a variety of interpretations of this ruling including leading some to only eat scaled fish.

27.5.3 Prohibition of blood

Both religions view blood as the 'life' fluid. Slaughter must be done in such a way that the animal bleeds out and dies in a humane manner. Taking the life of an animal is a privilege that the Supreme Being has given to humans. Slaughter needs to be done with respect for the animal and with technical proficiency. Jews assign this important task to specially trained religious slaughter men. Muslims permit all sane adult Muslims to slaughter.

Both Judaism and Islam provide far more humane guidelines for conducting slaughter than those alternate techniques used when the rules were promulgated. The slaughter in both cases requires a quick and deep horizontal cut across the neck of an animal that has not been stunned, cutting arteries and veins, as well as the trachea, but not severing the spine. The continued functioning of the nervous system permits the animal's heart to beat and to pump blood out.

Both religions require a prayer, i.e., the Jewish slaughter man says the prayer following a format that is used for all blessings in Judaism at the beginning of the slaughter session. The Muslim says a prayer, referring to the greatness of G-d, over each animal at the actual time of slaughter.

Religious slaughter has been drawn into a wider debate about what constitutes a humane method of slaughter. This debate includes the relative merits of first stunning the animal before slaughter. On a practical note, religious slaughter may in fact be more humane, as it leads to the release of endorphins, which allows the animals to die more peacefully. However, the relevant scientific research remains to be done. With poorly trained slaughter men and meat handlers, no slaughter, with or without stunning, will be humane. With well trained religious slaughter men using a very sharp, properly designed knife and equipment for holding a calm animal, religious slaughter may prove as humane as any other method of slaughter. Efforts to completely ban religious slaughter would prevent devout Jews and Muslims from keeping their faith.

Following slaughter, the Muslim returns the animal back to the secular authorities for further processing. Slaughter rules cover only meat and poultry, the warm blooded animals. Fish are not covered.

27.5.4 Additional rules unique to kosher

Additional steps at the time of slaughter

Jewish law involves a whole additional collection of mandated and extremely detailed practices following slaughter:

- an inspection of the internal organs focusing on the ruminant's lung,
- removal of specific veins and arteries (those with high levels of coagulated blood), and certain visceral fats,
- removal of the sciatic nerve to commemorate Jacob's bout with an angel,
- soaking the meat and poultry in cool water for half an hour, salting for an hour, and
- finally give the meat three rinses to remove additional blood.

These later steps of 'koshering' meat with salt were traditionally done at home, but are now generally done either at the initial processing plant or at the local butcher store, which is often now part of a regional brand supermarket. Meat that is not salted or rinsed within a 72-h period is not used as glatt kosher (a more stringent set of rules specifically for large domesticated ruminants (essentially cattle) adopted initially in America in the 20th century), but meat and poultry can be washed and used for regular kosher with up to three washes each done within 72 h of the previous wash.

Since many of the forbidden parts are located in the rear of the animal and since the process to remove them is extremely complicated and time-consuming, European Jews generally avoid the entire hindquarter.

Glatt ('smooth' in Yiddish) refers to an animal that upon examination is determined to have no lung adhesions or other defects on its lungs. Only a relatively small percentage of cattle prove to be truly glatt (a requirement of the Sephardic community for all animals, where it is referred to as 'Beis Yosef' meat), frequently as few as one in twenty for cattle. The practice among European Jews was that if small adhesions can be peeled from the surface of a cow's lung without perforating it so that it remains airtight, the animal is deemed glatt kosher; if there are more than two or three such defects, it is kosher although not glatt. Most Sephardim and Chassidim, on the other hand, only accepted glatt or Beis Yosef meat as kosher (although Sephardim continued to make use of the rear portion of the animal).

Prohibition of mixing milk and meat
This is based on the Biblical passage that appears three times in the first five books of Hebrew Scriptures ('Thou shalt not cook (seethe) the kid in the milk of its mother'.) Poultry, because it is slaughtered according to religious law, also may not be cooked or served with milk.

Not all foods fall into those two categories – so there is a third category, which is the 'all other' or 'neutral' category, generally referred to by its Yiddish name as '*pareve*', with many different English spellings. All items in the plant and mineral kingdoms that are permitted are pareve (almost all are), but a few animal items are also *pareve*, primarily fish, eggs and honey.

Fish are considered *pareve* and do not require religious slaughter. This for most Jews opens up interesting culinary opportunities – dishes that cannot be made with milk and 'kosher' meat, can often be made with fish. Although pareve, the custom is not to serve kosher fish directly with meat or poultry. The custom is to eat fish first at a meat meal, then rinse all the involved dishes and utensils, and then serve the meat course. One or two of the Orthodox sects also do not permit mixing milk and fish, e.g., the Lubavitch sect.

The stringency with which the separation of meat and milk is enforced often indicates where on the spectrum of kosher observance each person places him- or herself. Some Jewish homes essentially have two separate kitchens: meat and dairy with three sets of dishes: meat, milk and *pareve*, as do some food service establishments such as kosher hospitals and campus dining halls. Most kosher restaurants, on the other hand, simply function as either a dairy or meat establishment. The dairy restaurants will either feature traditional dairy dishes like blintzes (crepe suzette), latkes (potato pancakes), kugels etc. along with fish products (which are pareve). The meat restaurant is likely to be your typical meat, potato, vegetable type of place or Chinese food.

Special rules for Passover
The kosher laws provide specific rules for dealing with the week of Passover, a special holiday in late March or early April commemorating the Exodus from Egypt that freed the Hebrew people from bondage. For this holiday, special bread (Matzos) is baked and all other uses of the grains wheat, oat, rye, barley, and spelt are prohibited. Thus, all baked goods start with matzos as the raw material. (Some more observant Jews do not accept any such products for Passover other than the whole matzos.) In addition, the European rabbis of the late middle-ages added Passover prohibitions covering many other critical plant materials in the modern food supply, e.g., legumes (peas, beans, soy), rice and corn, and some seeds (poppy and sesame) and spices. These limitations have become significant. Many key food ingredients are made from rice and corn. For example, high fructose corn syrup (HFCS) is pervasive in the food supply and expressly prohibited by most European rabbis for Passover.

Other kosher rules
For certain foods to be considered kosher, at least one Jew must participate in the process of making the food to ascertain it is kosher. These include some cooked and baked products along with cheeses and, in some groups, milk (chalav Yisrael). Bread, to avoid confusion, cannot normally be made containing dairy ingredients, although making it in a special unique shape is permitted. A special 'rich' (egg) bread, called 'Challah' is used for the Sabbath and special holidays, although in America it has become widespread and certainly is available the rest of the week. Other kosher issues include special rules for grape juice-derived products, which has limited the development of fine quality wines until very recently; for the status of flour with respect to receiving the special blessings of Passover (the grain offering on the second day of Passover) before being used; the need to wait until the fourth year to harvest fruit; and the rules for functioning on the Jewish holidays, the most

important being the weekly Sabbath, where one is commanded according to the third of the Ten Commandments not to perform creative work including cooking. Some laws actually are more stringent if the food company, supermarket, restaurant or food store is owned by a Jew. For such Jewish companies, there are special rules for the purchasing of new equipment, and for baking bread, for example. These laws offer both challenges and opportunities to the food industry and to kosher consumers in a world where our food comes from around the globe and is often prepared in central facilities producing many different products in one food plant.

27.5.5   Additional rules unique to halal
Prohibition of alcohol
The Muslim community has a prohibition of drugs and intoxicants, most specifically drinking alcohol (ethanol, ethyl alcohol). Other chemical compounds that are scientifically classified and labeled as an alcohol are permitted. Obviously, the drinking of alcoholic beverages is prohibited, but this also affects the food industry and food preparation in other ways. Many flavors, e.g., vanilla, almond, and lemon, are extracted in alcohol and sold as alcoholic extracts. Most food processors and government regulators have simply accepted the assumption that any significant amount of alcohol is removed with boiling or baking. Many Muslims have been suspicious of this claim, and finally the USDA tested food processes to discover that significant amounts of the banned alcohol remain after cooking (Larsen, 1995). Other products may naturally contain alcohol, such as vinegar (which is derived from alcohol and is definitely permitted in Islam) and even fresh squeezed orange juice.

Prohibition of suspicious additives and drugs
The other aspect of this halal requirement is the 'drug' issue. Beyond the obvious 'recreational' drugs, various Muslim religious leaders will put out lists of substances that they rule are 'suspicious', such as phosphates and other additives, and, therefore, they recommend that Muslims not use these items. There is no consensus regarding the content of these lists.

27.6   Food, meal preparation and sharing in Muslim and Jewish homes

While substantial bodies of law shape what is understood to be good food, learning about good food, preparing food and sharing food with others occurs primarily within the home. In keeping with any cultural or religious minorities, when Muslims or Jews are minorities within larger societies, the emphasis on the role of home and family in learning becomes even greater. Food can be so much more than just something to satisfy our hunger, based on the perception of the person, depending on what sort of 'food values' we grew up with or learned to practice.

27.6.1   Concept of food in a Muslim home

Food and mealtimes are regarded with honor in a Muslim home. Food is considered a form of divine gift that exists not just to satisfy a Muslim's hunger but also to provide a lifestyle which creates and encourages gratitude, longevity, sustainability and contentment. The practice of taking the time to prepare meals for family or guests stems from appreciating the food as a gift in itself. This perception creates humility and allows the Muslim to take the time away from the demands of the day-to-day life, to honor such a respectful ritual of planning a meal and making meal choices. Eating 'on the go' is not widely practiced, let alone respected, way to get through the mealtimes as it definitely encourages the concept of merely fulfilling hunger.

Food is seen as a way to reach out to other people, a circle of family and friends or even strangers. Inviting others to share meals and joining together to prepare something exquisite can be the best medium to present and learn about the food practices, thoughts, and processes of a different culture. Working together opens a discussion about differences in cultural and religious practices. Such collaboration makes available a wealth of knowledge which may not have been as accessible before.

All in all, a life which consists of the body and the soul is considered a gift from G-d and keeping both of these in line with religious beliefs is the person's responsibility. This responsibility is first learned at home. Most Muslims believe that a person's body is a sanctuary and to protect it from 'bad' stuff is a person's responsibility. To accept and include something that is considered unholy, like alcohol or any pork items, would be considered a violation of a 'clean' body and soul. Eating practices nourish the body and the soul equally, hence the reason for ensuring that the guidelines continue to be met.

27.6.2   Food preparation, serving equipment and practical limits to sharing at home

In both Jewish and Muslim homes, food preparation and serving equipment has a religious designation. To keep either a kosher or halal home takes an economic commitment, a great deal of knowledge, a certain amount of extra time and a willingness to accept many restrictions when operating in a contemporary, secular society. Strict observers of halal practices want dishes and equipment thoroughly cleansed before use for halal cooking. There are, however, no especially dedicated dishes needed to keep a halal home. A Jewish home will need special meat and milk dishes for regular use and usually others for Passover. A kosher observant person needs to have at least four, and possibly six (fancy dishes for company) sets of dishes to operate throughout the year, not counting any *pareve* dishes or special items for the special meals often associated with the holidays.

The Jewish separation of milk and meat and the separate dishes for the Passover holiday leads to much more kitchen complexity in a Jewish home than in other homes. Thus, the investment in the management of the household is significant. The need to separate milk and meat means that all meals need to be defined as either meat or milk. There are varying amounts of waiting time between eating meat and then milk today ranging from the time it takes to rinse one's month with water up to one hour.

Usually, because of the shorter waiting period (when not using hard cheeses) for dairy to meat and the higher value in most societies of meat, the typical home will start with a dairy breakfast and usually have a dairy lunch and then switch to meat for supper. On the other hand, some will have a lunch of meat, which generally means any afternoon snack would have to be meat, or in practical terms, pareve. From meat to milk may be from 3 to 6 h.

For both Jews and Muslims, the bringing in of 'dish to pass/pot luck' food and/or participating in a 'dish to pass' elsewhere outside of the community with whom one's standards are not compatible presents a challenge. Those who bring food to these events must know the rules and may even be restricted to those of equal following of the law. In addition, dishes or plates taken out of the house may be 'removed' from the household kitchenware and are often not brought back for further use.

27.6.3   The Jewish home and the impact of prohibiting work on the Sabbath and any holy days

To the observant Jew, the Sabbath is a day of rest. No work whatsoever may be done, with the exception of heroic efforts to preserve life. On that day even food preparation is restricted because of the very extensive laws relating to the prohibition of any form of work (work is historically defined by activities associated with the building of the Holy Temple in Jerusalem). Every Sabbath or holiday requires special preparations with respect to the meal: the stove top is normally left on and is placed beneath a thin sheet of metal cover, i.e., the blech – that helps avoid the temptation to fiddle with the dials and also distributes heat across the entire surface, permitting more foods to be placed on the 'stove top'. A large hot plate can also be used to keep the foods warm. Refrigerators and ovens need to have their 'light bulbs' turned off. Thus, the regular meal is sometimes defined as the opposite – a meal that is easier to prepare and can use all the gadgets and be prepared last minute. Some kosher agencies even give supervision to appliances that are specifically designed to operate without violating the Sabbath laws, which prohibit actions that are religiously defined as work.

Within Jewish and Muslim food practices, the home is equally central, but the way that centrality functions is significantly different because of the complete prohibition of work on holy days as well as the extensive food laws pertaining to the separation of milk and meat, the origins of specific types of food and the special laws for Passover. Muslim devotion is a disciplined way of life that emanates from the home. Jewish devotion is a disciplined way of life that has many prescribed rules for worship that is grounded within the home. In the sense that a Jewish home is also a center of weekly and seasonally mandated rituals of religious worship, Judaism is closer to Hinduism and other more ancient Eastern religions, than to Islam. Within Islam the rituals of devotion are more closely associated with daily life than with holy days. Even though special activities for holy days do exist, it is more a matter of emphasis whether something is routine and sacred or special and sacred.

27.7 Routine meal etiquette, worship, and hospitality: eating, washing, prayers, blessings

Assuming that the meal meets halal or kosher standards, both Islam and Judaism have elaborate rituals associated with routine meals as occasions for gratitude and blessings. Nevertheless, the two religions differ in emphasis: Islam emphasizes more the way in which the meal is consumed; Judaism emphasizes more what is consumed.

27.7.1 Muslim practices

The Quran and the traditions of the Prophet Muhammad (PBUH: Peace Be Upon Him), offer Muslims fairly extensive rules concerning the eating of routine meals. Again, it can be seen that the act of eating is sanctified as both a reminder of the overall scope of the religion and as a way to sanctify the normal, the everyday acts such as eating (Husseini, 1993).

> Eat of the good things We have provided for your sustenance, but commit no excess therein. (Al-Qur'an 20: 81)

1. Never criticize any food.
2. Du'a (say a supplication) before each meal.

> The devil considers food lawful for him when Allah's name is not mentioned over it. (Muslim)
> O Allah! Bless the food You have bestowed upon us and protect us from the torment of hell. In the name of Allah we start. (An Nasa'i)

3. Du'a after each meal.

> Praise be to Allah the One Who gave us the food and the drinks. Praise be to Him Who made us Muslims. (Tirmidhi and Abu Dawud)

4. Eating less.

> Nothing is worse than a person who fills his stomach. It should be enough for the son of Adam to have a few bites to satisfy his hunger. If he wishes more, it should be: One-third for his food, one-third for his liquids, and one-third for his breath. (Tirmidhi and Ibn Majah)

5. Eating slowly.
6. Moderation.

> Eat of the good things We have provided for your sustenance, but commit no excess therein... (Al-Qur'an 20:81)

7. Sharing food with family, friends, and others such as the wayfarer and the less privileged.
8. Eating together.

> Eat together and not separately, for the blessing is associated with the company. (Ibn Majah)

In addition, Islam has developed a set of 'good eating habits' that modern society may think of as 'outdated' but in fact addresses many of the issues that have changed the social covenant within and between families (Husseini, 1993). These remain good reminders for all.

1. Sit down while eating and/or drinking.
2. Eat together and share food.
3. Serve others first, especially guests.
4. The host is the first to start eating and the last to finish.
5. Take smaller portions of food on the plate.
6. Eat from one's own side of the plate and not from the middle.
7. Finish the plate of food without leftovers. (It looks like sustainability is nothing new!)
8. Eat with the right hand.
9. Wait for everyone to finish before finally leaving the table.

Often the meal even today is eaten without benefit of table but rather as a meal on the floor. Often the men and women sit separately, and may even sit in separate rooms. In some Muslim cultures the use of one's right hand for eating is still common, while others may use the various 'breads' as the way to eat the food. Watching a Somali Muslim take spaghetti and meat sauce, mix it with one hand, create small balls of spaghetti and eat that with no mess or spillage is truly an art form. While the degree of observance about using the right versus the left hand is changing, historically only the right hand was designated 'clean' and would be used for greeting and eating; the left hand would be reserved for more modest activities of bathing and bodily functions.

27.7.2   Jewish practices

While both Jews and Muslims say a prayer before and after eating, for Jews the prayers depend on the actual foods consumed and whether, therefore, it is even defined as a meal or a snack. Jewish tradition is that 'bread' is special and its presence becomes the definition of a meal. This changes the nature of the blessings that are needed.

For Jews, plant foods have special specific prayers – the major food prayers cover the following. Notice that there is no special blessing for animal products, even those not evolved directly from slaughter – for which Noah and humans were given permission after the flood because of their desire.

All Jewish blessings start with the same six Hebrew words, 'Blessed are Thou our Lord, King of the Universe, who provided....'

Bread (Motzi)
From the ground (Adama)
From a tree (Etz)
From a vine (grapes) (Hagafen)
Grains other than bread (Mezonos)
Other foods not covered by the above categories (Shehakol)

Both bread and wine are foods processed from raw materials, the resulting product becoming much more valuable and useful than the original state, an apt description of personal growth as well. The unique status of bread and wine is reflected in the hierarchy of blessings. With all other foods, the degree of specificity of the blessing drops when they are processed, both bread and wine, however, actually acquire a blessing for themselves. Before eating a dish of bulgur wheat or barley, a person is required to recite the blessing for 'creating the types of grains', while bread calls for the higher form of blessing, '*ha-motzi lechem min ha-arertz*' ('brings forth bread from the earth'). Similarly, plain grapes require the generic 'created the fruit of the tree', while before drinking wine one recites the more specific '*borei peri ha-gafen*' ('created the fruit of the vine'). Any other type of juice requires the less specific blessing of '*shehakol*'.

However, if the event is defined as a meal, then only the prayer over bread needs to be said, but is coupled with the need to wash one's hands ritually with a blessing before the blessing over bread and also requires a longer closing 'Grace After Meals'. On the Sabbath and festivals, the washing and the prayer over the bread is usually preceded by the Kiddush, the blessing of the wine. In some communities the wine is shared with all (including the children) at the table.

In some cases, one even ends up with two blessings over one food, the best example being the cereal product 'Crispex', which has as its major component on one side corn and on the other side as its major component wheat. Since these require different prayers and each is totally dominant on one side, the rabbinical decision has been to go with two prayers, turning the product over to expose the side related to the current blessing.

The only Biblically-mandated benediction is after a meal (Deuteronomy 8:10), which according to most rabbis must include the consumption of bread, the Torah directed the *Birkhat ha'Mazon* (Grace after Meals), consisting of three blessings. In addition, the rabbis instituted an abridged version, called 'the blessing condensed from three', a special blessing recited after eating any of the Seven Species, 'a land of wheat and barley and vine and fig and pomegranate; a land of oil-olives and [date] honey' (Deuteronomy 8:8) with which the land of Israel is praised. After a 'snack' the short blessing is recited. But if Motzi was said, then the longer blessing needs to be said.

If at least three men according to Orthodox tradition are at the meal, the start is done communally with additional verses by both the leader of the group with responses by the other participants. On holidays or other ceremonial occasions, e.g., weddings, bar and bat mitzvahs (the rite of passage to adulthood) or the bris (circumcision of the male child on the 8th day which is done even on the Sabbath or a holiday), the Benching (Grace after Meals) may be sung – this can be a very beautiful and meaningful sense of community.

The handling and preparation of a kosher meal also depends on the placement of the meal in the religious cycle. Is it a 'weekday' meal, a Sabbath meal, a holiday meal or a meal before or after a fast? This may also affect the timing of the meal.

27.8   Special meals: the religious calendar, fasting schedules, special foods and guests

In this section we elaborate on special meals in Judaism and Islam, as well as the religiously mandated and voluntary fasting central to both religions.

27.8.1   Judaism and the concept of Seudah

There are two Hebrew words for meal: *Arucha* and *seudah*. An *arucha* derives from the root *arak* (to wander), which also gives the words *oreach* (guest), *orach* (path), and *orcha* (caravan). Hence the original intent of *arucha* was 'food for the journey' or 'simple food'. There are many types of meals, such as an *aruchat boker* (breakfast), *arucha mispachtit* (family meal*), arucha iskit* (business meal), and *arucha meshutaf* (joint meal/meal for everyone). It is not the same as an *aracha* (reception/ceremony). On the other hand, a *seudah*, from the root *sa'ad* (to support), denotes a 'feast' or 'banquet', implying more than simply the intake of food, but 'to strengthen, to refresh (by food)'.

Not all meals are created equal. The Talmud (Pesachim 49a) differentiated between two forms of *seudah* – *seudah shel* mitzvah (a meal associated with a religious purpose), including the Sabbath and festival meals as well as those connected to life cycle events and rituals, and *seudah shel reshoot* (a feast for a temporal purpose). Pointedly, the meal before the Fast of Tisha B'Av (the day commemorating the destruction of both the first and second Temple along with the exile from Spain), consisting of only one cooked food and bread, for Ashkenazim typically a cold hard-boiled egg and a roll, is still called a *Seudat Hamafseket* (Meal of Separation).

So the intent of seudah is not necessarily fancy food, but special food. As Ecclesiastes (3:4) expressed, there is 'a time to weep and a time to laugh; a time to wail and a time to dance'. There are occasions when a lot of food and fancy food is appropriate and other times calling for merely a hard-boiled egg and a bagel. Some occasions demand extravagant treats, while others require simple, but meaningful fare. A party commemorating a life cycle event calls for something special like wine, which would be most inappropriate at a meal to end a fast. Knowing when and what to serve is more than a matter of etiquette – it is a mark of understanding and commitment.

Following the destruction of the Temple in Jerusalem in 70 CE, the home table became a *Mikdash Ma'at* (a miniature Sanctuary), symbolically replacing the altar, as the Sabbath loaves replaced the *lechem ha'panim* (Showbread). Thus, the Jewish table, whether for an arucha or seudah, is more than mere physical nourishment, but also a source of spiritual sustenance and a place of worship. In Judaism, food, entertaining, and ritual are intertwined. Integral to every holiday and *simcha* (celebration) is a *seudat mitzvah*, establishing and enhancing the spirit of the occasion, many associated with traditional dishes.

Jewish celebrations are, by design, communal acts accompanied with shared meals. These feasts are much more than an opportunity to eat, although plenty of food is the general rule. Eating together builds relationships and reinforces the bonds of family and community. Among the most important roles of the seudat mitzvah is one of education. Indeed, the Sages structured the most well-known seudah, the Passover Seder, to serve as a pedagogic tool and its continuing popularity illustrates just how incredibly successful those ancient rabbis were at achieving their goal.

Although no other seudat mitzvah is as highly ritualized and structured as the Passover Seder, all of them offer unparalleled opportunities for learning in an informal atmosphere. After all, these meals provide occasions when various generations are represented and connected. There is an unrivaled opportunity to observe and ask questions. Children discover the appropriate customs, mood, and terminology as well as how to interpret nonverbal social cues (e.g., family roles, who sits at the head of the table, how are strangers treated, social humor, when it is time to be quiet). They learn how people interact. Indeed it is at such occasions that young children learn the power of speech as they observe diners breaking out in laughter in direct response to someone speaking or one person honoring another with deference. Perhaps the greatest lesson of a seudat mitzvah is the development of an emotional attachment to Judaism, which has been a key to its continuing survival.

One of the questions that comes up with the celebration of holiday meals is whether there is a requirement for meat. This is debated, but a fair proportion of Jews who are essentially vegetarian or even vegan at all other times, will eat meat (often poultry) on Friday night and the beginning of other major holidays.

27.8.2   The Jewish Sabbath

For Jews, the weekly Sabbath has a major impact on meals. For the Sabbath the meal has to be more than half cooked before the holiday starts and once a liquid food is removed from the heat source, it cannot be returned, while dry foods cannot be placed directly on the heat source. Since the Sabbath starts at sundown on Friday night, foods for the two religiously required meals on Saturday must be 'stable', i.e., they are often put into the oven and kept there for almost 24 h.

When holidays and the Sabbath occur on days next to each other preparation of meals can get very complicated. Meals for one event usually cannot be prepared on the other, so even longer preparation lead-times are needed. Special attention needs to be paid to getting the sequences to happen smoothly, to keep foods safe, and to still have enjoyable meals. Not all foods are able to meet all these requirements, so special recipes that are festive but have long hot holding times have been developed.

Further, Passover, with its eight days of special rules presents additional challenges that are made more interesting in some fervently Orthodox homes by the limitations that are imposed for only the first seven days, i.e., only dry matzos are used and any recipe calling for further processing of the matzos are avoided. This creates a unique singular culinary event on the eighth day when many foods eaten throughout Passover by less fervent Jews are permitted by these Jews in the diaspora, i.e., Jews in Israel only celebrate Passover for seven days and, therefore, miss this opportunity (e.g., the famous matzos balls for soup and matzos brei – scrambled eggs with matzos).

There is the further issue of the Sabbath year, where agriculture in the holy land is highly restricted as the land lays fallow. This may impact the available meal options for those who observe these rules. Many of the normally available foods, especially plant-based foods, may be unavailable.

27.8.3   Islam and the concept of the Eid feasts

The Muslim community has two major feast days. Eid ul Adha and Eid ul Fitr. The first, Eid ul Adha, the feast of the Sacrifice, commemorates the almost sacrifice of Ibrahim's (Abraham's) son Ishmael. It is celebrated during the period of the Haj to Mecca and Medina by those able to go; ideally one should go at least once in one's life if one can afford it. If one is on Haj, one goes into the dessert on the ninth day of Dhu al-Hijjah, the twelfth month, and the actual sacrifice will take place on the tenth day.

At the same time, Muslims around the world also slaughter an animal in preparation for this feast. This responsibility cannot be replaced by a cash or other in-kind donation. Ideally, the animal will be an intact animal that is at least a year old. Although all Muslim adults can slaughter an animal, this task is usually reserved for the men of the family. Once the animal has been properly killed and dressed, the meat is cut into relatively random pieces so that it can be rather 'accurately', and without preferential treatment, divided into three portions, one for the immediate family, one for friends, relatives and neighbors, and one for the poor.

Ideally, the poor will be given the food ahead of time although the transfer can happen on the first day of the Eid, which is normally three days long. A custom in some communities is to break the offering for the poor into six to eight portions, so all who are needy can be served. If all of the portions are not claimed by the poor, who may well come around to homes and knock on doors, one needs to search out the poor.

One concern in modern times with the slaughter of many animals during the actual Haj in Saudi Arabia concerns whether or not better methods can be developed to distribute the shares for the poor around the world where people really need the meat.

The flow of the day is also unique. There is a special prayer just after sunrise on both Eids. A person doing the actual slaughter would fast until the time of the slaughter. However, if they are slaughtering for someone else including someone who is deceased, then technically a fast is not required although generally they will still fast. To end the fast, one of the quickest cooking meats would be cooked – generally one of the internal organs. So as not to 'cheat' the poor, other items will be reserved for the poor, e.g., the head, the legs, and possibly the skin (which can be sold for cash to the hide collector).

By the time the slaughter and distribution are completed, it will probably be time for noon prayers. After prayers, the family will sit down to a full meal including the meat that has been prepared following the sacrifice.

Most people will do the sacrifice on the first day, but it can also be done on the second or third day and those with the means may sacrifice an animal on more than one day. The rest of the holiday is generally spent with family.

The second feast is the Eid ul Fitr, which occurs in the first three days after the end of Ramadan in the month of Shanwal. This festival centers around the Zakat (charity) of the Fitr offering. A certain amount of money is given to the poor for each member of the family. The amount should be equivalent to the cost of a meal (in lieu of the actual meal, which is also permitted) for each of the people in one's household. (In the USA, the sum is about five to seven dollars per person.) This contribution can be done during Ramadan, so that it is received before the Eid, which is ideal, as it allows the needy time to purchase food for the celebration. But it must be done before standing for the special prayers after sunrise for the Eid. This must be completed before G-d will accept the fasting of Ramadan.

In many countries people will shower and put on 'new' clothes to celebrate this Eid. After the special prayers they will have something sweet to eat, essentially a dessert item. They will then greet their family and go around and give money to the children and other non-wage earners. Also, some people also give candy to the children.

For this Eid, the heavier meal will generally occur before the noon prayer – it would be something like a brunch. There is also the custom of visiting other people before the noon prayers. Sometimes the men will go alone, sometimes the women alone, and sometimes the family will go together. The idea is to eat a little bit of 'sweet' or snack food in each house. The goal would be to visit around six to eight families. Those families would reciprocate by coming to the guests' houses who would, in turn, also serve snacks and desserts. One then returns to one's own home to have a real meal. This meal, unlike the Eid ul Adha is not centered on meat. Then noon prayers are said at home or at the Mosque. In the afternoon, it is time for a family celebration and socializing within the nuclear family and closer friends and family. The goal is spending time together with loved ones. Supper is then a regular evening meal. For those who are able, the second and third day is mainly taken up with socializing with the closer circle. Meals are festive but do not have a special format.

The importance of the giving of charity in both Islam and Judaism is extremely important. However, the specific giving of food, especially meat, is unique to Muslim practice, although Judaism does require that the poor be provided with food, especially for the Sabbath and holidays, and persons from outside the community appearing in the synagogue are expected to be offered home hospitality.

27.8.4   The impact of fasting on meals

While our focus is on the meal, Jews and Muslims also incorporate fasts of various sorts as part of their religious regimen. So that, too, needs to be considered. For Muslims, for whom fasting is much more common, special meals center more often on the beginning and end of the fast. The meals before and after the fast need to be 'more nutritious' but also help with beginning a fast and the often unrecognized greater difficulty of ending a fast without getting sick. For both Muslims and Jews, there are a number of meals that are defined by both holidays and fast days. Thus, a frequent or common meal may actually not be thought of as a special meal.

For Muslims, in particular, this entails dealing with the fast during the entire month of Ramadan, but also the many other fasts that a Muslim may self-impose. (The evening meal during Ramadan is more common as a communal meal than the Friday meal; a few even do breakfast as a communal meal.) These fasts have a major impact on both the nature of the meal and the timing of the meal. In addition, the communal meal following the communal Friday prayers may impact the weekly cycle of meals within the community, particularly within the Shia community. In other traditions this communal meal may be at other times during the weekend. Many Muslim communities do not have this tradition. (Of 100 mosques in the Chicago area, only three do the communal Friday meal.)

There are preferred food items for ending the fast (Husseini, 1993). Some may eat dates, sip water, and eat a pinch of salt. Others eat fruit cocktail, juice, fritter, or food traditional to a region – e.g., yogurt.

> The Length of the Fast Day: Dawn (approximately 1.5 h before sunrise) to sunset
> Suhur (predawn meal)

>> The difference between our fasting and that of the people of the Book [Jewish] is eating shortly before dawn. (Muslim)

> Iftar (ending the fast) Initially eat a light snack.

>> The people will continue to prosper as long as they hasten to break the fast (in time). (Bukhari and Muslim)

> The Maghrib Prayer is said (after sunset) and then a full course dinner may be eaten. This is a very substantial meal.

Part of the custom is for these meals to be a 'celebration', which leads to these becoming communal meals as part of a community fast.

In Judaism, there is one biblically mandated fast, Yom Kippur, a 25-h abstention from food and drink beginning at nightfall. Another major fast is Tisha b'Av (the ninth day of the month of Hebrew lunar month of Av) to commemorate a number of calamities that befell the Jewish people on this day, including the destruction of both Temples. In addition, the Jewish year contains a number of rabbinically ordained fast days that begin at dawn and last until nightfall, helping to commemorate various national tragedies of the past.

27.9   'Common' meals in Islam and 'special' meals in Judaism

27.9.1   A few Muslim meals
From south Asia
Lunch:

> Lamb Korma (tender lamb cooked in traditional Pakistani spices in a yogurt sauce)
> Plain Rice (boiled basmati rice)
> Dal (lentils cooked with turmeric and chili powder)
> Baigan Bhurta (roasted eggplant mixed with spices and cilantro)
> Kheer (rice pudding with almonds and raisins)

Dinner:

>   Matar Pulao (basmati rice cooked with green peas and whole spices)
>   Karahi Chicken (boneless chicken cooked with garlic, green peppers, onions and tomatoes with cilantro)
>   Bhindi ki Sabzi (okra lightly sautéed with dried red chilies and onions)
>   Paratha (pan fried flat layered bread)
>   Raita (yogurt side sauce with cumin and chopped cucumbers and cilantro)
>   Gulab Jamun (dried milk balls, deep fried and soaked in sugar syrup)

27.9.2   Jewish Sabbath and holiday meals

The Sabbath is ushered in with the lighting of candles and a blessing over a cup of wine (Kiddush) – wine symbolizes joy and fruitfulness – and departs with the lighting of a candle, a sniff of spice and a blessing over a cup of wine (the Havdallah ceremony). In between these ancient ceremonies, Jews pray, study and reflect and, in fulfillment of the commandment of *oneg Shabbat* ('enjoyment of the Sabbath'), Jews also socialize, sing and partake of three meals: the first meal on Friday night; the second on Saturday following morning prayer services; and, late on Saturday afternoon, *shalosh seudot* ('third meal').

Sabbath dinner possesses a singular ambience. The table is set with the family's finery. Candlesticks sprout dancing flames, casting a genial glow over the celebrants. Two loaves of bread, covered with a special cloth, dominate the table. After reciting the Kiddush and sampling the wine, the diners wash their hands, the bread is uncovered and the Hamotzi (blessing over bread) recited. A repast of favorite and traditional delicacies follows. Fish is a customary appetizer. Chicken or meat is usually served as the main course. Lively *zemirot* (traditional songs) are sung. Thus, a profoundly religious activity and an enjoyably gastronomic experience become one and the same. In the words of the writer Isaac Bashevis Singer, 'Our house was filled with the odor of burning wax, blessed spices and with the atmosphere of wonder and miracles'.

Ashkenazic-Polish Sabbath dinner

>   Egg challah
>   Gefilte fish with horseradish
>   Chicken soup with noodles
>   Roast chicken or brisket
>   Noodle or rice kugel
>   Carrot tzimmes
>   Babka or mandelbrot
>   Fruit compote

Sephardic-Greek Sabbath dinner

>   Pita bread

Bean soup or red lentil soup
Fish with plum, rhubarb, or tomato sauce
Eggplant-and-meat casserole or chicken in rice
Fresh fruit

Historically, morning prayer services were conducted very early in the day, even on the Sabbath and holidays, and since it is traditional not to eat before services, the first meal of the day is truly a break the fast. However, when worshippers returned home from synagogue, it was generally too early for lunch, typically the heaviest meal of the day, yet too late for a usual breakfast. Instead, following Sabbath and festival morning synagogue services and also on Sundays, Sephardim enjoyed a *desayuno*, a casual dairy meal consisting primarily of finger foods, including cheeses (kaskaval and feta), olives, fresh fruit, jams, yogurt, rice pudding, and *ouzo/raki* (anise liqueur). Ubiquitous to the *desayunos* of Turkish and Balkan Sephardim are a trio of pastries, the 'three Bs' – borekas, boyos, and bulemas. Since many of these items require assembly, what Ashkenazim call potchke, they are typically prepared by a group of relatives or friends, during which time there is much socializing. Some of the popular foods at a contemporary *desayuno* are:

Sephardic brown eggs (Huevos Haminados)
Egyptian slow-simmered fava beans (Ful Medames)
Greek marinated fried eggplant (Melitzanes Tiganites)
Turkish eggplant and cheese casserole (Almodrote de Berengena)
Middle Eastern stuffed grape leaves (Yaprakes Finos/Dolmas)
Middle Eastern chickpea spread (Hummus)
Sephardic egg casseroles (Fritadas)
Sephardic miniature pies (Pastelitos)
Sephardic small cheese pies (Quesadas)
Sephardic filled cheese pastries (Boyos)
Turkish filled phyllo coils (Bulemas)
Turkish turnovers (Borekas)
Sephardic egg cookies (Biscochos de Huevo)

The mood at the second meal is similar to the first, but the foods served are quite different. The prohibition against creative work on the Sabbath includes cooking and these restrictions play a role in the development of numerous Jewish dishes. Since most people prefer warm dishes, particularly on a chilly Saturday afternoon, slow-cooked dishes such as cholent/chamin (bean stews), kugel (baked or boiled puddings of noodles, potato, or matzos) and kishke/nakahoris [stuffed derma (i.e., the large intestine)] were devised to simmer over a very low fire until Sabbath lunch. These ancient dishes have remained an important part of Jewish cuisine through the ages as well as having inspired a host of slow-cooked dishes among non-Jews ranging from French cassoulet to Boston baked beans.

Beginning in the 8th century, the practice of serving hot food on the Sabbath took on special import with the emergence of the Karaites, a sect that denied the authority of the Talmud and rabbinic interpretation. One of the Karaite's principal doctrines literally interpreted the admonishment 'You shall kindle no fire throughout your habitations upon the Sabbath day (Exodus 35:3)' as forbidding any fire or hot food on the Sabbath. Thereafter, lighting the Sabbath candles to usher in the day and eating hot dishes for Sabbath lunch were no longer simply a matter of enjoyment, but an attestation of identification with rabbinical tradition.

Ashkenazic Sabbath lunch

Egg challah
Chopped liver
Marinated mushrooms or dill pickles
Cholent (Sabbath stew)
Potato kugel
Stuffed kishke
Babka or mandelbrot

Sephardic Sabbath lunch

Pita
Stuffed grape leaves
Eggplant salad
Hamin (Sabbath stew)
Huevos Haminados (long-cooked eggs)
Fresh fruit

The atmosphere at shalosh seudot, the participants saddened by the imminent departure of the Sabbath, contrasts with that of the rest of the day. The songs are slower and almost mournful. Shalosh seudot fare is generally simpler than that served at the other two Sabbath meals – consisting primarily of bread, cold fish and perhaps a few salads and some leftover dessert. This meal is often held in the synagogue between the afternoon and evening services.

On Saturday night following the Sabbath, many homes hold a party called *melaveh malkeh* (literally 'escorting the queen'). The Sabbath, being metaphorically viewed as a queen, is symbolically escorted away at this gathering. In the process the special feelings of the Sabbath are prolonged for a while longer. Food, of course, is an integral part of the event.

Each holiday has its own customs and traditional foods, differing among the various Jewish communities.

Ashkenazic Passover Dinner

Matza
Gefilte fish with horseradish
Chicken soup with matza balls
Stuffed veal breast
Matza kugel
Sweet potato tzimmes

Compote
Sponge cake or Nut torte
Sephardic Passover Dinner
  Matza
  Huevos Haminados (long-cooked eggs)
  Fish with egg-lemon sauce
  Mina (layered matza)
  Apio (sweet-and-soar celeriac)
  Keftes (vegetable or meat patties)
  Shoulder of lamb roast
  Artichokes
  Walnut crescent cookies or almond cookies
  Tishpishti (syrup-soaked nut cake)

27.10   Variations in meals within Judaism and Islam

We must note that for any one person, family or community group, the religious influence on a meal and on food choices is only one force that frames a meal. Individual and familial choice is always present and the ethnic/cultural heritage, discussed in many of the other chapters of this book, also help to frame a person's or group's food choices.

27.10.1   Culturally 'Jewish' food and meals

The history, economics, geography, and climate of each area in which the Jews settled determined what produce was available, what resources were at their disposal, and what societal influences affected them. Local fare was incorporated into the Jewish culinary repertoire while traditional recipes were refashioned or forgotten. Jewish food was all too often the product of poverty and scarcity, housewives created filling and flavorful fare from little. To make a cross-cultural connection, rural or southern cooking in the USA has often had this same quality.

Around the turn of the first millennium, the Jewish world dramatically changed. At this point in time, there were around one million Jews in the world, and only one-tenth of them lived in Europe outside of Spain. The vast majority lived primarily in the Muslim lands stretching from central Asia through North Africa. Then a series of events would begin a shift of demographics and influence. By the end of the eighteenth century, of the world's 2.5 million Jews, 1.5 million lived in Europe and a million in Muslim lands. In 1939, at the high point of Jewish demography, there were nearly 16.7 million Jews in the world with 9.5 million in Europe, another 4.8 million in America, the bulk of them on both continents Ashkenazim, i.e., people of European origin, and only one million in the Muslim lands, each group with drastically different forms of cooking and thinking about food.

As the authority and importance of the ancient Judean and then Babylonian communities waned, two important Jewish centers emerged in Western Europe – Iberia (called Sepharad after a city in Asia Minor), and the Rhine River Valley (called Ashkenaz after a kingdom in the upper Euphrates region). Although these two medieval Jewish communities were geographically close to each other, their experiences were light years apart: The Muslims of Iberia generated a relatively liberal, inclusive society; while Christian Franco-Germany developed into a restrictive, frequently hostile feudal world. As a result, Ashkenazim and Sephardim religious law, although both based on the Babylonian Talmud (there is also a Jerusalem Talmud covering the same areas of oral law, but which is considered less 'binding'), gradually developed differences in areas such as customs, law, liturgy, the pronunciation of Hebrew, mindsets, and foods.

The Jews of medieval Franco-Germany found themselves in an alien culinary environment. Their non-Jewish neighbors ate conspicuous quantities of pork and shellfish, used lard and butter for frying, freely mixed meat with dairy products, and possessed a completely different culture of food and hospitality. As a result, Ashkenazim required a great deal of ingenuity to adapt the local fare.

Medieval Jews and Muslims, on the other hand, possessed a similar Middle Eastern attitude toward food and hospitality. Both cultures, for example, attached a special importance to the spiritual cleanliness of food, using only the right hand to touch food and eating from a common serving platter (without using silverware or tablecloths), and placed a religious dimension on entertaining strangers as well as friends. In addition, Muslims eschew pork and blood, rarely eat shellfish, and only occasionally use dairy products in cooking. As a result, Sephardim easily adopted the dishes of their Muslim neighbors and vice versa. During the course of more then a millennium, Spanish Jews developed a highly sophisticated and distinctive cuisine – a synthesis of Iberian, Arabic, and Jewish influences – one that was much more refined and diverse than that of medieval Franco-Germany. The high regard for Sephardic cooking was reflected in the Arabic maxim 'wear Muslim clothes, sleep in a Christian bed, and enjoy Jewish food'.

Still, when the early Ashkenazim lived in France and Sephardim in Spain, there was frequent contact between them and Sephardic dishes frequently flowed north into Ashkenazic kitchens, such as cholent (a relative dry meat and bean stew that can be served for Sabbath lunch) and pashtida (a quiche-like meat pie). Then, beginning with the First Crusades as the bulk of the Ashkenazim began moving (or were driven) further and further east, these contacts became fewer and fewer, and Teutonic and later Slavic culinary influences grew increasingly strong. Meanwhile, folk traditions gradually emerged in these disparate Jewish communities, including everyday language, proverbs, clothing, synagogue melodies, and holiday foods. During the second millennium, the differences between Sephardim and Ashkenazim grew into a fully fledged division.

In actuality, a mosaic of enduring Jewish cultural communities of varying sizes and antiquity grew up across the globe, including in Afghanistan, Azerbaijan, Ethiopia, Franco-Germany, Georgia, Greece (Romaniots), India (Bombay, Calcutta, and Cochin), Iran (Persia), Italy, Kurdistan, Spain, Tajikistan, Turkmenistan, Uzbekistan (Bukhara), and Yemen, each possessing its own unique history, customs, and cuisine. To further complicate matters, Sephardic and Ashkenazic food developed differently in the various new homelands of the refugees from Sephard (Spain and Portugal), including the Levant, the Maghreb (Algeria, Morocco, and Tunisia), Egypt, Greece, Libya, Syria, Turkey, and western Europe (UK, Holland, and France), and Ashkenaz (Franco-Germany), including Alsace, Austria, the Baltic States (Latvia, Lithuania, and Estonia), Czechoslovakia, Germany, Hungary, Poland, Romania, and Ukraine. Since it was primarily in the Ottoman Empire that the Spanish–Portuguese refugees found haven, the synthesis of Iberian and Ottoman cuisines emerged as the most conspicuous form of Sephardic cooking. Since the ancestors of the majority of Ashkenazim came from the Slavic regions of eastern Europe (Poland, Latvia, Lithuania, and Ukraine), it is this form of Ashkenazic cooking that is most widespread and which most Americans associate with Jewish food. Some of these Eastern European foods are bagels, borchst (beet soup), gefilte fish (chopped carp and other fish that made it edible despite the floating 'Y' bones in the fillets), grebenis (rendered poultry skin), schmaltz (rendered poultry fat), kugel (a noodle, potato, or matzos baked sweet main meal starch), tzimmes (vegetables and some sweet fruits with cheap meat), and cholent (as mentioned earlier a bean and cheap meat (e.g., ribs) dish that can be kept through Sabbath lunch).

After all is said and done, what is Jewish food? It is food that evokes the spirit of a Jewish community as it celebrates the Sabbath, festivals, and life cycle events. It is a dish that possesses the power to nostalgically conjure up the joy of millions of Sabbath dinners or resounds with the memory of the myriad of ghettos, shtetlach (small, isolated rural communities), and mellahs (Jewish quarters in Morocco) in which for millennia Jews struggled to eke out a living and raise their children as Jews.

Thus, for those in the Jewish and Muslim communities, meals are a complex set of interactions reflecting history, custom, tradition, and law as modified by all the other factors that affect everyone's meal choices.

27.11 Meal occasion conflict and collaboration with outsiders

In general Muslim and Jewish dietary laws operate in their own realm. But in a country like Ethiopia, the basic framework common to kosher and halal seems to have spilled over into the local Christian population, who essentially use the same food rules, e.g., no pork and a live slaughter by a cut across the neck. And the growth of an unusual grain, tef, requires a different type of bread, i.e., injera. Breads often define a society as it is typically the staple around which meals are built. And these generally are not driven by the religious laws, although the kosher laws do put some restrictions on breads (see above for an explanation of why breads are required to be pareve).

Reflecting the difference between ancient agricultural Jewish and nomadic Arab cultures, the generic Hebrew term for bread, *lechem*, is also used as a synecdoche for food in general, while the similar Arabic *lahum* possesses the specific denotation of meat. Another important insight into the food of Ethiopia and many African countries is that in many cases livestock becomes a mobile bank. Thus, having herds of animals permits one to wander as necessary to obtain resources for them, but they also can be used in trade or eaten at times when additional resources are needed.

Foods can become the occasions for forming identities that include some while excluding others. This process of differentiation occurs both across religions and among diverse practitioner of a single religion who do not recognize each other's integrity. This creates an ongoing challenge over the issue of having meals together or in common. But it also creates an increased opportunity to collaborate in meal production, as well as to learn to offer hospitality to others who are not like us.

An interesting note is the role of certain animals in various societies. Ruminants are the great converter of grass and other high fiber materials into foods humans can consume. Pigs and poultry require a more balanced diet, but often can be 'scavengers' of the by-products of human food production, particularly at the household level, where by-product amounts are relatively small each day, so keeping a few chickens or a pig may provide a natural outlet for these materials.

In India we may find a Hindu avoiding all beef dishes, while a Muslim will require the meats to be halal slaughtered. However, cultural sensitivity generally precludes the Muslim in some circumstances from having beef as a meal in India. In India, pork is simply not used by either Hindus or Muslims. Notice as you go to various Indian restaurants that none that we are aware of ever have pork (whether or not the restaurant observes halal rules), while some do and others don't have beef dishes, not to mention the pareve and meat kosher Indian restaurants.

Many Christians and secularists do not find the actual food that is designated kosher or halal offensive. In fact, at times, they may even believe such food is handled in ways superior to non-certified food and this comes out through the fact that approximately 75 to 80% of intentional kosher purchasers are non-Jews. Muslims and Jews would be offended by pork or Jews would be offended by the mixing of dairy and meat. The highest form of hospitality might well be to welcome others with the food they believe they need, even if one would not be able to eat that food oneself. People feel most accepted when their specific needs are met.

The need for meals that are religiously 'good' and nutritionally sound, that fit the desires and needs of individuals and groups make food and meal choices intractably complicated.

Many Eastern Religions limit or restrict the use of animal products (please see the chapter on Chinese Food), i.e., they contain within their prescriptions rules that result in different degrees of vegetarian (no meat) or vegan (no animal products, usually including eggs, dairy and honey as prohibited) practice. Some even exclude any vegetable grown underground or beets, as it is red like blood.

Yet, eating different foods can actually lead to social and market differentiation that produces more efficient use of food products. For example, the hind quarters of beef are expensive to make kosher, but there is a ready non-kosher market of Christians and secularists for these preferred cuts, a practice that has existed in Europe for nearly a millennium.

Finally, the need for continuous production, especially dairy, may make it possible for Muslims, Jews and Christians to collectively own and operate dairies that can function without impinging on any of the workers' religious observances as well as prepare for the diverse holidays that usually do not fall on the same days.

While we have noted the antipathy with Christian reliance on pork, there are many other Christian ties to foods, while few, that are as fully developed as in Islam or Judaism. Some groups prohibit alcohol. Others, such as Mormons (Church of Jesus Christ of Latter-Day Saints) also prohibit caffeine. (They also require a year of stored food – a great driver of food preservation technology.) Some groups, such as the Seventh Day Adventists, follow aspects of the Hebrew Scriptural Code, although taking on different nuances than the Jewish version of these codes, i.e., kosher, which are generally not as technical and comprehensive as the Jewish laws. Some, such as the Roman Catholic community, places some limitations on food during certain periods of the year, e.g., Lent.

The world's religious and secular traditions are too broad for any adherent to fully understand the breadth and depth of even their own traditions, let alone those of others. While a third of the world's population is Christian, a fifth of the world's population is Muslim and only a fraction of a percent is Jewish, it is likely that high levels of ignorance of the details and lack of observance, except at the level of cultural practices, hold in all these groups. Food is as much an occasion for learning as it is for conflict over what is 'right' or 'wrong'.

Within Judaism and Islam, there are believed to be a minimal set of rules that are binding on all people outside the community. Only Jews (or Muslims) would be bound by the more specific laws of their religions. This creates an opportunity to consider what may be held as universal and what we may be free to differ about. The Jewish concept of the Noahide code is incumbent on all of humanity. This is the humans' part of the covenant with G-d at the end of the flood that gave us the rainbow, consisting of the minimum requirements for gentiles (non-Jews, 'goy' in Hebrew, i.e., people of the other nations) to conduct their lives. This covenant, like the others of Hebrew Scripture, remains in place to this day, although all of the others are only incumbent on Jews.

1. The required establishment of courts of justice.
2. The prohibition of blasphemy.
3. The prohibition of idolatry.
4. The prohibition of incest.
5. The prohibition of bloodshed.
6. The prohibition of robbery.
7. The prohibition of eating flesh cut from a living animal [treif = torn from a limb. Note: a food law makes the top 7!).

Thus, our brothers and sisters in Judaism (and Islam?) can ask of other faiths, especially the third Abrahamic faith, i.e., Christianity, to consider and reconsider slaughter practices. Again, for more details we recommend a recently published text (Weiner, 2008).

27.12   Meals in desperate circumstances, or among outsiders: the need for proficient knowledge and religious maturity

As we have seen, the subtle distinctions made by individuals, not to mention whole religious traditions, require intimate knowledge of diverse practices. Persons acting in faith are not dismayed by the inevitable institutional or individual hypocrisies that arise. Persons acting in faith become energized and empowered by the inevitable paradoxes and contradictions that frame daily life.

Recognizing boundaries, and discerning when to cross them, is part of the paradoxical nature of religious approaches to food. For example, vegetarianism or very low levels of meat consumption, is often a religious or scientific ideal. Muhammad (PBUH) ate very little meat. Vegetarian or vegan practices are often a least-common-denominator compromise that permits diverse peoples to eat together. However, as likely as not, vegetarianism can be compelled by poverty as well as devotion. Also, food processing technologies, which have existed in some forms for millennia, allow the affluent and even the poor of affluent nations, to consume or waste unprecedented amounts of food, especially animal products. Discerning holy and unholy desires as well as the good or evil results they may have is part of becoming mature in faith and religious practice.

Proficiency is a basic ability to communicate about food practices in another's religious tradition, with an openness to practices that diverge strongly from one's own and a willingness to accept what may seem paradoxical, contradictory or even hypocritical results in one's own and others practices, without being cynical or suspicious. Fluency is knowledge of the basic food practices and rules of another's religious tradition and the absence of fear when exploring unknown matters or dealing with unfamiliar dimensions of particular traditions. Below we offer several examples of personal discipline that exhibit fluency in core religious and spiritual teachings.

Within the Mishnah (commentaries), a part of the Talmud (with the Gemorrah being the actual law) compiled from around 70 to 200 CE of Jewish legal interpretations, the teaching about the importance of honoring the Sabbath begins with a lesson about hunger and food. It is forbidden to work on the Sabbath day of rest. An example of work is moving objects across the threshold of a house from the private residence to the public domain. On one Sabbath, the householder is greeted at the door by a pauper in need of food. The commandment to feed the hungry would seem to be in conflict with the commandment to not work on the Sabbath; it is prohibited to fulfill one commandment by breaking another. A 16th century rabbi noted that both commandments can be fulfilled if the outsider is invited into the house. Hence, the first and core teaching about keeping the Sabbath emphasizes that there should be no 'outsiders' or 'insiders' on the Sabbath. On the Sabbath, all are 'insiders' whose needs can be met by properly honoring G-d's law and inviting the pauper inside (Feigelson, 2008).

Within the Christian Gospels, Jesus is confronted by a woman who is identified as either a pagan worshipper of other gods or simply a gentile, a non-Jew. She seeks Jesus as one who can heal her daughter, but Jesus rebuffs her request. He explains that his first efforts are to serve the needs of Jews not outsiders. Jesus claims that 'It is not fair to take the children's food and throw it to the dogs', treating the woman as a dog. The woman confronts Jesus' claim stating, 'Yes, Lord, yet even the dogs eat the crumbs that fall from their masters' table'. In this challenge over who should be 'fed', her persistent faith compels Jesus to acknowledge that by the woman's actions her daughter has been healed. In this case, being fed in its many dimensions, spiritually, physically and politically, does the work of restoring oneself and others to wholeness.

Personal discipline is the cultivation of individual human judgment in the face of a reality that is larger than any person, tradition or practice can fully comprehend, let alone control. These incidents provoke deeply personal reactions that are also inherently public in nature. Often the most profoundly faithful action produces what may appear as a scandalous outcome to naïve or partisan observers. Welcoming outsiders that some insiders or 'authorities' deem as unworthy is often perceived as scandalous.

To some fervently Orthodox Jews, Christians or Muslims, to eat with others not like themselves is to render them suspect within their own community, with respect to their relationship to G-d, their personal ethics and their understanding of their religious faith. Yet we have seen the same reactions of disdain and exclusion from secularists or atheists who do not want their presence or behaviors among religionists to be seen as tacit endorsements of religious fervor.

Those who exercise authority within a community will often be protective of insiders and of their own positions. They will see conflict as a challenge to their authority. Yet honest and pointed conflict over how to meet real needs is not disrespectful. Religion, spirituality and faith are deeply personal, but not private, and conflict over them is almost always profound and often painful.

A dramatic example comes to mind. While the injunction against eating pork is absolute in Islam, even that injunction can be violated if starvation is imminent and there is no other food. G-d would not will for a creature to starve if she or he could live. The duty to live out of honor for Allah is more important than the rule to abstain from pork. The person of faith must judge for him or herself what the options are and act upon them, not with a cynical doubt regarding G-d's sovereignty or a wavering commitment to G-d, but rather with a completely open attitude to discover G-d's gift of abundance.

For nearly a fifth of the world's peoples, those who practice Hinduism, Buddhism and Jainism, vegetarianism is a religious ideal. Food – what one eats and what one shares with others, withholds from others or wastes – is a significant part of one's karma. Eating is a matter of both nutritional and spiritual attainment. One's karma is a simple expression of the living consequences of a person's actions and thoughts. These religions seek to minimize harm to sentient beings. Jains, Buddhists, higher caste Hindus and many Sikhs practice strict vegetarianism. While there are religious and spiritual rules and traditions to follow, the actual practice is interpreted and implemented by individuals, according to their sense of personal discipline, without being based in religious law.

According to Abraham Isaac Hacohen Kook (1865–1935), the first Ashkenazic Chief Rabbi of Israel and himself a vegetarian, the original state of man was vegetarian, a condition that will be restored in the time of the Messiah when 'the lion shall eat straw like an ox' (Isaiah 11:7). This is in accord with the many biblical commentators who explain that before the time of Noah it was forbidden to eat meat and only afterward was humanity allowed this concession called *basar ta'avah* ('meat of lust'). The Talmud (*Chulin* 84a), reflecting this vegetarian sentiment, declared, 'The Torah teaches a lesson in ethical conduct that man shall not eat meat unless he has a special craving for it… and even then he should eat it only occasionally and sparingly'. Also, many people are reducing or eliminating meat as a matter of religious and spiritual discipline, more or less supported by secular sciences, to reduce the perceived likelihood of developing some diseases and to reduce the impact of raising animals on the environment. Many peoples of all religious backgrounds may be too poor to purchase meat or raise their own.

In the above examples, we noted that the Sabbath law could be simplistically applied as an excuse for not helping someone who might be scandalous to associate with. Upon closer examination, a valid spiritual re-interpretation of the mandate forces an opening to those we might most wish to ignore. The concept of insiders and outsiders and boundaries between the two is important.

This is how groups define themselves and what they affirm, as well as how others define them. To be skillful practitioners of science or religious faith, we need to know when to honor those boundaries as well as when to violate them for a greater good. While the power of institutions is palpable, only individuals act on behalf of those institutions. One has choices about how to act in institutionally loyal ways without being a partisan. Those actions stem from a context specific knowledge.

27.13   Meals by degree of individual religious observance and the perceptions of others

Most of us care deeply about what we eat and do not eat and, for a host of reasons, some we can name and some we cannot, some conscious and others less so. Except for those few who never think about food, or who only eat on the run, food has a special and arguably sacred status. Care is given to what is put in the mouth and body, as human beings do not just graze upon anything or only upon what the individual wants. The care that is given is never wholly rational, as it is too complicated to rationally evaluate what we eat and its impact on our well-being, even though we do evaluate it by internal and external standards. The concept of a sacred foods movement is emerging among those who would like to better support the choices of individuals and families in the kinds of meals they prepare and consume. This section of our chapter explores some of the issues raised when we try to honor individual meal choices in a time of great changes in food markets and equally rapidly diversifying societies.

A symptom of our time is that people seek what knowledge they need from the diverse resources near at hand and claim it wherever they can find it – whether from science, a variant of their own religious tradition or a completely different religion. People may find religious order and structure in one set of spiritual practices, while finding the process of discernment more usefully embodied in another tradition. At different times in life people may find comfort in very strict allegiance to religious rules and at other times in adopting a stance of more openness to trying new things.

Religious fundamentalism and spiritual universalism are two variants of a similar phenomenon: the need of individuals to claim more certainty than a human really can. Such rigid adherence to either religious fundamentalism or spiritual or scientific universalism often draws the deep suspicion of outsiders. Youth often go through both extremes as they mature. What is needed as we accommodate the choices of individuals is to respect the ambiguity of human understanding of what is fundamental and what is universal.

Again, the point here is to neither endorse nor condemn a universalizing syncretism, or endorse or condemn religious institutions, but to note that both syncretism and fervent, exclusive devotion occur. Both are important processes for individuals and families to work through. The core issue is maintaining the space for individuals and groups to work independently as well as collaboratively, and this requires a deep knowledge of one's own traditions, as well as those of others. Imitation, as well as differentiation, is rampant among religious (and scientific) practices, as long as a practice can be borrowed without threatening the perceived integrity of the borrower or the perceived integrity of the originator.

The Quran speaks directly to this ambiguity in an important verse (5:5) – 'the food of the people of the book is acceptable to you and our food is acceptable to them'. This is the concept that people from different communities can (and should) sit down and share meals and discuss things over a meal. The implication being its importance as a practice for today, and this practice is endorsed even though our specific food rules are at times contradictory. In practical terms, though, there has always been some engagement across the three religions of the book, and now the growing intermingling of Eastern and Western religious traditions is prompting a reassessment of who may share meals and discuss matters of ultimate, religious importance together.

Some of these interactions and changes are relatively benign in terms of their effect on meals. The development of tofu burgers, tofu dogs and tofurkey allow vegetarians to eat foods that resemble meat without eating meat. Kosher shrimp or crab made from pollock is another example. Turkey bacon might be both kosher and halal. But the eating of religiously certified imitations of otherwise banned foods may raise eyebrows.

The emphasis of some foods in a religious tradition may come about as an effort to undermine syncretism and force greater differentiation. For example, scholars believe that the emphasis on pork in Christian diets, especially as a substitute for lamb at Easter, was a direct effort to distance Christianity from Judaism and Islam. The emphasis on blood sausage in many predominately Catholic cultures or nations has a similar effect. Differentiation can occur within groups, as when secular, anti-religious Jews take up eating pork as a matter of personal pride and a sign of freedom from religious dogma. The use of grape juice as an alternative to wine for communion among many Protestants was a result of an attack on alcoholism and Catholicism.

Sometimes dominant religious groups are able to control policy in such a way as to attempt to force assimilation or syncretistic practices upon religious minorities. The banning or attempted banning of religious slaughter and privileging stunning of animals, without being based on research about the actual killing practices, is an example of using policy to force conformity on divergent communities. In a similar manner, the failure to test the presence of residual alcohol in food products prevents Muslims from being confident consumers of foods containing flavor extracts.

Since religious practices are changing constantly, if at times very slowly, the fact of changing practices is not the problem. Rather, the problem is the contest over who decides what changes should be accepted that disrupts the internal management of the religion or community. What follows is a process social scientists call triangulation. This process of triangulation has led to fundamentalist reactions within all the major historic religious groups as well as the policies of liberal, secular states. Because the impact on food choices, meals and diet is invidious, we need to elaborate here upon the process.

Syncretism forces a battle for control within religious groups between the strong willed and strongly positioned. Leaders may want to prevent or foster changes that are out of step with their communities. Individual adherents to any religion may face decisions and actions that need quicker resolution than the leaders can give or the decisions and actions may be perceived to be within the realm of individual discretion. What follows is a competition to define authentic practice. At some point one of the contestants will blame the outside cultures or religions as being the source of 'the problem' for the community, but the conflict is, properly understood, an internal matter. Religious leaders and participants have not kept pace with one another in the elaboration of their traditions.

As for science, diets and dieting have assumed a quality of religious vocation with attendant rules, guilt and shame. This is more than a metaphor for religion; it is a version of religion. Syncretism occurs with respect to science in a similar fashion. The institutions that maintain and promote scientific knowledge do not keep pace with the public's need for reliable and context-specific knowledge creating avenues for many, and sometimes unsubstantiated, alternatives. In the process, all knowledge is discredited in the ensuing conflicts.

When an internal conflict pulls in outsiders, they will understandably feel attacked, but the attacks are not really about the outsiders; the attacks take the focus of the pain of the conflict away from the community, pushing the conflict over its boundaries. Outsiders can choose to remain outsiders. But outsiders become protagonists if they engage, usually in one of two ways. If the outsiders mock the provincialism or backwardness of the community in crisis as engaging in something petty, an unintelligible conflict, they will unify the community in crisis in a distraction of opposition to the evil, or merely questionable, ways of knowing in the wider cultural, religious, or even scientific world. If the outsiders try to take what they perceive as the moral high ground by asserting that the crisis would be resolved if only those in conflict could find some allegiance to 'universal' standards, such as values, human rights or modernity, this will provoke not just a feeling of opposition, but a feeling of the outside invading the integrity of the community in transition.

Curiosity and wonder drive religion, spirituality and science. As we are more dependent upon scientific knowledge, we have come to expect more sophistication in both science and spiritual practices. Yoga is a perfect example of a religious practice that involves study, meditation and movement that arose out of the Hindu religion, but is no longer inherently Hindu and has become a more universal, spiritual practice. This is an example of syncretism: there is some scientific support for these practices and many of the practices can be substantially found in ancient Jewish, Christian and Muslim disciplines.

There is no such thing as a generic person, a generically observant Jew or Muslim, or for that matter Christian or Hindu, Buddhist, atheist or secularist. We are specific persons, holding specific practices and attitudes about being careful and careless in ways that are our own. By habit or choice we fashion interpretations of institutional mandates and policies, whether they are religious, scientific or governmental; by habit or choice, we ignore some mandates. It both matters and does not matter how we are regarded by others and what we do that is seen by others. Our individual ways of being observant and the way we perceive others perceptions vary as enormously as what we may perceive G-d or reality demanding of us.

This creates a rather ironic situation. Given the wealth of nutritional studies, as well as the diversity of secular and religious prescriptive teachings about food, it is still unclear how individual people and families make choices about food as they consume their meals. We do not know empirically how people are implementing their religious or spiritual disciplines. No one really knows how 'scientific' information might support a 'rational' decision for individuals or small groups to make meal choices. Many crucial steps are missing from most food discussions because of the individual nature of food consumption, competing institutional mandates for what is good, the complex interactions of food and lifestyle, and the effects of both on individual bodies.

In this sea of complexity, people are trying to make choices for good food, good meals and good lives. These choices may only be partially, inadequately supported by the market and businesses selling provisions, government policies on labeling or restricting foods, and religiously affiliated certification bodies. Yet even if we had fully functioning institutions, individuals would make their own choices.

We are not aware of any studies that compare spiritual and religious practices with the quality of diet in the elderly or any other group. But surely fasting, moderate food consumption, avoidance or the embrace of some foods and other dietary practices could have measurable effects. These practices are nearly universal among spiritual disciplines and would be worth studying, in general, as well as assessing the more idiosyncratic practices of specific movements. This is a great opportunity to build respect among scientists and persons of faith. Such research would not threaten core religious claims, but only shape the development of less mandated practices.

In the coming era, food scientists and religious and spiritual leaders would do well to study what, if any, differences people may realize in their food choices and in their lives based upon the interplay of the religious, spiritual and scientific information individuals use and adapt in practice. While doctrinal disputes may be frightening or alien to some secularists, scientists and persons of different religious traditions, they are also fun. They can be helpful if they give us sharper tools for assessing the sorts of lives we want to live, and to decide for ourselves the sorts of foods we should eat, and to honestly and accurately describe our desires and experiences to one another. The special requirements suggested by nutrigenomics, i.e., gearing a person's nutrition to their specific genetic chemistry, will provide some new guidance to individuals; specific religious laws will have a new system of wisdom to interact with.

A critical question that merits much follow-up would address the role of animal-based foods in the diet. Some studies have led to inferences that plant-based, vegan diets would be substantially healthier and pose less of an impact on the environment than diets that included any animal-based foods. These inferences have fueled great speculation about plant-based diets substantially reducing energy consumption and slowing or reversing the environmental damage that has led to global warming. Religious groups and spiritual movements around the world are coming to address issues of over-consumption and environmental degradation and, the more accurate the science that can be brought to bear, the more chance such movements have for being effective. Spiritual practices evolve over time with respect to more fixed religious mandates. Spiritual practices also borrow from the other cultures and religions which surround practitioners. Food is a central occasion for teaching and learning about the ambiguities, paradoxes and vagaries of cultural, scientific and spiritual practices.

For many years, traditional Western, Eastern, Native American and other spiritual practices associated with health and the environment were not taken seriously by modern science. Practitioners carried on, preserving and adapting local and traditional knowledge about foods, medicines, stewardship of the land and animal husbandry in the shadows and on the sidelines of modern markets. It has become clear that some of the so called scientific practices were not based on good science, as well as that some traditional practices dismissed as archaic were truly beneficial. Vetting received knowledge, from any source may produce welcome changes or disturbing breaks with received wisdom.

Yet syncretism poses a specific kind of organizational threat that may be hard to discern for outsiders. Besides the financial stake in certain products, production techniques or research protocols from those of different cultures or religions from which practices are borrowed, outsiders are rarely aware of what may be causing internal strife. Initially, outsiders may be quite unaware of what is going on, or pleased, as imitation is the finest form of flattery.

A more useful response is for everyone to make a serious effort to address and resolve each other's problems, without imposing a solution on any one group or groups. Collaborative inquiry both respecting differences and boundaries, while crossing them at times is what is required.

27.14   Glossary of key terms

Note: There is no agreed upon spelling of words that have been transliterated from other languages, in this case, mainly from Hebrew and Arabic. The reader can usually recognize variants of the spelling used in this chapter.

Ashkenazim refers to descendents of Jews from the medieval Franco-German Jewish communities. The practices of the various Jewish communities sometimes differ from those of the others.

Bodek ('examiner') is a person specially trained to examine the insides of an animal following slaughter to determine if it was healthy and, therefore, kosher. Since the lungs are most commonly affected by defects, they must be examined before an animal can be certified as kosher. The act of examination is called bedikah.

Chaylev ('fat') are certain fats attached to the stomach and intestines of an animal that are prohibited. (See also Treiber.)

Eid-ul-Adha is the Muslim feast of the sacrifice.

Eid-ul-Fitr is the Muslim feast celebrating the end of the Ramadan fast.

El-Kitabi are People of the Book, i.e., essentially Jews and Christians.

Fleishig ('meat') refers to meat and meat utensils.

Glatt ('smooth') refers to an animal that upon examination are determined to have no more than two adhesions or other defects on its lungs.

Hadacha ('washing') is the custom of soaking meat before and after salting (see also Melicha) to remove any blood.

Hadiths are the traditions (sayings and actions) of the Prophet Muhammad (PBUH).

Halal is lawful for Muslims.

Haram is not lawful for Muslims.

Kashrut refers to the entirety of Jewish dietary practices.

Kasher is the act of making an item kosher.

Kosher ('fit') refers to the fitness of foods for the Jewish table.

Kosher Kosher, also called 'Bosor Kosher', a term referring to non-glatt kosher meat.

Makbooh is used to describe foods that are 'suspect', i.e., they are not prohibited but should be avoided by Muslims.

Mashgiach ('supervisor') is a person qualified to supervise the production of kosher foods in accordance with the Jewish dietary laws. A mashgiach does not need to be a rabbi, but when dealing with meat must be specially trained.

Milchig ('dairy') refers to dairy foods and utensils.

Melicha ('salting') refers to the practice of salting raw meat and poultry to extract the blood. (See Hadacha).

Mizrachim refers to Jews from an 'Eastern', i.e., Asian background.

Nevayla denotes an animal that dies of natural causes or becomes unfit through faulty slaughtering. Such an animal is not kosher.

Quran is the Muslim holy books revealed to the Prophet Muhammad (Peace Be Upon Him) in Mecca and Medina in the years 610 to 622.

Pareve ('neutral') refers to foods that are neither meat nor dairy.

Plumba is a tag affixed to a food, certifying that it is kosher.

Rav Hamachshir ('supervising rabbi') is the rabbi who supervises a mashgiach.

Sephardim refers to descendents of the Jewish community of Spain.

Shechita ('slaughter') is the act of ritual slaughter of animals and fowl. If an animal or fowl is killed other than by shechita, it is not kosher.

Sheriah is the Muslim set of laws.

Shia are a minority in the Islamic community that believe that the rightful ruler of the community should be a descendant of the Prophet.

Shochet ('slaughterer') is a person trained to perform ritual slaughter.

Sunna are the traditions of the Prophet.

Sunni are the majority in Islam who do not require that the leaders of the community be a descendant of the Prophet and is selected by the people through a democratic process.

Treif ('torn') means an animal killed by a predator, but also used to denote all foods that are not kosher.

Treiber is the process of removing veins, arteries, the sciatic nerve and forbidden fats from animals. Since this process is extremely exacting and tedious, Ashkenazim do not use the hindquarter of animals where most of the forbidden items are located.

27.15 References

Blech, Z. (2004). *Kosher food production*. Ames, IO: Blackwell Publishing.

Brillat-Savarin, J.A. (1825/1972). *The physiology of taste or meditations on transcendental gastronomy*. (1825 Trans. M.F.K. Fisher 1972): New York: Alfred A. Knopf.

Feigelson, M. (2008). (D'var Torah. Los Angeles: American Jewish University.

Grunfeld I. (1972). *The Jewish dietary laws*. London: The Soncino Press.

Husseini, M.M. (1993). *The Islamic dietary concepts and practices*. Chicago: Islamic Food and Nutrition Council.

Larsen J. (1995). *Ask the dietitian*. Hopkins, MN: Hopkins Technology, LLC. http://www.dietitian.com/alcohol.html.

Regenstein, J.M., Chaudry, M. and Regenstein, C.E. (2003). The kosher and halal food laws. *Comprehensive Reviews in Food Science and Food Safety*, **2**(3), 111–127.

Riaz, M.N. and Chaudry, M.M. (2004). *Halal food production*. Boca Raton, FL: CRC Press.

Weiner M. (Hebrew author); Touger, E., Schulman, Y., Schulman, M. (editor), (2008).*The Divine Code: The Guide to observing God's will for mankind, revealed from Mount Sinai in the Torah of Moses*, Volume 1, Pittsburgh: Ask Noah International.

28

Revisiting British meals

**D. Marshall, University of Edinburgh Business School, UK, and
C. Pettinger, University of Plymouth, UK**

Abstract: The idea of the British meal is revisited, and its historical definitions and the contribution that research on the family meal has made to this debate are discussed. The definition of what 'eating properly' means for some British households is studied and conclusions from contemporary research are drawn to comment on the time spent eating at home, changing meal patterns and the nature of domestic meals. Accounts based on time use diary data and market research on the status of the domestic family are cited. British meal patterns are described and the effects of acculturation, convenience, and concerns over health and indulgence on contemporary British meals are outlined.

Key words: British meals, meal patterns, proper meals, acculturation, menu pluralism, convenience foods, cooking, eating out, health.

Mealtimes have adjusted to new patterns of work. In Britain and America, they are vanishing from weekday lives. Lunch has disappeared in favour of daytime 'grazing'. People eat while they are doing other things, with eyes averted from company. They snack in the street, trailing litter, spreading smell pollution and dropping fodder for rats. Office workers forage for impersonal sandwiches, grab them ready-made from refrigerated shelves and bolt them down in isolation. Before leaving home in the morning they do not share breakfast with loved ones. Family breakfast has been crowded out of daily routines. In the evening there may be no meal to share – or, if there is, there may be a shortage of sharers. Latchkey kids come home alone and fall ravenously on instantly infused pot noodles or beans eaten straight from the tin… so the family mealtime looks irretrievably dead. The future, however, usually turns out to be surprisingly like the past. We are in a blip, not a trend. Cooking will revive, because it is inseparable from humanity: a future without it is impossible. Communal feeding is essential to social life: we shall come to value it more highly in awareness of the present threat. There is bound to be a reaction in favour of traditional eating habits, as nostalgia turns into fashion and evidence builds up of the deleterious effects of snacking. The advertisers are already beginning to re-romanticise family feeding. Some convenience foods can be adapted as friends of family values: fast preparation time can make fixed mealtimes possible.

Felipe Fernandez-Armesto is a professorial fellow at Queen Mary, University of London, and author of *Food: A History*. See also *The Guardian*, Saturday 14 September 2002 (at http://www.guardian.co.uk/society/2002/sep/14/ publichealth.comment).

28.1   Introduction

Meiselman's (2000) collection of essays in *Dimensions of the meal* underpins some of the complexity and latitude for interpretation in discussing meals. In the introduction, he acknowledges that the term 'meal' means different things to different people. Discussing the broader definition of the meal, for a food service audience, he notes that the '"meal" refers to both the event of eating as well as to what is eaten – meal is both an event and a product' (Meiselman 2008:13). There are some questions over what we recognise as a meal in contemporary British society and what this eating occasion represents for consumers. Addressing this question Marshall (2000) looked at the nature and structure of the British meal in relation to how eating occasions were organised in terms of time and space. Meal patterns, meal formats and the family meal all embodied social and ritualistic aspects of eating. The chapter ended with a speculative classification of British meals and cogitated on the shift towards more convenience reflecting greater individualisation of British eating and the implications for the sociability of the meal and communal feeding. At the end of the millennium it seemed that 'British meals remained relatively resilient to change, despite the proliferation of new products, the rise in eating out, and greater exposure, through the media, to new cuisine' (Marshall 2000:216). This chapter revisits some of those issues and considers what has happened to British meals in the intervening period – specifically drawing on new research data to ask has the British meal changed and in what direction?

The chapter revisits the idea of the British meal, its historical definitions and the contribution that research on the family meal has made to this debate. It considers some of the important social changes in British society and what eating properly means for some British households. It draws on empirical work using time diary data on the status of the domestic family meal and more contemporary accounts from market research. The chapter looks at British meal patterns and the ways in which trends towards more snacking, greater convenience, and concerns over health are accommodated within contemporary British meals.

28.2   Revisiting the 'proper' British meal

Much of the pioneering work on routine food practice such as that by social anthropologist Mary Douglas (1975) on the structure of the British meal, or Anne Murcott's (1982) description of cooked meals in South Wales households, or Charles and Kerr's (1988) accounts of 'proper meals' in the North East of England, provided a specifically British slant to the ongoing discussion around meals. Although it could be argued that in these accounts, meals were a convenient medium through which to study social relations, gender and class, this British work provides much of the theoretical underpinning in other Western cultures (see for example Bisogni *et al.* 2007, Bugge and Almas 2006, Mäkelä *et al.* 1999, and Mäkelä and Fjellstrom in Chapter 3). It is worth reiterating some of the points raised by these scholars in terms of what they reveal about the nature of meals and eating in Britain in that historical period, as it serves as a useful starting point from which to consider what has changed.

Douglas's work is particularly notable in identifying that, where food is taken as part of a structured event – a social occasion with prescribed time, place and sequence of actions – we have a 'meal'. As eating occasions, meals have a recognizable structure, are sequenced through the day with individual elements linked together in pre-determined combinations and successions of courses (Douglas and Nicod, 1974; Douglas 1975). Food choice, far from being arbitrary, is socially constructed and this structural analysis revealed a 'grammar of eating' reflected in organisation of courses within the meal around a tripartite structure with a stressed main course and two unstressed courses and individual dishes. This structure was repeated in individual dishes with a main centrepiece, usually meat, and staple and trimmings, usually potato and vegetables. The progression through courses and dishes reflected savory/sweet, hot/cold and liquid /dry binary oppositions (see also Mintz 1992; Koctürk 1995; Makela 1995, 1999, 2000). However, Lalonde (1992) argued that to define the meal purely by its structure did not adequately express its symbolic significance. He proposed the 'meal-as-event'; a lived experience drawing its meanings from a complex array of sensory, emotive and cognitive factors.

While this structural approach provides a useful framework to look at the organisation of eating over time, much of the debate has shifted away from sequencing of courses, e.g. British domestic meals of one to three courses (Marshall 2000, 2005) across the eating event to a consideration of the 'cooked dinner'. Murcott's (1982, 1983) research, which related specifically to the gender differentiation of cooking in the household, revealed the ubiquity of the domestic 'cooked dinner' and provides a valuable insight into the rules regarding what is considered appropriate and characteristically British. As Murcott notes:

> It is not a whole menu, but one course presented on a single plate. It centrally consists of meat – flesh, not offal (e.g. liver), fresh, not preserved

(e.g. sausages). Poultry is acceptable, fish is not. As essential to its definition are the accompanying vegetables, potatoes and at least one in addition that has to be green. It is not complete until gravy is added to the whole assemblage' (Murcott 1995: 228)

Moreover, specific cooking techniques are employed, roasting the meat and boiling the vegetables into a recognisable combination that is, literally and figuratively, bound together by the gravy. Proper meals are eaten at the table with all members of the family present. Epitomised by the Sunday roast dinner these eating occasions were symbolic of family life. Subsequent work with families in the North of England found the idea of eating 'properly' to be synonymous with 'meat and two veg' and the debate moved more towards the discussion of distinct individual dishes, rather than a series of courses (Charles and Kerr, 1988). This early work on meals is important not only in terms of what it tells us about British family life but in informing us about how foods are combined into recognisable formats that impact upon what is deemed acceptable (Marshall, 2005, 2000, 1993). It seems that while these ideas of 'proper eating' have traditionally been located in the family unit (see Chapters 3, 11 and 10 by Makela, Fjellstrom, and McIntosh *et al.*, respectively), there is less empirical evidence to illustrate how this pattern may be reflected in other types of household. Other work has shown the meal to symbolise a sense of 'togetherness' for young couples setting up home for the first time (Marshall and Anderson 2002, Kemmer *et al.* 1998, see Sobal *et al.* 2003 for a discussion of this with young American couples). Gender has continued to be a topic in meals, as evidenced by Ueland, and Jonsson and Ekstrom in Chapters 6 and 12.

28.2.1 Social change

If we accept the relationship between meals and family life then any question of whether British meals have changed cannot be separated from broader questions of social change in Britain. As noted elsewhere in this book, the terms 'family' and 'household' are not synonymous and the composition of households has changed (ONS 2007, Beardsworth and Keil 1997, Bell and Valentine 1997). In the last census, 2001, families accounted for 67% of households with single person households up to 30% of households. Family life remains important and while marriage is still the most common form of partnership, the proportion of unmarried and cohabiting partnerships has doubled in the last ten years. Of particular relevance is the increase in the number of people living alone in the UK; as a proportion of total households in the UK, one-person households increased from 18 to 29% between 1971 and 2006 (ONS, 2007) and most of this increase was made up of people below state pension age living alone (or from approximately 8.5% in 1981 to 14% in 2006 (by type of household) (ONS 2007) and this is mirrored by a decline in the number of traditional family/household units. While the total number of UK households grew to 24.9 million in 2006 most of the increase came from the growth in single-person households. We know much less about eating in non-family households or households without children (see Pliner and Bell's account of eating meals alone in Chapter 9).

The number of women in paid employment is up from just under 60 to 70% since the early 1970s bringing with it greater financial independence and contributing to the rise in dual-earner households. The most recent British figures show that household net wealth has more than doubled in real terms between 1987 and 2006. In the 50 years since data on British household expenditure was first collected, family expenditure on food and non-alcoholic drink has halved to around 15% of average weekly expenditure (FES 2008). In the mid-1970s, expenditure on food and drink still represented a quarter of average weekly household expenditure. Accompanying the social changes were a number of developments in types of food and technologies that impacted upon attitudes towards food, meal preparation, and healthy eating (TNS 2007).

Post-war food rationing ended in 1954 (Fig. 28.1) and the period in which Douglas was writing saw the expansion and development of basic convenience foods and innovative cooking methods. The microwave was introduced in 1974 and fast food outlets such as McDonalds appeared around this time. Foreign travel broadened tastes in the 1980s and contributed to increased demand for ethnic cuisine. In the 1990s, there was considerable product development in the area of ready meals and high value added products. The start of this century saw the establishment of specific products targeted at children and introduction of reduced fat versions of ready meals. More recent interest in organic and sustainable foods along with an explosion of cookery

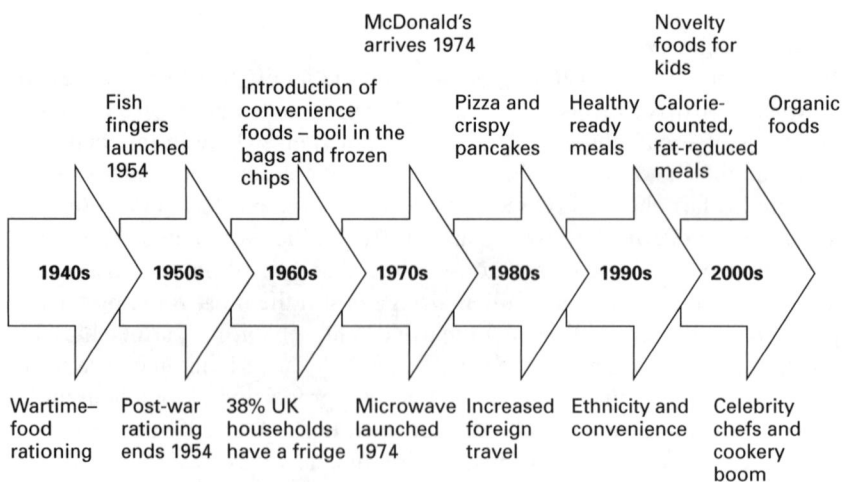

Fig. 28.1 Social trends and developments in food types and technologies in the UK over the past 50 years (Adapted from TNS UK Consumer Attitudes Survey 2007) Source: TNS Consumer Attitudes Survey 2007).

books and programmes has brought with it a changing food environment. Moreover, the food industry has not simply responded to these changes but shaped and redefined our ideas about eating, and meals, through product development aimed at tapping into some of these changes.

28.3 Redefining the proper meal and menu pluralism

Reflecting on British food habits at the turn of the century Beardsworth and Keil (1997) suggested that these social changes were manifest in what they call 'menu pluralism' – '(a) product of the processes which have combined to create the modern food system with its globalization of food supply, its industrialization of production and distribution' (Beardsworth and Keil 1997: 68). This described a situation in which there were a number of alternative schemes to structure food choice and eating patterns centred around issues such as health and dietary performance, minimizing time and effort, ethical considerations and pleasure. They proposed that individuals adjust their menu choice to suit, for example, their mood, economic circumstances or the setting in which the eating event takes place. The question is how and where are these changes accommodated within the British meal system?

One way to look at this is through a typology of British meals, that includes domestic eating events from the festive through to the informal. This suggests changes at the household level are more likely to impact on the number and type of meals eaten at home rather than fundamentally challenging the character of individual meals, for example fewer lunches at home in response to changing work patterns; a higher incidence of lighter meals in response to health concerns. Moreover, there is scope to accommodate some of these changes and still retain the structure and format of the meal as an event (Marshall 2005). Some evidence for this can be found in newly established households where young couples were interpreting proper meals to include a range of dishes beyond the traditional 'meat and two veg'. Accounts of main meals in these households included pasta and rice dishes, salad and some of the 'favourites' referred to above such as spaghetti bolognaise and stir fry dishes. Data on what the couples ate over a seven day period showed that one quarter (27%) of the 273 meals recorded across the sample, were categorised as 'meat and potatoes' (27% of meals) followed by light meals that included dishes like baked potatoes, salad, quiche, and beans on toast (19%). Pasta dishes, usually with a sauce or accompanied by some meat, proved popular (16%) as did rice-based dishes (7%) and these included things like chilli-con-carne, or rice with chicken. The tri-component structure was evident in many of these meals, confirming the structural model proposed by Douglas (1982), yet the incidence of rice and pasta meals, served in 23% of all the evening meals, reveals a repertoire of dishes that extend beyond the traditional meat and vegetables.

Further evidence of this change can be found in the most recent figures from the Expenditure and Food Survey that show a decline of 6.2% in the quantity of fresh and processed potato purchased (from 864 g/per person/ week in 2003/4 to 810 g/per person/week in 2006) and a increase of 4.3% in the quantity of pasta purchased (from 84g/per person/week to 87g/per person/ week) over the same period (DEFRA 2008). These figures reflect the trend in consumption of these staples reported in an analysis of the National Diet and Nutrition Survey between the late 1980s and turn of the century (Hoare and Henderson 2004). What is not clear is the extent to which pasta and rice are replacing potatoes as the staple component in the meal, while retaining the tripartite structure, or if we are seeing greater use of rice-based and pasta dishes, where the staple is a more integral part of the meal, as part of an expanding repertoire of 'proper' meals (Marshall and Anderson 2002). As well as the inclusion of different staple components there was some evidence of 'restructuration' of meals, in which meat is regarded as an ingredient rather than the most highly valued part of the meal (see Holm and Møhl 2000 for a discussion of this in relation to Danish meals). While household purchase of carcass meat actually rose by 5.7% between 2003/4 and 2006 (from 225 g/per person/week to 238 g/per person/week), a reversal of earlier trends, purchase of both fish and fruit and vegetables (excluding potatoes) showed respective increases of 9.1% and 8.2% over the period (DEFRA 2008).

What we don't know is how the food habits of this cohort of young people develop as they establish their own families? There is evidence that children are playing an ever increasing role in family food choice and what is served at mealtimes (Cook 2007; Romani 2005) but we have little data on British families. The influence of children on family meals is also discussed in this book by Fjellstrom and by McIntosh *et al.* in Chapters 11 and 10, respectively.

28.4   Time use and domestic dining

Cheng *et al.* (2007) offer an insightful analysis into the changing nature of food consumption in the UK using time use diary data from seven-day diaries collected from 1274 people in 1975 and a weekday and weekend day diary collected from 8552 people in 2000, recording activities in 30 min and 10 min slots, respectively. The data was weighted to account for over-sampling of specific sub-groups and non-response and to reflect distribution of gender and age in the national population (further details can be found in Cheng *et al.* 2007). They found that there was an overall decline in the mean amount of time devoted to eating and drinking from 105 min per day in 1975 to 98 min per day in 2000. Over this same period, the mean amount of time devoted to eating and drinking at home fell from 71 to 56 min per day, while the eating or drinking outside the home increased from 11 to 25 min per day. As the authors note, while this may suggest that the family meal at home is not as prevalent in 2000, the data on meal duration shows little change over the two time periods. If anything people spent slightly longer per episode at home. This leads them to suggest that '(a)pparently, although people devoted less time overall to eating and drinking at home, when they were at home episodes remained of sufficient duration to suggest that they were eating with others, an impression further supported by a tendency for episodes to be slightly longer at weekends in both years' (2007: 47). Those in full time employment spent less time eating and drinking at home than other employment groups. Women spent more time eating and drinking at home than men, as did non-single households compared with single households, although the differences diminished over the two periods. In 2000, those with young children spent less time eating and drinking at home than those without children, a reversal of the situation in 1975, whereas those with older children spent more time than those without children in 2000, again a reversal of the earlier period.

Further evidence for the continued importance of the family meal can be found in the UK Food Standards Agency (FSA) annual survey of British attitudes that has been running since 2000 and draws on around 3000 consumer interviews. In 2005, 57% of respondents claimed to sit down to a main meal once a day with family members, a further 14% participating more that once a day (corresponding figures for 2003 were 39 and 9%, respectively). In the last two years the question was changed and 76% (2007) and 73% (2006) agreed with the statement that they 'make time for proper meals' (FSA 2006, 2007, 2008). Although this research does not offer any definition of what constitutes a proper meal it does support previous findings by Cheng *et al.* (2007) suggesting overwhelming support for the idea of the family meal. In the Future Foundation report for Kellogg's 58% of respondents claimed to sit down to a family meal every day, despite 83% believing that they ate fewer family meals compared with five years ago (Future Foundation/Kellogg's 2008). Older respondents and households without children were more likely to believe the family meal was in decline, yet 82% claimed to have a family meal with their children all the time or most of the time and this rose to 92% in Scotland. In contrast, both lunch and breakfast were less likely to be eaten with the family with under 20% agreeing that they ate with the family all of the time on these occasions (Future Foundation/Kellogg's 2008). Market research data from TNS Worldpanel covering 2005 to 2008 shows the number of domestic meal occasions involving three or more people was greater than meals with one or two people lending further support to the importance of the family meal in homes with children. Yet solitary meals were up between 1995 and 2008 from 33% to 41% of all occasions reflecting the growth in single-person households and an increase in meal fragmentation where household members are eating the same meal but at different times. While last year saw a decline in meals eaten alone there are still a high proportion of meals being eaten in the company of others (TNS 2008).

28.5   British meal patterns

As we have seen most of the discussion around meals centres on the main meal of the day, usually a hot cooked dinner eaten by the family. Yet there may be a number of occasions when food and drinks are consumed throughout the day. Meal patterns are essentially focused on the distribution and timing of meals throughout the day and across the week. Indeed, the idea of a regular meal pattern of three meals a day, breakfast, lunch and dinner, is a relatively recent historical development (Burnett 1979, Lehmann 2002, Visser 1997, Wood 1995). Data from TNS Worldpanel shows that breakfast accounted for one quarter of all meal occasions recorded by their UK household panel, followed by dinner (21%) and lunch (17%). Together these three meals accounted for 63% of all meal occasions (includes in-home/lunchbox/out-of-home snacks but excluding foodservice) with teatime (10%) in-home snacks (13%) and out-of-home snacks and lunchboxes (15%) constituting the remaining occasions. Meals appear to be less bound to specific times as more people eat when it is convenient, consequently we are seeing a 'flattening' of the mealtime peaks (Fig. 28.2) of 8 am (breakfast), 1 pm (midday meal) and 5.30 pm (evening meal) that were evident around the time of Warren's account (Warren, 1958). In 2000, breakfast was eaten between 7 am and 11 am and the evening meal between 6 pm and 10 pm reflecting this temporal change (Vidal 2008, Cabinet Office 2008, Future Foundation/Kellogg's 2008). More recent market research data shows that British consumers are likely to eat at home before 9 am and not after 9 pm (TNS 2008). This probably

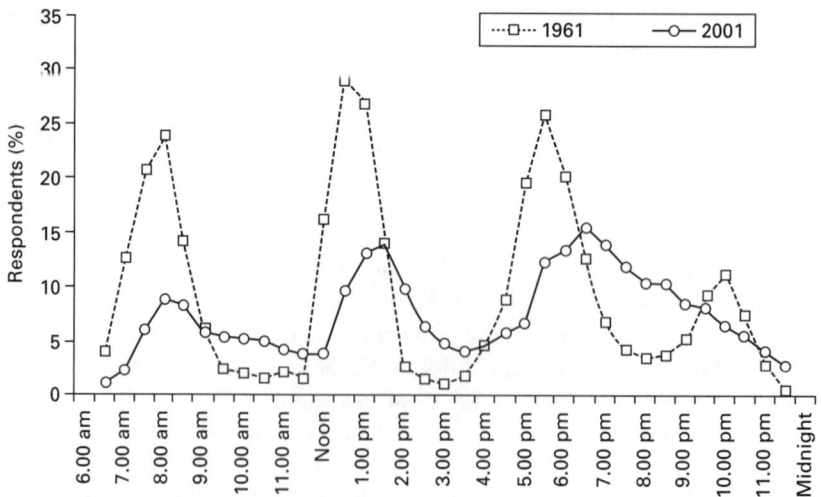

Fig. 28.2   Percentage respondents of eating and drinking, in or out of home, by time of day, all days, 6.00 am to midnight (Source: BBC/NationalStatistics/NVision/Future in Foundation/Kellogg's Base 545/654 adults in households with children).

reflects a changing lifework balance and a more relaxed attitude towards the timing, format and where meals are eaten (IGD 2005).

One notable change has been the shift in timing of the main hot cooked meal from the middle to the end of the day, a response to changing work patterns and demands in the labour market (in Warren's, 1958 account 64% described the midday meal as 'dinner' the remainder called it 'lunch' and 60% of men went home for their midday meal). This practice of eating the main meal in the evening, also evident in the USA, Denmark, Norway, South Africa and New Zealand stands in contrast with Southern Europe, Germany, Switzerland, Poland, Czech Republic, Sweden and Finland, where the main meal is still eaten in the middle of the day (Meiselman 2008) and the occasion is often still a long drawn out affair (Volatier, 1999).

There has been a suggestion that the pattern of three meals a day appears to be giving way to de-structured and even 'anarchic tendencies to ingest food on impulse rather than according to a ritualized socialized timetable' (Lehmann, 2002) as lunch becomes more of a 'continuation of work as opposed to a break' (Datamonitor 2003). This appears to lend some support to the grazing thesis and echoes Fischler's (1988) claims of 'gastro-anomie' as the breakdown of structured eating events in France brings with it moral connotations on contemporary family life. However, as we have seen above, data on the time spent eating at home in British households suggests that conviviality and spending time together as a family is still important. Moreover, evidence from the Nordic and low countries refutes this grazing thesis (Kjaernes *et al.* 2001, Mestdag 2005). The growth in eating out appears to be replacing lunch and entertaining at home rather than fundamentally threatening the practice of eating together as a household (Cheng *et al.* 2007).

In a study of people in Nottingham ($n = 826$), adults still had regular meal patterns but a sizable proportion were likely to skip breakfast (contradicting earlier market research by Taylor Nelson Sofres, 1998) and lunch (71.3% ate breakfast; 77.5% ate lunch and over 90% ate an evening meal) and this was most prevalent in younger adults, where 'snacking' habits were also more commonplace (Pettinger *et al.* 2006, 2004). Despite being considered an important meal, breakfast was most likely to be skipped particularly during the week owing to time constraints, although in some cases breakfast is being eaten at work or on the way to work. The incidence of cooked breakfast, which more closely resembles a meal than a snack, is more likely to be eaten at the weekend (IGD 2005). Given the increased focus on healthy eating there is something of a renewed interest in having a healthy breakfast, especially for children, and some evidence that this will reduce the demand for mid morning snacks (TNS 2008, Kellogg's 2008). Most weekday lunchtime meals occur outside of the the home and sandwiches dominate the menu, both outside and in the home, representing around one third of lunches (IGD 2005). Pettinger *et al.* (2006) found 51% of the English households in their sample reported eating together with their family on a daily basis compared with 65% of French households, yet when this is extended to eating together at least once per week the contrast is not so great with 38% of the English and 22% of the French report eating with their families; around 10% in both samples only ate with the family once a month. The English sample were also much more likely to purchase a takeaway meal, 23% purchasing one up to six times per week compared with 7% in French households, although it is not clear if this is eaten at home with the family which perhaps warrants further investigation. Despite the limitation in definition of 'household' in this research, and the distinction between eating daily and at least once per week, one difference may relate to the 'conviviality vector' (Volatier 1999, Rozin *et al.* 1999), There still remains a much stronger regularity of meal patterns in France (Chapter 16 by Grignon and Grignon). De Saint Pol's (2006) recent publication on the 'maintenance/synchronisation of French mealtimes' adds support to this thesis. In this study, UK data from Time Use Surveys (2000) were compared with French data from an ONS survey (1998–1999). Results showed fewer defined trends in temporal meal occasions in the UK (17.6% of those surveyed were involved in a food activity at 13.00) than in France (54.1% of French were involved in a food activity at 12.30) suggesting that UK mealtimes are far less 'synchronised' than French mealtimes (de Saint Pol, 2006).

Some market researchers claim that this breakdown of traditional structures and a ritualized timetable of eating, favours a shift to snacks and lighter meals, and convenience foods that require little or no preparation (Mintel, 2004). Yet this may reflect broader social changes in relation to household types, rather than any demise of importance attached to meals within British households. Despite this alleged demise of the family meal, and/or increase in snacking and grazing, part of a broader moral panic, the formal dinner is still regarded by some as a highly ritualised activity (Marshall 2005, Murcott 1997), especially if strict definitions of the meal and family are relaxed.

As we have seen above the family meal remains relatively resilient to change and what we are seeing is an increased incidence of shorter eating events, light meals and snacking, but not necessarily at the expense of the main meal (Cheng *et al.*, 2007). Further evidence from British market research reveals something of a general trend towards more lighter meals, originally among the 35–54 years age group, in response to health concerns. However, more recent figures show a reverse in this trend as the number of main meals grows (Fig. 28.3). The incidence of light meals, up from 4500 million meal occasions to 5500 million meal occasions between 2000 and 2006 has dropped slightly. In contrast, the number of main meal occasions, down from 7000 to 6000 million meal occasions in the same period has shown a recent increase (TNS 2008). One factor in all of this has been the increased sales of home-made and part-made food, up over 10% over the past four years, in a renewed interest in cooking meals, and growth in homemade savory food occasions (TNS 2008). This corroborates findings from France, some of Nordic and Low Countries, where there was no sign of 'disorganisation' of the meal (de Saint Pol 2006, Kjaernes *et al.* 2001, Mestdag, 2005).

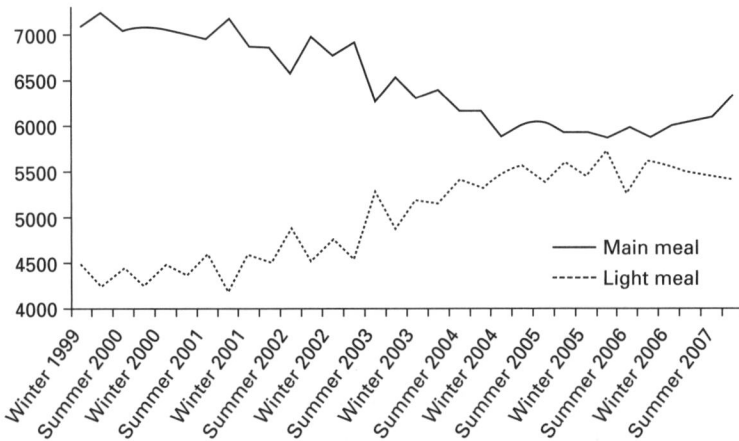

Fig. 28.3 In-home main meals versus light meals (millions of meal occasions) winter 1999 to summer 2007 (Source: TNS Worldpanel Usage).

28.5.1 Culinary diversity: Acculturation, cookbooks and favourite dishes

The 1990s saw evidence of 'Foodways in flux' (Goode *et al.* 1995) and British food habits continue to be influenced by the new immigrant population and exposure to these foods through cultural experiences at home and abroad (Mitchell 1999, 2006). The British foreign secretary Robin Cook's declared Chicken Tikka Masala to be a 'true British national dish' (*The Guardian* 2001) and a survey for the BBC (Hall *et al.* 2003) revealed spaghetti bolognaise and curry and rice as the nation's favourite dish along with pizza, crispy duck, chicken tikka masala, Chinese sweet and sour chicken, chow mein, rogan josh, chicken madras, thai green curry and chicken korma (Mitchell 2006). In the same survey, favourite convenience foods also affirmed the popularity of 'foreign' dishes; except for fish and chips, the rest were of ethnic origin. These dishes have, according to some writers, usurped the traditional 'sausage and your joints of beef... just your traditional meat and two veg... stews, pies... and probably the heavy puddings' (Bell and Valentine 1997, p. 170; *The Telegraph* 2008) to become mainstream British dishes (Mitchell, 2006). Ethnic food and acculturation promote 'cultural' health as they enhance variety of food and meal availability and permit us to embrace multiculturalism, described as a 'positive force for our economy and society and will have significant implications for our understanding of Britishness' (*The Guardian*, 2001).

Yet accounts of favourite dishes do not necessarily reflect what people are eating on a regular basis and ethnic cookbooks, the other part of Mitchell's argument around culinary acculturation, may have more to do with domestic entertainment and special cooking occasions than everyday eating. Indeed, other writers have explored the extent to which cookbooks, such as those by the celebrity chef Jamie Oliver, may be more about forms of social distinction, through the use and assemblage of traditional and exotic ingredients, and making life easier than cooking per se (Brownlie *et al.* 2005). The incorporation of these dishes into British domestic meals probably owes as much to the development of pre-prepared foods, meal components and dishes, and stocking 'exotic' ingredients in supermarkets. Yet, we know relatively little about the place of these foods in British domestic meals. In the next section, we look at the impact of convenience on our eating habits.

28.5.2   Convenience and cooking

The notion of eating for 'convenience' stems back to post-war Britain, where exponential change and developments occurred, all of which contributed to the evolution of British eating habits. In the 1940s, individual food choice was limited by (lack of) availability and budget after which the 'technological revolution' and growth of food processing and the food industry enhanced and influenced consumer choice (Foster and Lunn 2007). This, coupled with changing working patterns, as more women went out to work, created the prediction by Hollingsworth in 1960 that choice of food would in the future be influenced by 'lack of time and inclination to cook and ready availability of a wide choice of foods in supermarkets, which are likely to favour the trend towards increased consumption of precooked and other processed foods' (Hollingsworth 1960: 29).

Convenience is a complex concept; it is often conceived as 'time saving' or 'time buying' and reflects the problem of 'timing' and being in the wrong place. Others have related the concept to energy saving and the 'transference of culinary skills' reflecting social and emotional aspects of food preparation (Carrigan 2006, Brewis and Jack 2005, Candel 2001, Warde, 1999, Gofton 1995). There are no fixed boundaries in the definitions of convenience but a useful working definition is: 'transfer the time and activities of preparation from the household manager to the food processor' (Buckley *et al.* 2005: 7, cited in Carrigan 2006). A study by Carrigan *et al.* (2006) explored the meaning of convenience food for UK mothers, in the context of their family's food found that while some mothers are ambivalent towards convenience foods other 'embrace it enthusiastically'. They found multiple meanings surrounding the terms 'proper' and 'convenience' supporting the idea of reconceptualisation of 'proper' food especially around weekday meals that permitted the inclusion of convenient solutions into these meals. A hierarchy of acceptable convenience foods were outlined, whereby there were strong status distinctions, with price being an important factor as regards quality; cheap convenience food was considered to be of dubious origin. Yet they show the value and importance of convenience foods to these mothers in the context of everyday meals. As they note 'their (mothers) experience of family meals is one bounded by time-shortages and complex scheduling. Convenience food alleviates the difficulties circumstances create for mothers trying to deliver wholesome, nutritionally balanced meals by avoiding unpleasantness, and by saving and buying time on various stages in the food preparation and cooking' (Carrigan 2006: 382).

Convenience food was described as allowing them to cope with the demands of their complicated lives and family routines, for these young mothers cooking from scratch was no longer an essential ingredient in the family meal. However, they note that while convenience food allows mothers to accommodate different demands from individual family members, by effectively customising each serving, it represents a deviation from the idea of one meal for everyone. Families are eating together but they may be eating different things. They go on to add that time saved in the kitchen is replaced by time spent in the supermarket catering to individual family needs, although this was more evident in families with older children. This mirrors recent qualitative research undertaken in North Devon, in which a group of young mothers suggested that 'convenience' foods were normally a less healthy option, used mainly when they needed to feed their children quickly (Pettinger and Lankshear 2008,). This raises the issue of the common perception of 'convenience' being associated with sub-standard quality and perceived as less 'healthy', which is questionable. Moreover, it is important to stress that convenience and health are not always in opposition (Gofton and Ness 1999) and both trends are still prevalent in the British food market (Mintel 2008, Shiu *et al.* 2004).

This trend towards the incorporation of more convenience into mealtimes, for some households, is reflected in the retail provision of more convenience foods as well as the growth in takeaway and eating out (Cheng *et al.* 2007, Candel 2001, King 2007, Mintel 2008, Tansey and Worsley 1995, Warde and Martens 2000). The importance of convenience is revealed in the fact that for some consumers, particularly those under 35 years of age, 'convenience' seems to have become more important than the pleasure of eating (see Pettinger *et al.* 2004, 2006) and may be detrimental to health (Rozin *et al.* 1999). One needs only to observe the proportion of supermarket shelving dedicated to ready-made and instant chilled meals to appreciate their significance in the British diet. There was a 15% increase reported in the sale of ready meals in the UK between 1995 and 1999 (Eurofood, 1999, cited in Shiu *et al.* 2004). More recent data shows that the ready meals market has been slowing down following this period of strong growth with nearly seven in ten consumers using ready meals, mainly on an occasional basis, although 15.5% were reported as being 'heavy users'. One of the reasons for this slow down appears to be growing concerns over the fat, salt and other ingredients (Mintel 2008). Convenience orientation was seen as being more important in single-person households and was negatively related to cooking enjoyment, involvement with food products and variety seeking (Candel 2001) although as Carrigan's (2006) work has shown they appeal to a broader range of consumers. One report found that consumers were not only using ready meals but over half of British consumers had claimed pre-prepared supermarket foods or meals bought from takeaways or restaurants was their own cooking, with one in ten serving them to guests (IGD 2008). The same report also suggests that there are a group of British consumers who are shunning ready meals and indulging themselves in buying quality ingredients and cooking at home. Yet, we need to see convenience in the context of the meal, in certain situations it is perfectly acceptable to use convenience foods – frozen foods, cook in sauces, ready meals – in other circumstances it is inappropriate.

This growth in demand for convenience appears, in part, to be related to cooking skills transition (Caraher et al. 1999; Caraher and Lang 1999) which shows that cooking skills, like nutrition and the range of foods that we eat, have been undergoing significant change with considerable social, economic and health consequences. Increasingly, it appears that technology and the media are replacing the traditional role of the mother or grandmother as teacher, resulting in the 'modern gastronomic revolution' (Meades 2003). Convenience food has become so ingrained in food culture that some consumers can no longer distinguish between convenience cooking and 'cooking from scratch' (Mintel 2004, Short 2003). Cheng et al. (2007) found less time being devoted to food preparation in the home; down from 57 to 41 minutes per day between 1975 and 2000. The gendered division of labour in the household is becoming more equitable and men are doing more cooking, up from 11 to 23 minutes per day, but women continued to carry more of the responsibility despite the time devoted to cooking falling from 100 to 58 minutes per day. This reduction in time allocation to food preparation for women likely reflects their increased participation in the labour market and the growth in pre-prepared and convenience foods, as well as the introduction of new storage and processing technologies in the kitchen (Gershuny 2000, Shove and Southerton 2000). There remains a persistence of the gender differentiation within the domestic setting with regard to food-related tasks (preparation, shopping and deciding) (Mennell 1996, Murcott 1982), dispelling the idea of the decline of the traditional female role of 'housekeeper' as more women are engaged in employment outside the home (Tansey and Worsley 1995). Although some of these tasks appear to be more often shared in the UK, particularly food shopping (Marshall and Anderson 2000, Pettinger et al. 2006) women continue to carry much of the responsibility for food-related tasks. This is reflected in several chapters in this book.

One survey of British consumers found that 67% liked cooking, and 76% make time to cook proper meals, although we have no indication of the frequency with which they partake in either activity (FSA 2008) this may be every day or on special occasions. Moreover, enjoyment of cooking was higher for women that men, and women were also more likely to say they made time to cook proper meals. Making time to cook was less common among youngest age consumers and pre-nesters life stage (16–25 with no children). In the same survey around half of those surveyed enjoyed watching cooking programmes on television. Conversely, time use data shows people are spending less time cooking (Cheng et al. 2007) adding further weight to the convenience thesis yet the idea of cooking is subject to multiple interpretations with 'cooking from scratch' and 'cooking with pre-prepared foods' no longer clearly discernable concepts according to some accounts (Short 2003, Mintel 2004). Further distinctions can be drawn between homemade, or part homemade (using packet mix, sauce, or bases), convenience (includes ready meals, pizza) and assembled meals (bacon, eggs which represent around three quarters of home savoury foods (TNS 2008). Homemade and part homemade foods are more popular Monday through Thursday with convenience and assembled usage increasing at the weekend (TNS 2008). The British consumer attitudes survey reports almost three quarters in their survey claimed to be using raw/fresh ingredients once a day or more frequently (FSA 2006). Similar results are found in the Kellogg's report where around 60% of respondents in a UK survey ($n = 1000$) claim to cook using raw ingredients at least three times per week and UK adults were estimated to make an average of three meals per week from scratch, up from 2.5 meals per week four years ago (Future Foundation/Kellogg's 2008). This is partly explained by the increased interest in watching cookery programmes and looking through magazines for recipes. Yet, in another study less than a quarter of UK respondents cooked a meal from raw ingredients on a daily basis (22.2%) and this was explained by the samples' working patterns and their heavy reliance on convenience food in the form of ready-prepared meals, take away and snack food (Pettinger *et al.* 2006). Changes in work and leisure patterns are clearly having an impact on how and what British consumers are eating and one development has been the increase in eating outside of the home.

28.5.3 Eating out

The most recent expenditure and food survey indicated an increase in 'eating out' expenditure since 2003–4 (DEFRA, 2008) growing from £7.39 per person per week in 2003/4 to £8.00 per person per week in 2006 being spent on food and non-alcoholic drinks eaten outside the home. However, taking out the effect of a rise in the retail price index and using constant 2006 prices, we actually see a fall of 3.3% in eating out expenditure. This currently represents about one third of all household food expenditure; in 2006, average expenditure on food and drink (excluding alcoholic drinks) was £21.55 per person per week and expenditure on eating out (excluding alcoholic drinks) was £8.00 per person per week. Moreover, the data does not show any increase in consumption of take away foods brought home over the past two years.

It is important to consider why people eat out and Warde and Martens (2000) summarise reasons for eating out to include: experiencing something different from the everyday; getting a break from cooking and serving; relaxing; having a treat; socialising; celebrating; a liking for food; and preventing hunger. They suggest that eating out for pleasure is more widespread and household expenditure on food eaten outside the home has altered accordingly.

Indeed, this appears to be a new form of distinction as highly educated and single households are spending more time on this eating activity (Martens and Warde 1997, Cheng *et al.* 2007). Eating out, as well as being convenient, might be seen as convivial pastime, promoting variety and choice, although some research has also associated it with sedentary lifestyles and increased energy intake, particularly amongst the young (Orfanos *et al.* 2007) and lower socio-economic groups (Acheson 1998). This links to our previous discussion of changing meal patterns and the shift from more highly structured main meals to lighter meals. Perhaps ironically, given the emphasis on the family meal, eating out may present a better opportunity for family/household socialisation than the domestic table (Wilk 2006) so whether eaten in or out of the home, the social aspects of the proper British meal remain very much alive? One might also argue that eating out reinforces our ideas about eating 'properly', the vast majority of eating out involves traditional dishes that reflect ideas about proper meals.

28.5.4   Consumer attitudes towards meals and health

There is considerable public debate around health and increased media interest in the subject but what are the implications for British meals and the ways we eat? It is clear that health is a growing influence on what we eat and is more likely to be a consideration for meals eaten earlier in the day or at the weekend (TNS 2008). The 2007 Consumer Attitudes Survey based on British consumer attitudes towards food safety and food standards offers one insight into the relationship between health and meals. The report identified four attitudinal clusters (Table 28.1) (TNS 2007).

While primarily focused on health, the description of the segments reveals that two groups are involved in cooking proper meals but for different reasons. *Traditional Cooking Enthusiasts* (24% – women, married, white, from higher socio-economic groups, and rural consumers) try to cook and eat proper meals together as a family on a regular basis while *Concerned Health Advocates* (25% – married, with children, 36–65 age group and ethnic minorities) do so in the pursuit of health. The *Health Conscious Pragmatists* (22%) (more likely to include women, urban and retired individuals) have more limited cooking skills and lack of time, but are trying to cook from scratch and use fresh ingredients. The *Convenience Driven Health Rejectors* (29% more likely to include men, 16–25 from lower socioeconomic groups) have few formal eating occasions at home, except for the occasional Sunday roast dinner, and place minimal emphasis on meal preparation which they see as arduous. Consequently, they do not like cooking and use ready meals, takeaways and snacks. Eating for this group is much more erratic and they exhibit limited appreciation for cooking from scratch. The study reveals some of the subtleties in terms of attitudes towards meals that exist in the UK but also shows the extent to which the proper meal continues to feature in British ideas about eating healthily even among those who are convenience orientated.

Table 28.1 TNS attitudinal segments (Source: table created from TNS UK Consumer Attitude Survey 2007: p. 3)

Health Conscious Pragmatists	Represent 22% of the UK adult population and are characterised by a feeling that healthy eating is important but also that convenience foods are not necessarily bad for you. They are liberalists in their approach to eating; satisfied with their general diet, as well as their ability to adapt their eating habits in different circumstances. Health Conscious Pragmatists tend to eat a varied diet, embracing modern cuisines and demonstrating a willingness to experiment with different foods and move with the times.
Convenience Driven Health Rejectors	Represent 29% of the UK adult population, and are defined by a low enthusiasm for healthy eating and an endorsement of convenience foods. They tend to have a relatively short-term outlook in life, living more 'day to day, week to week', and notably, may also lack stability in one or more areas of their lives, such as relationships or work. Meal preparation is mostly perceived as laborious and their overall relationship with food is fairly functional. They do, however, often look to food for immediate gratification and stimulation, and enjoy eating out regularly.
Concerned Health Advocates	Represent 25% of the UK adult population, and are characterised by a high importance of healthy eating coupled with a concern and need for more information about food. This group is aware of the need to eat a balanced diet and have a level of anxiety regarding what they eat. This translates into a more principled and controlled approach to food, with knowledge of nutritional values and ingredients lists. Concerned Health Advocates enjoy indulgent/rich food usually in the form of occasional snacks. However, these are only digressions and are not allowed at the expense of healthy eating.
Traditional Cooking Enthusiasts	Comprise 24% of the adult UK population, which is defined by an enthusiasm for cooking and a negative view of convenience foods and eating habits. Traditional Cooking Enthusiasts tend to have a certain level of stability in their lives, with rather structured and conventional lifestyles. They are organised and relatively disciplined individuals. Food is an extremely important part of their lives, and they are passionate about cooking and making 'proper meals' the 'proper way'.

Moreover, it reflects contemporary views on meals and how this relates to British ideas about family life and healthy lifestyles. If anything, the health debate adds some context to previous discussions by reaffirming support for the family meal. The ubiquity of the family meal and its role in domestic life seems to be reflected in ideas that eating meals together is 'healthier', although there is little nutritional evidence for such claims (see the chapter by Mackintosh *et al.* in this book for an American perspective and the link at the end of the chapter to CASA).

28.6 Conclusions

The 'panic' over the breakdown in convivial eating at the end of the 20th century, and resulting anomie, was seen as symptomatic of the demise of the family and an increasingly individualised dining experience. This was best reflected in Fischler's accounts of de-structuration of the meal and the increase in grazing in France. As we have seen there are distinct differences between France and Britain but there is little evidence to suggest that we have become a nation of grazers or that the domestic family meal is dead. There remains what we might call a 3-2-1 pattern of eating, *three* meals per day, centred on breakfast, lunch and dinner, with snacks in between and tea still featuring in households with children. Of these meals *two*, breakfast and lunch, are increasingly likely to be eaten alone or outside the home, although there is a resurgence of interest in breakfast at home and weekends reveal different patterns from midweek. This leaves *one* main meal per day, usually at the end of the working day when most families try to eat a 'proper meal' together. As discussed this includes a broader repertoire of dishes and more ready prepared, cooked, or bought in foods. Rather than disappearing the family meal seems to have proved resilient and while the evidence shows that this is not an everyday event it has taken on renewed significance. Eating properly is not simply about the food but about the ritual around eating (as many chapters in this book note). While family meals are not, by any means, some haven of domesticity and conviviality, as depicted in the ads (see for example, the Bisto, the nation's favourite provider of instant gravy, (Hylton 2005) advertising campaign and website in the UK that promotes the idea of a proper home cooked 'family meal' (*The Observer* 2005), they remain important in providing an opportunity for people to get together and are seen as good for family 'health'.

Sticking with the family meal, we might ask what has happened to the British meal in the last ten years? On the one hand very little has changed; our ideas about what constitutes a proper British meal is still grounded in the menus and meal structures initially proposed by Douglas, Murcott and Charles and Kerr and exemplified by that British 'institution' of the Sunday roast dinner (Delia Smith 1995). While some traditional home-cooked dishes such as toad in the hole or steak and kidney pie are on the wane (*The Telegraph* 2008) these dishes can still be found on the supermarket shelves and on restaurant menus. On the other hand, a lot has changed. At a broad social level certain trends continue to impact on British meals: the emergence of new household structures, the increase in the number of women in fulltime employment and the rise of dual-income households has meant less time available in the home for cooking and preparation as well as some reprioritisation of how that time should be spent.

Exposure to new cuisines and our willingness, and ability, to incorporate these new dishes into the diurnal repertoire of family meals has led to an increasingly varied and diverse set of domestic meals that look very different to those served up 30 years ago. The same report that showed a decline in some traditional home-cooked dishes found spaghetti bolognaise and curry was popular among the under 40s reflecting an interest in 'foreign' cuisine (although there is a question about the extent to which this is seen as 'foreign'). This increasing variety has been encouraged by eating out experiences and increasingly facilitated by retailers and food manufacturers sourcing ingredients from further afield and developing innovative products that cater to the demands on modern life. But this growth in convenience is paralleled by increasing concerns over health (Gofton and Ness 1991) and greater awareness about what is good to eat. In terms of meals, this appears to be manifest initially in a shift towards more frequent lighter meals, but more latterly, we have seen a renewed interest in home cooking and eating together as a family (TNS 2008). However, the return to home cooking does not necessarily mean a more nutritionally 'healthy' diet in the same way that a shift towards using more convenience foods does not necessarily mean a less healthy diet. It is apparent that enjoyment and indulgence are also an important part of the dining experience and we are more likely to indulge later in the day and at the weekend. To fully understand this we need to reflect on how health and indulgence play out at the individual, rather than the aggregate, level. Health appears to be more prevalent midweek and for breakfast and lunch meals whereas indulgence is more prevalent at weekend and dinner meals (TNS 2008).

Within domestic meals, we have learnt to accommodate tradition and novelty; the everyday and the festive, the functional and pleasurable; the convenient and the healthy. Meals, allow us to live with this ambivalence, for example balancing indulgent weekends with healthy midweek eating, or convenient snacks with home-cooked meals. In looking at British meals we need to consider not only what is happening overall but consider changes in households as well as patterns of eating. Family meals are still prevalent but in households that have children. Some of the changes in terms of individualisation relate more to broader societal changes than any fundamental shift away from the ideal of the proper family meal. Overall, we can say that British meals are more 'flexible' in their timing, their structure and their content, to quote one market research report 'Eating in the UK today has become a more flexible exercise, reflecting a broader shift in the way that modern British consumers live their lives. Rigid 9 to 5 working structures are giving way to more flexible arrangements for many while commuting times, the school-run, growing leisure commitments and an ever-shrinking gender gap in the workplace place additional pressures on when, where (and with whom?) food appears in our lives' (Future Foundation/Kellogg's 2008:23). Amidst these changes meals remain an important part of the way we 'do' food.

28.7   Future trends

Not much time has passed since we last considered British meals but despite concerns over the demise of the family meal at the start of 2000 it continues to be an important part of British domestic life. Yet this resurgence of interest in family meals and cooking is not unrelated to ongoing concerns about health and obesity as well as reflecting a growing interest in indulgence and pleasure. Convenience will continue to be an integral part of our food culture. As manufacturers adapt products to reflect the health concerns these will become acceptable across a number of eating occasions as individual components for assembly or complete takeaway meals. We are increasingly likely to differentiate between meals that are about refuelling and meals that are about pleasure. Greater importance will be attached to proper family meals where eating together can be convenient, healthy and pleasurable especially when they incorporate the flexibility to cater for the different needs and time schedules of contemporary life.

28.8   Acknowledgements

Many thanks to Giles Quick and Jonathan Firth at TNS Worldpanel Usage (the UK's largest and only continuous monitor of food and drink consumption in the UK) for all their help with researching this chapter and adding some 'quantification' to the British meals debate.

28.9   Sources of further information

British Sociological Association; Food Studies Group http://www.britsoc.co.uk/
British Nutrition Foundation http://www.nutrition.org.uk/
Nutrition Society http://www.nutritionsociety.org/index.php
UK Public Health Association http://www.ukpha.org.uk/
Public Health Nutrition Journal http://journals.cambridge.org/action/ displayJournal?jid=PHN
TNS Video Insights http://www.tnsglobal.com/news/video-insights/ ?page=3&count=5

- 20.06.2008 A Focus on Long term consumption trends (UK) Dominic Brown, TNS Worldpanel, UK video (17.28 mb) http://www.tnsglobal.com/ _assets/video/TNS_Market_Research_DBrown_Hot_Topics.wmv (accessed 260808)
- 28.05.2008 TNS Market Research - Dispelling the Myth of Snacking (UK) Jonathan Firth, TNS Worldpanel Usage UK video (12.62 mb) http://www.tnsglobal.com/_assets/video/TNS_Market_Research_ Snacking_HT.wmv(accessed 260808)

http://www.oxfordsymposium.org.uk/
Bisto advertising campaign http://aahnight.co.uk

28.10   References

Acheson D (1998) *Independent enquiry into Inequalities in Health.* HMSO, London.
Beardsworth A and Keil T (1997) *Sociology on the menu: an invitation to the study of food and society.* Routledge, London
Bell D and Valentine G (eds) (1997) *Consuming geographies: we are where we eat.* Chapter 7 'Nation', (p. 170). Routledge, London.
Bisogni CA, Winter Falk L, Madore E, Blake CE, Jastran M, Sobal J and Devine CM (2007) Dimensions of everyday eating and drinking episodes, *Appetite*, **48**(2), 218–231.
Brownlie D, Hewer P and Horne S (2005) Culinary tourism: An exploratory reading of contemporary representations of cooking, *Consumption Markets and Culture*, **8**(1), 7–26.
Buckley M, Cowan C, McCarthy M and O'Sullivan C (2005) The convenience consumer and food related lifestyles in Great Britain. *Journal of Food Products Marketing*, **11**(3), 3–25.
Bugge AB and Almas R (2006) Domestic dinner: representations and practices of a proper meal among young suburban mothers, *Journal of Consumer Culture*, **6**(2), 203–228.
Burnett J (1979) Plenty and want: a social history of diet in England from 1815 to the present day, Scolar Press, London.
Cabinet Office (2008) *Food:an analysis of the issues.* Strategy Unit discussion paper, January.
Candel MJJM (2001) Consumers' convenience orientations towards meal preparation: conceptualisation and measurement, *Appetite*, **36**, 15–28.
Caraher M, Lang T, Dixon P and Carr-Hill R (1999) The state of cooking skills and their relevance to health promotion, *British Food Journal*, **101**(8), 590–609.
Caraher M and Lang T (1999) Can't cook won't cook: a review of cooking skills and their relevance to health promotion, *International Journal of Health Promotion and Education*, **37**(3), 89–100.
Carrigan M, Szmigin I and Leek S (2006) Managing routine food choices in UK families: The role of convenience consumption, *Appetite*, **47**, 372–383.
Charles N and Kerr M (1988) *Women, food and families.* Manchester, Manchester University Press.
Cheng SL, Olsen W, Southerton D and Warde A (2007) The changing practice of eating: evidence from UK time diaries, 1975 and 2000, *The British Journal of Sociology*, **58**(1), 39–61.
Cook DT (2007) Semantic Provisioning of Children's Food: Commerce, Care and Maternal Practice, ESRC Cultures of Consumption, Working Paper 036. http://www.consume.bbk.ac.uk/publications.html (accessed 26/08/08).
Datamonitor (2003) Workplace consumption, *A Datamonitor Report (summary).* www.datamonitor.com/consumer.
De Saint Paul T (2006) Le dîner Français: un synchronisme alimentaire qui se maintient, *Economique et Statistique*, **400**, 45–69.
DEFRA (2000) *National Food Survey*, HMSO, London.
DEFRA (2008) *Family Food in 2008.* Report on the Expenditure and Food Survey. http://statistics.defra.gov.uk/esg/publications/efs/default.asp (accessed 5/6/08).
Douglas M (1975) Deciphering a meal, *Daedalus* **101**(1), 61–81.

Douglas M (1982) *In the active voice*, London, Routledge and Kegan Paul.

Douglas M and Nicod M (1974) Taking the biscuit: the structure of British meals, *New Society*, 744–747.

Family Expenditure Survey (FES) (2008) Family Spending Survey celebrates 50 years of Family Spending, 28 January. Accessed at http://www.statistics.gov.uk/cci/ nugget.asp?id=1921

Fischler C (1988) Food, Self and Identity, *Social Science Information* **27**(2), 275–292.

Food Standards Agency (2006) Consumer Attitudes to Food Standards: Wave 6 UK Report Final, Prepared by TNS, February http://www.food.gov.uk/science/socsci/surveys/ foodsafety-nutrition-diet/cas2005 (accessed 1/9/08).

Food Standards Agency (2007) Consumer Attitudes to Food Standards: Wave 7 UK Report Final, Prepared by TNS, February http://www.food.gov.uk/science/socsci/surveys/ foodsafety-nutrition-diet/cas07 (accessed 010908).

Food Standards Agency (2008) Consumer Attitudes to Food Standards: Wave 8 UK Report Final, Prepared by TNS, January http://www.food.gov.uk/science/socsci/surveys/ foodsafety-nutrition-diet/eighthcas2007 (accessed 010908).

Foster R and Lunn J (2007) Food availability and our changing diet. 40th Anniversary briefing paper, *Nutrition Bulletin*, **32**(3), 187–249.

Gershuny J 2000 *Changing Times: Work and Leisure in Postindustrial Society*, Oxford: Oxford University Press. (cited in Cheng *et al.* 2007 *op. cit.*)

Gofton L (1995a) Convenience and the moral status of consumer practices. In D Marshall (ed), *Food choice and the consumer*, Blackie Academic and Professional.

Gofton L (1995b) Dollar rich and time poor? Some problems in interpreting changing food habits, *British Food Journal*, **97**(10), 11–16.

Gofton L and Ness M (1991) Twin Trends: Health and Convenience in Food Change or Who Killed the Lazy Housewife? *British Food Journal* **93**, 7.

Goode J, Beardsworth A, Haslam C, Keil T and Sherratt E (1995) Dietary Dilemmas: nutritional concerns of the 1990s, *British Food Journal*, **97**, 11, 3–12.

Guardian (2001) Robin Cook's chicken tikka masala speech. www.guardian.co.uk/world/ 2001/apr/19/race.britishidentity/print (accessed 19/6/08).

Hall C, Hayes J and Pratt J (2003) *Recipes for the Nation's Favourite Food: Britain's Top 100 Dishes*. BBC:London. (cited in Mitchell 1999)

Hoare J and Henderson L (2004) *NDNS adults aged 19–64, Summary Report* volume 5. London HMSO.

Hollingsworth DF (1960) The changing patterns in British Food Habits since the 1939– 45 war. Symposium proceedings, Food Habits in Britain: 135th scientific meeting, 59th Scottish meeting, Institute of Physiology, University of Glasgow.

Holm L and Mohl M (2000) The role of meat in everyday food culture: an analysis of an interview study in Copenhagen, *Appetite*, **34**, 277–283.

Hylton S (2005) The last word – Bisto: altogether now: 'Aah'. Archive report, *Ethical Corporation*. http://www.ethicalcorp.com/content.asp?ContentID=4019 (accessed 19/ 6/08).

IGD (2008) Delia's cookbook is just the tip of the cheating iceberg. Press office http:// www.igd.com/cir.asp?menuid=9&cirid=2670.

IGD (2007) Brits Shun Fast-Food for Healthy Home Cooking (http://www.igd.com/ cir.asp?menuid=9&cirid=2461).

Kemmer D, Anderson A and Marshall D (1998) Living together and eating together: changes in food choices and eating habits during the transition from single to married/ cohabiting, *The Sociological Review*, **46**(1), 48–72.

Kjaernes U, Ekström M P, Gronow J, Holm L and Mäkelä J (2001) Introduction. In U. Kjaernes, *et al.* (Ed.), *Eating patterns. A day in the lives of Nordic peoples* (pp. 25– 63). Lysaker: SIFO-National Institute for Consumer Research.

King D (2007) Foresight. Tackling Obesities: Future Choices. Project Report, 2nd edition, Government office for Science, London.

Koctürk T (1995) Structure and Change in Food Habits, *Scandanavian Journal of Nutrition/ Näringsforskning*, **39**(1), 2–4.

Lalonde MP (1992) Deciphering a meal again, or the anthropology of taste, *Social Science Information*, **31**(1), 69–86.

Lehmann G (2002) Meals and Mealtimes, 1600–1800. In *The Meal,* H Walker (ed) Proceedings of the Oxford Symposium of Food and Cookery 2001. Prospect Books, Totnes, Devon.

Mäkelä J, Roos E and Pratalla R (1995) Ideas of meals in Finland and Kentucky: Results from a collaborative research project. Paper presented at the 4th International Food Choice Conference, Birmingham, AL. In Abstracts of the Fourth Food Choice Conference (1995) *Appetite*, **24**, 272.

Mäkelä J, Kjaernes U, Ekström MP, L'Orange Fürst E, Gronow J and Holm L (1999) Nordic meals: Methodological notes on a comparative study, *Appetite*, **32**, 73–79.

Mäkelä J (2000) Cultural definitions of the meal. In HL Meiselman (ed) *Dimensions of the meal, the science, culture, business and art of eating.* Aspen, Maryland.

Marshall D (1993) Appropriate Meal Occasions: Understanding Conventions and Exploring Situational Influences on Food choice, *The International Review of Retail, Distribution and Consumer Research*, **3**(3), July, 279–301.

Marshall D (2000) British Meals and Food Choice. Chapter 13. In HL Meiselman (ed) *Dimensions of the meal, the science, culture, business and art of eating.* Aspen, Gaithersburg, Maryland.

Marshall D (2005) Food as Ritual, Routine or Convention, *Consumption, Markets and Culture*, **8**(1), 69–85.

Marshall D and Anderson A (2000) Who's Responsible for the Food Shopping. A Study of Young Scottish Couples in their 'Honeymoon' Period, *International Review of Retail, Distribution and Consumer Research*, **10**(1), January, 59–72.

Martens L and Warde A (1997) Urban pleasure? On the meaning of eating out in a northern city. In P. Caplan (ed) *Food, Health and Identity.* Chapter 7, Routledge, London.

Meades J (2003) *Meades eats.* 3 part TV series, March, BBC2. http://www.bbc.co.uk/food/meadeseats/20030303.shtml

Meiselman HL (2000) *Dimensions of the meal, the science, culture, business and art of eating.* Aspen, Gaithersburg, Maryland.

Meiselman HL (2008) Dimensions of the meal, *Journal of Food Service*, **19**(1), 13–21.

Mennell S (1996) *All manners of food: eating and taste in England and France from the Middle Ages to the present.* Chicago: University of Illinois Press.

Mestdag I (2005) Disappearance of the traditional meal: Temporal, social and spatial destructuration, *Appetite*, **45**, 62–74.

Mintel International Group Ltd (2003) *Snacks – Pan-European Overviews.* European Consumer Goods Intelligence Series. London.

Mintel International Group Ltd (2004) *The Evening Meal – UK, Market Intelligence,* Mintel, London

Mintel International Group Ltd (2008) *Chilled and frozen ready meals – UK.* Market Intelligence, Mintel, London.

Mintz S (1992) A Taste of History, *The Times Higher Education Supplement*, 8 May, 15–18.

Mitchell J (1999) The British main meal in the 1990s: has it changed its identity? *British Food Journal*, **101**(11), 871–883.

Mitchell J (2006) Food acceptance and acculturation, *Journal of Foodservice*, **17**, 77–83.

Murcott A (1982) On the social significance of the 'cooked dinner' in South Wales, *Social Science Information*, **21**(4/5), 677–695.

Murcott A (1983) Cooking and the cooked: a note on the domestic preparation of meals. In Murcott, A. (Eds), *The Sociology of Food and Eating*, Gower, Aldershot.

Murcott A (1995) It's such a pleasure to cook for him': food, mealtimes and gender in some South Wales households. In Jackson, S, Moores, S (Eds), *The Politics of Domestic Consumption*, Prentice Hall/Harvester Wheatsheaf, Hemel Hempstead.

Murcott A (1997) 'The Lost Supper' Perspective, *Times Higher Education Supplement*, 31 January, p 15.

Observer The (2005) Aah! Bisto wants us all to eat as a family again. Article by Kim Hunter Gordon, Journalist.

Oddy DJ and Burnett J (1992) 'British Diet Since Industrialization: A Bibliographic Study'. In HJ Teuteberg, *European Food History: A Research Review*, Leicester University Press.

Office for National Statistics (ONS) (2007) *Social Trends Number 37*, Edited by A Self and L Zealey. Office for National Statistics, HMSO, London.

Orfanos P *et al.* (2007) Eating out of home and its correlates in 10 European countries. The EPIC Study, *Public Health Nutrition*, **10**(12), 1515–1525.

Pettinger C, Holdsworth M and Gerber M (2004) Psycho-social influences on food choice is Southern France and Central England, *Appetite*, **42**, 307–316.

Pettinger C, Holdsworth M and Gerber M (2006) Meal patterns and cooking practices in Southern France and Central England *Public Health Nutrition*, **9**(8), 1020–1026.

Pettinger C and Lankshear G (2008) Food choice priorities in rural communities in North Devon: an exploratory study. Poster presented at British Sociological Association Food Study Group conference, July 2008, *Food, Society and Public Health*, Book of abstracts, p. 34. British Library, London.

Prentice AM and Jebb SA (2003) Fast foods, energy density and obesity: a possible mechanistic link. *Obesity Reviews*, **4**, 187–194.

Romani S (2005) Feeding post-modern families: food preparation and consumption practices in new family structures, *European Association for Consumer Research Conference*, Goteborg.

Rozin P, Fischler C, Imada S, Sarubin A and Wrzesniewski A (1999) Attitudes to food and the role of food in life in the USA, Japan, Flemish Belgium and France: possible implications for the diet-health debate, *Appetite*, **33**, 163–180.

Shiu ECC, Dawson JA and Marshall DW (2004) Segmenting the convenience and health trends in the British food market, *British Food Journal*, **106**(2), 106–127.

Short F (2003) Domestic cooking skills – what are they? *Journal of the Home Economics Institute of Australia*, **10**(3), 13–21.

Shove E and Southerton D (2000) Defrosting the Freezer: From Novelty to Convenience. A Story of Normalization, *Journal of Material Culture*, **5**(3), 301–19. (cited in Cheng *et al.* 2007 *op. cit.*)

Sobal J and Nelson MK (2003) Commensal eating patterns: A community study, *Appetite* **41**, 181–190.

Tansey G and Worsley T (1995) *The Food System: A Guide*. Earthscan Publications.

Telegraph (2008) British home cooking 'is becoming extinct', March 2008. http://www.telegraph.co.uk/news/uknews/1581090/British-home-cooking-'is-becoming-extinct?html 01/008 (accessed 01/10/08)

TNS/FSA (2007) *Eighth Consumer Attitudes Survey Segmentation report* October 2007 http://www.food.gov.uk/news/newsarchive/2008/jul/attitudes1507 (accessed 01/09/08).

TNS Worldpanel Usage (The UK's largest and only continuous monitor of food and drink consumption in the UK) September 2008 (Personal communication). TNS Worldpanel) http://www.tnsglobal.com/market-research/fmcg-research/consumer-panel/ (accessed 1/9/08).

Vidal J (2008) Way we eat now: later, faster, and increasingly in Asian restaurants. The Guardian, Saturday, 5 January 2008. http://www.guardian.co.uk/uk/2008/jan/05/lifeandhealth.foodanddrink (accessed 26/8/08)

Visser M (1991) *The rituals of dinner:the origins, evolution, eccentricities, and meaning of table manners*, Penguin, London.

Volatier JL (1999) Le repas traditionnel se porte encore bien. *CREDOC comsommations et modes de vie 1999* (132).

Warde A (1999) Convenience food: space and timing, *British Food Journal*, **101**(7) 518–527.

Warde A and Martens L (2000) *Eating out: social differentiation, consumption and pleasure.* Cambridge University Press.

Warren GC (ed) (1958) *The foods we eat.* Cassell, London.

Wilk R (2006) Power at the Table: Happy meals and food fights, *Appetite*, **47**(3), 401.

Wood R (1995) *The sociology of the meal*, Edinburgh University Press.

Index